LIFE
The Science of Biology
EIGHTH EDITION

 Sinauer Associates, Inc.

 W. H. Freeman and Company

EIGHTH EDITION LIFE The Science of Biology

DAVID SADAVA
The Claremont Colleges
Claremont, California

H. CRAIG HELLER
Stanford University
Stanford, California

GORDON H. ORIANS
Emeritus, University of Washington
Seattle, Washington

WILLIAM K. PURVES
Emeritus, Harvey Mudd College
Claremont, California

DAVID M. HILLIS
University of Texas
Austin, Texas

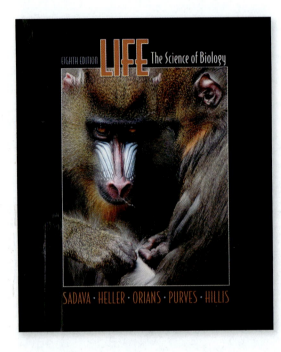

Cover Photograph
"Amor de Madre," photograph of mandrill (*Mandrillus sphinx*) mother and young. Copyright © Max Billder.

Frontispiece
Grizzly bears (*Ursus arctos horribilis*) hunt salmon in an Alaskan river. Copyright © Lynn M. Stone/Naturepl.com.

LIFE: The Science of Biology, Eighth Edition

Copyright © 2008 by Sinauer Associates, Inc. All rights reserved.
This book may not be reproduced in whole or in part without permission.

Address editorial correspondence to:
Sinauer Associates Inc., 23 Plumtree Road, Sunderland, MA 01375 U.S.A.
www.sinauer.com
publish@sinauer.com

Address orders to:
VHPS/W.H. Freeman & Co., Order Dpt., 16365 James Madison Highway,
U.S. Route 15, Gordonsville, VA 22942 U.S.A.
www.whfreeman.com

Examination copy information: 1-800-446-8923
Orders: 1-888-330-8477

Library of Congress Cataloging-in-Publication Data
Life: the science of biology / David Sadava ... [et al.]. — 8th ed.
        p. cm.
    Includes index.
    ISBN-13: 978-0-7167-7671-0 (hardcover) – ISBN 978-0-7167-7673-4 (Volume I) –
    ISBN 978-0-7167-7674-1 (Volume 2) – ISBN 978-0-7167-7675-8 (Volume 3)
    1. Biology. I. Sadava, David E.
QH308.2.L565 2007
570—dc22   2006031320

Printed in U.S.A.
First Printing December 2006
The Courier Companies, Inc.

*To our students, especially the more than 30,000
we have collectively instructed in introductory biology over the years.*

The Authors

Craig Heller   Gordon Orians   Bill Purves   David Sadava   David Hillis

David Sadava is the Pritzker Family Foundation Professor of Biology at the Keck Science Center of Claremont McKenna, Pitzer, and Scripps, three of The Claremont Colleges. Twice winner of the Huntoon Award for superior teaching, Dr. Sadava has taught courses on introductory biology, biotechnology, biochemistry, cell biology, molecular biology, plant biology, and cancer biology. He is a visiting scientist in medical oncology at the City of Hope Medical Center. He is the author or coauthor of five books on cell biology and on plants, genes, and crop biotechnology. His research has resulted in over 50 papers, many coauthored with undergraduates, on topics ranging from plant biochemistry to pharmacology of narcotic analgesics to human genetic diseases. For the past 15 years, he and his collaborators have investigated multi-drug resistance in human small-cell lung carcinoma cells with a view to understanding and overcoming this clinical challenge. Their current work focuses on new anti-cancer agents from plants.

Craig Heller is the Lorry I. Lokey/Business Wire Professor in Biological Sciences and Human Biology at Stanford University. He earned his Ph.D. from the Department of Biology at Yale University in 1970. Dr. Heller has taught in the core biology courses at Stanford since 1972 and served as Director of the Program in Human Biology, Chairman of the Biological Sciences Department, and Associate Dean of Research. Dr. Heller is a fellow of the American Association for the Advancement of Science and a recipient of the Walter J. Gores Award for excellence in teaching. His research is on the neurobiology of sleep and circadian rhythms, mammalian hibernation, the regulation of body temperature, and the physiology of human performance. Dr. Heller has done research on sleeping kangaroo rats, diving seals, hibernating bears, and exercising athletes. Some of his recent work on the effects of temperature on human performance is featured in the opener to Chapter 40.

Gordon Orians is Professor Emeritus of Biology at the University of Washington. He received his Ph.D. from the University of California, Berkeley in 1960 under Frank Pitelka. Dr. Orians has been elected to the National Academy of Sciences and the American Academy of Arts and Sciences, and is a Foreign Fellow of the Royal Netherlands Academy of Arts and Sciences. He was President of the Organization for Tropical Studies, 1988–1994, and President of the Ecological Society of America, 1995–1996. He is a recipient of the Distinguished Service Award of the American Institute of Biological Sciences. Dr. Orians is a leading authority in ecology, conservation biology, and evolution. His research on behavioral ecology, plant–herbivore interactions, community structure, and environmental policy has taken him to six continents. He now devotes full time to writing and to helping apply scientific information to environmental decision-making.

Bill Purves is Professor Emeritus of Biology as well as founder and former Chair of the Department of Biology at Harvey Mudd College in Claremont, California. He received his Ph.D. from Yale University in 1959 under Arthur Galston. A fellow of the American Association for the Advancement of Science, Dr. Purves has served as head of the Life Sciences Group at the University of Connecticut, Storrs, and as Chair of the Department of Biological Sciences, University of California, Santa Barbara, where he won the Harold J. Plous Award for teaching excellence. His research interests focused on the hormonal regulation of plant growth. Dr. Purves elected early retirement in 1995, after teaching introductory biology for 34 consecutive years, in order to concentrate entirely on research directed at learning and science education. He is currently participating in the development of a virtual technical high school, with responsibility for curriculum design in scientific reasoning and health science.

David Hillis is the Alfred W. Roark Centennial Professor in Integrative Biology and the Director of the Center for Computational Biology and Bioinformatics at the University of Texas at Austin, where he also has directed the School of Biological Sciences. Dr. Hillis has taught courses in introductory biology, genetics, evolution, systematics, and biodiversity. He has been elected into the membership of the American Academy of Arts and Sciences, awarded a John D. and Catherine T. MacArthur Fellowship, and has served as President of the Society for the Study of Evolution and of the Society of Systematic Biologists. His research interests span much of evolutionary biology, including experimental studies of evolving viruses, empirical studies of natural molecular evolution, applications of phylogenetics, analyses of biodiversity, and evolutionary modeling. He is particularly interested in teaching and research about the practical applications of evolutionary biology.

Preface

As active scientists working in a wide variety of both basic and applied biology, we are fortunate to be part of a field that is not only fascinating but also changes rapidly. It is apparent not just in the time span since we started our careers—we see it every day when we open a newspaper or a scientific journal. As educators of both introductory and advanced-level students, we desire to convey our excitement about biology's dynamic nature.

This new edition of *Life* looks, and is, quite different from its predecessors. In planning the Eighth Edition, we focused on three fundamental goals. The first was to maintain and enhance what has worked well in the past—an emphasis on not just what we know but how we came to know it; the incorporation of exciting new discoveries; an art program distinguished by its beauty and clarity; plus a unifying theme. As should be the case in any biology textbook, that theme is evolution by natural selection, a 150-year-old idea that more than ever ties together the living world. We have been greatly helped in this endeavor by the addition of a new author, David Hillis. His knowledge and insights have been invaluable in developing our chapters on evolution, phylogeny, and diversity, and they permeate the rest of the book as well.

Our second goal has been to make *Life* more pedagogically accessible. From the bold new design to the inclusion of numerous learning aids throughout each chapter (see New Pedagogical Features), we have worked to make our writing consistently easy to follow as well as engaging.

Third, between editions we asked seven distinguished ecologists—all of whom teach introductory biology—to provide detailed critiques of the Ecology unit. As a result of their extensive suggestions, Part Nine, Ecology, has a fresh organization (see The Nine Parts). And one of the seven, May Berenbaum, has agreed to join the *Life* author team for the Ninth Edition. The other six stalwarts are thanked in the "Reviewers of the Eighth Edition" section.

Enduring Features

As stated above, we are committed to a blending of a presentation of the core ideas of biology with an emphasis on introducing our readers to the process of scientific inquiry. Having pioneered the idea of depicting seminal experiments in specially designed figures, we continue to develop this here, with 96 EXPERIMENT figures (28 percent more than in the Seventh Edition). Each follows the structure: Hypothesis, Method, Result, and Conclusion. Many now include "Further Research," which asks students to conceive an experiment that explores a related question.

A related feature is the RESEARCH METHOD figures, depicting many laboratory and field methods used to do this research. All the Experiment and Research Method figures are listed in the endpapers at the back of the book.

Another much-praised feature—which we pioneered ten years ago in *Life*'s Fifth Edition—is the BALLOON CAPTIONS used in our figures. We know that many students are visual learners. The balloon captions bring explanations of intricate, complex processes directly into the illustration, allowing the reader to integrate the information without repeatedly going back and forth between the figure and its legend.

Life is the only introductory biology book for science majors that begins each chapter with a story. These OPENING STORIES, most of which are new to this edition, are meant to intrigue students while helping them see how the chapter's biological subject relates to the world around them.

New Pedagogical Features

There are several new elements in the Eighth Edition chapters. Each has been designed as a study tool to aid the student in mastering the material. In the opening page spread, IN THIS CHAPTER previews the chapter's content, and the CHAPTER OUTLINE gives the major section headings, all numbered and all framed as questions to emphasize the inquiry basis of science.

Each main section of a chapter now ends with a RECAP. This key element briefly summarizes the important concepts in the section, then provides two or three questions to stimulate immediate review. Each question includes reference to pertinent text or a figure or both.

The CHAPTER SUMMARY boldfaces key terms introduced and defined in the chapter. We have kept the highlighted references to key figures and to the Web tutorials and activities that support a topic in the chapter.

Another new element, BIOBITS, is not strictly a learning aid, but offers intriguing (occasionally amusing) supplemental information. BioBits, like the opening stories, are intended to help students appreciate the interface between biology and other aspects of life.

Redesigned WEB ICONS alert the reader to the tutorials and activities on *Life*'s companion website (www.thelifewire.com). Each of these study and review resources, many of which are new for the Eighth Edition, has been created specifically for *Life*. A full list, by chapter, is found in the front endpapers of the book.

The Nine Parts

We have reorganized the book into nine parts. Part One sets the stage for the entire book, with the opening chapter on biology as an exciting science, starting with a student project, and how evolution unites the living world. This is followed by chapters on the basic chemical building blocks that underlie life. We have tried to tie this material together by relating it to theories on the origin of life, with new discoveries of water in our solar system as an impetus.

In Part Two, Cells and Energy, we present an integrated view of the structure and biochemical functions of cells. The discussions of biochemistry are often challenging for students; thus we have reworked both the text and illustrations for greater clarity. These discussions are presented in the context of the latest discoveries on the origin of life and evolution of cells.

Part Three, Heredity and the Genome, begins with continuity at the cellular level, and then outlines the principles of genetics and the identification of DNA as the genetic material. New examples, such as the genetics of coat color in dogs, enliven these chapters. This is followed by chapters on gene expression and on the prokaryotic and eukaryotic genomes. Many new discoveries have been made in this new field of genomics, ranging from tracking down the bird flu virus to the genomes of wild cats, such as the cheetah.

Part Four, Molecular Biology: The Genome in Action, reinforces the basic principles of classical and molecular genetics by applying them to such diverse topics as cell signaling, biotechnology, and medicine. We use many new experiments and examples from applied biology to illustrate these concepts. These include the latest information on the human genome and the emerging field of systems biology. The chapter on natural defenses now includes a discussion of allergy.

Part Five, The Patterns and Processes of Evolution, has been updated in several important ways. We have emphasized the importance of evolutionary biology as a basis for comparing and understanding all aspects of biology, and have described numerous practical applications of evolutionary biology that will be familiar and relevant to the everyday lives of most students.

Recent experimental studies of evolution are described and explained, to help students understand that evolution is an ongoing, observable process. The chapters on phylogenetics and molecular evolution have been completely rewritten to reflect recent advances in those fields. Other changes reflect our growing knowledge of the history of life on Earth and the mechanisms of evolution that have given rise to all of biodiversity.

Part Six, The Evolution of Diversity, reflects the latest views on phylogeny. It continues to emphasize groups united by evolutionary history over classically defined taxa. This emphasis is now supported by an appendix on the Tree of Life that clearly maps out and describes all groups discussed in the text, so that students can quickly look up unfamiliar names and see how they fit into the larger context of life. We now discuss aspects of phylogeny that are still under study or debate (among major groups of eukaryotes, plants, and animals, for instance).

In Part Seven, Flowering Plants: Form and Function, we report on several exciting new discoveries. These include the receptors for auxin, gibberellins, and brassinosteroids as well as great progress on the florigen problem. We have updated our treatment of signal transduction pathways and of circadian rhythms in plants. The already strong treatment of environmental challenges to plants has been augmented by new Experiment figures on plant defenses against herbivores, one confirming that nicotine does help tobacco plants resist certain insects.

Part Eight, Animals: Form and Function, is about how animals work. Although we give major attention to human physiology, we embed it in a background of comparative animal physiology. Our focus is systems physiology but we also introduce the underlying cellular and molecular mechanisms. For example, our explanations of nervous system phenomena—whether they be action potentials, sensation, learning, or sleep—are discussed in terms of the properties of ion channels. The actions of hormones are explained in terms of the molecular mechanisms. Maximum athletic performance is explained in terms of the underlying cellular energy systems. Throughout Part Eight we try to help the student make the connections across all levels of biology, from molecular to behavioral, and to see the relevance of physiology to issues of health and disease. Of central importance in each chapter is mechanisms of control and regulation.

Part Nine, Ecology, begins with a new chapter that describes the scope of ecological research and discusses recent advances in our understanding of the broad patterns in the distribution of life on Earth. The next chapter, also new, combines Behavior and Behavioral Ecology. It shows how the decisions that organisms make during their lives influence both their survival and reproductive success, and also the dynamics of populations and the structure of ecological communities. The chapter on Population Ecology has new material that explains how ecologists are able to mark and follow individual organisms in the wild to determine their survival and reproductive success. Following a chapter on Community Ecology, another new chapter, Ecosystems and Global Ecology, shows how ecologists are expanding the scope of their studies to encompass the functioning of the global ecosystem. This discussion leads naturally to the final chapter in the book, Conservation Biology, which describes how ecologists and conservation biologists work to reduce the rate at which species are becoming extinct as a result of human activities.

Full Books, Paperbacks, or Loose-Leaf

We again provide *Life* both as the full book and as a cluster of paperbacks. Thus, instructors who want to use less than the whole book, or who want their students to have more portable units, can choose from these split volumes:

Volume I, The Cell and Heredity, includes: Part One, The Science and Building Blocks of Life (Chapters 1–3); Part Two, Cells and Energy (Chapters 4–8); Part Three, Heredity and the Genome (Chapters 9–14); and Part Four, Molecular Biology: The Genome in Action (Chapters 15–20).

Volume II, Evolution, Diversity, and Ecology, includes: Chapter 1, Studying Life; Part Five, The Patterns and Processes of Evolution (Chapters 21–25); Part Six, The Evolution of Diversity (Chapters 26–33); and Part Nine, Ecology (Chapters 52–57).

Volume III, Plants and Animals, includes: Chapter 1, Studying Life; Part Seven, Flowering Plants: Form and Function (Chapters 34–39); and Part Eight, Animals: Form and Function (Chapters 40–51).

Note that each volume also includes the book's front matter, Appendixes, Glossary, and Index.

Life is also available in a loose-leaf version. This shrink-wrapped, unbound, 3-hole punched version is designed to fit into a 3-ring binder. Students take only what they need to class and can easily integrate any instructor handouts or other resources.

Media and Supplements for the Eighth Edition

The media and supplements for *Life*, Eighth Edition have been assembled with two main goals in mind: (1) to provide students with a collection of tools that helps them effectively master the vast amount of new information that is being presented to them in the introductory biology course; and (2) to provide instructors with the richest possible collection of resources to aid in teaching the course—preparing, presenting the lecture, providing course materials online, and assessing student comprehension.

All of the *Life* media and supplemental resources have been developed specifically for this textbook. This gives the student the greatest degree of consistency when studying across different media. For example, the animated tutorials and activities found on the Companion Website were built using textbook art, so that the manner in which structures are illustrated, the colors used to identify objects, and the terms and abbreviations used are all consistent.

The rich collection of visual resources in the Instructor's Media Library provides instructors with a wide range of options for enhancing lectures, course websites, and assignments. Highlights include: layered art PowerPoint® presentations that break down complex figures into detailed, step-by-step presentations; a collection of approximately 200 video segments that can help capture the attention and imagination of students; and the new set of PowerPoint® slides of textbook art with editable labels and leaders that allow easy customization of the figures.

For a detailed description of all the media and supplements available to accompany the Eighth Edition, please turn to "Life's Media and Supplements package" on page xiii.

Many People to Thank

One of the wisest pieces of advice ever given to a textbook author is to "be passionate about your subject, but don't put your ego on the page." Considering all the people who looked over our shoulders throughout the process of creating this book, this advice could not be more apt. We are indebted to many people, who gave invaluable help to make this book what it is. First and foremost are our colleagues, biologists from over 100 institutions. Some were users of the previous edition, who suggested many improvements. Others reviewed our chapter drafts in detail, including advice on how to improve the illustrations. Still others acted as accuracy re-

viewers when the book was almost completed. Our publishers created an advisory group of introductory course coordinators. They advised us on a variety of issues, ranging from book content and design to elements of the print and media supplements. All of these biologists are listed in the Reviewer credits.

We needed a fresh editorial eye for this edition, and we were fortunate to work with Carol Pritchard-Martinez as development editor. With a level head that comes from years of experience, she was a major presence as we wrote and revised. Elizabeth Morales, our artist, was on her second edition with us. This time, she extensively revised almost all of the prior art and translated our crude sketches into beautiful new art. We hope you agree that our art program remains superbly clear and elegant. Once again, we were lucky to have Norma Roche as the copy editor. Her firm hand and encyclopedic recall of our book's many chapters made our prose sharper and more accurate. For this edition, Norma was joined by the capable and affable Maggie Brown. Susan McGlew coordinated the hundreds of reviews that we described above. David McIntyre was a truly proactive photo editor. Not only did he find over 500 new photographs, including many new ones of his own, that enrich the book's content and visual statement, but he set up, performed, and photographed the experiment shown in Figure 36.1. The elegant new interior design is the creation of Jeff Johnson. He also coordinated the book's layout and designed the cover. Carol Wigg, for the eighth time in eight editions, oversaw the editorial process. Her influence pervades the entire book—she created many BioBits, shaped and improved the chapter-opening stories, interacted with David McIntyre in conceiving many photo subjects, and kept an eagle eye on every detail of text, art, and photographs.

W. H. Freeman continues to bring *Life* to a wider audience. Associate Director of Marketing Debbie Clare, the Regional Specialists, Regional Managers, and experienced sales force are effective ambassadors and skillful transmitters of the features and unique strengths of our book. We depend on their expertise and energy to keep us in touch with how *Life* is perceived by its users.

Finally, we are indebted to Andy Sinauer. Like ours, his name is on the cover of the book, and he truly cares deeply about what goes into it.

DAVID SADAVA

CRAIG HELLER

GORDON ORIANS

BILL PURVES

DAVID HILLIS

Reviewers for the Eighth Edition

Between-Edition Reviewers (Ecology and Animal Parts)

May Berenbaum, University of Illinois, Urbana-Champaign

Carol Boggs, Stanford University

Judie Bronstein, University of Arizona

F. Lynn Carpenter, University of California, Irvine

Dan Doak, University of California, Santa Cruz

Jessica Gurevitch, SUNY, Stony Brook

Margaret Palmer, University of Maryland

Marty Shankland, University of Texas, Austin

Advisory Board Members

Heather Addy, University of Calgary

Art Buikema, Virginia Polytechnic Institute and State University

Jung Choi, Georgia Technical University

Rolf Christoffersen, University of California, Santa Barbara

Alison Cleveland, Florida Southern University

Mark Decker, University of Minnesota

Ernie Dubrul, University of Toledo

Richard Hallick, University of Arizona

John Merrill, Michigan State University

Melissa Michael, University of Illinois

Deb Pires, University of California, Los Angeles

Sharon Rogers, University of Nevada, Las Vegas

Marty Shankland, University of Texas, Austin

Manuscript Reviewers

John Alcock, Arizona State University

Charles Baer, University of Florida

Amy Baird, University of Texas, Austin

Patrice Boily, University of New Orleans

Thomas Boyle, University of Massachusetts, Amherst

Mirjana Brockett, Georgia Institute of Technology

Arthur Buikema, Virginia Polytechnic Institute and State University

Hilary Callahan, Barnard College

David Champlin, University of Southern Maine

Chris Chanway, University of British Columbia

Mike Chao, California State University, San Bernardino

Rhonda Clark, University of Calgary

Elizabeth Connor, University of Massachusetts, Amherst

Deborah A. Cook, Clark Atlanta University

Elizabeth A. Cowles, Eastern Connecticut State University

Joseph R. Cowles, Virginia Polytechnic Institute and State University

William L. Crepet, Cornell University

Martin Crozier, Wayne State University

Donald Dearborn, Bucknell University

Mark Decker, University of Minnesota

Michael Denbow, Virginia Polytechnic Institute and State University

Jean DeSaix, University of North Carolina, Chapel Hill

William Eldred, Boston University

Andy Ellington, University of Texas, Austin

Gordon L. Fain, University of California, Los Angeles

Kevin M. Folta, University of Florida

Miriam Goldbert, College of the Canyons

Kenneth M. Halanych, Auburn University

Susan Han, University of Massachusetts, Amherst

Tracy Heath, University of Texas, Austin

Shannon Hedtke, University of Texas, Austin

Mark Hens, University of North Carolina, Greensboro

Albert Herrera, University of Southern California

Barbara Hetrich, University of Northern Iowa

Erec Hillis, University of California, Berkeley

Jonathan Hillis, Austin, Texas

Hopi Hoekstra, University of California, San Diego

Kelly Hogan, University of North Carolina, Chapel Hill

Carl Hopkins, Cornell University

Andrew Jarosz, Michigan State University

Norman Johnson, University of Massachusetts, Amherst

Walter Judd, University of Florida

David Julian, University of Florida

Laura Katz, Smith College

Melissa Kosinski-Collins, Massachusetts Institute of Technology

William Kroll, Loyola University of Chicago

Marc Kubasak, University of California, Los Angeles

Josephine Kurdziel, University of Michigan

John Latto, University of California, Berkeley

Brian Leander, University of British Columbia

Jennifer Leavey, Georgia Institute of Technology

Arne Lekven, Texas A&M University

Don Levin, University of Texas, Austin

Rachel Levin, Amherst College

Thomas Lonergan, University of New Orleans

Blase Maffia, University of Miami

Meredith Mahoney, University of Texas, Austin

Charles Mallery, University of Miami

Ron Markle, Northern Arizona University

Mike Meighan, University of California, Berkeley

Melissa Michael, University of Illinois, Urbana-Champaign

Jill Miller, Amherst College

Subhash Minocha, University of New Hampshire

Thomas W. Moon, University of Ottawa

Richard Moore, Miami University of Ohio

John Morrissey, Hofstra University

Leonie Moyle, University of Indiana

Mary Anne Nelson, University of New Mexico

Dennis O'Connor, University of Maryland, College Park

Robert Osuna, SUNY, Albany

Cynthia Paszkowski, University of Alberta

Diane Pataki, University of California, Irvine

Ron Patterson, Michigan State University

Craig Peebles, University of Pittsburgh

Debra Pires, University of California, Los Angeles

Greg Podgorski, Utah State University

Chuck Polson, Florida Institute of Technology

Donald Potts, University of California, Santa Cruz

Jill Raymond, Rock Valley College

Ken Robinson, Purdue University

Sharon L. Rogers, University of Nevada, Las Vegas

Laura Romano, Denison University

Pete Ruben, Utah State University, Logan

Albert Ruesink, Indiana University

Walter Sakai, Santa Monica College

Mary Alice Schaeffer, Virginia Polytechnic Institute and State University

Daniel Scheirer, Northeastern University

Stylianos Scordilis, Smith College

Kevin Scott, University of Calgary

Jim Shinkle, Trinity University

Denise Signorelli, Community College of Southern Nevada

Thomas Silva, Cornell University

Jeffrey Tamplin, University of Northern Iowa

Steve Theg, University of California, Davis

Sharon Thoma, University of Wisconsin, Madison

Jeff Thomas, University of California, Los Angeles

Christopher Todd, University of Saskatchewan

John True, SUNY, Stony Brook

Mary Tyler, University of Maine

Fred Wasserman, Boston University

John Weishampel, University of Central Florida

Elizabeth Willott, University of Arizona

David Wilson, University of Miami

Heather Wilson-Ashworth, Utah Valley State College

Accuracy Reviewers

John Alcock, Arizona State University

John Anderson, University of Minnesota

Brian Bagatto, University of Akron

Lisa Baird, University of San Diego

May Berenbaum, University of Illinois, Urbana-Champaign

Gerald Bergtrom, University of Wisconsin, Milwaukee

Stewart Berlocher, University of Illinois, Urbana-Champaign

Mary Bisson, SUNY, Buffalo

Arnold Bloom, University of California, Davis

Judie Bronstein, University of Arizona

Jorge Busciglio, University of California, Irvine

Steve Carr, Memorial University of Newfoundland

Thomas Chen, Santa Monica College

Randy Cohen, California State University, Northridge

Reid Compton, University of Maryland, College Park

James Courtright, Marquette University

Jerry Coyne, University of Chicago

Joel Cracraft, American Museum of Natural History

Joseph Crivello, University of Connecticut, Storrs

Gerrit De Boer, University of Kansas, Lawrence

Arturo DeLozanne, University of Texas, Austin

Stephen Devoto, Wesleyan University

Laura DiCaprio, Ohio University

John Dighton, Rutgers Pinelands Field Station

Jocelyne DiRuggiero, University of Maryland, College Park

W. Ford Doolittle, Dalhousie University

Emanuel Epstein, University of California, Davis

Gordon L. Fain, University of California, Los Angeles

Lewis J. Feldman, University of California, Berkeley

James Ferraro, Southern Illinois University

Cole Gilbert, Cornell University

Elizabeth Godrick, Boston University

Martha Groom, University of Washington

Kenneth M. Halanych, Auburn University

Mike Hasegawa, Purdue University

Mark Hens, University of North Carolina, Greensboro

Richard Hill, Michigan State University

Franz Hoffman, University of California, Irvine

Sara Hoot, University of Wisconsin, Milwaukee

Carl Hopkins, Cornell University

Alfredo Huerta, Miami University

Michael Ibba, The Ohio State University

Walter Judd, University of Florida

Laura Katz, Smith College

Manfred D. Laubichler, Arizona State University

Brian Leander, University of British Columbia

Mark V. Lomolino, SUNY College of Environmental Science and Forestry

Jim Lorenzen, University of Idaho

Denis Maxwell, University of Western Ontario

Brad Mehrtens, University of Illinois, Urbana-Champaign

John Merrill, Michigan State University

Allison Miller, Saint Louis University

Clara Moore, Franklin and Marshall College

Julie Noor, Duke University

Mohamed Noor, Duke University

Theresa O'Halloran, University of Texas, Austin

Norman R. Pace, University of Colorado

Randall Packer, George Washington University

Walt Ream, Oregon State University

Eric Richards, Washington University

Steve Rissing, The Ohio State University

R. Michael Roberts, University of Missouri, Columbia

Pete Ruben, Simon Fraser University

David A. Sanders, Purdue University

Mike Sanderson, University of California, Davis

Marty Shankland, University of Texas, Austin

Jeff Silberman, University of Arkansas

Margaret Silliker, DePaul University

Dee Silverthorn, University of Texas, Austin

M. Suzanne Simon-Westendorf, Ohio University

Alastair G.B. Simpson, Dalhousie University

John Skillman, California State University, San Bernardino

Frederick W. Spiegel, University of Arkansas

John J. Stachowicz, University of California, Davis

Heven Sze, University of Maryland

E.G. Robert Turgeon, Cornell University

Mary Tyler, University of Maine

Mike Wade, Indiana University

Leslie Winemiller, Texas A&M University

Mimi Zolan, Indiana University

Tree of Life Appendix Reviewers

John Abbott, University of Texas, Austin

Joseph Bischoff, National Center for Biotechnology Information

Ruth Buskirk, University of Texas, Austin

David Cannatella, University of Texas

Joel Cracraft, American Museum of Natural History

Scott Federhen, National Center for Biotechnology Information

Carol Hotton, National Center for Biotechnology Information

Robert Jansen, University of Texas, Austin

Brian Leander, University of British Columbia

Detlef Leipe, National Center for Biotechnology Information

Beryl Simpson, University of Texas, Austin

Richard Sternberg, National Center for Biotechnology Information

Edward Theriot, University of Texas

Sean Turner, National Center for Biotechnology Information

Supplements Authors

Dany Adams, The Forsyth Institute

Erica Bergquist, Holyoke Community College

Ian Craine, University of Toronto

Ernest Dubrul, University of Toledo

Edward Dzialowski, University of North Texas

Donna Francis, University of Massachusetts, Amherst

Jon Glase, Cornell University

Lindsay Goodloe, Cornell University

Celine Muis Griffin, Queen's University

Nancy Guild, University of Colorado at Boulder

Norman Johnson, University of Massachusetts, Amherst

James Knapp, Holyoke Community College

Jennifer Knight, University of Colorado, Boulder

David Kurjiaka, University of Arizona

Richard McCarty, Johns Hopkins University

Betty McGuire, Cornell University

Nancy Murray, Evergreen State College

Deb Pires, University of California, Los Angeles

Catherine Ueckert, Northern Arizona University

Jerry Waldvogel, Clemson University

LIFE's Media and Supplements Package

For the Student

Companion Website www.thelifewire.com

(Also available as a CD, optionally packaged with the book)

The *Life*, Eighth Edition Companion Website is available free of charge to all students (no access code required). The site features a variety of study and review resources designed to help students master the wide range of material presented in the introductory biology course. Features of the site include:

- *Interactive Summaries.* These summaries combine a review of important concepts with links to all the key figures from the chapter as well as all of the relevant animated tutorials and activities.

- *Animated Tutorials.* Over 100 in-depth animated tutorials present complex topics in a clear, easy-to-follow format that combines a detailed animation with an introduction, conclusion, and quiz.

- *Activities.* Over 120 interactive activities help the student learn important facts and concepts through a wide range of activities, such as labeling steps in processes or parts of structures, building diagrams, and identifying different types of organisms.

- *Flashcards.* For each chapter of the book, there is a set of flashcards that allows the student to review all the key terminology from the chapter. Students can review the terms in the study mode, and then quiz themselves on a list of terms.

- *New! Experiment Links.* New for the Eighth Edition, each experiment featured in the textbook has a corresponding treatment on the companion website that links to further information about the experiment, further research that followed, and applications derived from the research.

- *Interactive Quizzes.* Every question includes an image taken from the textbook, thorough feedback on both right and wrong answer choices, references to textbook pages, and links to electronic versions of book pages, where the related material is highlighted.

- *Online Quizzes.* These quizzes test the student's comprehension of the chapter material, and the results are stored in the online gradebook.

- *Key Terms.* The key terminology introduced in each chapter is listed, with definitions and audio pronunciations from the Glossary.

- *Suggested Readings.* For each chapter of the book, a list of suggested readings is provided as a resource for further study.

- *Glossary.* The language of biology is often difficult for students taking introductory biology, so we have created a full glossary with audio pronunciations.

- *Math for Life* (Dany Adams, *The Forsyth Institute*). *Math for Life* is a collection of mathematical shortcuts and references to help students with the quantitative skills they need in the laboratory.

- *Survival Skills* (Jerry Waldvogel, *Clemson University*). *Survival Skills* is a guide to more effective study habits. Topics include time management, note-taking, effective highlighting, and exam preparation.

Study Guide (ISBN 978-0-7167-7893-6)

Edward M. Dzialowski, *University of North Texas*; Jon Glase, *Cornell University*; Lindsay Goodloe, *Cornell University*; Nancy Guild, *University of Colorado*; and Betty McGuire, *Smith College*

For each chapter of the textbook, the *Life* Study Guide offers a variety of study and review tools. The contents of each chapter are broken down into both a detailed review of the Important Concepts covered and a boiled-down Big Picture snapshot. In addition, Common Problem Areas and Study Strategies are highlighted. A set of study questions (both multiple-choice and short-answer) allows students to test their comprehension. All questions include answers and explanations.

Lecture Notebook (ISBN 978-0-7167-7894-3)

This invaluable printed resource consists of all the artwork from the textbook (more than 1,000 images with labels) presented in the order in which they appear in the text, with ample space for note-taking. Because the Notebook has already done the drawing, students can focus more of their attention on the concepts. They will absorb the material more efficiently during class, and their notes will be clearer, more accurate, and more useful when they study from them later.

MCAT® Practice Test (ISBN 0-7167-5907-1)

A complete printed MCAT exam, with answers, allows students to test their knowledge of the full range of introductory biology content as they prepare for the medical school entrance exams.

CatchUp Math & Stats

Michael Harris, Gordon Taylor, and Jacquelyn Taylor

This primer will help your students quickly brush up on the quantitative skills they need to succeed in biology. Presented in brief, accessible units, the book covers topics such as working with powers, logarithms, using and understanding graphs, calculating standard deviation, preparing a dilution series, choosing the right statistical test, analyzing enzyme kinetics, and many more.

Student Handbook for Writing in Biology, Second Edition

Karen Knisely (ISBN 0-7167-6709-0)

This book provides practical advice to students who are learning to write according to the conventions in biology. Using the standards of journal publication as a model, the author provides, in a user-friendly format, specific instructions on: using biology databases to locate references; paraphrasing for improved comprehension; preparing lab reports, scientific papers, posters; preparing oral presentations in PowerPoint®, and more.

Bioethics and the New Embryology: Springboards for Debate

Scott F. Gilbert, Anna Tyler, and Emily Zackin (ISBN 0-7167-7345-7)

Our ability to alter the course of human development ranks among the most significant changes in modern science and has brought embryology into the public domain. The question that must be asked is: Even if we *can* do such things, *should* we do such things?

BioStats Basics: A Student Handbook

James L. Gould and Grant F. Gould (ISBN 0-7167-3416-8)

BioStats Basics provides introductory-level biology students with a practical, accessible introduction to statistical research. Engaging and informal, the book avoids excessive theoretical and mathematical detail to focus on how core statistical methods are put to work in biology.

Laboratory Manuals

W. H. Freeman publishes a range of high-quality biology lab texts, all of which are available for bundling with *Life,* Eighth Edition. Our laboratory texts are available as complete paperback texts, or as Freeman Laboratory Separates.

- *Biology in the Laboratory,* Third Edition
  Doris R. Helms, Carl W. Helms, Robert J. Kosinski, and John C. Cummings (ISBN 0-7167-3146-0)
- *Laboratory Outlines in Biology-VI*
  Peter Abramoff and Robert G. Thomson (ISBN 0-7167-2633-5)
- *Anatomy and Dissection of the Frog,* Second Edition
  Warren F. Walker, Jr. (ISBN 0-7167-2636-X)
- *Anatomy and Dissection of the Rat,* Third Edition
  Warren F. Walker, Jr. and Dominique Homberger (ISBN 0-7167-2635-1)
- *Anatomy and Dissection of the Fetal Pig,* Fifth Edition
  Warren F. Walker, Jr. and Dominique Homberger (ISBN 0-7167-2637-8)
- *Atlas and Dissection Guide for Comparative Anatomy,* Sixth Edition
  Saul Wischnitzer and Edith Wischnitzer (ISBN 0-7167-6959-X)

Custom Publishing for Laboratory Manuals

http://custompub.whfreeman.com

Instructors can build and order customized lab manuals in just minutes, choosing material from Freeman's biology laboratory manuals, as well as their own material.

For the Instructor

Instructor's Media Library

In order to give you the widest possible range of resources to help engage students and better communicate the material, we have assembled an unparalleled collection of media resources. The Eighth Edition of *Life* features an expanded Instructor's Media Library (available on a set of CDs and DVDs) that includes:

- *Textbook Figures and Tables.* Every image from the textbook is provided in both JPEG (high- and low-resolution) and PDF formats.
- *Unlabeled Figures.* Every figure in the textbook is provided in an unlabeled format. These are useful for student quizzing and custom presentation development.
- *Supplemental Photos.* The supplemental photograph collection contains over 1,500 photographs (all in addition to those in the text), forming a rich resource of visual imagery.
- *Animations.* A collection of over 100 in-depth animations, all of which were created from the textbook's art program, and which can be viewed in either narrated or step-through mode.
- *Videos.* This collection of approximately 200 video segments covering topics across the entire textbook helps demonstrate the complexity and beauty of life.
- *PowerPoint® Resources.* For each chapter of the textbook, we have created several different types of PowerPoint® presentations. These give instructors the flexibility to build a presentation in the manner that best suits their needs. Included are:
  - Figures and Tables
  - Lecture Presentation
  - New! Editable Labels
  - Layered Art
  - Supplemental Photos
  - Videos, Animations
- *Clicker Questions.* A set of questions written specifically to be used with classroom personal response systems ("clickers") is provided for each chapter. These questions are designed to reinforce concepts, gauge student comprehension, and provide an outlet for active participation.
- *Chapter Outlines, Lecture Notes,* and the complete *Test File* are all available in Microsoft Word® format for easy use in lecture and exam preparation.
- An intuitive *Browser Interface* provides a quick and easy way to preview all of the content on the Instructor's Media Library.
- *Computerized Test Bank.* The entire printed Test File, plus the textbook end-of-chapter Self-Quizzes, the Companion Website Online Quizzes, and the Study Guide questions are all included in Brownstone's easy-to-use Diploma® software.
- *Instructor's Website.* A wealth of instructor's media, as well as electronic versions of other instructor supplements, are available online for instant access anytime.
- *Online Quizzes.* The Companion Website includes an Online Quiz for each chapter of the textbook. Instructors can choose to use these quizzes as assignments, and can view the results in the online gradebook.

■ *Course Management System Support.* As a service for adopters using WebCT, Blackboard, or ANGEL for their courses, full electronic course packs are available.

Instructor's Resource Kit

The *Life,* Eighth Edition Instructor's Resource Kit includes a wealth of information to help instructors in the planning and teaching of their course. The Kit includes:

■ *Instructor's Manual,* featuring:
  • A "What's New" guide to the Eighth Edition
  • A brief chapter overview
  • A key terms section with all the boldface terms from the text
  • Chapter outlines
■ *Lecture Notes*—detailed notes for each chapter that can serve as the basis for lectures, including references to figures and media resources.
■ *Media Guide*—A visual guide to the extensive media resources available with the Eighth Edition of *Life.* The guide includes thumbnails and descriptions of every video, animation, PowerPoint®, and supplemental photo in the Media Library, all organized by chapter.
■ *Lab manual* and custom lab manual information.

Overhead Transparencies

This set includes over 1,000 transparencies—all the four-color line art and all the tables from the text—along with convenient binders. Balloon captions have been removed and colors have been enhanced for clear projection in a wide range of conditions. Labels and images have been resized for improved readability.

Test File

Ernest Dubrul, *University of Toledo;* Jon Glase, *Cornell University;* Norman Johnson, *University of Massachusetts;* Catherine Ueckert, *Northern Arizona University*

The test file offers more than 5,000 questions, including fill-in-the-blank and multiple-choice test questions. The electronic version of the Test File also includes all of the textbook end-of-chapter Self-Quiz questions, all of the Student Website Online Quiz questions, and all of the Study Guide questions.

iclicker

Developed for educators by educators, iclicker is a hassle-free radio-frequency classroom response system that makes it easy for instructors to ask questions, record responses, take attendance, and direct students through lectures as active participants. For more information, visit www.iclicker.com.

The *Life* eBook

The *Life,* Eighth Edition eBook is a complete online version of the textbook that can be purchased online, or packaged with the printed textbook. This online version of *Life* is a substantially less expensive alternative that gives your students an efficient and rich learning experience by integrating all of the resources from the companion website directly into the eBook text. In addition, the eBook offers instructors unique opportunities to customize the text with their own content. Key features of the eBook include:

■ *For Students:*
  • Integration of all website **activities** and **animated tutorials**
  • In-text **self quiz questions**
  • Interactive **summary exercises**
  • Custom text **highlighting**
  • A **notes** feature that allows students to annotate the text
  • Complete **glossary**, **index**, and **full-text search** features

■ *For Instructors:*
  • A powerful notes feature that can incorporate text, Web links, and documents directly into the text
  • Easy integration of instructor media resources, including videos and supplemental photographs
  • Ability to link directly to any eBook page from other sites

BIO PORTAL

New for the Eighth Edition, BioPortal is the digital gateway to all of the teaching and learning resources that are available with *Life.* BioPortal integrates the *Life* textbook, all of the student and instructor media, extensive assessment resources, and course planning resources all into a powerful and easy-to-use learning management system. All of this means that your students get easy access to learning resources, presented in the proper context and at the proper time, and you get a complete learning management system, ready to use, without hours of prep work. Features of BioPortal include:

■ *eBook*
  • Completely integrated with all media resources
  • Customizable with notes, sections, images, and more

■ *Student Resources*
  • Animated Tutorials and Activities
  • Assessment: Online Quizzes and Interactive Quizzes

■ *Instructor Resources*
  • Complete Test Bank
  • All quizzes
  • Media resources: Videos, PowerPoints, Supplemental Photos, and more

■ *Assignments*
  • Quizzes and exams
  • Assignable textbook sections
  • Assignable animations and activities
  • Custom assignments

■ *Easy-to-Use Course Management*
  • Complete course customization
  • Custom resources/Document posting
  • Announcements/Calendar/Course Email/Discussions
  • Robust Gradebook

Contents in Brief

Contents

Part Two ■ Cells and Energy

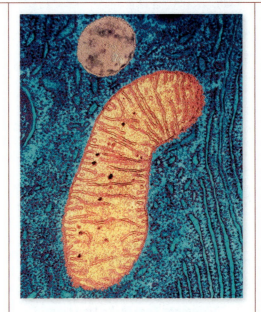

Part Three ■ Heredity and the Genome

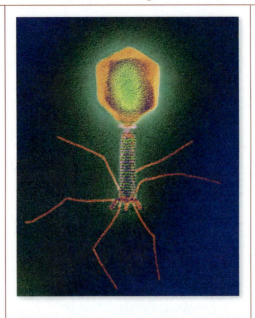

Part Four ■ Molecular Biology: The Genome in Action

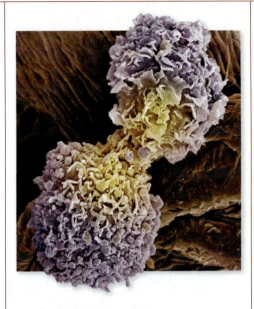

Part Five ▪ The Patterns and Processes of Evolution

Part Six ■ The Evolution of Diversity

Part Seven ■ Flowering Plants: Form and Function

Part Eight ▪ Animals: Form and Function

Part Nine ▪ Ecology

PART ONE
The Science and
Building Blocks of Life

CHAPTER 1

Studying Life

Why are frogs croaking?

In August of 1995, a group of Minnesota middle school students on a field trip were hiking through some local wetlands when they discovered a horde of young frogs, most of them with deformed, missing, or extra legs. The students' find made the national news and focused public attention on amphibian population declines, an issue already being studied by many scientists.

There are a number of possible reasons for the problems amphibians are facing. Water pollution is an obvious possibility, since these animals breed and spend their early lives in ponds and streams. Acidic rain resulting from air pollution could also affect their watery homes. Could ultraviolet radiation be causing "mutant" frogs? Is global warming adversely affecting amphibians? Is some disease attacking them? Evidence exists to support each of these possibilities, and there is no single answer. In one case, a college undergraduate came up with an answer, and in the process gave scientists a whole new perspective on the question.

In 1996, Stanford University sophomore Pieter Johnson was shown a collection of Pacific tree frogs with extra legs growing out of their bodies. He decided to focus his honors research project on finding out what caused these deformities. The frogs came from a pond in an agricultural region near abandoned mercury mines; thus two possible causes of the deformities were agricultural chemicals, and heavy metals from the old mines.

Pieter applied the *scientific method*. Based on what he knew and on his library research, he proposed a logical explanation for the monster frogs—environmental water pollution—and designed an experiment to test his idea. His experiment compared ponds where there were deformed frogs with ponds where the frogs were normal and tested for the presence or absence of pollutants. As frequently happens in science, his proposed explanation, or *hypothesis*, was disproved by his experiment. But his field work led to a new hypothesis: that the deformities are caused by a parasite. Pieter conducted laboratory experiments, the results of which supported the conclusion that a certain type of parasite is present in some ponds. These parasites burrow into newly hatched tadpoles and disrupt the development of the adult

Frogs Are Having Serious Problems
These preserved Pacific tree frogs (*Hyla regilla*) exhibit multiple deformities of the hind legs. Similar deformities have been found in frogs from different regions of the world.

A Biologist at Work As a college sophomore, Pieter Johnson studied numerous ponds that were home to Pacific tree frogs, trying to discover why some ponds had so many deformed frogs.

frog's hind legs. Pieter's research did not explain the global decline in amphibians, but it did illuminate one type of problem amphibians encounter. Science usually progresses in such small but solid steps.

Biologists use the scientific method to investigate the processes of life at all levels, from molecules to ecosystems. Some of these processes happen in millionths of seconds and others cover millions of years. The goals of biologists are to understand how organisms and groups of organisms function, and sometimes they use that knowledge in practical and beneficial ways.

IN THIS CHAPTER we examine the most common features of living organisms and put those features into the context of the major principles that underlie all biology. Next we offer a brief outline of how life evolved and how the different organisms on Earth are related. We then turn to the subjects of biological inquiry and the scientific method.

1.1 What Is Biology?

Biology is the scientific study of living things. Biologists define "living things" as all the diverse organisms descended from a single-celled ancestor that appeared almost 4 billion years ago. Because of their common ancestry, living organisms share many characteristics that are not found in the nonliving world. Most living organisms:

- consist of one or more cells
- contain genetic information
- use genetic information to reproduce themselves
- are genetically related and have evolved
- can convert molecules obtained from their environment into new biological molecules
- can extract energy from the environment and use it to do biological work
- can regulate their internal environment

This list can serve as a rough guide to the major themes and unifying principles of biology that you will encounter in this book. A simple list, however, belies the incredible complexity and diversity of life. Some forms of life may not display all of these characteristics all of the time. For example, the seed of a desert plant may go for many years without extracting energy from the environment, converting molecules, regulating its internal environment, or reproducing; yet the seed is alive.

And what about viruses? Although they do not consist of cells, viruses probably evolved from cellular organisms, and many biologists consider them to be living organisms. Viruses cannot carry out physiological functions on their own, but must parasitize the machinery of host cells to do those jobs for them—including reproduction. Yet viruses contain genetic information, and they certainly evolve (as we know because evolving flu viruses require annual changes in the vaccines we create to combat them). Are viruses alive? What do you think?

This book explores the characteristics of life, how these characteristics vary among organisms, how they evolved, and how they work together to enable organisms to survive and reproduce. *Evolution* is a central theme in biology, and therefore in this book. Through differential survival and reproduction, living systems evolve and become adapted to Earth's many environments. The processes of evolution have generated the enormous diversity that we see today as life on Earth (**Figure 1.1**).

Living organisms consist of cells

Cells and the chemical processes within them are the topic of Part Two of this book. Some organisms are *unicellular,* consisting of a single cell that carries out all the functions of life, while others are *multicellular,* made up of a number of cells that are specialized for different functions.

The discovery of cells was made possible by the invention of the microscope in the 1590s by father and son Dutch spectacle makers Zaccharias and Hans Janssen. The first biologists to improve on their technology and use it to study living organisms were Antony van Leeuwenhoek (Dutch) and Robert Hooke (English) in the middle to late 1600s. It was van Leeuwenhoek who discovered that drops of pond water teemed with single-celled organisms, and he made many other discoveries as he progressively improved his microscopes over a long lifetime of research. Hooke carried out similar studies. From observations of plant tissues (cork, to be specific) he concluded that the tissues were made up of repeated units he called *cells* (**Figure 1.2**).

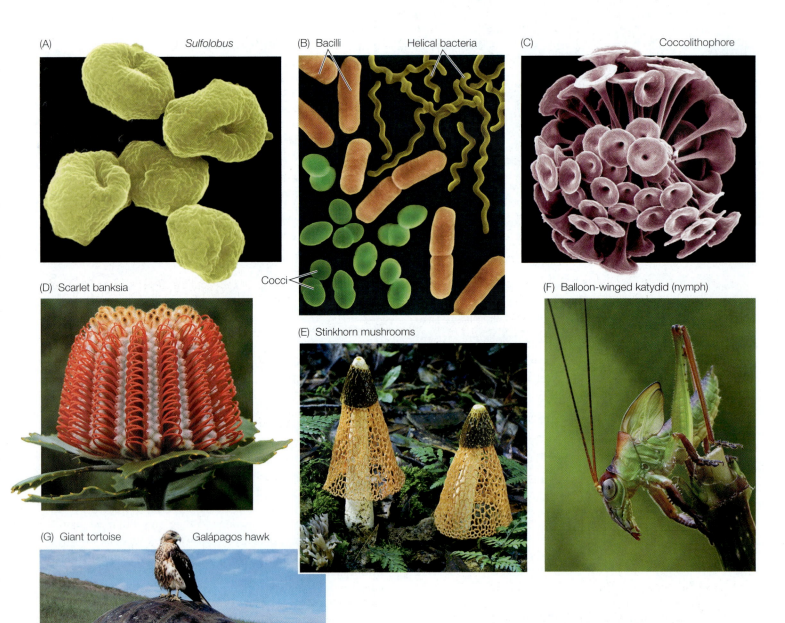

(A) *Sulfolobus*

(B) Bacilli   Helical bacteria   Cocci

(C) Coccolithophore

(D) Scarlet banksia

(E) Stinkhorn mushrooms

(F) Balloon-winged katydid (nymph)

(G) Giant tortoise   Galápagos hawk

1.1 The Many Faces of Life The processes of evolution have led to the millions of diverse organisms living on Earth today. Archaea (A) and bacteria (B) are single-celled organisms. The three different bacteria in (B) represent the three shapes (rods, or bacilli; helices; and spheres, or cocci) often seen in these organisms, which are described in Chapter 26. (C) Organisms whose single cells display more complexity are commonly known as protists. Protists are the subjects of Chapter 27. (D) Chapters 28 and 29 cover the multicellular green plants. The other broad groups of multicellular organisms are (E) the fungi, discussed in Chapter 30, and (F,G) the animals, covered in Chapters 31–33.

(A)

(B)

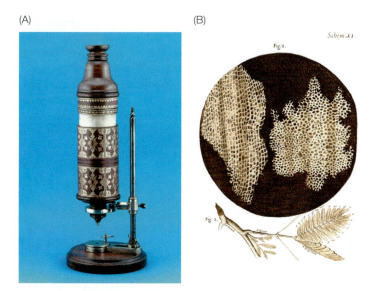

(C)

(D)

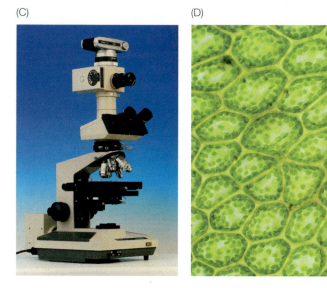

In 1676, Hooke wrote that van Leeuwenhoek had observed "a vast number of small animals in his Excrements which were most abounding when he was troubled with a Loose-nesse and very few or none when he was well." This simple observation represents the discovery of bacteria.

1.2 All Life Consists of Cells (A) The development of microscopes, such as this instrument of Robert Hooke's, revealed the microbial world to seventeenth-century scientists. (B) Hooke was the first to propose the concept of cells, based on his observations of thin slices of plant tissue (cork) under his microscope. (C) A modern version of the optical, or "light" microscope. (D) A modern light micrograph reveals the intricacies of cells in a leaf.

More than a hundred years passed before studies of cells advanced significantly. In 1838, Matthias Schleiden, a German biologist, and Theodor Schwann, from Belgium, were having dinner together and discussing their work on plant and animal tissues, respectively. They were struck by the similarities in their observations and came to the conclusion that the structural elements of plants and animals were essentially the same. They formulated their conclusion as the **cell theory**, which states that:

- Cells are the basic structural and physiological units of all living organisms.
- Cells are both distinct entities and building blocks of more complex organisms.

But Schleiden and Schwann did not understand the origin of cells. They thought cells emerged by the self-assembly of nonliving materials, much as crystals form in a solution of salt. This conclusion was in accordance with the prevailing view of the day that life arises from nonlife by spontaneous generation—mice from dirty clothes, maggots from dead meat, insects from mixtures of straw and pond water. The debate over whether or not life could arise from nonlife continued until 1859, when the French Academy of Sciences sponsored a contest for the best experiment to prove or disprove spontaneous generation. The prize was won by the great French scientist Louis Pasteur; his experiment proving that life must be present in order for life to be generated is described in Figure 3.30. Pasteur's vision of microorganisms led him to propose the germ theory of disease and explain the role of single-celled organisms in the fermentation of beer and wine. He also designed a method of preserving milk by heating it to kill microorganisms, a process we now know as pasteurization.

Today we readily accept the fact that all cells come from preexisting cells. In addition, we understand that the functional properties of organisms derive from the properties of their cells. We also understand that cells of all kinds share many essential mechanisms because they share a common ancestry that goes back billions of years. We therefore add a few more elements to the cell theory:

- All cells come from preexisting cells.
- All cells are similar in chemical composition.
- Most of the chemical reactions of life occur within cells.
- Complete sets of genetic information are replicated and passed on during cell division.

At the same time Schleiden and Schwann were building the foundation for the cell theory, Charles Darwin was beginning to understand how organisms undergo evolutionary change.

The diversity of life is due to evolution by natural selection

Evolution by **natural selection**, as proposed by Charles Darwin, is perhaps the major unifying principle of biology and is the topic of Part Five of this book.

Darwin proposed that living organisms are descended from common ancestors and are therefore related to one another. He did not have the advantage of understanding the mechanisms of genetic inheritance that you will learn about in Part Three, but even so he surmised that such mechanisms existed because offspring resembled their parents in so many different ways. That simple fact is the basis for the concept of a **species**. Although the precise definition of a species is complicated, in its most widespread usage it refers to a group of organisms that look alike ("are morphologically similar") and can breed successfully with one another.

But offspring also differ from their parents. Any population of a plant or animal species displays variation, and if you select breeding pairs on the basis of some particular trait, that trait is more likely to be present in their offspring than in the general population. Darwin himself bred pigeons, and was well aware of how pigeon fanciers selected for unusual feather patterns, beak shapes, and body sizes. He realized that if humans could select for specific traits, the same process could operate in nature; hence the term *natural selection*.

How would selection function in nature? Darwin postulated that different probabilities of survival and reproductive success would do the job. He reasoned that the reproductive capacity of plants and animals, if unchecked, would result in unlimited growth of populations, but we do not observe such unlimited population growth in nature; therefore only a small percentage of offspring must sur-

vive to reproduce. So, any trait that confers even a small increase in the probability that its possessor will survive and reproduce would be strongly favored and would spread in the population. Darwin called this phenomenon *natural selection*.

Because organisms with certain traits survive and reproduce best under specific sets of conditions, natural selection leads to **adaptations**: structural, physiological, or behavioral traits that enhance an organism's chances of survival and reproduction in its environment (**Figure 1.3**). The many different environments and ecological communities organisms have adapted to over evolutionary history have led to a remarkable amount of diversity, which we will survey in Part Six of this book.

If all cells come from preexisting cells, and if all the diverse species of organisms on Earth are related by descent with modification from a common ancestor, then what is the source of information that is passed from parent to daughter cells and from parental organisms to their offspring?

Many leaves are wide and flat, a configuration that presents a maximum of photosynthetic surface to the sun. Some trees, such as this sugar maple, lose their leaves in response to cold or dry weather.

The leaves of many evergreen conifers, such as spruce trees, are waxy-coated needles that resist water loss and are not shed on a yearly basis.

These water lilies are rooted in the pond bottom; their large leaves are flat "pads" that float on the surface.

The leaves of pitcher plants form a vessel that holds water. The plant receives extra nutrients from the decomposing bodies of insects that drown in the pitcher.

The ability to climb can be advantageous to a plant, enabling it to reach above other plants to obtain more sunlight. Some of the leaves of this climbing cucumber are tightly furled tendrils that wrap around a stake.

1.3 Adaptations to the Environment The leaves of all plants are specialized for photosynthesis—the sunlight-powered transformation of water and carbon dioxide into larger structural molecules called carbohydrates. The leaves of different plants, however, display many different adaptations to their individual environments.

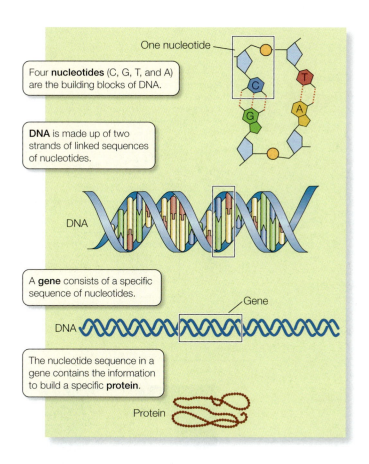

Four **nucleotides** (C, G, T, and A) are the building blocks of DNA.

DNA is made up of two strands of linked sequences of nucleotides.

A **gene** consists of a specific sequence of nucleotides.

The nucleotide sequence in a gene contains the information to build a specific **protein**.

1.4 The Genetic Code Is Life's Blueprint The instructions for life are contained in the sequences of nucleotides in DNA molecules. Specific DNA sequences comprise genes, and the information in each gene provides the cell with the information it needs to manufacture a specific protein. The average length of a single human gene is 16,000 nucleotides.

ism must express different parts of their genome. How the control of gene expression enables a complex organism to develop and function is a major focus of current biological research.

The genome of an organism consists of thousands of genes. If the nucleotide sequence of a gene is altered, it is likely that the protein that gene encodes will be altered. Alterations of genes are called *mutations*. Mutations occur spontaneously; they can also be induced by many outside factors, including chemicals and radiation. Most mutations are deleterious, but occasionally a change in the properties of a protein alters its function in a way that improves the functioning of the organism under the environmental conditions it encounters. Such beneficial mutations are the raw material of evolution.

Cells use nutrients to supply energy and to build new structures

Living organisms acquire substances called *nutrients* from the environment. Nutrients supply the organism with energy and raw materials for building biological structures. Cells take in nutrient molecules and break them down into smaller chemical units. In doing so, they can capture the energy contained in the chemical bonds of the nutrient molecules and use that energy to do different kinds of work. One kind of cellular work is the building, or *synthesis*, of new complex molecules and structures from smaller chemical units. For example, we are all familiar with the fact that carbohydrates eaten today may be deposited in the body as fat tomorrow (**Figure 1.5A**). Another kind of work cells do is mechanical work—for example, moving molecules from one cellular location to another, or even moving whole cells or tissues, as in the case of muscles (**Figure 1.5B**).

Living organisms control their internal environment

Life depends on thousands of biochemical reactions that occur inside cells. These reactions require materials to be moved into and out of cells in a controlled manner. Within the cell, those reactions are linked in that the products of one are the raw materials of the next. If this complex network of reactions is to be properly integrated, reaction rates within a cell must be precisely controlled. A large proportion of the activities of cells are directed toward the regulation of the multiple chemical reactions continuously in progress inside the cell.

Organisms that are made up of more than one cell have an *internal environment* that is not cellular. That is, their individual cells are bathed in extracellular fluids, from which they receive nutrients and into which they excrete wastes. The cells of multicellular organisms are specialized to contribute in some way to the maintenance of that internal environment. However, with the evolution of specialized

Biological information is contained in a genetic language common to all organisms

A cell's instructions—or "blueprints" for existence—are contained in its **genome**, which is the sum total of all the DNA molecules in the cell. **DNA** (deoxyribonucleic acid) molecules are long sequences of four different subunits called **nucleotides**. The sequence of the nucleotides contains genetic information. Specific segments of DNA called **genes** contain the information the cell uses to make proteins (**Figure 1.4**). **Proteins** make up much of an organism's structure and are the molecules that govern the chemical reactions within cells. By analogy with a book, the nucleotides of DNA are like the letters of an alphabet. Protein molecules are the sentences they spell. Combinations of proteins that constitute structures and control biochemical processes are the paragraphs. The structures and processes that are organized into different systems with specific tasks (such as digestion or transport) are the chapters of the book, and the complete book is the organism. Natural selection is the author and editor of all the books in the library of life.

If you were to write out your own genome using four letters to represent the four nucleotides, you would have to write a total of more than 3 billion letters. If you used the same size type as you are reading in this book, your genome would fill about a thousand books the size of this one.

All the cells of a multicellular organism contain the same genome, yet different cells have different functions and form different structures. Therefore, different types of cells in an organ-

(B)

(A)

1.5 Energy from Nutrients Can Be Stored or Used Immediately (A) The cells of this Arctic ground squirrel have broken down the complex carbohydrates in plants and converted their molecules into fats, which are stored in the animal's body to provide an energy supply for the cold months. (B) The cells of this kangaroo are breaking down food molecules and using the energy in their chemical bonds to do mechanical work—in this case, to jump.

functions, these cells lost many of the functions carried out by single-celled organisms, and therefore depend on the internal environment for essential services. The interdependence of the different kinds of cells in a multicellular organism can by expressed in the famous motto of the Three Musketeers: "One for all, and all for one!"

To accomplish their specialized tasks, assemblages of similar cells are organized into *tissues*. For example, a single muscle cell cannot generate much force, but when many of these cells combine to form the tissue of a working muscle, considerable force and movement can be generated (see Figure 1.5B). Different tissue types are organized to form *organs* that accomplish specific functions. Familiar organs include the heart, brain, and stomach. Organs whose functions are interrelated can be grouped into *organ systems*. The functions of cells, tissues, organs, and organ systems are all integral to the multicellular *organism* (**Figure 1.6**). The biology of organisms is the subject of Parts Seven and Eight of this book.

1.6 Biology Is Studied at Many Levels of Organization
Life's properties emerge when DNA and other molecules are organized in cells. Energy flows through all the biological levels shown here.

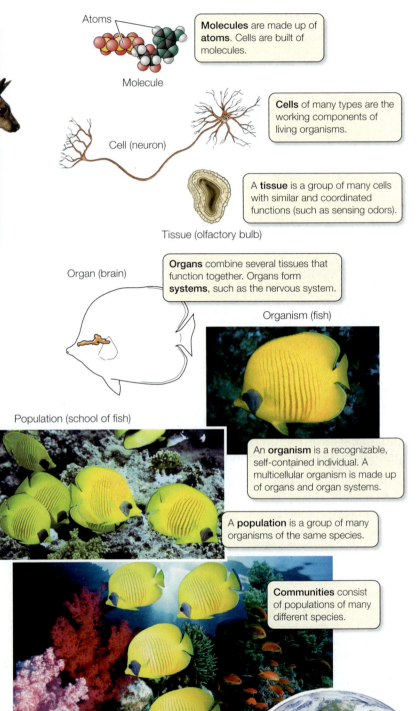

Atoms

Molecules are made up of **atoms**. Cells are built of molecules.

Molecule

Cells of many types are the working components of living organisms.

Cell (neuron)

A **tissue** is a group of many cells with similar and coordinated functions (such as sensing odors).

Tissue (olfactory bulb)

Organ (brain)

Organs combine several tissues that function together. Organs form **systems**, such as the nervous system.

Organism (fish)

An **organism** is a recognizable, self-contained individual. A multicellular organism is made up of organs and organ systems.

Population (school of fish)

A **population** is a group of many organisms of the same species.

Communities consist of populations of many different species.

Community (coral reef)

Biological communities in the same geographical location form **ecosystems**. Ecosystems exchange energy and create Earth's **biosphere**.

Biosphere

(A)

(B)

1.7 Conflict and Cooperation Organisms of the same species interact with one another in various ways. (A) Territorial elephant seal bulls defend stretches of beach from other males. The single male who controls a stretch of beach is able to mate with the many females (seen in the background) that live there. (B) Members of a meerkat colony are usually related to one another. Meerkats cooperate in many ways, such as watching out for predators and giving a warning bark if one appears.

As Figure 1.6 shows, individual organisms do not live in isolation, and the hierarchy of biology continues well beyond that level.

Living organisms interact with one another

Organisms interact with their external environments as well as the internal environment. Individual organisms are part of *populations* that interact among themselves and with populations of different organisms.

Organisms interact in many different ways. For example, some animals are *territorial* and will try to prevent other individuals of their species from exploiting the resource they are defending, whether it be food, nesting sites, or mates (**Figure 1.7A**). Animals may also *cooperate* with members of their species, forming social units such as a termite colony, a school of fish, or a meerkat colony (**Figure 1.7B**). Such interactions among individuals have resulted in the evolution of social behaviors such as communication.

The interaction of populations of many different species forms a *community*, and interactions between different species are a major evolutionary force. Adaptations that give an individual of one species an advantage in obtaining members of another species as food (and the converse, adaptations that lessen an individual's chances of becoming food) are paramount in evolutionary history. Organisms of different species may compete for the same resources, resulting in natural selection for specialized adaptations that allow some individuals to exploit those resources more efficiently than others can.

In any given geographic locality, the interacting communities form *ecosystems*. Organisms in the ecosystem can modify the environment in ways that affect other organisms. For example, in most terrestrial (land) environments, the dominant plants greatly modify the environmental conditions in which animals and other plants must live. The ways in which species interact with one another and with their environment is the subject of *ecology*, the topic of Part Nine of this book.

Discoveries in biology can be generalized

Because all life is related by descent from a common ancestor, shares a genetic code, and consists of similar building blocks—cells—knowledge gained from investigations of one type of organism can, with care, be generalized to other organisms. Therefore, biologists can use **model systems** for research, knowing that they can extend their findings to other organisms and to humans. For example, our basic understanding of the chemical reactions in cells came from research on bacteria, but is applicable to all cells, including those of humans. Similarly, the biochemistry of photosynthesis—the process by which plants use sunlight to produce biological molecules—was largely worked out from experiments on *Chlorella* (a type of pond scum; see Figure 8.12). We learned much of what we know about the genes that control plant development from work on a single species of plant (see Chapter 19). Knowledge about how animals develop has come from work on sea urchins, frogs, chickens, roundworms, and fruit flies. And recently, the discovery of a major gene controlling human skin color came from work on zebrafish. Being able to generalize from model systems is a powerful tool.

1.1 RECAP

Living organisms are made of cells, evolve by natural selection, contain genetic information, extract energy from their environment and use it to do biological work, control their internal environment, and interact with one another.

- Can you describe the relationship between evolution by natural selection and the genetic code? See pp. 5–7

- Do you understand why results of biological research on one species can be generalized to very different species? See p. 9

Now that we have an overview of the major features of life that will be explored in depth in this book, we can ask how and when life first emerged. In the next section we will describe the history of life from the earliest simple life forms to the complex and diverse organisms that inhabit our planet today.

1.2 How Is All Life on Earth Related?

What do biologists mean when they say that all organisms are *genetically related*? They mean that species on Earth share a *common ancestor*. If two species are similar, as dogs and wolves are, then they probably have a common ancestor in the fairly recent past. The common ancestor of two species that are more different—say, a dog and a deer—probably lived in the more distant past. And if two organisms are very different—such as a dog and a clam—then we must go back to the *very* distant past to find their common ancestor. How can we tell how far back in time the common ancestor of any two organisms lived? In other words, how do we discover the evolutionary relationships among organisms?

For many years, biologists investigated the history of life by studying the *fossil record*—the preserved remains of organisms that lived in the distant past (**Figure 1.8**). Geologists supplied knowledge about the ages of fossils and the nature of the environments in which they lived. Biologists then inferred the evolutionary relationships among living and fossil organisms by comparing their anatomical similarities and differences. The development of modern molecular methods for comparing genomes, described in Chapter 24, has enabled biologists to more accurately establish the degrees of relationship between living organisms and use that information to help us interpret the fossil record.

In general, the greater the differences between the genomes of two species, the more distant was their common ancestor. Using molecular techniques, biologists are exploring fundamental questions about the history of life on Earth. What were the earliest forms of life? How did those simple organisms give rise to the great diversity of organisms alive today? Can we reconstruct the family tree of all life?

Life arose from nonlife via chemical evolution

Geologists estimate that Earth is about 5 billion years old. For the first billion years, it was not a very hospitable place for life, and there was no life. If we picture the history of Earth as a 30-day calendar, life probably arose around the middle of the first week, or about 4 billion years ago (**Figure 1.9**).

When we consider how life might have arisen from nonliving matter, we must take into account the properties of the young Earth's atmosphere, oceans, and climate, all of which were very differ-

The presence of well-preserved ammonoid mollusks in a fossil bed often helps scientists date other fossils found there.

1.8 Fossils Give Us a View of Past Life The most prominent of the many fossilized organisms in this rock sample are ammonoids, an extinct mollusk group whose living relatives include squids and octopus. Ammonoids flourished between 200 million and 60 million years ago; this particular group of fossils is about 185 million years old.

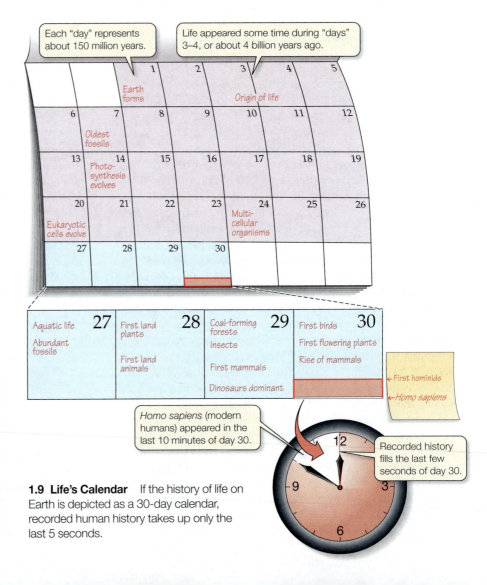

Each "day" represents about 150 million years.

Life appeared some time during "days" 3–4, or about 4 billion years ago.

1.9 Life's Calendar If the history of life on Earth is depicted as a 30-day calendar, recorded human history takes up only the last 5 seconds.

Homo sapiens (modern humans) appeared in the last 10 minutes of day 30.

Recorded history fills the last few seconds of day 30.

ent from today's. Biologists postulate that complex biological molecules first arose through the random physical association of chemicals in that environment. Experiments that simulate the conditions on early Earth have confirmed that the generation of complex molecules under such conditions is possible, even probable. The critical step, however, for the evolution of life had to be the appearance of molecules that could reproduce themselves and also serve as templates for the synthesis of large molecules with complex but stable shapes. The variation of the shapes of these large, stable molecules (described in Chapter 3) enabled them to participate in increasing numbers and kinds of interactions with other molecules: chemical reactions.

Biological evolution began when cells formed

The second critical step in the origin of life was the enclosure of complex biological molecules in *membranes*, which kept them close together and increased the frequency with which they interacted. Fat-like molecules were the critical ingredient: because these molecules are not soluble in water, they form membrane-like films. These films tend to form spherical *vesicles*, which could have enveloped assemblages of other biological molecules. Scientists postulate that about 3.8 billion years ago, this natural process of membrane formation resulted in the first cells with the ability to replicate themselves—an event that marked the beginning of biological evolution.

For 2 billion years after cells originated, all organisms consisted of only one cell. These first unicellular organisms were (and are, as multitudes of their descendants exist in similar form today) **prokaryotes**. Prokaryotic cell structure consists of DNA and other biochemicals enclosed in a membrane.

These early prokaryotes were confined to the oceans, where there was an abundance of complex molecules they could use as raw materials and sources of energy. The ocean shielded them from the damaging effects of ultraviolet light, which was intense at that time because there was no oxygen in the atmosphere, and hence no protective ozone layer.

Photosynthesis changed the course of evolution

The sum total of all the chemical reactions that go on inside a cell constitutes the cell's **metabolism**. To fuel their metabolism, the earliest prokaryotes took in molecules directly from their environment, breaking these small molecules down to release the energy contained in their chemical bonds. Many modern species of prokaryotes still function this way, and very successfully.

An extremely important step that would change the nature of life on Earth occurred about 2.5 billion years ago with the evolution of **photosynthesis**. The chemical reactions of photosynthesis (which are explained in Chapter 8) transform the energy of sunlight into a form of energy that can power the synthesis of large biological molecules. These large molecules become the building blocks of cells; they can also be broken down to provide metabolic energy. Because its energy-capturing processes provide food for other organisms, photosynthesis is the basis of much of life on Earth today.

Early photosynthetic cells were probably similar to present-day prokaryotes called *cyanobacteria* (**Figure 1.10**). Over time, photo-

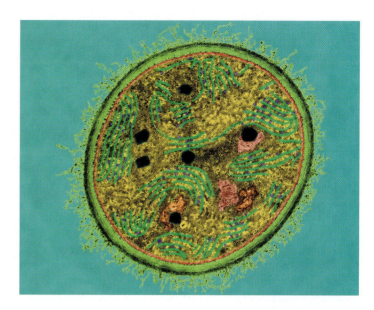

1.10 Photosynthetic Organisms Changed Earth's Atmosphere This modern cyanobacterium may be very similar to the early photosynthetic prokaryotes that introduced oxygen into Earth's atmosphere.

synthetic prokaryotes became so abundant that vast quantities of oxygen gas—O_2, which is a by-product of photosynthesis—began slowly to accumulate in the atmosphere. O_2 was poisonous to many of the prokaryotes that lived at that time. However, those organisms that tolerated O_2 were able to proliferate as the presence of oxygen opened up vast new avenues of evolution. Metabolism based on the use of O_2, called *aerobic metabolism*, is more efficient than the *anaerobic* (non-oxygen-using) *metabolism* that characterized earlier organisms. Aerobic metabolism allowed cells to grow larger, and today it is used by the majority of Earth's organisms.

Over millions of years, the vast quantities of oxygen released by photosynthesis formed a layer of ozone (O_3) in the upper atmosphere. As the ozone layer thickened, it intercepted more and more of the sun's deadly ultraviolet radiation. Only in the last 800 million years has the presence of a dense ozone layer allowed organisms to leave the protection of the ocean and live on land.

Eukaryotic cells evolved from prokaryotes

Another important step in the history of life was the evolution of cells with discrete intracellular compartments, called **organelles**, that were capable of taking on specialized cellular functions. This event happened about 3 weeks into our calendar of Earth's history (see Figure 1.9). One of these organelles, the *nucleus*, came to contain the cell's genetic information. The nucleus has the appearance of a dense kernel, giving these cells their name: **eukaryotes** (from the Greek *eu*, "true," and *karyon*, "kernel"), as distinguished from the cells of prokaryotes, which lack internal compartments (*pro*, "before").

Some organelles are hypothesized to have originated when cells ingested smaller cells (see Figure 4.26). For example, the organelle specialized to conduct photosynthesis, the *chloroplast*, could have originated as a photosynthetic prokaryote that was ingested by a larger eukaryote. If the larger cell failed to break down this intended food object, a partnership could have evolved in which the ingested prokaryote provided the products of photosynthesis and the host cell provided a good environment for its smaller partner.

Multicellularity arose and cells became specialized

Until slightly more than 1 billion years ago, all the organisms that existed—whether prokaryotic or eukaryotic—were unicellular. Yet another important evolutionary step occurred when some eukaryotes failed to separate after cell division, remaining attached to each other. The permanent association of cells made it possible for some cells to specialize in certain functions, such as reproduction, while other cells specialized in other functions, such as absorbing nutrients and distributing them to neighboring cells. This **cellular specialization** enabled multicellular eukaryotes to increase in size and become more efficient at gathering resources and adapting to specific environments.

Biologists can trace the evolutionary Tree of Life

If all the species of organisms on Earth today are the descendants of a single kind of unicellular organism that lived almost 4 billion years ago, how have they become so different? And why are there so many species?

As long as individuals within a population mate at random, structural and functional changes may evolve within that popula-

tion, but the population will remain one species. However, if some event isolates some members of a population from the others, structural and functional differences between them may accumulate over time. In short, the evolutionary paths of the two groups may diverge to the point where their members can no longer reproduce with each other. They have evolved into different species. This evolutionary process, called *speciation*, is detailed in Chapters 22 and 23.

Biologists give each species a distinct scientific name formed from two Latinized names (a *binomial*). The first name identifies the species' *genus*—a group of species that share a recent common ancestor. The second is the name of the species. For example, the scientific name of the human species is *Homo sapiens*: *Homo* is our genus and *sapiens* is our species. Scientific names usually refer to some characteristic of the species. *Homo* is derived from the Latin word for "man," and *sapiens* is derived from the Latin word for "wise" or "rational."

As many as 30 million species of organisms may exist on Earth today. Many times that number lived in the past but are now extinct. Many millions of speciation events created this vast diversity, and the unfolding of these events can be diagrammed as an evolutionary "tree" showing the order in which populations split and eventually evolved into new species. An evolutionary tree traces the descendants of ancestors that lived at different times in the past. The organisms on any one branch share a common ancestor at the base of that branch. The most closely related groups are placed together on the same branch; more distantly related organisms are on different branches. In this book, we adopt the convention that time flows from left to right, so the tree in Figure 1.11 (and other trees in this book) lies on its side, with its root—the ancestor of all life—at the left. Although many details remain to be clarified, the broad outlines of the Tree of Life have been determined. Its branching patterns are based on a rich array of evidence from fossils, structures, metabolic processes, behavior, and molecular analyses of genomes.

No fossils exist to help us determine the earliest divisions in the lineage of life because those unicellular organisms had no parts that could be preserved as fossils. However, molecular evidence has been used to separate all living organisms into three major **domains**: Archaea, Bacteria, and Eukarya (**Figure 1.11**). The organisms of each domain have been evolving separately from organisms in the other domains for more than a billion years.

Organisms in the domains **Archaea** and **Bacteria** are all prokaryotes. Archaea and Bacteria differ so fundamentally from each other in their metabolic processes that they are believed to have separated into distinct evolutionary lineages very early.

Members of the third domain—**Eukarya**— have eukaryotic cells. Three major groups of multicellular eukaryotes—plants, fungi, and animals—all evolved from unicellular *microbial eukaryotes*, more generally referred to as *protists*. The photosynthetic protist that gave rise to plants was completely distinct from the protist that was ancestral to both animals and fungi, as can be seen from the branching pattern of Figure 1.11.

Common ancestor of all organisms

BACTERIA

ARCHAEA

Archaea and Eukarya share a common ancestor not shared by the bacteria.

The eukaryotic cell probably evolved only once. Many different microbial eukaryote (protist) groups arose from this common ancestor.

Three major groups of multicellular eukaryotes evolved from different groups of microbial eukaryotes.

Plants

Fungi

EUKARYA

Animals

Ancient →→→→→ Present
Time

1.11 The Tree of Life The classification system used in this book divides Earth's organisms into three domains: Bacteria, Archaea, and Eukarya. The unlabeled blue branches within the Eukarya represent various groups of microbial eukaryotes, more commonly known as "protists."

Some bacteria, some archaea, some protists, and most plants are capable of photosynthesis. These organisms are called *autotrophs* ("self-feeders"). The biological molecules they produce are the primary food for nearly all other living organisms.

Fungi include molds, mushrooms, yeasts, and other similar organisms, all of which are *heterotrophs* ("other-feeders")—that is, they require a source of molecules synthesized by other organisms, which they then break down to obtain energy for their own metabolic processes. Fungi break down energy-rich food molecules in their environment and then absorb the breakdown products into their cells. Some fungi are important as decomposers of the waste products and dead bodies of other organisms.

Like fungi, animals are heterotrophs, but unlike fungi they ingest their food source, then break down the food in a digestive tract. Animals eat other forms of life, including plants, fungi, and other animals. Their cells absorb the breakdown products and obtain energy from them.

1.2 RECAP

The first cellular life on Earth was prokaryotic and arose about 4 billion years ago. The complexity of the organisms that exist today is the result of several important evolutionary events, including the evolution of photosynthesis, eukaryotic cells, and multicellularity. The genetic relationships of all organisms can be shown as a branching Tree of Life.

- Can you explain the evolutionary significance of photosynthesis? See p. 11

- What do the domains of life represent? What are the major groups of eukaryotes? See p. 12 and Figure 1.11

In February of 1676, Robert Hooke received a letter from the physicist Sir Isaac Newton. In this letter Newton famously remarked to Hooke, "If I have seen a little further, it is by standing on the shoulders of giants." We all stand on the shoulders of giants, building on the research of earlier scientists. By the end of this course, you will know more about evolution than Darwin ever could have, and you will know infinitely more about cells than Schleiden and Schwann did. Let's look at the methods biologists use to expand our knowledge of life.

1.3 How Do Biologists Investigate Life?

Biologists use many tools and methods in their research, but regardless of the methods they use, biologists take two basic approaches to their investigations of life: they observe and they conduct experiments.

Observation is an important skill

Biologists have always observed the world around them, but today their abilities to observe are greatly enhanced by many sophisticated technologies, such as electron microscopes, DNA chips,

magnetic resonance imaging, and global positioning satellites. Advances in technology have been responsible for most major advances in biology. For example, not too long ago it was extremely difficult and time-consuming to decipher the nucleotide sequence that makes up a single gene. New technologies enabled biologists to sequence the entire human genome in only 13 years (1990–2003). Scientists now use these methods routinely, sequencing the genomes of organisms (including organisms that cause serious diseases) in only days. We will explore some of these technologies and what we have learned from them in Part Four of this book.

Our ability to observe the distributions of organisms, such as fish in the world's oceans, has also improved dramatically. A short time ago, researchers could put physical tags on fish and then only hope that someday a fisherman would catch one and send back the tag, which would at least reveal where the fish ended up. Today, electronic recording devices attached to fish can continuously record not only where the fish is, but also how deep it swims at different times of day and the temperature and salinity of the water around it (**Figure 1.12**). At set intervals, these tags download their information to a satellite, which relays it back to researchers. Suddenly we are acquiring a great deal of new knowledge about the distribution of life in the oceans.

The scientific method combines observation and logic

Observations lead to questions, and scientists make additional observations and do experiments to answer those questions. The conceptual approach that underlies the design and conduct of most modern scientific investigations is called the **scientific method**. This

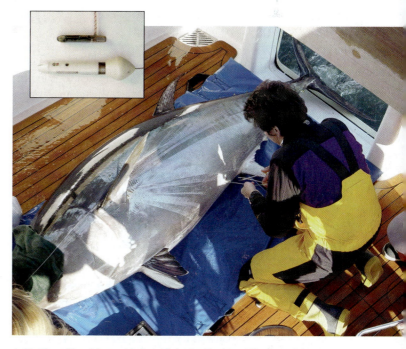

1.12 Tuna Tracking Marine biologist Barbara Block attaches a computerized data recording tag (inset) to a bluefin tuna. The use of such tags makes it possible to track individual tuna wherever they travel in the world's oceans.

powerful tool, also called the *hypothesis–prediction (H–P) method*, is a conceptual approach that provides a strong foundation for making advances in biological knowledge. The scientific method has five steps:

- Making *observations*
- Asking *questions*
- Forming *hypotheses*, or tentative answers to the questions
- Making *predictions* based on the hypotheses
- *Testing* the predictions by making additional observations or conducting experiments

Once a question has been posed, a scientist uses *inductive logic* to propose a tentative answer to the question. That tentative answer is called a **hypothesis**. For example, at the opening of this chapter, you learned that Pieter Johnson was shown abnormal frogs gathered in certain ponds. The first question stimulated by this observation was, is there something in these ponds that caused frogs to develop such extreme anatomical abnormalities?

In formulating a hypothesis, scientists put together the facts they already know to formulate one or more possible answers to the question. Pieter knew that there were likely to be contaminants in the ponds where deformed frogs were found because agricultural pesticides were used heavily in the region. In addition, mercury had once been mined nearby, and the abandoned mines could be a source of heavy metals in the water. He also knew that there were nearby ponds in which the frogs were normal. His first hypothesis, therefore, was that contaminants in the water caused mutations in the frog eggs.

The next step in the scientific method is to apply a different form of logic—*deductive logic*—to make predictions based on the hypothesis. Based on his hypothesis, Pieter predicted (1) that he would find contaminants in the ponds with the abnormal frogs, and (2) that eggs from those ponds would produce abnormal frogs when they were hatched in the laboratory.

Good experiments have the potential of falsifying hypotheses

Once predictions are made from a hypothesis, **experiments** can be designed to test those predictions. The most informative experiments are those that have the ability to show that the prediction is wrong. If the prediction is wrong, the hypothesis must be questioned, modified, or rejected.

Both of Pieter Johnson's initial predictions proved to be wrong. He counted frogs and other organisms in 35 ponds in the region where the deformed frogs had been found and measured chem-

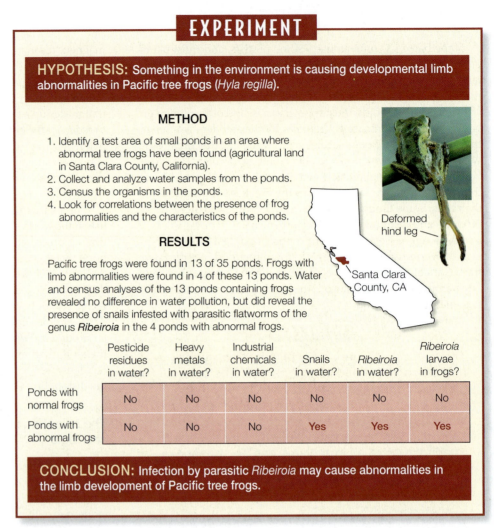

EXPERIMENT

HYPOTHESIS: Something in the environment is causing developmental limb abnormalities in Pacific tree frogs (*Hyla regilla*).

METHOD

1. Identify a test area of small ponds in an area where abnormal tree frogs have been found (agricultural land in Santa Clara County, California).
2. Collect and analyze water samples from the ponds.
3. Census the organisms in the ponds.
4. Look for correlations between the presence of frog abnormalities and the characteristics of the ponds.

Deformed hind leg

Santa Clara County, CA

RESULTS

Pacific tree frogs were found in 13 of 35 ponds. Frogs with limb abnormalities were found in 4 of these 13 ponds. Water and census analyses of the 13 ponds containing frogs revealed no difference in water pollution, but did reveal the presence of snails infested with parasitic flatworms of the genus *Ribeiroia* in the 4 ponds with abnormal frogs.

	Pesticide residues in water?	Heavy metals in water?	Industrial chemicals in water?	Snails in water?	*Ribeiroia* in water?	*Ribeiroia* larvae in frogs?
Ponds with normal frogs	No	No	No	No	No	No
Ponds with abnormal frogs	No	No	No	Yes	Yes	Yes

CONCLUSION: Infection by parasitic *Ribeiroia* may cause abnormalities in the limb development of Pacific tree frogs.

1.13 Comparative Experiments Look for Differences between Groups Pieter Johnson analyzed the differences between ponds in which deformed frogs were present versus nearby ponds in which there were no deformed frogs. Such comparisons can result in valuable insights.

icals in the water. Thirteen of the ponds were home to Pacific tree frogs, but he found deformed frogs in only four ponds. To Pieter's surprise, analysis of the water samples failed to reveal higher amounts of pesticides, industrial chemicals, or heavy metals in the ponds with deformed frogs. Also surprisingly, when he collected eggs from those ponds and hatched them in the laboratory, he always got normal frogs. The original hypothesis that contaminants caused mutations in the frog eggs had to be rejected. A new hypothesis had to be formulated and new experiments had to be conducted.

There are two general types of experiments and Pieter used both:

- In a **comparative experiment**, we predict that there will be a difference between samples or groups based on our hypothesis. We then test whether or not the predicted difference exists.

- In a **controlled experiment**, we also compare samples or groups, but in this case we start the experiment with groups that are as similar as possible. We predict on the basis of our hypothesis that some factor, or *variable*, plays a role in the phenomenon we are investigating. We then use some method to manipulate that variable in an "experimental" group while leaving the "control" group unaltered. We then test to see if the manipulation created the predicted difference between the experimental and control groups.

COMPARATIVE EXPERIMENTS Comparative experiments are valuable when we do not know or cannot control the critical variables. Pieter Johnson performed a comparative experiment when he tested the water in the ponds (**Figure 1.13**). His challenge was to find some variable that differed between the ponds with normal and abnormal frogs. Finding no differences in the water chemistry of the two types of ponds, he had to reject his hypothesis that environmental contaminants were causing mutations in the frogs. So he compared the two types of ponds to see what variables *were* different between them.

Pieter found that a species of freshwater snail was present in the ponds with abnormal frogs, but absent from the ponds with normal ones. Freshwater snails are hosts for many parasites. His new hypothesis was that a parasite infecting the snail was in some way responsible for the frogs' deformities. To test that hypothesis, he performed controlled experiments.

CONTROLLED EXPERIMENTS In controlled experiments, one variable is manipulated while others are held constant. The variable that is manipulated is called the *independent variable* and the response that is measured is the *dependent variable*. A good controlled experiment is not easy to design because biological variables are so interrelated that it is difficult to alter just one.

Many parasites go through complex life cycles with several stages, each of which requires a specific host animal. Pieter focused on the possibility that some parasite that used freshwater snails as one of its hosts was infecting the frogs and causing their deformities. Pieter found a candidate parasite with this type of life cycle: a small flatworm called *Ribeiroia*, which was present in the ponds where the deformed frogs were found.

In Pieter's controlled experiment, the independent variable was the presence or absence of the snail and the parasite (**Figure 1.14**). He controlled all other variables by collecting frog eggs from ponds where there were no snails or flatworm parasites and hatching them in the laboratory. He divided the resulting tadpoles into two groups and placed them in separate tanks. He introduced snails and parasites into half of the tanks (the experimental group) and left the other tanks (the control group) free of snails and parasites. His dependent variable was the frequency of abnormalities in the frogs that developed under the different sets of conditions. He found that 85 percent of the frogs in the experimental tanks with *Ribeiroia* alone, but none of the frogs in the other tanks, developed abnormalities. Thus Pieter's results supported his hypothesis, and he could go on to investigate how the parasites caused abnormalities in the developing frogs.

Ribeiroia uses three hosts in California ponds: snails, frogs, and predatory birds such as herons. For the parasite to complete its life cycle and reproduce, it must be able to move from a frog to a bird. The limb deformities *Ribeiroia* causes may actually make infected frogs easier for predatory birds to capture and eat.

EXPERIMENT

HYPOTHESIS: Infection of Pacific tree frog tadpoles by the parasite *Ribeiroia* causes developmental limb abnormalities.

METHOD

1. Collect *Hyla regilla* eggs from a site with no record of abnormal frogs.
2. Allow eggs to hatch in laboratory aquaria. Randomly divide equal numbers of the resulting tadpoles into control and experimental groups.
3. Allow the control group to develop normally. Subject the experimental groups to infection with *Ribeiroia*, a different parasite (*Alaria*), and a combination of both parasites.
4. Follow tadpole development. Count and assess the resulting adult frogs.

Control (no parasites) Experiment 1 (with *Alaria*) Experiment 2 (with *Ribeiroia*) Experiment 3 (with *Alaria* and *Ribeiroia*)

RESULTS

- Survivorship (percent of tadpoles reaching adulthood)
- Abnormality rate (percent of adults with limb abnormalities)

CONCLUSION: *Ribeiroia* causes developmental limb abnormalities in Pacific tree frogs.

1.14 Controlled Experiments Manipulate a Variable The variable Johnson manipulated was the presence or absence of two species of parasitic flatworm. Other conditions of the experiment remained constant.

Statistical methods are essential scientific tools

Whether we are doing comparative or controlled experiments, at the end we have to decide whether there is a difference between the samples, individuals, groups, or populations in the study. How do we decide whether a measured difference is enough to support or falsify a hypothesis? In other words, how do we decide in an unbiased, objective way that the measured difference is significant?

Significance can be measured with statistical methods. Scientists use statistics because they recognize that variation is ubiquitous. Statistical tests analyze that variation and calculate the probability that the differences observed could be due to random variation. The results of statistical tests are therefore probabilities. A statistical test starts with a **null hypothesis**—the premise that no difference exists. When quantified observations, or **data**, are collected, statistical methods are applied to those data to calculate the likelihood that the null hypothesis is correct.

More specifically, statistical methods tell us the probability of obtaining the same results by chance even if the null hypothesis were true. Put another way, we need to eliminate insofar as possible the chance that any differences showing up in the data are merely the result of random variation in the samples tested. Scientists generally conclude that the differences they measure are significant if the statistical tests show that the *probability of error* (the probability that the results can be explained by chance) is 5 percent or lower. In particularly critical experiments, such as tests of the safety of a new drug, scientists require much lower probabilities of error, such as 1 percent or even 0.1 percent.

Not all forms of inquiry are scientific

Science is a unique human endeavor that is bounded by certain standards of practice. Other areas of scholarship share with science the practice of making observations and asking questions, but scientists are distinguished by what they do with their observations and how they answer their questions. Data, subjected to appropriate statistical analysis, are critical in the testing of hypotheses. The scientific method is the most powerful way humans have devised for learning about the world and how it works. Scientific explanations for natural processes are objective and reliable because the hypotheses proposed *must be testable* and *must have the potential of being rejected* by direct observations and experiments. Scientists clearly describe the methods they have used to test hypotheses so that other scientists can repeat their observations or experiments. Not all experiments are repeated, but surprising or controversial results are always subjected to independent verification. All scientists worldwide share this built-in process of testing and rejecting hypotheses, so they all contribute to a common body of scientific knowledge.

If you understand the methods of science, you can distinguish science from non-science. Art, music, and literature are activities that contribute to the quality of human life, but they are not science. They do not use the scientific method to establish what is fact. Religion is not science, although religions have historically purported to explain natural events ranging from unusual weather patterns to crop failures to human diseases and mental afflictions. Many such phenomena that at one time were mysterious are now explicable in terms of scientific principles.

The power of science derives from the uncompromising objectivity and absolute dependence on evidence that comes from *reproducible and quantifiable observations*. A religious or spiritual explanation of a natural phenomenon may be coherent and satisfying for the person or group holding that view, but it is not testable, and therefore it is not science. To invoke a supernatural explanation (such as an "intelligent designer" with no known bounds) is to depart from the world of science.

Science describes the facts about how the world works, not how it "ought to be." Many of the recent scientific advances that have contributed so much to human welfare also raise major ethical issues. Developments in genetics and developmental biology, for example, now enable us to select the sex of our children, to use stem cells to repair our bodies, and to modify the human genome. Although scientific knowledge allows us to do these things, science cannot tell us whether or not we should do them, or if we choose to do so, how we should regulate them.

Making wise decisions about such issues requires a clear understanding of the implications of available scientific information. Success in surgery depends on an accurate diagnosis. So does success in environmental management. However, to make wise decisions about public policy, we also need to employ the best possible ethical reasoning in deciding which outcomes we should strive for. For a bright future, society needs both good science and good ethics, as well as an educated public that understands the importance of both and the critical differences between them.

1.3 RECAP

The scientific method of inquiry starts with the formulation of hypotheses based on observations and data. Comparative and controlled experiments are carried out to test hypotheses.

- Can you explain the relationship between a hypothesis and an experiment? See p. 14

- What features characterize questions that can be answered only by using a comparative approach? See p. 14 and Figure 1.13

- What is controlled in a controlled experiment? See p. 15 and Figure 1.14

- Do you understand why arguments must be supported by quantifiable and reproducible data in order to be considered scientific? See p. 16

The vast amount of scientific knowledge accumulated over centuries of human civilization allows us to understand and manipulate aspects of the natural world in ways that no other species can. These abilities present us with challenges, opportunities, and, above all, responsibilities. Let's look at how knowledge of biology can affect the formulation of public policy.

1.4 How Does Biology Influence Public Policy?

The study of biology has long had major implications for human life. Agriculture and medicine are two important human activities that depend on biological knowledge. Our ancestors unknowingly applied the principles of evolutionary biology when they domesticated plants and animals. People have also been speculating about the causes of diseases and searching for methods of combating them since ancient times. Long before the causes of diseases were known, people recognized that diseases could be passed from one person to another. Isolation of infected persons has been practiced as long as written records have been available, but most so-called cures were not effective until scientists found out what caused diseases.

Today, thanks to the deciphering of genomes and the ability to manipulate them, vast new possibilities exist for improvements in the control of human diseases and agricultural productivity. At the same time, these capabilities have raised important ethical and policy issues. How much and in what ways should we tinker with the genetics of humans and other species? Does it matter whether our crops and domesticated animals are changed by traditional breeding experiments or by gene transfers? What rules should govern the release of genetically modified organisms into the environment? Science alone cannot provide answers to those questions, but wise policy decisions must be based on accurate scientific information.

Another reason for studying biology is to understand the effects of the vastly increased human population on its environment. Our use of natural resources is putting stress on the ability of Earth's ecosystems to continue to produce the goods and services on which our society depends. Human activities are changing global climates, causing the extinctions of a large number of species, and spreading new diseases while facilitating the resurgence of old ones. The rapid spread of the SARS and West Nile viruses, for example, was facilitated by modern modes of transportation, and the recent resurgence of tuberculosis is the result of the evolution of bacteria that are resistant to antibiotics. Biological knowledge is vital for determining the causes of these changes and for devising wise policies to deal with them. An understanding of biology also helps people appreciate the marvelous diversity of living organisms that provides goods and services for humankind and also enriches our lives aesthetically and spiritually.

Biologists are increasingly called on to advise government agencies concerning the laws, rules, and regulations by which society deals with the increasing number of problems and challenges that have at least a partial biological basis. As an example of the value of scientific knowledge for the assessment and formulation of public policy, let's return to the tracking study of bluefin tuna introduced in Section 1.3. Prior to this study, both scientists and fishermen knew that bluefins had a western Atlantic breeding ground in the Gulf of Mexico and an eastern Atlantic breeding ground in the Mediterranean Sea. Overfishing was endangering the western breeding population. Everyone assumed that the fish from the two breeding populations had geographically separate feeding grounds as well as separate breeding grounds, so an international commission drew a line down the middle of the Atlantic Ocean and established stricter fishing quotas on the western side of the line. The intent was to allow the western population to recover. However, new data revealed that in fact the eastern and western bluefin populations mix freely on the feeding (and hence fishing) grounds across the entire North Atlantic (**Figure 1.15**). Thus a fish caught on the eastern side of the line could be from the western breeding population, so the established policy was not appropriate for achieving its intended goal.

Throughout this book we will share with you the excitement of studying living things and illustrate the rich array of methods that biologists use to determine why the world of living things looks and functions as it does. The most important motivator of most biologists is curiosity. People are fascinated by the richness and diversity of life and want to learn more about organisms and how they interact with one another. The trait of human curiosity

1.15 Bluefin Tuna Do Not Recognize the Lines Drawn on Maps by International Commissions   Because it was assumed that western (red dots) and eastern (gold dots) breeding populations of bluefin tuna also fed on their respective sides of the Atlantic Ocean, separate fishing quotas were established to either side of 45°W longitude (dashed line). It was believed this would allow the endangered western population to recover. However, tracking data showed that the two populations mix freely, especially in the heavily fished waters of the northernmost Atlantic (blue circle); so in fact the established policy does not protect the western population.

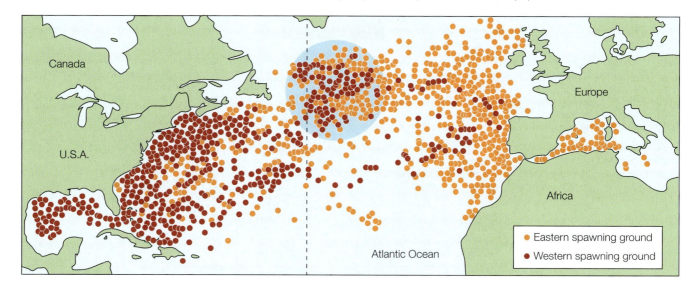

might even be seen as adaptive, and could have been selected for if individuals who were motivated to learn about their surroundings were likely to have survived and reproduced better, on average, than their less curious relatives!

There are vast numbers of questions for which we do not yet have answers, and new discoveries usually engender questions no one thought to ask before. Perhaps you will eventually pose and answer one or more of those questions.

CHAPTER SUMMARY

1.1 What is biology?

Biology is the study of life at all levels of organization, ranging from molecules to the biosphere.

The **cell theory** states that all life consists of cells, and all cells come from preexisting cells.

All living organisms are related to one another through descent with modification. **Evolution** by **natural selection** is responsible for the diversity of **adaptations** found in living organisms.

The instructions for a cell are contained in its **genome**, which consists of DNA molecules made up of sequences of **nucleotides**. Specific segments of DNA called **genes** contain the information the cell uses to make **proteins**. Review Figure 1.4

Cells are the basic structural and physiological units of life. Most of the chemical reactions of life take place in cells. Living organisms control their internal environment. They also interact with other organisms of the same and different species. Biologists study life at all these levels of organization. Review Figure 1.6, Web/CD Activity 1.1

Biological knowledge obtained from a **model system** may be generalized to other species.

1.2 How is all life on Earth related?

Biologists use fossils, anatomical similarities and differences, and molecular comparisons of genomes to reconstruct the history of life. Review Figure 1.9

Life first arose by chemical evolution. Biological evolution began with the formation of cells.

Photosynthesis was an important evolutionary step because it changed Earth's atmosphere and provided a means of capturing energy from sunlight.

The earliest organisms were **prokaryotes**; organisms with more complex cells, called **eukaryotes**, arose later. Eukaryotic cells have discrete intracellular compartments, called **organelles**, including a **nucleus** that contains the cell's genetic material.

The genetic relationships of **species** can be represented as an evolutionary tree. Species are grouped into three **domains**: **Archaea**, **Bacteria**, and **Eukarya**. The domains Archaea and Bacteria consist of unicellular prokaryotes. The domain Eukarya contains the microbial eukaryotes (protists), plants, fungi, and animals. Review Figure 1.11, Web/CD Activity 1.2

1.3 How do biologists investigate life?

The **scientific method** used in most biological investigations involves five steps: making observations, asking questions, forming hypotheses, making predictions, and testing those predictions.

Hypotheses are tentative answers to questions. Predictions made on the basis of a hypothesis are tested with additional observations and two kinds of **experiments**: **comparative** and **controlled experiments**. Review Figures 1.13 and 1.14

Statistical methods are applied to **data** to establish whether or not the differences observed are significant or whether they could be expected by chance. These methods start with the **null hypothesis** that there are no differences.

Science can tell us how the world works, but it cannot tell us what we should or should not do.

1.4 How does biology influence public policy?

Wise public policy decisions must be based on accurate scientific information. Biologists are often called on to advise governmental agencies on the solution of important problems that have a biological component.

FOR DISCUSSION

1. Even if we knew the sequences of all of the genes of a single-celled organism and could cause those genes to be expressed in a test tube, we still could not create one of those organisms in the test tube. Why do you think this is so? In light of this fact, what do you think of the statement that the genome contains all of the information for a species?

2. If someone told you that giraffes developed long necks because they stretched their necks to reach leaves higher and higher on trees, how would you help that person think about giraffes more accurately, in terms of evolution by natural selection?

3. In a recent discovery of the genes that control skin color in zebrafish, why did the biologists assume that the same genes might be responsible for skin color in humans?

4. Why is it so important in science that we design and perform tests capable of falsifying a hypothesis?

5. What features characterize questions that can be answered only by using a comparative approach?

FOR INVESTIGATION

1. The abnormalities of frogs in Pieter Johnsons's study were associated with the presence of a parasite. How would you investigate how the parasites induced the formation of monster frogs? Hint: When tadpoles were exposed to the parasites *after* they began to develop legs, they did not show abnormalities.

2. Just as all cells come from preexisting cells, mitochondria—the cell organelles that convert energy in food to a form of energy that can do biological work—all come from preexisting mitochondria. Cells do not synthesize mitochondria from the genetic information in their nuclei. What investigations would you carry out to understand the nature of mitochondria?

Where there is water, there can be life

On July 14, 2005, the spacecraft *Cassini*, launched 8 years earlier by NASA, flew 168 kilometers above the south pole of Enceladus, one of Saturn's moons. As it approached this small, frigid body, *Cassini*'s instruments relayed information on chemicals in the atmosphere while its camera snapped photos. Back on Earth, scientists looking at this data were in for a big surprise. Photographs showed an immense plume of water vapor, ice particles, and liquid water that spewed from the moon like the geysers of Earth's Yellowstone Park. Since the temperature on Enceladus is about −200°C, surface water would freeze solid; thus the source of the plume must have been a pool of liquid water below the surface, probably heated by molten rocks.

Meanwhile, a little closer to home, two robotic vehicles were looking for water on Mars. One of them landed on what photographs indicated was a huge dry lake bed. With instructions from scientists on Earth, the vehicle dug up and analyzed Martian soil samples, finding a mineral called dry hematite that is often a chemical signature of the presence of water. In addition, there was salt residue, suggesting that water had evaporated. These discoveries by geologists sparked the interest of biologists, because where there is water, there can be life.

Put another way, there is good reason to believe that life as we know it cannot exist without water. Animals and plants could not survive on Earth's land masses until adaptations evolved that allowed them to retain the water that makes up some 70 percent of their bodies. Aquatic organisms, of course, do not need such water-retention mechanisms, which leads biologists to conclude that life originated in a watery environment. That environment need not have been the lakes, rivers, and oceans with which we are familiar. Living organisms have been found in hot springs at temperatures above the usual boiling point of water, in a lake beneath the frozen Antarctic ice, in water trapped 2 miles below Earth's surface, in water 3 miles below the surface of the sea, in extremely acidic and extremely salty water, and even in the water that cools the interiors of nuclear reactors.

Biologists' attention focuses not only on the presence of water, but on what is dissolved in that water. A major discovery of biology is that living things are composed of the same chemical elements as the vast nonliving por-

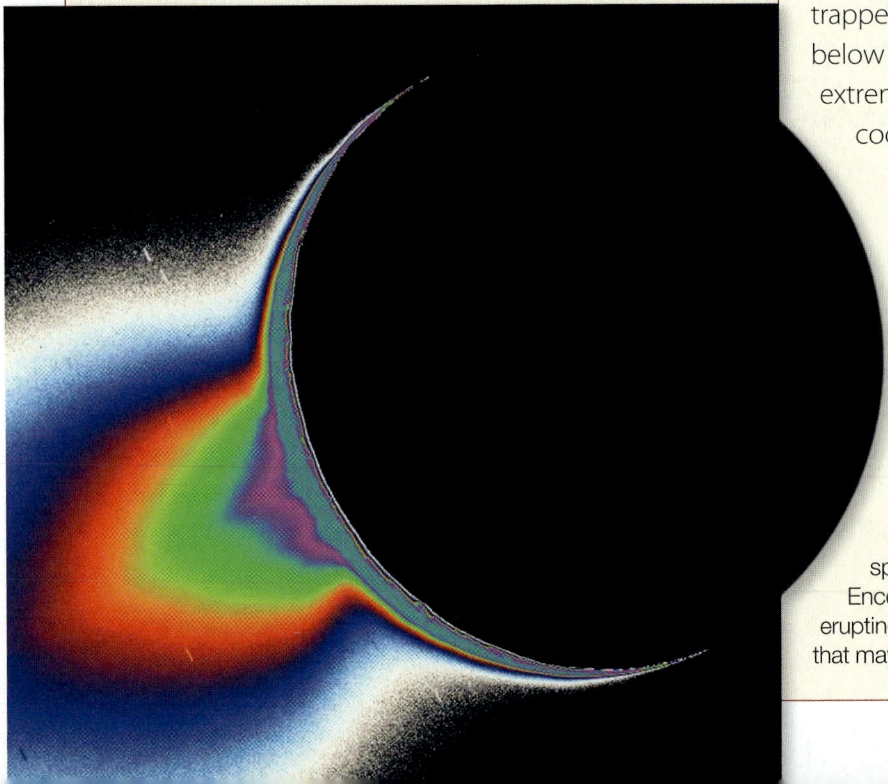

Geysers on a Frozen Moon Images from the spacecraft *Cassini*, enhanced with computer-generated color, show huge plumes of water and water vapor being sprayed from the south polar surface of Saturn's moon Enceladus. Scientists believe that these jets are geysers erupting from pools of liquid water, heated by volcanic activity, that may lie just below the frozen surface of this moon.

Looking for Water The robotic rover vehicle *Opportunity*, shown here in a NASA mock-up, was sent to Mars to look for evidence of water. The rover's instruments found salt residue and other evidence of evaporated water in the rock and sand of a region that appears to have once been at the bottom of a huge lake.

tion of the universe. This *mechanistic* view—that life is chemically based and obeys universal laws of chemistry and physics—is relatively new in human history. Until the nineteenth century, a "vital force" (from the Latin *vitalis*, "of life"), distinct from the mechanistic forces governing physics and chemistry, was presumed to be responsible for life. Many people still assume that a vital force exists, but the mechanistic view of life has led to great advances in biological science and is the cornerstone of modern medicine and agriculture. The search for life on Mars and the description of life here on Earth both begin with chemistry.

IN THIS CHAPTER we will introduce the constituents of matter: atoms. We will examine their variety, their properties, and their capacity to combine with other atoms. Then we will consider how matter changes. In addition to changes in state (solid to liquid to gas), substances undergo chemical reactions that transform both their composition and their characteristic properties. Finally, we will take a closer look at the structure and properties of water and its relationship to chemical acids and bases.

2.1 What Are the Chemical Elements That Make Up Living Organisms?

All matter is composed of **atoms**. Atoms are tiny—more than a trillion (10^{12}) of them could fit on top of the period at the end of this sentence. Each atom consists of a dense, positively charged **nucleus**, around which one or more negatively charged **electrons** move (**Figure 2.1**). The nucleus contains one or more **protons** and may contain one or more **neutrons**. Atoms and their component particles have volume and mass, which are properties of all matter. *Mass* measures the quantity of matter present; the greater the mass, the greater the quantity of matter.

The mass of a proton serves as a standard unit of measure called the **atomic mass unit** (**amu**), or *dalton* (named after the English chemist John Dalton). A single proton or neutron has a mass of about 1 dalton (Da), which is 1.7×10^{-24} grams (0.0000000000000000000000017 g). The mass of an electron is 9×10^{-28} g (0.0005 Da). Because the mass of an electron is negligible compared with the mass of a proton or a neutron, the contribution of electrons to the mass of an atom can usually be ignored when measurements and calculations are made. It is electrons, however, that determine how atoms will interact in chemical reactions, and we will discuss them extensively later in this chapter.

Each proton has a positive electric charge, defined as +1 unit of charge. An electron has a negative charge equal and opposite to that of a proton; thus the charge of an electron is –1 unit. The neutron, as its name suggests, is electrically neutral, so its charge is 0. Charges that are not alike (+/–) attract each other, whereas charges that are alike (+/+, –1–1) repel each other. Atoms are electrically neutral because the number of electrons in an atom equals the number of protons.

An element consists of only one kind of atom

An **element** is a pure substance that contains only one kind of atom. The element hydrogen consists only of hydrogen atoms; the element iron consists only of iron atoms. The atoms of each element have certain characteristics or properties that distinguish them from the atoms of other elements. The more than 100 elements found in the universe are arranged in the *periodic table* (**Figure 2.2**). These elements are not found in equal amounts. Stars have abundant hydrogen and helium. Earth's crust, and the surfaces of the neighboring planets, are almost

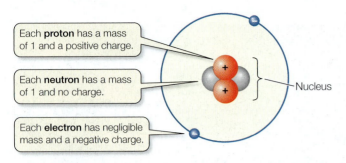

Each **proton** has a mass of 1 and a positive charge.

Each **neutron** has a mass of 1 and no charge.

Each **electron** has negligible mass and a negative charge.

Nucleus

2.1 The Helium Atom This representation of a helium atom is called a Bohr model. It exaggerates the space occupied by the nucleus. In reality, although the nucleus accounts for virtually all of the atomic mass, it occupies only about 1/10,000 of the atom's volume.

half oxygen, 28 percent silicon, 8 percent aluminum, and 2–5 percent each of sodium, magnesium, potassium, calcium, and iron; they contain much smaller amounts of the other elements.

About 98 percent of the mass of every living organism (bacterium, turnip, or human) is composed of just six elements: carbon, hydrogen, nitrogen, oxygen, phosphorus, and sulfur. The chemistry of these six elements will be our primary concern, but

other elements found in living things are important as well. Sodium and potassium, for example, are essential for nerve function; calcium can act as a biological signal; iodine is a component of a vital hormone; and magnesium and molybdenum are essential to plants (magnesium as part of their chlorophyll pigment, and molybdenum for incorporating nitrogen into biologically useful substances).

Protons: Their number identifies an element

An element differs from other elements by the number of protons in each of its atoms. This **atomic number** is unique to each element and does not change. An atom of helium always has 2 protons, and an atom of oxygen always has 8 protons; the atomic numbers of helium and oxygen are thus 2 and 8, respectively.

Atomic number (number of protons)

2

He

Chemical symbol (for helium)

4.003

Atomic mass (number of protons plus number of neutrons averaged over all isotopes)

2.2 The Periodic Table The periodic table groups the elements according to their physical and chemical properties. Elements 1–92 occur in nature; elements with atomic numbers above 92 were created in the laboratory.

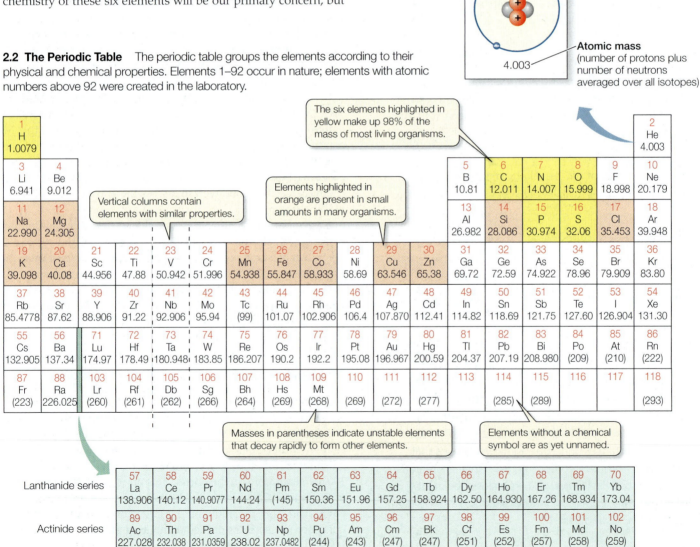

The six elements highlighted in yellow make up 98% of the mass of most living organisms.

Vertical columns contain elements with similar properties.

Elements highlighted in orange are present in small amounts in many organisms.

Masses in parentheses indicate unstable elements that decay rapidly to form other elements.

Elements without a chemical symbol are as yet unnamed.

1 H 1.0079																	2 He 4.003
3 Li 6.941	4 Be 9.012											5 B 10.81	6 C 12.011	7 N 14.007	8 O 15.999	9 F 18.998	10 Ne 20.179
11 Na 22.990	12 Mg 24.305											13 Al 26.982	14 Si 28.086	15 P 30.974	16 S 32.06	17 Cl 35.453	18 Ar 39.948
19 K 39.098	20 Ca 40.08	21 Sc 44.956	22 Ti 47.88	23 V 50.942	24 Cr 51.996	25 Mn 54.938	26 Fe 55.847	27 Co 58.933	28 Ni 58.69	29 Cu 63.546	30 Zn 65.38	31 Ga 69.72	32 Ge 72.59	33 As 74.922	34 Se 78.96	35 Br 79.909	36 Kr 83.80
37 Rb 85.4778	38 Sr 87.62	39 Y 88.906	40 Zr 91.22	41 Nb 92.906	42 Mo 95.94	43 Tc (99)	44 Ru 101.07	45 Rh 102.906	46 Pd 106.4	47 Ag 107.870	48 Cd 112.41	49 In 114.82	50 Sn 118.69	51 Sb 121.75	52 Te 127.60	53 I 126.904	54 Xe 131.30
55 Cs 132.905	56 Ba 137.34	71 Lu 174.97	72 Hf 178.49	73 Ta 180.948	74 W 183.85	75 Re 186.207	76 Os 190.2	77 Ir 192.2	78 Pt 195.08	79 Au 196.967	80 Hg 200.59	81 Tl 204.37	82 Pb 207.19	83 Bi 208.980	84 Po (209)	85 At (210)	86 Rn (222)
87 Fr (223)	88 Ra 226.025	103 Lr (260)	104 Rf (261)	105 Db (262)	106 Sg (266)	107 Bh (264)	108 Hs (269)	109 Mt (268)	110 (269)	111 (272)	112 (277)	113	114 (285)	115 (289)	116	117	118 (293)

Lanthanide series

57 La 138.906	58 Ce 140.12	59 Pr 140.9077	60 Nd 144.24	61 Pm (145)	62 Sm 150.36	63 Eu 151.96	64 Gd 157.25	65 Tb 158.924	66 Dy 162.50	67 Ho 164.930	68 Er 167.26	69 Tm 168.934	70 Yb 173.04

Actinide series

89 Ac 227.028	90 Th 232.038	91 Pa 231.0359	92 U 238.02	93 Np 237.0482	94 Pu (244)	95 Am (243)	96 Cm (247)	97 Bk (247)	98 Cf (251)	99 Es (252)	100 Fm (257)	101 Md (258)	102 No (259)

Along with a definitive number of protons, every element except hydrogen has one or more neutrons in its nucleus. The **mass number** of an atom is the total number of protons and neutrons in its nucleus. The nucleus of a carbon atom contains 6 protons and 6 neutrons, and has a mass number of 12. Oxygen has 8 protons and 8 neutrons, and has a mass number of 16. The mass number is essentially the mass of the atom in daltons.

Each element has its own one- or two-letter chemical symbol. For example, H stands for hydrogen, C for carbon, and O for oxygen. Some symbols come from other languages: Fe (from the Latin, *ferrum*) stands for iron, Na (Latin, *natrium*) for sodium, and W (German, *wolfram*) for tungsten.

In text, immediately preceding the symbol for an element, the atomic number is written at the lower left and the mass number at the upper left. Thus hydrogen, carbon, and oxygen are written as $_1^1H$, $_6^{12}C$, and $_8^{16}O$, respectively.

Neutrons: Their number differs among isotopes

In some elements, the number of neutrons in the atomic nucleus is not constant. **Isotopes** of the same element all have the same, definitive, number of protons, but differ in their number of neutrons. Many elements have several isotopes. The isotopes of hydrogen shown in **Figure 2.3** have special names, but the isotopes of most elements do not have distinct names. The natural isotopes of carbon, for example, are ^{12}C (6 neutrons in the nucleus), ^{13}C (7 neutrons), and ^{14}C (8 neutrons). Note that all three (pronounced "carbon-12," "carbon-13," and "carbon-14") have 6 protons, so they are all carbon. Most carbon atoms are ^{12}C, about 1.1 percent are ^{13}C, and a tiny fraction are ^{14}C. An element's atomic mass, or **atomic weight**, is the average of the mass numbers of a representative sample of atoms of the element, with all isotopes in their normally occurring proportions. The atomic weight of carbon, taking into account all of its isotopes and their abundances, is thus calculated to be 12.011.

Most isotopes are stable. But some, called **radioisotopes**, are unstable and spontaneously give off energy in the form of α (alpha), β (beta), or γ (gamma) radiation from the atomic nucleus.

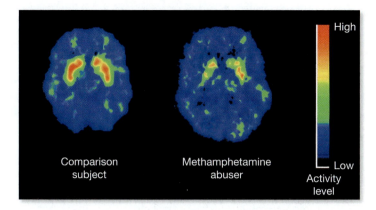

2.4 Tagging the Brain A radioactively labeled sugar detects differences between the brain activity of a healthy person and that of a person abusing methamphetamines. The more active a brain region is, the more sugar it takes up. The healthy brain (left) shows more activity in the region involved in memory (the red area) than the drug abuser's brain does.

Known as *radioactive decay*, this release of energy transforms the original atom. These transformations can extend even to a change in the number of protons, so that the original atom is now a different element. The energy released by radioactive decay can interact with surrounding substances. Scientists can incorporate such radiation-sensitive substances into instruments that allow them to detect the presence of radioisotopes. For instance, if an earthworm is given food to which a radioisotope has been added, its path through the soil can be followed by a simple detector called a Geiger counter. If a radioisotope is incorporated into a molecule, it acts as a tag or label, allowing researchers and physicians to trace that molecule and to identify any changes the molecule undergoes inside the body (**Figure 2.4**). Radioisotopes are also used to date fossils, an application described in Section 21.1.

Although radioisotopes are useful in research and in medicine, even a low dose of the radiation they emit has the potential to damage molecules and cells. However, these damaging effects are sometimes used to our advantage; for example, the γ radiation from ^{60}Co (cobalt-60) is used in medical practice to kill cancer cells.

Electrons: Their behavior determines chemical bonding

The characteristic number of electrons in each atom of an element determines how its atoms will react with other atoms. Biologists are interested in how chemical changes take place in living cells. When considering atoms, they are concerned primarily with electrons because the behavior of electrons explains how chemical *reactions* occur. These reactions often cause the atomic composition of substances to be altered. Reactions usually involve changes in the distribution of electrons between atoms.

The location of a given electron in an atom at any given time is impossible to determine. We can only describe a volume of space within the atom where the electron is likely to be. The region of space where the electron is found at least 90 percent of the time is the electron's **orbital**. Orbitals have characteristic shapes and ori-

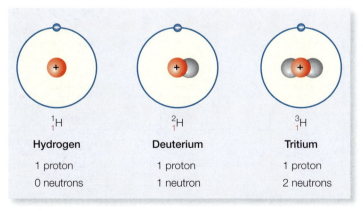

$_1^1H$	$_1^2H$	$_1^3H$
Hydrogen	**Deuterium**	**Tritium**
1 proton	1 proton	1 proton
0 neutrons	1 neutron	2 neutrons

2.3 Isotopes Have Different Numbers of Neutrons The isotopes of hydrogen all have one proton in the nucleus, which defines them as hydrogen. Their differing mass numbers are due to different numbers of neutrons.

entations, and a given orbital can be occupied by a maximum of two electrons. Thus any atom larger than helium (atomic number 2) must have electrons in two or more orbitals. The orbitals are filled in a specific sequence, in a series of what are known as **electron shells**, or *energy levels*, around the nucleus (**Figure 2.5**).

■ *First shell*: The innermost electron shell consists of just one orbital, called an *s* orbital. A hydrogen atom has one electron in its first shell ($_1$H); helium has two ($_2$He). Atoms of all other elements have two or more shells to accommodate orbitals for additional electrons.

■ *Second shell*: The second shell contains four orbitals (an *s* orbital and three *p* orbitals), and hence holds up to eight electrons.

■ *Additional shells*: Elements with more than ten electrons have three or more electron shells. The farther a shell is from the nucleus, the higher the energy level is for an electron occupying that shell. (That is, the negatively charged electron needs to absorb a greater amount of energy to overcome the "pull" of the positively charged nucleus and remain in that shell.)

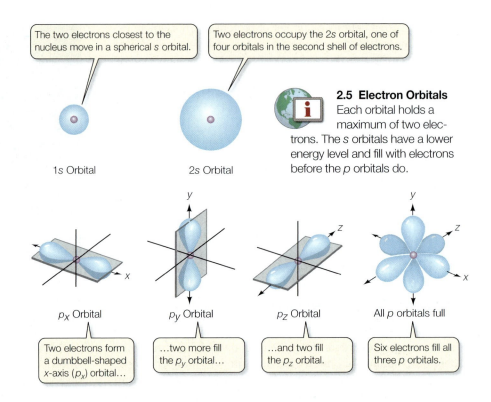

The two electrons closest to the nucleus move in a spherical *s* orbital.

Two electrons occupy the 2*s* orbital, one of four orbitals in the second shell of electrons.

1*s* Orbital

2*s* Orbital

2.5 Electron Orbitals Each orbital holds a maximum of two electrons. The *s* orbitals have a lower energy level and fill with electrons before the *p* orbitals do.

p$_x$ Orbital

p$_y$ Orbital

p$_z$ Orbital

All *p* orbitals full

Two electrons form a dumbbell-shaped *x*-axis (*p*$_x$) orbital…

…two more fill the *p*$_y$ orbital…

…and two fill the *p*$_z$ orbital.

Six electrons fill all three *p* orbitals.

2.6 Electron Shells Determine the Reactivity of Atoms Each orbital holds a maximum of two electrons, and each shell can hold a specific maximum number of electrons. Each shell must be filled before electrons move into the next shell. The energy level of electrons is higher in shells farther from the nucleus. An atom with unpaired electrons in its outermost shell can react (bond) with other atoms.

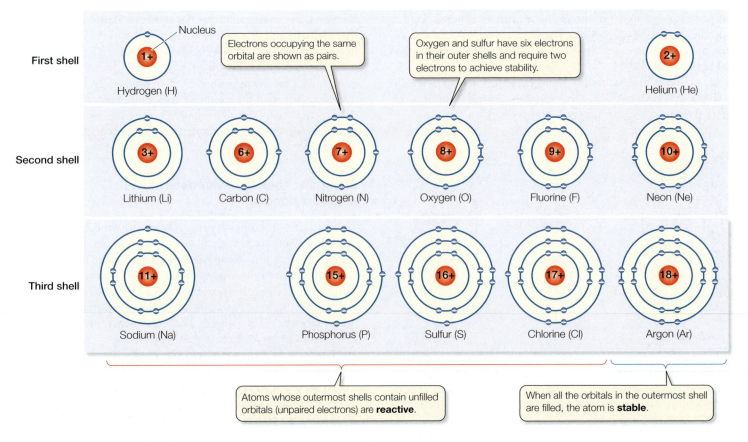

First shell

Nucleus

Hydrogen (H) 1+

Electrons occupying the same orbital are shown as pairs.

Oxygen and sulfur have six electrons in their outer shells and require two electrons to achieve stability.

Helium (He) 2+

Second shell

Lithium (Li) 3+

Carbon (C) 6+

Nitrogen (N) 7+

Oxygen (O) 8+

Fluorine (F) 9+

Neon (Ne) 10+

Third shell

Sodium (Na) 11+

Phosphorus (P) 15+

Sulfur (S) 16+

Chlorine (Cl) 17+

Argon (Ar) 18+

Atoms whose outermost shells contain unfilled orbitals (unpaired electrons) are **reactive**.

When all the orbitals in the outermost shell are filled, the atom is **stable**.

The *s* orbitals fill with electrons first, and their electrons have the lowest energy level. Subsequent shells have different numbers of orbitals, but the outermost shells usually hold only eight electrons. In any atom, the outermost electron shell (the *valence shell*) determines how the atom combines with other atoms—that is, how the atom behaves chemically. When a valence shell with four orbitals contains eight electrons, there are no unpaired electrons, and the atom is *stable*—it will not react with other atoms (**Figure 2.6**). Examples of chemically stable elements are helium, neon, and argon.

> Life on Earth is based on carbon chemistry. Because silicon shares many biochemical properties with carbon, scientists and science fiction writers alike have speculated about the possibility and probable nature of a silicon-based life form.

Non-stable, or *reactive*, atoms have unpaired electrons in their outermost shells. Reactive atoms can attain stability either by sharing electrons with other atoms or by losing or gaining one or more electrons. In either case, the atoms involved are *bonded* together into stable associations called **molecules**. The tendency of atoms in stable molecules to have eight electrons in their outermost shells is known as the *octet rule*. Many atoms in biologically important molecules—for example, carbon (C) and nitrogen (N)—follow this rule. An important exception is hydrogen (H), which attains stability when two electrons occupy its single shell (consisting of just one *s* orbital).

2.1 RECAP

The living world is composed of the same set of chemical elements as the rest of the universe. The structure of an atom—with its nucleus of protons and neutrons and its characteristic configuration of electrons in orbitals around the nucleus—determines its properties.

- Can you describe the arrangement of protons, neutrons, and electrons in an atom? See Figure 2.1

- Can you use the periodic table to identify some of the differences and similarities in atomic structure among different elements (for example, oxygen, carbon, and helium)? Do you understand how the configuration of the valence shell influences the placement of an element in the periodic table? See pp. 22–24 and Figures 2.2 and 2.6

- Do you understand how bonding can help a reactive atom achieve stability? See p. 25 and Figure 2.6

We have introduced the individual players on the biochemical stage—the atoms. We have shown how the energy levels of electrons drive an atomic "quest for stability." Next we will describe the different types of chemical bonds that can lead to stability, joining atoms together into molecular structures with hosts of different properties.

2.2 How Do Atoms Bond to Form Molecules?

A **chemical bond** is an attractive force that links two atoms together in a molecule. There are several kinds of chemical bonds (**Table 2.1**). In this section we will begin with *covalent bonds*, the strong bonds that result from the sharing of electrons. Next we will examine *ionic bonds*, which form when an atom gains or loses electrons to achieve stability. We will then consider other, weaker, kinds of interactions, including hydrogen bonds, that are enormously important to biology.

Covalent bonds consist of shared pairs of electrons

A **covalent bond** forms when two atoms attain stable electron numbers in their outermost shells by *sharing* one or more pairs of electrons. Consider two hydrogen atoms coming into close proximity, each with a single unpaired electron in its single shell (**Figure 2.7**). Each positively charged nucleus *attracts the other atom's* unpaired negatively charged electron, but this attraction is countered by each electron's *attraction to its own nucleus*. Thus the two unpaired electrons become shared, filling the shells of both atoms. The shared attraction links the two hydrogen atoms in a covalent bond, and a stable hydrogen gas molecule (H_2) is formed.

A **compound** is a molecule made up of atoms of two or more elements bonded together in a fixed ratio. Methane gas (CH_4), water (H_2O), and table sugar (sucrose, $C_{12}H_{22}O_{11}$) are examples of compounds. The chemical symbols identify the different elements in a compound, and the subscript numbers indicate how many atoms of each element are present. Every compound has a **molecular weight** (molecular mass) that is the sum of the atomic

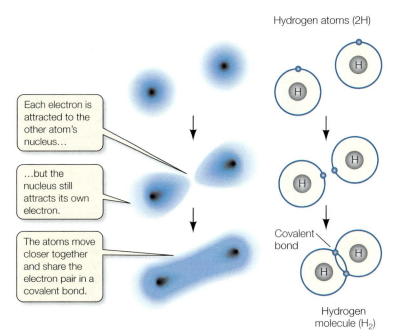

Hydrogen atoms (2H)

Each electron is attracted to the other atom's nucleus…

…but the nucleus still attracts its own electron.

The atoms move closer together and share the electron pair in a covalent bond.

Covalent bond

Hydrogen molecule (H_2)

2.7 Electrons Are Shared in Covalent Bonds Two hydrogen atoms can combine to form a hydrogen molecule. A covalent bond forms when the electron orbitals of the two atoms overlap.

TABLE 2.1

Chemical Bonds and Interactions

NAME	BASIS OF INTERACTION	STRUCTURE	BOND ENERGY^a (KCAL/MOL)
Covalent bond	Sharing of electron pairs		50–110
Ionic bond	Attraction of opposite changes		3–7
Hydrogen bond	Sharing of H atom		3–7
Hydrophobic interaction	Interaction of nonpolar substances in the presence of polar substances (especially water)		1–2
van der Waals interaction	Interaction of electrons of nonpolar substances		1

^{a}*Bond energy* is the amount of energy needed to separate two bonded or interacting atoms under physiological conditions.

weights of all atoms in the molecule. Looking at the periodic table in Figure 2.2, you can calculate the molecular weights of these three compounds to be 16, 18, and 342, respectively. Molecular weights are usually related to a molecule's size.

Consider the electrons involved in covalent bonding in methane gas. The carbon atom in this compound has six electrons:

two electrons fill its inner shell, and four travel in its outer shell. Because its outer shell can hold up to eight electrons, carbon can share electrons with up to four other atoms—*it can form four covalent bonds* (**Figure 2.8A**). When an atom of carbon reacts with four hydrogen atoms, methane forms. Thanks to electron sharing, the outer shell of methane's carbon atom is now filled with eight electrons; the outer shell of each of the four hydrogen atoms is also filled. Four covalent bonds—four shared electron pairs—hold methane together.

Figure 2.8B shows several different ways to represent the molecular structure of methane. **Table 2.2** shows the covalent bonding capacities of some biologically significant elements.

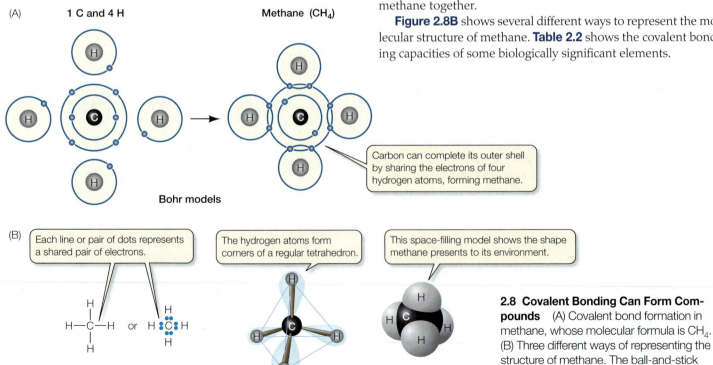

(A) 1 C and 4 H   Methane (CH_4)

Carbon can complete its outer shell by sharing the electrons of four hydrogen atoms, forming methane.

Bohr models

(B) Each line or pair of dots represents a shared pair of electrons.

The hydrogen atoms form corners of a regular tetrahedron.

This space-filling model shows the shape methane presents to its environment.

Structural formulas   Ball-and-stick model   Space-filling model

2.8 Covalent Bonding Can Form Compounds (A) Covalent bond formation in methane, whose molecular formula is CH_4. (B) Three different ways of representing the structure of methane. The ball-and-stick model and the space-filling model show the spatial orientation of the bonds.

TABLE 2.2

Covalent Bonding Capabilities of Some Biologically Important Elements

ELEMENT	USUAL NUMBER OF COVALENT BONDS
Hydrogen (H)	1
Oxygen (O)	2
Sulfur (S)	2
Nitrogen (N)	3
Carbon (C)	4
Phosphorus (P)	5

STRENGTH AND STABILITY Covalent bonds are very strong, meaning that it takes a lot of energy to break them. The thermal energy that biological molecules ordinarily have under physiological conditions (that is, the chemical and physical environment found in living tissues) is less than 1 percent of that needed to break covalent bonds. So biological molecules, most of which are put together with covalent bonds, are quite stable, as are their three-dimensional structures and the spaces they occupy. Most biochemical structures, ranging from the muscles you are using to move your eyes as you read this book to the paper it is printed on, are based on the stability of covalent bonds.

ORIENTATION For a given pair of elements—for example, carbon bonded to hydrogen—the length, angle, and direction of the covalent bonds are consistently the same, regardless of the larger molecule of which the particular bond is a part. The four filled orbitals around the carbon nucleus of methane, for example, are always distributed in space so that the bonded hydrogens define the corners of a regular tetrahedron, with carbon in the center (see Figure 2.8B). The three-dimensional structure formed by carbon and four hydrogens is the same in a large, complicated protein as it is in a simple methane molecule. This property of covalent bonds makes the prediction of biological structure possible. The shapes of molecules contribute to their biological functions, as we will see in Section 3.1.

Multiple covalent bonds

A covalent bond can be represented by a line between the chemical symbols for the linked atoms:

- A *single bond* involves the sharing of a single pair of electrons (for example, H—H or C—H).
- A *double bond* involves the sharing of four electrons (two pairs) (C=C).
- *Triple bonds*—six shared electrons—are rare, but there is one in nitrogen gas (N≡N), the major component of the air we breathe.

UNEQUAL SHARING OF ELECTRONS If two atoms of the same element are covalently bonded, there is an equal sharing of the pair(s) of electrons in the outermost shell. However, when the two atoms are of different elements, the sharing is not necessarily equal. One nucleus may exert a greater attractive force on the electron pair than the other nucleus, so that the pair tends to be closer to that atom.

The attractive force that an atomic nucleus exerts on electrons is its **electronegativity**. The electronegativity of a nucleus depends on how many positive charges it has (nuclei with more protons are more positive and thus more attractive to electrons) and on the distance between an electron and the nucleus (the closer the electron, the greater the pull of the electronegativity). The closer two atoms are in electronegativity, the more equal their sharing of electrons will be. **Table 2.3** shows the electronegativities of some elements important in biological systems.

If two atoms are close to each other in electronegativity, they will share electrons equally in what is called a *nonpolar covalent bond*. Two oxygen atoms, for example, both with electronegativity of 3.5, will share electrons equally. So will two hydrogen atoms (both with electronegativity of 2.1). But when hydrogen bonds with oxygen to form water, the electrons involved are unequally shared: they tend to be nearer to the oxygen nucleus because it is the more electronegative of the two. When electrons are drawn to one nucleus more than to the other, the result is a *polar covalent bond* (**Figure 2.9**).

Because of this unequal sharing of electrons, the oxygen end of the hydrogen–oxygen bond has a slightly negative charge (symbolized δ^- and spoken as "delta negative," meaning a partial unit of charge), and the hydrogen end has a slightly positive charge (δ^+). The bond is **polar** because these opposite charges are separated at the two ends, or poles, of the bond. The partial charges that result from polar covalent bonds produce polar molecules or polar regions of large molecules. Polar bonds greatly influence the interactions between molecules that contain them.

Ionic bonds form by electrical attraction

When one interacting atom is much more electronegative than the other, a complete transfer of one or more electrons may take place. Consider sodium (electronegativity 0.9) and chlorine (3.1). A sodium atom has only one electron in its outermost shell; this condition is unstable. A chlorine atom has seven electrons in its out-

TABLE 2.3

Some Electronegativities

ELEMENT	ELECTRONEGATIVITY
Oxygen (O)	3.5
Chlorine (Cl)	3.1
Nitrogen (N)	3.0
Carbon (C)	2.5
Phosphorus (P)	2.1
Hydrogen (H)	2.1
Sodium (Na)	0.9
Potassium (K)	0.8

Bohr model

Space-filling model

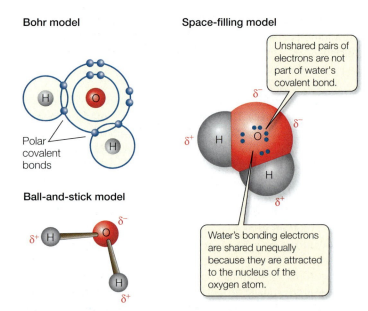

Polar covalent bonds

Unshared pairs of electrons are not part of water's covalent bond.

Ball-and-stick model

Water's bonding electrons are shared unequally because they are attracted to the nucleus of the oxygen atom.

2.9 Water's Covalent Bonds are Polar   These three representations all illustrate polar covalent bonding in water (H_2O). When atoms with different electronegativities, such as oxygen and hydrogen, form a covalent bond, the electrons are drawn to one nucleus more than to the other. A molecule held together by such a polar covalent bond has partial (δ) charges at the ends. In water, the shared electrons are displaced toward the oxygen atom's nucleus.

ermost shell—another unstable condition. Since the electronegativity of chlorine is so much greater than that of sodium, any electrons involved in bonding will tend to be much nearer to the chlorine nucleus—so near, in fact, that a complete transfer of the outermost electron from sodium to chlorine's outermost shell takes place (**Figure 2.10**). This reaction between sodium and chlorine makes the resulting atoms more stable. The result is two *ions*.

Ions are electrically charged particles that form when atoms gain or lose one or more electrons:

- The sodium ion (Na^+) in our example has a +1 unit of charge because it has one less electron than it has protons. The outermost electron shell of the sodium ion is full, with eight electrons, so the ion is stable. Positively charged ions are called **cations**.

- The chloride ion (Cl^-) has a –1 unit of charge because it has one more electron than it has protons. This additional electron gives Cl^- a stable outermost shell with eight electrons. Negatively charged ions are called **anions**.

Some elements can form ions with multiple charges by losing or gaining *more than one* electron. Examples are Ca^{2+} (calcium ion, a calcium atom that has lost two electrons) and Mg^{2+} (magnesium ion). Two biologically important elements can each yield more than one stable ion: iron yields Fe^{2+} (ferrous ion) and Fe^{3+} (ferric ion), and copper yields Cu^+ (cuprous ion) and Cu^{2+} (cupric ion). Groups of covalently bonded atoms that carry an electric charge are called *complex ions*; examples include NH_4^+ (ammonium ion), SO_4^{2-} (sulfate ion), and PO_4^{3-} (phosphate ion). The charge from an ion radi-

ates in all directions. Once formed, ions are usually stable and no more electrons are lost or gained.

Ionic bonds are bonds formed by electrical attraction between ions bearing opposite charges. Ions can form bonds that result in stable solid compounds (referred to by the general term *salts*) such as sodium chloride (NaCl) and potassium phosphate (K_3PO_4). In sodium chloride—familiar to us as table salt—cations and anions are held together by ionic bonds. In solids, the ionic bonds are strong because the ions are close together. However, when ions are dispersed in water, the distance between them can be large; the strength of their attraction is thus greatly reduced. Under the conditions in living cells, an ionic attraction is usually less than one-tenth as strong as a nonpolar covalent bond (see Table 2.1).

Not surprisingly, ions can interact with polar molecules, since they both carry electric charges. Such an interaction results when a salt such as NaCl dissolves in water. Water molecules surround the individual ions, separating them (**Figure 2.11**). The negatively charged chloride ions attract the positive pole of the water molecules; the negative pole of the water molecules is, in contrast, oriented toward the positively charged sodium ions. The polarity of water gives it other special properties, as we will see.

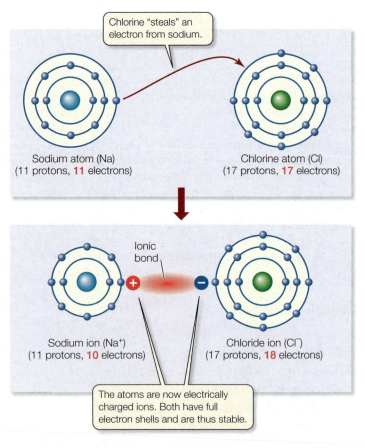

Chlorine "steals" an electron from sodium.

Sodium atom (Na)
(11 protons, **11** electrons)

Chlorine atom (Cl)
(17 protons, **17** electrons)

Ionic bond

Sodium ion (Na^+)
(11 protons, **10** electrons)

Chloride ion (Cl^-)
(17 protons, **18** electrons)

The atoms are now electrically charged ions. Both have full electron shells and are thus stable.

2.10 Formation of Sodium and Chloride Ions   When a sodium atom reacts with a chlorine atom, the more electronegative chlorine fills its outermost shell by "stealing" an electron from the sodium. In so doing, the chlorine atom becomes a negatively charged chloride ion (Cl^-). The sodium atom, upon losing the electron, becomes a positively charged sodium ion (Na^+).

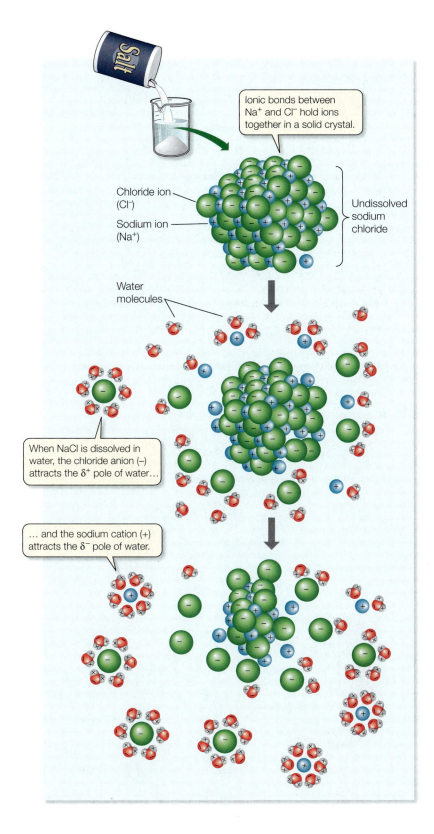

Ionic bonds between Na⁺ and Cl⁻ hold ions together in a solid crystal.

Chloride ion (Cl⁻)

Sodium ion (Na⁺)

Undissolved sodium chloride

Water molecules

When NaCl is dissolved in water, the chloride anion (−) attracts the δ⁺ pole of water…

… and the sodium cation (+) attracts the δ⁻ pole of water.

2.11 Water Molecules Surround Ions When an ionic solid dissolves in water, polar water molecules cluster around cations or anions, blocking their reassociation into a solid and forming a solution.

tween a strongly electronegative atom and a hydrogen atom covalently bonded to a different electronegative atom, as shown in **Figure 2.12B**. A hydrogen bond is weaker than most ionic bonds because it is formed by partial charges (δ^+ and δ^-). It has about one-tenth (10 percent) the strength of a covalent bond between a hydrogen atom and an oxygen atom (see Table 2.1). However, where many hydrogen bonds form, they have considerable strength and greatly influence the structure and properties of substances. Later in this chapter we'll see how hydrogen bonding between water molecules contributes to many of the properties that make water so significant for living systems. Hydrogen bonds also play important roles in determining and maintaining the three-dimensional shapes of giant molecules such as DNA and proteins (see Section 3.5).

Polar and nonpolar substances: Each interacts best with its own kind

Just as water molecules can interact with one another through polarity-induced hydrogen bonds, any molecule that is polar can interact with other polar molecules through the weak (δ^+ to δ^-) attractions of hydrogen bonds. If a polar molecule interacts with water in this way, it is called **hydrophilic** ("water-loving").

What about nonpolar molecules? Nonpolar molecules tend to interact with other nonpolar molecules. For example, carbon (electronegativity 2.5) forms nonpolar bonds with hydro-

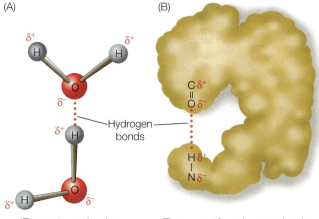

(A)

(B)

Hydrogen bonds

Two water molecules

Two parts of one large molecule (or two large molecules)

2.12 Hydrogen Bonds Can Form Between or Within Molecules (A) A hydrogen bond between two molecules is an attraction between negative charges on one molecule and positive charges on the hydrogen atoms of a second molecule. (B) Hydrogen bonds can form between large molecules, as well as between different parts of the same large molecule.

Hydrogen bonds may form within or between molecules with polar covalent bonds

In liquid water, the negatively charged oxygen (δ^-) atom of one water molecule is attracted to the positively charged hydrogen (δ^+) atoms of another water molecule (**Figure 2.12A**). The bond resulting from this attraction is called a **hydrogen bond**. Hydrogen bonds are not restricted to water molecules; they may also form be-gen (electronegativity 2.1). The resulting *hydrocarbon molecule*—that is, a molecule containing only hydrogen and carbon atoms—is nonpolar, and in water it tends to aggregate with other nonpolar molecules rather than with polar water. Such nonpolar molecules are known as **hydrophobic** ("water-hating"), and the interactions between them are called *hydrophobic interactions*.

Of course, hydrophobic substances do not really "hate" water; they can form weak interactions with it, since the electronegativities of carbon and hydrogen are not exactly the same. But these interactions are far weaker than the hydrogen bonds between the water molecules, so the nonpolar substances keep to themselves.

Hydrophobic interactions between nonpolar substances are enhanced by **van der Waals forces**, which result when two atoms of a nonpolar molecule are in close proximity. These brief interactions are a result of random variations in the electron distribution in one molecule, which create an opposite charge distribution in the adjacent molecule. Although a single van der Waals interaction is brief and weak at any given site, the sum of many such interactions over the entire span of a large nonpolar molecule can produce substantial attraction.

2.2 RECAP

Chemical bonds link atoms together to form molecules. Strong covalent bonds are important in establishing biological structures, as are ionic bonds in salt formation. Polarity and weak hydrophobic interactions allow molecules to interact.

- How do two atoms' electronegativities result in the unequal sharing of electrons in polar molecules? See p. 27 and Figure 2.9

- Can you explain why a covalent bond is stronger than an ionic bond? See pp. 27–28 and Table 2.1

- What is a hydrogen bond and how is it important in biological systems? See p. 29 and Figure 2.12

The bonding of atoms into molecules is not necessarily a permanent affair. The dynamic of life involves constant change, even at the molecular level. Let's look at how molecules interact with one another—how they break up, how they find new partners, and what the consequences of those changes can be.

2.3 How Do Atoms Change Partners in Chemical Reactions?

A **chemical reaction** occurs when atoms combine or change their bonding partners. Consider the combustion reaction that takes place in the flame of a propane stove. When propane (C_3H_8) reacts with oxygen gas (O_2), the carbon atoms become bonded to oxygen atoms instead of hydrogen atoms, and the hydrogen atoms become bonded to oxygen instead of carbon (**Figure 2.13**). As the covalently bonded atoms change partners, the composition of the

matter changes, and propane and oxygen gas become carbon dioxide and water. This chemical reaction can be represented by the equation

$$C_3H_8 + 5\,O_2 \rightarrow 3\,CO_2 + 4\,H_2O + \text{Energy}$$

$$\text{Reactants} \quad \rightarrow \quad \text{Products}$$

In this equation, the propane and oxygen are the **reactants**, and the carbon dioxide and water are the **products**. In this case, the reaction is *complete*: all the propane and oxygen are used up in forming the two products. The arrow symbolizes the direction of the chemical reaction. The numbers preceding the molecular formulas indicate how many molecules are used or produced.

Note that in this and all other chemical reactions, *matter is neither created nor destroyed*. The total number of carbon atoms on the left (3) equals the total number of carbon atoms on the right (3). In other words, the equation is *balanced*. However, there is another aspect of this reaction: the heat and light of the stove's flame reveal that the reaction of propane and oxygen released a great deal of energy. **Energy** is defined as the capacity to do work, but on a more intuitive level, it can be thought of as the capacity for change. Chemical reactions do not create or destroy energy, but *changes in the form of energy* usually accompany chemical reactions.

In the reaction between propane and oxygen, a large amount of heat energy was released. This energy was present in the covalent bonds of the propane and oxygen gases in another form, called *potential chemical energy*. Not all reactions release energy; indeed, many chemical reactions require that energy be supplied from the environment, and some of this supplied energy is stored as potential chemical energy in the bonds formed in the products. We will see in future chapters how reactions that release energy and reactions that require energy can be linked together.

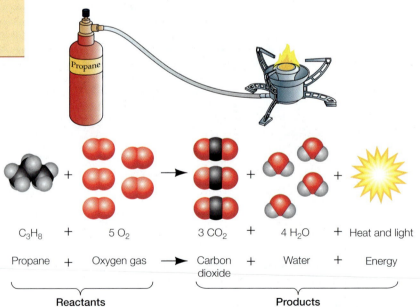

C_3H_8 + 5 O_2 → 3 CO_2 + 4 H_2O + Heat and light

Propane + Oxygen gas → Carbon dioxide + Water + Energy

Reactants — Products

2.13 Bonding Partners and Energy May Change in a Chemical Reaction
One molecule of propane reacts with five molecules of oxygen gas to give three molecules of carbon dioxide and four molecules of water. This reaction releases energy in the form of heat and light.

Many reactions take place in living cells, and some of these have much in common with the combustion of propane. The fuel is different—it is the sugar glucose, rather than propane—and the reactions proceed by many intermediate steps that permit the energy released from the glucose to be harvested and put to use by the cell. But the products are the same: carbon dioxide and water. These reactions, which we will study in detail in Chapter 7, are a key to the origin of life from simpler molecules.

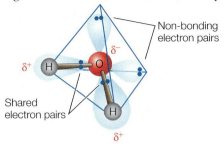

Non-bonding electron pairs

Shared electron pairs

2.3 RECAP

Chemical reactions involve the making and breaking of bonds between atoms. Many chemical reactions involve changes in the form of energy.

- ■ What are the components of a chemical reaction? See Figure 2.13

- ■ Do you understand how the form of energy can change during a chemical reaction?

We will present and discuss energy changes, oxidation–reduction reactions, and several other types of chemical reactions that are prevalent in living systems in Part Two of this book. First, however, we must understand the unique properties of the substance in which most biochemical reactions take place: water.

pairs of electrons in the outer shell of the oxygen atom repel one another, giving the water molecule a tetrahedral shape:

These chemical features explain some of the interesting properties of water, such as the ability of ice to float, the melting and freezing temperatures of water, the ability of water to store heat, and the ability of water droplets to form.

ICE FLOATS   In ice, water's solid state, individual water molecules are held in place by hydrogen bonds. Each water molecule is hydrogen-bonded to four other water molecules in a rigid, crystalline structure (**Figure 2.14**). Although the molecules are held firmly in place, they are not as tightly packed as they are in liquid water. In other words, *solid water is less dense than liquid water*, which is why ice floats.

Think of the biological consequences if ice were to sink in water. A pond would freeze from the bottom up, becoming a solid

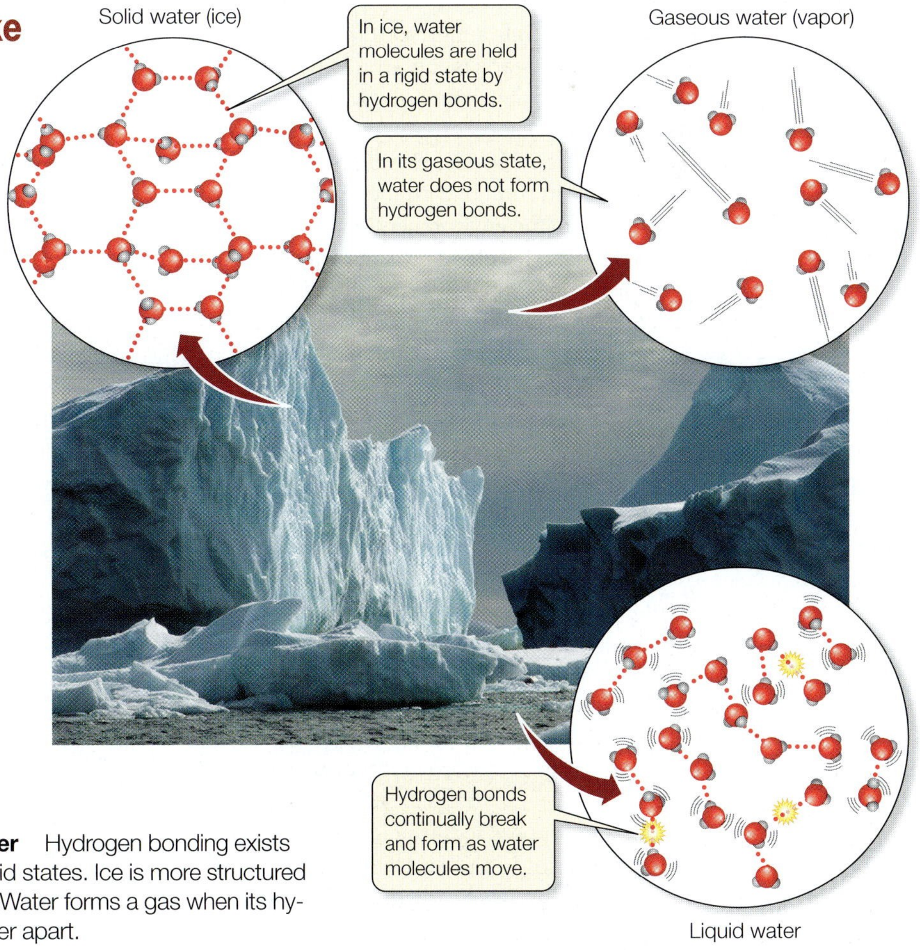

2.4 What Properties of Water Make It So Important in Biology?

Like most matter, water can exist in three states: solid (ice), liquid, or gas (vapor). As we mentioned at the start of this chapter, liquid water is probably the medium in which life originated, and it is in water that life on Earth evolved for its first billion years. In this section we will explore how the structure and interactions of water molecules make water essential to life.

Water has a unique structure and special properties

The molecule H_2O has unique chemical features. As we have already learned, water is a polar molecule that can form hydrogen bonds. In addition, the four

Solid water (ice)

In ice, water molecules are held in a rigid state by hydrogen bonds.

In its gaseous state, water does not form hydrogen bonds.

Gaseous water (vapor)

Hydrogen bonds continually break and form as water molecules move.

Liquid water

2.14 Hydrogen Bonds Hold Water Molecules Together   Hydrogen bonding exists between the molecules of water in both its liquid and solid states. Ice is more structured but less dense than liquid water, which is why ice floats. Water forms a gas when its hydrogen bonds are broken and the molecules move farther apart.

block of ice in winter and killing most of the organisms living there. Once the whole pond was frozen, its temperature could drop well below the freezing point of water. But in fact ice floats, forming a protective insulating layer on the top of the pond, reducing heat flow to the cold air above. Thus fish, plants, and other organisms in the pond are not subjected to temperatures lower than 0°C, the freezing point of pure water. The recent discovery by the *Global Surveyor* satellite of liquid water beneath the Martian polar ice has created speculation that life might exist there.

MELTING, FREEZING, AND HEAT CAPACITY   Compared with many other substances of the same molecular size, ice requires a great deal of heat energy to melt, because hydrogen bonds must be broken in order for water to change from solid to liquid. In the opposite process—freezing—a great deal of energy is lost when water is transformed from liquid to solid.

Water contributes to the surprising constancy of the temperatures found in the oceans and other large bodies of water throughout the year. The temperature changes of coastal land masses are also moderated by large bodies of water. Indeed, water helps minimize variations in atmospheric temperature across the planet. This moderating ability is a result of the high *heat capacity* of liquid water, which is in turn a result of its high specific heat. The **specific heat** of a substance is the amount of heat energy required to raise the temperature of 1 gram of that substance by 1°C. Raising the temperature of liquid water takes a relatively large amount of heat because much of the heat energy is used to break the hydrogen bonds that hold the liquid together. Compared with other small molecules that are liquids, water has a high specific heat.

Water also has a high **heat of vaporization**, which means that a lot of heat is required to change water from its liquid to its gaseous state (the process of *evaporation*). Once again, much of the heat energy is used to break hydrogen bonds. This heat must be absorbed from the environment in contact with the water. Evaporation thus has a cooling effect on the environment—whether a leaf, a forest, or an entire land mass. This effect explains why sweating cools the human body: as sweat evaporates from the skin, it uses up some of the adjacent body heat.

COHESION AND SURFACE TENSION   In liquid water, individual water molecules are free to move about. The hydrogen bonds between the molecules continually form and break. Chemists estimate that this occurs about a trillion times a minute in a single water molecule, making it a truly dynamic structure. At any given time, a water molecule will form an average of 3.4 hydrogen bonds with other water molecules. These hydrogen bonds explain the *cohesive strength* of liquid water. This cohesive strength, or **cohesion**, is defined as the capacity of water molecules to resist coming apart from one another when placed under tension. Water's cohesive strength permits narrow columns of liquid water to move from the roots to the leaves of trees more than 100 meters high. When water evaporates from the leaves, the entire column moves upward in response to the pull of the molecules at the top.

The surface of liquid water exposed to the air is difficult to puncture because the water molecules in this surface layer are hydrogen-bonded to other water molecules below them (**Figure 2.15**).

2.15 Surface Tension   Water droplets form "beads" on the surface of a leaf because hydrogen bonds keep the water molecules together.

This *surface tension* of water permits a container to be filled slightly above its rim without overflowing, and it permits insects to walk on the surface of a pond.

There are at least 20 trillion galaxies in the universe and an average of 100 billion stars in each galaxy. Probability analyses of these numbers suggest that many undiscovered planets must exist, and that some of those planets probably have water—and thus the possibility of life.

Water is the solvent of life

A human body is over 70 percent water by weight, excluding the minerals contained in bones. Water is the dominant component of virtually all living organisms, and most biochemical reactions take place in this watery environment.

A **solution** is produced when a substance (the **solute**) is dissolved in a liquid (the **solvent**). If the solvent is water, then the solution is an *aqueous solution*. Many of the important molecules in biological systems are polar, and therefore soluble in water. Many important biochemical reactions occur in aqueous solutions. Biologists study what happens in these reactions, both in terms of the nature of the reactants and products and in terms of their amounts:

■ *Qualitative analysis* deals with substances dissolved in water and the chemical reactions that occur there. For example, it would investigate the steps involved, and the products formed, during the combustion of glucose in living tissues. Qualitative analysis is the subject of much of the next few chapters.

■ *Quantitative analysis* measures concentrations, or the amount of a substance in a given amount of solution. For example, it would seek to describe *how much* of a certain product is formed during the combustion of a given amount of glucose. What follows is a brief introduction to some of the quantitative chemical terms you will see in this book.

Fundamental to quantitative thinking in chemistry and biology is the mole concept. A **mole** is the amount of a substance (in grams) the mass of which is numerically equal to its molecular weight. So a mole of table sugar ($C_{12}H_{22}O_{11}$) weighs 342 grams; a mole of sodium ion (Na^+) weighs 23 grams; and a mole of hydrogen gas (H_2) weighs 2 grams.

Quantitative analysis does not yield direct counts of molecules. Chemists use a constant that relates the weight of any substance to the number of molecules of that substance. This constant is called **Avogadro's number**, which is 6.02×10^{23} molecules per mole. It allows chemists to work with moles of substances (which can be weighed out in the laboratory) instead of actual molecules (which are too numerous to be counted). Consider 34.2 grams (just over 1 ounce) of table sugar, $C_{12}H_{22}O_{11}$. This is one-tenth of a mole, or as Avogadro puts it, 6.02×10^{22} molecules.

If you have trouble grasping the idea of a mole, think of the mole concept the way you think of the concept of a dozen: we buy a dozen eggs or a dozen doughnuts, knowing that we will get 12 of whichever we buy, even though they don't weigh the same or take up the same amount of space.

When a physician injects a certain molar concentration of a drug into the bloodstream of a patient, a rough calculation can be made of the actual number of drug molecules that will interact with the patient's cells. In the same way, a chemist can dissolve a mole of sugar in water to make 1 liter of solution, knowing that the mole contains 6.02×10^{23} individual sugar molecules. This solution—1 mole of a substance dissolved in water to make 1 liter—is called a 1 molar (1 M) solution.

The many molecules dissolved in the water in living tissues are not present at anything close to 1 molar concentrations. Most are in the micromolar (millionths of a mole per liter of solution; μM) to millimolar (thousandths of a mole per liter; mM) range. Some, such as hormone molecules, are even less concentrated than that. While these molarities seem to indicate very low concentrations, remember that even a 1 μM solution has 6.02×10^{17} molecules of the solute per liter.

Aqueous solutions may be acidic or basic

When some substances dissolve in water, they release *hydrogen ions* (H^+), which are actually single, positively charged protons. Hydrogen ions can attach to other molecules and change their properties. For example, the protons in "acid rain" can damage plants, and you probably have experienced the excess of hydrogen ions we know as "acid indigestion."

Here we will examine the properties of **acids**, which release H^+, and **bases**, which accept H^+. We will distinguish between strong and weak acids and bases and provide a quantitative means for stating the concentration of H^+ in solutions: the pH scale.

ACIDS RELEASE H$^+$, BASES ACCEPT H$^+$　When hydrochloric acid (HCl) is added to water, it dissolves, releasing the ions H^+ and Cl^-:

$$HCl \rightarrow H^+ + Cl^-$$

Because its H^+ concentration has increased, such a solution is *acidic*. Just like the combustion reaction of propane and oxygen shown in Figure 2.13, the dissolution of HCl to form its ions is a *complete* reaction. HCl is therefore called a *strong acid*.

Acids are substances that *release* H^+ ions in solution. HCl is an acid, as is H_2SO_4 (sulfuric acid). One molecule of sulfuric acid may ionize to yield two H^+ and one SO_4^{2-}. Biological compounds that contain —COOH (the carboxyl group) are also acids because

$$—COOH \rightarrow —COO^- + H^+$$

However, not all acids ionize fully in water. For example, if acetic acid is added to water, at the end of the reaction, there are two ions, but some of the original acetic acid remains as well. Because the reaction is *not complete*, acetic acid is a *weak acid*.

Bases are substances that *accept* H^+ in solution. Just as with acids, there are strong and weak bases. If NaOH (sodium hydroxide) is added to water, it dissolves and ionizes, releasing OH^- and Na^+ ions:

$$NaOH \rightarrow Na^+ + OH^-$$

Because the concentration of OH^- increases and OH^- absorbs H^+ to form water, such a solution is *basic*. Because this reaction is complete, NaOH is a *strong base*.

Weak bases include the bicarbonate ion (HCO_3^-), which can accept a H^+ ion and become carbonic acid (H_2CO_3), and ammonia (NH_3), which can accept a H^+ and become an ammonium ion (NH_4^+). Biological compounds that contain —NH_2 (the amino group) are also bases because

$$—NH_2 + H^+ \rightarrow —NH_3^+$$

ACID–BASE REACTIONS MAY BE REVERSIBLE　When acetic acid is dissolved in water, two reactions happen. First, the acetic acid forms its ions:

$$CH_3COOH \rightarrow CH_3COO^- + H^+$$

Then, once the ions are formed, they re-form acetic acid:

$$CH_3COO^- + H^+ \rightarrow CH_3COOH$$

This pair of reactions is reversible. A **reversible reaction** can proceed in either direction—left to right or right to left—depending on the relative starting concentrations of the reactants and products. The formula for a reversible reaction can be written using a double arrow:

$$CH_3COOH \rightleftharpoons CH_3COO^- + H^+$$

In terms of acids and bases, there are two types of reactions, depending on the extent of reversibility:

■ The ionization of strong acids and bases is virtually irreversible.

■ The ionization of weak acids and bases is somewhat reversible.

WATER IS A WEAK ACID   The water molecule has a slight but significant tendency to ionize into a hydroxide ion (OH⁻) and a hydrogen ion (H⁺). Actually, two water molecules participate in this reaction. One of the two molecules "captures" a hydrogen ion from the other, forming a hydroxide ion and a hydronium ion:

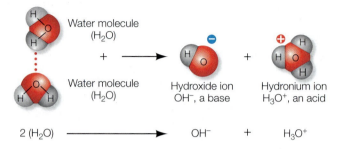

Water molecule (H₂O)

Water molecule (H₂O)

Hydroxide ion OH⁻, a base

Hydronium ion H₃O⁺, an acid

$$2\ (H_2O) \longrightarrow OH^- + H_3O^+$$

The hydronium ion is, in effect, a hydrogen ion bound to a water molecule. For simplicity, biochemists tend to use a modified representation of the ionization of water:

$$H_2O \rightarrow H^+ + OH^-$$

The ionization of water is important to all living creatures. This fact may seem surprising, since only about one water molecule in 500 million is ionized at any given time. But we will be less surprised if we focus on the abundance of water in living systems and the reactive nature of the H⁺ produced by ionization.

pH is the measure of hydrogen ion concentration

Solutions are acidic or basic; compounds or ions can be acids or bases. We can measure how acidic or basic a solution is by measuring the concentration of H⁺ in moles per liter, or *molarity* (see page 33). Here are some examples:

- Pure water has a H⁺ concentration of $10^{-7}\ M$.

- A 1 M HCl solution has a H⁺ concentration of 1 M (recall that all the HCl dissolves into its ions).

- A 1 M NaOH solution has a H⁺ concentration of $10^{-14}\ M$.

This is a very wide range of numbers to work with (think about the decimals!). It is easier to work with the *logarithm* of the H⁺ concentration, because logarithms compress this range: the $\log_{10}$ of 100, for example is 2, and the $\log_{10}$ of 0.01 is –2. Because most H⁺ concentrations in living systems are less than 1, we convert these negative numbers into positive ones by using the *negative* of the logarithm of the H⁺ molar concentration (designated by square brackets: [H⁺]). This number is called the **pH** of the solution.

Since the H⁺ concentration of pure water is $10^{-7}\ M$, its pH is $-\log(10^{-7}) = -(-7)$, or 7. A smaller negative logarithm means a larger number. In practical terms, a lower pH means a higher H⁺ concentration, or greater acidity. In 1 M HCl, the H⁺ concentration is 1 M, so the pH is the negative logarithm of 1 ($-\log 10^0$), or 0. The pH of 1 M NaOH is the negative logarithm of 10^{-14}, or 14.

A solution with a pH of less than 7 is acidic—it contains more H⁺ ions than OH⁻ ions. A solution with a pH of 7 is neutral, and a solution with a pH value greater than 7 is basic. **Figure 2.16** shows the pH values of some common substances.

Buffers minimize pH change

The maintenance of internal constancy—*homeostasis*— is a hallmark of all living things and extends to pH. As we mentioned earlier, when H⁺ is added to many molecules, they change their properties, thus upsetting constancy. Maintaining internal constancy can be achieved by buffers: solutions that maintain a relatively constant pH even when substantial amounts of an acid or base are added to a system. How can this work?

A **buffer** is a solution of a weak acid and its corresponding base—for example, carbonic acid (H_2CO_3) and bicarbonate ions (HCO_3^-). If an acid is added to a solution containing this buffer, not all the H⁺ ions from that acid stay in solution. Instead, many of them combine with the bicarbonate ions to produce more carbonic acid. This reaction uses up some of the H⁺ ions in the solution and decreases the acidifying effect of the added acid:

$$HCO_3^- + H^+ \rightarrow H_2CO_3$$

If a base is added, the reaction essentially reverses. Some of the carbonic acid ionizes to produce bicarbonate ions and more H⁺, which counteracts some of the added base. In this way, the buffer minimizes the effects of an added acid or base on pH. This is what happens in the blood, where this buffering system is important in preventing significant changes in pH that could disrupt the ability of the blood to carry vital oxygen to tissues. A given amount of acid or base causes a smaller change in pH in a buffered solution than in an unbuffered one (**Figure 2.17**).

How do you spell relief? The lining of the stomach constantly secretes hydrochloric acid, making the stomach contents acidic. Excessive stomach acid inhibits digestion and causes discomfort, but it can be relieved by ingesting a salt such as NaHCO₃ ("bicarbonate of soda") that acts as a buffer.

Buffers illustrate an important chemical principle of reversible reactions, called the *law of mass action*. Addition of a reactant on one side of a reversible system drives the reaction in the direction that uses up that compound. In the case of buffers, addition of an acid drives the reaction in one direction; addition of a base drives the reaction in the other direction.

Life's chemistry began in water

As we have emphasized throughout this chapter, the presence of water on a planet—Mars, Earth, or any other—is a necessary prerequisite for life as we know it. Astronomers believe our solar system began forming about 4.6 billion years ago, when a star exploded and collapsed to form the sun and 500 or so bodies, called planetesimals. These planetesimals collided with one another to form the inner planets, including Earth and Mars. The first chemical signatures indicating the presence of life on Earth now appear to be about 4 billion years old. So it took 600 million years, during a geological time frame called the Hadean, for the chemical conditions on Earth to become just right for life. Key among those conditions was the presence of water.

2.16 pH Values of Some Familiar Substances
An electronic instrument can be used to measure the pH of a solution.

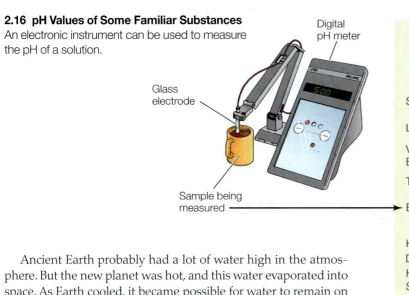

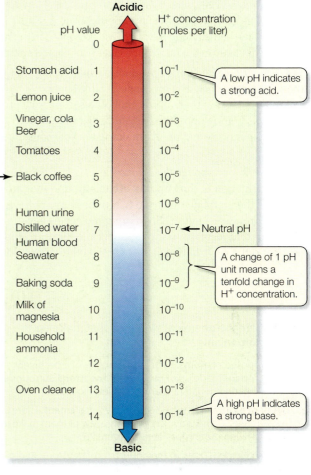

Ancient Earth probably had a lot of water high in the atmosphere. But the new planet was hot, and this water evaporated into space. As Earth cooled, it became possible for water to remain on its surface, but where did that water come from? One current view is that comets—loose agglomerations of dust and ice that have orbited the sun since the planets formed—struck Earth and Mars repeatedly and brought not only water but other chemical components of life, such as nitrogen. As the planets cooled, chemicals from their crusts dissolved in the water, and simple chemical reactions would have taken place. Some of these reactions could have led to life, but impacts by large comets and rocky meteorites would have released enough energy to heat the developing oceans almost to boiling, thus destroying any early life. On Earth, these large impacts eventually subsided, and life gained a foothold about 3.8 to 4 billion years ago. The prebiotic Hadean was over. The Archean had begun, and there has been life on Earth ever since.

In Section 3.6 we will return to the question of how the first life could have arisen from inanimate chemicals.

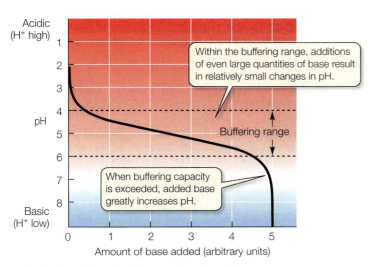

2.17 Buffers Minimize Changes in pH With increasing amounts of added base, the overall slope of a graph of pH is downward. Inside the buffering range, however, the slope is shallow. At very high and very low values of pH, where the buffer is ineffective, the slopes are much steeper.

2.4 RECAP

Most of the chemistry of life occurs in water, which has molecular properties that make it suitable for its important biochemical roles. A special property of water is its ability to ionize (release hydrogen ions). The presence of hydrogen ions in solution can change the properties of biological molecules.

- What are some biologically important properties of water arising from its molecular structure? See pp. 31–32 and Figure 2.14

- Do you understand what a solution is, and why we call water "the solvent of life"? See pp. 32–33

- Do you see the relationship between hydrogen ions, acids, and bases? Can you explain what the pH scale measures? See pp. 33–34 and Figure 2.16

- Can you explain how a buffer works, and why buffering is important to living systems? See p. 34 and Figure 2.17

An Overview and a Preview

Now that we have covered the major properties of atoms and molecules, let's review them and see how they will be applied in the next chapter, which focuses on the major molecules of biological systems.

■ *Molecules vary in size.* Some are small, such as H_2 and CH_4. Others are larger, such as a molecule of table sugar ($C_{12}H_{22}O_{11}$), which has 45 atoms. Still others, especially proteins such as hemoglobin (the oxygen carrier in red blood cells), are gigantic, sometimes containing tens of thousands of atoms.

■ *All molecules have a specific three-dimensional shape.* For example, the orientation of the bonding orbitals around the carbon atom gives the methane molecule (CH_4) the shape of a regular tetrahedron (see Figure 2.8B). Larger molecules have complex shapes that result from the numbers and kinds of atoms present and the ways in which they are linked together. Some large molecules, such as hemoglobin, have compact, ball-like shapes. Others, such as the protein called keratin that makes up your hair, have long, thin, ropelike structures. Their shapes relate to the roles these molecules play in living cells.

■ *Molecules are characterized by certain chemical properties* that determine their biological roles. Chemists use the characteristics of composition, structure (three-dimensional shape), reactivity, and solubility to distinguish a pure sample of one molecule from a sample of a different molecule. The presence of certain groups of atoms can impart distinctive chemical properties to a molecule.

Between the small molecules discussed in this chapter and the world of the living cell stand the macromolecules. These huge molecules—proteins, lipids, carbohydrates, and nucleic acids—are the subject of the next chapter.

CHAPTER SUMMARY

2.1 What are the chemical elements that make up living organisms?

Matter is composed of atoms. Each **atom** consists of a positively charged **nucleus** made up of **protons** and **neutrons**, surrounded by **electrons** bearing negative charges. Review Figure 2.1

The number of protons in the nucleus defines an **element**. There are many elements in the universe, but only a few of them make up the bulk of living organisms. Review Figure 2.2

Isotopes of an element differ in their numbers of neutrons. **Radioisotopes** are radioactive, emitting radiation as they decay.

Electrons are distributed in **shells**, which are volumes of space defined by specific numbers of orbitals. Each **orbital** contains a maximum of two electrons. Review Figure 2.6, Web/CD Activity 2.1

In losing, gaining, or sharing electrons to become more stable, an atom can combine with other atoms to form **molecules**.

2.2 How do atoms bond to form molecules?

See Web/CD Tutorial 2.1

A **chemical bond** is an attractive force that links two atoms together in a molecule. Review Table 2.1

A **compound** is a molecule made up of atoms of two or more elements bonded together in a fixed ratio, such as water (H_2O) or table sugar ($C_6H_{12}O_6$).

Covalent bonds are strong bonds formed when two atoms share one or more pairs of electrons. Review Figure 2.7

When two atoms of unequal electronegativity bond with each other, a **polar** covalent bond is formed. The two ends, or poles, of the bond have partial charges (δ^+ or δ^-). Review Figure 2.9

Ions are electrically charged bodies that form when an atom gains or loses one or more electrons. **Anions** and **cations** are negatively and positively charged ions, respectively.

Ionic bonds are electrical attractions between oppositely charged ions. Ionic bonds are strong in solids (salts), but weaken when the ions are separated from one another in solution. Review Figure 2.10

Hydrogen bonds are weak electrical attractions that form between a δ^+ hydrogen atom in one molecule and a δ^- atom in another molecule (or in another part of a large molecule). Hydrogen bonds are abundant in water.

Nonpolar molecules interact very little with polar molecules, including water. Nonpolar molecules are attracted to one another by very weak bonds called **van der Waals forces**.

2.3 How do atoms change partners in chemical reactions?

In **chemical reactions**, atoms combine or change their bonding partners. **Reactants** are converted into **products**.

Some chemical reactions release **energy** as one of their products; other reactions can occur only if energy is provided to the reactants.

Neither matter nor energy is created or destroyed in a chemical reaction, but both change form. Review Figure 2.13

Some chemical reactions, especially in biology, are reversible. That is, the products formed may be converted back to the reactants.

In living cells, chemical reactions take place in multiple steps so that the released energy can be harvested for cellular activities.

2.4 What properties of water make it so important in biology?

Water's molecular structure and its capacity to form hydrogen bonds give it unique properties that are significant for life. Review Figure 2.14

The high **specific heat** of water means that water gains or loses a great deal of heat when it changes state. Water's high **heat of vaporization** ensures effective cooling when water evaporates.

The **cohesion** of water molecules refers to their capacity to resist coming apart from one another.

Solutions are produced when solid substances (**solutes**) dissolve in a liquid (the **solvent**). Water is the critically important solvent for life.

Acids are solutes that release hydrogen ions in aqueous solution. **Bases** accept hydrogen ions.

The **pH** of a solution is the negative logarithm of its hydrogen ion concentration. Values lower than pH 7 indicate a solution is acidic; values above pH 7 indicate a basic solution. Review Figure 2.16

Buffers are mixtures of weak acids and bases that limit the change in the pH of a solution when acids or bases are added.

SELF-QUIZ

1. The atomic number of an element
   a. equals the number of neutrons in an atom.
   b. equals the number of protons in an atom.
   c. equals the number of protons minus the number of neutrons.
   d. equals the number of neutrons plus the number of protons.
   e. depends on the isotope.

2. The atomic weight (atomic mass) of an element
   a. equals the number of neutrons in an atom.
   b. equals the number of protons in an atom.
   c. equals the number of electrons in an atom.
   d. equals the number of neutrons plus the number of protons.
   e. depends on the relative abundances of its electrons and neutrons.

3. Which of the following statements about the isotopes of an element is *not* true?
   a. They all have the same atomic number.
   b. They all have the same number of protons.
   c. They all have the same number of neutrons.
   d. They all have the same number of electrons.
   e. They all have identical chemical properties.

4. Which of the following statements about covalent bonds is *not* true?
   a. A covalent bond is stronger than a hydrogen bond.
   b. A covalent bond can form between atoms of the same element.
   c. Only a single covalent bond can form between two atoms.
   d. A covalent bond results from the sharing of electrons by two atoms.
   e. A covalent bond can form between atoms of different elements.

5. Hydrophobic interactions
   a. are stronger than hydrogen bonds.
   b. are stronger than covalent bonds.
   c. can hold two ions together.
   d. can hold two nonpolar molecules together.
   e. are responsible for the surface tension of water.

6. Which of the following statements about water is *not* true?
   a. It releases a large amount of heat when changing from liquid into vapor.
   b. Its solid form is less dense than its liquid form.
   c. It is the most effective solvent of polar molecules.
   d. It is typically the most abundant substance in a living organism.
   e. It takes part in some important chemical reactions.

7. The reaction $HCl \rightarrow H^+ + Cl^-$ in the human stomach is an example of the
   a. cleavage of a hydrophobic bond.
   b. formation of a hydrogen bond.
   c. elevation of the pH of the stomach.
   d. formation of ions by dissolving an acid.
   e. formation of polar covalent bonds.

8. The hydrogen bond between two water molecules arises because water is
   a. polar.
   b. nonpolar.
   c. a liquid.
   d. small.
   e. hydrophobic.

9. When table salt (NaCl) is added to water,
   a. a covalent bond is broken.
   b. an acidic solution is formed.
   c. Na^+ and Cl^- ions are separated.
   d. Na^+ is attracted to the hydrogen atoms of water.
   e. water molecules surround Na (but not Cl) atoms.

10. The three most abundant elements in a human skin cell are
    a. calcium, carbon, and oxygen.
    b. carbon, hydrogen, and oxygen.
    c. carbon, hydrogen, and sodium.
    d. carbon, nitrogen, and potassium.
    e. nitrogen, hydrogen, and argon.

FOR DISCUSSION

1. Using the information in the periodic table (Figure 2.2), draw a Bohr model (see Figure 2.8) of silicon dioxide showing electrons shared in covalent bonds.

2. Compare a covalent bond between two hydrogen atoms and a hydrogen bond between hydrogen and oxygen atoms with regard to the electrons involved, the role of polarity, and the strength of the bond.

3. Write an equation describing the combustion of glucose ($C_6H_{12}O_6$) to produce carbon dioxide and water.

4. The pH of the human stomach is about 2.0, while the pH of the small intestine is about 10.0. What are the hydrogen ion (H^+) concentrations inside these two organs?

FOR INVESTIGATION

Would you expect the elemental composition of Earth's crust to be the same as that of the human body? How could you find out?

CHAPTER 3

Macromolecules and the Origin of Life

How sweet it is

Sometimes a scientist gets lucky. In 1989, a British sugar company was looking for nontraditional uses for its product, and wondered if sugar might be chemically modified to make other useful substances. As often happens in industry, the company sought the expertise of university scientists. A professor at the University of London knew how to modify table sugar (sucrose, $C_{12}H_{22}O_{11}$) by replacing hydrogen and oxygen atoms with other elements. The professor assigned a graduate student the job of making sugar with chlorine atoms in place of some of the OH (hydroxyl) groups.

The professor's idea was to test each new molecule for possible use as a starting material for other compounds. But the graduate student, Shahikant Phadnis, responded to his professor's request for "testing" one molecule by "tasting." To his astonishment, the new molecule—$C_{12}H_{19}O_8Cl_3$—was sweet. Indeed, it was many times sweeter than sugar. But unlike sugar, it could not be used by the body as food energy. Phadnis had discovered a new artificial sweetener, now known as sucralose.

The commercial demand for artificial sweeteners is huge. About one-third of all people in Europe, the Americas, and Australia consume low-calorie foods and beverages. Some people, such as diabetics, must restrict their sugar intake or suffer serious health consequences. People fighting obesity want to reduce their intake of sugar, which is a major calorie source. Still others are prone to tooth decay from bacteria in the mouth that feed on sugar. With sucralose, all these people can have their cake and eat it too. It is used in hundreds of products, and is sold by itself as a sugar substitute, packaged in yellow under the brand name "Splenda."

Ordinary sugar has two important chemical properties that affect people who eat it. First, it binds noncovalently to certain proteins found in taste buds on the tongue. Sucralose binds to the taste buds even better than sugar does, hence its intense sweet taste. Second, sugar molecules have a three-dimensional structure that allows them to bind to digestive molecules in the body, which break sugar down for energy and conversion to other substances (including fat). It is in this second chemical property that sucralose differs strikingly from ordinary sugar. Sucralose does not bind to digestive molecules, but passes through the body largely unchanged.

Sucrose and its artificial cousin sucralose are carbohydrates, one of the four major kinds of large molecules that characterize living systems. These *macromolecules*,

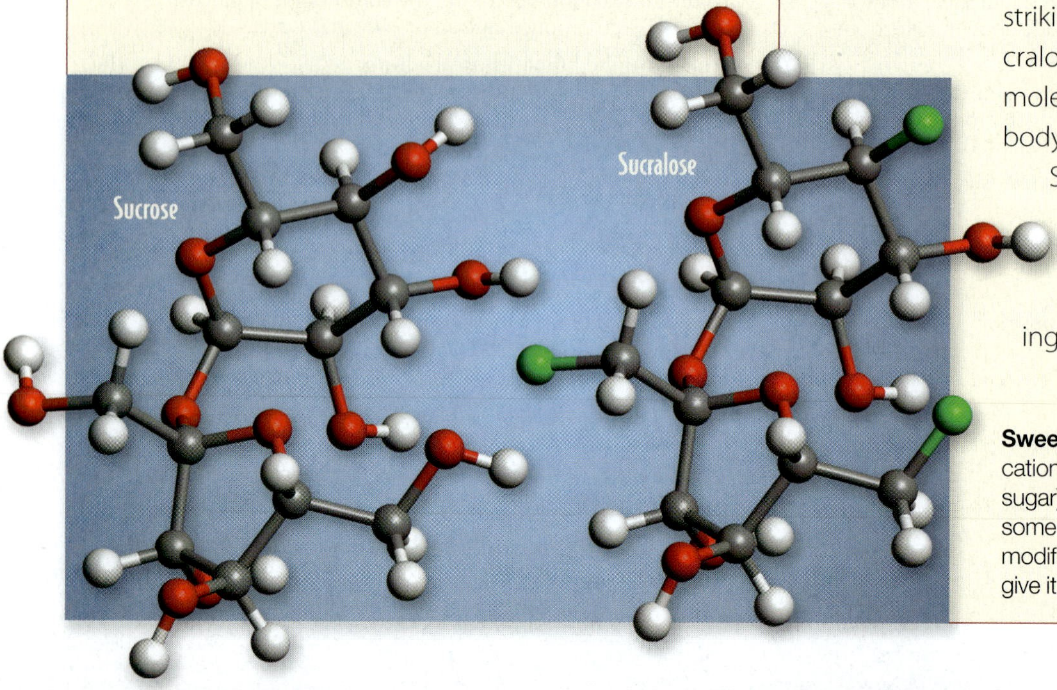

Sucrose

Sucralose

Sweeter Than Sugar Sucralose, a modification of the biological molecule sucrose (table sugar), has chlorine atoms (green) in place of some of the hydroxyl groups of sucrose. This modification results in chemical properties that give it a sweet taste, but no nutritional value.

A Substitute for Sugar Sucralose can be used in place of sugar in baked goods, allowing dieters and people with diabetes to enjoy sweet treats.

which also include proteins, lipids, and nucleic acids, differ in several significant ways from the small molecules and ions described in Chapter 2. First—not a surprise—they are larger, with molecular weights ranging from hundreds of Daltons (sucrose) to billions (some nucleic acids).

Second, these molecules all contain carbon atoms, and so belong to a group of what are known as *organic* chemicals. Third, they are held together largely by covalent bonds, which gives them important structural stability and forms the basis of some of their functions. And finally, carbohydrates, proteins, lipids, and nucleic acids are all unique to the living world. These molecular classes do not occur in inanimate nature. You won't find proteins in rocks—and if you do, you can be sure they came from some living organism.

> **IN THIS CHAPTER** we will describe the chemical and biological properties of proteins, lipids, carbohydrates, and nucleic acids. We will see what these building blocks of life are made of, how they are assembled, and what roles they play in living organisms. We end the chapter with some ideas on how these molecules originated when life first began.

3.1 What Kinds of Molecules Characterize Living Things?

Four kinds of molecules are characteristic of living things: proteins, carbohydrates, lipids, and nucleic acids. Most of these *biological molecules* are large **polymers** (*poly*, "many"; *mer*, "unit") constructed by the covalent bonding of smaller molecules called **monomers** (**Table 3.1**). The monomers that make up each kind of biological molecule have similar chemical structures. Thus chains of chemically similar sugar monomers (*saccharides*) form the different carbohydrates; the thousands of different proteins are formed from combinations of a mere 20 *amino acids*, all of which share chemical similarities.

Polymers with molecular weights exceeding 1,000 are usually considered to be **macromolecules**. The proteins, polysaccharides (large carbohydrates), and nucleic acids of living systems certainly fall into this category. How these macromolecules function and interact with other molecules depends on properties of chemical groups in their monomers, the *functional groups*.

Functional groups give specific properties to molecules

Certain small groups of atoms, called **functional groups**, are consistently found together in a variety of different biological molecules. You will encounter several functional groups repeatedly in your study of biology (**Figure 3.1**). Each functional group has specific chemical properties, and when it is attached to a larger molecule, it confers those properties on the larger molecule. For example, the *hydroxyl group* is polar and attracts water molecules. Small molecules containing the hydroxyl group usually dissolve easily in water. The oxygen atom in the *keto group* is highly electronegative and can attract the hydrogen atom held by another electronegative atom (such as nitrogen), forming a hydrogen bond. The consistent chemical behavior of these functional groups helps us understand the properties of the molecules that contain them.

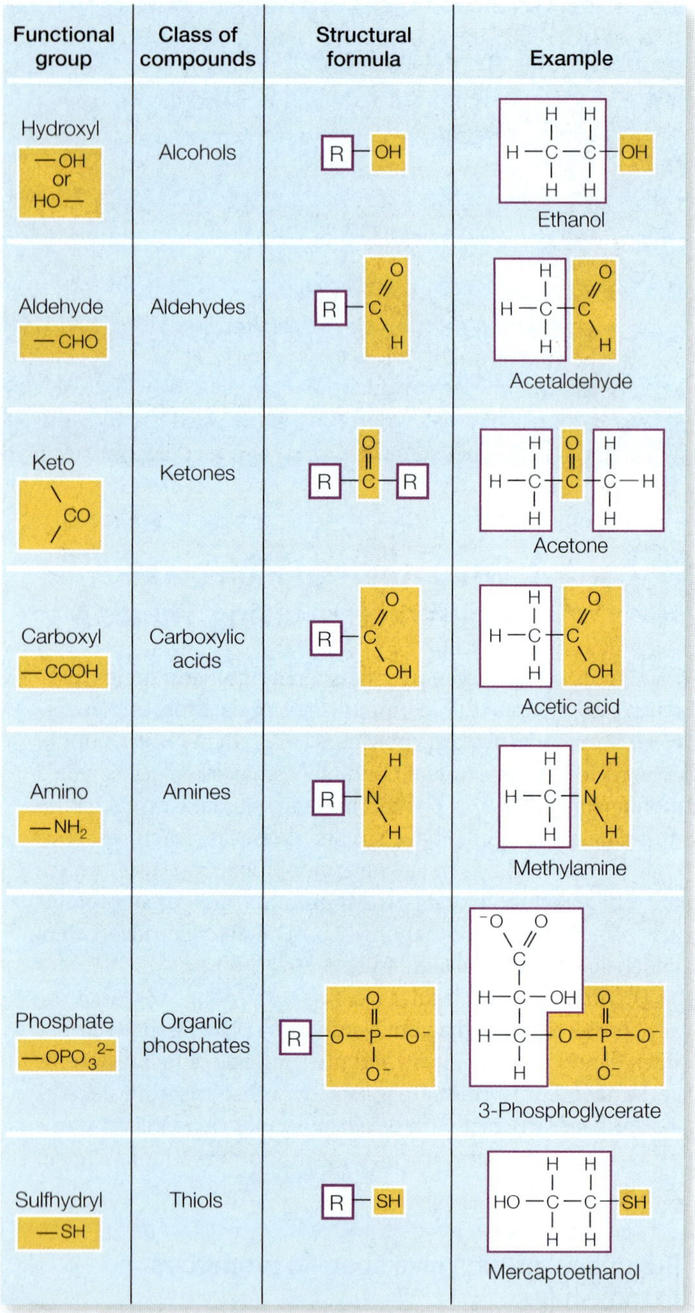

Functional group	Class of compounds	Structural formula	Example
Hydroxyl — OH or HO—	Alcohols	R—OH	Ethanol
Aldehyde — CHO	Aldehydes	R—C(=O)H	Acetaldehyde
Keto \CO/	Ketones	R—C(=O)—R	Acetone
Carboxyl — COOH	Carboxylic acids	R—C(=O)OH	Acetic acid
Amino — NH₂	Amines	R—N(H)H	Methylamine
Phosphate — OPO₃²⁻	Organic phosphates	R—O—P(=O)(O⁻)O⁻	3-Phosphoglycerate
Sulfhydryl — SH	Thiols	R—SH	Mercaptoethanol

3.1 Some Functional Groups Important to Living Systems
The seven functional groups shown here (highlighted in yellow) are the most common ones found in biologically important molecules. R represents the "remainder" of the molecule, which may be any of a large number of carbon skeletons or other chemical groups.

TABLE 3.1

The Building Blocks of Organisms

MONOMER	COMPLEX POLYMER (MACROMOLECULE)
Amino acid	Polypeptide (protein)
Monosaccharide (sugar)	Polysaccharide (carbohydrate)
Nucleotide	Nucleic acid

Isomers have different arrangements of the same atoms

Isomers are molecules that have the same chemical formula—the same kinds and numbers of atoms—but the atoms are arranged differently. (The prefix *iso-*, meaning "same," is encountered in many biological terms.) Of the different kinds of isomers, we will consider two: structural isomers and optical isomers.

Structural isomers differ in how their atoms are joined together. Consider two simple molecules, each composed of four carbon and ten hydrogen atoms bonded covalently, both with the formula C_4H_{10}. These atoms can be linked in two different ways, resulting in two different forms of the molecule:

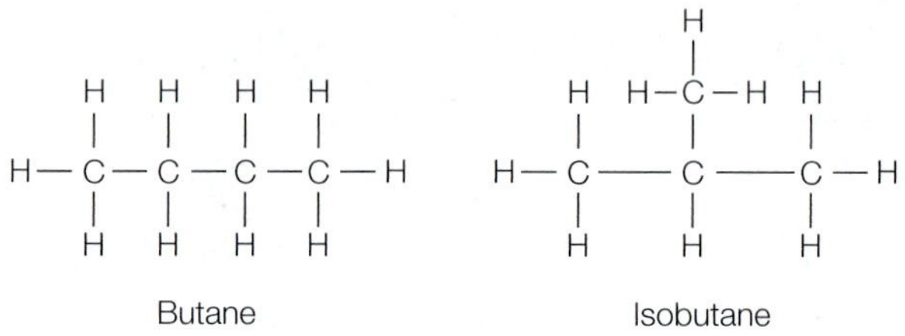

Butane   Isobutane

The different bonding relationships in butane and isobutane are distinguished by their structural formulas, and the two molecules have different chemical properties.

Optical isomers occur when a carbon atom has four different atoms or groups of atoms attached to it. This pattern allows two different ways of making the attachments, each the mirror image of the other (**Figure 3.2**). Such a carbon atom is called an *asymmetrical carbon*, and the two resulting molecules are optical isomers of each other. You can envision your right and left hands as optical isomers. Just as a glove is specific for a particular hand, some biochemical molecules that can interact with one optical isomer of a carbon compound are unable to "fit" the other.

The structures of macromolecules reflect their functions

The four kinds of biological macromolecules are present in roughly the same proportions in all living organisms (**Figure 3.3**). A protein that has a certain function in an apple tree probably has a similar function in a human being because its chemistry is the same wherever it is found. One important advantage of this *biochemical unity* is that organisms can acquire needed biochemicals by eating other organisms. When you eat an apple, the molecules you take in include carbohydrates, lipids, and proteins that can be refashioned into the special varieties of those molecules needed by humans.

Each type of macromolecule performs some combination of functions, such as energy storage, structural support, protection, catalysis (speeding up a chemical reaction), transport, defense, regulation, movement, and information storage. These roles are not necessarily exclusive; for example, both carbohydrates and proteins can play structural roles, supporting and protecting tissues and organs. However, only the nucleic acids specialize in information storage. These macromolecules function as hereditary material, carrying the traits of both species and individuals from generation to generation.

The functions of macromolecules are directly related to their three-dimensional shapes and to the sequences and chemical prop-

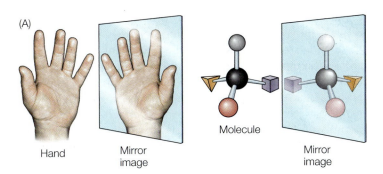

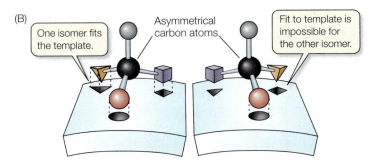

3.2 Optical Isomers (A) Optical isomers are mirror images of each other. (B) Molecular optical isomers result when four different atoms or groups are attached to a single carbon atom. If a template (representing a larger biological molecule) is laid out to match the groups on one carbon atom, the groups on that molecule's mirror-image isomer cannot be rotated to fit the same template.

erties of their monomers. Some macromolecules fold into compact spherical forms with surface features that make them water-soluble and capable of intimate interaction with other molecules. Some proteins and carbohydrates form long, fibrous systems that provide strength and rigidity to cells and organisms. Still other long, thin assemblies of proteins can contract and cause movement.

Because macromolecules are so large, they contain many different functional groups (see Figure 3.1). For example, a single large

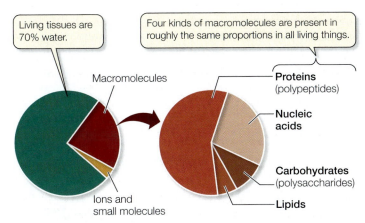

3.3 Substances Found in Living Tissues The substances shown here make up the nonmineral components of living tissue (bone would be an example of a mineral component).

protein may contain hydrophobic, polar, and charged functional groups, each of which gives different specific properties to local sites on the macromolecule. As we will see, this diversity of properties determines the shapes of macromolecules and their interactions both with other macromolecules and with smaller molecules.

Most macromolecules are formed by condensation and broken down by hydrolysis

Polymers are constructed from monomers by a series of reactions called **condensation reactions** (sometimes called *dehydration* reactions; both terms refer to the loss of water). Condensation reactions result in covalently bonded monomers. These reactions release a molecule of water for each covalent bond formed (**Figure 3.4A**). The condensation reactions that produce the different kinds of polymers differ in detail, but in all cases, polymers form only if energy is added to the system. In living systems, specific energy-rich molecules supply this energy.

The reverse of a condensation reaction is a **hydrolysis reaction** (*hydro*, "water"; *lysis*, "break"). Hydrolysis reactions digest polymers and produce monomers. Water reacts with the covalent bonds that link the polymer together, and the products are free

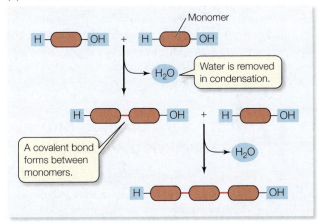

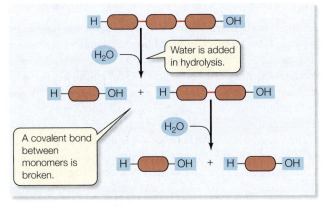

3.4 Condensation and Hydrolysis of Polymers (A) A condensation reaction links monomers into polymers. (B) A hydrolysis reaction breaks polymers into individual monomers.

monomers. The elements (H and O) of H_2O become part of the products (**Figure 3.4B**). The linkages between monomers thus can be formed and broken inside living tissues.

Condensation and hydrolysis reactions are universal in living things and, as we will see at the end of this chapter, their role in assembling polymers was an important step in the origin of life in an aqueous environment. We begin our study of the building blocks of life with a diverse group of polymers, the proteins.

3.2 What Are the Chemical Structures and Functions of Proteins?

The functions of **proteins** include structural support, protection, transport, catalysis, defense, regulation, and movement. The monomeric subunits of proteins are the 20 **amino acids**. Among the functions of macromolecules listed earlier, only two—energy storage and information storage—are not usually performed by proteins.

Proteins range in size from small ones such as the hormone insulin, which has a molecular weight of 5,733 and 51 amino acids, to huge molecules such as the muscle protein titin, which has a molecular weight of 2,993,451 and 26,926 amino acids. Each of these proteins consists of a single unbranched polymer of amino acids (a *polypeptide chain*), which is folded into a specific three-dimensional shape.

Many proteins are made up of more than one polypeptide chain. For example, the oxygen-carrying protein hemoglobin has four chains that are folded separately and come together to make up the functional protein. As we will see later in this book, proteins can associate with one another, forming "multi-protein machines" to carry out complex tasks such as DNA synthesis.

The *composition* of a protein refers to the relative amounts of the different amino acids it contains. Variation in the precise *sequence*

of the amino acids in each polypeptide chain is the source of the diversity in protein structures and functions.

To understand the many functions of proteins that will be described throughout this book, we must first explore protein structure. First we will examine the properties of amino acids and see how they link together to form polypeptide chains. Then we will systematically examine protein structure and look at how a linear chain of amino acids is consistently folded into a specific, compact, three-dimensional shape. Finally, we will see how this three-dimensional structure provides a definitive physical and chemical environment that influences how other molecules can interact with the protein.

Amino acids are the building blocks of proteins

The amino acids have both a carboxyl functional group and an amino functional group (see Figure 3.1) attached to the same carbon atom, called the α carbon. At the pH values commonly found in cells, both of these groups are ionized: the carboxyl group has lost a hydrogen ion, and the amino group has gained one. Because they possess both carboxyl and amino groups, *amino acids are simultaneously acids and bases*.

Also attached to the α carbon atom are a hydrogen atom and a **side chain**, or **R group**, designated by the letter R:

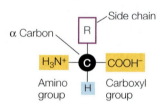

The α carbon in an amino acid is an asymmetrical carbon because it is bonded to four different atoms or groups or atoms. Therefore, amino acids exist in two isomeric forms, called D-amino acids and L-amino acids. D and L are abbreviations for the Latin terms for right (*dextro*) and left (*levo*). Only L-amino acids are commonly found in proteins in most organisms, and their presence is an important chemical "signature" of life.

The side chains of amino acids also contain functional groups, which are important in determining the three-dimensional structure and function of the protein macromolecule. The side chains are highlighted in white in **Table 3.2**. As the table shows, the 20 amino acids found in living organisms are grouped and distinguished by their side chains.

- The five amino acids that have electrically charged side chains (+1, −1) attract water (are hydrophilic) and oppositely charged ions of all sorts.

- The five amino acids that have polar side chains (δ^+, δ^-) tend to form hydrogen bonds with water and with other polar or charged substances. These amino acids are hydrophilic.

- Seven amino acids have side chains that are nonpolar hydrocarbons or very slightly modified hydrocarbons. In the watery environment of the cell, these hydrophobic side chains may cluster together in the interior of the protein. These amino acids are hydrophobic.

Three amino acids—cysteine, glycine, and proline—are special cases, although their side chains are generally hydrophobic.

TABLE 3.2

The Twenty Amino Acids

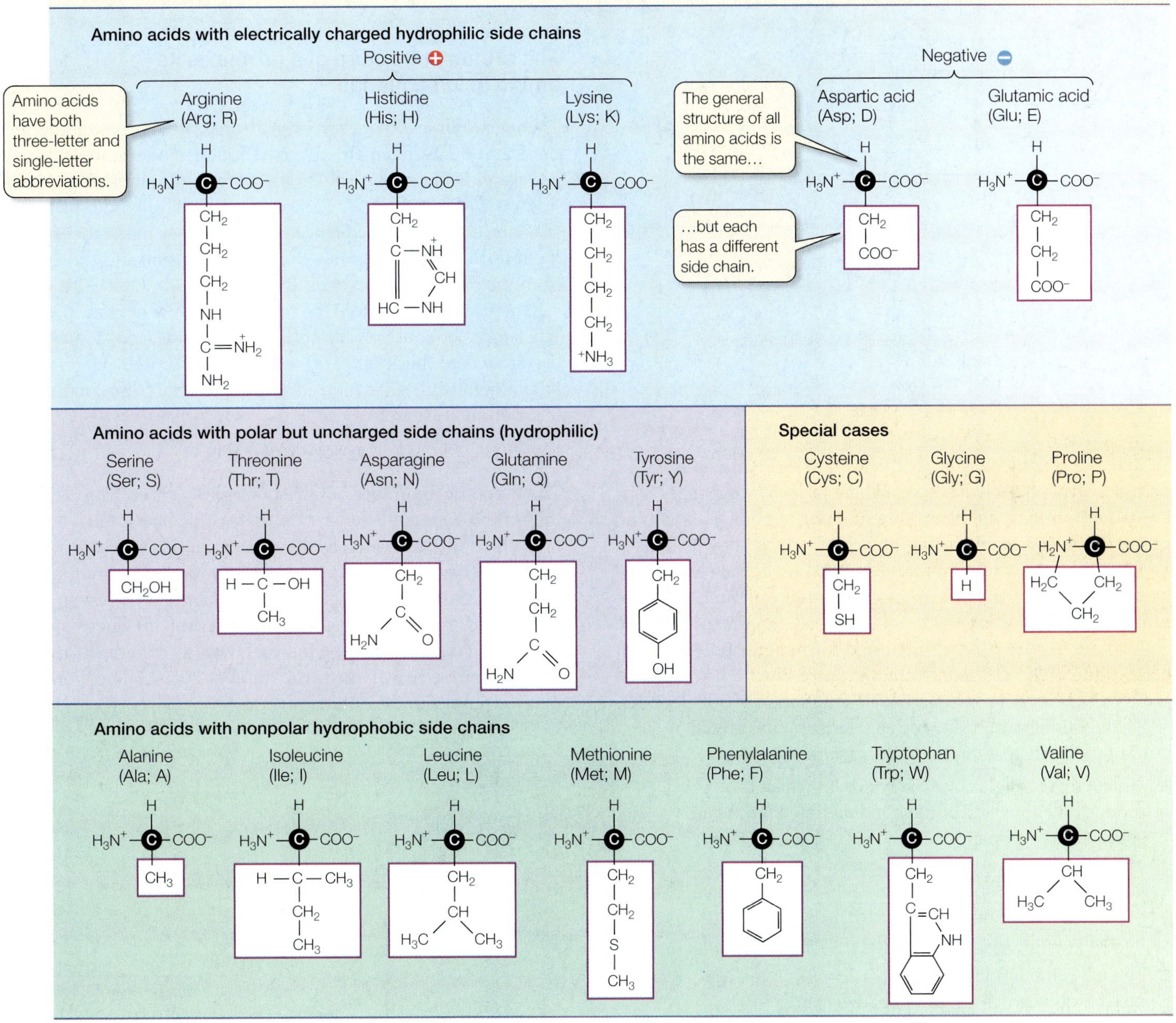

Amino acids with electrically charged hydrophilic side chains

Positive ⊕

Negative ⊖

Amino acids have both three-letter and single-letter abbreviations.

The general structure of all amino acids is the same…

…but each has a different side chain.

Arginine (Arg; R) — side chain: CH_2–CH_2–CH_2–NH–C=NH_2^+–NH_2

Histidine (His; H) — side chain: CH_2–C–NH⁺, CH, HC–NH

Lysine (Lys; K) — side chain: CH_2–CH_2–CH_2–CH_2–$^+NH_3$

Aspartic acid (Asp; D) — side chain: CH_2–COO^-

Glutamic acid (Glu; E) — side chain: CH_2–CH_2–COO^-

Amino acids with polar but uncharged side chains (hydrophilic)

Serine (Ser; S) — side chain: CH_2OH

Threonine (Thr; T) — side chain: H–C–OH, CH_3

Asparagine (Asn; N) — side chain: CH_2–C(=O)–H_2N

Glutamine (Gln; Q) — side chain: CH_2–CH_2–C(=O)–H_2N

Tyrosine (Tyr; Y) — side chain: CH_2–(benzene ring)–OH

Special cases

Cysteine (Cys; C) — side chain: CH_2–SH

Glycine (Gly; G) — side chain: H

Proline (Pro; P) — side chain: H_2C–CH_2–CH_2 (ring)

Amino acids with nonpolar hydrophobic side chains

Alanine (Ala; A) — side chain: CH_3

Isoleucine (Ile; I) — side chain: H–C–CH_3, CH_2, CH_3

Leucine (Leu; L) — side chain: CH_2–CH–H_3C, CH_3

Methionine (Met; M) — side chain: CH_2–CH_2–S–CH_3

Phenylalanine (Phe; F) — side chain: CH_2–(benzene ring)

Tryptophan (Trp; W) — side chain: CH_2–C=CH–NH (indole ring)

Valine (Val; V) — side chain: CH–H_3C, CH_3

(General amino acid structure shown for all: H_3N^+–C(H)–COO^-; proline shown as H_2N^+–C(H)–COO^-)

- The *cysteine* side chain, which has a terminal —SH group, can react with another cysteine side chain to form a covalent bond called a **disulfide bridge** (—S—S—) (**Figure 3.5**). Disulfide bridges help determine how a polypeptide chain folds. When cysteine is not part of a disulfide bridge, its side chain is hydrophobic.

- The *glycine* side chain consists of a single hydrogen atom and is small enough to fit into tight corners in the interior of a protein molecule, where a larger side chain could not fit.

- *Proline* differs from other amino acids because it possesses a modified amino group lacking a hydrogen on its nitrogen, which limits its hydrogen-bonding ability. In addition, the ring structure of proline limits rotation about its α carbon, so proline is often found at bends or loops in a protein.

Peptide bonds form the backbone of a protein

When amino acids polymerize, the carboxyl and amino groups attached to the α carbon are the reactive groups. The carboxyl group of one amino acid reacts with the amino group of another, undergoing a condensation reaction that forms a **peptide bond** (also called a *peptide linkage*). **Figure 3.6** gives a simplified description of this reaction. (In living systems, other molecules must activate the amino acids before this reaction proceeds, and there are intermediate steps in the process as well. We will examine these steps in Section 12.4.)

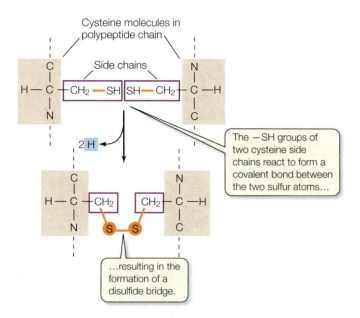

Cysteine molecules in polypeptide chain

Side chains

The —SH groups of two cysteine side chains react to form a covalent bond between the two sulfur atoms…

…resulting in the formation of a disulfide bridge.

3.5 A Disulfide Bridge Two cysteine molecules in a polypeptide chain can form a disulfide bridge (—S—S—).

The primary structure of a protein is its amino acid sequence

There are four levels of protein structure: primary, secondary, tertiary, and quaternary. The precise sequence of amino acids in a polypeptide chain constitutes the **primary structure** of a protein (**Figure 3.7A**). The peptide backbone of the polypeptide chain consists of the repeating sequence —N—C—C—, made up of the N atom from the amino group, the α carbon atom, and the C atom from the carboxyl group of each amino acid.

Scientists have deduced the primary structure of many proteins. The single-letter abbreviations for amino acids (see Table 3.2) are used to record the amino acid sequence of a protein. Here, for example, are the first 20 amino acids (out of a total of 124) in the protein ribonuclease from a cow:

$$\text{KETAAAKFERQHMDSSTSAA}$$

The theoretical number of different proteins is enormous. Since there are 20 different amino acids, there could be $20 \times 20 = 400$ distinct dipeptides (two linked amino acids), and $20 \times 20 \times 20 = 8,000$ different tripeptides (three linked amino acids). Imagine this process of multiplying by 20 extended to a protein made up of 100 amino acids (which is considered a small protein). There could be 20^{100} (10^{130}) such small proteins, each with its own distinctive primary structure. How large is the number 20^{100}? There aren't that many electrons in the entire universe!

Just as a sentence begins with a capital letter and ends with a period, polypeptide chains have a linear order. The chemical "capital letter" marking the beginning of a polypeptide is the amino group of the first amino acid in the chain and is known as the *N terminus*. The "punctuation mark" for the end of the chain is the carboxyl group of the last amino acid—the *C terminus*. All the other amino and carboxyl groups in the chain (except those in side chains) are involved in peptide bond formation, so they do not exist in the chain as "free," intact groups. Biochemists refer to the "N → C," or "amino-to-carboxyl," orientation of polypeptides.

The peptide bond has two characteristics that are important in the three-dimensional structure of proteins:

■ Unlike many single covalent bonds, in which the groups on either side of the bond are free to rotate in space, the C—N peptide linkage is relatively inflexible. The adjacent atoms (the α carbons of the two adjacent amino acids, shown in black in Figure 3.6) in a peptide bond are not free to rotate fully, and this inflexibility limits the folding of the polypeptide chain.

■ The oxygen bound to the carbon (C=O) in the carboxyl group carries a slight negative charge (δ^-), whereas the hydrogen bound to the nitrogen (N—H) in the amino group is slightly positive (δ^+). This asymmetry of charge favors hydrogen bonding within the protein molecule itself and with other molecules, contributing to both the structure and the function of many proteins.

Before we explore the significance of these characteristics, however, we need to examine the importance of the order of amino acids.

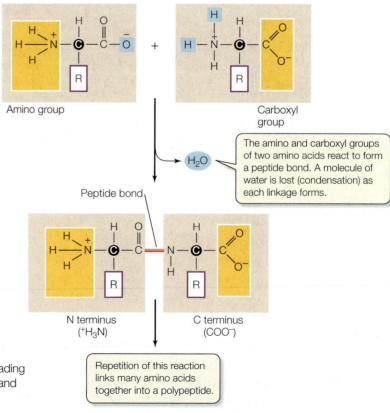

Amino group

Carboxyl group

The amino and carboxyl groups of two amino acids react to form a peptide bond. A molecule of water is lost (condensation) as each linkage forms.

Peptide bond

N terminus (^+H_3N)

C terminus (COO^-)

Repetition of this reaction links many amino acids together into a polypeptide.

3.6 Formation of Peptide Bonds In living things, the reaction leading to a peptide bond has many intermediate steps, but the reactants and products are the same as those shown in this simplified diagram.

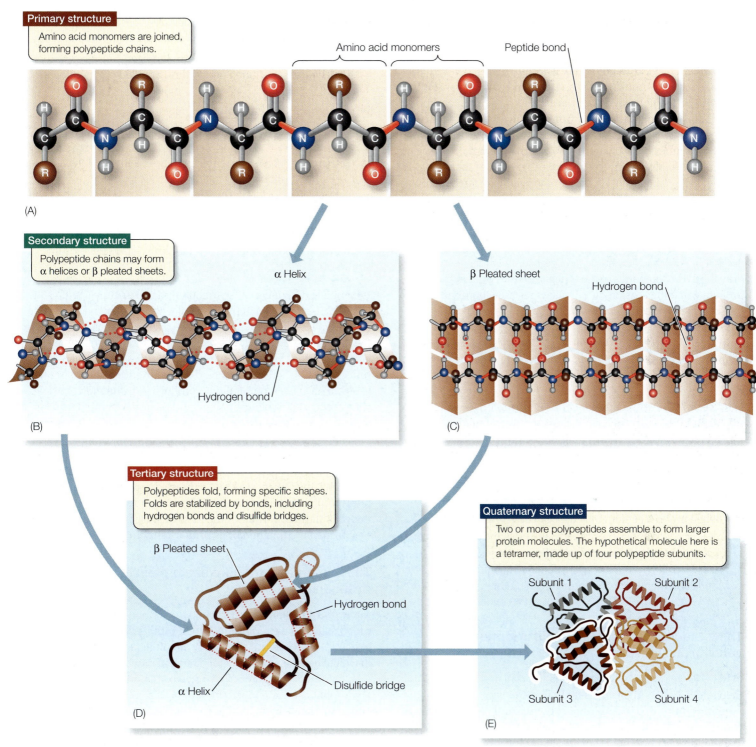

Primary structure
Amino acid monomers are joined, forming polypeptide chains.

Amino acid monomers

Peptide bond

(A)

Secondary structure
Polypeptide chains may form α helices or β pleated sheets.

α Helix

Hydrogen bond

(B)

β Pleated sheet

Hydrogen bond

(C)

Tertiary structure
Polypeptides fold, forming specific shapes. Folds are stabilized by bonds, including hydrogen bonds and disulfide bridges.

β Pleated sheet

Hydrogen bond

α Helix

Disulfide bridge

(D)

Quaternary structure
Two or more polypeptides assemble to form larger protein molecules. The hypothetical molecule here is a tetramer, made up of four polypeptide subunits.

Subunit 1   Subunit 2

Subunit 3   Subunit 4

(E)

3.7 The Four Levels of Protein Structure   Secondary, tertiary, and quaternary structure all arise from the primary structure of the protein.

At the higher levels of protein structure, local coiling and folding give the molecule its final functional shape, but all of these levels derive from the primary structure—that is, the precise location of specific amino acids in the polypeptide chain. The properties associated with a precise sequence of amino acids determine how the protein can twist and fold, thus adopting a specific stable structure that distinguishes it from every other protein.

Primary structure is determined by covalent bonds. The next level of protein structure is determined by weaker hydrogen bonds.

The secondary structure of a protein requires hydrogen bonding

A protein's **secondary structure** consists of regular, repeated spatial patterns in different regions of a polypeptide chain. There are two basic types of secondary structure, both of them determined by hydrogen bonding between the amino acids that make up the primary structure.

THE α HELIX The α (alpha) helix is a right-handed coil that is "threaded" in the same direction as a standard wood screw (**Figure 3.7B**). The R groups extend outward from the peptide backbone of the helix. The coiling results from hydrogen bonds that form between the δ^+ hydrogen of the N—H of one amino acid and the δ^- oxygen of the C=O of another. When this pattern of hydrogen bonding is established repeatedly over a segment of the protein, it stabilizes the coil. The presence of amino acids with large R groups that distort the coil or otherwise prevent the formation of the necessary hydrogen bonds will keep an α helix from forming.

The α-helical secondary structure is common in the fibrous structural proteins called keratins, which make up hair, hooves, and feathers. Hair can be stretched because stretching requires that only the hydrogen bonds of the α helix, not the covalent bonds, be broken; when the tension on the hair is released, both the hydrogen bonds and the helix re-form.

THE β PLEATED SHEET A β (beta) pleated sheet is formed from two or more polypeptide chains that are almost completely extended and aligned. The sheet is stabilized by hydrogen bonds between the N—H groups on one chain and the C=O groups on the other (**Figure 3.7C**). A β pleated sheet may form between separate polypeptide chains, as in spider silk, or between different regions of a single polypeptide chain that is bent back on itself. Many proteins contain regions of both α helix and β pleated sheet in the same polypeptide chain.

When proteins get together, they can be very strong. Spider silk, for example, is stronger than Kevlar, the artificial polymer used in bulletproof vests. Potential uses for spider silk (if it could be purified in large enough quantities) include automobile airbags and surgical sutures.

The tertiary structure of a protein is formed by bending and folding

In many proteins, the polypeptide chain is bent at specific sites and then folded back and forth, resulting in the **tertiary structure** of the protein (**Figure 3.7D**). Although α helices and β pleated sheets contribute to the tertiary structure, usually only portions of the macromolecule have these secondary structures, and large regions consist of tertiary structure unique to a particular protein. The tertiary structure results in a macromolecule with a definite three-dimensional shape. The outer surfaces of the macromolecule present functional groups capable of chemically interacting with other molecules in the cell. These molecules might be other proteins (as in quaternary structure, as we will see below) or small molecules (as in enzymes; see Section 6.4).

While hydrogen bonding between the N—H and C=O groups within and between chains is responsible for secondary structure, the interactions between R groups—the amino acid side chains—determine tertiary structure. We described the various strong and weak interactions between atoms in Section 2.2. Many of these interactions are involved in determining tertiary structure.

- Covalent *disulfide bridges* can form between specific cysteine side chains (see Figure 3.5), holding a folded polypeptide in place.

- *Hydrophobic* side chains can aggregate together in the interior of the protein, away from water, folding the polypeptide in the process.

- *van der Waals forces* can stabilize the close interactions between hydrophobic side chains.

- *Ionic bonds* can form between positively and negatively charged side chains buried deep within a protein, away from water, forming a *salt bridge*.

- *Hydrogen bonds* between side chains also stabilize folds in proteins.

A complete description of a protein's tertiary structure specifies the location of every atom in the molecule in three-dimensional space in relation to all the other atoms. Such a description is available for the protein lysozyme (**Figure 3.8**). The first tertiary structures to be determined took years to figure out, but today dozens of new structures are published every week. The major advances making this possible have been our ability to produce large quantities of specific proteins using biotechnology, and the use of computers to analyze the atomic data.

Bear in mind that both tertiary structure and secondary structure derive from a protein's primary structure. If lysozyme is heated slowly, the heat energy will disrupt only the weak interactions and cause only the tertiary structure to break down. But the protein will return to its normal tertiary structure when it cools, demonstrating that all the information needed to specify the unique shape of a protein is contained in its primary structure.

The quaternary structure of a protein consists of subunits

Many functional proteins contain two or more polypeptide chains, called *subunits*, each of them folded into its own unique tertiary structure. The protein's **quaternary structure** results from the ways in which these subunits bind together and interact (see Figure 3.7E).

Quaternary structure is illustrated by hemoglobin (**Figure 3.9**). Hydrophobic interactions, van der Waals forces, hydrogen bonds, and ionic bonds all help hold the four subunits together to form the hemoglobin molecule. The function of hemoglobin is to carry oxygen in red blood cells. As hemoglobin binds one O_2 molecule, the four subunits shift their relative positions slightly, changing the quaternary structure. Ionic bonds are broken, exposing buried side chains that enhance the binding of additional O_2 molecules. The quaternary structure changes again when hemoglobin releases its O_2 molecules to the cells of the body.

Both shape and surface chemistry contribute to protein specificity

The specific shape and structure of a protein allows it to bind *noncovalently* to another molecule (often called a **ligand**), which in turn

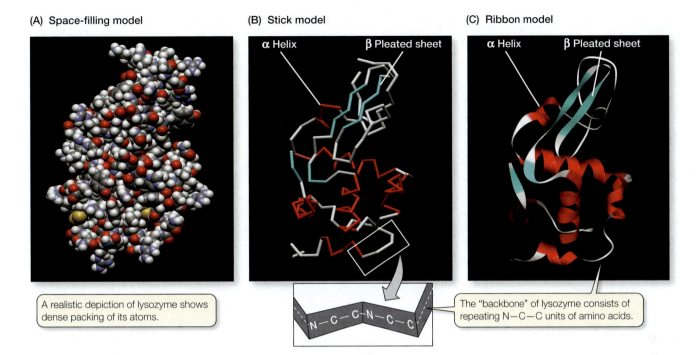

(A) Space-filling model

A realistic depiction of lysozyme shows dense packing of its atoms.

(B) Stick model

α Helix   β Pleated sheet

—N—C—C—N—C—C—

(C) Ribbon model

α Helix   β Pleated sheet

The "backbone" of lysozyme consists of repeating N—C—C units of amino acids.

3.8 Three Representations of Lysozyme Different molecular representations of a protein emphasize different aspects of its tertiary structure. These three representations of lysozyme are similarly oriented.

allows other important biological events to occur. Here are just a few examples of how protein functions follow from protein structure:

- Two adjacent cells can stick together because proteins protruding from each of the cells interact with each other (cell junctions; see Section 5.2).

- A substance can enter a cell by binding to a carrier protein in the cell surface membrane (membrane transport; see Section 5.3).

- A chemical reaction can be speeded up when a type of protein called an *enzyme* binds to the reactants (enzyme–substrate interactions; see Section 6.3).

- Chemical signals such as hormones can bind to proteins on a cell's outer surface (receptor proteins; see Section 15.1 and Chapter 41).

- Defensive proteins called antibodies can recognize the shape of a virus coat and bind to it (immune response; see Chapter 18).

The biological specificity of protein function depends on two general properties of the protein: its shape, and the chemistry of its exposed surface groups.

- *Shape.* When a small molecule collides with and binds to a much larger protein, it is like a baseball being caught by a catcher's mitt: the mitt has a shape that binds to the ball and fits around it. Just as a hockey puck or a ping-pong ball is not designed to fit a baseball catcher's mitt, a given molecule will bind to a protein only if there is a general "fit" between their two three-dimensional shapes. The need for the right "fit" becomes even more specific after initial binding.

- *Chemistry.* The functional groups on the surface of a protein promote chemical interactions with other substances (**Figure 3.10**). These groups are the side chains of the exposed amino acids, and are therefore a property of the protein's primary structure.

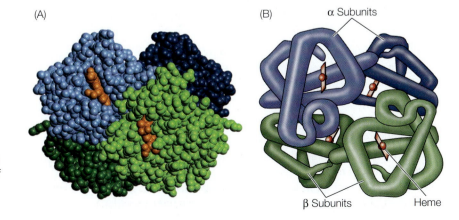

3.9 Quaternary Structure of a Protein Hemoglobin consists of four folded polypeptide subunits that assemble themselves into the quaternary structure shown here. In these two graphic representations, each type of subunit is a different color. The heme groups contain iron and are the oxygen-carrying sites.

(A)

(B)

α Subunits

β Subunits   Heme

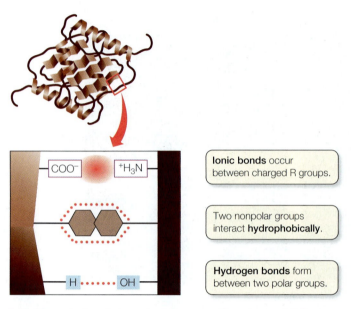

3.10 Noncovalent Interactions between Proteins and Other Molecules Noncovalent interactions allow a protein to bind tightly to another molecule with specific properties, or allow regions within a protein to interact with one another.

Examine the structures of the 20 amino acids in Table 3.2, noting the properties of the side chains. Exposed hydrophobic groups can bind to similarly nonpolar groups in the ligand. Charged R groups can bind to oppositely charged groups on the ligand. Polar R groups containing a hydroxyl (—OH) group can form a hydrogen bond with the ligand. These three types of interactions—hydrophobic, ionic, and hydrogen bonding—are each weak by themselves, but strong when all of them act together. So the exposure of appropriate amino acid R groups on the protein surface allows the binding of a specific ligand to occur.

Environmental conditions affect protein structure

Because it is determined by weak forces, the three-dimensional structure of proteins is sensitive to environmental conditions. Conditions that would not break covalent bonds, but can upset the weaker noncovalent interactions that determine secondary and tertiary structure, may affect a protein's shape and thus its function.

■ *Increases in temperature* cause more rapid molecular movements and thus can break hydrogen bonds and hydrophobic interactions.

■ *Alterations in pH* can change the pattern of ionization of carboxyl and amino groups in the R groups of amino acids, thus disrupting the pattern of ionic attractions and repulsions.

■ *High concentrations of polar substances* such as urea can disrupt the hydrogen bonding that is crucial to protein structure. Nonpolar substances may also disrupt normal protein structure.

The loss of a protein's normal three-dimensional structure is called **denaturation**, and it is always accompanied by a loss of the normal biological function of the protein (**Figure 3.11**). Denaturation is usually irreversible because amino acids that were buried in

the interior of the protein may now be exposed at the surface, and vice versa, causing a new structure to form or different molecules to bind to the protein. The boiling of an egg denatures its proteins and is, as you know, not reversible. However, as we saw earlier in the case of lysozyme, denaturation may be reversible in the laboratory. If the protein is allowed to cool or the denaturing chemicals are removed, the protein may return to its "native" shape and normal function.

Chaperonins help shape proteins

There are two occasions when a polypeptide chain is in danger of binding the wrong ligand. The first, as we have just seen, is following denaturation. In addition, when a protein has just been made and has not yet folded completely, it can present a surface that binds the wrong molecule. In the cell, a protein can sometimes fold incorrectly as it is made, with serious consequences. For example, in Alzheimer's disease, misfolded proteins accumulate in the brain and bind to one another, forming fibers in the areas of the brain that control memory, mood, and spatial awareness.

Living systems limit inappropriate protein interactions by making a class of proteins called, appropriately, **chaperonins** (recall the chaperones at school dances who tried to prevent "inappropriate interactions" among the students). Chaperonins were first identified in fruit flies as "heat shock" proteins, which prevented denaturing proteins from clumping together when the flies' temperatures were raised.

Some chaperonins work by trapping proteins that are in danger of inappropriate binding inside a molecular "cage" (**Figure 3.12**). This cage has several identical subunits, and is itself a good example of quaternary protein structure. Inside the cage, the targeted protein folds into the correct shape, and then is released at the appropriate time and place.

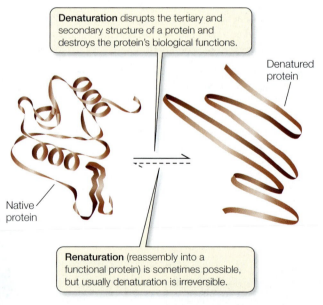

3.11 Denaturation Is the Loss of Tertiary Protein Structure and Function Agents that can cause denaturation include high temperatures and certain chemicals.

3.12 Chaperonins Protect Proteins from Inappropriate Binding Chaperonins surround new or denatured proteins and prevent them from binding to the wrong ligand.

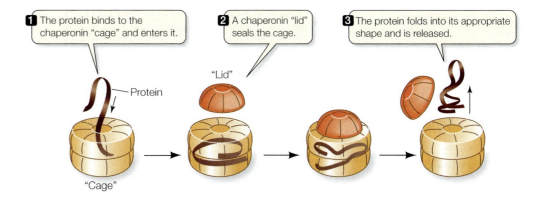

1 The protein binds to the chaperonin "cage" and enters it.

2 A chaperonin "lid" seals the cage.

3 The protein folds into its appropriate shape and is released.

Protein

"Lid"

"Cage"

3.2 RECAP

Proteins are polymers of amino acids. The sequence of amino acids in a protein determines its primary structure; secondary, tertiary, and quaternary structures arise through interactions between the amino acids. A protein's three-dimensional shape ultimately determines its function.

■ Can you describe the attributes of an amino acid's "R" group that would make it hydrophobic? Hydrophilic? See p. 42 and Table 3.2

■ Can you sketch and explain how two amino acids link together to form a peptide bond? See p. 43 and Figure 3.6

■ Can you describe the four levels of protein structure and how they are all ultimately determined by the protein's primary structure (i.e., its amino acid sequence)? See pp. 44–46 and Figure 3.7

■ Do you understand the different ways environmental factors such as temperature and pH can affect the weak interactions that give a protein its specific shape and function? See p. 48

3.3 What Are the Chemical Structures and Functions of Carbohydrates?

Carbohydrates are molecules containing carbon atoms flanked by hydrogen atoms and hydroxyl groups (H—C—OH). They have two major biochemical roles:

■ Carbohydrates are a source of energy that can be released in a form usable by body tissues.

■ Carbohydrates serve as *carbon skeletons* that can be rearranged to form new molecules that are essential for biological structures and functions.

Some carbohydrates are relatively small, with molecular weights of less than 100. Others are true macromolecules, with molecular weights in the hundreds of thousands.

There are four categories of biologically important carbohydrates, which we will discuss in turn:

■ **Monosaccharides** (*mono*, "one"; *saccharide*, "sugar"), such as glucose, ribose, and fructose, are simple sugars. They are the monomers from which the larger carbohydrates are constructed.

■ **Disaccharides** (*di*, "two") consist of two monosaccharides linked together by covalent bonds.

■ **Oligosaccharides** (*oligo*, "several") are made up of several (3–20) monosaccharides.

■ **Polysaccharides** (*poly*, "many"), such as starch, glycogen, and cellulose, are large polymers composed of hundreds or thousands of monosaccharides.

The general formula for carbohydrates, CH_2O, gives the relative proportions of carbon, hydrogen, and oxygen in a monosaccharide (i.e., the proportions of these atoms are 1:2:1). In disaccharides, oligosaccharides, and polysaccharides, these proportions differ slightly from the general formula because two hydrogens and an oxygen are lost during each of the condensation reactions that form them.

Monosaccharides are simple sugars

Green plants produce monosaccharides through photosynthesis, and animals acquire them directly or indirectly from plants. All living cells contain the monosaccharide **glucose**. Cells use glucose as an energy source, breaking it down through a series of reactions that release stored energy and produce water and carbon dioxide. Glucose exists in two forms, the straight chain form and the ring form. The ring form predominates in more than 99 percent of biological circumstances because it is more stable under physiological conditions. There are two versions of the ring form, called α- and β-glucose, which differ only in the orientation of the —H and —OH attached to carbon 1 (**Figure 3.13**). The α and β forms interconvert and exist in equilibrium when dissolved in water.

Different monosaccharides contain different numbers of carbons. (The standard convention for numbering carbons in carbohydrates shown in Figure 3.13 is used throughout this book.) Most of the monosaccharides found in living systems belong to the D series of optical isomers. (Recall also that only L-amino acids occur in proteins—there is amazing specificity in biology!) Some monosaccharides are structural isomers, with the same kinds and numbers of atoms, but in different arrangements. For example, the **hexoses** (*hex*, "six"), a group of structural isomers, all have the for-

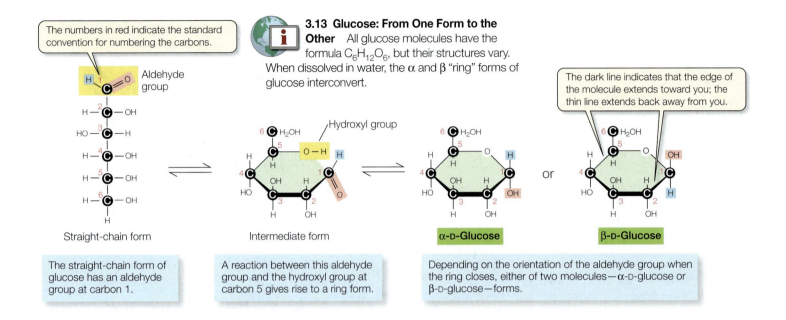

3.13 Glucose: From One Form to the Other All glucose molecules have the formula $C_6H_{12}O_6$, but their structures vary. When dissolved in water, the α and β "ring" forms of glucose interconvert.

The numbers in red indicate the standard convention for numbering the carbons.

The dark line indicates that the edge of the molecule extends toward you; the thin line extends back away from you.

Straight-chain form — The straight-chain form of glucose has an aldehyde group at carbon 1.

Intermediate form — A reaction between this aldehyde group and the hydroxyl group at carbon 5 gives rise to a ring form.

α-D-Glucose or **β-D-Glucose** — Depending on the orientation of the aldehyde group when the ring closes, either of two molecules—α-D-glucose or β-D-glucose—forms.

mula $C_6H_{12}O_6$. Included among the hexoses are glucose, fructose (so named because it was first found in fruits), mannose, and galactose (**Figure 3.14**).

Pentoses (*pente*, "five") are five-carbon sugars. Two pentoses are of particular biological importance: ribose and deoxyribose form part of the backbones of the nucleic acids RNA and DNA, respectively (see Section 3.5). These two pentoses are not isomers; rather, one oxygen atom is missing from carbon 2 in deoxyribose (*de-*, "absent"). The absence of this oxygen atom is an important distinction between RNA and DNA.

Glycosidic linkages bond monosaccharides

The disaccharides, oligosaccharides, and polysaccharides are all constructed from monosaccharides that are covalently bonded together by condensation reactions that form **glycosidic linkages**. A single glycosidic linkage between two monosaccharides forms a disaccharide. For example, a molecule of sucrose is a disaccharide formed from a glucose molecule and a fructose molecule, while lactose (milk sugar) contains glucose and galactose. Sucrose, which is common table sugar in the human diet, is a major disaccharide in plants.

The disaccharide maltose contains two glucose molecules, but it is not the only disaccharide that can be made from two glucoses. When glucose molecules form a glycosidic linkage, the linkage will be one of two types, α or β, depending on whether the molecule that bonds its carbon 1 is α-D-glucose or β-D-glucose (see Figure 3.13). An α linkage with carbon 4 of a second glucose molecule produces maltose, whereas a β linkage gives cellobiose (**Figure 3.15**). Maltose and cellobiose are structural isomers, both having the formula $C_{12}H_{22}O_{11}$. However, they are different compounds with different properties. They undergo different chemical reactions and are recognized by different enzymes. For example, maltose can be hydrolyzed into its monosaccharides in the human

Three-carbon sugar

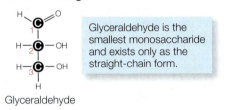

Glyceraldehyde is the smallest monosaccharide and exists only as the straight-chain form.

Glyceraldehyde

Five-carbon sugars (pentoses)

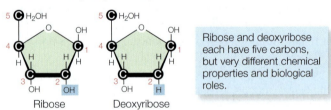

Ribose and deoxyribose each have five carbons, but very different chemical properties and biological roles.

Ribose   Deoxyribose

Six-carbon sugars (hexoses)

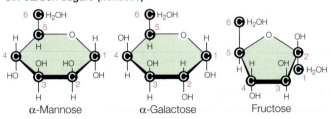

α-Mannose   α-Galactose   Fructose

These hexoses are structural isomers. All have the formula $C_6H_{12}O_6$, but each has distinct biochemical properties.

3.14 Monosaccharides Are Simple Sugars Monosaccharides are made up of varying numbers of carbons. Some hexoses are structural isomers that have the same kind and number of atoms, but the atoms are arranged differently. Fructose, for example, is a hexose, but forms a five-membered ring like the pentoses.

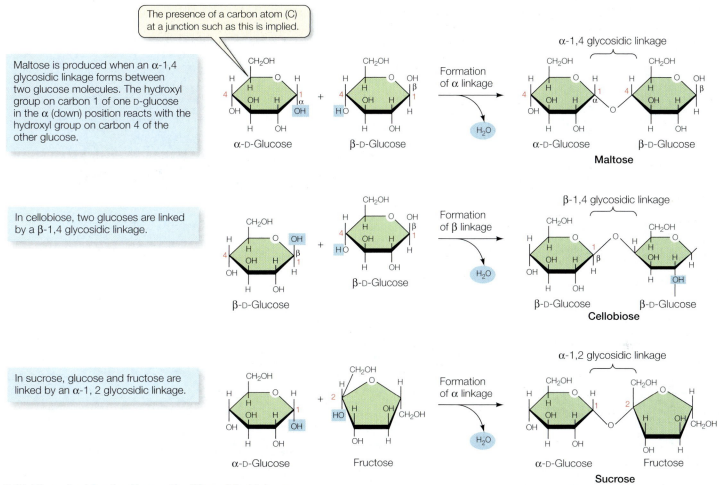

3.15 Disaccharides Are Formed by Glycosidic Linkages
Glycosidic linkages between two monosaccharides can create many different disaccharides. Which disaccharide is formed depends on which monosaccharides are linked, and on the site (which carbon atom is linked) and form (α or β) of the linkage.

body, whereas cellobiose cannot. Certain microorganisms have the chemistry needed to break down cellobiose.

Oligosaccharides contain several monosaccharides bound by glycosidic linkages at various sites. Many oligosaccharides have additional functional groups, which give them special properties. Oligosaccharides are often covalently bonded to proteins and lipids on the outer cell surface, where they serve as recognition signals. The different human blood groups (such as the ABO blood types) get their specificity from oligosaccharide chains.

Polysaccharides store energy and provide structural materials

Polysaccharides are giant polymers of monosaccharides connected by glycosidic linkages (**Figure 3.16**):

- *Starch* is a polysaccharide of glucose with α-glycosidic linkages.
- *Glycogen* is a highly branched polysaccharide of glucose.
- *Cellulose* is also a polysaccharide of glucose, but its individual monosaccharides are connected by β-glycosidic linkages.

Starch comprises a family of giant molecules of broadly similar structure. While all starches are large polymers of glucose with α linkages (see Figure 3.16A), the different starches can be distinguished by the amount of branching that occurs at carbons 1 and 6 (see Figure 3.16B). Some plant starches are unbranched, such as plant amylose; others are moderately branched, such as plant amylopectin. Starch readily binds water, and when that water is removed, unbranched starch tends to form hydrogen bonds between the polysaccharide chains, which then aggregate.

Bread becomes hard and stale because when it dries out, the polysaccharide chains in starch aggregate. Adding water and gentle heat separates the chains and the bread becomes softer.

Glycogen stores glucose in animal livers and muscles. Starch and glycogen serve as energy storage compounds for plants and animals, respectively. Both of these polysaccharides are readily hydrolyzed into glucose monomers, which in turn can be further degraded to liberate their stored energy. But if it is glucose that is actually needed for fuel, why must it be stored as a polymer? The reason is that 1,000 glucose molecules would exert 1,000 times the osmotic pressure of a single glycogen molecule, causing water to

(A) Molecular structure

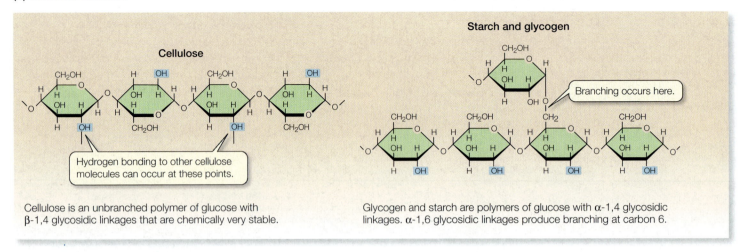

Cellulose

Starch and glycogen

Branching occurs here.

Hydrogen bonding to other cellulose molecules can occur at these points.

Cellulose is an unbranched polymer of glucose with β-1,4 glycosidic linkages that are chemically very stable.

Glycogen and starch are polymers of glucose with α-1,4 glycosidic linkages. α-1,6 glycosidic linkages produce branching at carbon 6.

(B) Macromolecular structure

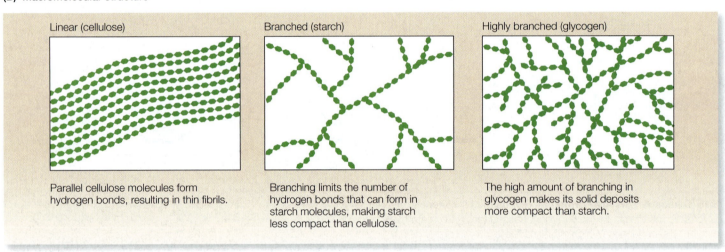

Linear (cellulose)

Parallel cellulose molecules form hydrogen bonds, resulting in thin fibrils.

Branched (starch)

Branching limits the number of hydrogen bonds that can form in starch molecules, making starch less compact than cellulose.

Highly branched (glycogen)

The high amount of branching in glycogen makes its solid deposits more compact than starch.

(C) Polysaccharides in cells

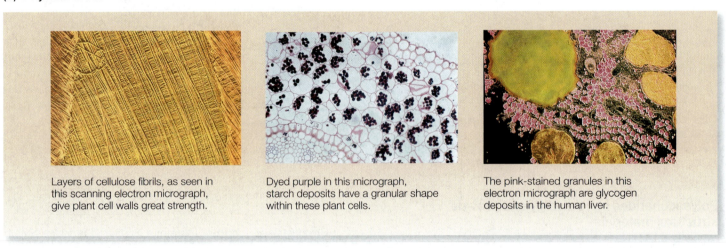

Layers of cellulose fibrils, as seen in this scanning electron micrograph, give plant cell walls great strength.

Dyed purple in this micrograph, starch deposits have a granular shape within these plant cells.

The pink-stained granules in this electron micrograph are glycogen deposits in the human liver.

3.16 Representative Polysaccharides Cellulose, starch, and glycogen demonstrate different levels of branching and compaction in polysaccharides.

enter the cells (see Section 5.3). If it were not for polysaccharides, many organisms would expend a lot of time and energy expelling excess water from their cells.

Cellulose is the predominant component of plant cell walls, and is by far the most abundant organic compound on Earth. Starch can be easily degraded by the actions of chemicals or enzymes. Cellulose, however, is chemically more stable because of its β-glycosidic linkages (see Figure 3.16A). Thus starch is a good storage medium that can be easily broken down to supply glucose for energy-producing reactions, while cellulose is an excellent structural material that can withstand harsh environmental conditions without changing.

Chemically modified carbohydrates contain additional functional groups

Some carbohydrates are chemically modified by the addition of functional groups, such as phosphate and amino groups (**Figure 3.17**). For example, carbon 6 in glucose may be oxidized from —CH_2OH to a carboxyl group (—COOH), producing glucuronic acid. Or a phosphate group may be added to one or more of the —OH sites. Some of the resulting *sugar phosphates*, such as fructose 1,6-bisphosphate, are important intermediates in cellular energy reactions, which will be discussed in Chapter 7.

When an amino group is substituted for an —OH group, *amino sugars* such as glucosamine and galactosamine, are produced. These compounds are important in the extracellular matrix (see Section 4.4), where they form parts of glycoproteins involved in keeping tissues together. Galactosamine is a major component of cartilage, the material that forms caps on the ends of bones and stiffens the ears and nose. A derivative of glucosamine produces the polymer *chitin*, which is the principal structural polysaccharide in the skeletons of insects and many crustaceans (e.g., crabs and lobsters), as well as in the cell walls of fungi.

Taken together, fungi, insects, and crustaceans account for more than 80 percent of the species ever described. Thus the chitin that supports their bodies is one of the most abundant substances on Earth.

3.17 Chemically Modified Carbohydrates Added functional groups can modify the form and properties of a carbohydrate.

(A) Sugar phosphate

Fructose 1,6 bisphosphate is involved in the reactions that liberate energy from glucose. (The numbers in its name refer to the carbon sites of phosphate bonding; *bis-* indicates that two phosphates are present.)

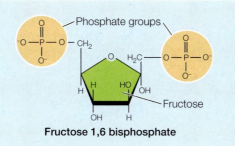

Phosphate groups

Fructose

Fructose 1,6 bisphosphate

(B) Amino sugars

The monosaccharides glucosamine and galactosamine are amino sugars with an amino group in place of a hydroxyl group.

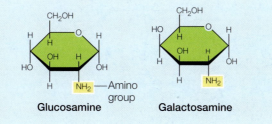

Amino group

Glucosamine   **Galactosamine**

Galactosamine is an important component of cartilage, a connective tissue in vertebrates.

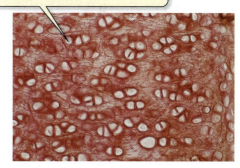

(C) Chitin

Chitin is a polymer of *N*-acetylglucosamine; *N*-acetyl groups provide additional sites for hydrogen bonding between the polymers.

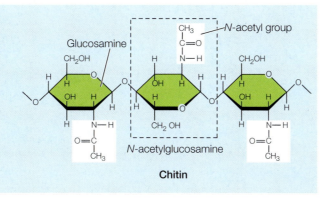

Glucosamine

N-acetyl group

N-acetylglucosamine

Chitin

The external skeletons of insects are made up of chitin.

3.3 RECAP

Carbohydrates are composed of carbon, hydrogen, and oxygen in the general ratio of 1:2:1. They provide energy and structure to cells and are precursors of numerous important biological molecules. Monosaccharide monomers can be connected by glycosidic linkages to form disaccharides, oligosaccharides, and polysaccharides.

- Can you draw the chemical structure of a disaccharide formed by two monosaccharides? See Figure 3.15

- What qualities of the polysaccharides starch and glycogen make them useful for energy storage? See p. 51 and Figure 3.16

- From looking at the cellulose molecules in Figure 3.16A, can you see where a large number of hydrogen bonds are present in the linear structure of cellulose shown in Figure 3.16B? Do you understand why this structure is so strong?

We have now seen how amino acid monomers form protein polymers and how saccharide monomers form the polymers of carbohydrates. Now we will look at the lipids, which are unique among the four classes of large biological molecules in that they are not, strictly speaking, polymers.

3.4 What Are the Chemical Structures and Functions of Lipids?

Lipids are hydrocarbons that are insoluble in water because of their many nonpolar covalent bonds. As we saw in Section 2.2, nonpolar hydrocarbon molecules are hydrophobic and preferentially aggregate among themselves, away from water, which is polar. When these nonpolar hydrocarbons are sufficiently close together, weak but additive van der Waals forces hold them together. These huge macromolecular aggregations are not polymers in a strict chemical sense, because the individual lipid molecules are not covalently bonded. However, they can be considered to be polymers of individual lipid subunits.

There are several different types of lipids, and they play a number of roles in living organisms:

- Fats and oils store energy.
- Phospholipids play important structural roles in cell membranes.
- The carotenoids help plants capture light energy.
- Steroids and modified fatty acids play regulatory roles as hormones and vitamins.
- The fat in animal bodies serves as thermal insulation.
- A lipid coating around nerves provides electrical insulation.
- Oil or wax on the surfaces of skin, fur, and feathers repels water.

3.18 Synthesis of a Triglyceride In living things, the reaction that forms triglycerides is more complex, but the end result is as shown here.

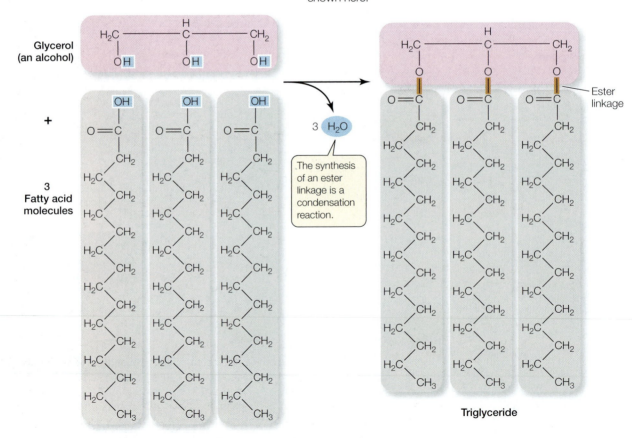

Fats and oils store energy

Chemically, fats and oils are *triglycerides*, also known as *simple lipids*. Triglycerides that are solid at room temperature (20°C) are called **fats**; those that are liquid at room temperature are called **oils**. Triglycerides are composed of two types of building blocks: *fatty acids* and *glycerol*. Glycerol is a small molecule with three hydroxyl (—OH) groups (thus it is an alcohol). A **fatty acid** is made up of a long nonpolar hydrocarbon chain and a polar carboxyl group (—COOH). A **triglyceride** contains three fatty acid molecules and one molecule of glycerol. The carboxyl group of a fatty acid can bond with the hydroxyl group of glycerol, resulting in a covalent bond called an **ester linkage** and water (**Figure 3.18**).

The three fatty acids in a triglyceride molecule need not all have the same hydrocarbon chain length or structure:

- In **saturated** fatty acids, all the bonds between the carbon atoms in the hydrocarbon chain are single bonds—there are no double bonds. That is, all the bonds are saturated with hydrogen atoms (**Figure 3.19A**). These fatty acid molecules are relatively rigid and straight, and they pack together tightly, like pencils in a box.

- In **unsaturated** fatty acids, the hydrocarbon chain contains one or more double bonds. Oleic acid, for example, is a *monounsaturated* fatty acid that has one double bond near the middle of the hydrocarbon chain, which causes a kink in the molecule (**Figure 3.19B**). Some fatty acids have more than one double bond and have multiple kinks; they are *polyunsaturated* fatty acids. Kinks prevent the unsaturated fat molecules from packing together tightly.

The kinks in fatty acid molecules are important in determining the fluidity and melting point of a lipid. The triglycerides of animal fats tend to have many long-chain saturated fatty acids, packed tightly together; these fats are usually solids at room temperature and have a high melting point. The triglycerides of plants, such as corn oil, tend to have short or unsaturated fatty acids. Because of their kinks, these fatty acids pack together poorly and have a low melting point, and these triglycerides are usually liquids at room temperature.

Fats and oils are marvelous energy storehouses. In 1900, the German engineer Rudolf Diesel used peanut oil to power one of his early automobile engines, and there has been a recent resurgence of biodiesel technology. One successful application converts used cooking oil into "French fry fuel."

Phospholipids form biological membranes

Because lipids and water do not interact, when water and lipids are mixed together two distinct layers form. Many biologically important substances—such as ions, sugars, and free amino acids—that are soluble in water are insoluble in lipids.

Like triglycerides, **phospholipids** contain fatty acids bound to glycerol by ester linkages. In phospholipids, however, any one of several phosphate-containing compounds replaces one of the fatty acids (**Figure 3.20A**). The phosphate functional group has a neg-

ative electric charge, so this portion of the molecule is hydrophilic, attracting polar water molecules. But the two fatty acids are hydrophobic, so they tend to aggregate away from water.

In an aqueous environment, phospholipids line up in such a way that the nonpolar, hydrophobic "tails" pack tightly together and the phosphate-containing "heads" face outward, where they interact with water. The phospholipids thus form a **bilayer**: a sheet two molecules thick, with water excluded from the core (**Figure 3.20B**). Biological membranes have this kind of phospholipid bilayer structure, and we will devote all of Chapter 5 to their biological functions.

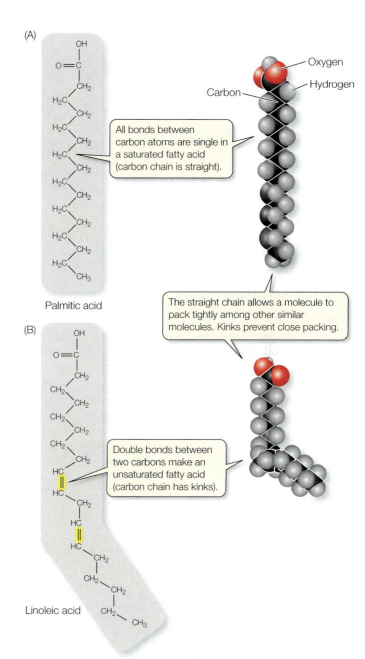

3.19 Saturated and Unsaturated Fatty Acids (A) The straight hydrocarbon chain of a saturated fatty acid allows the molecule to pack tightly among other similar molecules. (B) In unsaturated fatty acids, kinks in the chain prevent close packing. Note the convention in the space-filling molecular models shown here: gray, H; red, O; black, C.

(A) Phosphatidylcholine

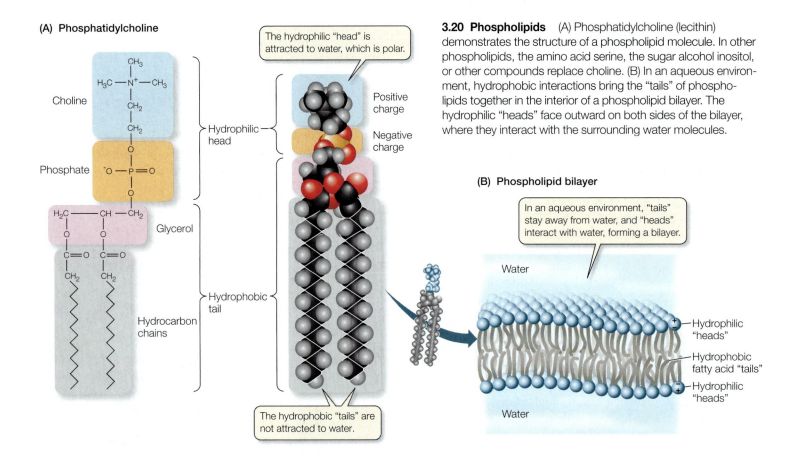

The hydrophilic "head" is attracted to water, which is polar.

Choline

Phosphate

Glycerol

Hydrocarbon chains

Hydrophilic head

Hydrophobic tail

Positive charge

Negative charge

The hydrophobic "tails" are not attracted to water.

3.20 Phospholipids (A) Phosphatidylcholine (lecithin) demonstrates the structure of a phospholipid molecule. In other phospholipids, the amino acid serine, the sugar alcohol inositol, or other compounds replace choline. (B) In an aqueous environment, hydrophobic interactions bring the "tails" of phospholipids together in the interior of a phospholipid bilayer. The hydrophilic "heads" face outward on both sides of the bilayer, where they interact with the surrounding water molecules.

(B) Phospholipid bilayer

In an aqueous environment, "tails" stay away from water, and "heads" interact with water, forming a bilayer.

Water

Water

Hydrophilic "heads"

Hydrophobic fatty acid "tails"

Hydrophilic "heads"

Not all lipids are triglycerides

A number of classes of lipids are not based on the glycerol-fatty acid structure we have just described. Yet because they are made up largely of carbon and hydrogen and are nonpolar, these molecules are still classified as lipids.

CAROTENOIDS The *carotenoids* are a family of light-absorbing pigments found in plants and animals. Beta-carotene (β-carotene) is one of the pigments that traps light energy in leaves during photosynthesis. In humans, a molecule of β-carotene can be broken down into two vitamin A molecules (**Figure 3.21**), from which we

make the pigment rhodopsin, which is required for vision. Carotenoids are responsible for the colors of carrots, tomatoes, pumpkins, egg yolks, and butter.

STEROIDS The *steroids* are a family of organic compounds whose multiple rings share carbons (**Figure 3.22**). The steroid cholesterol is an important constituent of membranes. Other steroids function as hormones, chemical signals that carry messages from one part of the body to another (see Chapter 41). Cholesterol is synthesized in the liver and is the starting material for making testosterone and other steroid hormones, as well as the bile salts that help break down dietary fats so that they can be digested. Cholesterol is absorbed from foods such as milk, butter, and animal fats.

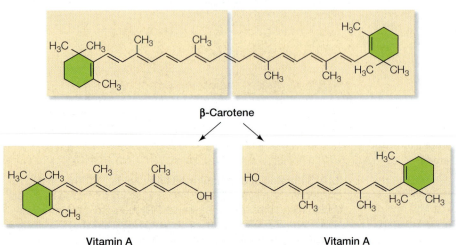

β-Carotene

Vitamin A

Vitamin A

3.21 β-Carotene is the Source of Vitamin A The carotenoid β-carotene is symmetrical around its central double bond; when that bond is broken, the result is two vitamin A molecules. The simplified structural formula used here is standard chemical shorthand for large organic molecules with many carbon atoms. Structural formulas are simplified by omitting the C (indicating a carbon atom) at the intersections of the lines representing covalent bonds. Hydrogen atoms (H) to fill all the available bonding sites on each C are assumed.

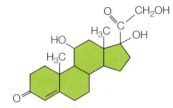

Cholesterol is a constituent of membranes and is the source of steroid hormones.

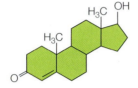

Vitamin D₂ can be produced in the skin by the action of light on a cholesterol derivative.

Cortisol is a hormone secreted by the adrenal glands.

Testosterone is a male sex hormone.

3.22 All Steroids Have the Same Ring Structure The steroids shown here, all important in vertebrates, are composed of carbon and hydrogen and are highly hydrophobic. However, small chemical variations, such as the presence or absence of a hydroxyl group, can produce enormous functional differences among these molecules.

VITAMINS **Vitamins** are small molecules that are not synthesized by the human body and so must be acquired from the diet (see Chapter 50). For example, *vitamin A* is formed from the β-carotene found in green and yellow vegetables (see Figure 3.21). In humans, a deficiency of vitamin A leads to dry skin, eyes, and internal body surfaces, retarded growth and development, and night blindness, which is a diagnostic symptom for the deficiency. Vitamins D, E, and K are also lipids.

WAXES The sheen on human hair is more than cosmetic. Glands in the skin secrete a waxy coating that repels water and keeps the hair pliable. Birds that live near water have a similar waxy coating on their feathers. The shiny leaves of holly plants, familiar during winter holidays, also have a waxy coating. Finally, bees make their honeycombs out of wax. All waxes have the same basic structure: they are formed by an ester linkage between a saturated, long-chain fatty acid and a saturated, long-chain alcohol. The result is a very long molecule, with 40–60 CH_2 groups. For example, here is the structure of beeswax:

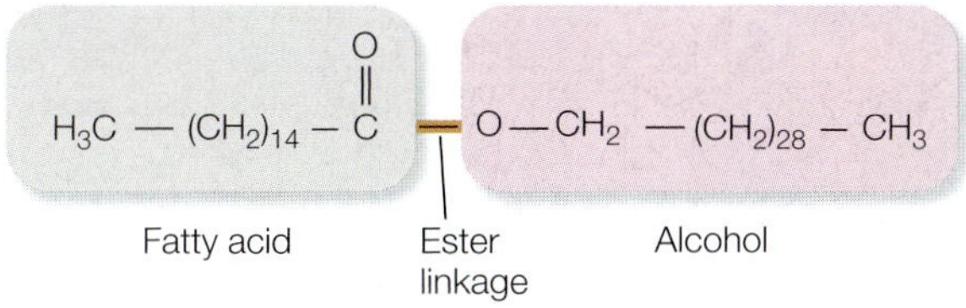

$$H_3C-(CH_2)_{14}-\overset{\overset{\displaystyle O}{\|}}{C}-O-CH_2-(CH_2)_{28}-CH_3$$

Fatty acid   Ester linkage   Alcohol

This highly nonpolar structure accounts for the impermeability of wax to water.

Although all the large molecules discussed in this chapter are found only in living organisms, the final class of biological molecules we will discuss is especially linked to the living world. The function of the nucleic acids is nothing less than the transmission of life's "blueprint" to every new organism.

3.5 What Are the Chemical Structures and Functions of Nucleic Acids?

The **nucleic acids** are polymers specialized for the storage, transmission, and use of genetic information. There are two types of nucleic acids: **DNA** (deoxyribonucleic acid) and **RNA** (ribonucleic acid). DNA is a macromolecule that encodes hereditary information and passes it from generation to generation. Through an RNA intermediate, the information encoded in DNA is used to specify the amino acid sequence of proteins. Information flows from DNA to DNA in reproduction, but in the nonreproductive activities of the cell, information flows from DNA to RNA to proteins, which ultimately carry out life's functions.

Nucleotides are the building blocks of nucleic acids

Nucleic acids are composed of monomers called **nucleotides**, each of which consists of a pentose sugar, a phosphate group, and a nitrogen-containing **base**. The bases of the nucleic acids take one of two chemical forms: a single-ring structure called a **pyrimidine** or a fused-ring structure called a **purine** (**Figure 3.23**). (Molecules consisting of a pentose sugar and a nitrogenous base—but no phosphate group—are called *nucleosides*.) In DNA, the pentose sugar is **deoxyribose**, which differs from the **ribose** found in RNA by one oxygen atom (see Figure 3.14).

3.23 Nucleotides Have Three Components

A nucleotide consists of a phosphate group, a pentose sugar (ribose or deoxyribose), and a nitrogen-containing base, all linked together by covalent bonds. The nitrogenous bases have two different chemical forms: purines have two fused rings, and the smaller pyrimidines have a single ring.

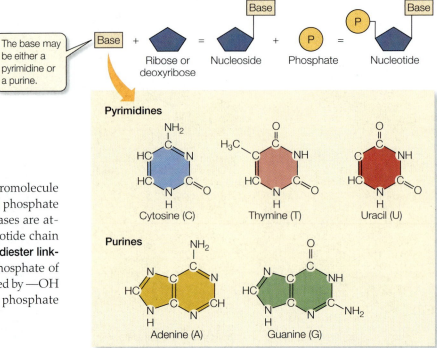

The base may be either a pyrimidine or a purine.

Base + Ribose or deoxyribose = Nucleoside + Phosphate = Nucleotide

Pyrimidines

Cytosine (C) Thymine (T) Uracil (U)

Purines

Adenine (A) Guanine (G)

In both RNA and DNA, the backbone of the macromolecule consists of a chain of alternating pentose sugars and phosphate groups (sugar–phosphate–sugar–phosphate). The bases are attached to the sugars and project from the polynucleotide chain (**Figure 3.24**). The nucleotides are joined by **phosphodiester linkages** between the sugar of one nucleotide and the phosphate of the next (*diester* refers to the two covalent bonds formed by —OH groups reacting with acidic phosphate groups). The phosphate

3.24 Distinguishing Characteristics of DNA and RNA

RNA is usually a single strand. DNA usually consists of two strands running in opposite directions (antiparallel).

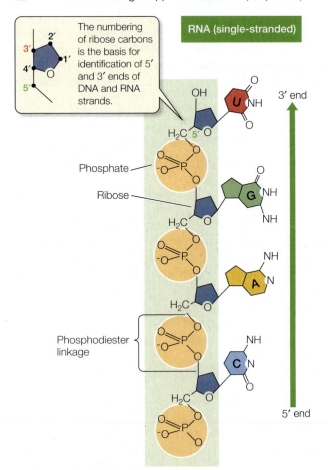

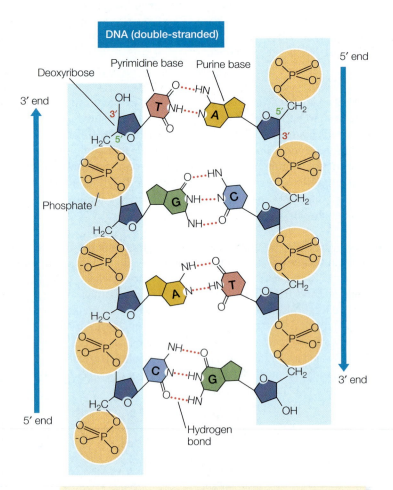

The numbering of ribose carbons is the basis for identification of 5′ and 3′ ends of DNA and RNA strands.

RNA (single-stranded)

3′ end

Phosphate

Ribose

Phosphodiester linkage

5′ end

DNA (double-stranded)

Deoxyribose

Pyrimidine base Purine base

Phosphate

5′ end

3′ end

3′ end

5′ end

Hydrogen bond

In RNA, the bases are attached to ribose. The bases in RNA are the purines adenine (A) and guanine (G) and the pyrimidines cytosine (C) and uracil (U).

In DNA, the bases are attached to deoxyribose, and the base thymine (T) is found instead of uracil. Hydrogen bonds between purines and pyrimidines hold the two strands of DNA together.

groups link carbon 3′ in one pentose sugar to carbon 5′ in the adjacent sugar.

Most RNA molecules consist of only one polynucleotide chain. DNA, however, is usually double-stranded; its two polynucleotide chains are held together by hydrogen bonding between their nitrogenous bases. The two strands of DNA run in opposite directions. You can see what this means by drawing an arrow through a phosphate group from carbon 5 to carbon 3 in the next ribose. If you do this for both strands of the DNA in Figure 3.24, the arrows will point in opposite directions. This *antiparallel* orientation allows the strands to fit together in three-dimensional space.

The uniqueness of a nucleic acid resides in its nucleotide sequence

Only four nitrogenous bases—and thus only four nucleotides—are found in DNA. The DNA bases and their abbreviations are **adenine** (A), **cytosine** (C), **guanine** (G), and **thymine** (T). A key to understanding the structure and function of nucleic acids is the principle of **complementary base pairing**. In double-stranded DNA, adenine and thymine always pair (A-T), and cytosine and guanine always pair (C-G).

Base pairing is complementary because of three factors: the sites for hydrogen bonding on each base; the geometry of the sugar–phosphate backbone, which brings opposite bases near each other; and the molecular sizes of the paired bases. Adenine and guanine are both purines; thymine and cytosine are both pyrimidines. The pairing of a large purine with a smaller pyrimidine ensures stability and consistency in the double-stranded molecule of DNA.

RNA is also made up of four different monomers, but its nucleotides differ from those of DNA. In RNA the nucleotides are termed *ribonucleotides* (the ones in DNA are *deoxyribonucleotides*). They contain ribose rather than deoxyribose, and instead of the base thymine, RNA uses the base **uracil** (U). The other three bases are the same as those in DNA (**Table 3.3**).

Although RNA is generally single-stranded, complementary hydrogen bonding between ribonucleotides plays important roles in determining the three-dimensional shapes of some types of RNA molecules (**Figure 3.25**). Complementary base pairing can

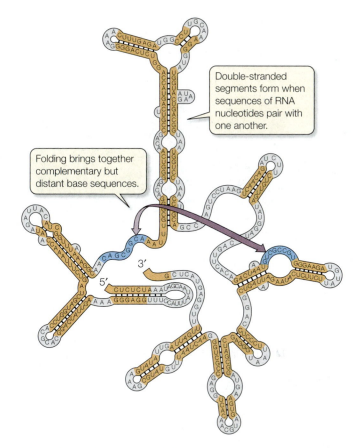

3.25 Hydrogen Bonding in RNA When a single-stranded RNA folds in on itself, hydrogen bonds can stabilize it into a three-dimensional shape.

also take place between ribonucleotides and deoxyribonucleotides. In RNA, guanine and cytosine pair (G-C), as in DNA, but adenine pairs with uracil (A-U). Adenine in an RNA strand can pair either with uracil (in another RNA strand) or with thymine (in a DNA strand).

DNA is a purely *informational* molecule. The information in DNA is encoded in the sequence of bases carried in its strands—the information encoded in the sequence TCAG is different from the information in the sequence CCAG. RNA uses the information carried in the nucleotide sequence of a DNA molecule to specify the amino acid sequence that in turn dictates the primary structure of a protein. This information can be read easily and reliably, in a specific order, as we will see in Chapter 12.

The three-dimensional physical appearance of DNA is strikingly uniform. The segment shown in **Figure 3.26** could be from any DNA molecule. The variations in DNA—the different sequences of bases—are strictly internal. Through hydrogen bonding, the two complementary polynucleotide strands pair and twist to form a **double helix**. When compared with the complex and varied tertiary structures of proteins, this uniformity is surprising. But this structural contrast makes sense in terms of the functions of these two classes of macromolecules. As we saw in Section 3.2, the different and unique shapes of proteins permit these macromolecules to recognize specific "target" molecules. The unique three-dimensional form of each protein matches at least a portion of the

TABLE 3.3		
Distinguishing RNA from DNA		
NUCLEIC ACID	**SUGAR**	**BASES**
RNA	Ribose	Adenine
		Cytosine
		Guanine
		Uracil
DNA	Deoxyribose	Adenine
		Cytosine
		Guanine
		Thymine

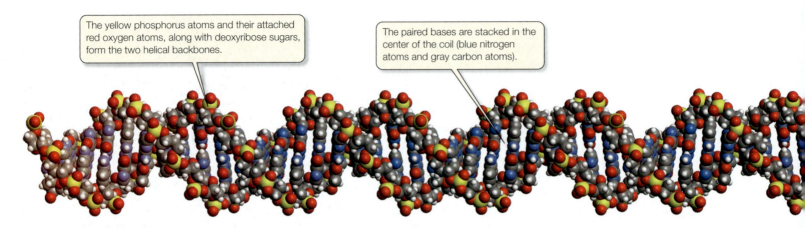

The yellow phosphorus atoms and their attached red oxygen atoms, along with deoxyribose sugars, form the two helical backbones.

The paired bases are stacked in the center of the coil (blue nitrogen atoms and gray carbon atoms).

3.26 The Double Helix of DNA The backbones of the two strands in a DNA molecule are coiled in a double helix. In this model, the small white atoms represent hydrogen.

surface of the target molecule. In other words, structural diversity in the molecules to which proteins bind requires corresponding diversity in the structure of the proteins themselves. Structural diversity is necessary in DNA as well. However, the diversity of DNA is found in the structure of its base sequence rather than in the physical shape of the molecule.

DNA reveals evolutionary relationships

Because DNA carries hereditary information between generations, a theoretical series of DNA molecules, with changes in base sequences, stretches back through the lineages of all organisms to the beginning of evolutionary time. Closely related living species should have more similar base sequences than species judged by other criteria to be more distantly related. The details of how scientists use this information are covered in Chapter 25.

The elucidation and examination of DNA base sequences has confirmed many of the evolutionary relationships that have been inferred from the more traditional comparisons of body structures, biochemistry, and physiology. For example, the closest living relative of humans (*Homo sapiens*) has usually been considered to be the chimpanzee (genus *Pan*), and in fact chimpanzee DNA shares more than 98 percent of its DNA base sequence with human DNA. Increasingly, scientists turn to DNA analyses to elucidate evolutionary relationships when studies of structure are not possible or are not conclusive. For example, DNA studies revealed a close relationship between starlings and mockingbirds that was not expected on the basis of their anatomy or behavior.

Nucleotides have other important roles

Nucleotides are more than just the building blocks of nucleic acids. As we will see in later chapters, there are several nucleotides with other functions:

■ ATP (adenosine triphosphate) acts as an energy transducer in many biochemical reactions (see Section 6.2).

■ GTP (guanosine triphosphate) serves as an energy source, especially in protein synthesis. It also has a role in the transfer of information from the environment to cells (see Section 15.2).

■ cAMP (cyclic adenosine monophosphate), a special nucleotide in which an additional bond forms between the sugar and phosphate groups, is essential in many processes, including the actions of hormones and the transmission of information by the nervous system (see Section 15.3).

3.5 RECAP

The nucleic acids DNA and RNA are polymers of nucleotide monomers. The sequence of nucleotides in DNA carries the information used by RNA to specify primary protein structure. The genetic information in DNA is passed from generation to generation and can be studied to understand evolutionary relationships.

■ Can you describe the key differences between DNA and RNA? Between purines and pyrimidines? See pp. 57–59, Figure 3.23, and Table 3.3

■ Do you understand how purines and pyrimidines pair up in complementary bonding between nucleotides? See p. 59 and Figure 3.24

■ Do you see how there can be immense diversity among DNA molecules even though they all appear to be structurally similar? See pp. 59–60

We have seen that the nucleic acids RNA and DNA carry the blueprint of life, and that the heritage of these macromolecules reaches back to the beginning of evolutionary time. But where did the nucleic acids come from? How did the building blocks of life originally arise?

3.6 How Did Life on Earth Begin?

As we saw in Chapter 2, living things are composed of the same atomic elements as the inanimate universe—the 92 naturally occurring elements of the periodic table (see Figure 2.2). But the arrangements of these atoms into molecules are unique in biological systems. You won't find biological molecules in inanimate matter (unless they came from a once-living organism).

How life began on Earth, or anywhere else, is impossible to know for certain. There are two prevailing scientific theories for the origin of life on Earth:

- The molecules of life arrived on Earth from extraterrestrial sources.
- Life is the result of chemical evolution on Earth.

Could life have come from outside Earth?

As mentioned in Chapter 2, comets are thought to have brought Earth most of its life-engendering water. Recently, it has become apparent that several meteorites from Mars have landed on Earth, and that some of these meteors carry some molecules that are possibly characteristic of life.

In 1984, a softball-sized rock was found on the ice in the Allan Hills region of Antarctica. ALH 84001, as it came to be called, was a meteorite that came from Mars (**Figure 3.27**). We know this because the composition of the gases trapped within the rock was identical to the Martian atmosphere, which is quite different from Earth's atmosphere. Radioactive dating and mineral analyses determined that ALH 84001 was 4.5 billion years old and was blasted off the Martian surface 16 million years ago, landing on Earth fairly recently, about 13,000 years ago.

Scientists found water trapped below the Martian meteorite's surface. This discovery was not surprising, considering that surface observations had shown that liquid water was once abundant on Mars (see Chapter 2). Because water is essential for life, scientists wondered whether the meteorite might contain other signs of life as well. Their analysis revealed two substances related to living systems. First, simple carbon-containing molecules called polycyclic aromatic hydrocarbons were present in small but unmistakable amounts; these substances can be formed by living organisms. Second, crystals of magnetite, an iron oxide mineral made by many living things on Earth, were found in the interior of the rock.

ALH 84001 is not the only visitor from outer space that has been shown to contain chemical signatures of life. Fragments of a meteorite that fell around the town of Murchison, Australia, in 1969 were found to contain molecules that are unique to life, including purines, pyrimidines, and amino acids. Although the presence of such molecules in rocks may suggest that those rocks once harbored life, it does not prove that there were living things in the rocks when they landed on Earth.

Most scientists find it hard to believe that an organism in a meteorite could survive thousands of years of traveling through space, followed by intense heat as the meteorite passed through Earth's atmosphere. But there is some evidence that the heat inside some meteorites may not have been severe. When weakly magnetized rock is heated, it reorients its magnetic field to align with the magnetic field around it. In the case of ALH 84001, this would have been Earth's powerful magnetic field, which would have affected the meteorite as it approached our planet. Careful measurements indicate that, while reorientation did occur at the surface of the rock, it did not occur in the inside. The scientists who took these measurements, Benjamin Weiss and Joseph Kirschvink at the California Institute of Technology, concluded that the inside of ALH 84001 was never heated over 40°C on its trip to Antarctica. This evidence makes a long interplanetary trip by living organisms more plausible.

Did life originate on Earth?

Both Earth and Mars once had the water and other simple molecules that could, under the right conditions, form the large molecules unique to life. The second theory of the origin of life on Earth, **chemical evolution**, holds that conditions on primitive Earth led to the emergence of these molecules. Scientists have sought to reconstruct those primitive conditions.

Early in the twentieth century, researchers proposed that there was little oxygen gas (O_2) in Earth's first atmosphere. Oxygen gas is thought to have accumulated in quantity about 2.5 billion years ago as the by-product of photosynthesis by single-celled life forms (today, it constitutes 21 percent of our atmosphere). In the 1950s, Stanley Miller and Harold Urey set up an experimental "atmosphere" containing the gases they believed to have been present in Earth's early atmosphere: hydrogen gas, ammonia, methane gas, and water vapor. Through these gases, they passed a spark to simulate lightning, then cooled the system so the gases would condense and collect in a watery solution, or "ocean" (**Figure 3.28**). Within days, the system contained numerous complex molecules, including amino acids, purines, and pyrimidines—some of the building blocks of life.

In science, an experiment and its results must be constantly reinterpreted, repeated, and refined as more knowledge accumulates. The results of the Miller–Urey experiments have undergone several such refinements:

3.27 Was Life Once Here? The meteorite ALH 84001, which came from Mars and landed in Antarctica, contains chemical signatures of life.

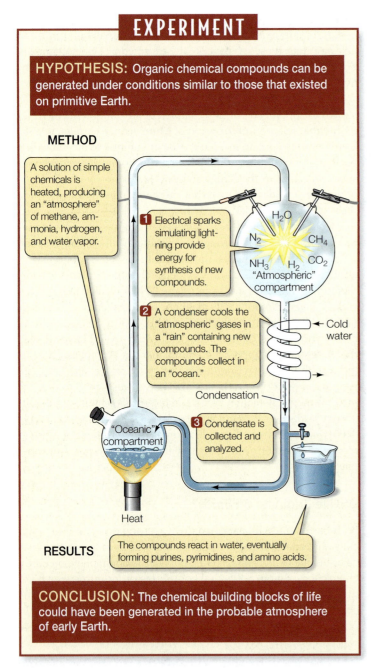

EXPERIMENT

HYPOTHESIS: Organic chemical compounds can be generated under conditions similar to those that existed on primitive Earth.

METHOD

A solution of simple chemicals is heated, producing an "atmosphere" of methane, ammonia, hydrogen, and water vapor.

1 Electrical sparks simulating lightning provide energy for synthesis of new compounds.

H_2O N_2 CH_4 NH_3 H_2 CO_2
"Atmospheric" compartment

2 A condenser cools the "atmospheric" gases in a "rain" containing new compounds. The compounds collect in an "ocean."

Cold water

Condensation

3 Condensate is collected and analyzed.

"Oceanic" compartment

Heat

RESULTS   The compounds react in water, eventually forming purines, pyrimidines, and amino acids.

CONCLUSION: The chemical building blocks of life could have been generated in the probable atmosphere of early Earth.

3.28 Synthesis of Prebiotic Molecules in an Experimental Atmosphere The Miller–Urey experiment simulated possible atmospheric conditions on primitive Earth and obtained some of the molecular building blocks of biological systems.
FURTHER RESEARCH: If O_2 were present in the "atmosphere" in this experiment, what results would you predict?

■ The amino acids in living things are always L-isomers (see Figure 3.2 and p. 40). But a mixture of D- and L-isomers appeared in the amino acids formed in the Miller–Urey experiments. Recent experiments show that natural processes could have selected the L-amino acids from the mixture. Some minerals, especially calcite-based rocks, have unique crystal struc-

tures that selectively bind to D- or L-amino acids, separating the two. Such rocks were abundant on early Earth.

■ Scientists' views of Earth's original atmosphere have changed since Miller and Urey did their experiment. There is abundant evidence of major volcanic eruptions 4 billion years ago, which would have released carbon dioxide (CO_2), nitrogen (N_2), hydrogen sulfide (H_2S), and sulfur dioxide (SO_2) into the atmosphere. Experiments using these gases in addition to the ones in the original experiment have produced more diverse molecules.

Chemical evolution may have led to polymerization

The Miller–Urey experiment and other experiments that followed it provided a plausible scenario for the formation of the building blocks of life. The next step would be the condensation of these monomers into polymers (see Figure 3.4). This poses a major problem, because in water, short polymers tend to hydrolyze back into monomers.

Scientists have used model systems to try to simulate conditions (most of them with a low water content) under which polymers could be made:

■ *Solid mineral surfaces*, such as finely divided clays, have a large surface area, and the silicates within the minerals may have been catalytic (speeded up the reactions) for the early carbon-based molecules.

■ *Hydrothermal vents* deep in the ocean, where hot water emerges from beneath Earth's crust, contain metals such as iron and nickel. These metals have been shown in laboratory experiments to catalyze polymerization of amino acids in the absence of oxygen.

■ *Hot pools* at the edges of oceans may, through evaporation, have concentrated monomers to the point where polymerization was favored (the "primordial soup" hypothesis).

In whatever ways the earliest stages of chemical evolution occurred, they resulted in the emergence of monomers and polymers that have probably remained unchanged in their general structure and function for 3.8 billion years.

RNA may have been the first biological catalyst

The three-dimensional structure of a folded RNA molecule presents a unique surface to the external environment (see Figure 3.25). These surfaces are every bit as specific as those of proteins. Just as the shapes of proteins allow them to function as catalysts, speeding up reactions that would ordinarily take place too slowly to be biologically useful, the three-dimensional shapes and other chemical properties of certain RNA molecules allow them to function as catalysts.

The Miller–Urey experiment and other such experiments in prebiotic chemistry yielded both amino acids and nucleotides. Organisms can synthesize RNA and proteins from these monomers. But if protein synthesis requires DNA and RNA, and nucleic acid synthesis requires proteins (enzymes), we are confronted with a chicken-and-egg type of question: when life originated, which came first, the proteins or the nucleic acids? The discovery of cat-

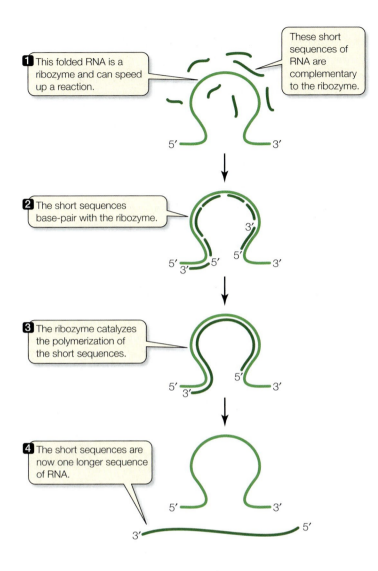

① This folded RNA is a ribozyme and can speed up a reaction.

These short sequences of RNA are complementary to the ribozyme.

② The short sequences base-pair with the ribozyme.

③ The ribozyme catalyzes the polymerization of the short sequences.

④ The short sequences are now one longer sequence of RNA.

3.29 An Early Catalyst for Life? This laboratory reconstruction shows that a ribozyme (a folded RNA molecule) can catalyze the polymerization of several short RNA strands into a longer molecule. Such a process could be a precursor for the copying of nucleic acids, which is essential for their duplication and expression.

- In certain viruses called retroviruses, there is an enzyme called reverse transcriptase that catalyzes the synthesis of DNA from RNA.

While this evidence suggests that RNA could have been the first polymer, scientists are a long way from a plausible explanation for the origins of the other large molecules characteristic of life, such as polysaccharides, proteins, and lipids.

Experiments disproved spontaneous generation of life

The idea that life originated from nonliving matter is not new. In fact, many cultures and religions have descriptions of such events. During the Renaissance (a period from about 1450 to 1700 AD, marked by the birth of modern science), most people thought that at least some forms of life arose repeatedly and directly from inanimate or decaying matter by *spontaneous generation*. For instance, it was suggested that mice arose from sweaty clothes placed in dim light, frogs came from moist soil, and flies were produced from meat. Scientists such as the Italian physician and poet Francesco Redi, however, doubted the assumptions of the time. Redi proposed that flies arose not by some mysterious transformation of decaying meat, but from other flies that laid their eggs on the meat. In 1668, Redi performed a scientific experiment—a relatively new concept at the time—to test his hypothesis. He set out several jars containing chunks of meat.

- One jar contained meat exposed both to air and to flies.
- A second jar contained meat in a container wrapped in a fine cloth so that the meat was exposed to air, but not to flies.
- The meat in the third jar was in a sealed container and thus was not exposed to either air or flies.

As he had hypothesized, Redi found maggots, which then hatched into flies, only in the first container. This finding demonstrated that maggots could occur only where flies were present. The idea that a complex organism like a fly could appear de novo from a nonliving substance in the meat, or from "something in the air," was laid to rest.

With the advent of the Leeuwenhook microscope in the 1660s, a vast new biological world was unveiled. Under microscopic observation, virtually every environment on Earth was found to be teeming with tiny organisms such as bacteria. Some scientists believed that these organisms arose spontaneously from their rich chemical environment. Experiments by the great French scientist Louis Pasteur disproved this idea, showing that microorganisms come only from other microorganisms, and that an environment without life remains lifeless unless contaminated by living creatures (**Figure 3.30**).

alytic RNAs provided a solution to this dilemma. Catalytic RNAs, called **ribozymes**, can catalyze reactions on their own nucleotides as well as in other cellular substances.

Given that RNA can be both informational (in its nucleotide sequence) and catalytic (due to its ability to form unique three-dimensional shapes; see Figure 3.25), it has been hypothesized that early life existed in an "RNA world"—a world before DNA. It is thought that when RNA was first made, it could have acted as a catalyst for its own replication as well as for the synthesis of proteins. DNA could eventually have evolved from RNA. Some laboratory evidence supports this scenario:

- RNAs of different sequences have been put in a test tube and made to replicate on their own. Such self-replicating ribozymes speed up the synthesis of RNA 7 million times.

- In the test tube, a ribozyme can catalyze the assembly of short RNAs into a longer molecule—the beginning of priming of RNA for protein synthesis (**Figure 3.29**).

- In living organisms today, the formation of peptide bonds (see Figure 3.6) is catalyzed by a ribozyme.

3.30 Disproving the Spontaneous Generation of Life
Louis Pasteur's classic experiments showed that, under conditions existing on Earth today, an inanimate solution remains lifeless unless a living organism contaminates it.

Pasteur's experiments proved that life cannot arise from nonliving materials, but the Miller–Urey experiments indicate that it could have—at least at the molecular level. How can we reconcile the results of these two experiments? Bear in mind that the atmospheric and planetary constituents of Earth today are vastly different from those of prebiotic Earth. The oxygen in today's atmosphere would break down many molecules as they were formed, and the energy sources that might have propelled those early chemical reactions no longer dominate our planet.

3.6 RECAP

The chemicals of life could have originated elsewhere in the universe, or they could have evolved on Earth. Chemical signatures of life on meteorites from Mars make the first idea plausible. Laboratory experiments support the second hypothesis.

- What experimental evidence indicates that the chemical evolution of life could have taken place on Earth? See pp. 61–62 and Figure 3.28

- Do you understand what polymerization is? Reviewing the macromolecules described in this chapter, do you understand why this chemical process is so important? See p. 62 and Figure 3.29

- Francesco Redi's experiment, described on page 63, is one of the earliest known examples of the scientific method. Do you see all the elements of the method, as described in Section 1.3, in this experiment? Are these elements present in the experiments shown in Figures 3.28 and 3.30?

The diverse life forms of Earth are composed of chemical building blocks including atoms, small molecules, and large biological molecules. But from this molecular foundation emerges the cell, a structure that is the basis of life as we know it today. It is the cell that perpetuates life, and the cell is the basis for Pasteur's truism "all life from life." The next chapter describes the structure and function of living cells.

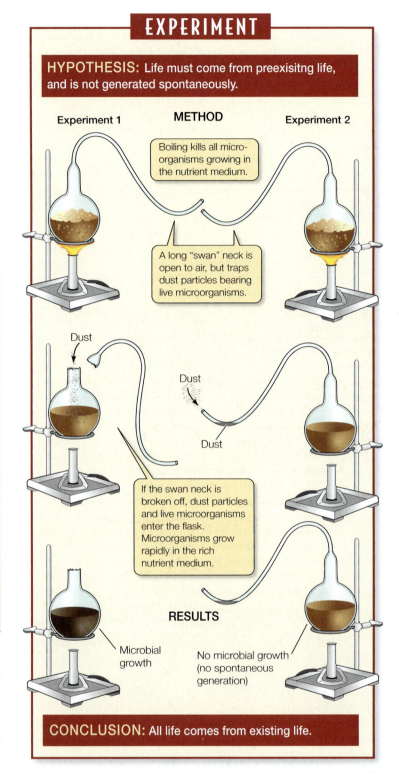

EXPERIMENT

HYPOTHESIS: Life must come from preexisitng life, and is not generated spontaneously.

METHOD

Experiment 1   Experiment 2

Boiling kills all microorganisms growing in the nutrient medium.

A long "swan" neck is open to air, but traps dust particles bearing live microorganisms.

Dust

Dust

Dust

If the swan neck is broken off, dust particles and live microorganisms enter the flask. Microorganisms grow rapidly in the rich nutrient medium.

RESULTS

Microbial growth

No microbial growth (no spontaneous generation)

CONCLUSION: All life comes from existing life.

CHAPTER SUMMARY

3.1 What kinds of molecules characterize living things?

See Web/CD Tutorial 3.1

Macromolecules are **polymers** constructed by the formation of covalent bonds between smaller molecules called **monomers**. Macromolecules in living organisms include polysaccharides, proteins, and nucleic acids.

Functional groups are small groups of atoms that are consistently found together in a variety of different macromolecules. Functional groups have particular chemical properties that they confer on any larger molecule of which they are a part. Review Figure 3.1, Web/CD Activity 3.1

Structural and optical **isomers** have the same kinds and numbers of atoms, but differ in their structures and properties. Review Figure 3.2

The many functions of macromolecules are directly related to their three-dimensional shapes, which in turn is the result of the sequences and chemical properties of their monomers.

Monomers are joined by **condensation reactions**, which release a molecule of water for each bond formed. **Hydrolysis reactions** use water to break polymers into monomers. Review Figure 3.4

3.2 What are the chemical structures and functions of proteins?

See Web/CD Activity 3.2

The functions of proteins include support, protection, catalysis, transport, defense, regulation, and movement.

Amino acids are the monomers from which proteins are constructed. The properties of the amino acids depend on their **side chains**, or **R groups**, which may be charged, polar, or hydrophobic. Review Table 3.2

Peptide bonds covalently link amino acids into polypeptide chains. These bonds form by condensation reactions between the carboxyl and amino groups. Review Figure 3.6

The **primary structure** of a protein is the order of amino acids in the chain. This chain is folded into a **secondary structure**, which in different parts of the protein may be an α **helix** or a β **pleated sheet**. Review Figure 3.7A–C

Disulfide bonds and noncovalent interactions between amino acids cause the polypeptide chain to fold into a three-dimensional **tertiary structure** and allow multiple chains to interact in a **quaternary structure**. Review Figure 3.7D,E

The specific shape and structure of a protein allows it to bind noncovalently to other molecules, often called **ligands.**

Heat, alterations in pH, or certain chemicals can all result in protein **denaturation**, which involves the loss of tertiary and/or secondary structure as well as biological function. Review Figure 3.11

Chaperonins assist protein folding by preventing binding to inappropriate ligands. Review Figure 3.12

3.3 What are the chemical structures and functions of carbohydrates?

Carbohydrates contain carbon bonded to hydrogen and oxygen atoms in a ratio of 1:2:1, or $(CH_2O)_n$.

Monosaccharides are the monomers that make up carbohydrates. **Hexoses** such as **glucose** are six-carbon monosaccharides; **pentoses** have five carbons. Review Figure 3.14, Web/CD Activity 3.3

Glycosidic linkages, which have either an α or a β orientation in space, covalently link monosaccharides into larger units such as **disaccharides**, **oligosaccharides**, and **polysaccharides**. Review Figure 3.15

Starch stores energy in plants. Starch and **glycogen** are formed by α-glycosidic linkages between glucose monomers and are distinguished by the amount of branching they exhibit. They can be easily broken down to release stored energy.

Cellulose, a very stable glucose polymer, is the principal component of the cell walls of plants.

3.4 What are the chemical structures and functions of lipids?

Fats and **oils** are **triglycerides**, composed of three **fatty acids** covalently bonded to a molecule of **glycerol** by **ester linkages**. Review Figure 3.18

Saturated fatty acids have a hydrocarbon chain with no double bonds. The hydrocarbon chains of **unsaturated** fatty acids have one or more double bonds that bend the chain, making close packing less possible. Review Figure 3.19

Phospholipids have a hydrophobic hydrocarbon "tail" and a hydrophilic phosphate "head." In water, the interactions of the hydrophobic tails and hydrophilic heads of phospholipids generate a **phospholipid bilayer** that is two molecules thick. The heads are directed outward, where they interact with the surrounding water. The tails are packed together in the interior of the bilayer. Review Figure 3.20

3.5 What are the chemical structures and functions of nucleic acids?

The unique function of the **nucleic acids**—**DNA** and **RNA**—is information storage; they form the hereditary material that passes genetic information to the next generation.

Nucleic acids are polymers of nucleotides. A **nucleotide** consists of a phosphate group, a pentose sugar (**ribose** in RNA and **deoxyribose** in DNA), and a nitrogen-containing **base**. Review Figure 3.23, Web/CD Activity 3.4

In DNA the nucleotide bases are **adenine**, **guanine**, **cytosine**, and **thymine**. **Uracil** substitutes for thymine in RNA. The nucleotides are joined by **phosphodiester linkages** between the sugar of one nucleotide and the phosphate of the next.

RNA is single-stranded. DNA is a **double helix** in which there is **complementary base pairing** based on hydrogen bonds between adenine and thymine (A-T) and between guanine and cytosine (G-C). The two strands of the DNA double helix run in opposite directions. Review Figures 3.24 and 3.26, Web/CD Activity 3.5

The information content of DNA and RNA resides in their **base sequences**.

3.6 How did life on Earth begin?

Chemical evolution proposes that conditions on early Earth could have produced the macromolecules that distinguish living things. Review Figure 3.28, Web/CD Tutorial 3.2

Because it can form a three-dimensional shape, RNA could act as a **ribozyme**, an RNA surface on which chemical reactions proceed at a faster rate. Review Figure 3.29

Experiments have ruled out the continuous spontaneous generation of life. Review Figure 3.30, Web/CD Tutorial 3.3

SELF-QUIZ

1. The most abundant molecule in the cell is
   a. a carbohydrate.
   b. a lipid.
   c. a nucleic acid.
   d. a protein.
   e. water.

2. All lipids are
   a. triglycerides.
   b. polar.
   c. hydrophilic.
   d. polymers of fatty acids.
   e. more soluble in nonpolar solvents than in water.

3. All carbohydrates
   a. are polymers.
   b. are simple sugars.
   c. consist of one or more simple sugars.
   d. are found in biological membranes.
   e. are more soluble in nonpolar solvents than in water.

4. Which of the following is not a carbohydrate?
   a. Glucose
   b. Starch
   c. Cellulose
   d. Hemoglobin
   e. Deoxyribose

5. All proteins
   a. are enzymes.
   b. consist of one or more polypeptide chains.
   c. are amino acids.
   d. have quaternary structures.
   e. are more soluble in nonpolar solvents than in water.

6. Which of the following statements about the primary structure of a protein is not true?
   a. It may be branched.
   b. It is determined by the structure of the corresponding DNA.
   c. It is unique to that protein.
   d. It determines the tertiary structure of the protein.
   e. It is the sequence of amino acids in the protein.

7. The amino acid leucine
   a. is found in all proteins.
   b. cannot form peptide linkages.
   c. is hydrophobic.
   d. is hydrophilic.
   e. is identical to the amino acid lysine.

8. The quaternary structure of a protein
   a. consists of four subunits—hence the name quaternary.
   b. is unrelated to the function of the protein.
   c. may be either alpha or beta.
   d. depends on covalent bonding among the subunits.
   e. depends on the primary structures of the subunits.

9. All nucleic acids
   a. are polymers of nucleotides.
   b. are polymers of amino acids.
   c. are double-stranded.
   d. are double-helical.
   e. contain deoxyribose.

10. Which of the following statements about condensation reactions is *not* true?
    a. Protein synthesis results from them.
    b. Polysaccharide synthesis results from them.
    c. Nucleic acid synthesis results from them.
    d. They consume water as a reactant.
    e. Different condensation reactions produce different kinds of macromolecules.

FOR DISCUSSION

1. Suppose that, in a given protein, one lysine is replaced by aspartic acid (see Table 3.2). Does this change occur in the primary structure or in the secondary structure? How might it result in a change in tertiary structure? In quaternary structure?

2. If there are 20 different amino acids commonly found in proteins, how many different dipeptides are there? How many different tripeptides? How many different trinucleotides? How many different single-stranded RNAs composed of 200 nucleotides?

3. Why might RNA have preceded proteins in the evolution of biological macromolecules?

FOR INVESTIGATION

1. The Miller–Urey experiment (see Figure 3.28) showed that it was possible for amino acids to be formed from gases hypothesized to have been in Earth's early atmosphere. These amino acids were dissolved in water. Knowing what you do about the polymerization of amino acids into proteins (see Figure 3.6), how would you set up experiments to show that proteins can form under the conditions of early Earth? What properties would you expect of those proteins?

2. The interpretation of Pasteur's experiment (see Figure 3.30) depended on the inactivation of microorganisms by heat. We now know that some microorganisms can survive at very high temperatures. How would this change the interpretation of Pasteur's experiment? What experiments would you do to inactivate such microbes?

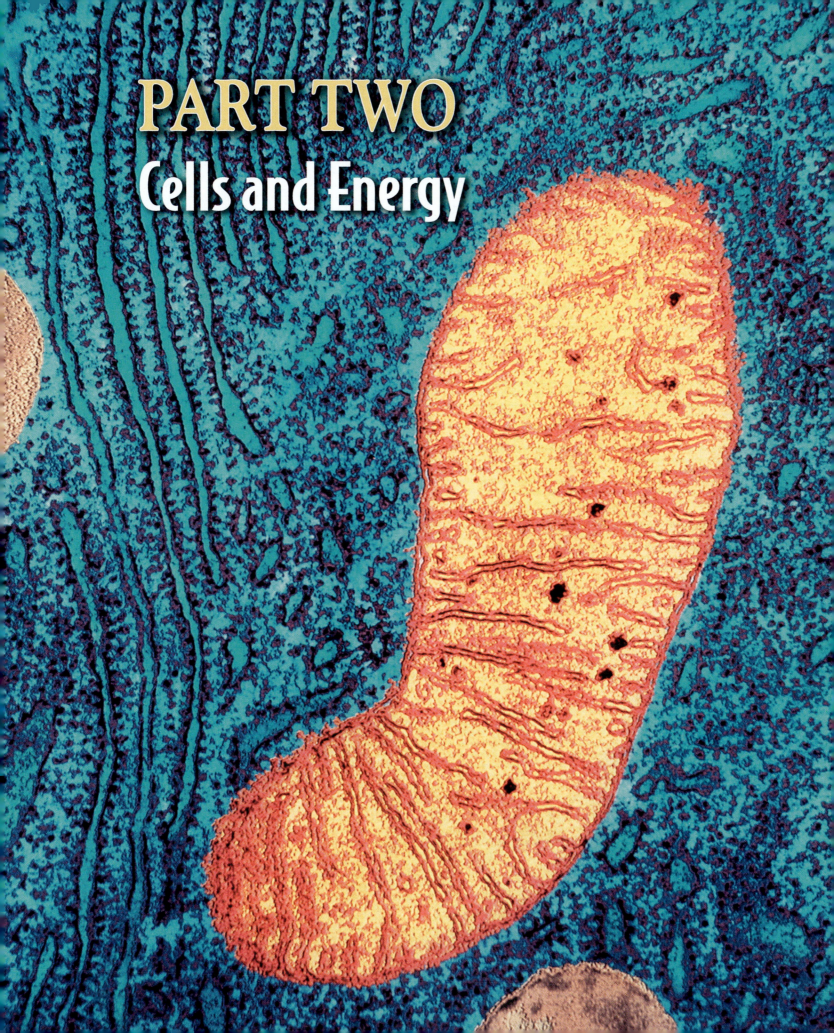

PART TWO
Cells and Energy

The oldest evidence of life?

Charles Darwin faced a dilemma. In his great book, *The Origin of Species*, he proposed the theory of natural selection to explain the gradual appearance and disappearance of different forms of organisms. But he realized that the fossil record, on which he based his theory, was incomplete, especially for the beginning of life. In Darwin's time—the middle of the nineteenth century—the oldest known fossils were complex organisms found in rocks dated at about 550 million years ago (the Cambrian period). Did simpler organisms exist before that time? If so, where were their fossils? These would surely provide a link to the origin of life.

Conditions on Earth were probably suitable for the emergence of life by 4 billion years ago, about 600 million years after Earth began to form. But at the turn of the twentieth century, the oldest known fossils were clumps of algae (simple aquatic photosynthetic organisms) close to 1 billion years old—still far short of the origin of life. Would it even be possible to find older fossils? Most ancient rocks are igneous—formed by high-temperature processes such as volcanic eruptions—and geologic action has altered them drastically over the millenia. It is unlikely that any cellular fossil could survive these extreme conditions.

It took until the 1990s to find older evidence of life. It was then that, in a few places on Earth's surface, scientists discovered a phenomenon—some relatively unchanged 3.5-billion-year-old rocks. In one of these rock samples from what is now Australia, geologist J. William Schopf saw chains and clumps of what looked tantalizingly like contemporary cyanobacteria ("blue-green" bacteria). Schopf needed to prove that these chains were once alive, not just the results of simple chemical reactions. He and his colleagues looked for chemical evidence of photosynthesis.

The use of carbon dioxide in photosynthesis is a hallmark of life and leaves a unique chemical signature—a specific ratio of isotopes of carbon (C^{13}: C^{12}) in the resulting carbohydrates. Schopf showed that the Australian material had this signature. Microscopic examination of the chains revealed substructures characteristic of living systems that were not likely to be the result of simple chemical reactions. Schopf's evidence suggests that the Australian sample is indeed the remains of a truly ancient living organism.

In addition to confirming the presence of the earliest life, Schopf's fossils suggest that life requires not just the right collection of macromolecules, but compartmentalization. The macromolecules of life perform unique functions because they are enclosed in

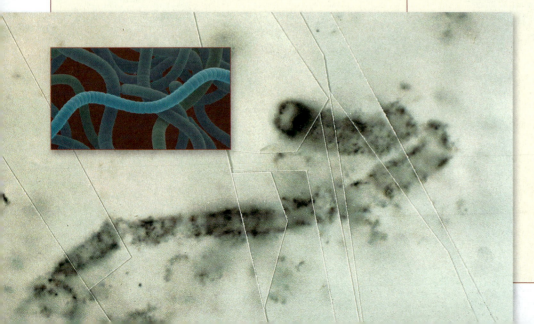

The Earliest Evidence of Life? This fossil from Western Australia is 3.5 billion years old. Its form is similar to that of modern filamentous cyanobacteria (inset).

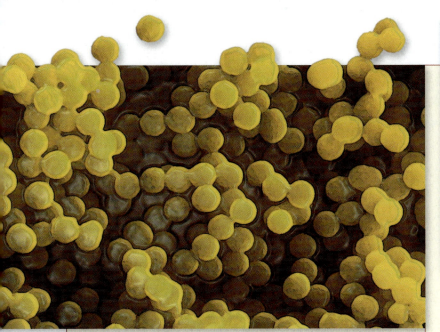

An Early "Cell" from the Laboratory   Mixing amino acids and lipids formed in experiments simulating prebiotic conditions on Earth results in cell-like structures called proteinoids. They are surrounded by a membrane bilayer and can carry out some chemical reactions.

structures that separate them from one another and from the external environment. This *compartmentalization* takes the form of cells and is seen in organelles that characterize eukaryotic cells. How did such compartmentalization arise?

Scientists have tried to model the origin of cells in the laboratory. In such modeling experiments, aggregates of molecules form round structures similar to cells. These chemical compartments can perform some biochemical reactions and can exchange materials with their environment. Taken together with the results of prebiotic chemistry experiments and the "RNA world" hypothesis described in Section 3.6, these experiments suggest that something similar to these aggregates might have been the first cells.

IN THIS CHAPTER we examine the structure and some of the functions of the "living compartment" known as the cell. We begin with the cell theory (the basis of cell biology), and then examine the simple cells of the single-celled organisms known as prokaryotes. We then tour the more complex eukaryotic cell and its various internal compartments, each of which performs specific functions for the cell.

4.1 What Features of Cells Make Them the Fundamental Unit of Life?

Just as atoms are the building blocks of chemistry, *cells are the building blocks of life*. The **cell theory** was described in Section 1.1 as the first unifying principle of biology. Recall the critical three tenets of the cell theory:

- Cells are the fundamental units of life.
- All organisms are composed of cells.
- All cells come from preexisting cells.

Cells contain water and the other small and large molecules we examined in the previous two chapters. Each cell contains at least 10,000 different types of molecules, most of them present in many copies. Cells use these molecules to transform matter and energy, to respond to their environment, and to reproduce themselves.

The cell theory has three important implications:

- Studying cell biology is in some sense the same as studying life. The principles that underlie the functions of the single cell of a bacterium are similar to those governing the approximately 60 trillion cells of your body.
- Life is continuous. All those cells in your body came from a single cell, a fertilized egg, which came from the fusion of two cells, a sperm and an egg from your parents, whose cells also came from fertilized eggs, and so on.
- The origin of life on Earth was marked by the origin of the first cells.

Cell size is limited by the surface area-to-volume ratio

Most cells are tiny. The volumes of cells range from 1 to 1,000 cubic micrometers (μm^3) (**Figure 4.1**). There are some exceptions: the eggs of birds are, relatively speaking, enormous, and individual cells of several types of algae and bacteria are large enough to be viewed with the unaided eye. And although neurons (nerve cells) have a volume that is within the "normal" range, they often have fine projections that may extend for meters, carrying signals from one part of a large animal to another.

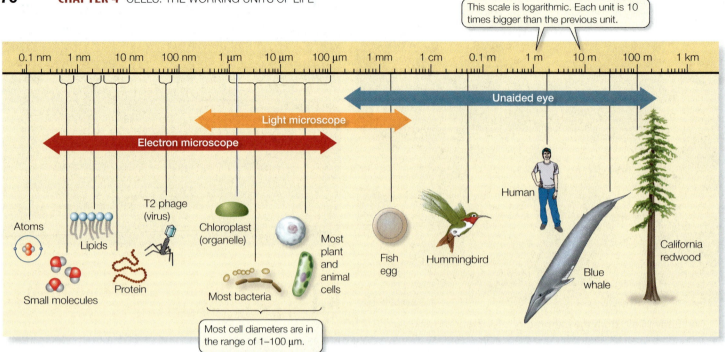

This scale is logarithmic. Each unit is 10 times bigger than the previous unit.

Most cell diameters are in the range of 1–100 µm.

4.1 The Scale of Life This logarithmic scale shows the relative sizes of molecules, cells, and multicellular organisms.

Most cells are miniscule. About 2,000 human skin cells lined up in a row would fit across this page.

Small cell size is a practical necessity arising from the change in the **surface area-to-volume ratio** of any object as it increases in size. As an object increases in volume, its surface area also increases, but not to the same extent (**Figure 4.2**). This phenomenon has great biological significance for two reasons:

- The *volume* of a cell determines the amount of chemical activity it carries out per unit of time.

- The *surface area* of a cell determines the amount of substances the cell can take in from the outside environment and the amount of waste products it can release to the environment.

As a living cell grows larger, its chemical activity, and thus its rate of waste production and its need for resources, increase faster than its surface area. In addition, cells must often distribute substances from one place to another within the cell; the smaller the cell, the more easily this is accomplished. This explains why large organisms must consist of many small cells: cells must be small in volume in order to maintain a large enough surface area-to-volume ratio and an ideal internal volume. The large surface area represented by the myriad small cells of a multicellular organism enables it to carry out the many different functions required for survival.

4.2 Why Cells Are Small Whether it is cuboid (A) or spheroid (B), as an object grows larger its volume increases more rapidly than its surface area. Cells must maintain a large surface area-to-volume ratio in order to function. This fact explains why large organisms must be composed of many small cells rather than a few huge ones.

Microscopes are needed to visualize cells

The smallest object a person can typically discern is about 0.2 mm (200 µm) in size. We refer to this measure as **resolution**, the distance apart two objects must be in order for the eye to distinguish them as separate; if they are closer together, they appear as a single blur. Most cells are much smaller than 200 µm, and thus are invisible to

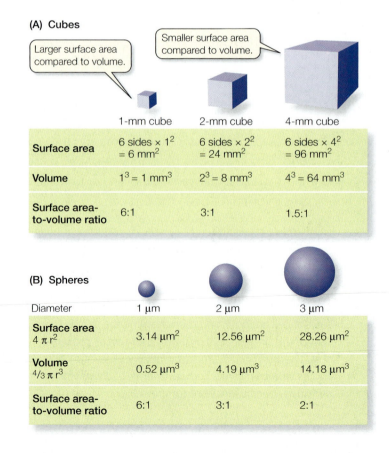

(A) Cubes

Larger surface area compared to volume.

Smaller surface area compared to volume.

	1-mm cube	2-mm cube	4-mm cube
Surface area	6 sides × 1^2 = 6 mm^2	6 sides × 2^2 = 24 mm^2	6 sides × 4^2 = 96 mm^2
Volume	1^3 = 1 mm^3	2^3 = 8 mm^3	4^3 = 64 mm^3
Surface area-to-volume ratio	6:1	3:1	1.5:1

(B) Spheres

Diameter	1 µm	2 µm	3 µm
Surface area $4\pi r^2$	3.14 µm^2	12.56 µm^2	28.26 µm^2
Volume $^4/_3 \pi r^3$	0.52 µm^3	4.19 µm^3	14.18 µm^3
Surface area-to-volume ratio	6:1	3:1	2:1

RESEARCH METHOD

140 μm

In **bright-field microscopy**, light passes directly through these human cells. Unless natural pigments are present, there is little contrast and details are not distinguished.

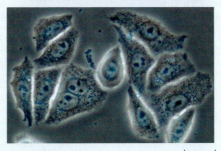

30 μm

In **phase-contrast microscopy**, contrast in the image is increased by emphasizing differences in refractive index (the capacity to bend light), thereby enhancing light and dark regions in the cell.

30 μm

Differential interference-contrast microscopy uses two beams of polarized light. The combined images look as if the cell is casting a shadow on one side.

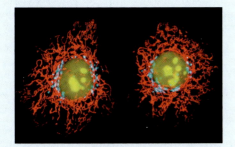

20 μm

In **fluorescence microscopy**, a natural substance in the cell or a fluorescent dye that binds to a specific cell material is stimulated by a beam of light, and the longer-wavelength fluorescent light is observed coming directly from the dye.

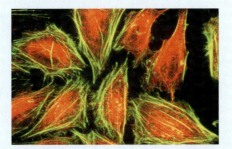

20 μm

Confocal microscopy uses fluorescent materials but adds a system of focusing both the stimulating and emitted light so that a single plane through the cell is seen. The result is a sharper two-dimensional image than with standard fluorescence microscopy.

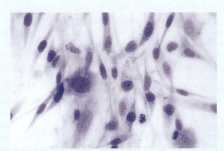

30 μm

In **stained bright-field microscopy**, a stain added to preserve cells enhances contrast and reveals details not otherwise visible. Stains differ greatly in their chemistry and their capacity to bind to cell materials, so many choices are available.

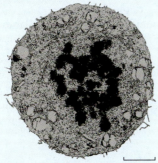

10 μm

In **transmission electron microscopy** (TEM), a beam of electrons is focused on the object by magnets. Objects appear darker if they absorb the electrons. If the electrons pass through they are detected on a fluorescent screen.

20 μm

Scanning electron microscopy (SEM) directs electrons to the surface of the sample, where they cause other electrons to be emitted. These electrons are viewed on a screen. The three-dimensional surface of the object can be visualized.

0.1 μm

In **freeze-fracture microscopy**, cells are frozen and then a knife is used to crack them open. The crack often passes through the interior of plasma and internal membranes. The "bumps" that appear are usually large proteins embedded in the interior of the membrane.

4.3 Looking at Cells The top two rows show some techniques used in light microscopy. The lower three images were created using electron microscopes. All of these images are of a particular type of cultured cell known as HeLa cells. The story of these cells and their widespread use in research is told at the start of Chapter 9.

the human eye. *Microscopes* improve resolution so that cells and their internal structures can be seen (**Figure 4.3**).

There are two basic types of microscopes:

- A *light microscope* uses glass lenses and visible light to form a magnified image of an object. It has a resolution of about 0.2 μm, which is 1,000 times that of the human eye. It allows visualization of cell sizes and shapes and some internal cell structures. Internal structures are hard to see under visible light, so cells are often chemically treated and their components stained with various dyes to make certain structures stand out.

■ An *electron microscope* uses electromagnets to focus an electron beam, much as a light microscope uses glass lenses to focus a beam of light. Since we cannot see electrons, the electron microscope directs them at a fluorescent screen or photographic film to create a visible image. The resolution of electron microscopes is about 0.2 nm, about 1,000,000 times greater than the human eye. This resolution permits the details of many subcellular structures to be distinguished.

Many techniques have been developed to enhance the views of cells we see under the light and electron microscopes.

Cells are surrounded by a plasma membrane

Each cell is surrounded by a membrane that separates it from its environment, creating a segregated (but not isolated) compartment. This **plasma membrane** is composed of a phospholipid bilayer, with the hydrophilic "heads" of the lipids facing the cell's aqueous interior on one side of the membrane and the extracellular environment on the other (see Figure 3.20). Proteins and other molecules are embedded in the lipids. We will devote most of Chapter 5 to detailing the structure and functions of the plasma membrane, but summarize its roles here:

■ The plasma membrane allows the cell to maintain a more or less *constant internal environment*. A self-maintaining, constant internal environment (the state known as *homeostasis*) is a key characteristic of life that will be discussed in detail in Chapter 40.

■ The plasma membrane acts as a *selectively permeable barrier*, preventing some substances from crossing it while permitting other substances to enter and leave the cell.

■ As the cell's boundary with the outside environment, the plasma membrane is important in *communicating with adjacent cells* and *receiving signals from the environment*. We will describe this function in Chapter 15.

■ The plasma membrane often has proteins protruding from it that are responsible for *binding and adhering to adjacent cells*.

Cells are prokaryotic or eukaryotic

As we learned in Section 1.2, biologists classify all living things into three domains: Archaea, Bacteria, and Eukarya. The organisms in Archaea and Bacteria are collectively called **prokaryotes** because they have in common a *prokaryotic* cell organization. Prokaryotic cells do not typically have membrane-enclosed internal compartments. The first cells were probably similar in organization to modern prokaryotes.

Eukaryotic cell organization, on the other hand, is particular to members of the domain Eukarya (**eukaryotes**), which includes the protists, plants, fungi, and animals. The genetic material (DNA) of eukaryotic cells is contained in a special membrane-enclosed compartment called the **nucleus**. Eukaryotic cells also contain other membrane-enclosed compartments in which specific chemical reactions take place.

4.1 RECAP

The cell theory is a unifying principle of biology. Surface area-to-volume ratios limit the size of cells. Both prokaryotic and eukaryotic cells are enclosed within a plasma membrane, but prokaryotic cells lack membrane-enclosed internal compartments.

■ Why can we say that cell biology embodies all the principles of life? See p. 69

■ Can you explain why cells are small? See p. 70 and Figure 4.2

Prokaryotes and eukaryotes alike have prospered for many hundreds of millions of years, and many evolutionary success stories have arisen from both types of cell organization. Let's look first at the organization of prokaryotic cells.

4.2 What Are the Characteristics of Prokaryotic Cells?

Prokaryotes can live on more different and diverse energy sources than any other living organisms, and they can inhabit greater environmental extremes, such as very hot springs and very salty water. As we examine prokaryotic cells in this section, bear in mind that there are vast numbers of prokaryotic species, and that the Bacteria and Archaea are distinguished in numerous ways. These differences, and the vast diversity of organisms in these two domains, will be the subject of Chapter 26.

Prokaryotic cells are generally smaller than eukaryotic cells, ranging from 0.25×1.2 μm to 1.5×4 μm. Each individual prokaryote is a single cell, but many types of prokaryotes are usually seen in chains, small clusters, or even clusters containing hundreds of individuals. In this section we will first consider the features that cells in the domains Bacteria and Archaea have in common. Then we will examine structural features that are found in some, but not all, prokaryotes.

Prokaryotic cells share certain features

All prokaryotic cells have the same basic structure (**Figure 4.4**):

■ The plasma membrane encloses the cell, regulating the traffic of materials into and out of the cell and separating it from its environment.

■ The **nucleoid** contains the hereditary material (DNA) of the cell.

The rest of the material enclosed in the plasma membrane is called the **cytoplasm**. The cytoplasm is composed of two components: the more fluid cytosol and insoluble suspended particles, including ribosomes.

■ The **cytosol** consists mostly of water that contains dissolved ions, small molecules, and soluble macromolecules such as proteins.

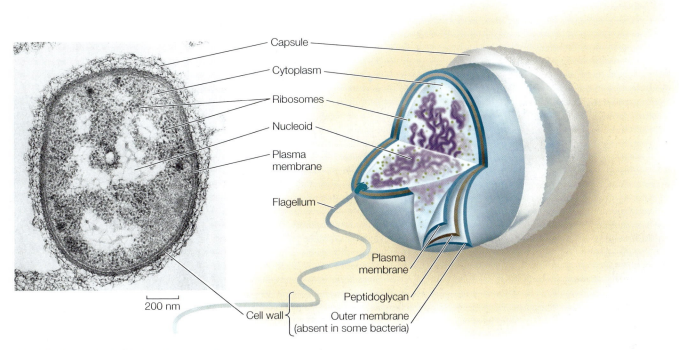

Capsule
Cytoplasm
Ribosomes
Nucleoid
Plasma membrane
Flagellum
Plasma membrane
Peptidoglycan
Cell wall
Outer membrane (absent in some bacteria)

200 nm

4.4 A Prokaryotic Cell The bacterium *Pseudomonas aeruginosa* illustrates the typical structures shared by all prokaryotic cells. This bacterium also has protective structures in an outer membrane that some, but not all, prokaryotes have. The flagellum and capsule are also structures found in some, but not all, prokaryotic cells.

■ **Ribosomes** are complexes of RNA and proteins about 25 nm in diameter. They are the sites of protein synthesis.

The cytoplasm is not a static region. Rather, the substances in this environment are in constant motion. For example, a typical protein moves around the entire cell within a minute, and it encounters many molecules along the way.

Although structurally less complicated than eukaryotic cells, prokaryotic cells are functionally complex, carrying out thousands of biochemical reactions.

Some prokaryotic cells have specialized features

As they evolved, some prokaryotes developed specialized structures that gave a selective advantage to those cells that had them. These structures include a protective cell wall, an internal membrane for compartmentalization of some chemical reactions, and flagella for cell movement through the watery environment. These features are shown in Figures 4.4 and 4.5.

CELL WALLS Most prokaryotes have a cell wall located *outside* the plasma membrane. The rigidity of the cell wall supports the cell and determines its shape. The cell walls of most bacteria, but not archaea, contain *peptidoglycan*, a polymer of amino sugars, cross-linked by covalent bonds to form a single giant molecule around the entire cell. In some bacteria, another layer—the outer membrane (a polysaccharide-rich phospholipid membrane)—encloses the peptidoglycan layer. Unlike the plasma membrane, this outer membrane is not a major permeability barrier.

When pathogenic bacteria such *Salmonella*, *Shigella*, and *Neisseria* infect a human, lipopolysaccharide fragments from their membranes, called endotoxins, are released into the bloodstream. Among other unfortunate effects, these molecules cause fever and interfere with blood clotting, leading to excessive bleeding (hemorrhaging).

Enclosing the cell wall in some bacteria is a layer of slime, composed mostly of polysaccharides and referred to as a *capsule*. The capsules of some bacteria may protect them from attack by white blood cells in the animals they infect. The capsule helps keep the cell from drying out, and sometimes it helps the bacterium attach to other cells. Many prokaryotes produce no capsule, and those that do have capsules can survive even if they lose them, so the capsule is not essential to prokaryotic life.

As you will see later in this chapter, eukaryotic plant cells also have a cell wall, but it differs in composition and structure from the cell walls of prokaryotes.

INTERNAL MEMBRANES Some groups of bacteria—the cyanobacteria and some others—carry on photosynthesis. In these photosynthetic bacteria, the plasma membrane folds into the cytoplasm to form an internal membrane system that contains compounds needed for photosynthesis. The development of photosynthesis, which requires membranes, was an important event in the early evolution of life on Earth. Other prokaryotes have internal membrane folds that remain attached to the plasma membrane. These folds may function in cell division or in various energy-releasing reactions.

FLAGELLA AND PILI Some prokaryotes swim by using appendages called **flagella** (**Figure 4.5A**). A single flagellum, made of a protein called *flagellin*, looks at times like a tiny corkscrew. A complex mo-

(A)

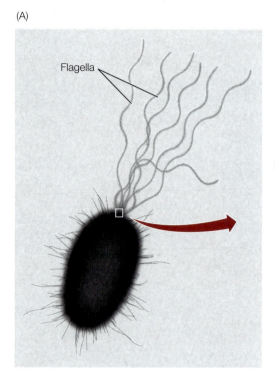

(B)

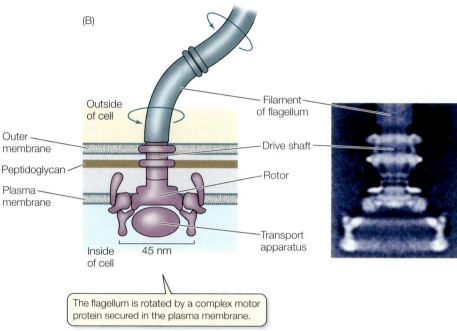

The flagellum is rotated by a complex motor protein secured in the plasma membrane.

4.5 Prokaryotic Flagella   (A) Flagella contribute to the movement and adhesion of prokaryotic cells. (B) Complex protein ring structures anchored in the plasma membrane form a motor unit that rotates the flagellum and propels the cell.

tor protein spins the flagellum on its axis like a propeller, driving the cell along. The motor protein is anchored to the plasma membrane and, in some bacteria, to the outer membrane of the cell wall (**Figure 4.5B**). We know that the flagella cause the motion of the cell because if they are removed, the cell does not move.

Pili project from the surfaces of some groups of bacteria. Shorter than flagella, these hairlike structures help bacteria adhere to one another when they exchange genetic material, as well as to animal cells for protection and food.

CYTOSKELETON   Some prokaryotes, especially rod-shaped bacteria, have an internal filamentous helical structure just inside the plasma membrane. The proteins that make up this structure are similar in amino acid sequence to actin in eukaryotic cells, and since actin is part of the cytoskeleton in those cells (see Section 4.3), it has been suggested that the helical filaments in prokaryotes play a role in maintaining cell shape.

4.2 RECAP

Prokaryotic organisms can live on diverse energy sources and in extreme environments. Unlike eukaryotic cells, prokaryotic cells do not have internal compartments.

- **What structures are present in all prokaryotic cells? See pp. 72–73 and Figure 4.4**

- **Describe the structure and function of a specialized prokaryotic cell feature, such as the cell wall, capsule, or flagellum. See pp. 73–74**

The cells of animals, plants, fungi, and protists are usually larger than those of prokaryotic organisms, and they are structurally more complex. What specific structures characterize the eukaryotic cell?

4.3 What Are the Characteristics of Eukaryotic Cells?

Eukaryotic cells generally have dimensions up to 10 times greater than those of prokaryotes; for example, the spherical yeast cell has a diameter of 8 μm. Like prokaryotic cells, eukaryotic cells have a plasma membrane, cytoplasm, and ribosomes. But in addition to this shared organization, eukaryotic cells have compartments within the cytoplasm whose interiors are separated from the cytosol by a membrane.

Compartmentalization is the key to eukaryotic cell function

Some of the compartments in eukaryotic cells have been characterized as factories that make specific products. Others are like power plants that take in energy in one form and convert it into a more useful form. These membranous compartments, as well as other structures (such as ribosomes) that lack membranes but possess distinctive shapes and functions, are called **organelles**. Each of these organelles has specific roles in its particular cell. These roles are defined by the chemical reactions each organelle can carry out.

■ The *nucleus* contains most of the cell's genetic material (DNA). The replication of the genetic material and the first steps in decoding genetic information take place in the nucleus.

■ The *mitochondrion* is a power plant and industrial park, where energy stored in the bonds of carbohydrates and fatty acids is converted into a form more useful to the cell (ATP).

■ The *endoplasmic reticulum* and *Golgi apparatus* are compartments in which some proteins synthesized by the ribosomes are packaged and sent to appropriate locations in the cell.

■ *Lysosomes* and *vacuoles* are cellular digestive systems in which large molecules are hydrolyzed into usable monomers.

■ *Chloroplasts* (found in only some cells) perform photosynthesis. The membrane surrounding each organelle does two essential things. First, it keeps the organelle's molecules away from other molecules in the cell with which they might react inappropriately. Second, it acts as a traffic regulator, letting important raw materials into the organelle and releasing its products to the cytoplasm. The evolution of compartmentalization was an important development in the ability of eukaryotic cells to specialize, forming the organs and tissues of a complex multicellular body.

Organelles can be studied by microscopy or isolated for chemical analysis

Cell organelles were first detected by light and electron microscopy. The use of stains targeted to specific macromolecules has allowed cell biologists to determine the chemical compositions of organelles (see Figure 4.20, which shows a single cell stained for three different proteins).

Another way to look at cells is to take them apart. *Cell fractionation* begins with the destruction of the plasma membrane, which allows the cytoplasmic components to flow out into a test tube. The various organelles can then be separated from one another on the basis of size or density (**Figure 4.6**). Biochemical analyses can then be done on the isolated organelles. Microscopy and cell fractionation have complemented each other, giving us a complete picture of the structure and function of each organelle.

Microscopy of plant and animal cells reveals that many of the organelles are identical in both cell types. This can be seen in **Figure 4.7**, which also illustrates the prominent differences between eukaryotic cells and the prokaryotic cell diagrammed in Figure 4.4.

Some organelles process information

Living things depend on accurate, appropriate information—internal signals, environmental cues, and stored instructions—to respond appropriately to changing conditions and maintain a constant internal environment. In the cell, information is stored in the sequence of nucleotides in DNA molecules. Most of the DNA in eukaryotic cells resides in the nucleus. Information is translated from the language of DNA into the language of proteins at the ribosomes. This process is described in detail in Chapter 12.

NUCLEUS The single **nucleus** is usually the largest organelle in a cell (**Figure 4.8**; see also Figure 4.6). The nucleus of most animal

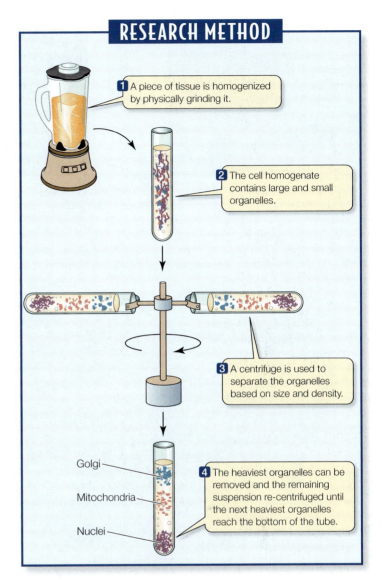

RESEARCH METHOD

1 A piece of tissue is homogenized by physically grinding it.

2 The cell homogenate contains large and small organelles.

3 A centrifuge is used to separate the organelles based on size and density.

Golgi

Mitochondria

Nuclei

4 The heaviest organelles can be removed and the remaining suspension re-centrifuged until the next heaviest organelles reach the bottom of the tube.

4.6 Cell Fractionation The organelles of cells can be separated from one another after the cells are broken open and centrifuged.

cells is approximately 5 μm in diameter—substantially larger than most prokaryotic cells. The nucleus has several roles in the cell:

■ It is the site of DNA replication.

■ It is the site of genetic control of the cell's activities.

■ A region within the nucleus, the **nucleolus**, begins the assembly of ribosomes from RNA and specific proteins.

The nucleus is surrounded by two membranes, which together form the *nuclear envelope*. The two membranes of the nuclear envelope are separated by 10–20 nm and are perforated by about 3,500 *nuclear pores* approximately 9 nm in diameter, which connect the interior of the nucleus with the cytoplasm. At these pores, the outer membrane of the nuclear envelope is continuous with the inner membrane. The pores are composed of over 100 different proteins, interacting hydrophobically. Each pore is surrounded by a pore complex made up of eight large protein aggregates arranged

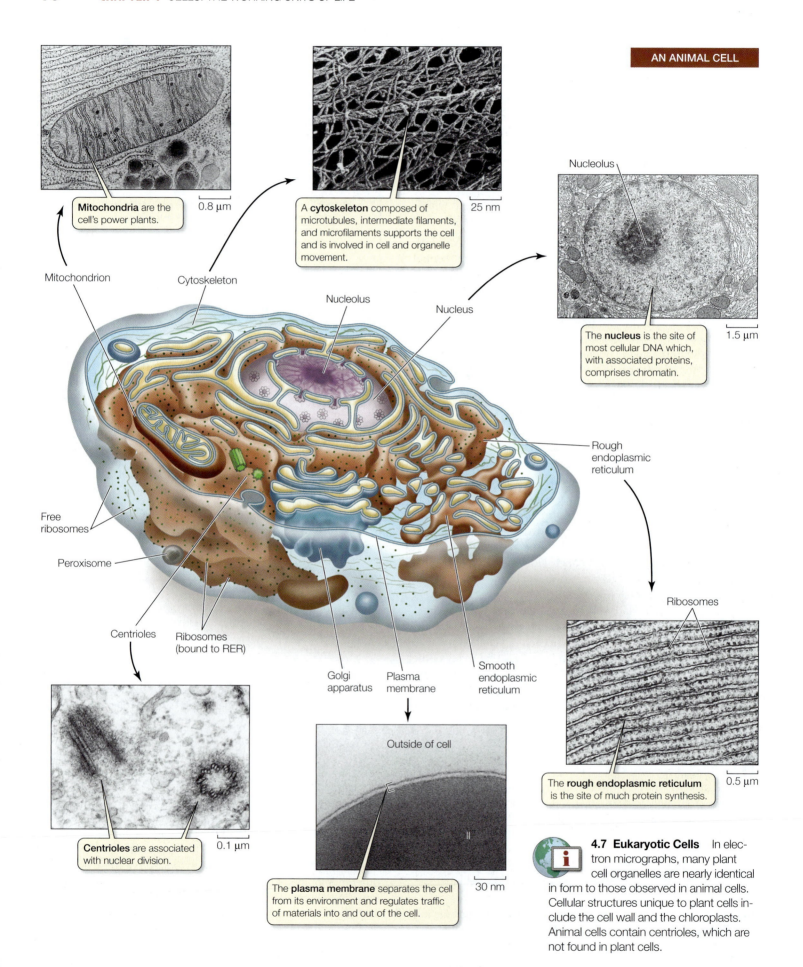

AN ANIMAL CELL

Mitochondria are the cell's power plants.

0.8 µm

A **cytoskeleton** composed of microtubules, intermediate filaments, and microfilaments supports the cell and is involved in cell and organelle movement.

25 nm

Nucleolus

The **nucleus** is the site of most cellular DNA which, with associated proteins, comprises chromatin.

1.5 µm

Mitochondrion

Cytoskeleton

Nucleolus

Nucleus

Rough endoplasmic reticulum

Free ribosomes

Peroxisome

Centrioles

Ribosomes (bound to RER)

Golgi apparatus

Plasma membrane

Smooth endoplasmic reticulum

Ribosomes

Centrioles are associated with nuclear division.

0.1 µm

Outside of cell

The **plasma membrane** separates the cell from its environment and regulates traffic of materials into and out of the cell.

30 nm

The **rough endoplasmic reticulum** is the site of much protein synthesis.

0.5 µm

4.7 Eukaryotic Cells In electron micrographs, many plant cell organelles are nearly identical in form to those observed in animal cells. Cellular structures unique to plant cells include the cell wall and the chloroplasts. Animal cells contain centrioles, which are not found in plant cells.

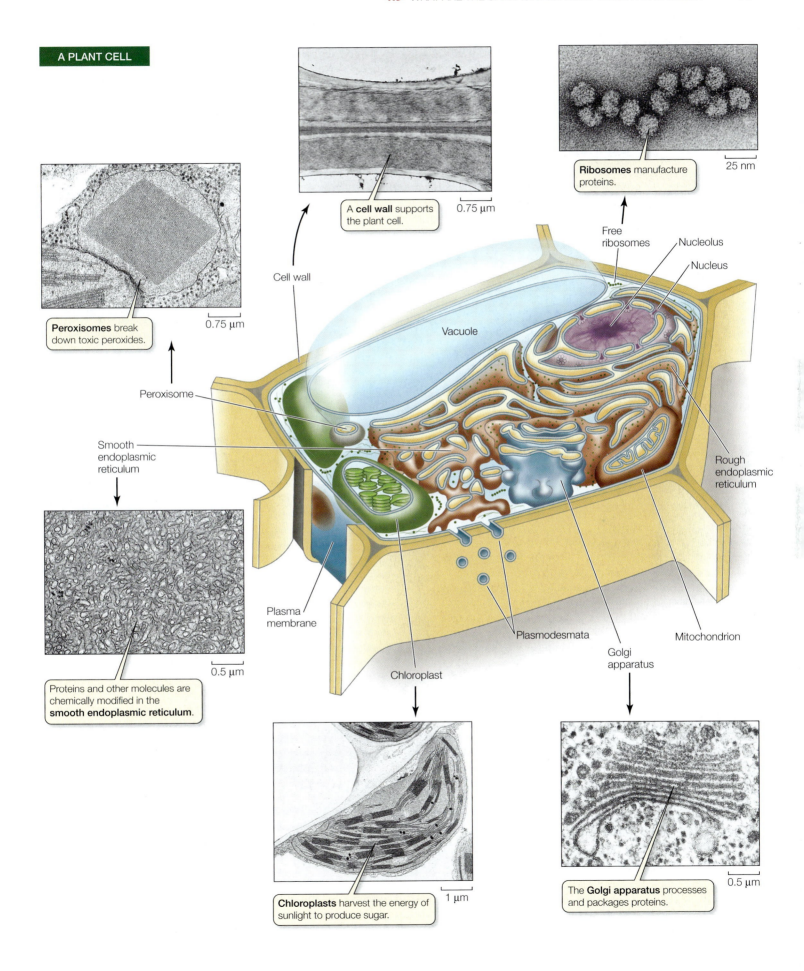

A PLANT CELL

A **cell wall** supports the plant cell.

0.75 μm

Ribosomes manufacture proteins.

25 nm

Peroxisomes break down toxic peroxides.

0.75 μm

Free ribosomes

Nucleolus

Nucleus

Cell wall

Vacuole

Peroxisome

Smooth endoplasmic reticulum

Rough endoplasmic reticulum

Plasma membrane

Plasmodesmata

Mitochondrion

Golgi apparatus

Chloroplast

Proteins and other molecules are chemically modified in the **smooth endoplasmic reticulum**.

0.5 μm

Chloroplasts harvest the energy of sunlight to produce sugar.

1 μm

The **Golgi apparatus** processes and packages proteins.

0.5 μm

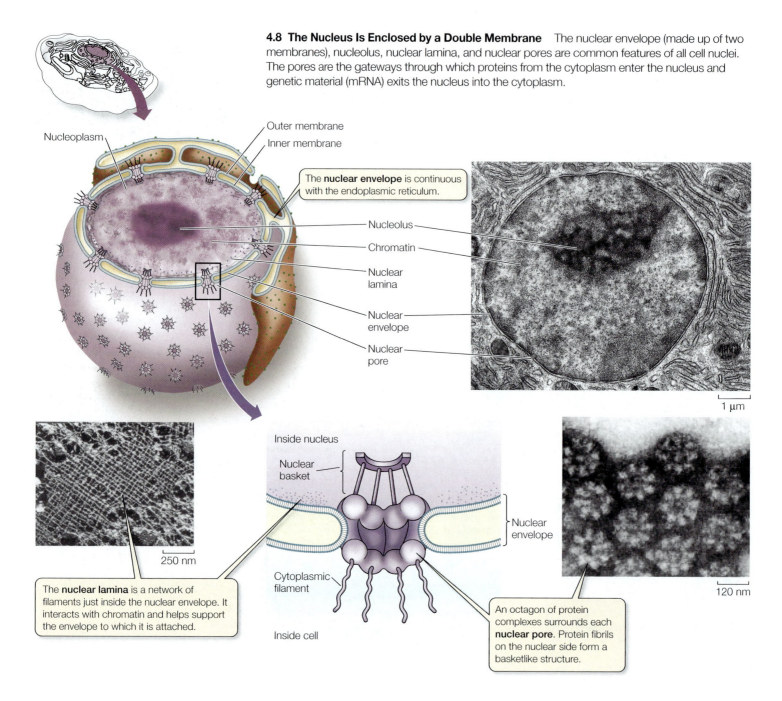

4.8 The Nucleus Is Enclosed by a Double Membrane The nuclear envelope (made up of two membranes), nucleolus, nuclear lamina, and nuclear pores are common features of all cell nuclei. The pores are the gateways through which proteins from the cytoplasm enter the nucleus and genetic material (mRNA) exits the nucleus into the cytoplasm.

Nucleoplasm

Outer membrane
Inner membrane

The **nuclear envelope** is continuous with the endoplasmic reticulum.

Nucleolus
Chromatin
Nuclear lamina
Nuclear envelope
Nuclear pore

1 μm

Inside nucleus
Nuclear basket
Nuclear envelope
Cytoplasmic filament
Inside cell

250 nm

The **nuclear lamina** is a network of filaments just inside the nuclear envelope. It interacts with chromatin and helps support the envelope to which it is attached.

An octagon of protein complexes surrounds each **nuclear pore**. Protein fibrils on the nuclear side form a basketlike structure.

120 nm

in an octagon where the inner and outer membranes merge (see Figure 4.8).

The nuclear pore works much like a turnstile gate at a sports event. Just as children can pass under the gate, small substances, such as ions and molecules with a molecular weight of less than 10,000 daltons, freely diffuse through the pore. Larger molecules (up to 50,000 Da) can also diffuse through, but they take longer. Still larger molecules, such as many proteins that are made in the cytoplasm and imported into the nucleus, are like adults at the gate: they cannot get through without a "ticket." In the case of imported proteins, the "ticket" is a short sequence of amino acids that is part of the protein. We know that this sequence is the *nuclear localization signal* from several lines of evidence:

- The signal sequence occurs in most nuclear proteins, but not in proteins that remain in the cytoplasm.

- If the signal sequence is removed from a protein, it stays in the cytoplasm.

- If the signal sequence is added to a protein that normally stays in the cytoplasm, that protein moves into the nucleus.

- Some viruses have a signal sequence that allows them to enter the nucleus; viruses without the signal sequence do not enter the nucleus as virus particles.

How does the signal sequence result in passage through the nuclear pore? Apparently, it has a three-dimensional structure that allows it to bind noncovalently to one of the proteins in the pore that acts as a *receptor*. Binding results in the receptor changing its three-dimensional shape so that it stretches the pore to let the large, imported protein pass through.

4.9 Chromatin and Chromosomes (A) When a cell is not dividing, the nuclear DNA is aggregated with proteins to form chromatin, which is dispersed throughout the nucleus. (B) The chromatin in a dividing cell is packed into dense bodies called chromosomes.

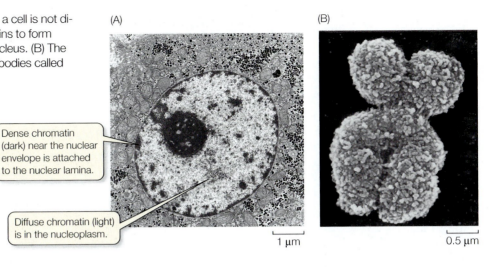

(A)

Dense chromatin (dark) near the nuclear envelope is attached to the nuclear lamina.

Diffuse chromatin (light) is in the nucleoplasm.

1 μm

(B)

0.5 μm

At certain sites, the outer membrane of the nuclear envelope folds outward into the cytoplasm and is continuous with the membrane of another organelle, the endoplasmic reticulum (which we will describe shortly). Inside the nucleus, DNA combines with proteins to form a fibrous complex called *chromatin*. Chromatin consists of exceedingly long, thin threads. Prior to cell division, the chromatin aggregates to form discrete, readily visible structures called *chromosomes* (**Figure 4.9**).

Surrounding the chromatin are water and dissolved substances collectively referred to as the *nucleoplasm*. Within the nucleoplasm, a network of apparently structural proteins called the *nuclear matrix* organizes the chromatin. At the periphery of the nucleus, the chromatin is attached to a protein meshwork, called the *nuclear lamina*, which is formed by the polymerization of proteins called *lamins* into filaments. The nuclear lamina maintains the shape of the nucleus by its attachment to both the chromatin and the nuclear envelope.

During most of a cell's life cycle, the nuclear envelope is a stable structure. When the cell reproduces, however, the nuclear envelope breaks down into pieces of membrane with attached pore complexes. The envelope re-forms when distribution of the replicated DNA to the daughter cells is completed (see Section 9.3). How this amazing self-assembly occurs is not known.

RIBOSOMES In prokaryotic cells, **ribosomes** float freely in the cytoplasm. In eukaryotic cells they are found in two places: in the cytoplasm, where they may be free or attached to the surface of the endoplasmic reticulum; and inside mitochondria and chloroplasts. In each of these locations, the ribosomes are the sites where proteins are synthesized under the direction of nucleic acids. Although they seem small in comparison to the cell in which they are contained, ribosomes are huge machines made up of several dozen kinds of molecules.

The ribosomes of prokaryotes and eukaryotes are similar in that both consist of two different-sized subunits. Eukaryotic ribosomes are somewhat larger, but the structure of prokaryotic ribosomes is better understood. Chemically, ribosomes consist of a special type of RNA called *ribosomal RNA* (*rRNA*), to which more than 50 different protein molecules are noncovalently bound.

The endomembrane system is a group of interrelated organelles

Much of the volume of some eukaryotic cells is taken up by an extensive **endomembrane system**. This system includes two main components, the endoplasmic reticulum and the Golgi apparatus. Con-tinuities between the nuclear envelope and the endomembrane system are visible under the electron microscope. Tiny, membrane-surrounded droplets called *vesicles* appear to shuttle between the various components of the endomembrane system. This system has various structures, but all of them are essentially compartments, closed off by their membranes from the cytoplasm. The membranes and materials in this system move between its various organelles.

ENDOPLASMIC RETICULUM Electron micrographs reveal a network of interconnected membranes branching throughout the cytoplasm of a eukaryotic cell, forming tubes and flattened sacs. These membranes are collectively called the **endoplasmic reticulum**, or **ER**. The interior compartment of the ER, referred to as the *lumen*, is separate and distinct from the surrounding cytoplasm (**Figure 4.10**). The ER can enclose up to 10 percent of the interior volume of the cell, and its foldings result in a surface area many times greater than that of the plasma membrane.

Rough endoplasmic reticulum (**RER**) is ER that is studded with ribosomes, which are temporarily attached to the outer faces of its flattened sacs.

- As a compartment, the RER segregates certain newly synthesized proteins away from the cytoplasm and transports them to other locations in the cell.

- While inside the RER, proteins can be chemically modified so as to alter their function and eventual destination.

The attached ribosomes are sites for the synthesis of proteins that function outside the cytosol—that is, proteins that are to be exported from the cell, incorporated into membranes, or moved into the organelles of the endomembrane system. These proteins enter the lumen of the RER as they are synthesized. As in nuclear localization, this is accomplished by a sequence of amino acids on the protein that acts as a RER localization signal. Once in the lumen of the RER, these proteins undergo several changes, including the formation of disulfide bridges and folding into their tertiary structures (see Figure 3.7D).

Some proteins gain carbohydrate groups in the RER, thus becoming *glycoproteins*. In the case of proteins directed to the lysosomes, the carbohydrate groups are part of an "addressing" system that ensures that the right proteins are directed to those organelles.

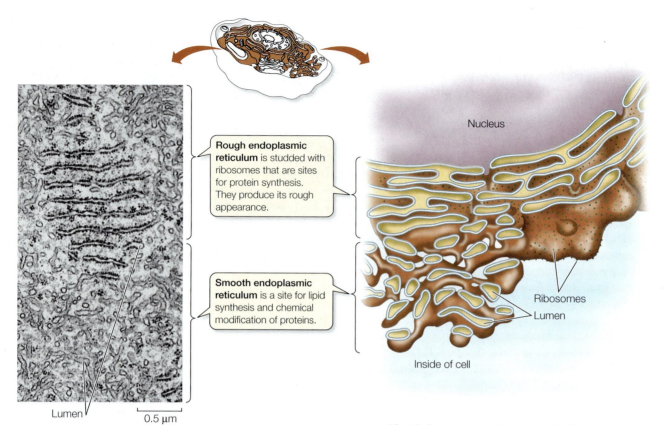

Rough endoplasmic reticulum is studded with ribosomes that are sites for protein synthesis. They produce its rough appearance.

Smooth endoplasmic reticulum is a site for lipid synthesis and chemical modification of proteins.

Nucleus

Ribosomes

Lumen

Inside of cell

Lumen

0.5 μm

4.10 Endoplasmic Reticulum The transmission electron micrograph on the left shows a two-dimensional slice through the three-dimensional structures depicted in the drawing. In normal living cells, the membranes of the ER never have open ends; they define closed compartments set off from the surrounding cytoplasm.

Smooth endoplasmic reticulum (SER) is more tubular (less like flattened sacs) than the RER and lacks ribosomes (see Figure 4.10). Within the lumen of the SER, some proteins that have been synthesized on the RER are chemically modified. In addition, the SER has three other important roles:

■ It is responsible for chemically modifying small molecules taken in by the cell, especially drugs and pesticides.

■ It is the site for the hydrolysis of glycogen in animal cells.

■ It is the site for the synthesis of lipids and steroids.

Cells that synthesize a lot of protein for export are usually packed with endoplasmic reticulum. Examples include glandular cells that secrete digestive enzymes and white blood cells that secrete antibodies. In contrast, cells that carry out less protein synthesis (such as storage cells) contain less ER. Liver cells, which modify molecules that enter the body from the digestive system, have abundant smooth ER.

GOLGI APPARATUS The exact appearance of the **Golgi apparatus** (named for its discoverer, Camillo Golgi) varies from species to species, but it almost always consists of flattened membranous sacs called *cisternae* (singular *cisterna*), piled up like saucers, and small membrane-enclosed vesicles (**Figure 4.11**). The entire apparatus is about 1 μm long.

The Golgi apparatus has several roles:

■ It receives proteins from the ER and may further modify them.

■ It concentrates, packages, and sorts proteins before they are sent to their cellular or extracellular destinations.

■ It is where some polysaccharides for the plant cell wall are synthesized.

In the cells of plants, protists, fungi, and many invertebrate animals, the stacks of cisternae are individual units scattered throughout the cytoplasm. In vertebrate cells, a few such stacks usually form a larger, single, more complex Golgi apparatus.

The Golgi apparatus appears to have three functionally distinct parts: a bottom, a middle, and a top. The bottom cisternae, constituting the *cis* region of the Golgi apparatus, lie nearest to the nucleus or a patch of RER (see Figure 4.11). The top cisternae, constituting the *trans* region, lie closest to the surface of the cell. The cisternae in the middle make up the *medial* region. (The terms *cis*, *trans*, and medial derive from Latin words meaning, respectively, "on the same side," "on the opposite side," and "in the middle.") These three parts of the Golgi apparatus contain different enzymes and perform different functions.

The Golgi apparatus receives proteins from the ER, packages them, and sends them on their way. Since there is often no direct membrane continuity between ER and Golgi apparatus, how does a protein get from one organelle to the other? The protein could simply leave the ER, travel across the cytoplasm, and enter the Golgi apparatus. But that would expose the protein to interactions with other molecules in the cytoplasm. On the other hand, segregation from the cytoplasm could be maintained if a piece of the ER could "bud off," forming a membranous vesicle that contains the protein—and that is exactly what happens (see Figure 4.11). As we will see in Section 5.1, many of the molecules in a mem-

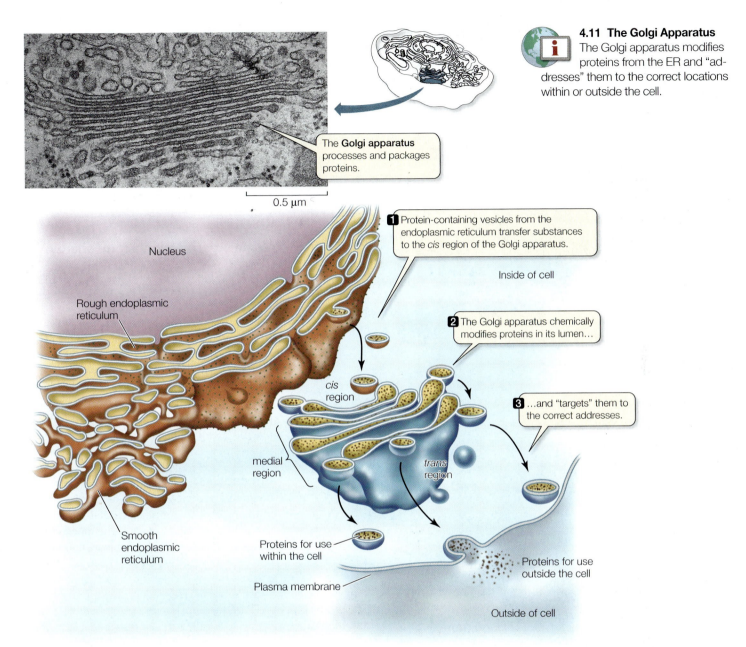

The **Golgi apparatus** processes and packages proteins.

0.5 µm

4.11 The Golgi Apparatus
The Golgi apparatus modifies proteins from the ER and "addresses" them to the correct locations within or outside the cell.

Nucleus

Rough endoplasmic reticulum

1 Protein-containing vesicles from the endoplasmic reticulum transfer substances to the *cis* region of the Golgi apparatus.

Inside of cell

cis region

2 The Golgi apparatus chemically modifies proteins in its lumen…

medial region

trans region

3 …and "targets" them to the correct addresses.

Smooth endoplasmic reticulum

Proteins for use within the cell

Plasma membrane

Proteins for use outside the cell

Outside of cell

brane are not bonded to one another (they "rub shoulders" rather than "hold hands"). Thus a biological membrane is not an ordered, solid structure, but a flexible, fluid boundary. Pieces of a membrane can easily bud off from or fuse with another membrane.

Proteins make the passage from ER to Golgi apparatus safely enclosed in a vesicle. Once it arrives, the vesicle fuses with the membrane of the Golgi apparatus, releasing its cargo. Other small vesicles may move between the cisternae, transporting proteins, and it appears that some proteins move from one cisterna to the next through tiny channels. Vesicles budding off from the *trans* region carry their contents away from the Golgi apparatus (see Figure 4.11).

LYSOSOMES Originating from the Golgi apparatus are organelles called **lysosomes**. They contain digestive enzymes, and they are the sites where macromolecules—proteins, polysaccharides, nucleic acids, and lipids—are hydrolyzed into their monomers (see Figure 3.4). Lysosomes are about 1 µm in diameter, are surrounded by a single membrane, and have a densely staining, featureless interior

(**Figure 4.12**). There may be dozens of lysosomes in a cell, depending on its needs.

Lysosomes are sites for the breakdown of food and foreign objects taken up by the cell. These materials get into the cell by a process called *phagocytosis* (*phago*, "eat"; *cytosis*, "cellular"), in which a pocket forms in the plasma membrane and eventually deepens and encloses material from outside the cell. This pocket becomes a small vesicle that breaks free of the plasma membrane to move into the cytoplasm as a *phagosome* containing the food or other material (see Figure 4.12). The phagosome fuses with a *primary lysosome*, forming a *secondary lysosome*, in which digestion occurs.

The effect of this fusion is rather like releasing hungry foxes into a chicken coop: the enzymes in the secondary lysosome quickly hydrolyze the food particles. These reactions are enhanced by the mild acidity of the lysosome's interior, where the pH is lower than in the surrounding cytoplasm. The products of digestion diffuse through the membrane of the lysosome, providing raw materials for other cell processes. The "used" secondary lyso-

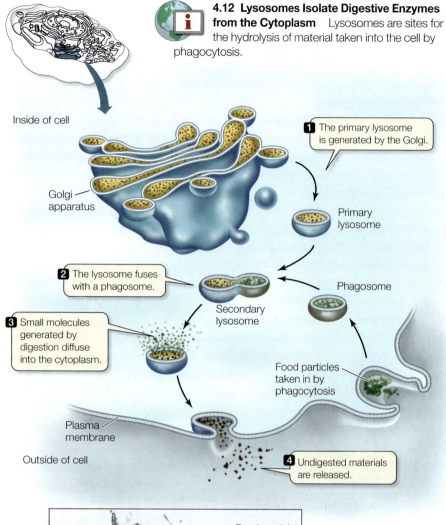

4.12 Lysosomes Isolate Digestive Enzymes from the Cytoplasm Lysosomes are sites for the hydrolysis of material taken into the cell by phagocytosis.

Inside of cell

Golgi apparatus

1 The primary lysosome is generated by the Golgi.

Primary lysosome

2 The lysosome fuses with a phagosome.

Secondary lysosome

Phagosome

3 Small molecules generated by digestion diffuse into the cytoplasm.

Food particles taken in by phagocytosis

Plasma membrane

Outside of cell

4 Undigested materials are released.

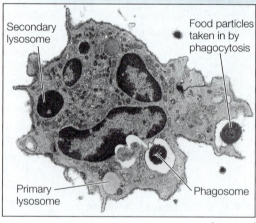

Secondary lysosome

Food particles taken in by phagocytosis

Primary lysosome

Phagosome

1 μm

Plant cells do not appear to contain lysosomes, but the central vacuole of a plant cell (which we will describe below) may function in an equivalent capacity because it, like lysosomes, contains many digestive enzymes.

Fatal lysosomal storage diseases can occur when lysosomes malfunction and molecules that should get broken down accumulate inside them instead.

Some organelles transform energy

A cell uses energy to synthesize the materials it needs for activities such as growth, reproduction, and movement. Energy is transformed from one form to another in mitochondria (found in eukaryotic cells) and in chloroplasts (found in eukaryotic cells that harvest energy from sunlight). In contrast, energy transformations in prokaryotic cells are associated with enzymes attached to the inner surface of the plasma membrane or to extensions of the plasma membrane that protrude into the cytoplasm.

MITOCHONDRIA In eukaryotic cells, the breakdown of fuel molecules such as glucose begins in the cytosol. The molecules that result from this partial degradation enter the **mitochondria** (singular *mitochondrion*), whose primary function is to convert the potential chemical energy of those fuel molecules into a form that the cell can use: the energy-rich molecule ATP (adenosine triphosphate). The production of ATP in the mitochondria using fuel molecules and molecular oxygen (O_2) is called *cellular respiration*.

Typical mitochondria are somewhat less than 1.5 μm in diameter and 2–8 μm in length—about the size of many bacteria. The number of mitochondria per cell ranges from one contorted giant in some unicellular protists to a few hundred thousand in large egg cells. An average human liver cell contains more than a thousand mitochondria. Cells that require the most chemical energy tend to have the most mitochondria per unit of volume.

Mitochondria have two membranes. The *outer membrane* is smooth and protective, and it offers little resistance to the movement of substances into and out of the mitochondrion. Immediately inside the outer membrane is an *inner membrane*, which folds inward in many places, and thus has a surface area much greater than that of the outer membrane (**Figure 4.13**). The folds tend to be quite regular, giving rise to shelflike structures called *cristae*.

Embedded in the inner mitochondrial membrane are many large protein complexes that participate in cellular respiration. The inner membrane exerts much more control over what enters and leaves the space it encloses than does the outer membrane. The

some, now containing undigested particles, then moves to the plasma membrane, fuses with it, and releases the undigested contents to the environment.

Lysosomes are also where the cell digests its own material in a process called *autophagy*. Autophagy is an ongoing process in which organelles such as mitochondria are engulfed by lysosomes and hydrolyzed into monomers, which pass out of the lysosome through its membrane into the cytoplasm for reuse.

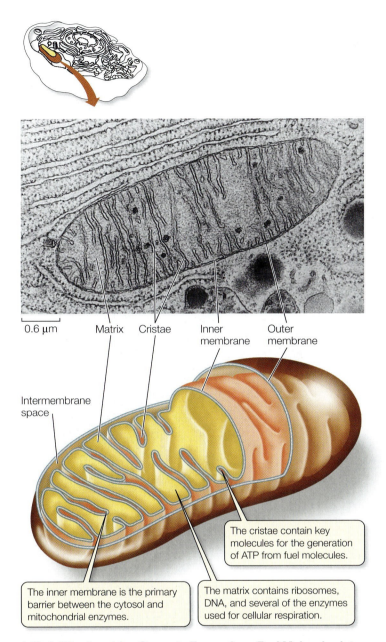

0.6 μm   Matrix   Cristae   Inner membrane   Outer membrane

Intermembrane space

The cristae contain key molecules for the generation of ATP from fuel molecules.

The inner membrane is the primary barrier between the cytosol and mitochondrial enzymes.

The matrix contains ribosomes, DNA, and several of the enzymes used for cellular respiration.

4.13 A Mitochondrion Converts Energy from Fuel Molecules into ATP The electron micrograph is a two-dimensional slice through a three-dimensional organelle. As the drawing emphasizes, the cristae are extensions of the inner mitochondrial membrane.

space enclosed by the inner membrane is referred to as the *mitochondrial matrix*. In addition to many enzymes, the matrix contains some ribosomes and DNA that are used to make some of the proteins needed for cellular respiration. We will see in Chapter 7 how the different parts of the mitochondrion work together in cellular respiration.

PLASTIDS One class of organelles—the *plastids*—is produced only in the cells of plants and certain protists. There are several types of plastids, with different functions.

Chloroplasts contain the green pigment *chlorophyll* and are the sites of photosynthesis (**Figure 4.14**). In photosynthesis, light

energy is converted into the chemical energy of bonds between atoms. The molecules formed by photosynthesis provide food for the photosynthetic organisms, as well as for other organisms that eat them. Directly or indirectly, photosynthesis is the energy source for most of the living world.

Chloroplasts are variable in size and shape (**Figure 4.15**). Like a mitochondrion, a chloroplast is surrounded by two membranes. In addition, there is a series of internal membranes whose structure and arrangement vary from one group of photosynthetic organisms to another. Here we concentrate on the chloroplasts of the flowering plants. Even those chloroplasts show some variation, but the pattern shown in Figure 4.14 is typical.

The internal membranes of chloroplasts look like stacks of flat, hollow pita bread. These stacks, called *grana* (singular *granum*), consist of a series of flat, closely packed, circular compartments called **thylakoids**. In addition to phospholipids and proteins, the membranes of the thylakoids contain chlorophyll and other pigments that harvest light for photosynthesis (we will see how they do this in Section 8.2). The thylakoids of one granum may be connected to those of other grana, making the interior of the chloroplast a highly developed network of membranes, much like the ER.

The fluid in which the grana are suspended is called the *stroma*. Like the mitochondrial matrix, the chloroplast stroma contains ribosomes and DNA, which are used to synthesize some, but not all, of the proteins that make up the chloroplast.

Animal cells do not produce chloroplasts, but some do contain functional chloroplasts. These chloroplasts are either taken up from the partial digestion of green plants or contained within unicellular algae that live within the animal's tissues. The green color of some corals and sea anemones comes from chloroplasts in algae that live within those animals (see Figure 4.15C). The animals derive some of their nutrition from the photosynthesis that their chloroplast-containing "guests" carry out. Such an intimate relationship between two different organisms is called *symbiosis*.

Other types of plastids have functions different from those of chloroplasts:

- *Chromoplasts* contain red, orange, and/or yellow pigments and give color to plant organs such as flowers (**Figure 4.16A**). The chromoplasts have no known chemical function in the cell, but the colors they give to some petals and fruits probably encourage animals to visit flowers and thus aid in pollination, or to eat fruits and thus aid in seed dispersal. (On the other hand, carrot roots gain no apparent advantage from being orange.)

- *Leucoplasts* are storage depots for starches and fats (**Figure 4.16B**).

Several other organelles are surrounded by a membrane

Peroxisomes are organelles that collect the toxic peroxides (such as hydrogen peroxide, H_2O_2) that are the unavoidable by-products of cellular chemical reactions. These peroxides can be safely broken down inside the peroxisomes without mixing with other parts of the cell. Peroxisomes are small organelles, about 0.2 to 1.7 μm

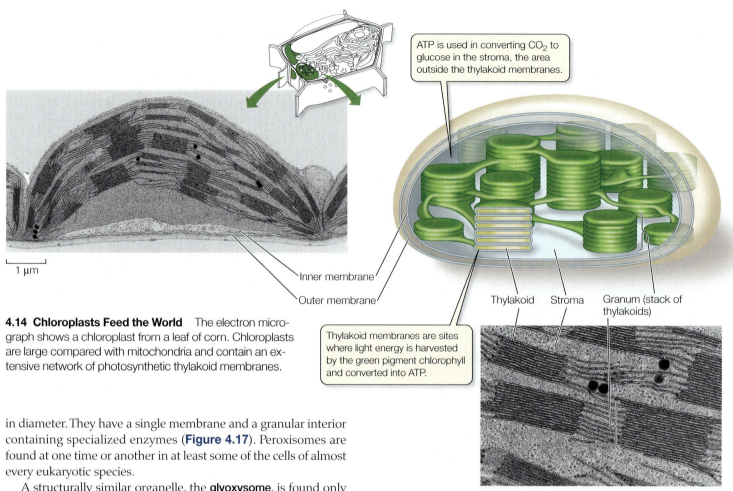

ATP is used in converting CO_2 to glucose in the stroma, the area outside the thylakoid membranes.

Inner membrane

Outer membrane

Thylakoid   Stroma   Granum (stack of thylakoids)

1 μm

Thylakoid membranes are sites where light energy is harvested by the green pigment chlorophyll and converted into ATP.

4.14 Chloroplasts Feed the World The electron micrograph shows a chloroplast from a leaf of corn. Chloroplasts are large compared with mitochondria and contain an extensive network of photosynthetic thylakoid membranes.

0.5 μm

in diameter. They have a single membrane and a granular interior containing specialized enzymes (**Figure 4.17**). Peroxisomes are found at one time or another in at least some of the cells of almost every eukaryotic species.

A structurally similar organelle, the **glyoxysome**, is found only in plants. Glyoxysomes, which are most prominent in young plants, are the sites where stored lipids are converted into carbohydrates for transport to growing cells.

Many eukaryotic cells, but particularly those of plants and protists, contain membrane-enclosed **vacuoles** filled with aqueous

4.15 Being Green (A) In green plants, chloroplasts are concentrated in the leaf cells. (B) Green algae are photosynthetic and filled with chloroplasts. (C) No animal species produces its own chloroplasts, but in this symbiotic arrangement, a sea anemone is nourished by the chloroplasts of unicellular green algae living within its tissues.

(A) Chloroplasts   Leaf cell

(B)

(C)

75 μm

250 μm

The chloroplasts in these single-celled green algae have assembled into spirals.

Chloroplast-filled green algae live in the tissues of this sea anemone.

(A)

Chromoplast

5 μm

(B)

Leucoplast

Starch grains

1 μm

4.16 Chromoplasts and Leucoplasts (A) Colorful pigments stored in the chromoplasts of flowers like this poppy may help attract pollinating insects. (B) Leucoplasts in the cells of a potato are filled with white starch grains.

solutions containing many dissolved substances (**Figure 4.18**). Plant vacuoles have several functions:

■ *Storage*: Plant cells produce a number of toxic by-products and waste products, many of which are simply stored within vacuoles. Because they are poisonous or distasteful, these stored materials deter some animals from eating the plants, and may thus contribute to plant survival.

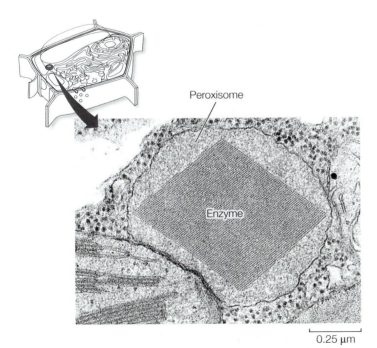

Peroxisome

Enzyme

0.25 μm

4.17 A Peroxisome A diamond-shaped crystal, composed of an enzyme, almost entirely fills this rounded peroxisome in a leaf cell. The enzyme catalyzes one of the reactions that breaks down toxic peroxides in the peroxisome.

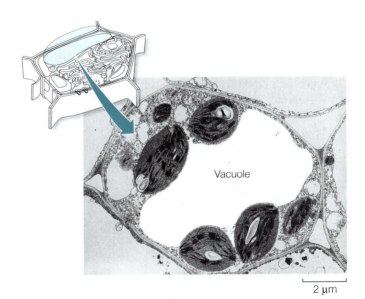

Vacuole

2 μm

4.18 Vacuoles in Plant Cells Are Usually Large The large central vacuole in this cell is typical of mature plant cells. Smaller vacuoles are visible toward each end of the cell.

- *Structure*: In many plant cells, enormous vacuoles take up more than 90 percent of the cell volume and grow as the cell grows. The presence of dissolved substances in the vacuole causes water to enter, making the vacuole swell like a balloon. Plant cells have a rigid cell wall, which resists the swelling of the vacuole, providing *turgor*, or stiffness, that helps to support the plant.

- *Reproduction*: Some pigments (especially blue and pink ones) in the petals and fruits of flowering plants are contained in vacuoles. These pigments—the *anthocyanins*—are visual cues that help attract the animals that assist in pollination or seed dispersal.

- *Digestion*: In some plants, vacuoles in seeds contain enzymes that hydrolyze seed proteins into monomers that a developing plant embryo can use as food.

Food vacuoles are found in some simple and evolutionarily ancient groups of eukaryotes—single-celled protists and simple multicellular organisms such as sponges, for example—that have no digestive system. In these organisms, the cells engulf food particles by phagocytosis, generating a food vacuole. Fusion of this vacuole with a lysosome results in digestion, and small molecules leave the vacuole and enter the cytoplasm for use or distribution to other organelles.

Contractile vacuoles are found in many freshwater protists. Their function is to get rid of the excess water that rushes into the cell because of the imbalance in solute concentration between the interior of the cell and its freshwater environment. The contractile vacuole enlarges as water enters, then abruptly contracts, forcing the water out of the cell through a special pore structure.

The cytoskeleton is important in cell structure

In addition to its many membrane-enclosed organelles, the eukaryotic cytoplasm contains a set of long, thin fibers called the **cytoskeleton**. The cytoskeleton fills several important roles:

- It supports the cell and maintains its shape.
- It provides for various types of cellular movement.
- It positions organelles within the cell.
- Some of its fibers act as tracks or supports for motor proteins, which move organelles within the cell.
- It interacts with extracellular structures, helping to anchor the cell in place.

How do we know that the structural fibers of the cytoskeleton can achieve all these dynamic functions? Observing an individual structure under the microscope and observing a function in a living cell that contains that structure may be suggestive, but in science *mere correlation does not show cause and effect*. In cell biology, there are two ways to show that a structure or process "A" causes function "B":

- *Inhibition*: Use a drug that inhibits A and see if B still occurs. If it does not, then A is probably a causative factor for B. **Figure 4.19** shows an experiment with such a drug (an *inhibitor*) that demonstrates cause and effect in the case of the cytoskeleton and cell movement.

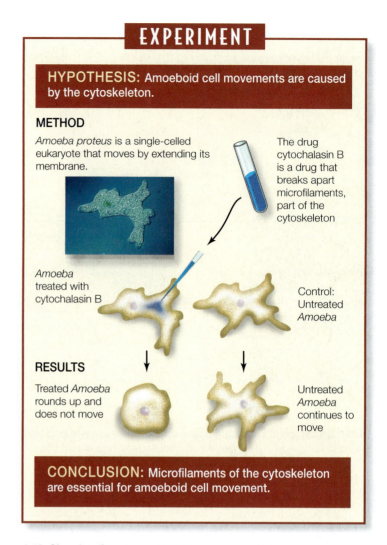

EXPERIMENT

HYPOTHESIS: Amoeboid cell movements are caused by the cytoskeleton.

METHOD

Amoeba proteus is a single-celled eukaryote that moves by extending its membrane.

The drug cytochalasin B is a drug that breaks apart microfilaments, part of the cytoskeleton

Amoeba treated with cytochalasin B

Control: Untreated *Amoeba*

RESULTS

Treated *Amoeba* rounds up and does not move

Untreated *Amoeba* continues to move

CONCLUSION: Microfilaments of the cytoskeleton are essential for amoeboid cell movement.

4.19 Showing Cause and Effect in Biology A substance known to inhibit a structure (in this case, cytochalasin B, a drug that inhibits microfilament formation) is applied in order to discover whether it also inhibits a function (in this case, cell movement in *Amoeba*). FURTHER RESEARCH: The drug colchicine breaks apart microtubules. How would you show that these components of the cytoskeleton are *not* involved in cell movement in *Amoeba*?

- *Mutation:* Look at a cell that lacks the gene (or genes) for A and see if B still occurs. If it does not, then A is probably a causative factor for B. Part Three of this book describes many experiments using this genetic approach.

MICROFILAMENTS Microfilaments can exist as single filaments, in bundles, or in networks. They are about 7 nm in diameter and up to several micrometers long. Microfilaments have two major roles:

- They help the entire cell or parts of the cell to move.
- They determine and stabilize cell shape.

Microfilaments are assembled from *actin*, a protein that exists in several forms and has many functions, especially in animals. The actin found in microfilaments (which are also known as *actin filaments*) is extensively folded and has distinct "plus" and "minus"

4.20 The Cytoskeleton Three highly visible and important structural components of the cytoskeleton are shown here in detail. These structures maintain and reinforce cell shape and contribute to cell movement.

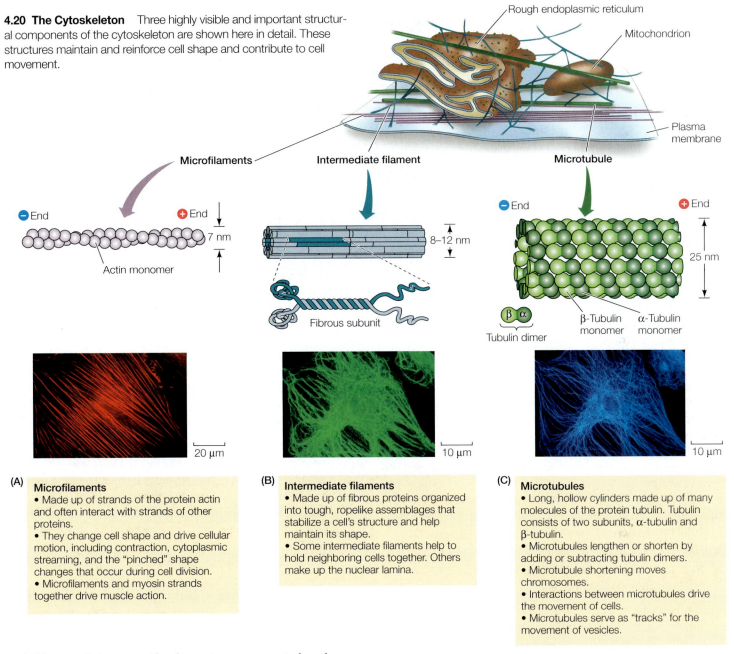

(A) Microfilaments
• Made up of strands of the protein actin and often interact with strands of other proteins.
• They change cell shape and drive cellular motion, including contraction, cytoplasmic streaming, and the "pinched" shape changes that occur during cell division.
• Microfilaments and myosin strands together drive muscle action.

(B) Intermediate filaments
• Made up of fibrous proteins organized into tough, ropelike assemblages that stabilize a cell's structure and help maintain its shape.
• Some intermediate filaments help to hold neighboring cells together. Others make up the nuclear lamina.

(C) Microtubules
• Long, hollow cylinders made up of many molecules of the protein tubulin. Tubulin consists of two subunits, α-tubulin and β-tubulin.
• Microtubules lengthen or shorten by adding or subtracting tubulin dimers.
• Microtubule shortening moves chromosomes.
• Interactions between microtubules drive the movement of cells.
• Microtubules serve as "tracks" for the movement of vesicles.

ends. These ends interact with other actin monomers to form long, double helical chains (**Figure 4.20A**). The polymerization of actin into microfilaments is reversible, and they can disappear from cells, breaking down into monomers of free actin.

In the muscle cells of animals, actin filaments are associated with another protein, the "motor protein" *myosin*, and the interactions of these two proteins account for the contraction of muscles (described in Section 47.1). In non-muscle cells, actin filaments are associated with localized changes of shape in the cell. For example, microfilaments are involved in a flowing movement of the cytoplasm called *cytoplasmic streaming* and in the "pinching" contractions that divide an animal cell into two daughter cells. Microfilaments are also involved in the formation of cellular extensions called *pseudopodia* (*pseudo*, "false;" *podia*, "feet") that enable some cells to move.

In some cell types, microfilaments form a meshwork just inside the plasma membrane. Actin-binding proteins then cross-link the microfilaments to form a rigid structure that supports the cell. For

example, microfilaments support the tiny microvilli that line the human intestine, giving it a larger surface area through which to absorb nutrients (**Figure 4.21**).

INTERMEDIATE FILAMENTS There are at least 50 different kinds of **intermediate filaments**, many of them specific to a few cell types. They generally fall into six molecular classes (based on amino acid sequence) that share the same general structure, being composed of fibrous proteins of the keratin family, similar to the protein that makes up hair and fingernails. These proteins are organized into tough, ropelike assemblages 8 to 12 nm in diameter (**Figure 4.20B**). Intermediate filaments have two major structural functions:

■ They stabilize cell structure.

■ They resist tension.

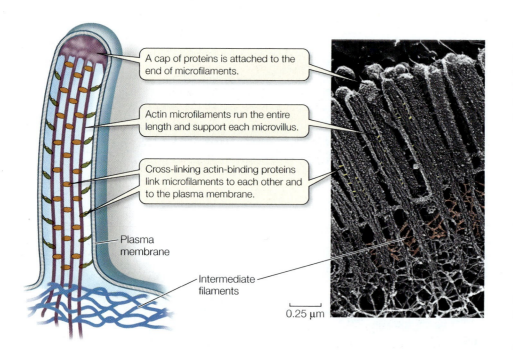

A cap of proteins is attached to the end of microfilaments.

Actin microfilaments run the entire length and support each microvillus.

Cross-linking actin-binding proteins link microfilaments to each other and to the plasma membrane.

Plasma membrane

Intermediate filaments

0.25 μm

4.21 Microfilaments for Support Cells that line the intestine are folded into tiny projections called microvilli, which are supported by microfilaments. The microvilli increase the surface area of the cells, facilitating their absorption of small molecules.

In some cells, intermediate filaments radiate from the nuclear envelope and may maintain the positions of the nucleus and other organelles in the cell. The lamins of the nuclear lamina are intermediate filaments. Other kinds of intermediate filaments help hold a complex apparatus of microfilaments in the microvilli in intestinal cells (see Figure 4.21). Still other kinds stabilize and help maintain rigidity in body surface tissues by connecting "spot welds" called *desmosomes* between adjacent cells (see Figure 5.7B).

MICROTUBULES Microtubules are long, hollow, unbranched cylinders about 25 nm in diameter and up to several micrometers long. Microtubules have two roles in the cell:

- They form a rigid internal skeleton for some cells.

- They act as a framework along which motor proteins can move structures within the cell.

Microtubules are assembled from molecules of the protein *tubulin*. Tubulin is a *dimer*—a molecule made up of two monomers. The polypeptide monomers that make up a tubulin dimer are known as α-tubulin and β-tubulin. Thirteen chains of tubulin dimers surround the central cavity of the microtubule (**Figure 4.20C**). The two ends of a microtubule are different: one end is designated the plus (+) end, the other the minus (–) end. Tubulin dimers can be rapidly added or subtracted, mainly at the plus end, lengthening or shortening the microtubule. This capacity to change length rapidly makes microtubules *dynamic structures*. This dynamic property of microtubules can be seen in animal cells, where they are often found in parts of the cell that are changing shape.

Many microtubules radiate from a region of the cell called the *microtubule organizing center*. Tubule polymerization results in a rigid structure, and tubule depolymerization leads to its collapse.

In plants, microtubules help control the arrangement of the cellulose fibers of the cell wall. Electron micrographs of plants frequently show microtubules lying just inside the plasma membrane of cells that are forming or extending their cell walls. Experimental alteration of the orientation of these microtubules leads to a similar change in the cell wall and a new shape for the cell.

In many cells, microtubules serve as tracks for **motor proteins**, specialized molecules that use energy to change their shape and move. Motor proteins bond to and move along the microtubules, carrying materials from one part of the cell to another. Microtubules are also essential in distributing chromosomes to daughter cells during cell division. They are also intimately associated with movable cell appendages: the cilia and flagella.

Because microtubules are crucial for proper chromosome alignment during cell division, drugs such as colchicine and taxol that disrupt microtubule dynamics also disrupt cell division. If these drugs can be targeted to cancer cells—whose division is rampant—they can be useful in treating the disease.

CILIA AND FLAGELLA Many eukaryotic cells possess flagella or cilia, or both. These whiplike organelles may push or pull the cell through its aqueous environment, or they may move surrounding fluid over the surface of the cell. Cilia and eukaryotic flagella (which are quite distinct from prokaryotic flagella) are both assembled from specialized microtubules and have identical internal structures, but differ in their length and pattern of beating:

- Cilia (singular *cilium*) are shorter than flagella and are usually present in great numbers (**Figure 4.22A**). They beat stiffly in one direction and recover flexibly in the other direction (like a swimmer's arm), so that the recovery stroke does not undo the work of the power stroke.

- Eukaryotic *flagella* are longer than cilia and are usually found singly or in pairs. Waves of bending propagate from one end of a flagellum to the other in snakelike undulation.

In cross section, a typical cilium or eukaryotic flagellum is surrounded by the plasma membrane and contains a "9 + 2" array of microtubules. As **Figure 4.22B** shows, nine fused pairs of microtubules—called *doublets*—form an outer cylinder, and one pair of unfused microtubules runs up the center. A spoke radiates from

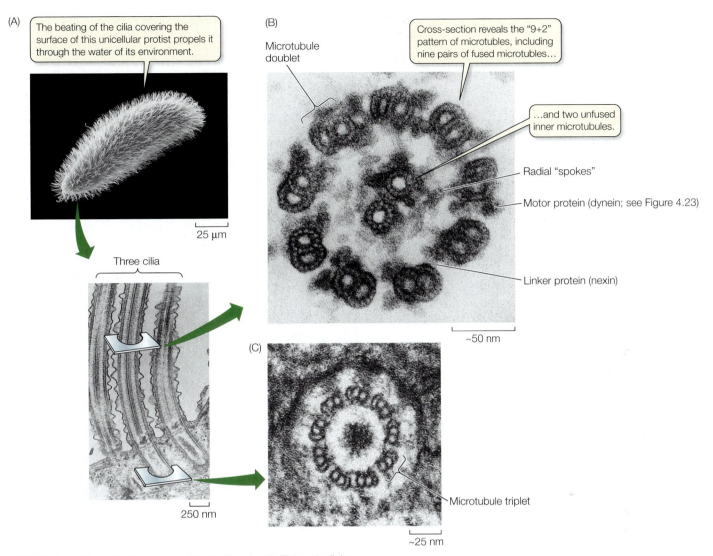

(A) The beating of the cilia covering the surface of this unicellular protist propels it through the water of its environment.

25 μm

Three cilia

250 nm

(B) Microtubule doublet

Cross-section reveals the "9+2" pattern of microtubules, including nine pairs of fused microtubles...

...and two unfused inner microtubules.

Radial "spokes"

Motor protein (dynein; see Figure 4.23)

Linker protein (nexin)

~50 nm

(C)

Microtubule triplet

~25 nm

4.22 Sliding Microtubules Cause Cilia to Bend (A) This unicellular eukaryotic organism (a ciliate protist) can coordinate the beating of its cilia, allowing rapid movement. (B) A cross section of a single cilium shows the arrangement of the microtubules and proteins. (C) Cross section of a basal body.

one microtubule of each doublet and connects the doublet to the center of the structure.

In the cytoplasm at the base of every eukaryotic flagellum or cilium is an organelle called a *basal body*. The nine microtubule doublets extend into the basal body. In the basal body, each doublet is accompanied by another microtubule, making nine sets of three microtubules. The central, unfused microtubules in the cilium do not extend into the basal body (**Figure 4.22C**).

Centrioles are almost identical to the basal bodies of cilia and flagella. Centrioles are found in the microtubule organizing centers of all eukaryotes except the seed plants and some protists. Under the light microscope, a centriole looks like a small, featureless particle, but the electron microscope reveals that it is made up of a precise bundle of microtubules arranged as nine sets of three fused microtubules each. Centrioles are involved in the forma-

tion of the mitotic spindle, to which the chromosomes attach (see Figure 9.9).

MOTOR PROTEINS The nine microtubule doublets of cilia and flagella are linked by proteins. The motion of cilia and flagella results from the sliding of the microtubule doublets past each other. This sliding is driven by a motor protein called *dynein*, which can change its three-dimensional shape. All motor proteins work by undergoing reversible shape changes powered by energy from ATP. Dynein molecules attached to one microtubule doublet bind to a neighboring doublet. As the dynein molecules change shape, they move the microtubule doublet past its neighbor (**Figure 4.23A**).

Another motor protein, *kinesin*, carries protein-laden vesicles from one part of the cell to another. Kinesin and similar motor proteins bind to a vesicle or other organelle, then "walk" it along a microtubule by changing their shape. This process is driven by a protein region of about 350 amino acids that is similar to myosin, the motor protein that binds to and moves actin microfilaments in muscle. Recall that microtubules have a plus end and a minus end. Dynein moves attached organelles toward the minus end, while kinesin moves them toward the plus end (**Figure 4.23B**).

(A)

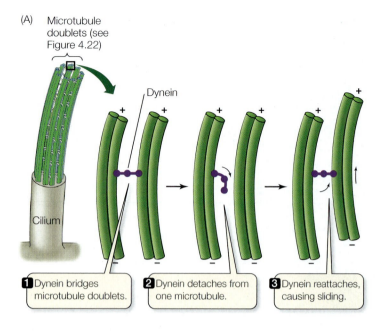

Microtubule doublets (see Figure 4.22)

Dynein

Cilium

1 Dynein bridges microtubule doublets.

2 Dynein detaches from one microtubule.

3 Dynein reattaches, causing sliding.

(B)

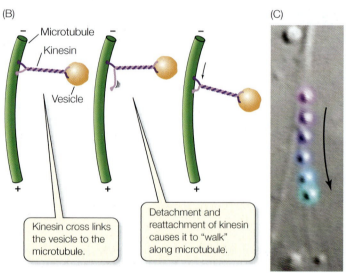

Microtubule

Kinesin

Vesicle

Kinesin cross links the vesicle to the microtubule.

Detachment and reattachment of kinesin causes it to "walk" along microtubule.

(C)

4.23 Motor Proteins Drive Vesicles along Microtubules (A) The motor protein dynein allows microtubule doublets to slide past one another in a cilium or flagellum. (B) Kinesin delivers vesicles or organelles to various parts of the cell by moving along microtubule "railroad tracks." Kinesin moves things from the minus toward the plus end of a microtubule; dynein works similarly, but moves from the plus toward the minus end. (C) Powered by kinesin, a vesicle (color) moves along a microtubule track in the protist *Dictyostelium*. The time sequence (time-lapse micrography at half-second intervals) is shown by the color gradient of purple to blue.

All cells interact with their environment, and many eukaryotic cells are part of a multicellular organism and must interact with other cells. The plasma membrane plays a crucial role in these interactions, but other structures outside that membrane are involved as well.

4.4 What Are the Roles of Extracellular Structures?

Although the plasma membrane is the functional barrier between the inside and the outside of a cell, many structures are produced by cells and secreted to the outside of the plasma membrane, where they play essential roles in protecting, supporting, or attaching cells. Because they are outside the plasma membrane, these structures are said to be *extracellular*. The peptidoglycan cell wall of bacteria is an example of such an extracellular structure. In eukaryotes, other extracellular structures—the cell walls of plants and the extracellular matrices found between the cells of animals—play similar roles. Both of these structures are made up of two components: a prominent fibrous macromolecule, and a gel-like medium.

The plant cell wall is an extracellular structure

The **cell wall** of plant cells is a semirigid structure outside the plasma membrane (**Figure 4.24**). It consists of cellulose fibers (see Figure 3.16) embedded in other complex polysaccharides and proteins. The cell wall has three major roles in plants:

- It provides support for the cell and limits its volume by remaining rigid.
- It acts as a barrier to infection by fungi and other organisms that can cause plant diseases.
- It contributes to plant form by growing as plant cells expand.

Because of their thick cell walls, plant cells viewed under a light microscope appear to be entirely isolated from one another. But electron microscopy reveals that this is not the case. The cytoplasm of adjacent plant cells is connected by numerous plasma membrane–lined channels, called **plasmodesmata**, that are about 20–40 nm in diameter and extend through the walls of adjoining cells (see Figure 15.20). Plasmodesmata permit the diffusion of water, ions, small molecules, and RNA and proteins between connected cells, ensuring uniform concentrations of these substances.

4.3 RECAP

The hallmark of eukaryotic cells is compartmentalization. Membrane-enclosed organelles process information, transform energy, form internal compartments for transporting proteins, and carry out intracellular digestion. An internal cytoskeleton plays several structural roles.

- What are some advantages of organelle compartmentalization? See pp. 74–75
- What are the differences between rough and smooth endoplasmic reticulum? See pp. 79–80 and Figure 4.10
- Describe how motor proteins and microtubules move materials within the cell. See pp. 88–90 and Figure 4.23

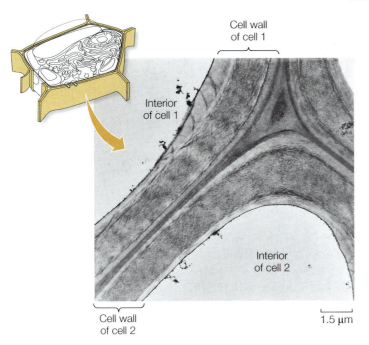

Cell wall
of cell 1

Interior
of cell 1

Interior
of cell 2

Cell wall
of cell 2

1.5 µm

4.24 The Plant Cell Wall The semirigid cell wall provides support for plant cells.

The extracellular matrix supports tissue functions in animals

The cells of animals lack the semirigid cell wall that is characteristic of plant cells, but many animal cells are surrounded by, or are in contact with, an **extracellular matrix**. This matrix is composed of fibrous proteins such as **collagen** (the most abundant protein in mammals, constituting 25 percent of the protein in the human body), a matrix of glycoproteins termed **proteoglycans**, consisting primarily of sugars, and a third group of proteins that link the fibrous proteins and gel-like proteoglycan matrix together (**Figure 4.25**). These proteins, along with other substances that are specific to certain body tissues, are secreted by cells that are present in or near the matrix.

The functions of the extracellular matrix are many:

- It holds cells together in tissues.

- It contributes to the physical properties of cartilage, skin, and other tissues.

- It helps filter materials passing between different tissues.

- It helps orient cell movements during embryonic development and during tissue repair.

- It plays a role in chemical signaling from one cell to another.

In the human body, some tissues, such as those in the brain, have very little extracellular matrix; other tissues, such as bone and cartilage, have large amounts. Bone cells are embedded in an extracellular matrix that consists primarily of collagen and calcium phosphate. This matrix gives bone its familiar rigidity. Epithelial cells, which line body cavities, lie together in a sheet spread over a *basal lamina*, or *basement membrane*, a form of extracellular matrix (see Figure 4.25).

Some extracellular matrices are made up, in part, of an enormous proteoglycan. A single molecule of this proteoglycan consists of many hundreds of polysaccharide chains covalently attached to about a hundred proteins, all of which are attached in turn to one enormous polysaccharide. The molecular weight of this proteoglycan can exceed 100 million Da; the molecule takes up as much space as an entire prokaryotic cell.

4.4 RECAP

Extracellular structures are produced by cells and secreted outside the plasma membrane. Most are composed of a fibrous component in a gel-like medium.

- Do you understand the functions of the cell wall in plants and the extracellular matrix in animals?

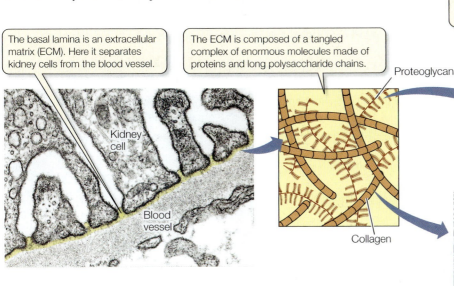

The basal lamina is an extracellular matrix (ECM). Here it separates kidney cells from the blood vessel.

The ECM is composed of a tangled complex of enormous molecules made of proteins and long polysaccharide chains.

Kidney cell

Blood vessel

Proteoglycan

Collagen

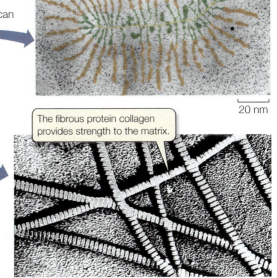

Proteoglycans have long polysaccharide chains that provide a viscous medium for filtering.

20 nm

The fibrous protein collagen provides strength to the matrix.

100 nm

4.25 An Extracellular Matrix Cells in the kidney secrete a basal lamina, an extracellular matrix that separates them from a nearby blood vessel and is also involved in filtering materials that pass between the kidney and the blood.

4.5 How Did Eukaryotic Cells Originate?

We began this chapter by describing some of the earliest evidence of prokaryotic cells on Earth, dating to about 3.5 billion years ago. The living world was entirely prokaryotic until about 1.5 billion years ago, when eukaryotic cells first appeared. This was a major event in the history of life, as the compartmentalization that is the hallmark of eukaryotic cells permitted new biochemical functions to evolve much more readily.

The endosymbiosis theory suggests how eukaryotes evolved

The first prokaryotes probably absorbed their food directly from the environment. Then some became photosynthetic (see Figure 1.10). Others, however, fed on smaller prokaryotes by engulfing them (**Figure 4.26**). Now suppose that a small, photosynthetic prokaryote was ingested by a larger one, but was not digested. Instead, the photosynthetic organism survived, trapped within the cytoplasm of the larger cell. Suppose further that the ingested prokaryote divided at about the same rate as the larger one, so successive generations of the larger cell contained the offspring of the smaller one. Such **endosymbiosis** (*endo*, "within"; *symbiosis*, "living together") would have benefited both partners: the smaller cell would have provided the larger one with monosaccharides made by photosynthesis, and the larger cell would have protected the smaller one. Over evolutionary time, the photosynthetic prokaryote might have evolved into the modern chloroplast.

Similar evidence and arguments support the proposition that mitochondria are the descendants of respiring prokaryotes engulfed by larger prokaryotes. The benefit of such an endosymbiotic relationship might have been the capacity of the engulfed prokaryote to detoxify molecular oxygen (O_2), which was increasing in Earth's atmosphere as a result of photosynthesis.

This **endosymbiosis theory** of eukaryote evolution was first suggested as early as the nineteenth century, but it was expounded and given substance and credence by the work of Lynn Margulis during the 1980s. The chloroplasts and mitochondria of eukaryotes are about the size of prokaryotic cells. These organelles also contain their own DNA and ribosomes and synthesize some of their own components. They are not independent of control by the nucleus, however. The vast majority of their proteins are encoded by nuclear DNA and synthesized in the cell's cytoplasm, then imported into the organelle. It is possible that over time the smaller prokaryotes that gave rise to organelles gradually lost much of their DNA to the nucleus of the larger cell.

Much circumstantial evidence favors the endosymbiosis theory:

- On an evolutionary time scale (i.e., millions of years), there is evidence of the movement of DNA between organelles in the eukaryotic cell.
- There are many biochemical similarities between chloroplasts and photosynthetic bacteria.
- DNA sequencing shows strong similarities between modern chloroplast DNA and that of a photosynthetic prokaryote.

Both prokaryotes and eukaryotes continue to evolve

Today's prokaryotes contain a number of the structures that are present in eukaryotic cells, such as a cytoskeleton, ribosomes, and a plasma membrane. It is probable that these structures evolved gradually. The shared chemistry of all living things, described in Chapter 3, suggests that eukaryotic cells evolved from prokaryotes. For example, both prokaryotes and eukaryotes:

- use nucleic acids as their genetic material;
- use the same 20 amino acids in their proteins; and
- use D sugars and L amino acids.

As described on page 73, many modern prokaryotes have folds in their internal membranes. It is believed that over evolutionary time such folds gradually closed off to form compartments, leading to the formation of the endoplasmic reticulum, Golgi apparatus, and lysosomes. Today's prokaryotic cells have their genetic material in a central region called the nucleoid (see Figure 4.4). During cell reproduction, this DNA becomes attached to the plasma membrane. If the membrane, aided by microfilaments, had folded over the DNA and sealed itself, the cell would have successfully compartmentalized its DNA in a nucleus.

4.5 RECAP

Eukaryotic cells arose long after prokaryotic cells. The endosymbiosis theory suggests how eukaryotic cells may have evolved from prokaryotic ancestors.

- Can you describe some of the evidence that suggests eukaryotic cells evolved from prokaryotic cells?
- What evidence suggests that mitochondria and chloroplasts were once independent prokaryotic cells?

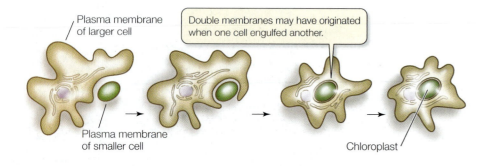

Plasma membrane of larger cell

Double membranes may have originated when one cell engulfed another.

Plasma membrane of smaller cell

Chloroplast

4.26 The Endosymbiosis Theory Chloroplasts may be descended from a small photosynthetic prokaryote that was engulfed by another, larger cell.

The fluidity of the plasma membrane undoubtedly played a major role in the compartmentalization of eukaryotic cells. Microfilaments of the cytoskeleton form a scaffold supporting the membrane, but recall that these microfilaments can contract. Contracting microfilaments attached to a membrane can deform and reshape it, like a rubber band. The double membrane that encloses mitochon-dria and chloroplasts could also have arisen through endosymbiosis. The outer membrane may have come from the engulfing cell's plasma membrane and the inner membrane from the engulfed cell's plasma membrane. This structure—the plasma membrane that surrounds and contains the cell—is so interesting and important that we devote all of Chapter 5 to explaining it.

CHAPTER SUMMARY

4.1 What features of cells make them the fundamental unit of life?

See Web/CD Activities 4.1 and 4.2

All cells come from preexisting cells.

Cells are small because a cell's **surface area** must be large compared with its **volume** to accommodate exchanges with its environment. Review Figure 4.2

All cells are enclosed by a selectively permeable plasma membrane that separates them from the external environment.

While certain biochemical processes, molecules, and structures are shared by all kinds of cells, two major categories of cells—**prokaryotes** and **eukaryotes**—are easily distinguished.

4.2 What are the characteristics of prokaryotic cells?

Prokaryotic cells have no internal compartments, but have a **nucleoid** region containing DNA and cytoplasm containing **cytosol**, **ribosomes**, proteins, and small molecules. Some prokaryotes have additional protective structures, including a cell wall, an outer membrane, and a capsule. Review Figure 4.4

Some prokaryotes have folded membranes that may be photosynthetic membranes, and some have flagella or pili. Review Figure 4.5

4.3 What are the characteristics of eukaryotic cells?

See Web/CD Tutorial 4.1

Eukaryotic cells are larger than prokaryotic cells and contain many membrane-enclosed **organelles**. The membranes that envelop organelles ensure compartmentalization of their functions. Review Figure 4.6

The **nucleus** contains most of the cell's DNA and controls protein synthesis. **Ribosomes** are sites of protein synthesis. Review Figure 4.8

The **endomembrane system**—consisting of the **endoplasmic reticulum** and **Golgi apparatus**—is a series of interrelated compartments enclosed by membranes. It segregates proteins and modifies them. **Lysosomes** contain many digestive enzymes. Review Figures 4.10, 4.11, and 4.12, Web/CD Activity 4.3, Web/CD Tutorial 4.2

Mitochondria and chloroplasts process energy. **Mitochondria** are present in most eukaryotic organisms and contain the enzymes needed for cellular respiration. The cells of photosynthetic eukaryotes contain **chloroplasts** that harvest light energy for photosynthesis. Review Figures 4.13 and 4.14

Vacuoles are prominent in many plant cells and consist of a membrane-enclosed compartment full of water and dissolved substances.

The **microfilaments**, **intermediate filaments**, and **microtubules** of the **cytoskeleton** provide the cell with shape, strength, and movement.

4.4 What are the roles of extracellular structures?

The plant **cell wall** consists principally of cellulose. Cell walls are pierced by **plasmodesmata** that join the cytoplasms of adjacent cells.

In animals, the **extracellular matrix** consists of different kinds of proteins, including **collagen** and **proteoglycans**. Review Figure 4.25

4.5 How did eukaryotic cells originate?

The **endosymbiosis theory** states that organelles such as mitochondria and chloroplasts originated when larger prokaryotes engulfed, but did not digest, smaller prokaryotes. Mutual benefits permitted this symbiotic relationship to be maintained, allowing the smaller cells to evolve into the eukaryotic organelles observed today. Review Figure 4.26

Infoldings of the plasma membrane could have led to the formation of some membrane-enclosed organelles, such as the endomembrane system and nucleus.

SELF-QUIZ

1. Which structure is present in both prokaryotic cells and eukaryotic plant cells?
   a. Chloroplasts
   b. Cell wall
   c. Nucleus
   d. Mitochondria
   e. Microtubules

2. The major factor limiting cell size is the
   a. concentration of water in the cytoplasm.
   b. need for energy.
   c. presence of membrane-enclosed organelles.
   d. ratio of surface area to volume.
   e. composition of the plasma membrane.

3. Which statement about mitochondria is *not* true?
   a. The inner mitochondrial membrane folds to form cristae.
   b. Mitochondria are usually 1 μm or smaller in diameter.
   c. Mitochondria are green because they contain chlorophyll.
   d. Fuel molecules from the cytosol are oxidized in mitochondria.
   e. ATP is synthesized in mitochondria.

4. Which statement about plastids is *true*?
   a. They are found in prokaryotes.
   b. They are surrounded by a single membrane.
   c. They are the sites of cellular respiration.
   d. They are found only in fungi.
   e. They contain several types of pigments or polysaccharides.

5. If all the lysosomes within a cell suddenly ruptured, what would be the most likely result?
   a. The macromolecules in the cytosol would begin to break down.
   b. More proteins would be made.
   c. The DNA within mitochondria would break down.
   d. The mitochondria and chloroplasts would divide.
   e. There would be no change in cell function.

6. The Golgi apparatus
   a. is found only in animals.
   b. is found in prokaryotes.
   c. is the appendage that moves a cell around in its environment.
   d. is a site of rapid ATP production.
   e. modifies and packages proteins.

7. Which organelle is *not* surrounded by one or more membranes?
   a. Ribosome
   b. Chloroplast
   c. Mitochondrion
   d. Peroxisome
   e. Vacuole

8. The cytoskeleton consists of
   a. cilia, flagella, and microfilaments.
   b. cilia, microtubules, and microfilaments.
   c. internal cell walls.
   d. microtubules, intermediate filaments, and microfilaments.
   e. calcified microtubules.

9. Microfilaments
   a. are composed of polysaccharides.
   b. are composed of actin.
   c. allow cilia and flagella to move.
   d. make up the spindle that aids the movement of chromosomes.
   e. maintain the position of the chloroplast in the cell.

10. Which statement about the plant cell wall is *not* true?
   a. Its principal chemical components are polysaccharides.
   b. It lies outside the plasma membrane.
   c. It provides support for the cell.
   d. It completely isolates adjacent cells from one another.
   e. It is semirigid.

FOR DISCUSSION

1. The drug vincristine is used to treat many cancers. It apparently works by causing microtubules to depolymerize. Vincristine use has many side effects, including loss of dividing cells and nerve problems. Explain why this might be so.

2. Through how many membranes would a molecule have to pass in moving from the interior of a chloroplast to the interior of a mitochondrion? From the interior of a lysosome to the outside of a cell? From one ribosome to another?

3. How does the possession of double membranes by chloroplasts and mitochondria relate to the endosymbiosis theory of the origins of these organelles? What other evidence supports the theory?

4. Compare the extracellular matrix of the animal cell with the plant cell wall with respect to composition of the fibrous and nonfibrous components, rigidity, and connectivity of cells.

FOR INVESTIGATION

The pathway of newly synthesized proteins through the cell can be followed by a "pulse-chase" experiment. Proteins are tagged with a radioactive isotope (the "pulse"), and the cell is allowed to process them for varying periods of time. The locations of the radioactive proteins are then determined by isolating cell organelles and quantifying their radioactivity. How would you use this method, and what results would you expect, for (A) a lysosomal enzyme? (B) a protein that is released from the cell?

CHAPTER 5

The Dynamic Cell Membrane

Disaster at the plasma membrane

On their first date, Paul and Ann shared an exotic lunch that included canned palm fruit. The next morning, they were suddenly stricken with the worst diarrhea, nausea, and vomiting either had ever experienced. They arrived at the hospital in shock, clinging to life with very low blood pressure and irregular heartbeat. Analysis of their stool samples quickly revealed the reason for their illness: *Vibrio cholerae*, the bacterium that causes cholera. Their infection was traced to the palm fruit, which was canned in El Salvador—a known location for the bacterium ever since a cholera epidemic devastated South and Central America in the early 1990s. Cholera spreads when an uninfected person ingests the bacterium, usually in water contaminated with the feces of infected individuals. In this case, contaminated water may have been used to irrigate the palm fruit.

Until late in the nineteenth century, people had no idea what caused this dreaded disease, nor did they understand exactly how it spread. Then, in 1854, the physician John Snow traced a cholera outbreak in central London to a single water pump. Once he removed the pump's handle, the outbreak subsided. Some 30 years later, in Berlin, Robert Koch examined cholera-contaminated water under a microscope and isolated *Vibrio cholerae*. These seemingly simple discoveries linking a disease to a specific cause helped usher in the medical science of *epidemiology*, the study of how diseases spread among populations and how they can be controlled.

Our current knowledge allows us to explain even more about cholera, including the physical events that occurred after Ann and Paul ate the contaminated palm fruit. Most of the bacteria they ingested died in the acidic environment of the stomach. But a few survived, bound themselves to the plasma membranes of cells in the small intestine, and released a toxic protein. This toxin entered healthy cells and did two things to their plasma membranes. It inactivated certain membrane proteins that "pump" sodium ions into the cell and, at the same time, it opened channels in the membrane that are normally closed, allowing chloride ions to escape from inside the cell into the intestine.

Healthy plasma membranes enclosing intestinal cells act as "gatekeepers," maintaining higher concentrations of Na$^+$ and Cl$^-$ *inside* the cells than *outside*, in the intestine. This necessary imbalance was overturned, and Na$^+$ and Cl$^-$ left the cells and accumulated in the intestine. Through a process called osmosis, this reversed

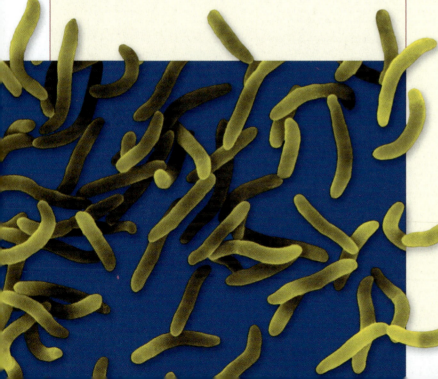

Vibrio cholerae This bacterium causes alterations in the membranes of cells in the human intestine, leading to cholera, a life-threatening but treatable disease.

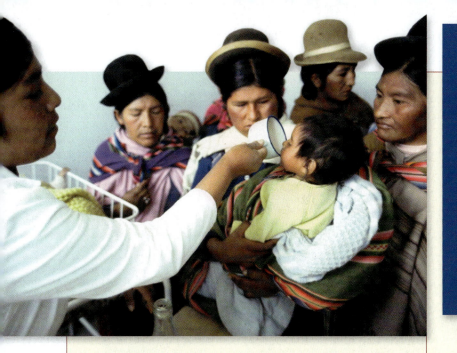

Treating Cholera In the early 1990s, a cholera epidemic broke out in Peru and spread to other countries in Central and South America, leading to over a million cases in 3 years. Here a medic from the World Health Organization administers oral rehydration therapy to a Peruvian child.

ion imbalance caused water to be pulled out of other body cells and into the intestine, resulting in severe diarrhea and potentially fatal dehydration. Fortunately, treating cholera is straightforward: doctors administer oral rehydration therapy to replace the lost ions and water. Paul and Ann were given quantities of a specially balanced solution of NaCl (salt) and glucose (sugar), and both recovered fully.

Cholera is a serious threat in regions where sanitation is inadequate, and whenever a disaster (such as a hurricane or an earthquake) disrupts water supplies. But rehydration therapy is inexpensive, safe, rapid, and effective—and it is all the result of knowing how biological membranes work.

IN THIS CHAPTER we focus on the structure and functions of biological membranes. Membranes are dynamic structures that perform their vital physiological roles by allowing cells to interact with other cells and with molecules in the environment. We describe the structural aspects of those interactions here. Membranes also regulate which molecules and ions enter and leave the cell. The selective permeability of membranes is an important characteristic of life.

5.1 What Is the Structure of a Biological Membrane?

The physical organization and functioning of all biological membranes depend on their constituents: lipids, proteins, and carbohydrates. The lipids establish the physical integrity of the membrane and create an effective barrier to the rapid passage of hydrophilic materials such as water and ions. In addition, the phospholipid bilayer serves as a lipid "lake" in which a variety of proteins "float" (**Figure 5.1**). This general design is known as the **fluid mosaic model**.

The proteins embedded in the phospholipid bilayer have a number of functions, including moving materials through the membrane and receiving chemical signals from the cell's external environment. Each membrane has a set of proteins suitable to the specialized function of the cell or organelle it surrounds.

The carbohydrates associated with membranes are attached either to the lipids or to protein molecules. In plasma membranes, they are located on the outside, where they protrude into the environment, away from the cell. Like some of the proteins, carbohydrates are crucial in recognizing specific molecules.

Lipids constitute the bulk of a membrane

The lipids in biological membranes are usually phospholipids. Recall from Section 2.2 that some compounds are hydrophilic ("water-loving") and others are hydrophobic ("water-hating"), and from Section 3.4 that a phospholipid molecule has regions of both kinds:

- *Hydrophilic regions*: The phosphorus-containing "head" of the phospholipid is electrically charged and hence associates with polar water molecules.
- *Hydrophobic regions*: The long, nonpolar fatty acid "tails" of the phospholipid associate with other nonpolar materials, but they do not dissolve in water or associate with hydrophilic substances.

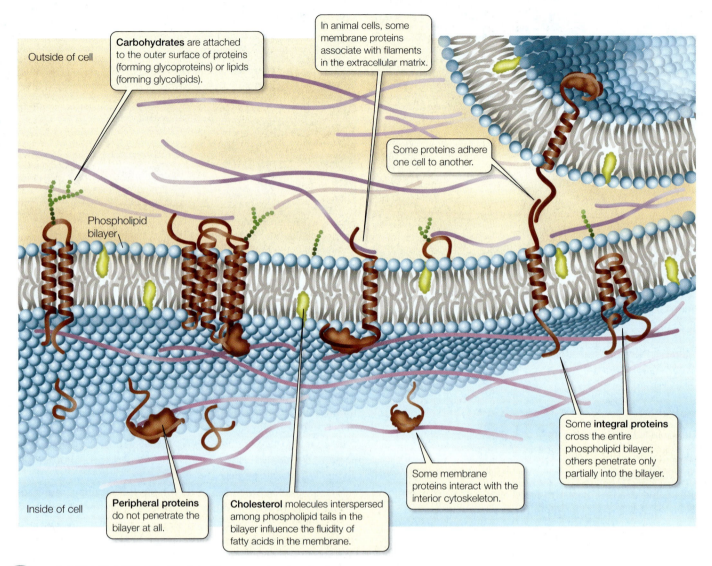

Outside of cell

Carbohydrates are attached to the outer surface of proteins (forming glycoproteins) or lipids (forming glycolipids).

In animal cells, some membrane proteins associate with filaments in the extracellular matrix.

Some proteins adhere one cell to another.

Phospholipid bilayer

Some **integral proteins** cross the entire phospholipid bilayer; others penetrate only partially into the bilayer.

Some membrane proteins interact with the interior cytoskeleton.

Peripheral proteins do not penetrate the bilayer at all.

Cholesterol molecules interspersed among phospholipid tails in the bilayer influence the fluidity of fatty acids in the membrane.

Inside of cell

5.1 The Fluid Mosaic Model The general molecular structure of biological membranes is a continuous phospholipid bilayer in which proteins are embedded. (The purple "ribbons" represent structural elements of the cell and extracellular matrix.)

Because of these properties, one way in which phospholipids can coexist with water is to form a *bilayer*, with the fatty acid "tails" of the two layers interacting with each other and the polar "heads" facing the outside aqueous environment (**Figure 5.2**).

In the laboratory, it is easy to make artificial bilayers with the same organization as natural membranes. In addition, small holes in a phospholipid bilayer seal themselves spontaneously. This capacity of lipids to associate with one another and maintain a bilayer organization helps biological membranes fuse during vesicle formation, phagocytosis, and related processes.

All biological membranes have a similar structure, but membranes from different cells or organelles may differ greatly in their lipid *composition*:

■ Phospholipids can differ in terms of fatty acid chain length, degree of unsaturation (double bonds) in the fatty acids, and the polar (phosphate-containing) groups present.

■ Up to 25 percent of the lipid content of a membrane may be cholesterol. When present, cholesterol is important to membrane integrity, and most cholesterol in membranes is not hazardous to your health. A molecule of cholesterol is commonly situated next to an unsaturated fatty acid (see Figure 5.1).

You can rinse mud off your hands with plain water, but you can't get rid of grease that way because grease isn't water-soluble. Soap molecules have one end that is water-soluble and one end that is fat-soluble, making it possible to wash grease off with soap and water.

The phospholipid bilayer stabilizes the entire membrane structure, but leaves it flexible, not rigid. At the same time, the fatty acids of the phospholipids make the hydrophobic interior of the membrane somewhat *fluid*—about as fluid as lightweight machine oil. This fluidity permits some molecules to move laterally (side to side) within the plane of the membrane. A given phospholipid molecule in the plasma membrane can travel from one end of the cell to the

5.2 A Phospholipid Bilayer Separates Two Aqueous Regions The eight phospholipid molecules shown here represent a small cross section of a membrane bilayer.

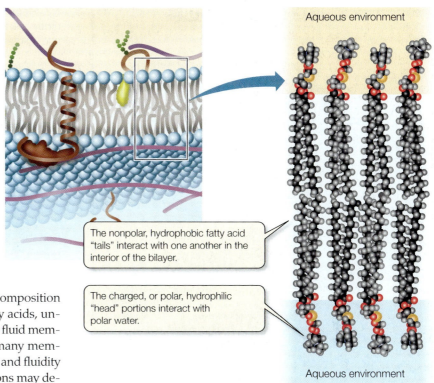

Aqueous environment

The nonpolar, hydrophobic fatty acid "tails" interact with one another in the interior of the bilayer.

The charged, or polar, hydrophilic "head" portions interact with polar water.

Aqueous environment

other in a little more than a second! On the other hand, seldom does a phospholipid molecule in one half of the bilayer flip over to the other side and trade places with another phospholipid molecule. For such a swap to happen, the polar part of each molecule would have to move through the hydrophobic interior of the membrane. Since phospholipid flip-flops are rare, the inner and outer halves of the bilayer may be quite different in the kinds of phospholipids they contain.

The fluidity of a membrane is affected by its lipid composition and by its temperature. In general, shorter-chain fatty acids, unsaturated fatty acids, and less cholesterol result in more fluid membranes. Adequate membrane fluidity is essential for many membrane functions. Because molecules move more slowly and fluidity decreases at reduced temperatures, membrane functions may decline in organisms that cannot keep their bodies warm. To address this problem, some organisms simply change the lipid composition of their membranes under cold conditions, replacing saturated with unsaturated fatty acids and using fatty acids with shorter tails. Such changes play a part in the survival of plants and hibernating animals and bacteria during the winter.

Membrane proteins are asymmetrically distributed

All biological membranes contain proteins. Typically, plasma membranes have one protein molecule for every 25 phospholipid molecules. This ratio varies, however, depending on membrane function. In the inner membrane of the mitochondrion, which is specialized for energy processing, there is one protein for every 15 lipids. On the other hand, myelin, a membrane that encloses some neurons (nerve cells) and uses the properties of lipids to act as an electrical insulator, has only one protein per 70 lipids.

Many membrane proteins are embedded in, or extend across, the phospholipid bilayer (see Figure 5.1). Like phospholipids, these proteins have both hydrophilic and hydrophobic regions.

- *Hydrophilic regions*: Stretches of amino acids with hydrophilic side chains (see Table 3.2) give certain regions of the protein a polar character. Those regions, or *domains*, interact with water, sticking out into the aqueous extracellular environment or cytoplasm.

- *Hydrophobic regions*: Stretches of amino acids with hydrophobic side chains give other regions of the protein a nonpolar character. These domains interact with the fatty acid chains in the interior of the phospholipid bilayer, away from water.

A special preparation method for electron microscopy, called **freeze-fracturing**, reveals proteins embedded in the phospholipid

bilayer of cellular membranes (**Figure 5.3**). The bumps that can be seen protruding from the interior of these membranes are not observed in pure lipid bilayers.

According to the fluid mosaic model, the proteins and lipids in a membrane are independent of each other and *interact only noncovalently*. The polar ends of proteins can interact with the polar ends of lipids, and the nonpolar regions of both molecules can interact hydrophobically.

There are two general types of membrane proteins:

- **Integral membrane proteins** have hydrophobic domains and penetrate the phospholipid bilayer. Many of these proteins have long, hydrophobic α-helical regions (see Section 3.2) that span the core of the bilayer. Their hydrophilic ends protrude into the aqueous environments on either side of the membrane (**Figure 5.4**).

- **Peripheral membrane proteins** lack hydrophobic domains and are not embedded in the bilayer. Instead, they have polar or charged regions that interact with similar regions on exposed parts of integral membrane proteins or with the polar heads of phospholipid molecules (see Figure 5.1).

Some membrane proteins are covalently attached to fatty acids or other lipid groups. These proteins can be classified as a special type of integral protein, as their hydrophobic lipid component allows them to insert themselves into the phospholipid bilayer.

Proteins are *asymmetrically distributed* on the inner and outer surfaces of a membrane. Integral membrane proteins that protrude on both sides of the membrane, known as **transmembrane proteins**, show different "faces" on the two membrane surfaces. Such proteins have certain specific domains on the outer side of the mem-

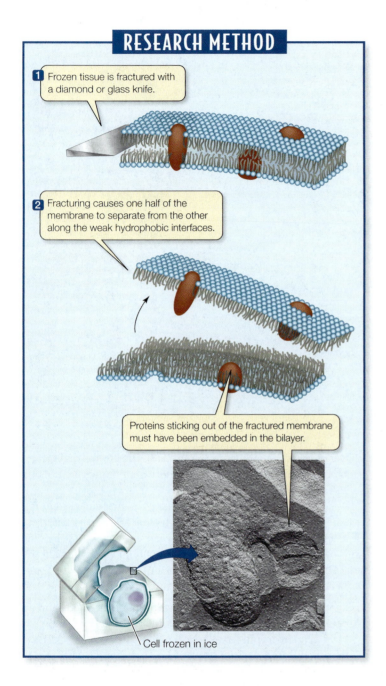

RESEARCH METHOD

1 Frozen tissue is fractured with a diamond or glass knife.

2 Fracturing causes one half of the membrane to separate from the other along the weak hydrophobic interfaces.

Proteins sticking out of the fractured membrane must have been embedded in the bilayer.

Cell frozen in ice

Hydrophilic R groups in exposed parts of the protein interact with aqueous environments.

Outside of cell (aqueous)

Hydrophobic interior of bilayer

Hydrophobic R groups interact with the hydrophobic core of the membrane, away from water.

Inside of cell (aqueous)

5.3 Membrane Proteins Revealed by the Freeze-Fracture Technique This membrane from a spinach chloroplast was first frozen and then separated so that the bilayer was split open.

brane, other domains within the membrane, and still other domains on the inner side of the membrane. Peripheral membrane proteins are localized on one side of the membrane or the other, but not both. This arrangement gives the two surfaces of the membrane different properties. As we will soon see, these differences have great functional significance.

Like lipids, many membrane proteins move around relatively freely within the phospholipid bilayer. Experiments using the technique of *cell fusion* illustrate this migration dramatically. When two cells are fused, a single continuous membrane forms and surrounds both cells, and some proteins from each cell distribute themselves uniformly around this membrane.

Although many proteins are free to migrate in the membrane, others are not, but rather appear to be "anchored" to a specific region of the membrane. These membrane regions are like a corral of horses on a farm: the horses are free to move around within the fenced area, but cannot get outside of it. For instance, the protein in the plasma membrane of a muscle cell that recognizes a chemical signal from neurons is normally found only at the site where a neuron meets the muscle cell. There are two ways in which the movement of proteins within a membrane can be restricted:

■ The cytoskeleton may have components just below the inner face of the membrane that are attached to membrane proteins protruding into the cytoplasm.

■ *Lipid rafts*, which are groups of lipids in a semisolid (not quite fluid) state, may trap proteins within a region. These lipids have a different composition than the surrounding phospholipids; for example, they may have very long fatty acid chains.

Membranes are dynamic

Membranes are constantly forming, transforming from one type to another, fusing with one another, and breaking down (**Figure 5.5**).

■ In eukaryotes, *phospholipids* are synthesized on the surface of the smooth endoplasmic reticulum and are rapidly distributed to membranes throughout the cell.

■ *Membrane proteins* are inserted into the rough endoplasmic reticulum as they form on ribosomes.

■ *Functioning membranes* also move about within eukaryotic cells. Portions of the rough ER bud away as vesicles and join the *cis* face of the Golgi apparatus. Rapidly—often in less than an hour—these segments of membrane find themselves in

5.4 Interactions of Integral Membrane Proteins An integral membrane protein is held in the membrane by the distribution of the hydrophilic and hydrophobic side chains of its amino acids. The hydrophilic ends extend into the aqueous cell exterior and internal cytoplasm; the lipid core of the membrane is hydrophobic.

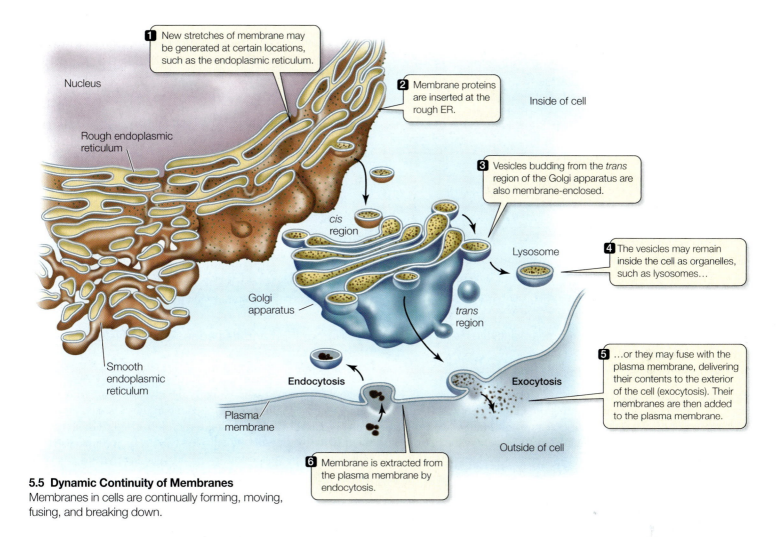

1 New stretches of membrane may be generated at certain locations, such as the endoplasmic reticulum.

Nucleus

2 Membrane proteins are inserted at the rough ER.

Inside of cell

Rough endoplasmic reticulum

3 Vesicles budding from the *trans* region of the Golgi apparatus are also membrane-enclosed.

cis region

Lysosome

4 The vesicles may remain inside the cell as organelles, such as lysosomes…

Golgi apparatus

trans region

Smooth endoplasmic reticulum

Endocytosis

Exocytosis

5 …or they may fuse with the plasma membrane, delivering their contents to the exterior of the cell (exocytosis). Their membranes are then added to the plasma membrane.

Plasma membrane

Outside of cell

6 Membrane is extracted from the plasma membrane by endocytosis.

5.5 Dynamic Continuity of Membranes
Membranes in cells are continually forming, moving, fusing, and breaking down.

the *trans* regions of the Golgi apparatus, from which they bud away to join the plasma membrane.

- Additions to the plasma membrane from fusion with *vesicles* derived from the Golgi apparatus are largely balanced by the removal of membrane in processes such as phagocytosis, affording a recovery path by which internal membranes are replenished.

Because all membranes appear similar under the electron microscope, and because they interconvert readily, we might expect all subcellular membranes to be chemically identical. However, that is not the case, for there are major chemical differences among the membranes of even a single cell. Membranes are changed chemically when they form parts of certain organelles. In the Golgi apparatus, for example, the membranes of the *cis* face closely resemble those of the endoplasmic reticulum in chemical composition, but those of the *trans* face are more similar to the plasma membrane. As a vesicle is formed, the mix of proteins and lipids in its membrane is selected, just as its internal contents are selected, to correspond with the vesicle's target membrane.

Membrane carbohydrates are recognition sites

In addition to lipids and proteins, many membranes contain significant amounts of carbohydrates. The carbohydrates are located on the outer surface of the membrane and serve as recognition sites for other cells and molecules (see Figure 5.1).

Membrane-associated carbohydrates may be covalently bonded to lipids or to proteins:

- **Glycolipids** consist of a carbohydrate covalently bonded to a lipid. The carbohydrate units of glycolipids often extend to the outside of the plasma membrane, where they serve as recognition signals for interactions between cells. For example, the carbohydrate of some glycolipids changes when a cell becomes cancerous. This change may allow white blood cells to target cancer cells for destruction.

- **Glycoproteins** consist of a carbohydrate covalently bonded to a protein. The bound carbohydrates are oligosaccharide chains, usually not exceeding 15 monosaccharide units in length. Glycoproteins enable a cell to be recognized by other cells and proteins.

An "alphabet" of monosaccharides on membranes can be used to generate a diversity of messages. Recall from Section 3.3 that sugar molecules can be formed from three to seven carbons attached at different sites to one another, forming linear or branched oligosaccharides with many different three-dimensional shapes. An oligosaccharide of a specific shape on one cell can bind to a mirror-image shape on an adjacent cell. This binding is the basis of cell–cell adhesion.

Now that we understand the structure of biological membranes, let's see how their components function. In the rest of this chapter we'll focus on the membrane that surrounds individual cells: the plasma membrane. We'll begin with a look at how the plasma membrane allows individual cells to be grouped together into tissues.

5.2 How Is the Plasma Membrane Involved in Cell Adhesion and Recognition?

Some organisms, such as bacteria, are *unicellular*; that is, the entire organism is a single cell. Others, such as plants and animals, are *multicellular*—composed of many cells. Often these cells exist in specialized blocks of cells with similar functions, called *tissues*. Your body has about 60 trillion cells, arranged in different kinds of tissues (such as muscle, nerve, or skin).

Two processes allow cells to arrange themselves in groups:

- **Cell recognition**, in which one cell specifically binds to another cell of a certain type

- **Cell adhesion**, in which the connection between the two cells is strengthened

Both processes involve the plasma membrane. They are most easily studied if the cells in a tissue are separated into individual cells, then allowed to adhere to one another again. Simple organisms provide a good model for the complex tissues of larger species. Studies of sponges, for example, have revealed how cells associate with one another.

A sponge is a multicellular marine animal with a simple body plan (see Section 31.5). The cells of the sponge are connected, but can be separated mechanically by passing the animal several times through a fine wire screen (**Figure 5.6A**). Through this process, what was a single animal becomes hundreds of individual cells suspended in seawater. Remarkably, if the cell suspension is shaken

for a few hours, the cells bump into one another and stick together in the same shape as the original sponge! The *cells recognize and adhere to one another*.

There are many different species of sponges. If disaggregated sponge cells from two different species are placed in the same container, the cells float around and bump into one another, but the cells will stick only to other cells of the same species. Two different sponges form, just like the ones at the start of the experiment.

Such tissue-specific and species-specific cell recognition and cell adhesion are essential in the formation and maintenance of tissues and multicellular organisms. Think of your own body. What keeps muscle cells bound to muscle cells and skin to skin? Specific cell adhesion is so obvious a characteristic of complex organisms that it is easy to overlook. You will see many examples of specific cell adhesion throughout this book; here, we describe its general principles. As you will see, cell recognition and cell adhesion depend on plasma membrane proteins.

Cell recognition and cell adhesion involve proteins at the cell surface

The molecule responsible for cell recognition and adhesion in sponges is a huge integral membrane glycoprotein (which is 80 percent sugar) that is partly embedded in the plasma membrane, with the carbohydrate sticking out and exposed to the environment (and to other sponge cells). As we saw in Section 3.2, a protein not only has a specific shape, but also has specific chemical groups exposed on its surface where they can interact with other substances, including other proteins. Both of these features allow binding to other specific molecules. The cells of the ground-up sponge in Figure 5.6A find one another again through recognition of exposed chemical groups on their membrane glycoproteins. In the majority of plant cells, the plasma membrane is covered with a thick cell wall, but this structure, too, has adhesion proteins that allow cells to bind to one another.

In most cases, the binding of cells in a tissue is **homotypic**; that is, the same molecule sticks out of both cells, and the exposed surfaces bind to each other. But **heterotypic** binding (of cells with different proteins) can also occur. In this case, *different* chemical groups on *different* surface molecules have an affinity for one another. For example, when the mammalian sperm meets the egg, different proteins on the two types of cells have complementary binding surfaces. Similarly, some algae form similar-appearing male and female reproductive cells (analogous to sperm and eggs) that have flagella to propel them toward each other. Male and female cells can recognize each other by heterotypic proteins on their flagella (**Figure 5.6B**).

Three types of cell junctions connect adjacent cells

In a complex multicellular organism, cell recognition proteins allow specific types of cells to bind to one another. Often, both cells contribute material to additional membrane structures that connect them to one another. These specialized structures, called **cell junctions**, are most evident in electron micrographs of *epithelial tissues*, which are layers of cells that line body cavities or cover body surfaces. We will examine three types of cell junctions that enable

(A) Homotypic binding

1 Tissue from a red sponge contains similar cells bound to each other.

2 The sponge tissue can be separated into single cells by passing it through a fine mesh screen.

3 Exposed regions of membrane glycoproteins bind to each other causing cells to adhere.

(B) Heterotypic binding

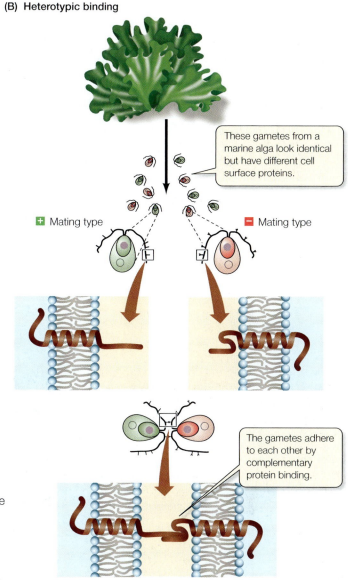

These gametes from a marine alga look identical but have different cell surface proteins.

+ Mating type

− Mating type

The gametes adhere to each other by complementary protein binding.

5.6 Cell Recognition and Adhesion (A) In most cases (including the aggregation of animal cells into tissues), protein binding is homotypic: molecules of the same protein occur on the surfaces of two cells of the same type and adhere to each other. (B) Heterotypic binding occurs between two different but complementary proteins.

cells to make direct physical contact and connect to one another: tight junctions, desmosomes, and gap junctions.

TIGHT JUNCTIONS SEAL TISSUES **Tight junctions** are specialized structures that link adjacent epithelial cells. They result from the mutual binding of specific proteins in the plasma membranes of two epithelial cells, which form a series of joints encircling each cell (**Figure 5.7A**). Found in the lining of the lumen (cavity) of organs such as the intestine, tight junctions have two functions:

■ They prevent substances from moving through the spaces between cells. Thus any substance entering the body from the lumen of the intestine must pass through the epithelial cells that form the tight junction.

■ They define different functional regions of the membrane by restricting the migration of membrane proteins and phospholipids from one region of the cell to another. Thus the membrane proteins and phospholipids in the *apical* (tip) region of the cell (facing the lumen) can be different from those in the *basolateral* (*basal*, bottom; *lateral*, side) regions of the junction cells (facing the body cavity outside the lumen).

By forcing materials to enter certain cells, and by allowing different areas of the same cell to express different membrane proteins with different functions, tight junctions help ensure the directional movement of materials into the body.

DESMOSOMES HOLD CELLS TOGETHER **Desmosomes** connect adjacent plasma membranes. Desmosomes hold adjacent cells firmly together, acting like spot welds or rivets (**Figure 5.7B**). Each desmosome has a dense structure called a *plaque* on the cytoplasmic side of the plasma membrane. To this plaque are attached special cell adhesion molecules (CAMs) that stretch from the plaque through the plasma membrane of one cell, across the intercellular space, and through the plasma membrane of the adjacent cell, where they bind to the plaque proteins in that cell.

The plaque is also attached to fibers in the cytoplasm. These fibers, which are intermediate filaments of the cytoskeleton (see Figure 4.20), are made of a protein called *keratin*. They stretch from

(A)

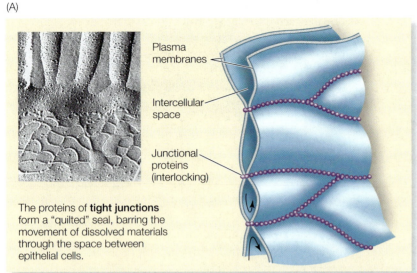

Plasma
membranes

Intercellular
space

Junctional
proteins
(interlocking)

The proteins of **tight junctions**
form a "quilted" seal, barring the
movement of dissolved materials
through the space between
epithelial cells.

(B)

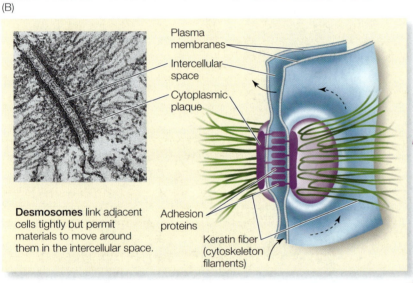

Plasma
membranes

Intercellular
space

Cytoplasmic
plaque

Adhesion
proteins

Keratin fiber
(cytoskeleton
filaments)

Desmosomes link adjacent
cells tightly but permit
materials to move around
them in the intercellular space.

(C)

Plasma
membranes

Intercellular
space

Hydrophilic
channel

Molecules
pass between
cells

Connexons
(channel proteins)

Gap junctions let adjacent
cells communicate.

5.7 Junctions Link Animal Cells Together
Tight junctions (A) and desmosomes (B) are
abundant in epithelial tissues. Gap junctions
(C) are also found in some muscle and nerve tissues, in
which rapid communication between cells is important.

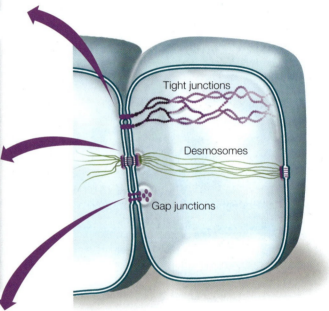

Tight junctions

Desmosomes

Gap junctions

stability to epithelial tissues, which often receive rough
wear while protecting the integrity of the organism's
body surface.

GAP JUNCTIONS ARE A MEANS OF COMMUNICATION
Whereas tight junctions and desmosomes have mechanical roles, **gap junctions** facilitate communication
between cells. Each gap junction is made up of specialized channel proteins, called *connexons*, that span
the plasma membranes of two adjacent cells and the
intercellular space between them (**Figure 5.7C**). Dissolved small molecules and ions can pass from cell to

one cytoplasmic plaque across the cell to connect with another
plaque on the other side of the cell. Anchored thus on both sides
of the cell, these extremely strong fibers provide great mechanical

cell through these junctions. We will describe their role in more
detail, as well as that of *plasmodesmata*, which perform a similar
role in plants, when we discuss cell communication in Chapter 15.

5.2 RECAP

In multicellular organisms, cells arrange themselves in groups by two processes, cell recognition and cell adhesion. Both processes are mediated by integral proteins in the plasma membrane.

- Can you explain the difference between cell recognition and cell adhesion? See p. 102

- The three types of cell junctions described here all have different effects for the passage of materials between cells and through intercellular space. Can you describe how each type deals with this molecular "traffic"? See pp. 103–104 and Figure 5.7

We have just examined how membrane structure accommodates one major membrane function: the binding of one cell to another. Now we turn to the second major function of membranes: regulating the substances that enter or leave a cell.

5.3 What Are the Passive Processes of Membrane Transport?

Biological membranes allow some substances, but not others, to pass through them. This characteristic of membranes is called **selective permeability**. Selective permeability allows the membrane to determine what substances enter or leave a cell or organelle.

There are two fundamentally different processes by which substances cross biological membranes:

- The processes of *passive transport* do not require any input of outside energy to drive them.

- The processes of *active transport* require the input of chemical energy from an outside source.

This section focuses on the passive processes by which substances enter and leave the cell. The energy for these processes is found in the substances themselves, and in the motive force generated by the difference in their concentrations on the two sides of the membrane. Passive transport processes include two types of *diffusion*: *simple diffusion* through the phospholipid bilayer, and *facilitated diffusion* through channel proteins or by means of carrier proteins.

Diffusion is the process of random movement toward a state of equilibrium

Nothing in this world is ever absolutely at rest. Everything is in motion, although the motions may be very small. An important consequence of all this random jiggling of molecules is that all the components of a solution tend eventually to become evenly distributed throughout the system. For example, if a drop of ink is allowed to fall into a container of water, the pigment molecules of the ink are initially very concentrated. Without human intervention, such as stirring, the pigment molecules of the ink move about at random, spreading slowly through the water until eventually the concentration of pigment—and thus the intensity of color—is

exactly the same in every drop of liquid in the container. A solution in which the solute particles are uniformly distributed is said to be at *equilibrium* because there will be no future net change in their concentration. Equilibrium does not mean that the particles have stopped moving; it just means that they are moving in such a way that their overall distribution does not change.

Diffusion is the process of random movement toward a state of equilibrium. Although the motion of each individual particle is absolutely random, the *net* movement of particles is directional until equilibrium is reached. Diffusion is thus *net movement from regions of greater concentration to regions of lesser concentration* (**Figure 5.8**).

In a complex solution (one with many different solutes), the diffusion of each solute is independent of that of the others. How fast a substance diffuses depends on four factors:

- The *diameter* of the molecules or ions: smaller molecules diffuse faster

- The *temperature* of the solution: higher temperatures lead to faster diffusion because ions or molecules have more energy, and thus move more rapidly

- The *electric charge*, if any, of the diffusing material

- The *concentration gradient* in the system—that is, the change in solute concentration with distance in a given direction: the greater the concentration gradient, the more rapidly a substance diffuses

DIFFUSION WITHIN CELLS AND TISSUES Within cells, or wherever distances are very short, solutes distribute themselves rapidly by diffusion. Small molecules and ions may move from one end of an organelle to another in a millisecond (10^{-3} s, or one-thousandth of a second). However, the usefulness of diffusion as a transport mechanism declines drastically as distances become greater. In the absence of mechanical stirring, diffusion across more than a centimeter may take an hour or more, and diffusion across meters may take years! Diffusion would not be adequate to distribute materials over the length of the human body (much less that of larger organisms), but within our cells or across layers of one or two cells, diffusion is rapid enough to distribute small molecules and ions almost instantaneously.

DIFFUSION ACROSS MEMBRANES In a solution without barriers, all the solutes diffuse at rates determined by temperature, their physical properties, and the concentration gradient of each solute. If a biological membrane divides the solution into separate compartments, then the movement of the different solutes can be affected by the properties of the membrane. The membrane is said to be *permeable* to solutes that can cross it more or less easily, but *impermeable* to substances that cannot move across it.

Molecules to which the membrane is impermeable remain in separate compartments, and their concentrations are apt to be different on the two sides of the membrane. Molecules to which the membrane is permeable diffuse from one compartment to the other until their concentrations are equal on both sides of the membrane. When the concentrations of the diffusing substance on the two sides of the permeable membrane are identical, equilibrium is reached. Individual molecules continue to pass through the membrane after equilibrium is established, but equal numbers

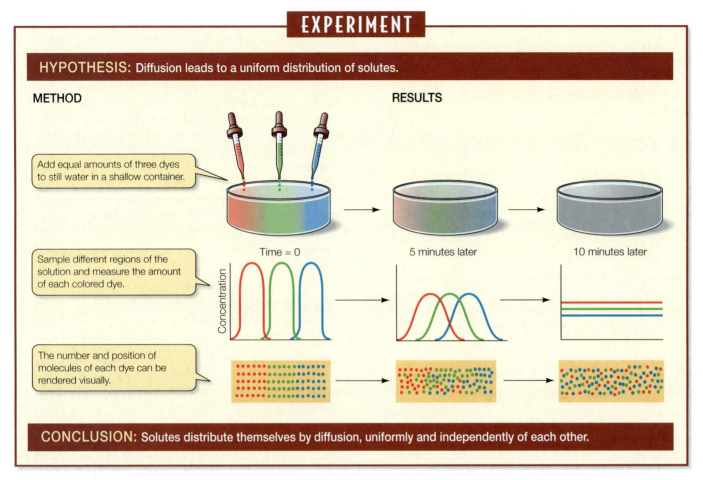

EXPERIMENT

HYPOTHESIS: Diffusion leads to a uniform distribution of solutes.

METHOD

Add equal amounts of three dyes to still water in a shallow container.

Sample different regions of the solution and measure the amount of each colored dye.

The number and position of molecules of each dye can be rendered visually.

RESULTS

Time = 0   5 minutes later   10 minutes later

Concentration

CONCLUSION: Solutes distribute themselves by diffusion, uniformly and independently of each other.

5.8 Diffusion Leads to Uniform Distribution of Solutes A simple experiment demonstrates that solutes move from regions of greater concentration to regions of lesser concentration until equilibrium is reached.

of molecules move in each direction, so there is *no net change* in concentration.

Simple diffusion takes place through the phospholipid bilayer

In **simple diffusion**, small molecules pass through the phospholipid bilayer of the membrane. A molecule that is itself hydrophobic, and hence is soluble in lipids, enters the membrane readily and is thus able to pass through it. The more lipid-soluble the molecule is, the more rapidly it diffuses through the membrane bilayer. This statement holds true over a wide range of molecular weights. Only water itself and the smallest of molecules seem to deviate from this rule, passing through bilayers much more rapidly than their lipid solubilities would predict.

On the other hand, electrically charged or polar molecules, such as amino acids, sugars, and ions, do not pass readily through a membrane, for two reasons:

- Cells are made up of, and exist in, water; polar substances form many hydrogen bonds with water and ions are surrounded by water molecules, thus preventing their "escape" to the membrane.

- The interior of the membrane is hydrophobic, and hydrophilic substances tend to be excluded from it.

Consider two types of molecules: a small protein made up of a few amino acids, and a cholesterol-based steroid of equivalent size. The protein, being polar, will diffuse slowly through the membrane, while the nonpolar steroid will diffuse through it readily.

Osmosis is the diffusion of water across membranes

Water molecules are abundant enough and small enough that they move through membranes by a diffusion process called **osmosis**. This completely passive process uses no metabolic energy and can be understood in terms of solute concentrations. Osmosis depends on the *number* of solute particles present, not on the kinds of particles. We will describe osmosis using red blood cells and plant cells as examples.

Red blood cells are normally suspended in a fluid called *plasma*, which contains salts, proteins, and other solutes. Examining a drop of blood under the light microscope reveals that these red cells have a characteristic doughnut shape. If pure water is added to the drop of blood, the red cells quickly swell and burst. Similarly, if slightly wilted lettuce is placed in pure water, it soon becomes crisp; by weighing it before and after, we can show that it has taken up water. If, on the other hand, red blood cells or crisp lettuce leaves are placed in a relatively concentrated solution of salt or sugar, the

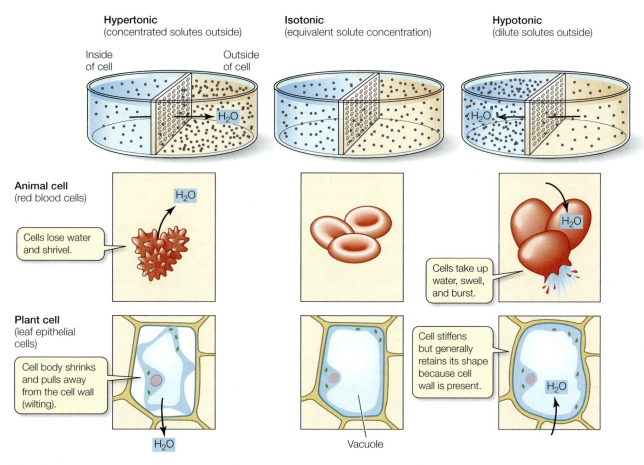

Hypertonic
(concentrated solutes outside)

Isotonic
(equivalent solute concentration)

Hypotonic
(dilute solutes outside)

Inside of cell

Outside of cell

H_2O

H_2O

H_2O

Animal cell
(red blood cells)

H_2O

Cells lose water and shrivel.

H_2O

Cells take up water, swell, and burst.

Plant cell
(leaf epithelial cells)

Cell body shrinks and pulls away from the cell wall (wilting).

Cell stiffens but generally retains its shape because cell wall is present.

H_2O

H_2O

Vacuole

5.9 Osmosis Can Modify the Shapes of Cells In an isotonic solution (center column), plant and animal cells maintain consistent, characteristic shapes. In a solution that is hypotonic to the cells (right), water enters the cells; an environment that is hypertonic to the cells (left) draws water out of the cells.

leaves become limp (they wilt), and the red blood cells pucker and shrink (see Figure 5.9, left).

From analyses of such observations, we know that the difference in solute concentration between a cell and its surrounding environment determines whether water will move from the environment into the cell or out of the cell into the environment. Other things being equal, if two different solutions are separated by a membrane that allows water, but not solutes, to pass through, water molecules will move across the membrane toward the solution with a higher solute concentration. In other words, *water will diffuse from a region of its higher concentration* (with a lower concentration of solutes) *to a region of its lower concentration* (with a higher concentration of solutes).

Three terms are used to compare the solute concentrations of two solutions separated by a membrane (**Figure 5.9**):

■ **Isotonic** solutions have equal solute concentrations.

■ A **hypertonic** solution has a higher solute concentration than the other solution with which it is being compared.

■ A **hypotonic** solution has a lower solute concentration than the other solution with which it is being compared.

Water moves from a hypotonic solution across a membrane to a hypertonic solution.

When we say that "water moves," bear in mind that we are referring to the *net* movement of water. Since it is so abundant, water is constantly moving across the plasma membrane into and out of cells. What concerns us here is whether the overall movement is greater in one direction or the other.

The concentration of solutes in the environment determines the direction of osmosis in all animal cells. A red blood cell takes up water from a solution that is hypotonic to the cell's contents. The cell bursts because its plasma membrane cannot withstand the swelling of the cell. The integrity of red blood cells (and other blood cells) is absolutely dependent on the maintenance of a constant solute concentration in the plasma in which they are suspended: the plasma must be isotonic to the blood cells if the cells are not to burst or shrink. Regulation of the solute concentration of body fluids is thus an important process for organisms without cell walls.

In contrast to animal cells, the cells of plants, archaea, bacteria, fungi, and some protists have cell walls that limit the volume of the cells and keep them from bursting. Cells with sturdy walls take up a limited amount of water, and in so doing they build up an internal pressure against the cell wall that prevents further water from entering. This pressure within the cell is called **turgor pressure**. Turgor pressure keeps plants upright and is the driving force for the enlargement of plant cells. It is a normal and essential component of plant growth. If enough water leaves the cells, turgor pressure drops and the plant wilts.

Turgor pressure in plant cells reaches about 100 pounds per square inch—several times greater than the pressure in automobile tires. This pressure is so great that the cells would slip by and detach from one another, were it not for adhesive molecules called pectins in the plant cell wall.

Diffusion may be aided by channel proteins

As we saw earlier, polar substances such as amino acids and sugars and charged substances such as ions do not readily diffuse across membranes. But they do cross the hydrophobic phospholipid bilayer passively (that is, without the input of energy) in two ways:

- Integral membrane proteins may form *channels* through which these substances can pass.
- Binding to a membrane protein called a *carrier protein* can speed up the diffusion of these substances.

Both of these processes are forms of **facilitated diffusion**.

Membrane **channel proteins** have a central pore lined with polar amino acids and water (to bind to polar or charged substances and allow them to pass through) and nonpolar amino acids on the outside of the protein (to keep them embedded in the phospholipid bilayer). The central pore can open when stimulated, allowing hydrophilic polar substances to pass through (**Figure 5.10**).

ION CHANNELS AND THE MEMBRANE POTENTIAL   The best-studied channel proteins are the **ion channels**. As you will see in later chapters, the movement of ions into and out of cells is important in many biological processes, ranging from the electrical activity of the nervous system to the opening of the pores in leaves that allow gas exchange with the environment. Hundreds of ion channels have been identified, each of them specific for a particular ion. All of them show the same basic structure of a hydrophilic pore that allows a particular ion to move through it.

Just as a fence may have a gate that can be opened or closed, most ion channels are *gated*: they can be closed to ion passage or opened. A **gated channel** opens when something happens to change the three-dimensional shape of the protein. Depending on the channel, this stimulus can range from the binding of a chemical signal (a *ligand-gated* channel; see Figure 5.10) to an electrical charge caused by an imbalance of ions (a *voltage-gated* channel).

Once a voltage-gated channel opens, millions of ions can rush through it per second. How fast the ions move, and in which direction (into or out of the cell), depends on two factors:

- The *concentration gradient* of the ion between the inside and the outside of the cell. For example, in animal cells, because of active transport (discussed below), the concentration of potassium (K^+) ions is usually much higher inside the cell than outside, so K^+ will tend to *diffuse out of the cell* through an open potassium channel.

- An *electrochemical gradient* forms when there is an overall imbalance in the charged substances between the outside and inside of the cell—that is, across the plasma membrane. In animal cells, there is a higher concentration of Cl^- and other negatively charged ions (anions) inside the cell than outside. These negatively charged substances cannot get through the membrane, so there is a tendency for K^+ to *stay inside the cell* to balance out the negative charges and maintain neutrality.

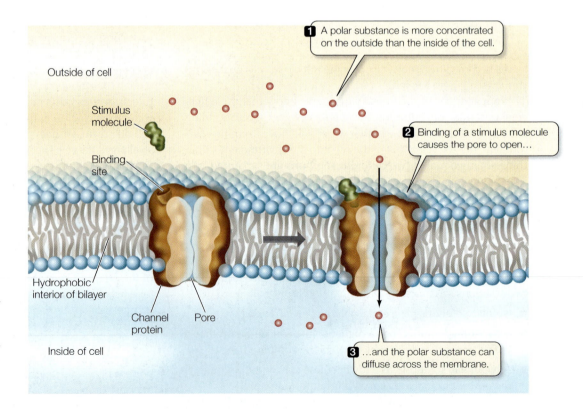

5.10 A Gated Channel Protein Opens in Response to a Stimulus
The channel protein has a pore of polar amino acids and water. It is anchored in the hydrophobic bilayer interior by its outer coating of nonpolar amino acids. The protein changes its three-dimensional shape when a stimulus molecule binds to it, opening the pore so that hydrophilic polar substances can pass through.

Outside of cell

Stimulus molecule

Binding site

Hydrophobic interior of bilayer

Channel protein

Pore

Inside of cell

1 A polar substance is more concentrated on the outside than the inside of the cell.

2 Binding of a stimulus molecule causes the pore to open…

3 …and the polar substance can diffuse across the membrane.

It is the totality of these two forces—simple diffusion based on concentration gradients, and electrochemical imbalances—that determines the direction in which an ion such as K^+ flows through a channel protein. Eventually, an equilibrium is reached at which the ion's rate of diffusion through the channel and out of the cell is balanced by the rate of entry through the channel due to electrical attraction. Obviously, the relative concentrations of K^+ on both sides of the membrane will not be equal, as we would expect if diffusion were the only force involved. Instead, the attraction of electrical charges keeps some extra K^+ inside the cell. This sets up a charge imbalance across the plasma membrane, as there is more K^+ on the inside than on the outside. This charge imbalance is called the **membrane potential**.

The membrane potential is related to the concentration imbalance of K^+ by the *Nernst equation*:

$$E_K = 2.3 \frac{RT}{zF} \log \frac{[K]_o}{[K]_i}$$

where R is the gas constant, F is the Faraday constant (both familiar to chemistry students), T is the temperature, and z is the charge on the ion (+1). Solving for $2.3\,RT/zF$ at 20°C ("room temperature"), the equation becomes much simpler:

$$E_K = 58 \log \frac{[K]_o}{[K]_i}$$

where E_K is the membrane potential (in millivolts, mV) that results from the ratio of K^+ concentrations outside the cell $[K]_o$ and inside the cell $[K]_i$.

Actual measurements from animal cells give a total membrane potential of approximately –70 mV across the membrane, where the inside is negative with respect to the outside (see Figure 44.7). Cells have a tremendous amount of potential energy stored in their membrane potentials. In fact, the brain cells you are using to read this book have more potential energy—about 200,000 volts per centimeter—than the high-voltage electric lines powering your reading light, which carry about 2 volts per centimeter.

As you will see when you study plant and animal physiology, the energy stored in the membrane potential of cells is the basis of many important biological processes. We will return to the Nernst equation and its applications to biology in Chapter 44.

THE SPECIFICITY OF ION CHANNELS How does an ion channel allow one ion, but not another, to pass through? It is not simply a matter of charge or size. For example, a sodium ion (Na^+), with a radius of 0.095 nanometers, is smaller than K^+, at 0.130 nm; both carry the same positive charge. Yet the potassium channel lets only K^+ pass through the membrane, and not the smaller Na^+. The elegant explanation was recently discovered when Roderick MacKinnon determined the structure of a potassium channel from a bacterium (**Figure 5.11**).

Being charged, both Na^+ and K^+ are attracted to water molecules. They have water "shells" in solution, held by the attraction of their positive charges to the negatively charged oxygen atoms on the water molecules (see Figure 2.11). To get through a membrane channel, K^+ must let go of its water. The "naked" K^+ is now attracted to the oxygen atoms on the pore of the channel protein. In the potassium channel, oxygen atoms are located at the stem of a funnel-shaped protein region. The K^+ ion just fits the stem, and so can get into a position where it is more strongly attracted to the oxygen atoms there than to those of water. The smaller Na^+ ion, on the other hand, is kept a bit more distant from the oxygen atoms on the stem of the channel, and prefers to be surrounded by water. So Na^+ does not enter the potassium channel.

As we mentioned, water crosses the plasma membrane at a rate far in excess of expectations, given its polarity. One way that water can do this is by hydrating some ions as they pass through ion channels. Up to 12 water molecules may coat an ion as it traverses a channel. Another way that water enters cells rapidly is through water channels, called **aquaporins**. Channel proteins that allow water to pass through them have been characterized in many membranes, from the membrane of plant vacuoles, where they are important in maintaining turgor pressure, to the mammalian kidney, where they function to retain water that would otherwise be lost through urine.

Many plants, like some humans, follow the sun. When ion channels in cells joining the leaf to the stem open in response to sunlight, K^+ and Cl^- enter the cells by diffusion. Water follows by osmosis and the stem cells swell, causing the leaf to tilt toward the sun.

(A) Side view

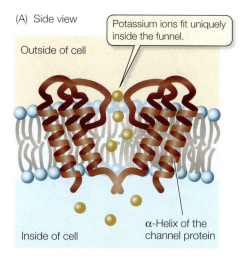

Potassium ions fit uniquely inside the funnel.

Outside of cell

Inside of cell

α-Helix of the channel protein

(B) "Top down" view

K^+

5.11 The Potassium Channel Roderick MacKinnon determined this structure for the selective K^+ channel in the bacterium *Streptomyces lividans*. (A) Potassium ions funnel through the channel in this side view, attracted by oxygen atoms on the α-helices of the channel protein. It is a "custom fit" for K^+; other ions do not pass through this channel. (B) View looking down from the top of the channel.

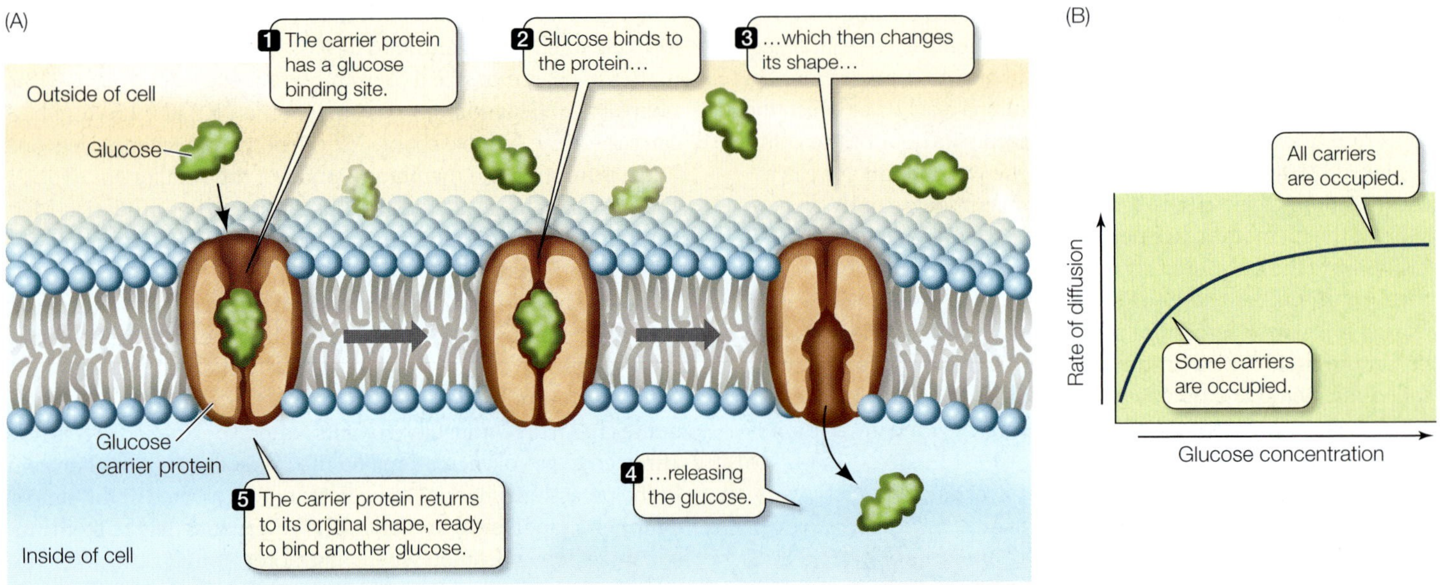

5.12 A Carrier Protein Facilitates Diffusion The glucose transporter is a carrier protein that allows glucose to enter the cell at a faster rate than would be possible by simple diffusion. (A) The transporter binds to glucose, brings it into the membrane interior, then changes shape, releasing glucose into the cell cytoplasm. (B) At lower glucose concentrations, not all the transporters are occupied and an increase in the number of glucose molecules increases the rate of diffusion. At high concentrations, all the transporter proteins are in use (the system is saturated), and the rate of diffusion plateaus.

Carrier proteins aid diffusion by binding substances

Another kind of facilitated diffusion involves not just the opening of a channel, but the actual binding of the transported substance to membrane proteins. These proteins are called **carrier proteins** and, like channel proteins, they allow diffusion both into and out of the cell. Carrier proteins transport polar molecules such as sugars and amino acids.

For example, glucose is the major energy source for most mammalian cells. The membranes of those cells contain a carrier protein—the glucose transporter—that facilitates glucose uptake into the cell. Binding of glucose to a specific three-dimensional site on the transporter protein causes the protein to change its shape and release glucose on the cytoplasmic side of the membrane (**Figure 5.12A**). Since glucose is broken down almost as soon as it enters a cell, there is almost always a strong concentration gradient favoring glucose entry, with a higher concentration outside the cell than inside. The transporter allows glucose molecules to cross the membrane and enter the cell much faster than they would by simple diffusion.

Transport by carrier proteins is different from simple diffusion. In both processes, the rate of movement depends on the concentration gradient across the membrane. However, in carrier-mediated transport, a point is reached at which increases in the concentration gradient are not accompanied by an increased rate of diffusion. At this point, the facilitated diffusion system is said to be *saturated* (**Figure 5.12B**). Because there are only a limited number of carrier protein molecules per unit of membrane area, the rate of diffusion reaches a maximum when all the carrier molecules are fully loaded with solute molecules. In other words, when the difference in solute concentration across the membrane is sufficiently high, not enough carrier molecules are free at a given moment to handle all the solute molecules.

5.3 RECAP

Diffusion is the movement of ions or molecules from a region of greater concentration to a region of lesser concentration. Water can diffuse through cell membranes by a process called osmosis. Channel proteins, which can be open or closed, and carrier proteins facilitate diffusion of charged and polar substances.

- What properties of a substance determine whether, and how fast, it will diffuse across a membrane? See p. 105

- Can you describe osmosis and explain the terms hypertonic, hypotonic, and isotonic? See p. 107 and Figure 5.9

- Do you understand how a channel protein facilitates diffusion? See pp. 108–109 and Figure 5.10

Passive transport allows substances to enter cells from the environment in such a way that, when equilibrium is reached, the concentrations of a substance inside the cell and just outside the cell are equal. But one hallmark of living things is that they can have a composition quite different from that of their environment. To achieve this they must sometimes move substances against their natural tendencies to diffuse. This process requires work—the input of energy—and is known as *active transport*.

5.4 How Do Substances Cross Membranes against a Concentration Gradient?

In many biological situations, an ion or small molecule must be moved across a membrane from a region of lower concentration to a region of higher concentration. In these cases, the substance cannot rush into or out of cells by diffusion. The movement of a substance across a biological membrane *against* a concentration gradient—called **active transport**—requires the expenditure of chemical energy. The differences between diffusion and active transport are summarized in **Table 5.1**.

TABLE 5.1

Membrane Transport Mechanisms

TRANSPORT MECHANISM	EXTERNAL ENERGY REQUIRED?	DRIVING FORCE	MEMBRANE PROTEIN REQUIRED?	SPECIFICITY
Simple diffusion	No	With concentration gradient	No	Not specific
Facilitated diffusion	No	With concentration gradient	Yes	Specific
Active transport	Yes	ATP hydrolysis (against concentration gradient)	Yes	Specific

Active transport is directional

Three types of membrane proteins are involved in active transport (**Figure 5.13**):

- **Uniports** move a single substance in one direction. For example, a calcium-binding protein found in the plasma membrane and endoplasmic reticulum of many cells actively transports Ca^{2+} to regions of its higher concentration either outside the cell or inside the ER.

- **Symports** move two substances in the same direction. For example, the uptake of amino acids from the intestine by the cells that line it requires the simultaneous binding of Na^+ and an amino acid to the same transport protein.

- **Antiports** move two substances in opposite directions, one into the cell and the other out of the cell. For example, many cells have a sodium–potassium pump that moves Na^+ out of the cell and K^+ into it.

Symports and antiports are known as *coupled transporters* because they move two substances at once.

Primary and secondary active transport rely on different energy sources

There are two basic types of active transport:

- **Primary active transport** requires the direct participation of the energy-rich molecule ATP.

- **Secondary active transport** does not use ATP directly; rather, its energy is supplied by an ion concentration gradient established by primary active transport.

In primary active transport, energy released by the hydrolysis of ATP drives the movement of specific ions against a concentration gradient. (We give the details of how ATP provides energy to cells in Section 6.2.) For example, we saw earlier that concentrations of potassium ions (K^+) inside a neuron are much higher than in the fluid bathing the nerve, whereas the concentration of sodium ions (Na^+) is much higher in the fluid outside. Nevertheless, a protein in the nerve cell membrane continues to pump Na^+ out of the neuron and K^+ in against these concentration gradients, ensuring that the gradients are maintained. This **sodium–potassium (Na^+–K^+) pump** is found in all animal cells. The pump is an integral membrane glycoprotein. It breaks down a molecule of ATP and uses the energy released to bring two K^+ ions into the cell and export three Na^+ ions (**Figure 5.14**). The Na^+–K^+ pump is thus an antiport.

In secondary active transport, the movement of a substance against its concentration gradient is accomplished using energy "regained" by letting ions move across the membrane *with* their concentration gradient. For example, once the sodium–potassium pump establishes a concentration gradient of sodium ions, the passive diffusion of some Na^+ back into the cell can provide energy for the secondary active transport of glucose into the cell (**Figure**

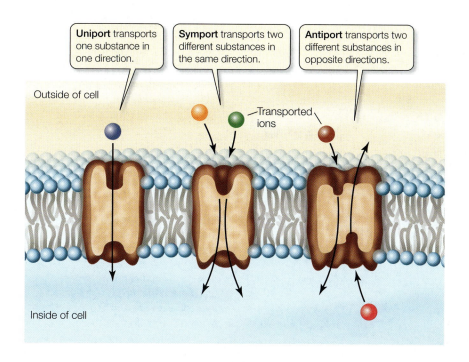

Uniport transports one substance in one direction.

Symport transports two different substances in the same direction.

Antiport transports two different substances in opposite directions.

Outside of cell

Transported ions

Inside of cell

5.13 Three Types of Proteins for Active Transport Note that in each of the three cases, transport is directional. Symport and antiport are examples of coupled transport.

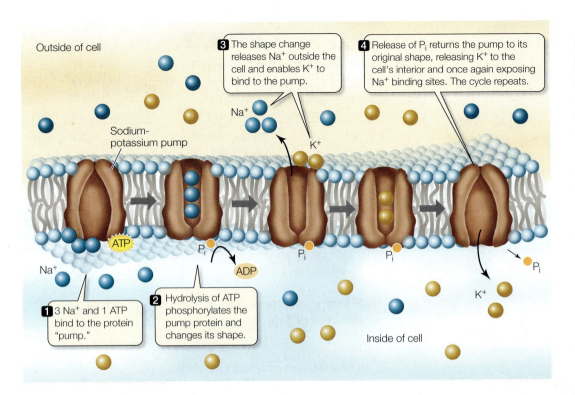

3 The shape change releases Na⁺ outside the cell and enables K⁺ to bind to the pump.

4 Release of Pᵢ returns the pump to its original shape, releasing K⁺ to the cell's interior and once again exposing Na⁺ binding sites. The cycle repeats.

Outside of cell

Sodium-potassium pump

Na⁺

K⁺

ATP

Pᵢ

ADP

Na⁺

1 3 Na⁺ and 1 ATP bind to the protein "pump."

2 Hydrolysis of ATP phosphorylates the pump protein and changes its shape.

Pᵢ

Pᵢ

Pᵢ

K⁺

Inside of cell

5.14 Primary Active Transport: The Sodium–Potassium Pump In active transport, energy is used to move a solute against its concentration gradient. Even though the Na⁺ concentration is higher outside the cell and the K⁺ concentration is higher inside the cell, for each molecule of ATP hydrolyzed, two K⁺ are pumped into the cell and three Na⁺ are pumped out of the cell. (The hydrolysis of ATP releases energy and splits the ATP molecule into a molecule of ADP and an inorganic phosphate ion, Pᵢ; see Section 6.2.)

5.15). Secondary active transport aids in the uptake of amino acids and sugars, which are essential raw materials for cell maintenance and growth. Both types of coupled transport proteins—symports and antiports—are used for secondary active transport.

Primary active transport
The sodium–potassium pump moves Na⁺, using the energy of ATP hydrolysis to establish a concentration gradient of Na⁺.

Secondary active transport
Na⁺, moving with the concentration gradient established by the sodium–potassium pump, drives the transport of glucose against its concentration gradient.

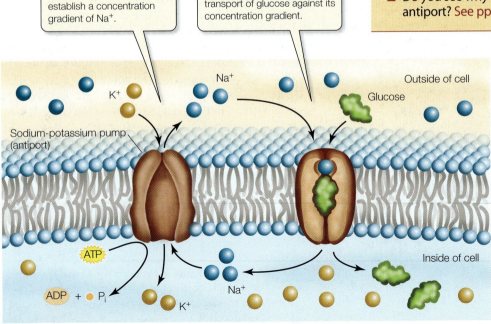

Sodium-potassium pump (antiport)

K⁺

Na⁺

Glucose

Outside of cell

ATP

ADP + Pᵢ

Na⁺

K⁺

Na⁺

Inside of cell

5.15 Secondary Active Transport The Na⁺ concentration gradient established by primary active transport (left) powers the secondary active transport of glucose (right). The movement of glucose across the membrane against its concentration gradient is coupled by a symport protein to the movement of Na⁺ into the cell.

We have examined a number of ways in which ions and small molecules can enter and leave cells. But what about large molecules? Their size means they diffuse very slowly, and their bulk makes it difficult for them to pass through the membrane directly. It takes a totally different mechanism to move large molecules across membranes.

5.5 How Do Large Molecules Enter and Leave a Cell?

Macromolecules such as proteins, polysaccharides, and nucleic acids are simply too large and too charged or polar to pass through biological membranes. This is actually a fortunate property—think of the consequences if such molecules diffused out of cells. A red blood cell would not retain its hemoglobin! On the other hand, cells must sometimes take up or secrete intact large molecules. As we saw in Section 4.3, this can be done by means of vesicles that either pinch off from the plasma membrane and enter the cell (*endocytosis*) or fuse with the plasma membrane and release their contents (*exocytosis*).

Macromolecules and particles enter the cell by endocytosis

Endocytosis is a general term for a group of processes that bring small molecules, macromolecules, large particles, and even small cells into the eukaryotic cell (**Figure 5.16A**). There are three types of endocytosis: *phagocytosis*, *pinocytosis*, and *receptor-mediated en-*

docytosis. In all three, the plasma membrane invaginates (folds inward) around materials from the environment, forming a small pocket. The pocket deepens, forming a vesicle. This vesicle separates from the plasma membrane and migrates with its contents to the cell's interior.

- In **phagocytosis** ("cellular eating"), part of the plasma membrane engulfs large particles or even entire cells. Phagocytosis is used by unicellular protists as a cellular feeding process and by some white blood cells that defend the body by engulfing foreign cells and substances. The food vacuole or phagosome that forms usually fuses with a lysosome, where its contents are digested (see Figure 4.12).

- In **pinocytosis** ("cellular drinking"), vesicles also form. However, these vesicles are smaller, and the process operates to bring small dissolved substances or fluids into the cell. Like phagocytosis, pinocytosis is relatively nonspecific as to what it brings into the cell. For example, pinocytosis goes on constantly in the *endothelium*, the single layer of cells that separates a tiny blood capillary from the surrounding tissue, allowing the cells to rapidly acquire fluids from the blood.

- In **receptor-mediated endocytosis**, specific reactions at the cell surface trigger the uptake of specific materials.

Let's take a closer look at this last process.

Receptor-mediated endocytosis is highly specific

Receptor-mediated endocytosis is used by animal cells to capture specific macromolecules from the cell's environment. This process depends on **receptor proteins**, integral membrane proteins that can bind to a specific molecule in the cell's environment. The uptake process is similar to that in nonspecific endocytosis. However, in receptor-mediated endocytosis, receptor proteins at particular sites on the extracellular surface of the plasma membrane bind to specific substances. These sites are called *coated pits* because they form a slight depression in the plasma membrane, and their cytoplasmic surfaces are coated by other proteins, such as *clathrin*.

When a receptor protein binds to its specific ligand, its coated pit invaginates and forms a *coated vesicle* around the bound macromolecule. Strengthened and stabilized by clathrin molecules, this vesicle carries the macromolecule into the cell (**Figure 5.17**). Once inside, the vesicle loses its clathrin coat and may fuse with a lysosome, where the engulfed material is processed and released into the cytoplasm. Because of its specificity for particular macromolecules, receptor-mediated endocytosis is a rapid and efficient method of taking up substances that may exist at low concentrations in the cell's environment.

Receptor-mediated endocytosis is the method by which cholesterol is taken up by most mammalian cells. Water-insoluble cholesterol and triglycerides are packaged by liver

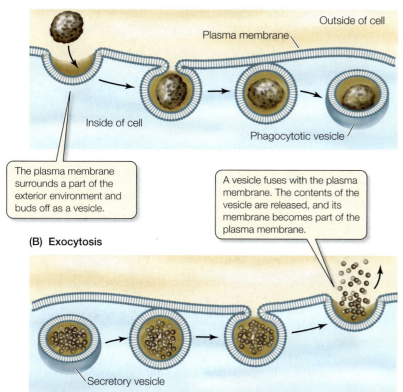

(A) Endocytosis

Outside of cell

Plasma membrane

Inside of cell

Phagocytotic vesicle

The plasma membrane surrounds a part of the exterior environment and buds off as a vesicle.

A vesicle fuses with the plasma membrane. The contents of the vesicle are released, and its membrane becomes part of the plasma membrane.

(B) Exocytosis

Secretory vesicle

5.16 Endocytosis and Exocytosis (A) Endocytosis and (B) exocytosis are used by eukaryotic cells to take up substances from and release substances to the outside environment.

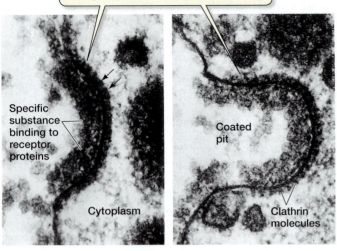

The protein clathrin coats the cytoplasmic side of the plasma membrane at a coated pit.

Specific substance binding to receptor proteins

Cytoplasm

Coated pit

Clathrin molecules

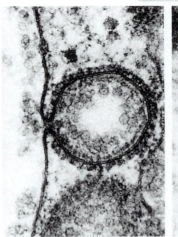

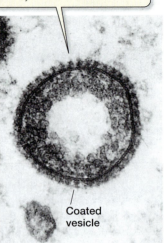

The endocytosed contents are surrounded by a clathrin-coated vesicle.

Coated vesicle

5.17 Formation of a Coated Vesicle In receptor-mediated endocytosis, the receptor proteins in a coated pit bind specific macromolecules, which are then carried into the cell by a coated vesicle.

cells into lipoprotein particles, which are then secreted into the bloodstream to provide body tissues with lipids. One type of lipoprotein particles, called *low-density lipoproteins* or LDLs, must be taken up by the liver for recycling. This uptake also occurs via receptor-mediated endocytosis, beginning with the binding of LDLs to specific receptor proteins on the liver cell surface. Once engulfed by endocytosis, the LDL particle is freed from the receptors. The receptors segregate to a region of the vesicle that buds off and forms a new vesicle, which is recycled to the plasma membrane. The freed LDL particle remains in the original vesicle, which fuses with a lysosome in which the LDL is digested and the cholesterol made available for cell use.

Persons with the inherited disease *familial hypercholesterolemia* have dangerously high levels of cholesterol in their blood because a deficient LDL receptor protein prevents receptor-mediated endocytosis of LDL in their liver.

Exocytosis moves materials out of the cell

Exocytosis is the process by which materials packaged in vesicles are secreted from a cell when the vesicle membrane fuses with the plasma membrane (**Figure 5.16B**). The initial event in this process is the binding of a membrane protein protruding from the cytoplasmic side of the vesicle with a membrane protein on the cytoplasmic side of the target site on the plasma membrane. The phospholipid bilayers of the two membranes merge, and an opening to the outside of the cell develops. The contents of the vesicle are released into the environment, and the vesicle membrane is smoothly incorporated into the plasma membrane.

In Chapter 4, we encountered exocytosis as the last step in the processing of material engulfed by phagocytosis: the secretion of indigestible materials to the environment. Exocytosis is also important in the secretion of many different substances, including digestive enzymes from the pancreas, neurotransmitters from neurons, and materials for the construction of the plant cell wall.

5.5 RECAP

Endocytosis and exocytosis transport molecules that are too large, charged, or polar to be transported through a membrane by passive or active transport. Endocytosis may be specific, mediated by a receptor protein in the membrane.

- Do you understand the difference between phagocytosis and pinocytosis? See p. 113

- Can you describe an example of receptor-mediated endocytosis? See p. 114 and Figure 5.17

We have now examined the structure of biological membranes and seen how macromolecules on the membrane surface allow cells to recognize and adhere to each other, allowing tissues and organs to form. We have also seen how the traffic of substances into and out of the cell is selectively regulated by the membrane. These are crucial functions, but they are not the only aspects of biological membranes. Let's take a brief look at a few other membrane functions.

5.6 What Are Some Other Functions of Membranes?

One important function of membranes is to *keep different materials separate* from one another. Recall from Section 4.3 that the membrane of the rough endoplasmic reticulum is a site for ribosome attachment. Newly formed proteins pass from the ribosomes through the membrane of the ER and into its interior, where the proteins are modified and then delivered to other parts of the cell.

This system requires a separate compartment to segregate these proteins from the rest of the cell.

The plasma membranes of certain types of cells, such as neurons, muscle cells, and some eggs, respond in different ways to the electric charges carried on ions. These membranes are thus *electrically excitable*, which gives them important special properties. For example, in neurons, the plasma membrane conducts nerve impulses from one end of the cell to the other.

Other biological activities and properties associated with membranes are discussed in the chapters that follow. These activities have been essential to the specialization of cells, tissues, and organisms throughout evolution. Three of these activities are especially important:

■ *Some organelle membranes help transform energy.* The membranes of certain organelles are specialized for processing energy (**Figure 5.18A**). For example, the inner mitochondrial membrane helps convert the energy of fuel molecules to the energy of ATP, and the thylakoid membranes of chloroplasts participate in the conversion of light energy to the energy of chemical bonds. These important processes, vital to the life of most eukaryotic organisms, are discussed in detail in Chapters 7 and 8.

■ *Some membrane proteins organize chemical reactions.* Many cellular processes depend on a series of enzyme-catalyzed reactions in which the products of one reaction serve as the reactants for the next. For such a series of reactions to occur, all the necessary molecules must come together. In a solution, for example, reactant and enzyme molecules are randomly distributed and collisions among them are random; thus a complete series of chemical reactions may occur only very slowly. However, if the different enzymes are bound to a membrane in sequential order, the product of one reaction can be released close to the enzyme for the next reaction. Such an "assembly line" allows reactions to proceed rapidly and efficiently (**Figure 5.18B**).

■ *Some membrane proteins process information.* As we have seen, biological membranes may have protruding integral membrane proteins or attached carbohydrates that can bind to specific substances in the environment. The binding of a specific ligand can serve as a signal to initiate, modify, or turn off a cell function (**Figure 5.18C**). In this type of information processing, specificity in binding is essential.

We have seen the informational role of a specific receptor protein in the endocytosis of LDL and its cargo of cholesterol. Another example is the binding of a hormone, such as insulin, to specific receptors on a target cell, such as a liver cell, to elicit a response

in that cell—in this case, the uptake of glucose. Chapter 15 will discuss many other examples of the role of membrane proteins in information processing, but one good example brings us back to the story that opened this chapter—that of the disease cholera.

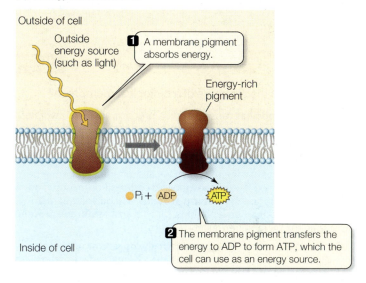

(A) Energy transformation

Outside of cell

Outside energy source (such as light)

1 A membrane pigment absorbs energy.

Energy-rich pigment

P_i + ADP   ATP

2 The membrane pigment transfers the energy to ADP to form ATP, which the cell can use as an energy source.

Inside of cell

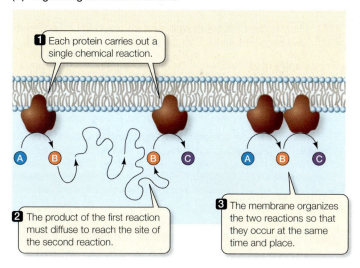

(B) Organizing chemical reactions

1 Each protein carries out a single chemical reaction.

A   B   B   C   A   B   C

2 The product of the first reaction must diffuse to reach the site of the second reaction.

3 The membrane organizes the two reactions so that they occur at the same time and place.

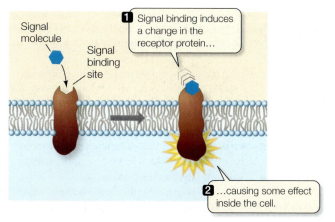

(C) Information processing

Signal molecule

Signal binding site

1 Signal binding induces a change in the receptor protein…

2 …causing some effect inside the cell.

5.18 More Membrane Functions (A) The membranes of organelles such as mitochondria and chloroplasts are specialized for the transformation of energy. (B) When a series of biochemical reactions must take place in sequence, the membrane can sometimes arrange the needed enzymes in an "assembly line" to ensure that the reactions occur in proximity to one another. (C) Membrane proteins have an information-processing role. Receptors on membranes conduct signals from outside the cell that trigger changes inside the cell.

Cholera toxin disrupts cellular information processing. The toxin protein has two subunits, one of which binds to a cell surface glycolipid, causing a change in the three-dimensional shape of the toxin protein so that the second subunit can enter the cell. This subunit acts as an enzyme to modify a peripheral protein called adenylate cyclase on the inner surface of the cell's plasma membrane. The modified adenylate cyclase opens chloride channels in the membrane. Resulting accumulation of Cl⁻ and Na⁺ in the intestine is followed by osmotic loss of water.

Whereas cholera was once almost invariably lethal, even in the developed world, it is now much less so. In 1997, a cholera epidemic broke out among 90,000 Rwandan refugees living in camps in the Congo. The majority of the subsequent 1,521 deaths occurred among people who were outside the health care network serving the camps. Everyone else was saved.

CHAPTER SUMMARY

5.1 What is the structure of a biological membrane?

Biological membranes consist of lipids, proteins, and carbohydrates. The **fluid mosaic model** of membrane structure describes a phospholipid bilayer in which proteins can move about laterally within the membrane. See Web/CD Activity 5.1

The two surfaces of a membrane may have different properties because of their different phospholipid composition, exposed domains of **integral membrane proteins**, and **peripheral membrane proteins**. Some proteins, called **transmembrane proteins**, span the membrane. Review Figure 5.1

Carbohydrates attached to proteins in **glycoproteins** or phospholipids in **glycolipids** project from the external surface of the plasma membrane and function as recognition signals.

Membranes are not static structures, but are constantly forming, exchanging and breaking down. Review Figure 5.5

5.2 How is the plasma membrane involved in cell adhesion and recognition?

The assembly of cells into tissues requires that they recognize and adhere to one another. **Cell recognition** and **cell adhesion** depend on integral membrane proteins that protrude from the cell surface. Binding can be between the same proteins from two cells (**homotypic**) or different proteins (**heterotypic**). Review Figure 5.6

Cell junctions connect adjacent cells. **Tight junctions** prevent the passage of molecules through the spaces between cells and restrict the migration of membrane proteins over the cell surface. **Desmosomes** allow cells to adhere firmly to one another. **Gap junctions** provide channels for communication between adjacent cells. Review Figure 5.7, Web/CD Activity 5.2

5.3 What are the passive processes of membrane transport?

See Web/CD Tutorial 5.1

Membranes exhibit **selective permeability**, regulating which substances pass through them.

Substances can diffuse passively across a membrane by two processes: **simple diffusion** through the phospholipid bilayer and **facilitated diffusion** through either **channel proteins** or by means of a carrier protein.

A solute diffuses across a membrane from a region with a greater concentration of that solute to a region with a lesser concentration of that solute. Equilibrium is reached when the concentrations of the solute are identical on both sides of the membrane. Review Figure 5.8

In **osmosis**, water diffuses from regions of higher water concentration to regions of lower water concentration. **Aquaporins** are membrane channels for water diffusion.

Most cells are in an **isotonic** environment, where solute concentrations on both sides of the plasma membrane are equal. If the solution surrounding a cell is **hypotonic** to the cell interior, more water enters the cell than leaves it. In plant cells, this leads to **turgor pressure**. In a **hypertonic** solution, more water leaves the cell than enters it. Review Figure 5.9

Ion channels are membrane proteins that allow rapid facilitated diffusion of ions through membranes. These can be **gated channels**, where the channel can be opened or closed by certain conditions of chemicals. The opening or closing of channels can set up an **electrochemical gradient** with unequal charged species on different sides of a membrane. Review Figure 5.10

Carrier proteins bind to polar molecules such as sugars and amino acids and transport them across the membrane. The maximum rate of this type of facilitated diffusion is limited by the number of carrier (transporter) proteins in the membrane; once all carrier proteins are in use, the system is saturated and increases in solute concentration will not increase the rate of diffusion. Review Figure 5.12

5.4 How do substances cross membranes against a concentration gradient?

See Web/CD Tutorial 5.2

Active transport requires the use of chemical energy to move substances across a membrane against a concentration gradient. Active transport proteins may be uniports, symports, or antiports. Review Figure 5.13

In **primary active transport**, energy from the hydrolysis of ATP is used to move ions into or out of cells against their concentration gradients. The **sodium-potassium pump** is an important example. Review Figure 5.14

Secondary active transport couples the passive movement of one substance with its concentration gradient to the movement of another substance against its concentration gradient. Energy from ATP is used indirectly to establish the concentration gradient that results in the movement of the first substance. Review Figure 5.15

5.5 How do large molecules enter and leave a cell?

See Web/CD Tutorial 5.3

Endocytosis is the transport of macromolecules, large particles, and small cells into eukaryotic cells by means of engulfment by and vesicle formation from the plasma membrane. **Phagocytosis** and **pinocytosis** are types of endocytosis. Review Figure 5.16A

In **receptor-mediated endocytosis**, a specific **receptor protein** on the plasma membrane binds to a particular macromolecule.

In **exocytosis**, materials in vesicles are secreted from the cell when the vesicles fuse with the plasma membrane. Review Figure 5.16B

5.6 What are some other functions of membranes?

Membranes function as sites for energy transformations, for organizing chemical reactions and for recognition and initial processing of extracellular signals. Review Figure 5.18

SELF-QUIZ

1. Which statement about membrane phospholipids is *not* true?
   a. They associate to form bilayers.
   b. They have hydrophobic "tails."
   c. They have hydrophilic "heads."
   d. They give the membrane fluidity.
   e. They flip-flop readily from one side of the membrane to the other.

2. When a hormone molecule binds to a specific protein on the plasma membrane, the protein it binds to is called a
   a. ligand.
   b. clathrin.
   c. receptor protein.
   d. hydrophobic protein.
   e. cell adhesion molecule.

3. Which statement about membrane proteins is *not* true?
   a. They all extend from one side of the membrane to the other.
   b. Some serve as channels for ions to cross the membrane.
   c. Many are free to migrate laterally within the membrane.
   d. Their position in the membrane is determined by their tertiary structure.
   e. Some play roles in photosynthesis.

4. Which statement about membrane carbohydrates is *not* true?
   a. Most are bound to proteins.
   b. Some are bound to lipids.
   c. They are added to proteins in the Golgi apparatus.
   d. They show little diversity.
   e. They are important in recognition reactions at the cell surface.

5. Which statement about animal cell junctions is *not* true?
   a. Tight junctions are barriers to the passage of molecules between cells.
   b. Desmosomes allow cells to adhere firmly to one another.
   c. Gap junctions block communication between adjacent cells.
   d. Connexons are made of protein.
   e. The fibers associated with desmosomes are made of protein.

6. You are studying how the protein transferrin enters cells. When you examine cells that have taken up transferrin, you find it inside clathrin-coated vesicles. Therefore, the most likely mechanism for uptake of transferrin is
   a. facilitated diffusion.
   b. an antiport.
   c. receptor-mediated endocytosis.
   d. gap junctions.
   e. ion channels.

7. Which statement about ion channels is *not* true?
   a. They form pores in the membrane.
   b. They are proteins.
   c. All ions pass through the same type of channel.
   d. Movement through them is from high concentrations to low concentrations.
   e. Movement through them is by simple diffusion.

8. Facilitated diffusion and active transport both
   a. require ATP.
   b. require the use of proteins as carriers.
   c. carry solutes in only one direction.
   d. increase without limit as the concentration gradient increases.
   e. depend on the solubility of the solute in lipids.

9. Primary and secondary active transport both
   a. generate ATP.
   b. are based on passive movement of Na^+ ions.
   c. include the passive movement of glucose molecules.
   d. use ATP directly.
   e. can move solutes against their concentration gradients.

10. Which statement about osmosis is *not* true?
    a. It obeys the laws of diffusion.
    b. In animal tissues, water moves into cells if they are hypertonic to their environment.
    c. Red blood cells must be kept in a plasma that is hypotonic to the cells.
    d. Two cells with identical solute concentrations are isotonic to each other.
    e. Solute concentration is the principal factor in osmosis.

FOR DISCUSSION

1. Muscle function requires calcium ions (Ca^{2+}) to be pumped into a subcellular compartment against a concentration gradient. What types of molecules are required for this to happen?

2. Section 27.5 will describe the diatoms, protists that have complex glassy structures in their cell walls (see Figure 27.1). These structures form within the Golgi apparatus. How do these structures reach the cell wall without having to pass through a membrane?

3. Organisms that live in fresh water are almost always hypertonic to their environment. In what way is this a serious problem? How do some organisms cope with this problem?

4. Contrast nonspecific endocytosis and receptor-mediated endocytosis.

5. The emergence of the phospholipid membrane was important to the origin of cells. Describe the most important properties of membranes that might have allowed cells containing them to thrive in comparison with molecular aggregates without membranes.

FOR INVESTIGATION

Under certain laboratory conditions, cells of two different species can be induced to fuse, combining their two plasma membranes in much the same way that vesicles fuse with the plasma membrane in exocytosis. Antibodies labeled with fluorescent dyes can be used to stain specific proteins on the cell surface. How would you use cell fusion to investigate whether membrane proteins diffuse in the plane of the membrane?

CHAPTER 6

Energy, Enzymes, and Metabolism

Sensitivity to alcohol

The guests were at their tables as the bride and groom entered the room to enthusiastic applause. Champagne glasses were raised in the first of many toasts. But no sooner had Frank drunk his first small glass of bubbly than he started feeling ill. With his face flushed and his heart beating rapidly, he excused himself and left the reception. It was hours before he felt normal again. The same thing had happened to him before, also when he drank a small amount of an alcoholic beverage.

There are people who can drink a lot of alcohol without getting sick, and there are others, like Frank, who become ill after consuming only a small amount. These individual differences are a function of the biochemistry of alcohol once it enters the body. Typically, a series of two chemical reactions transforms ethyl alcohol (the alcohol in beverages) into acetate in the body: first ethyl alcohol is transformed into acetaldehyde, then acetaldehyde is transformed into acetate.

These reactions constitute a *metabolic pathway*, in which the product of the first reaction, acetaldehyde, serves as the raw material for the second. Neither reaction can occur to any appreciable extent without help. Each reaction needs to be "speeded up" by a different *catalyst*—a specific *enzyme*.

Enzymes are proteins, and as such, each has a specific amino acid sequence. The enzyme that speeds up the second reaction in the metabolism of alcohol is aldehyde dehydrogenase (ALDH), which has 517 amino acids. Most of the guests at Frank's table had the same amino acid sequence in their ALDH. In this "typical" sequence, the 487th amino acid is glutamic acid, whose side chain carries a negative charge (see Table 3.1). The other guests' ALDH folded into the correct three-dimensional structure, and the enzyme did its job—speeding up the conversion of acetaldehyde into acetate. In Frank's cells, however, the amino acid at position 487 is lysine instead of glutamic acid; lysine has a *positive* charge. This seemingly tiny difference is enough to change the three-dimensional structure of ALDH and interfere with its catalytic function.

Because of its structural defect, Frank's ALDH works, but only slowly. When he drank that small glass of champagne, the first metabolic reaction (the conversion of alcohol to acetaldehyde) happened quickly, but the second reaction (the conversion of acetaldehyde to acetate) did not. And, whereas alcohol has some pleasant effects on the body, the effects produced by acetaldehyde are not so pleasant. As

A Catalyst for Alcohol The enzyme aldehyde dehydrogenase catalyzes an important step in the breakdown of alcohol.

A Precursor to Trouble   Some people have a weakly active form of the aldehyde dehydrogenase enzyme. When they drink alcohol, acetaldehyde accumulates in their bodies, with ill effects.

acetaldehyde accumulated in his body, Frank became ill. His facial flushing and accelerated heartbeat were typical short-term effects of acetaldehyde; in the longer term, it can cause brain abnormalities and even some cancers.

Frank's atypical ALDH sequence shows how individual differences in alcohol tolerance are not necessarily behavioral, but can be due to differences in basic biochemistry. (So think twice before kidding a person because he or she can't "hold his liquor.")

IN THIS CHAPTER we begin our study of biochemical transformations, focusing on the role of energy. After describing the physical principles that underlie energy transformations and how these principles apply to biology, we will see how the energy carrier ATP plays an important role in the cell. The rest of the chapter deals with the nature, activities, and regulation of enzymes, which speed up biochemical transformations and without which life would not be possible.

6.1 What Physical Principles Underlie Biological Energy Transformations?

Metabolic reactions and catalysts are essential to the biochemical transformation of energy by living things. Whether it is a plant using light energy to produce carbohydrates or a cat transforming food energy so it can leap to a countertop (where it hopes to find food so it can obtain more energy), the transformation of energy is a hallmark of life.

Physicists define **energy** as the capacity to do work. *Work* is what occurs when a force operates on an object over a distance. In biochemistry, it is more useful to consider energy as *the capacity for change*. No cell creates energy; all living things must obtain energy from the environment. Indeed, one of the fundamental physical laws is that energy can neither be created nor destroyed. However, energy can be *transformed* from one type into another, and living cells carry out many such energy transformations. Energy transformations are linked to the chemical transformations that occur in cells—the breaking of chemical bonds, the movement of substances across membranes, and so forth.

There are two basic types of energy and of metabolism

Energy comes in many forms: chemical, electric, heat, light, and mechanical. But all forms of energy can be considered as one of two basic types:

- **Potential energy** is the energy of state or position—that is, stored energy. It can be stored in many forms: in chemical bonds, or as a concentration gradient, or even as an electric charge imbalance (as in the membrane potential we described in Section 5.3). Think of a crouching cat, muscles hunched taut, holding still as it prepares to pounce.

- **Kinetic energy** is the energy of movement—that is, the type of energy that does work. Think of the cat leaping as some of the potential energy stored in its taut muscles is converted into muscle contractions.

Just as potential and kinetic energy can interconvert, the *form* of that energy can also interconvert. The potential energy in the cat's muscles is in covalent bonds (chemical energy), while

Potential chemical energy is converted into kinetic mechanical energy when the cat leaps.

Potential chemical energy is stored in the muscles of the cat.

6.1 Energy Conversions and Work   A leaping cat illustrates both the conversion between potential and kinetic energy and the conversion of energy from one form (chemical) to another (mechanical).

the kinetic energy of the pouncing cat is mechanical (**Figure 6.1**). You can think of many other such interconversions: reading this book, for example, involves light energy being converted into chemical energy in your eyes, and the chemical energy being converted into electric energy in the nerve cells that carry messages to your brain.

In any living organism, chemical reactions are occurring continuously. **Metabolism** is defined as the sum total of these reactions. Two types of metabolic reactions occur in all cells of all organisms:

- **Anabolic reactions** (*anabolism*) link simple molecules to form more complex molecules. The synthesis of a protein from amino acids is an anabolic reaction. Anabolic reactions require an input of energy and capture it in the chemical bonds that are formed.

- **Catabolic reactions** (*catabolism*) break down complex molecules into simpler ones and release the energy stored in chemical bonds.

Catabolic and anabolic reactions are often linked. The energy released in catabolic reactions is often used to drive anabolic reactions—that is, to do biological work, such as the contraction of a leaping cat's muscles or cellular activities such as active transport across a membrane.

Because living organisms are part of the physical universe, the **laws of thermodynamics** (*thermo*, "energy"; *dynamics*, "change") that operate in stars also apply to energy transformations in cells.

The first law of thermodynamics: Energy is neither created nor destroyed

The first law of thermodynamics states that in any conversion of energy, it is neither created nor destroyed. Another way of saying this is that *in any conversion of energy from one form to another, the total energy before and after the conversion is the same* (**Figure 6.2A**). As you will see in the next two chapters, potential energy in the chemical bonds of carbohydrates can be converted into potential energy in the form of ATP. This energy can then be converted into kinetic energy and used to do mechanical work, such as muscle contraction.

The second law of thermodynamics: Disorder tends to increase

The second law of thermodynamics states that, although energy cannot be created or destroyed, *when energy is converted from one form into another, some of that energy becomes unavailable to do work* (**Figure 6.2B**). In other words, no physical process or chemical reaction is 100 percent efficient, and not all the energy released can be converted into work. Some energy is lost to a form associated with disorder. Think of disorder as a kind of randomness. It takes energy to impose order on a system. Unless energy is applied to impose order on a system, that system will be randomly arranged or disordered. Your bedroom or dorm room is probably a good analogy. If you are like we were as students, your room is rather disordered, and it takes energy to organize it. The second law applies to all energy transformations, but we will focus here on chemical reactions in living systems.

NOT ALL ENERGY CAN BE USED   In any system, the total energy includes the usable energy that can do work *and* the unusable energy that is lost to disorder:

total energy = usable energy + unusable energy

In biological systems, the total energy is called **enthalpy** (**H**). The usable energy that can do work is called **free energy** (**G**). Free energy is what cells require for all the chemical reactions of cell growth, cell division, and the maintenance of cell health. The unusable energy is represented by **entropy** (**S**), which is a measure of the disorder of the system, multiplied by the *absolute temperature* (*T*). Thus we can rewrite the word equation above more precisely as

$$H = G + TS$$

Because we are interested in usable energy, we rearrange this expression:

$$G = H - TS$$

Why do you get hot when you exercise? Your muscles convert the chemical energy you obtain from food to the mechanical energy of muscle contraction. Less than 20 percent of this energy conversion does the work of exercise; the rest is lost as heat.

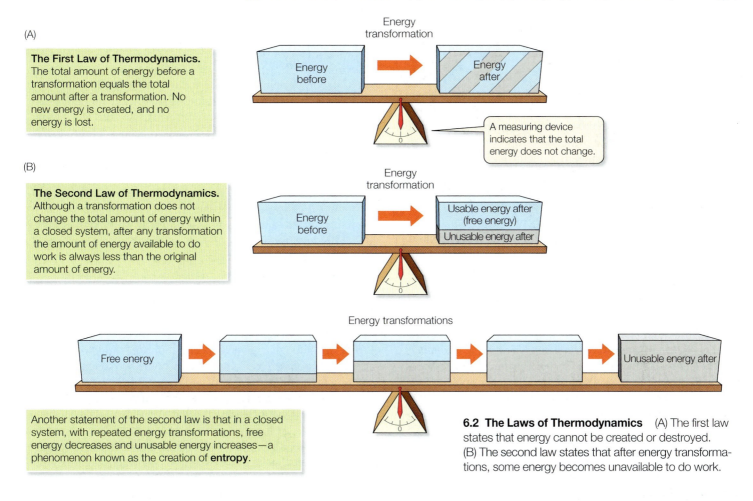

(A)

The First Law of Thermodynamics.
The total amount of energy before a transformation equals the total amount after a transformation. No new energy is created, and no energy is lost.

A measuring device indicates that the total energy does not change.

(B)

The Second Law of Thermodynamics.
Although a transformation does not change the total amount of energy within a closed system, after any transformation the amount of energy available to do work is always less than the original amount of energy.

Another statement of the second law is that in a closed system, with repeated energy transformations, free energy decreases and unusable energy increases—a phenomenon known as the creation of **entropy**.

6.2 The Laws of Thermodynamics (A) The first law states that energy cannot be created or destroyed. (B) The second law states that after energy transformations, some energy becomes unavailable to do work.

Although we cannot measure *G*, *H*, or *S* absolutely, we can determine the *change* in each at a constant temperature. Such energy changes are measured in calories (cal) or joules (J).* A change in energy is represented by the Greek letter *delta* (Δ). The **change in free energy (ΔG)** of any chemical reaction is equal to the difference in free energy between the products and the reactants,

$$\Delta G_{\text{reaction}} = G_{\text{products}} - G_{\text{reactants}}$$

Such a change can be either positive or negative; that is, the free energy of the products can be more or less than the free energy of the reactants. If the products have more free energy, then there must have been some input of energy into the reaction (remember, energy cannot be created, so some must somehow have been added from an external source).

At a constant temperature, ΔG is defined in terms of the change in total energy (ΔH) and the change in entropy (ΔS):

$$\Delta G = \Delta H - T\Delta S$$

This equation tells us whether free energy is released or consumed by a chemical reaction:

- If ΔG is negative (ΔG < 0), free energy is released.
- If ΔG is positive (ΔG > 0), free energy is required (consumed).

If the necessary free energy is not available, the reaction does not occur. The sign and magnitude of ΔG depend on the two factors on the right of the equation:

- ΔH: In a chemical reaction, ΔH is the total amount of energy added to the system (ΔH > 0) or released (ΔH < 0).
- ΔS: Depending on the sign and magnitude of ΔS, the entire term, TΔS, may be negative or positive, large or small. In other words, in living systems at a constant temperature (no change in T), the magnitude and sign of ΔG can depend a lot on changes in entropy. Large changes in entropy make ΔG more negative in value, as shown by the minus sign in front of the TΔS term.

If a chemical reaction increases entropy, its products are more disordered or random than its reactants. If there are more products than reactants, as in the hydrolysis of a protein to its amino acids, the products have considerable freedom to move around. The disorder in a solution of amino acids will be large compared with that in the protein, in which peptide bonds and other forces prevent free movement. So in hydrolysis, the change in entropy (ΔS) will be positive.

If there are fewer products, and they are more restrained in their movements than the reactants, ΔS will be negative. For example, a large protein linked by peptide bonds is less free in its move-

*A *calorie* is the amount of heat energy needed to raise the temperature of 1 gram of pure water from 14.5°C to 15.5°C. A *joule* is an energy measure in the commonly used SI system. 1 J = 0.239 cal; conversely, 1 cal = 4.184 J. Thus, for example, 486 cal = 2,033 J, or 2.033 kJ. Although defined in terms of heat, the calorie and the joule are measures of any form of energy—mechanical, electrical, or chemical. When you compare data on energy, always compare joules to joules and calories to calories.

ments than a solution of the hundreds or thousands of amino acids from which it was synthesized.

DISORDER TENDS TO INCREASE The second law of thermodynamics also predicts that, *as a result of energy transformations, disorder tends to increase.* Chemical changes, physical changes, and biological processes all tend to increase entropy and therefore tend toward disorder or randomness (see Figure 6.2B). This tendency for disorder to increase gives a directionality to physical processes and chemical reactions. It explains why some reactions proceed in one direction rather than another.

How does the second law apply to organisms? Consider the human body, with its highly complex structures constructed of simple molecules. This increase in complexity appears to be in conflict with the second law. But this is not the case! Constructing 1 kg of a human body requires that about 10 kg of biological materials be metabolized and in the process converted into CO_2, H_2O, and other simple molecules. This metabolism creates far more disorder than the order in 1 kg of flesh, and the conversions require a lot of energy. *Life requires a constant input of energy to maintain order.* There is no conflict with the second law of thermodynamics.

Having seen that the laws of thermodynamics apply to living things, we will now turn to a consideration of how these laws apply to biochemical reactions.

Chemical reactions release or consume energy

Recall that anabolic reactions tend to increase complexity (order) in the cell, while catabolic reactions break down complexity (cre-

ate disorder). Accordingly, anabolic reactions require energy, while catabolic reactions release energy:

- Catabolic reactions may break down an ordered reactant, such as a protein molecule, into smaller, more randomly distributed products, such as amino acids. *Reactions that release free energy* ($-\Delta G$) *are called* **exergonic reactions** (**Figure 6.3A**). For example:

  complex molecules → free energy + small molecules

- Anabolic reactions may make a single product, such as a protein (a highly ordered substance), out of many smaller reactants, such as amino acids (less ordered). *Reactions that require or consume free energy* ($+\Delta G$) *are called* **endergonic reactions** (**Figure 6.3B**). For example:

  free energy + small molecules → complex molecules

In principle, chemical reactions can run both forward and backward. For example, if compound A can be converted into compound B (A → B), then B, in principle, can be converted into A (B → A), although at given concentrations of A and B, only one of these directions will be favored. Think of the overall reaction as resulting from competition between forward and reverse reactions (A ⇌ B). Increasing the concentration of A speeds up the forward reaction, and increasing the concentration of B favors the reverse reaction. At some concentration of A and B, the forward and reverse reactions take place at the same rate. At this concentration, no further net change in the system is observable, although individual molecules are still forming and breaking apart. This balance between forward and reverse reactions is known as **chemical equilibrium**. Chemical equilibrium is a static state, a state of no net change, and a state in which $\Delta G = 0$.

(A) Exergonic reaction

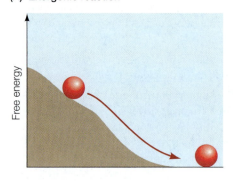
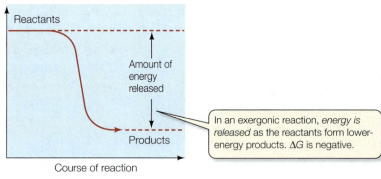

In an exergonic reaction, *energy is released* as the reactants form lower-energy products. ΔG is negative.

(B) Endergonic reaction

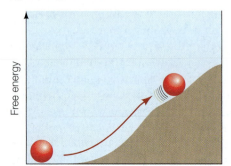
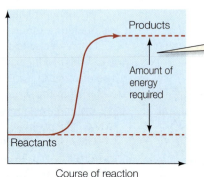

Energy must be added for an endergonic reaction, in which reactants are converted to products with a higher energy level. ΔG is positive.

6.3 Exergonic and Endergonic Reactions (A) In an exergonic reaction, the reactants behave like a ball rolling down a hill, and energy is released. (B) A ball will not roll uphill by itself. Driving an endergonic reaction, like moving a ball uphill, requires adding free energy.

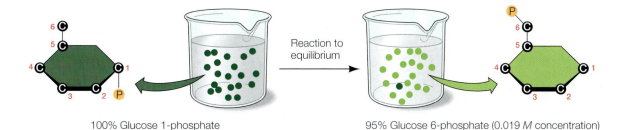

6.4 Chemical Reactions Run to Equilibrium No matter what quantities of glucose 1-phosphate and glucose 6-phosphate are dissolved in water, when equilibrium is attained, there will always be 95 percent glucose 6-phosphate and 5 percent glucose 1-phosphate.

Chemical equilibrium and free energy are related

Every chemical reaction proceeds to a certain extent, but not necessarily to completion. In other words, all the reactants present are not necessarily converted into products. Each reaction has a specific equilibrium point, and that equilibrium point is related to the free energy released by the reaction under specified conditions. To understand the principle of equilibrium, consider the following example.

Most cells contain glucose 1-phosphate, which is converted in the cell into glucose 6-phosphate. Imagine that we start out with an aqueous solution of glucose 1-phosphate that has a concentration of 0.02 *M*. (*M* stands for molar concentration; see Section 2.4). The solution is maintained under constant environmental conditions (25°C and pH 7). As the reaction proceeds slowly to equilibrium, the concentration of the product, glucose 6-phosphate, rises from 0 to 0.019 *M*, while the concentration of the reactant, glucose 1-phosphate, falls to 0.001 *M*. At this point, equilibrium is reached (**Figure 6.4**). From then on, the reverse reaction, from glucose 6-phosphate to glucose 1-phosphate, progresses at the same rate as the forward reaction.

At equilibrium, then, this reaction has a product-to-reactant ratio of 19:1 (0.019/0.001), so the forward reaction has gone 95 percent of the way to completion ("to the right," as written). Therefore, the forward reaction is an exergonic reaction. This result is obtained every time the experiment is run under the same conditions. The reaction is described by the equation

glucose 1-phosphate $\rightleftharpoons$ glucose 6-phosphate

The change in free energy (ΔG) for any reaction is related directly to its point of equilibrium. The further toward completion the point of equilibrium lies, the more free energy is released. In an exergonic reaction, such as the conversion of glucose 1-phosphate to glucose 6-phosphate, ΔG is a *negative* number (in this example, $\Delta G = -1.7$ kcal/mol, or -7.1 kJ/mol).

A large, *positive* ΔG for a reaction means that it proceeds hardly at all to the right (A $\rightarrow$ B). But if the product is present, such a reaction runs backward, or "to the left" (A $\leftarrow$ B) (nearly all B is converted into A). A ΔG value near zero is characteristic of a readily reversible reaction: reactants and products have almost the same free energies.

The principles of thermodynamics that we have been discussing apply to all energy transformations in the universe, so they are very powerful and useful. Next, we'll apply them to reactions in cells that involve the biological energy currency, ATP.

6.2 What Is the Role of ATP in Biochemical Energetics?

Cells rely on **adenosine triphosphate**, or **ATP**, for the capture and transfer of the free energy they need to do chemical work. ATP operates as a kind of "energy currency." That is, just as you may earn money from a job and then spend it on a meal, some of the free energy that is released by certain exergonic reactions is captured in ATP, which can then release free energy to drive endergonic reactions.

ATP is produced by cells in a number of ways (which we will describe in the next two chapters), and it is used in many ways. ATP is not an unusual molecule. In fact, it has another important use in the cell: it is a nucleotide that can be converted into a building block for nucleic acids. But two things about ATP make it especially useful to cells: it releases a relatively large amount of energy when hydrolyzed, and it can *phosphorylate* (donate a phosphate group to) many different molecules. We will examine these two properties in the discussion that follows.

ATP hydrolysis releases energy

An ATP molecule consists of the nitrogenous base adenine bonded to ribose (a sugar), which is attached to a sequence of three phosphate groups (**Figure 6.5A**). The hydrolysis of ATP yields free energy, as well as ADP (adenosine *di*phosphate) and an inorganic phosphate ion; the ion, HPO_4^{2-}, is commonly abbreviated P_i. Thus:

$$ATP + H_2O \rightarrow ADP + P_i + \text{free energy}$$

The important property of this reaction is that it is exergonic, releasing free energy. The change in free energy (ΔG) is about –7.3 kcal/mol (–30 kJ/mol) at a defined temperature, pH, and solute concentration.

Two characteristics of ATP account for the free energy released by the loss of one of its phosphate groups:

- The free energy of the P—O bond between phosphate groups (called a phosphoric acid anhydride bond) is much higher than the energy of the H—O bond that forms after hydrolysis. So some usable energy is released by hydrolysis.

- Because phosphates are negatively charged and so repel each other, it takes energy to get phosphates near enough to each other to make the covalent bond that links them together (e.g., to add a phosphate to ADP to make ATP).

Bioluminescence—the production of light by living organisms (**Figure 6.5B**)—is an example of an endergonic reaction driven by ATP hydrolysis as well as of the interconversion of energy forms (chemical to light). The chemical that becomes luminescent is called luciferin (after the devil, Lucifer):

$$\text{luciferin} + O_2 + ATP \xrightarrow{\text{luciferase}} \text{oxyluciferin} + AMP + PP_i + \text{light}$$

This reaction and the enzyme (luciferase) that catalyzes it occur in a wide variety of organisms besides the familiar firefly, ranging from marine organisms to mushrooms. The light is generally used to avoid predators or signal to mates.

Soft-drink companies use the firefly proteins luciferin and luciferase to detect bacterial contamination. Where there are living cells, there is ATP, and when the firefly proteins encounter ATP and oxygen, they give off light. Thus, soda that lights up is contaminated.

ATP couples exergonic and endergonic reactions

As we have just seen, the *hydrolysis of ATP is exergonic* and yields ADP, P_i, and free energy. The reverse reaction, the *formation of ATP from ADP and P_i, is endergonic* and consumes as much free energy as is released by the hydrolysis of ATP:

$$ADP + P_i + \text{free energy} \rightarrow ATP + H_2O$$

Many different exergonic reactions in the cell can provide the energy to convert ADP into ATP. In eukaryotes, the most important of these reactions is cellular respiration, in which some of the en-

(A)

ATP (space-filling model)   ATP (structural formula)

6.5 ATP (A) ATP is richer in energy than its relatives ADP and AMP. The hydrolysis of ATP releases the energy stored in the P—O bond between the second and third phosphates. (B) Fireflies use ATP to initiate the oxidation of luciferin. This process converts chemical energy into light energy, emitting rhythmic flashes that signal the insect's readiness to mate. Very little of the energy in this conversion is lost as heat.

ergy released from fuel molecules is captured in ATP. The formation and hydrolysis of ATP constitute what might be called an "energy-coupling cycle," in which ADP picks up energy from exergonic reactions to become ATP, which donates energy to endergonic reactions.

How does this ATP cycle trap and release energy? An exergonic reaction is coupled to the endergonic reaction that forms ATP from ADP and P_i (**Figure 6.6**). *Coupling of exergonic and endergonic reac-*

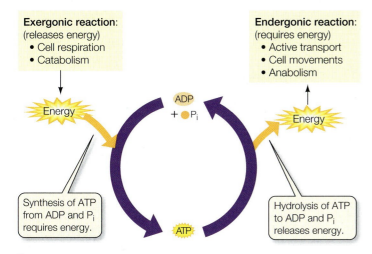

6.6 Coupling of Reactions Exergonic cellular reactions release the energy needed to make ATP from ADP. The energy released from the conversion of ATP back to ADP can be used to fuel endergonic reactions.

tions is very common in metabolism. When it forms, ATP captures free energy and retains it in the form of its P—O bond. ATP then diffuses to another site in the cell, where its hydrolysis releases free energy to drive an endergonic reaction.

A specific example of this energy-coupling cycle is shown in **Figure 6.7**. The formation of the amino acid glutamine has a pos-

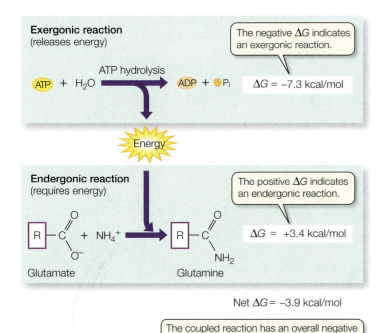

6.7 Coupling of ATP Hydrolysis to an Endergonic Reaction The synthesis of the amino acid glutamine from glutamate and an ammonium ion is an endergonic reaction that must be coupled with the exergonic hydrolysis of ATP.

itive ΔG (is endergonic) and will not proceed without the input of free energy from ATP hydrolysis, which has a negative ΔG (is exergonic). The overall ΔG for the coupled reactions (when the two ΔGs are added together) is negative. Hence the reactions proceed exergonically when they are coupled, and glutamine is synthesized.

An active cell requires millions of molecules of ATP per second to drive its biochemical machinery. An ATP molecule is consumed within a second of its formation, on average. At rest, an average person produces and hydrolyzes about 40 kg of ATP per day—as much as some people weigh. This means that each ATP molecule undergoes about 10,000 cycles of synthesis and hydrolysis every day.

6.2 RECAP

ATP is the "energy currency" of cells. ATP captures some of the free energy released by exergonic reactions, which can then be released by ATP hydrolysis and used to drive endergonic reactions.

■ How does ATP store energy? See p. 124

■ Can you explain the concept of coupled reactions? See pp. 124–125 and Figure 6.7

ATP is synthesized and used up very rapidly. But these biochemical reactions could not proceed so rapidly without the help of catalytic proteins called enzymes.

6.3 What Are Enzymes?

When we know the change in free energy (ΔG) of a reaction, we know where the equilibrium point of the reaction lies: the more negative ΔG is, the further the reaction proceeds toward completion. However, ΔG tells us nothing about the **rate** of a reaction—the speed at which it moves toward equilibrium. The reactions that occur in cells are so slow that they could not contribute to life unless the cells did something to speed them up. That is the role of **catalysts**: substances that speed up a reaction without being permanently altered by that reaction. A catalyst does not cause a reaction that would not take place eventually without it, but merely speeds up the rates of both forward and backward reactions, allowing equilibrium to be approached more rapidly.

Most biological catalysts are proteins called **enzymes**. Although we will focus here on proteins, some catalysts—perhaps the earliest ones in the origin of life—are RNA molecules called ribozymes (see Section 3.6). *A biological catalyst, whether protein or RNA, is a framework or scaffold in which chemical catalysis takes place.* Over time, proteins have evolved as catalysts, probably because of their great diversity in three-dimensional structure and variety of chemical functions.

In this section we will identify the energy barrier that controls the rate of chemical reactions. Then we'll focus on the role of enzymes: how they interact with specific reactants, how they lower the energy barrier, and how they permit reactions to proceed more quickly. In Section 6.4 we'll look at how enzymes contribute to the coupling of reactions.

For a reaction to proceed, an energy barrier must be overcome

An exergonic reaction may release a great deal of free energy, but the reaction may take place very slowly. Some reactions are slow because there is an *energy barrier* between reactants and products. Think about the propane stove we described in Section 2.3. The burning of propane ($C_3H_8 + 5 O_2 \rightarrow 3 CO_2 + 4 H_2O$ + energy) is an exergonic reaction—energy is released in the form of heat and light. Once started, the reaction goes to completion: all of the propane reacts with oxygen to form carbon dioxide and water vapor.

Because burning propane liberates so much energy, you might expect this reaction to proceed rapidly whenever propane is exposed to oxygen. But this does not happen. Simply mixing propane with air produces no reaction. Propane will start burning only if a spark—an input of energy—is provided. (In the case of the stove, this energy is supplied by a match.) The need for this spark to start the reaction shows that there is an energy barrier between the reactants and the products.

In general, exergonic reactions proceed only after the reactants are pushed over the energy barrier by a small amount of added energy. The energy barrier thus represents the amount of energy needed to start the reaction, known as the **activation energy (E_a)** (**Figure 6.8A**). Recall the ball rolling down the hill in Figure 6.3. The ball has a lot of potential energy at the top of the hill. However, if the ball is stuck in a small depression, it will not roll down the hill, even though that action is exergonic. To start the ball rolling, a small amount of energy (activation energy) is needed to get the ball out of the depression (**Figure 6.8B**).

In a chemical reaction, the activation energy is the energy needed to change the reactants into unstable molecular forms called **transition-state species**. Transition-state species have higher free energies than either the reactants or the products. Their bonds may be stretched and hence unstable. Although the amount of activation energy needed for different reactions varies, it is often small compared with the change in free energy of the reaction. The activation energy that starts a reaction is recovered during the ensuing "downhill" phase of the reaction, so it is not a part of the net free energy change, ΔG (see Figure 6.8A).

Where does the activation energy come from? In any collection of reactants at room or body temperatures, some molecules are moving around, and they could use their kinetic energy of motion to overcome the energy barrier, enter the transition state, and react. However, at these temperatures, only a few molecules have enough energy to do this; most have insufficient kinetic energy for activation, so the reaction takes place slowly. If the system were heated, all the reactant molecules would move faster and have more kinetic energy. Since more molecules would have energy exceeding the required activation energy, the reaction would speed up.

However, adding enough heat to increase the average kinetic energy of the molecules would not work in living systems. Such a nonspecific approach would accelerate all reactions, including destructive ones, such as the denaturation of proteins (see Figure 3.11). A more effective way to speed up a reaction in a living system is to

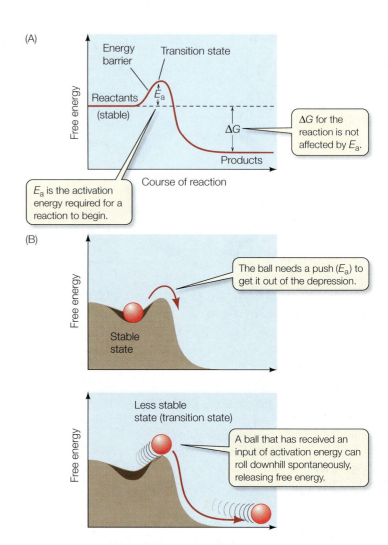

6.8 Activation Energy Initiates Reactions (A) In any chemical reaction, an initial stable state must become less stable before change is possible. (B) A ball on a hillside provides a physical analogy to the biochemical principle graphed in (A).

lower the energy barrier by bringing the reactants into close proximity to each other. In living cells, enzymes accomplish this task.

Enzymes bind specific reactant molecules

Catalysts increase the rate of chemical reactions. Most *nonbiological* catalysts are *nonspecific*. For example, powdered platinum catalyzes virtually any reaction in which molecular hydrogen (H_2) is a reactant. In contrast, most *biological* catalysts are *highly specific*. These complex molecules of protein (enzymes) or RNA (ribozymes) catalyze relatively simple chemical reactions. An enzyme or ribozyme usually recognizes and binds to only one or a few closely related reactants, and it catalyzes only a single chemical reaction. In the discussion that follows, we focus on enzymes, but remember that similar rules of chemical behavior apply to ribozymes as well.

In an enzyme-catalyzed reaction, the reactants are called **substrates**. Substrate molecules bind to a particular site on the en-

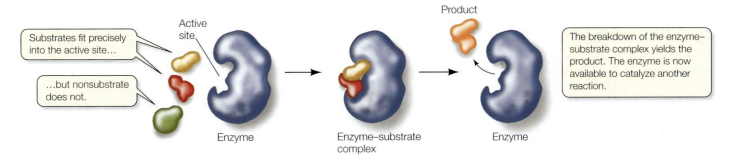

6.9 Enzyme and Substrate An enzyme is a protein catalyst with an active site capable of binding one or more substrate molecules.

zyme, called the **active site**, where catalysis takes place (**Figure 6.9**). The specificity of an enzyme results from the exact three-dimensional shape and structure of its active site, into which only a narrow range of substrates can fit. Other molecules—with different shapes, different functional groups, and different properties—cannot properly fit and bind to the active site.

The names of enzymes reflect the specificity of their functions and often end with the suffix "-ase." For example, the enzyme RNA polymerase catalyzes the formation of RNA, but not DNA, and the enzyme hexokinase accelerates the phosphorylation of hexose sugars, but not pentose sugars.

The binding of a substrate to the active site of an enzyme produces an **enzyme–substrate complex** (ES) held together by one or more means, such as hydrogen bonding, electrical attraction, or covalent bonding. The enzyme–substrate complex gives rise to product and free enzyme:

$$E + S \rightarrow ES \rightarrow E + P$$

where E is the enzyme, S is the substrate, P is the product, and ES is the enzyme–substrate complex. The free enzyme (E) is in the same chemical form at the end of the reaction as at the beginning. While bound to the substrate, it may change chemically, but by the end of the reaction it has been restored to its initial form.

Enzymes lower the energy barrier but do not affect equilibrium

When reactants are part of an enzyme–substrate complex, they require less activation energy than the transition-state species of the corresponding uncatalyzed reaction (**Figure 6.10**). Thus the enzyme lowers the energy barrier for the reaction—it offers the reaction an easier path. When an enzyme lowers the energy barrier, both the forward and the reverse reactions speed up, so the enzyme-catalyzed overall reaction proceeds toward equilibrium more rapidly than the uncatalyzed reaction. The final equilibrium is the same with or without the enzyme. Similarly, adding an enzyme to a reaction does not change the difference in free energy (ΔG) between the reactants and the products.

Each year over 500 tons of protease enzymes are added to laundry detergents to break down proteins such as the pizza stains on your shirt. These enzymes are produced by bacteria grown in huge stainless steel tanks.

Enzymes can change the rate of a reaction substantially. For example, if 600 molecules of a protein with arginine as its terminal amino acid just sit in solution, the protein molecules tend toward disorder, and the terminal peptide bonds break, releasing the arginines (ΔS increases). After 7 years, about half (300) of the proteins will have undergone this reaction. With the enzyme carboxypeptidase A catalyzing the reaction, however, the 300 arginines are released in half a second!

6.3 RECAP

A chemical reaction requires a "push" over the energy barrier to get started. Enzymes supply this activation energy by binding specific reactants (substrates).

- How does the structure of enzymes makes them specific? See pp. 126–127 and Figure 6.9
- Do you understand the relationship between enzymes and the equilibrium point of a reaction? See p. 127

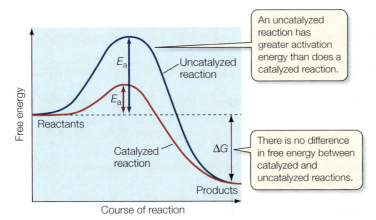

6.10 Enzymes Lower the Energy Barrier Although the activation energy is lower in an enzyme-catalyzed reaction than in an uncatalyzed reaction, the energy released is the same with or without catalysis. In other words, E_a is lower, but ΔG is unchanged.

Now that we understand the structure and specificity of enzymes, let's see how they work to speed up chemical reactions between the substrate molecules.

6.4 How Do Enzymes Work?

After formation of the enzyme–substrate complex, chemical interactions occur. These interactions contribute directly to the breaking of old bonds and the formation of new ones. In catalyzing a reaction, an enzyme may use one or more of the following mechanisms:

■ *Enzymes orient substrates.* While free in solution, substrates are rotating and tumbling around and may not have the proper orientation to interact when they collide. Part of the activation energy needed to start a reaction is used to bring together the specific atoms between which bonds are to form (**Figure 6.11A**). For example, if acetyl CoA and oxaloacetate are to form citrate (a reaction that is part of the metabolism of glucose, as we will see in Section 7.2), the two substrates must be oriented so that the carbon atom of the methyl group of acetyl CoA can form a covalent bond with the carbon atom of the carbonyl group of oxaloacetate. The active site of the enzyme citrate synthase has just the right shape to bind these two molecules so that these atoms are adjacent.

■ *Enzymes induce strain in the substrate.* Once a substrate has bound to its active site, an enzyme can cause bonds in the substrate to stretch, putting it in an unstable transition state (**Figure 6.11B**). For example, lysozyme is a protective enzyme that destroys invading bacteria by cleaving polysaccharide chains in their cell walls. Lysozyme's active site "stretches" the bonds of the bacterial polysaccharide—one of lysozyme's substrates. The stretching renders the polysaccharide bonds unstable and more reactive to lysozyme's other substrate, water.

■ *Enzymes temporarily add chemical groups to substrates.* The side chains (R groups) of an enzyme's amino acids may be direct participants in making its substrates more chemically reactive (**Figure 6.11C**).

■ In *acid–base catalysis*, the acidic or basic side chains of the amino acids forming the active site may transfer H^+ to or from the substrate, destabilizing a covalent bond in the substrate and permitting it to break.

■ In *covalent catalysis*, a functional group in a side chain forms a temporary covalent bond with a portion of the substrate.

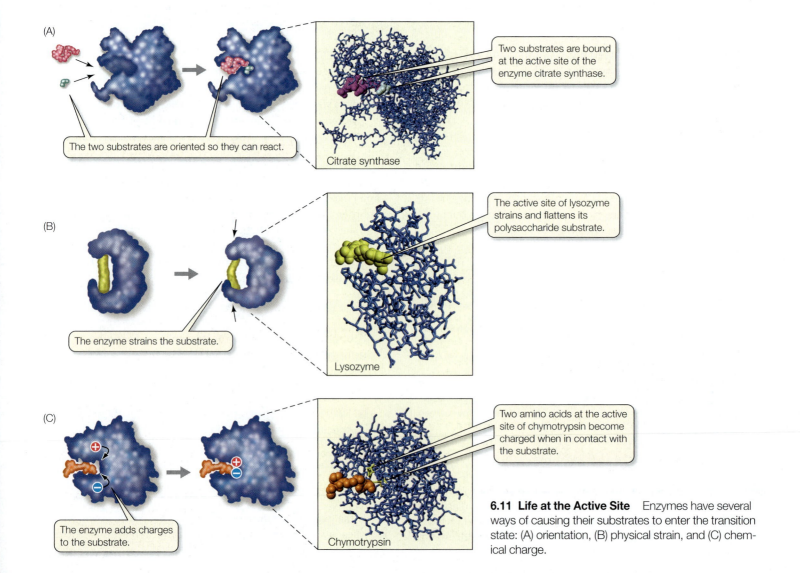

(A) The two substrates are oriented so they can react.

Two substrates are bound at the active site of the enzyme citrate synthase.

Citrate synthase

(B) The enzyme strains the substrate.

The active site of lysozyme strains and flattens its polysaccharide substrate.

Lysozyme

(C) The enzyme adds charges to the substrate.

Two amino acids at the active site of chymotrypsin become charged when in contact with the substrate.

Chymotrypsin

6.11 Life at the Active Site Enzymes have several ways of causing their substrates to enter the transition state: (A) orientation, (B) physical strain, and (C) chemical charge.

■ In *metal ion catalysis*, metal ions such as copper, iron, and manganese, which are firmly bound to side chains of the enzyme, can lose or gain electrons without detaching from the enzyme. This ability makes them important participants in oxidation–reduction reactions, which involve loss or gain of electrons.

Molecular structure determines enzyme function

Most enzymes are much larger than their substrates. An enzyme is typically a protein containing hundreds of amino acids and may consist of a single folded polypeptide chain or of several subunits. Its substrate is generally a small molecule. The active site of the enzyme is usually quite small, not more than 6–12 amino acids. Two questions arise from these observations:

■ What features of the active site allow it to recognize and bind the substrate?

■ What is the role of the rest of the huge protein?

THE ACTIVE SITE IS SPECIFIC TO THE SUBSTRATE The remarkable ability of an enzyme to select exactly the right substrate depends on a precise interlocking of molecular shapes and interactions of chemical groups at the active site. The binding of the substrate to the active site depends on the same kinds of forces that maintain the tertiary structure of the enzyme: hydrogen bonds, the attraction and repulsion of electrically charged groups, and hydrophobic interactions.

In 1894, the German chemist Emil Fischer compared the fit between an enzyme and its substrate to that of a lock and key. Fischer's model persisted for more than half a century with only indirect evidence to support it. The first direct evidence came in 1965, when David Phillips and his colleagues at the Royal Institution in London succeeded in crystallizing the enzyme lysozyme and determined its tertiary structure using the techniques of X-ray crystallography (described in Section 11.2). They observed a pocket in lysozyme that neatly fits its substrate (see Figure 6.11B).

AN ENZYME CHANGES SHAPE WHEN IT BINDS A SUBSTRATE As proteins, enzymes are not immutable structures. The three-dimensional shapes of many enzymes change when they bind to their substrates. These shape changes expose the active site (or sites) of the enzyme. A change in enzyme shape caused by substrate binding is called **induced fit**.

An example of induced fit can be seen in the enzyme hexokinase, which catalyzes the reaction

$$\text{glucose} + \text{ATP} \rightarrow \text{glucose 6-phosphate} + \text{ADP}$$

Induced fit brings reactive side chains from the hexokinase active site into alignment with the substrates (**Figure 6.12**), facilitating its catalytic mechanisms. Equally important, the folding of hexokinase to fit around the glucose substrate excludes water from the active site. This is essential, because the two molecules binding to the active site are glucose and ATP. If water were present, ATP could be hydrolyzed to ADP and phosphate. But since water is absent, the transfer of a phosphate from ATP to glucose is favored.

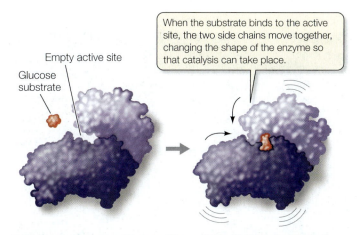

Glucose substrate

Empty active site

When the substrate binds to the active site, the two side chains move together, changing the shape of the enzyme so that catalysis can take place.

6.12 Some Enzymes Change Shape When Substrate Binds to Them Shape changes result in an induced fit between enzyme and substrate, improving the catalytic ability of the enzyme. Induced fit can be observed in the enzyme hexokinase seen with and without one of its substrates, glucose (its other substrate is ATP).

Induced fit at least partly explains why enzymes are so large. The rest of the macromolecule may have two roles:

■ It provides a framework so that the amino acids of the active site are properly positioned in relation to the substrate.

■ It participates in the small but significant changes in protein shape and structure that result in induced fit.

Some enzymes require other molecules in order to function

As large and complex as enzymes are, many of them require the presence of other, nonprotein molecular "partners" in order to function (**Table 6.1**):

TABLE 6.1

Some Examples of Nonprotein "Partners" of Enzymes

TYPE OF MOLECULE	ROLE IN CATALYZED REACTIONS
COFACTORS	
Iron (Fe^{2+} or Fe^{3+})	Oxidation/reduction
Copper (Cu^+ or Cu^{2+})	Oxidation/reduction
Zinc (Zn^{2+})	Helps bind NAD
COENZYMES	
Biotin	Carries $-COO^-$
Coenzyme A	Carries $-CO-CH_3$
NAD	Carries electrons
FAD	Carries electrons
ATP	Provides/extracts energy
PROSTHETIC GROUPS	
Heme	Binds ions, O_2, and electrons; contains iron cofactor
Flavin	Binds electrons
Retinal	Converts light energy

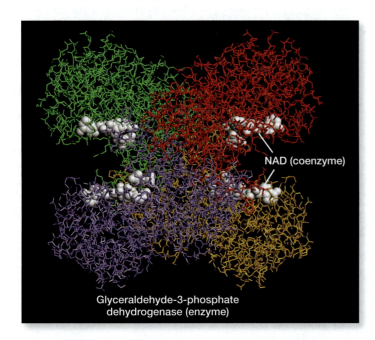

6.13 An Enzyme with a Coenzyme Some enzymes require coenzymes in order to function. This illustration shows the relative sizes of the four subunits (red, yellow, green, and purple) of the enzyme triose phosphate dehydrogenase and its coenzyme, NAD (white).

- **Prosthetic groups** are distinctive, non-amino acid atoms or molecular groupings that are permanently bound to their enzymes.

- **Cofactors** are inorganic ions such as copper, zinc, or iron that bind to certain enzymes and are essential to their function.

- **Coenzymes** are carbon-containing molecules that are required for the action of one or more enzymes. Coenzymes are usually relatively small compared with the enzyme to which they temporarily bind (**Figure 6.13**).

Prosthetic groups include, for example, the flavin nucleotides that are bound to the mitochondrial enzyme succinate dehydrogenase, which plays an important role in cellular respiration (see Section 7.2). Although it is not an enzyme, hemoglobin provides another example of a protein bound to a prosthetic group—in this case, heme (see Figure 3.9).

Coenzymes move from enzyme molecule to enzyme molecule, adding or removing chemical groups from the substrate. Coenzymes are like substrates in that they are not permanently bound to the enzyme, but must collide with the enzyme and bind to its active site. In addition, a coenzyme changes chemically during the reaction and then separates from the enzyme to participate in other reactions.

ATP and ADP can be considered coenzymes because they are necessary for some reactions, are changed by those reactions, and bind to and detach from the enzymes that catalyze those reactions. In the next chapter we will encounter other coenzymes that function in energy-harvesting reactions by accepting or donating electrons or hydrogen atoms. In animals, some coenzymes are produced from *vitamins*—substances that must be obtained from food because they cannot be synthesized by the body. For example, the B vitamin niacin is used to make the coenzyme NAD.

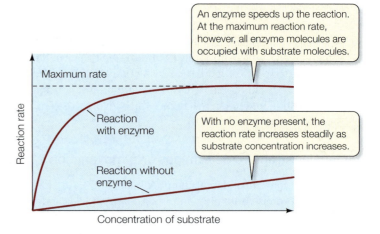

6.14 Catalyzed Reactions Reach a Maximum Rate Because there is usually less enzyme than substrate present, the reaction rate levels off when the enzyme becomes saturated.

Substrate concentration affects reaction rate

For a reaction of the type A → B, the rate of the uncatalyzed reaction is directly proportional to the concentration of A. The higher the concentration of substrate, the more reactions per unit of time. Addition of the appropriate enzyme speeds up the reaction, of course, but it also changes the shape of a plot of rate versus substrate concentration (**Figure 6.14**). At first, the rate of the enzyme-catalyzed reaction increases as the substrate concentration increases, but then it levels off. When further increases in the substrate concentration do not significantly increase the reaction rate, the maximum rate is attained.

Since the concentration of an enzyme is usually much lower than that of its substrate, what we are seeing is a *saturation* phenomenon like the one that occurs in facilitated diffusion (see Figure 5.12B). When all the enzyme molecules are bound to substrate molecules, the enzyme is working as fast as it can—at its maximum rate. Nothing is gained by adding more substrate, because no free enzyme molecules are left to act as catalysts.

The maximum rate of a catalyzed reaction can be used to measure how efficient the enzyme can be—that is, how many molecules of substrate are converted into product per unit of time when there is an excess of substrate present. This turnover number ranges from 1 molecule every 2 seconds for lysozyme to an amazing 40 million molecules per second for the liver enzyme catalase.

6.4 RECAP

Enzymes change during the reaction they catalyze, often participating in the reaction itself by temporarily changing shape or donating electrons. Some enzymes require cofactors, coenzymes, or prosthetic groups in order to function.

- Can you describe three mechanisms of enzyme catalysis? See p. 128 and Figure 3.11

- Do you understand the chemical role of coenzymes in enzymatic reactions? See p. 130

Now that we understand more about how enzymes work, let's see what makes it possible for the myriad enzymes in a complex organism, each specific to just one or a few reactions, to work together.

6.5 How Are Enzyme Activities Regulated?

A major characteristic of life is *homeostasis*, the maintenance of stable internal conditions (see Chapter 40). How a cell maintains a relatively constant internal environment while thousands of chemical reactions are going on at once is an amazing story. These chemical reactions are organized into *metabolic pathways* in which the product of one reaction is a reactant for the next. The pathway for the metabolism of alcohol in humans described at the opening of this chapter is just one of many such pathways that regulate the internal environment and provide for such diverse functions as the catabolism of glucose and the anabolism of amino acids. These pathways do not exist in isolation, but interact extensively. And *each* reaction in each pathway is catalyzed by a specific enzyme. Within a cell or organism, enzymes control whether and how much a metabolic pathway functions. If even one enzyme in the pathway is inactive, that step, and all steps subsequent to it, shut down.

The "flow" of chemicals (for example, a carbon atom) through these interacting pathways can be studied, but gets complicated quickly, since each pathway influences the others. Mathematical algorithms for the computer are used to model these pathways and show how they mesh in an interdependent system (**Figure 6.15**); such models can help predict what will happen if the concentration of one molecule or another is altered. This new field of biology, which has numerous applications, is called **systems biology**.

Regulation of the rates at which thousands of different enzymes operate contributes to homeostasis within an organism. In this section, we will investigate the role of enzymes in organizing and regulating metabolic pathways. In living cells, enzymes can be activated or inhibited in various ways, so the presence of an enzyme alone does not ensure that it is functioning. There are also mechanisms that alter the rate at which some enzymes catalyze reactions, making enzymes the target points at which entire sequences of chemical reactions can be regulated. Finally, we will examine how the environment—particularly temperature and pH—affects enzyme activity.

Enzymes can be regulated by inhibitors

Various **inhibitors** can bind to enzymes, slowing down the rates of enzyme-catalyzed reactions. Some inhibitors occur naturally in cells; others are artificial. Naturally occurring inhibitors regulate metabolism; artificial ones can be used to treat disease, to kill pests, or in the laboratory to study how enzymes work. Some inhibitors irreversibly inhibit the enzyme by permanently binding to it. Others have reversible effects; that is, they can become unbound from the enzyme. The removal of a natural reversible inhibitor increases an enzyme's rate of catalysis.

IRREVERSIBLE INHIBITION Some inhibitors covalently bond to certain side chains at the active site of an enzyme, thereby perma-

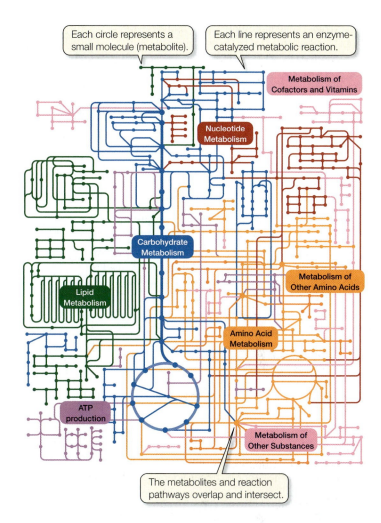

Each circle represents a small molecule (metabolite).

Each line represents an enzyme-catalyzed metabolic reaction.

Metabolism of Cofactors and Vitamins

Nucleotide Metabolism

Carbohydrate Metabolism

Lipid Metabolism

Metabolism of Other Amino Acids

Amino Acid Metabolism

ATP production

Metabolism of Other Substances

The metabolites and reaction pathways overlap and intersect.

6.15 Metabolic Pathways The complex interactions of metabolic pathways can be modeled by the tools of systems biology. In cells, the main elements controlling these pathways are enzymes.

nently inactivating the enzyme by destroying its capacity to interact with its normal substrate. An example of an irreversible inhibitor is DIPF (diisopropylphosphorofluoridate), which reacts with serine (**Figure 6.16**). DIPF is an irreversible inhibitor of acetylcholinesterase, whose operation is essential to the normal function of the nervous system. Because of their effect on acetylcholinesterase, DIPF and other similar compounds are classified as *nerve gases*. One of them, Sarin, was used in an attack on the Tokyo subway in 1995, resulting in a dozen deaths and hundreds hospitalized. The widely used insecticide malathion is a derivative of DIPF that inhibits only insect acetylcholinesterase, not the mammalian enzyme.

REVERSIBLE INHIBITION Not all inhibition is irreversible. Some inhibitors are similar enough to a particular enzyme's natural substrate to bind noncovalently to its active site, yet different enough that the enzyme catalyzes no chemical reaction. While such a molecule is bound to the enzyme, the natural substrate cannot enter the active site; thus the inhibitor effectively wastes the enzyme's time, preventing its catalytic action. Such molecules are called **com-**

6.16 Irreversible Inhibition DIPF forms a stable covalent bond with the side chain of the amino acid serine at the active site of the enzyme trypsin, thus irreversibly disabling the enzyme.

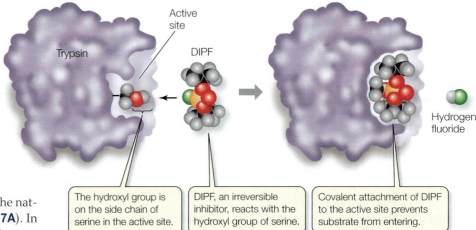

The hydroxyl group is on the side chain of serine in the active site.

DIPF, an irreversible inhibitor, reacts with the hydroxyl group of serine.

Covalent attachment of DIPF to the active site prevents substrate from entering.

petitive inhibitors because they compete with the natural substrate for the active site (**Figure 6.17A**). In these cases, the inhibition is reversible. When the concentration of the competitive inhibitor is reduced, it detaches from the active site, and the enzyme is active again.

Noncompetitive inhibitors bind to the enzyme at a site distinct from the active site. Their binding can cause a change in the shape of the enzyme that alters the active site (**Figure 6.17B**). In this case, the active site may still bind substrate molecules, but the rate of product formation may be reduced. Noncompetitive inhibitors, like competitive inhibitors, can become unbound, so their effects are reversible.

Frank (see p. 118) got sick from a single glass of champagne because he has nonfunctional aldehyde dehydrogenase (ALDH). A drug called Antabuse is a competitive inhibitor of ALDH. Antabuse is used to treat alcoholism, since it causes alcoholics to get sick if they consume alcohol.

Allosteric enzymes control their activity by changing their shape

The change in enzyme shape due to noncompetitive inhibitor binding is an example of **allostery** (*allo*, "different"; *stery*, "shape"). In this case, the binding of the inhibitor *induces* the protein to change its shape. More common are enzymes that already exist in the cell in more than one possible shape (**Figure 6.18**):

■ The *active* form of the enzyme has the proper shape for substrate binding.

■ The *inactive* form of the enzyme has a shape that cannot bind the substrate but can bind an inhibitor. Binding of an inhibitor to a site separate from the active site (i.e., the site where substrate binds) stabilizes the inactive form, and it becomes less likely to convert to the active form.

6.17 Reversible Inhibition (A) A competitive inhibitor binds temporarily to the active site of an enzyme. (B) A noncompetitive inhibitor binds temporarily to the enzyme at a site away from the active site. In both instances, the enzyme's function is disabled for only as long as the inhibitor remains bound.

■ The active form can be stabilized by binding of an *activator* to a third site on the enzyme.

Like substrate binding, the binding of inhibitors and activators is highly specific.

Most (but not all) enzymes that are allosterically regulated are proteins with quaternary structure; that is, they are made up of multiple polypeptide subunits. The active site is present on one such subunit, called the **catalytic subunit**, while the regulatory site(s)

(A) Competitive inhibition

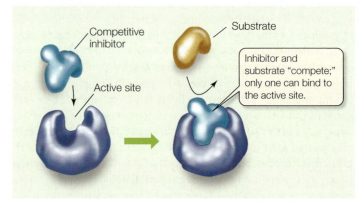

Competitive inhibitor

Substrate

Active site

Inhibitor and substrate "compete;" only one can bind to the active site.

(B) Noncompetitive inhibition

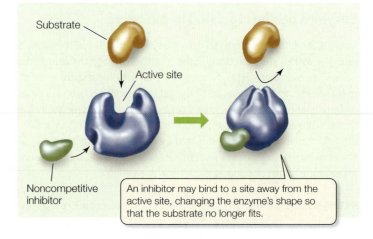

Substrate

Active site

Noncompetitive inhibitor

An inhibitor may bind to a site away from the active site, changing the enzyme's shape so that the substrate no longer fits.

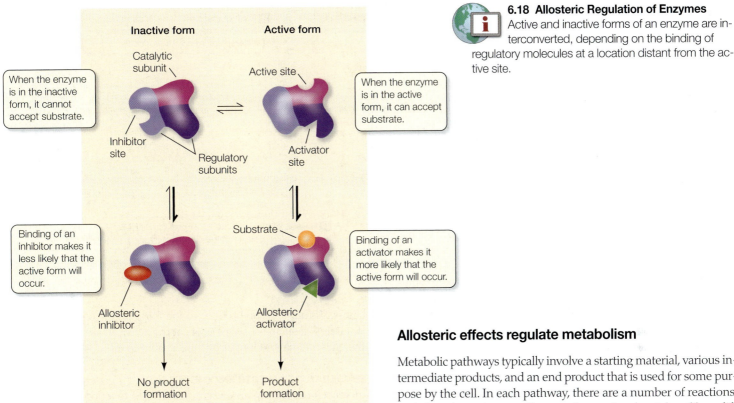

Inactive form   **Active form**

When the enzyme is in the inactive form, it cannot accept substrate.

When the enzyme is in the active form, it can accept substrate.

Binding of an inhibitor makes it less likely that the active form will occur.

Binding of an activator makes it more likely that the active form will occur.

No product formation   Product formation

6.18 Allosteric Regulation of Enzymes Active and inactive forms of an enzyme are interconverted, depending on the binding of regulatory molecules at a location distant from the active site.

for activators and/or inhibitors are on different polypeptide sequences, called **regulatory subunits**.

Allosteric enzymes and nonallosteric enzymes differ greatly in their reaction rates when the substrate concentration is low. Graphs of reaction rate plotted against substrate concentration show this relationship. For a nonallosteric enzyme, the plot looks like that in **Figure 6.19A**. The reaction rate first increases very sharply with increasing substrate concentration, then tapers off to a constant maximum rate as the supply of enzyme becomes saturated with substrate. The plot for many allosteric enzymes is radically different, having a *sigmoid* (S-shaped) appearance (**Figure 6.19B**). The increase in reaction rate with increasing substrate concentration is slight at low substrate concentrations, but within a certain range, the reaction rate is extremely sensitive to relatively small changes in substrate concentration. Because of this sensitivity, allosteric enzymes are important in regulating entire metabolic pathways.

Allosteric effects regulate metabolism

Metabolic pathways typically involve a starting material, various intermediate products, and an end product that is used for some purpose by the cell. In each pathway, there are a number of reactions, each forming an intermediate product and each catalyzed by a different enzyme. The first step in a pathway is called the **commitment step**, meaning that once this enzyme-catalyzed reaction occurs, the "ball is rolling," and the other reactions happen in sequence, leading to the end product. But what if the cell has no need for that product—for example, if that product is available from its environment in adequate amounts? It would be energetically wasteful for the cell to continue making something it does not need.

One way that cells solve this problem is to shut down the metabolic pathway by having the final product allosterically inhibit the enzyme that catalyzes the commitment step (**Figure 6.20**). This mechanism is known as **feedback inhibition** or **end-product inhibition**. When the end product is present at a high concentration, some of it binds to an allosteric site on the commitment step enzyme, thereby causing it to become inactive. We will describe many other examples of such allosteric interactions in later chapters.

Enzymes are affected by their environment

Enzymes enable cells to perform chemical reactions and carry out complex processes rapidly without using the extremes of temperature and pH employed by chemists in the laboratory. However, because of their three-dimensional structures and the chemistry of the side chains in their active sites, enzymes are highly sensitive to temperature and pH. We described the general effects of these environmental factors on proteins in Section 3.2. Here

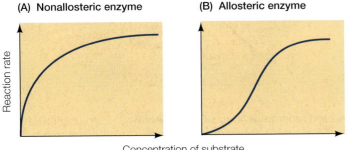

(A) Nonallosteric enzyme   **(B) Allosteric enzyme**

Reaction rate

Concentration of substrate

6.19 Allostery and Reaction Rate How the rate of an enzyme-catalyzed reaction changes with increasing substrate concentration depends on whether the enzyme is allosterically regulated.

6.20 Feedback Inhibition of Metabolic Pathways

The commitment step is catalyzed by an allosteric enzyme that can be inhibited by the end product of the pathway. The specific pathway shown here is the synthesis of isoleucine, an amino acid, from threonine in bacteria. This pathway is typical of many enzyme-catalyzed biological reactions.

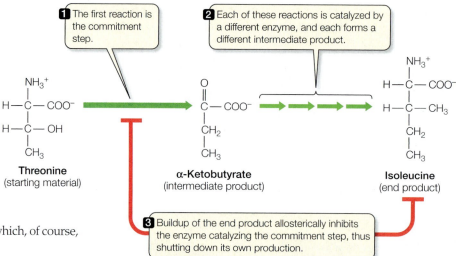

1 The first reaction is the commitment step.

2 Each of these reactions is catalyzed by a different enzyme, and each forms a different intermediate product.

Threonine
(starting material)

α-Ketobutyrate
(intermediate product)

Isoleucine
(end product)

3 Buildup of the end product allosterically inhibits the enzyme catalyzing the commitment step, thus shutting down its own production.

we will examine their effects on enzyme function, which, of course, depends on enzyme structure and chemistry.

pH AFFECTS ENZYME ACTIVITY The rates of most enzyme-catalyzed reactions depend on the pH of the solution in which they occur. Each enzyme is most active at a particular pH; its activity decreases as the solution is made more acidic or more basic than its "ideal" (optimal) pH (**Figure 6.21**).

Several factors contribute to this effect. One is the ionization of carboxyl, amino, and other groups on either the substrate or the enzyme. In neutral or basic solutions, carboxyl groups (—COOH) release H^+ to become negatively charged carboxylate groups (—COO⁻). Similarly, amino groups (—NH₂) accept H^+ in neutral or acidic solutions, becoming positively charged —NH_3^+ groups (see the discussion of acids and bases in Section 2.4). Thus, in a neutral solution, a molecule with an amino group is attracted electrically to another molecule that has a carboxyl group, because both groups are ionized and the two groups have opposite charges. If the pH changes, however, the ionization of these groups may change. For example, at a low pH (high H^+ concentration), the excess H^+ may react with —COO⁻ to form COOH. If this happens, the group is no longer charged and cannot interact with other

charged groups in the protein, so the folding of the protein may be altered. If such a change occurs at the active site of an enzyme, the enzyme may no longer be able to bind to its substrate.

TEMPERATURE AFFECTS ENZYME ACTIVITY In general, warming increases the rate of an enzyme-catalyzed reaction because at higher temperatures, a greater proportion of the reactant molecules have enough kinetic energy to provide the activation energy for the reaction (**Figure 6.22**). Temperatures that are too high, however, inactivate enzymes, because at high temperatures enzyme molecules vibrate and twist so rapidly that some of their noncovalent bonds break. When heat changes their tertiary structure, enzymes become denatured and lose their function. Some enzymes denature at temperatures only slightly above that of the human body, but a few are stable even at the boiling or freezing points of water. All enzymes, however, have an optimal temperature for activity.

Individual organisms adapt to changes in the environment in many ways, one of which is based on groups of enzymes, called **isozymes**, that catalyze the same reaction but have different chemical compositions and physical properties. Different isozymes

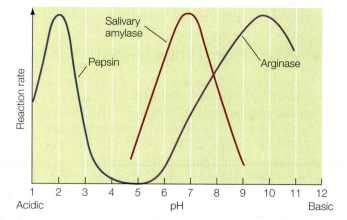

6.21 pH Affects Enzyme Activity Each enzyme catalyzes its reaction at a maximum rate at a particular pH. The activity curves peak at the pH at which each enzyme is most effective. For example, pepsin is a protease active in the acidic environment of the stomach.

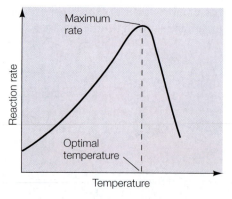

6.22 Temperature Affects Enzyme Activity Each enzyme is most active at a particular optimal temperature. At higher temperatures, denaturation reduces the enzyme's activity.

within a given group may have different optimal temperatures. The rainbow trout, for example, has several isozymes of the enzyme acetylcholinesterase. If a rainbow trout is transferred from warm water to near-freezing water (2°C), the fish produces an isozyme of acetylcholinesterase that is different from the one it produces at the higher temperature. The new isozyme has a lower optimal temperature, allowing the fish's nervous system to perform normally in the colder water.

In general, enzymes adapted to warm temperatures fail to denature at those temperatures because their tertiary structures are held together largely by covalent bonds, such as disulfide bridges, instead of the more heat-sensitive weak chemical interactions. Most enzymes in humans are more stable at high temperatures than those of the bacteria that infect us, so that a moderate fever tends to denature bacterial enzymes, but not our own.

6.5 RECAP

The rates of most enzyme-catalyzed reactions are affected by chemicals (such as inhibitors) and by environmental factors (such as temperature and pH).

- Can you describe the difference between reversible and irreversible inhibitors of enzymes? See p. 131

- How are allosteric enzymes regulated? See p. 131–133 and Figure 6.18

- Can you explain the concept of feedback inhibition? Can you envision how the reactions shown in Figure 6.20 might fit into a systems diagram such as Figure 6.15?

CHAPTER SUMMARY

6.1 What physical principles underlie biological energy transformations?

Energy is the capacity to do work. In a biological system, the usable energy is called **free energy** (**G**). The unusable energy is **entropy**, a measure of the disorder of the system.

Potential energy is the energy of state or position; it includes the energy stored in chemical bonds. **Kinetic energy** is the energy of motion; it is the type of energy that can do work.

The **laws of thermodynamics** apply to living organisms. The first law states that energy cannot be created or destroyed. The second law states that energy transformations decrease the amount of energy available to do work (free energy) and increase disorder. Review Figure 6.2

The **change in free energy** (**ΔG**) of a reaction determines its point of chemical equilibrium, at which the forward and reverse reactions proceed at the same rate.

Exergonic reactions release free energy and have a negative ΔG. **Endergonic reactions** consume or require free energy and have a positive ΔG. Endergonic reactions proceed only if free energy is provided. Review Figure 6.3

Metabolism is the sum total of the biochemical (metabolic) reactions in an organism. **Catabolic reactions** are associated with breakdown of complex molecules and release energy (are exergonic). **Anabolic reactions** build complexity in the cell and are endergonic.

6.2 What is the role of ATP in biochemical energetics?

ATP (**adenosine triphosphate**) serves as an energy currency in cells. Hydrolysis of ATP releases a relatively large amount of free energy.

The ATP cycle couples exergonic and endergonic reactions, transferring free energy from the exergonic to the endergonic reaction. Review Figure 6.6, Web/CD Activity 6.1

6.3 What are enzymes?

The **rate** of a chemical reaction is independent of ΔG, but is determined by the **energy barrier**. **Enzymes** are protein catalysts that affect the rates of biological reactions by lowering the energy barrier, supplying the **activation energy** needed to initiate a reaction.

Substrates bind to the enzyme's **active site**—the site of **catalysis**—forming an **enzyme–substrate complex**. Enzymes are highly specific for their substrates.

At the active site, a substrate can be oriented correctly, chemically modified, or strained. As a result, the substrate readily forms its **transition state**, and the reaction proceeds. Review Figure 6.9

6.4 How do enzymes work?

Binding substrate causes many enzymes to change shape, exposing their active site(s) and allowing catalysis. The change in enzyme shape caused by substrate binding is known as **induced fit**. Review Figure 6.10, Web/CD Activity 6.2

Some enzymes require other substances, known as **cofactors**, to carry out catalysis. **Prosthetic groups** are permanently bound to the enzyme; **coenzymes** are not. Coenzymes can be considered substrates, as they are changed by the reaction and then released from the enzyme.

Substrate concentration affects the rate of an enzyme-catalyzed reaction.

6.5 How are enzyme activities regulated?

Metabolism is organized into pathways in which the product of one reaction is a reactant for the next reaction. Each reaction in the pathway is catalyzed by an enzyme.

Enzyme activity is subject to regulation. Some inhibitors react irreversibly with enzymes. Others react reversibly. See Web/CD Tutorial 6.1

Allosteric regulators bind to a site different from the active site and stabilize the active or inactive form of an enzyme. Review Figure 6.18, Web/CD Tutorial 6.2

The end product of a metabolic pathway may inhibit the allosteric enzyme that catalyzes the **commitment step** of that pathway. Review Figure 6.20

Enzymes are sensitive to their environment. Both pH and temperature affect enzyme activity. Review Figures 6.21 and 6.22

SELF-QUIZ

1. Coenzymes differ from enzymes in that coenzymes are
   a. only active outside the cell.
   b. polymers of amino acids.
   c. smaller molecules, such as vitamins.
   d. specific for one reaction.
   e. always carriers of high-energy phosphate.

2. Which statement about thermodynamics is true?
   a. Free energy is used up in an exergonic reaction.
   b. Free energy cannot be used to do work.
   c. The total amount of energy can change after a chemical transformation.
   d. Free energy can be kinetic but not potential energy.
   e. Entropy has a tendency to increase.

3. In a chemical reaction,
   a. the rate depends on the value of ΔG.
   b. the rate depends on the activation energy.
   c. the entropy change depends on the activation energy.
   d. the activation energy depends on the value of ΔG.
   e. the change in free energy depends on the activation energy.

4. Which statement about enzymes is *not* true?
   a. They usually consist of proteins.
   b. They change the rate of the catalyzed reaction.
   c. They change the ΔG of the reaction.
   d. They are sensitive to heat.
   e. They are sensitive to pH.

5. The active site of an enzyme
   a. never changes shape.
   b. forms no chemical bonds with substrates.
   c. determines, by its structure, the specificity of the enzyme.
   d. looks like a lump projecting from the surface of the enzyme.
   e. changes the ΔG of the reaction.

6. The molecule ATP is
   a. a component of most proteins.
   b. high in energy because of the presence of adenine.
   c. required for many energy-producing biochemical reactions.
   d. a catalyst.
   e. used in some endergonic reactions to provide energy.

7. In an enzyme-catalyzed reaction,
   a. a substrate does not change.
   b. the rate decreases as substrate concentration increases.
   c. the enzyme can be permanently changed.
   d. strain may be added to a substrate.
   e. the rate is not affected by substrate concentration.

8. Which statement about enzyme inhibitors is *not* true?
   a. A competitive inhibitor binds the active site of the enzyme.
   b. An allosteric inhibitor binds a site on the active form of the enzyme.
   c. A noncompetitive inhibitor binds a site other than the active site.
   d. Noncompetitive inhibition cannot be completely overcome by the addition of more substrate.
   e. Competitive inhibition can be completely overcome by the addition of more substrate.

9. Which statement about feedback inhibition of enzymes is *not* true?
   a. It is exerted through allosteric effects.
   b. It is directed at the enzyme that catalyzes the first committed step in a metabolic pathway.
   c. It affects the rate of reaction, not the concentration of enzyme.
   d. It acts very slowly.
   e. It is an example of irreversible inhibition.

10. Which statement about temperature effects is *not* true?
    a. Raising the temperature may reduce the activity of an enzyme.
    b. Raising the temperature may increase the activity of an enzyme.
    c. Raising the temperature may denature an enzyme.
    d. Some enzymes are stable at the boiling point of water.
    e. All enzymes have the same optimal temperature.

FOR DISCUSSION

1. What makes it possible for endergonic reactions to proceed in organisms?

2. Consider two proteins: one is an enzyme dissolved in the cytosol of a cell, the other is an ion channel in its plasma membrane. Contrast the structures of the two proteins, indicating at least two important differences.

3. Plot free energy versus the course of an endergonic reaction and that of an exergonic reaction. Include the activation energy in both plots. Label E_a and ΔG on both graphs.

4. Consider an enzyme that is subject to allosteric regulation. If a competitive inhibitor (not an allosteric inhibitor) is added to a solution containing such an enzyme, the ratio of enzyme molecules in the active form to those in the inactive form increases. Explain this observation.

FOR INVESTIGATION

In humans, hydrogen peroxide (H_2O_2) is a dangerous toxin produced as a by-product of several metabolic pathways. The accumulation of H_2O_2 is prevented by its conversion to harmless H_2O, a reaction catalyzed by the appropriately named enzyme, catalase. Air pollutants can inhibit this enzyme and leave individuals susceptible to tissue damage by H_2O_2. How would you investigate whether catalase has an allosteric or a nonallosteric mechanism, and whether the pollutants are acting as competitive or noncompetitive inhibitors?

7 Pathways That Harvest Chemical Energy

Of mice and marathons

Like success in your biology course, winning a prestigious marathon comes only after a lot of hard work. The leg muscles of elite distance runners contain more mitochondria than most of us have. The chemical energy released by the hydrolysis of ATP in those mitochondria can be converted into mechanical energy to move the muscles.

The cells of muscle tissue associate into two types of muscle fibers. Most people have about equal proportions of each type. But in top marathon racers, 90 percent of the body's muscle is made up of so-called *slow-twitch* fibers. Cells of these fibers have lots of mitochondria and use oxygen to break down fats and carbohydrates, forming ATP. In contrast, the muscles of sprinters are about 80 percent *fast-twitch* fibers, which have fewer mitochondria. Fast-twitch fibers generate short bursts of ATP in the absence of O_2, but the ATP is soon used up. Extensive research with athletes has shown that training can improve the efficiency of the blood circulation to the muscle fibers, providing more oxygen, and even change the ratio of fast-twitch to slow-twitch fibers.

Now enter Marathon Mouse. No, this is not a cartoon character or a computer game, but a very real mouse that was genetically programmed by Ron Evans at the Salk Institute to express high levels of the protein PPARδ in its muscles. This protein normally controls the breakdown of fat in fat tissues, and it is also present in slow-twitch muscles, where it stimulates controlled fat breakdown to yield ATP. Evans's mouse was supposed to break down fats better, and thus be leaner—but there was an unexpected bonus. With high levels of PPARδ came an increase in slow-twitch fibers and a decrease in fast-twitch ones. It was as if the mouse had been in marathon training for a long time!

Marathon mice are leaner and meaner than ordinary mice. Leaner, because they are good at burning fat; and meaner in terms of their ability to run long distances. On an exercise wheel, a normal mouse can run for 90 minutes and about a half-mile (900 meters) before it gets tired. PPARδ-enhanced mice can run almost twice as long and twice as far—marks of true distance runners. Could we also manipulate genes to enhance performance (and fat burning) in humans?

The genetic engineering of people, if it is feasible, is probably far in the future. But implanting genetically altered muscle tissue is actually not such a farfetched idea, and has already raised concerns over improper athletic en-

Marathon Men It takes a lot of training to run a marathon. One of the results of all that training is that the leg muscles become packed with slow-twitch muscle fibers with cells rich in energy-metabolizing mitochondria.

Marathon Mouse This mouse can run for much longer than a normal mouse because its energy metabolism has been genetically altered.

hancement. More likely in the near term is the use of an experimental drug called GW501516 (developed by the drug company Glaxo-Smith/Kline), which activates the PPARδ protein. When Evans and colleagues gave the drug to normal mice, they achieved the same results as with the genetically modified mice. Because this drug stimulates fat breakdown, it is being tested to treat obesity.

The energy stored in ATP is the energy you use all the time to fuel both conscious actions, like running a marathon or turning the pages of this book, and your body's automatic actions, such as breathing or contracting your heart muscles.

IN THIS CHAPTER we will discover how cells produce usable energy, usually in the form of ATP. We will describe the metabolic pathway by which glucose is oxidized both in the presence and the absence of O_2. We close the chapter with an overview of the relationships between the metabolic pathways that use and produce the four biologically important classes of molecules—carbohydrates, fats, proteins, and nucleic acids.

7.1 How Does Glucose Oxidation Release Chemical Energy?

Fuels are molecules whose stored energy can be released for use. Wood burning in a campfire releases energy as heat and light. In cells, chemical fuels release chemical energy that is used to make ATP, which in turn can be used to drive endergonic reactions.

Photosynthetic organisms use energy from sunlight to synthesize their own fuels, as we will describe in Chapter 8. In nonphotosynthesizers, however, the most common chemical fuel is the sugar glucose ($C_6H_{12}O_6$). Other molecules, such as fats or proteins, can also supply energy, but to release it they must be converted either into glucose or into intermediate compounds in the various pathways of glucose metabolism.

In this section we explore how cells obtain energy from glucose by the chemical process of oxidation, which is carried out through a series of metabolic pathways. Several principles, some of which we noted in Section 6.5, govern metabolic pathways:

- Complex chemical transformations in the cell occur in a series of separate reactions that form a metabolic pathway.
- Each reaction in a metabolic pathway is catalyzed by a specific enzyme.
- Metabolic pathways are similar in all organisms, from bacteria to humans.
- Many metabolic pathways are compartmentalized in eukaryotes, with certain reactions occurring inside a specific organelle.
- Each metabolic pathway is regulated by key enzymes that can be inhibited or activated, thereby determining how fast the reactions will go.

Cells trap free energy while metabolizing glucose

As we saw in Section 2.3, the familiar process of combustion (burning) is very similar to the chemical processes that release energy in cells. If glucose is burned in a flame, it reacts with oxygen gas (O_2), forming carbon dioxide and water and releasing energy in the form of heat. The balanced equation for this combustion reaction is

$$C_6H_{12}O_6 + 6\,O_2 \rightarrow 6\,CO_2 + 6\,H_2O + \text{free energy}$$

The same equation applies to the metabolism of glucose in cells. The principles of metabolism cited above also apply to this process: the metabolism of glucose is a multistep pathway; each step is catalyzed by an enzyme; the process is compartmentalized; and the pathway is under enzymatic control.

The glucose metabolism pathway "traps" the stored energy of glucose in ATP molecules via the reaction

$$ADP + P_i + \text{free energy} \rightarrow ATP$$

The energy trapped in ATP can be used to do cellular work, such as movement of muscles or active transport across a membrane, just as heat energy captured from combustion can be used to do work.

The change in free energy (ΔG) resulting from the complete conversion of glucose and O_2 to CO_2 and water, whether by combustion or by metabolism, is −686 kcal/mol (−2,870 kJ/mol). Thus the overall reaction is highly exergonic and can drive the endergonic formation of a great deal of ATP from ADP and phosphate. The many steps of glucose metabolism allow this energy to be captured in ATP.

Three metabolic processes play important roles in the harvesting of energy from glucose: *glycolysis, cellular respiration,* and *fermentation* (**Figure 7.1**). All three involve metabolic pathways made up of many distinct chemical reactions.

- **Glycolysis** begins glucose metabolism in all cells and produces two molecules of the three-carbon product *pyruvate*. A small amount of the energy stored in glucose is captured in usable forms. Glycolysis does not use O_2.

- **Cellular respiration** uses O_2 (thus it is **aerobic**) from the environment and completely converts each pyruvate molecule into three molecules of CO_2 through a set of metabolic pathways. In the process, a great deal of the energy stored in the covalent bonds of pyruvate is released and transferred to ADP and phosphate to form ATP.

- **Fermentation** does not involve O_2 (it is **anaerobic**). Fermentation converts pyruvate into lactic acid or ethyl alcohol (ethanol), which are still relatively energy-rich molecules. Because the breakdown of glucose is incomplete, much less energy is released by fermentation than by cellular respiration.

An overview: Harvesting energy from glucose

The energy-harvesting processes in cells use different combinations of metabolic pathways depending on the presence or absence of O_2:

- When O_2 is available as the final electron acceptor, four pathways operate (**Figure 7.2A**). Glycolysis takes place first, and is followed by the three pathways of cellular respiration: **pyruvate oxidation**, the **citric acid cycle** (also called the *Krebs cycle* or the *tricarboxylic acid cycle*), and the **electron transport chain** (also called the *respiratory chain*).

- When O_2 is unavailable, pyruvate oxidation, the citric acid cycle, and the electron transport chain do not function, and the

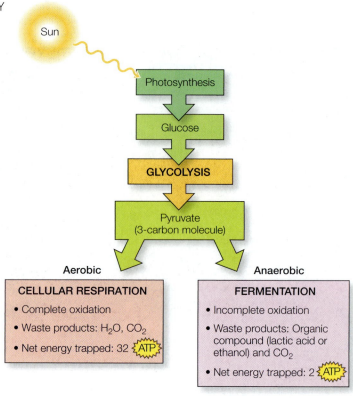

7.1 Energy for Life Living organisms obtain their energy from the food compounds produced by photosynthesis. They convert these compounds into glucose, which they metabolize by glycolysis to produce the 3-carbon compound pyruvate. Pyruvate molecules are then metabolized by either anaerobic fermentation or aerobic cellular respiration. The net result is energy "trapped" in ATP molecules that power the activities of living cells.

pyruvate produced by glycolysis is further metabolized by fermentation (**Figure 7.2B**).

These five metabolic pathways, which we will consider one at a time, occur in different locations in the cell (**Table 7.1**).

Redox reactions transfer electrons and energy

As Section 6.2 described, the addition of a phosphate group to ADP to make ATP is an endergonic reaction that can extract and store energy from exergonic reactions. Another way of transferring energy is to transfer electrons. A reaction in which one substance transfers one or more electrons to another substance is called an *oxidation–reduction reaction*, or **redox reaction**.

- **Reduction** is the gain of one or more electrons by an atom, ion, or molecule.

- **Oxidation** is the loss of one or more electrons.

Although oxidation and reduction are always defined in terms of traffic in electrons, we may also think in these terms when hydrogen atoms (*not* hydrogen ions) are gained or lost, because transfers of hydrogen atoms involve transfers of electrons (H = H⁺ + e⁻). Thus, when a molecule loses hydrogen atoms, it becomes oxidized.

Oxidation and reduction *always occur together*: as one material is oxidized, the electrons it loses are transferred to another material, reducing that material. In a redox reaction, we call the reactant that becomes reduced an *oxidizing agent* and the one that be-

(A) Glycolysis and cellular respiration

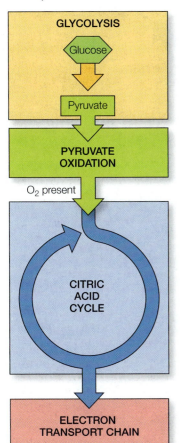

GLYCOLYSIS

Glucose

Pyruvate

PYRUVATE OXIDATION

O₂ present

CITRIC ACID CYCLE

ELECTRON TRANSPORT CHAIN

(B) Glycolysis and fermentation

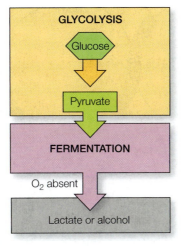

GLYCOLYSIS

Glucose

Pyruvate

FERMENTATION

O₂ absent

Lactate or alcohol

 7.2 Energy-Producing Metabolic Pathways Energy-producing reactions can be grouped into five metabolic pathways: glycolysis, pyruvate oxidation, the citric acid cycle, the electron transport chain, and fermentation. (A) The three lower pathways occur only in the presence of O₂ and are collectively referred to as cellular respiration. (B) When O₂ is unavailable, glycolysis is followed by fermentation.

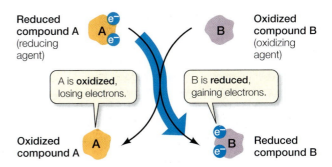

7.3 Oxidation and Reduction Are Coupled In a redox reaction, reactant A is oxidized and reactant B is reduced. In the process, A loses electrons and B gains electrons. Protons may be transferred along with electrons so that what is actually transferred (wide blue arrow) are hydrogen atoms: $AH_2 + B \rightarrow A + BH_2$.

comes oxidized a *reducing agent* (**Figure 7.3**). In both the combustion and the metabolism of glucose, glucose is the reducing agent (electron donor) and O₂ is the oxidizing agent (electron acceptor).

In a redox reaction, energy is transferred. Much of the energy originally present in the reducing agent becomes associated with the reduced product. (The rest remains in the reducing agent or

is lost.) As we will see, some of the key reactions of glycolysis and cellular respiration are highly exergonic redox reactions.

The coenzyme NAD is a key electron carrier in redox reactions

Section 6.4 described the role of coenzymes, small molecules that assist in enzyme-catalyzed reactions. ADP acts as a coenzyme when it picks up energy released in an exergonic reaction and uses it to form ATP (an endergonic reaction). In a similar fashion, the coenzyme **NAD** (**nicotinamide adenine dinucleotide**) acts as an electron carrier, in this case in redox reactions (**Figure 7.4A**).

NAD exists in two chemically distinct forms, one oxidized (NAD⁺) and the other reduced (NADH + H⁺) (**Figure 7.4B**). Both forms participate in biological redox reactions. The reduction reaction

$$NAD^+ + 2\,H \rightarrow NADH + H^+$$

is formally equivalent to the transfer of two hydrogen atoms (2 H⁺ + 2 e⁻). However, what is actually transferred is a *hydride ion* (H⁻, which is a proton and two electrons), leaving a free proton (H⁺).

 TABLE 7.1

Cellular Locations for Energy Pathways in Eukaryotes and Prokaryotes

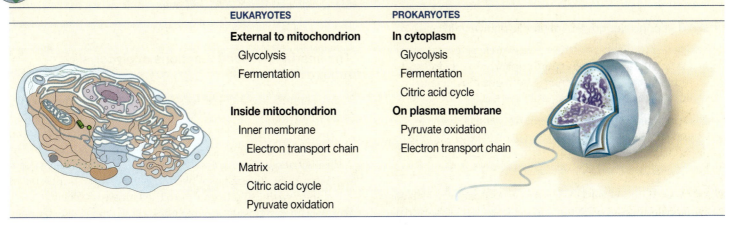

EUKARYOTES	PROKARYOTES
External to mitochondrion	**In cytoplasm**
Glycolysis	Glycolysis
Fermentation	Fermentation
	Citric acid cycle
Inside mitochondrion	**On plasma membrane**
Inner membrane	Pyruvate oxidation
Electron transport chain	Electron transport chain
Matrix	
Citric acid cycle	
Pyruvate oxidation	

7.4 NAD Is an Energy Carrier in Redox Reactions

Thanks to its ability to carry free energy and electrons, NAD is a major energy carrier in redox reactions and a universal energy intermediary in cells. (A) Each curved black arrow represents either an oxidation or reduction reaction. The wide blue arrow shows the path of transferred electrons (compare with Figure 7.2). (B) NAD$^+$ is the oxidized form and NADH the reduced form of NAD$^+$. The unshaded portion of the molecule (left) remains unchanged by the redox reaction.

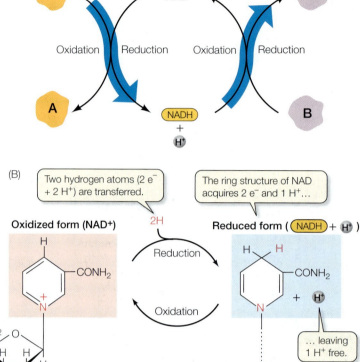

This notation emphasizes that reduction is accomplished by the addition of electrons.

Oxygen is highly electronegative and readily accepts electrons from NADH. The oxidation of NADH + H$^+$ by O$_2$,

$$NADH + H^+ + \tfrac{1}{2} O_2 \rightarrow NAD^+ + H_2O$$

is highly exergonic, with a ΔG of –52.4 kcal/mol (–219 kJ/mol). Note that the oxidizing agent appears here as "½ O$_2$" instead of "O." This notation emphasizes that it is molecular oxygen, O$_2$, that acts as the oxidizing agent.

Just as a molecule of ATP can be thought of as bundling or packaging about 12 kcal/mol (50 kJ/mol) of free energy, NAD can be thought of as packaging free energy in larger bundles (approximately 50 kcal/mol, or 200 kJ/mol). NAD is a common, but not the only, electron carrier in cells. As we will see, another carrier, **FAD (flavin adenine dinucleotide)**, also transfers electrons during glucose metabolism.

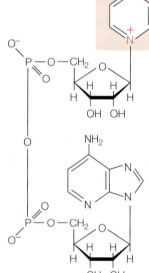

tion of those pathways with a look at the three that begin the process when O$_2$ is available as an electron acceptor: glycolysis, pyruvate oxidation, and the citric acid cycle.

7.2 What Are the Aerobic Pathways of Glucose Metabolism?

Glycolysis takes place in the cytosol of cells. It converts glucose into pyruvate, produces a small amount of energy, and generates no CO$_2$. In glycolysis, some of the covalent bonds between carbon and hydrogen in the glucose molecule are oxidized, releasing some of the energy stored in this carbohydrate. After 10 enzyme-catalyzed reactions, the end products of glycolysis are 2 molecules of **pyruvate** (pyruvic acid), 4 molecules of ATP, and 2 molecules of NADH. Glycolysis can be divided into two stages: energy-investing reactions that use ATP, and energy-harvesting reactions that produce ATP (**Figure 7.5**).

The energy-investing reactions of glycolysis require ATP

Using Figure 7.5 as a guide, let's work our way through the glycolytic pathway.

The first five reactions of glycolysis are *endergonic*; that is, the cell is investing free energy in the glucose molecule, rather than releasing energy from it. In two separate reactions (*reactions 1 and 3* in Figure 7.5), the energy of two molecules of ATP is invested in attaching two phosphate groups to the glucose molecule to form fructose 1,6-bisphosphate, which has a free energy substantially

---

7.1 RECAP

When glucose is oxidized, the free energy released is trapped in ATP. The five metabolic pathways of the cell combine in different ways to produce ATP, which supplies the energy for myriad other reactions in living cells.

- What principles govern metabolic pathways in cells? See p. 139

- Do you understand how the coupling of oxidation and reduction can transfer energy from one molecule to another? See pp. 140–141 and Figure 7.2

- Do you understand the roles of NAD and O$_2$ with respect to electrons in a redox reaction? See pp. 141–142 and Figure 7.3

Now that we have an overview of the metabolic pathways that harvest energy from glucose, let's begin our more detailed examina-

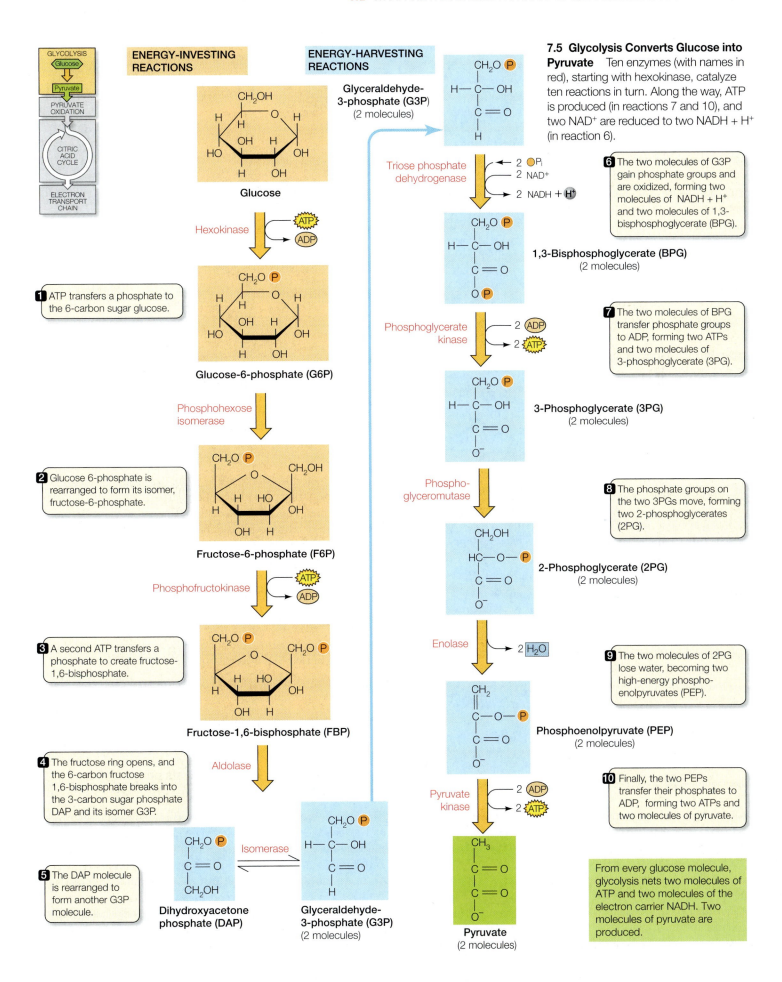

7.5 Glycolysis Converts Glucose into Pyruvate Ten enzymes (with names in red), starting with hexokinase, catalyze ten reactions in turn. Along the way, ATP is produced (in reactions 7 and 10), and two NAD$^+$ are reduced to two NADH + H$^+$ (in reaction 6).

ENERGY-INVESTING REACTIONS

ENERGY-HARVESTING REACTIONS

GLYCOLYSIS
Glucose
Pyruvate
PYRUVATE OXIDATION
CITRIC ACID CYCLE
ELECTRON TRANSPORT CHAIN

Glucose

Hexokinase

1 ATP transfers a phosphate to the 6-carbon sugar glucose.

Glucose-6-phosphate (G6P)

Phosphohexose isomerase

2 Glucose 6-phosphate is rearranged to form its isomer, fructose-6-phosphate.

Fructose-6-phosphate (F6P)

Phosphofructokinase

3 A second ATP transfers a phosphate to create fructose-1,6-bisphosphate.

Fructose-1,6-bisphosphate (FBP)

Aldolase

4 The fructose ring opens, and the 6-carbon fructose 1,6-bisphosphate breaks into the 3-carbon sugar phosphate DAP and its isomer G3P.

5 The DAP molecule is rearranged to form another G3P molecule.

Dihydroxyacetone phosphate (DAP)

Isomerase

Glyceraldehyde-3-phosphate (G3P)
(2 molecules)

Glyceraldehyde-3-phosphate (G3P)
(2 molecules)

Triose phosphate dehydrogenase
2 P$_i$
2 NAD$^+$
2 NADH + H$^+$

6 The two molecules of G3P gain phosphate groups and are oxidized, forming two molecules of NADH + H$^+$ and two molecules of 1,3-bisphosphoglycerate (BPG).

1,3-Bisphosphoglycerate (BPG)
(2 molecules)

Phosphoglycerate kinase
2 ADP
2 ATP

7 The two molecules of BPG transfer phosphate groups to ADP, forming two ATPs and two molecules of 3-phosphoglycerate (3PG).

3-Phosphoglycerate (3PG)
(2 molecules)

Phospho-glyceromutase

8 The phosphate groups on the two 3PGs move, forming two 2-phosphoglycerates (2PG).

2-Phosphoglycerate (2PG)
(2 molecules)

Enolase
2 H$_2$O

9 The two molecules of 2PG lose water, becoming two high-energy phospho-enolpyruvates (PEP).

Phosphoenolpyruvate (PEP)
(2 molecules)

Pyruvate kinase
2 ADP
2 ATP

10 Finally, the two PEPs transfer their phosphates to ADP, forming two ATPs and two molecules of pyruvate.

Pyruvate
(2 molecules)

From every glucose molecule, glycolysis nets two molecules of ATP and two molecules of the electron carrier NADH. Two molecules of pyruvate are produced.

higher than that of glucose. Later, these phosphate groups will be transferred to ADP to make new molecules of ATP.

Although both of these steps of glycolysis use ATP as one of their substrates, each is catalyzed by a different, specific enzyme. The enzyme hexokinase catalyzes *reaction 1*, in which a phosphate group from ATP is attached to the six-carbon glucose molecule, forming glucose 6-phosphate. (A *kinase* is any enzyme that catalyzes the transfer of a phosphate group from ATP to another substrate.) In *reaction 2*, the six-membered glucose ring is rearranged into a five-membered fructose ring. In *reaction 3*, the enzyme phosphofructokinase adds a second phosphate (taken from another ATP) to the fructose ring, forming a six-carbon sugar, fructose 1,6-bisphosphate.

Reaction 4 opens up the six-carbon sugar ring and cleaves it into two different three-carbon sugar phosphates: dihydroxyacetone phosphate and glyceraldehyde 3-phosphate. In *reaction 5*, one of those products, dihydroxyacetone phosphate, is converted into a second molecule of the other one, glyceraldehyde 3-phosphate (G3P, a triose phosphate). In summary, by the halfway point of the glycolytic pathway, two things have happened:

- Two molecules of ATP have been invested.

- The six-carbon glucose molecule has been converted into two molecules of a three-carbon sugar phosphate, glyceraldehyde 3-phosphate (G3P).

The energy-harvesting reactions of glycolysis yield NADH + H+ and ATP

In the discussion that follows, remember that *each reaction occurs twice for each glucose molecule* because each glucose molecule has been split into two molecules of G3P. It is the fate of G3P that now concerns us—its transformation will generate both NADH + H+ and ATP.

PRODUCING NADH + H+ *Reaction 6* is catalyzed by the enzyme triose phosphate dehydrogenase, and its end product is a phosphate ester, 1,3-bisphosphoglycerate (BPG). Reaction 6 is an *oxidation*, and it is accompanied by a large drop in free energy—more than 100 kcal of energy per mole of glucose is released in this exergonic reaction (**Figure 7.6**). If this big energy drop were simply a loss of heat, glycolysis would not provide useful energy to the cell. However, rather than being lost as heat, this energy is *stored* as chemical energy by reducing two molecules of NAD+ to make two molecules of NADH + H+.

Because NAD+ is present in small amounts in the cell, it must be recycled to allow glycolysis to continue; if none of the NADH is oxidized back to NAD+, glycolysis comes to a halt. The metabolic pathways that follow glycolysis carry out this oxidation, as we will see.

PRODUCING ATP In *reactions 7–10* of glycolysis, the two phosphate groups of BPG are transferred one at a time to molecules of ADP, with a rearrangement in between. More than 20 kcal (83.6 kJ/mol) of free energy is stored in ATP for every mole of BPG broken down. Finally, we are left with two moles of pyruvate for every mole of glucose that entered glycolysis.

The enzyme-catalyzed transfer of phosphate groups from donor molecules to ADP molecules to form ATP is called **substrate-level phosphorylation**. (*Phosphorylation* is the addition of a phosphate

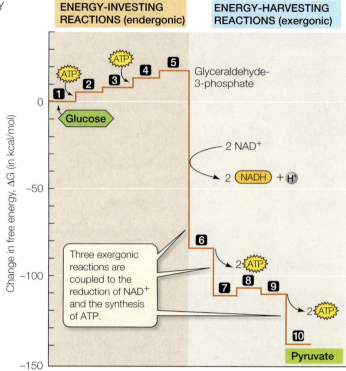

7.6 Changes in Free Energy During Glycolysis Each reaction of glycolysis results in a change in free energy. Refer to Figure 7.5 for the numbered reactions (e.g., reaction 1 is catalyzed by hexokinase).

Each glucose yields:
2 Pyruvate
2 NADH + 2 H+
2 ATP

group to a molecule. *Substrate-level phosphorylation* is distinguished from the oxidative phosphorylation carried out by the electron transport chain, which we will discuss later in this chapter.) In glycolysis, energy released by oxidation is used to form ATP. An example of substrate-level phosphorylation occurs in *reaction 7*, in which phosphoglycerate kinase catalyzes the transfer of a phosphate group from BPG to ADP, forming ATP. Both reactions 6 and 7 are exergonic, even though a substantial amount of energy is consumed in the formation of ATP.

To summarize:

- The energy-investing steps of glycolysis use the energy of hydrolysis of two ATP molecules per glucose molecule.

- The energy-releasing steps of glycolysis produce four ATP molecules per glucose molecule.

If O_2 is present, glycolysis is followed by the three pathways of cellular respiration.

Pyruvate oxidation links glycolysis and the citric acid cycle

The oxidation of pyruvate to acetate and its subsequent conversion to **acetyl CoA** is the link between glycolysis and all the other reactions of cellular respiration (see Figure 7.7). Coenzyme A (CoA) is a complex molecule responsible for binding the two-carbon acetate molecule. Acetyl CoA formation is a multi-step reaction catalyzed by the *pyruvate dehydrogenase complex*, an enormous multi-enzyme complex that is attached to the inner mitochondrial membrane. Pyruvate diffuses into the mitochondrion, where a series of coupled reactions takes place:

1. Pyruvate is oxidized to a two-carbon acetyl group (acetate), and CO_2 is released.
2. Part of the energy from this oxidation is captured by the reduction of NAD^+ to $NADH + H^+$.
3. Some of the remaining energy is stored temporarily by the combining of the acetyl group with CoA, forming acetyl CoA:

pyruvate + NAD^+ + CoA → acetyl CoA + $NADH + H^+ + CO_2$

Acetyl CoA has 7.5 kcal/mol (31.4 kJ/mol) more energy than simple acetate. Acetyl CoA can donate the acetyl group to acceptor molecules, much as ATP can donate phosphate groups to various acceptors. Acetyl CoA donates its acetyl group to the four-carbon compound oxaloacetate to form the six-carbon citrate, the compound that initiates the citric acid cycle, one of life's most important energy-harvesting pathways.

Arsenic, the classic poison of rodent exterminators and murder mysteries, acts by inhibiting pyruvate dehydrogenase, thus decreasing acetyl CoA production. The lack of acetyl CoA stops the citric acid cycle and cells eventually "starve to death" for lack of ATP.

The citric acid cycle completes the oxidation of glucose to CO_2

Acetyl CoA is the starting point for the citric acid cycle. This pathway of eight reactions completely oxidizes the two-carbon acetyl group to two molecules of carbon dioxide. The free energy released from these reactions is captured by ADP and the electron carriers NAD and FAD. **Figure 7.7** shows the energetics of the pathway. In addition, recall that energy is released upon oxidation and stored in either ATP, $FADH_2$, or $NADH + H^+$.

The citric acid cycle is maintained in a *steady state*—that is, although the intermediate compounds in the cycle enter and leave it, the concentrations of those intermediates do not change much. Pay close attention to the numbered reactions in **Figure 7.8** as you read the next several paragraphs.

The energy temporarily stored in acetyl CoA drives the formation of citrate from oxaloacetate (*reaction 1*). During this reaction, the coenzyme A molecule is removed and can be reused. In *reaction 2*, the citrate molecule is rearranged to form isocitrate. In *reaction 3*, a CO_2 molecule and two hydrogen atoms are removed, converting isocitrate into α-ketoglutarate. This reaction produces a large drop in free energy, some of which is stored in $NADH + H^+$.

Reaction 4 of the citric acid cycle, in which α-ketoglutarate is oxidized to succinyl CoA, is similar to oxidation of pyruvate to acetyl CoA and, like that reaction, is catalyzed by a multi-enzyme complex. In *reaction 5*, some of the energy in succinyl CoA is harvested to make GTP (guanosine triphosphate) from GDP and P_i, which is another example of substrate-level phosphorylation. GTP is then used to make ATP from ADP.

Free energy is released in *reaction 6*, in which the succinate released from succinyl CoA in reaction 5 is oxidized to fumarate. In the process, two hydrogens are transferred to an enzyme that contains the carrier FAD. After a molecular rearrangement (*reaction 7*), one more NAD^+ reduction occurs, producing oxaloacetate from malate (*reaction 8*). These two reactions illustrate a common biochemical mechanism: water (H_2O) is added in reaction 7 to form an —OH group, and then the H from that —OH group is removed in reaction 8 to reduce NAD^+ to $NADH + H^+$. The final product, oxaloacetate, is ready to combine with another acetyl group from acetyl CoA and go around the cycle again. The citric acid cycle operates twice for each glucose molecule that enters glycolysis (once for each pyruvate that enters the mitochondrion).

To summarize:

■ The *inputs* to the citric acid cycle are acetate (in the form of acetyl CoA), water, and oxidized electron carriers (NAD^+ and FAD).

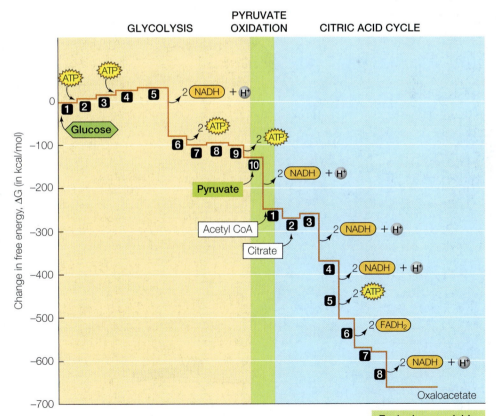

7.7 The Citric Acid Cycle Releases Much More Free Energy Than Glycolysis Does Electron carriers (NAD in glycolysis; NAD and FAD in the citric acid cycle) are reduced and ATP is generated in reactions coupled to other reactions, producing major drops in free energy as metabolism proceeds.

Each glucose yields:
3 CO_2
10 $NADH + H^+$
2 $FADH_2$
4 ATP

7.8 Pyruvate Oxidation and the Citric Acid Cycle

Pyruvate diffuses into the mitochondrion and is oxidized to acetyl CoA, which enters the citric acid cycle. Reactions 3, 4, 6, and 8 accomplish the major overall effects of the cycle—the trapping of energy—by passing electrons to NAD or FAD. Reaction 5 traps energy directly in ATP. Each reaction is catalyzed by a specific enzyme, although the enzymes are not shown in this figure.

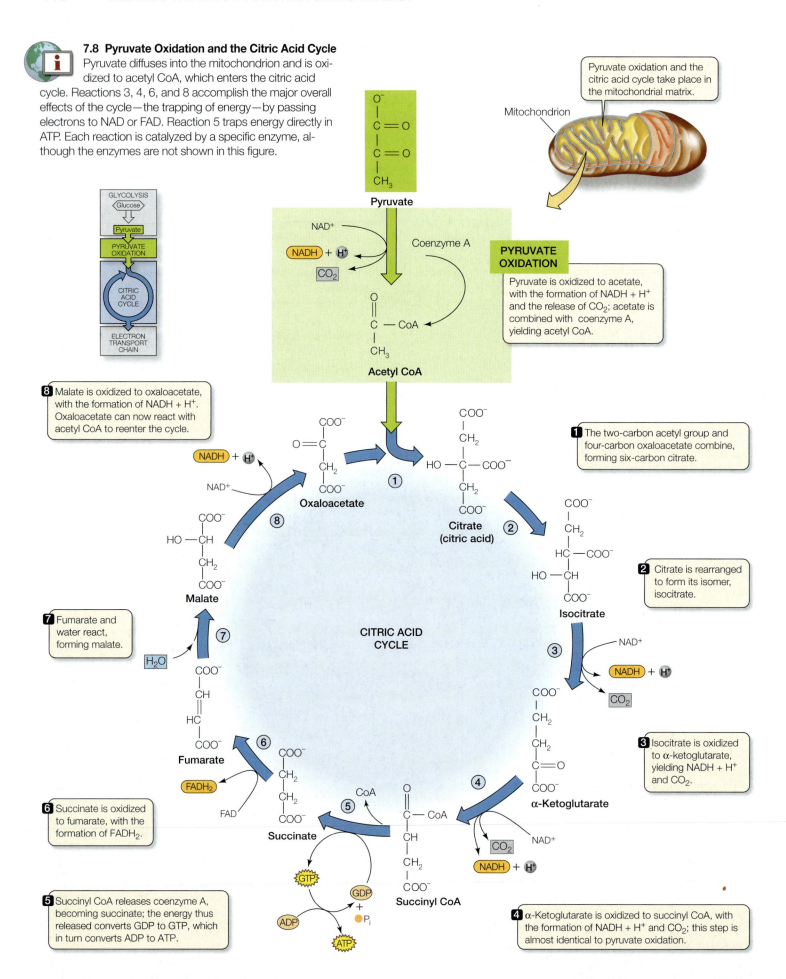

Pyruvate oxidation and the citric acid cycle take place in the mitochondrial matrix.

Mitochondrion

PYRUVATE OXIDATION

Pyruvate is oxidized to acetate, with the formation of NADH + H$^+$ and the release of CO$_2$; acetate is combined with coenzyme A, yielding acetyl CoA.

8 Malate is oxidized to oxaloacetate, with the formation of NADH + H$^+$. Oxaloacetate can now react with acetyl CoA to reenter the cycle.

1 The two-carbon acetyl group and four-carbon oxaloacetate combine, forming six-carbon citrate.

2 Citrate is rearranged to form its isomer, isocitrate.

3 Isocitrate is oxidized to α-ketoglutarate, yielding NADH + H$^+$ and CO$_2$.

7 Fumarate and water react, forming malate.

6 Succinate is oxidized to fumarate, with the formation of FADH$_2$.

5 Succinyl CoA releases coenzyme A, becoming succinate; the energy thus released converts GDP to GTP, which in turn converts ADP to ATP.

4 α-Ketoglutarate is oxidized to succinyl CoA, with the formation of NADH + H$^+$ and CO$_2$; this step is almost identical to pyruvate oxidation.

■ The *outputs* are carbon dioxide, reduced electron carriers ($NADH + H^+$ and $FADH_2$), and a small amount of ATP. Overall, for each acetyl group, the citric acid cycle removes two carbons as CO_2 and uses four pairs of hydrogen atoms to reduce electron carriers.

The citric acid cycle is regulated by concentrations of starting materials

We have seen how pyruvate, a three-carbon molecule, is completely oxidized to CO_2 by pyruvate dehydrogenase and the citric acid cycle. For the cycle to start anew, the starting molecules—acetyl CoA from oxidation of glucose, and oxidized electron carriers—must again all be in place. Because the electron carriers were reduced during the cycle (as well as in glycolysis—see reaction 6 in Figure 7.5), they are not able to accept electrons from the citric acid cycle reactions until they are *reoxidized*:

$$NADH \rightarrow NAD^+ + H^+ + e^-$$

$$FADH_2 \rightarrow FAD + 2H^+ + 2e^-$$

The oxidations of these electron carriers take place as coupled redox reactions, so that some other molecule (call it "X") gets reduced:

■ NAD gets oxidized: $NADH \rightarrow NAD^+ + H^+ + e^-$

■ An electron acceptor gets reduced: $X + H^+ + e^- \rightarrow XH$

Cells have two chemical ways of accomplishing this:

■ *Fermentation*: If O_2 is absent, "X" is pyruvate, the end product of glycolysis. In the process of reoxidizing NADH formed by glycolysis, pyruvate is reduced to lactate or ethyl alcohol. These are the final products, so the citric acid cycle cannot occur.

■ *Oxidative phosphorylation*: If O_2 is present, "X" is O_2. Pyruvate is fully oxidized to CO_2, and all the NADH and $FADH_2$ of the citric acid cycle is reoxidized. Energy is released by the oxidation of these electron carriers, and this energy is tapped to form ATP.

7.2 RECAP

The oxidation of glucose in the presence of O_2 involves glycolysis, pyruvate oxidation, and the citric acid cycle. In glycolysis, glucose is converted into pyruvate with some resulting energy capture. In the citric acid cycle, pyruvate is fully oxidized to CO_2, and more energy is captured in the form of reduced electron carriers.

■ Can you describe the *net energy yield* of glycolysis in terms of energy invested and energy harvested? See p. 144 and Figure 7.6

■ What role does pyruvate oxidation play in the citric acid cycle? See pp. 144–145 and Figure 7.8

■ Why is reoxidation of NADH crucial for continuation of the citric acid cycle? See p. 147

The oxidation of pyruvate and the citric acid cycle require the presence of O_2 as an electron receptor. We will look now at fermentation, the metabolic pathway that processes pyruvate in the absence of oxygen. Section 7.4 will cover oxidative phosphorylation, which completes the electron transfers of aerobic cellular respiration

7.3 How Is Energy Harvested from Glucose in the Absence of Oxygen?

Fermentation, like glycolysis, occurs in the cytosol. With no O_2 present to serve as an electron acceptor, fermentation uses NADH + H^+ formed by glycolysis to reduce pyruvate (or one of its metabolic derivatives) and regenerate NAD^+. As we saw in Figure 7.5, NAD^+ is required for *reaction 6* of glycolysis, so once the cell has replenished its NAD^+ supply by fermentation, it can metabolize more glucose.

Prokaryotic organisms are known to use many different fermentation pathways. Best understood, however, are two short fermentation pathways found in many different cells: *lactic acid fermentation*, whose end product is lactic acid (lactate); and *alcoholic fermentation*, whose end product is ethyl alcohol.

Pyruvate serves as the electron acceptor in **lactic acid fermentation** (**Figure 7.9**), which takes place in many microorganisms. It can also take place in some muscle cells, especially during high activity. As you will see in Chapter 47, muscular contraction requires a lot of energy, which must be supplied by ATP. But the flow of

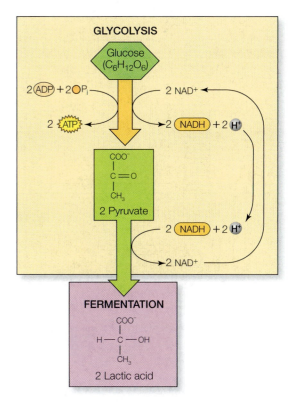

7.9 Lactic Acid Fermentation Glycolysis produces pyruvate, as well as ATP and NADH + H^+, from glucose. Lactic acid fermentation, using NADH + H^+ as a reducing agent, reduces pyruvate to lactic acid (lactate).

blood cannot supply enough O_2 during active contractions. To allow muscle tissue to continue to function in an anaerobic state, the muscle cells "switch" to lactic acid fermentation. Lactic acid buildup becomes a problem after prolonged periods because the acid ionizes, forming H^+ and lowering the pH of the cell, which reduces cellular activities (and causes muscle cramps).

Alcoholic fermentation takes place in certain yeasts and some plant cells under anaerobic conditions. This process requires two enzymes to metabolize pyruvate (**Figure 7.10**). First, carbon dioxide is removed from pyruvate, leaving the compound acetaldehyde. Second, the acetaldehyde is reduced by $NADH + H^+$, producing NAD^+ and ethyl alcohol (ethanol). Alcoholic beverages are made by the anaerobic fermentation of yeast cells using glucose from plant sources, such as grapes (wine) and barley (beer).

Billions and billions of prokaryotic microorganisms live by fermentation in the anaerobic environments of mammalian digestive organs such as the small intestine and the rumen (a specialized "stomach" found in cows and their relatives). These microorganisms may make crucial contributions to the "host" mammal's health and survival.

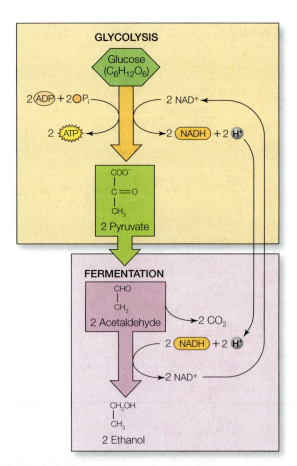

7.10 Alcoholic Fermentation In alcoholic fermentation, pyruvate from glycolysis is converted into acetaldehyde, and CO_2 is released. The $NADH + H^+$ from glycolysis acts as a reducing agent, reducing acetaldehyde to ethanol.

Fermentation enables glycolysis to produce a small amount of ATP through substrate-level phosphorylation. The net yield of two ATPs per glucose molecule is a minimal amount of usable energy, so it should not surprise you to learn that most organisms existing in anaerobic environments are small microbes that grow relatively slowly.

7.3 RECAP

In the absence of O_2, fermentation pathways use $NADH + H^+$ formed by glycolysis to reduce pyruvate or its derivatives and regenerate NAD^+. The energy yield of fermentation is low because glucose is only partially oxidized.

- Do you understand why replenishing NAD^+ is crucial to cellular metabolism? See p. 11

Fermentation pathways were the source of energy for much of life's early evolution, since Earth's early atmosphere lacked free oxygen. But when O_2 is present to accept electrons, the reoxidation of NADH and $FADH_2$ results in the synthesis of large amounts of ATP. Let's see how this process works.

7.4 How Does the Oxidation of Glucose Form ATP?

The overall process of ATP synthesis resulting from the reoxidation of electron carriers in the presence of O_2 is called **oxidative phosphorylation**. Two stages of the process can be distinguished:

1. *The electron transport chain.* The electrons from NADH and $FADH_2$ pass through a series of membrane-associated electron carriers. The flow of electrons along this electron transport chain accomplishes the active transport of protons across the inner mitochondrial membrane, out of the matrix, creating a proton concentration gradient.

2. *Chemiosmosis.* The protons diffuse back into the mitochondrial matrix through a proton channel, which couples this diffusion to the synthesis of ATP.

Before we proceed with the details of these two pathways, let's reflect on an important question: Why should the electron transport chain have so many components and complex processes? Why, for example, don't cells use the following single step?

$$NADH + H^+ + \tfrac{1}{2}O_2 \rightarrow NAD^+ + H_2O$$

The answer is that this reaction would be fundamentally untamable. It would be extremely exergonic—rather like setting off a stick of dynamite in the cell. There is no biochemical way to harvest that burst of energy efficiently and put it to physiological use (that is, no metabolic reaction is so endergonic as to consume a significant fraction of that energy in a single step). To control the release of energy during the oxidation of glucose in a cell, evolution has resulted in the lengthy electron transport chain: a series of reactions, each of which releases a small, manageable amount of energy.

The electron transport chain shuttles electrons and releases energy

The electron transport chain contains large integral proteins, smaller mobile proteins, and even a smaller lipid molecule (**Figure 7.11**):

- Four large protein complexes (I, II, III, and IV) containing electron carriers and associated enzymes are integral proteins of the inner mitochondrial membrane in eukaryotes (see Figure 4.13). Three of them are transmembrane proteins (extend through both sides of the membrane).

- **Cytochrome c** is a small peripheral protein that lies in the intermembrane space. It is loosely attached to the inner mitochondrial membrane.

- A nonprotein component called **ubiquinone** (abbreviated **Q**) is a small, nonpolar molecule that moves freely within the hydrophobic interior of the phospholipid bilayer of the inner mitochondrial membrane.

As illustrated in **Figure 7.12**, NADH + H⁺ passes electrons to the first large protein complex (I), called NADH-Q reductase,

which in turn passes the electrons to Q. The second complex (II), succinate dehydrogenase, passes electrons to Q from $FADH_2$ during the formation of fumarate from succinate in reaction 6 of the citric acid cycle (see Figure 7.8). These electrons enter the chain later than those from NADH.

The third complex (III), cytochrome c reductase, receives electrons from Q and passes them to cytochrome c. The fourth complex (IV), cytochrome c oxidase, receives electrons from cytochrome c and passes them to oxygen, which, along with these extra electrons ($\frac{1}{2} O_2^-$) picks up two hydrogen ions (H⁺) to form H_2O.

The electron carriers of the electron transport chain (including those contained in the three transmembrane protein complexes) differ as to how they change when they become reduced. NAD⁺, for example, accepts H⁻ (a hydride ion—one proton and two electrons), leaving the proton from the other hydrogen atom to float free: that is, the result is NADH + H⁺ (see Figure 7.3B). Other carriers, including Q, bind both protons and both electrons, becoming, for example, QH_2. Beyond Q, the chain transports electrons only. Electrons, not protons, are passed from Q to cytochrome c. An electron from QH_2 reduces the cytochrome's Fe^{3+} to Fe^{2+}.

What happens to the protons? Electron transport within each of the three transmembrane protein complexes results, as we'll see, in the pumping of protons across the inner mitochondrial membrane, and the return of those protons across the membrane is coupled to the formation of ATP. Thus the energy originally contained

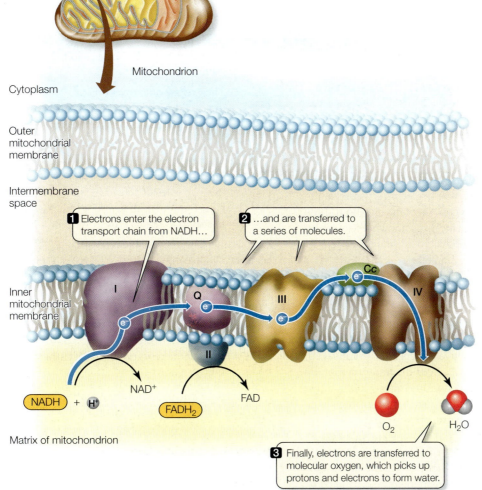

7.11 The Oxidation of NADH + H⁺ Electrons from NADH + H⁺ are passed through the electron transport chain, a series of protein complexes in the inner mitochondrial membrane containing electron carriers and enzymes. The carriers gain free energy when they become reduced and release free energy when they are oxidized.

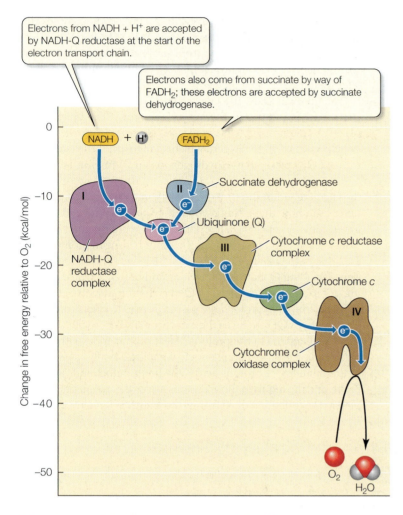

Electrons from NADH + H$^+$ are accepted by NADH-Q reductase at the start of the electron transport chain.

Electrons also come from succinate by way of FADH$_2$; these electrons are accepted by succinate dehydrogenase.

7.12 The Complete Electron Transport Chain   Electrons enter the chain from two sources, but they follow the same pathway from Q onward.

in glucose and other fuel molecules is finally captured in the cellular energy currency, ATP. For each pair of electrons passed along the chain from NADH + H$^+$ to oxygen, theoretically about three molecules of ATP are formed. Actually, it is somewhat less, about 2.5 ATP per NADH oxidized and 1.5 ATP per FADH oxidized.

Proton diffusion is coupled to ATP synthesis

As shown in Figure 7.11, all the electron carriers and enzymes of the electron transport chain except cytochrome *c* are embedded in the inner mitochondrial membrane. The operation of the electron transport chain results in the active transport of protons (H$^+$), against their concentration gradient, across the inner membrane of the mitochondrion from the mitochondrial matrix to the intermembrane space. This occurs because the electron carriers contained in the three large transmembrane protein complexes (I, III, and IV) are arranged in such a way that protons are taken up on one side of the membrane (in the mitochondrial matrix) and transported, along with electrons, to the other side (to the intermembrane space) (**Figure 7.13**). Thus the transmembrane protein com-

plexes of the electron transport chain act as *proton pumps*. Because of the positive charge on the protons (H$^+$), this pumping creates not only a difference in proton concentration, but also a difference in electric charge, across the inner mitochondrial membrane, making the mitochondrial matrix more negative than the intermembrane space. (In other words, the intermembrane space becomes more acidic than the mitochondrial matrix.)

Together, the proton concentration gradient and the charge difference constitute a source of potential energy called the **proton-motive force**. This force tends to drive the protons back across the membrane, just as the charge on a battery drives the flow of electrons, discharging the battery.

The conversion of the proton-motive force into kinetic energy is prevented by the fact that protons cannot cross the hydrophobic phospholipid bilayer of the inner mitochondrial membrane by simple diffusion. However, they can diffuse across the membrane by passing through a specific proton channel, called **ATP synthase**, which couples proton movement to the synthesis of ATP. This coupling of proton-motive force and ATP synthesis is called the *chemiosmotic mechanism*, or **chemiosmosis**.

THE CHEMIOSMOTIC MECHANISM FOR ATP SYNTHESIS   The chemiosmotic mechanism uses ATP synthase to couple proton diffusion to ATP synthesis. This mechanism has three parts:

1. The flow of electrons from one electron carrier to another in the electron transport chain is a series of exergonic reactions that occurs in the inner mitochondrial membrane.

2. These exergonic reactions drive the endergonic pumping of H$^+$ out of the mitochondrial matrix and across the inner membrane into the intermembrane space. This pumping establishes and maintains a H$^+$ gradient.

3. The potential energy of the H$^+$ gradient, or proton-motive force, is harnessed by ATP synthase. This protein has two roles: it acts as a channel allowing H$^+$ to diffuse back into the matrix, and it uses the energy of that diffusion to make ATP from ADP and P$_i$.

ATP synthesis is a reversible reaction, and ATP synthase can also act as an ATPase, hydrolyzing ATP to ADP and P$_i$:

$$ATP \rightleftharpoons ADP + P_i + \text{free energy}$$

If the reaction goes to the right, free energy is released and is used to pump H$^+$ out of the mitochondrial matrix. If the reaction goes to the left, it uses free energy from H$^+$ diffusion into the matrix to make ATP. What makes it prefer ATP synthesis? There are two answers to this question:

■ ATP leaves the mitochondrial matrix for use elsewhere in the cell as soon as it is made, keeping the ATP concentration in the matrix low and driving the reaction toward the left.

■ The H$^+$ gradient is maintained by electron transport and proton pumping. (The electrons, you will recall, come from the oxidation of NADH and FADH$_2$, which are themselves re-

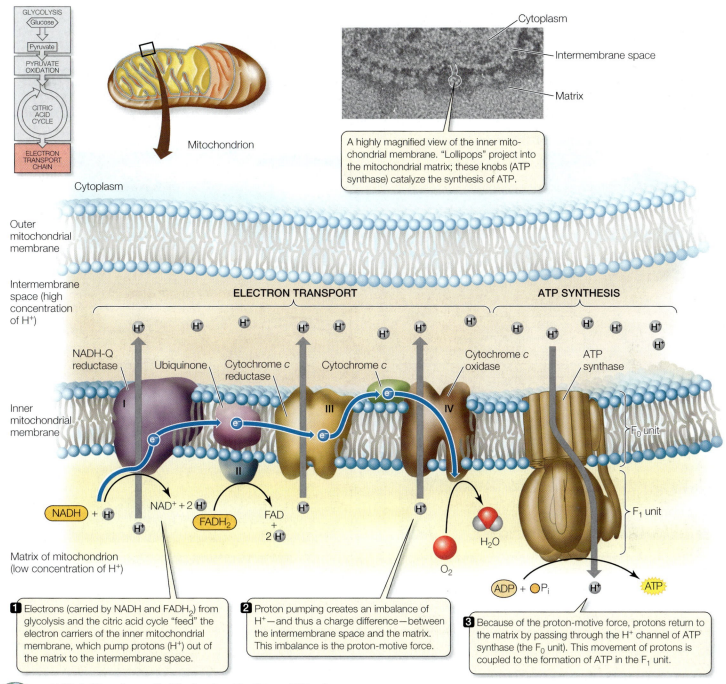

A highly magnified view of the inner mitochondrial membrane. "Lollipops" project into the mitochondrial matrix; these knobs (ATP synthase) catalyze the synthesis of ATP.

1 Electrons (carried by NADH and FADH₂) from glycolysis and the citric acid cycle "feed" the electron carriers of the inner mitochondrial membrane, which pump protons (H⁺) out of the matrix to the intermembrane space.

2 Proton pumping creates an imbalance of H⁺—and thus a charge difference—between the intermembrane space and the matrix. This imbalance is the proton-motive force.

3 Because of the proton-motive force, protons return to the matrix by passing through the H⁺ channel of ATP synthase (the F_0 unit). This movement of protons is coupled to the formation of ATP in the F_1 unit.

7.13 A Chemiosmotic Mechanism Produces ATP   As electrons pass through the transmembrane protein complexes in the electron transport chain, protons are pumped from the mitochondrial matrix into the intermembrane space. As the protons return to the matrix through ATP synthase, ATP is formed.

duced by the oxidations of glycolysis and the citric acid cycle. So, one reason you eat food is to replenish the H⁺ gradient!)

Every day a person hydrolyzes about 10^{25} ATP molecules to ADP. The vast majority of this ADP is "recycled"—converted back to ATP—using free energy from the oxidation of glucose.

EXPERIMENTS DEMONSTRATE CHEMIOSMOSIS   Two key experiments demonstrated (1) that a proton (H⁺) gradient across a membrane can drive ATP synthesis; and (2) that the enzyme ATP synthase is the catalyst for this reaction (**Figure 7.14**):

■ *Experiment 1* tested the hypothesis that ATP synthesis is driven by the H⁺ gradient across the inner mitochondrial membrane. In this experiment, mitochondria without a food source were "fooled" into making ATP by raising the H⁺ concentration in their environment. Isolated mitochondria were exposed to a low H⁺ concentration, then suddenly placed in a medium with a high H⁺ concentration. The outer mitochondrial membrane, unlike the inner one, is freely permeable to H⁺, so H⁺ rapidly diffused into the intermembrane space. This

EXPERIMENT 1

HYPOTHESIS: An H⁺ gradient can drive ATP synthesis by isolated mitochondria.

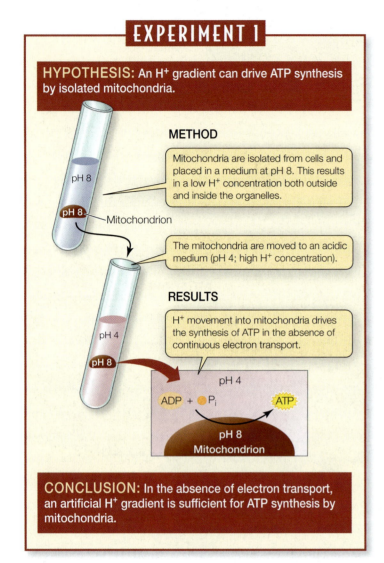

METHOD

Mitochondria are isolated from cells and placed in a medium at pH 8. This results in a low H⁺ concentration both outside and inside the organelles.

pH 8

pH 8 — Mitochondrion

The mitochondria are moved to an acidic medium (pH 4; high H⁺ concentration).

RESULTS

pH 4

pH 8

H⁺ movement into mitochondria drives the synthesis of ATP in the absence of continuous electron transport.

pH 4

ADP + P$_i$ → ATP

pH 8
Mitochondrion

CONCLUSION: In the absence of electron transport, an artificial H⁺ gradient is sufficient for ATP synthesis by mitochondria.

EXPERIMENT 2

HYPOTHESIS: ATP synthase is needed for ATP synthesis.

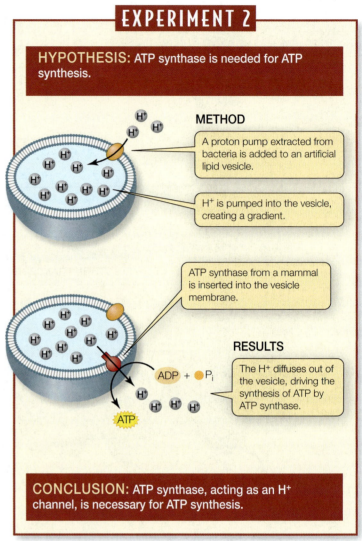

METHOD

A proton pump extracted from bacteria is added to an artificial lipid vesicle.

H⁺ is pumped into the vesicle, creating a gradient.

ATP synthase from a mammal is inserted into the vesicle membrane.

RESULTS

ADP + P$_i$

ATP

The H⁺ diffuses out of the vesicle, driving the synthesis of ATP by ATP synthase.

CONCLUSION: ATP synthase, acting as an H⁺ channel, is necessary for ATP synthesis.

7.14 Two Experiments Demonstrate the Chemiosmotic Mechanism An H⁺ gradient across a membrane is all that is needed to drive the synthesis of ATP by the enzyme ATP synthase. It doesn't matter whether the H⁺ gradient is produced artificially, as in these experiments, or by the electron transport chain found in nature. FURTHER RESEARCH: What would happen in Experiment 2 if a second ATP synthase, oriented in the opposite way to the one originally inserted in the membrane, was added?

created an artificial gradient across the inner membrane, which the mitochondria used to make ATP from ADP and P$_i$. This result supported the hypothesis, providing strong evidence for the chemiosmotic mechanism.

■ *Experiment 2* tested the hypothesis that the enzyme ATPase couples a proton gradient to ATP synthesis. A proton pump isolated from a bacterium was added to artificial membrane vesicles. When an appropriate energy source was provided, H⁺ was pumped into the vesicles, creating a proton gradient. If mammalian ATP synthase was then inserted into the membranes of these vesicles and the energy source removed, the vesicles made ATP, even in the absence of the usual electron carriers. Again, this result supported the hypothesis, showing that ATP synthase is the coupling factor.

UNCOUPLING PROTON DIFFUSION FROM ATP PRODUCTION Another way to demonstrate the chemiosmotic mechanism is to show that the diffusion of H⁺ and the formation of ATP must be *tightly coupled*; that is, that protons must pass only through the ATP synthase channel in order to move into the mitochondrial matrix. If a second type of H⁺ diffusion channel (not ATP synthase) is inserted into the mitochondrial membrane, the energy of the H⁺ gradient is released as heat, rather than being coupled to the synthesis of ATP. Such uncoupling molecules are deliberately used by some organisms to generate heat instead of ATP. For example, the natural uncoupling protein thermogenin plays an important role in regulating the temperature of some mammals under some circumstances, especially newborn human infants, who lack the hair to keep warm, and hibernating animals. We will describe this process in more detail in Section 40.4.

HOW ATP SYNTHASE WORKS: A MOLECULAR MOTOR   Now that we have established that the H⁺ gradient is needed for ATP synthesis, a question remains: How does the enzyme actually make ATP from ADP and P_i? This is certainly a fundamental question in biology, as it underlies energy harvesting in most cells. Look at the structure of ATP synthase in Figure 7.13. It is composed of two parts: the F_0 unit, a transmembrane region that is the H⁺ channel, and the F_1 unit, the "lollipop" of interacting subunits that constitute the active site for ATP synthesis. It is believed that ATP synthase turns potential energy from the H⁺ gradient into the kinetic energy of movement. The subunits of F_1 rotate, exposing the active site for ATP synthesis.

> ### 7.4 RECAP
>
> **Energy from the reoxidation of electron carriers is released in small steps by the electron transport chain. Through chemiosmosis, this energy is used to synthesize ATP.**
>
> - Can you describe roles of oxidation and reduction in the electron transport chain? See Figures 7.11–7.13
>
> - Do you understand the proton motive force and how it drives chemiosmosis? See p. 150 and Figure 7.13
>
> - Do you see how the two experiments described in Figure 7.14 demonstrate the chemiosmotic mechanism? See pp. 151–152

Oxidative phosphorylation captures a great deal of energy in ATP. What makes it so much more efficient than fermentation?

7.5 Why Does Cellular Respiration Yield So Much More Energy Than Fermentation?

The total net energy yield from glycolysis via fermentation is 2 molecules of ATP per molecule of glucose oxidized. The maximum yield of ATP that can be harvested from a molecule of glucose through glycolysis followed by cellular respiration is much greater—about 32 molecules of ATP (**Figure 7.15**). (See Figures 7.5, 7.8, and 7.13 to review where the ATP molecules come from.)

Why is so much more ATP produced by the metabolic pathways that operate in the presence of O_2? Recall that glycolysis is only a partial oxidation of glucose, as is fermentation. Much more energy remains in the end products of fermentation, lactic acid and ethanol than in the end product of cellular respiration, CO_2. In cellular respiration, carriers (mostly NAD⁺) are reduced in pyruvate oxidation and the citric acid cycle, then oxidized by the electron transport chain, with the accompanying production of ATP by chemiosmosis (2.5 ATP for each NADH + H⁺ and 1.5 ATP for each FADH₂). In an aerobic environment, a cell or organism capable of aerobic metabolism will be at an advantage (in terms of energy availability per glucose molecule) over one limited to fermentation.

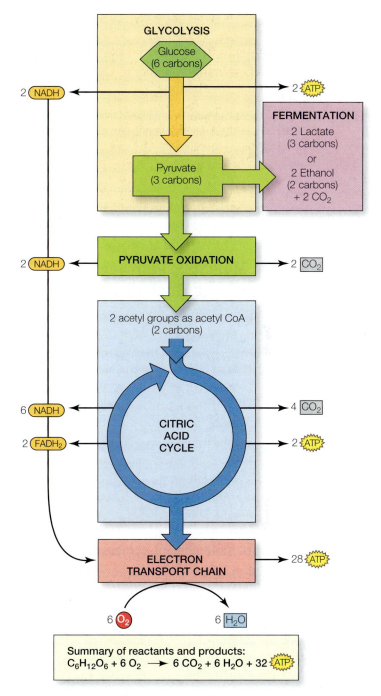

7.15 Cellular Respiration Yields More Energy Than Glycolysis Does   Electron carriers are reduced in pyruvate oxidation and the citric acid cycle, then oxidized by the electron transport chain. These reactions produce ATP via chemiosmosis.

The total gross yield of ATP from one molecule of glucose processed through glycolysis and cellular respiration is 32. However, we may subtract two from that gross—for a net yield of 30 ATP—because in some animal cells the inner mitochondrial membrane is impermeable to NADH, and a "toll" of one ATP must be paid for each NADH produced in glycolysis that is shuttled into the mitochondrial matrix.

The electron carriers of cellular respiration allow for the full oxidation of glucose, so the energy yield from glucose is much higher in the presence of O_2 than in its absence.

- Can you calculate the total energy yield from glucose in human cells in the presence versus the absence of O_2? See p. 153 and Figure 7.15

Now that we've seen how cells harvest energy, let's see how that energy moves through the interconnections among metabolic pathways in the cell.

7.6 How Are Metabolic Pathways Interrelated and Controlled?

Glycolysis and the pathways of cellular respiration do not operate in isolation from the rest of metabolism. Rather, there is an interchange, with biochemical traffic flowing both into these pathways and out of them, to and from the synthesis and breakdown of amino acids, nucleotides, fatty acids, and other building blocks of life. Carbon skeletons enter these pathways from other molecules that are broken down to release their energy (catabolism), and carbon skeletons leave these pathways to form the major macromolecular constituents of the cell (anabolism). These relationships are summarized in **Figure 7.16**.

Catabolism and anabolism involve interconversions of biological monomers

A hamburger or veggie burger contains three major sources of carbon skeletons for the person who eats it: carbohydrates, mostly as starch (a polysaccharide); lipids, mostly as triglycerides (three fatty acids attached to glycerol); and proteins (polymers of amino acids). Looking at Figure 7.16, you can see how each of these three types of macromolecules can be used in catabolism or anabolism.

CATABOLIC INTERCONVERSIONS Polysaccharides, lipids, and proteins can all be broken down to provide energy:

- *Polysaccharides* are hydrolyzed to glucose. Glucose then passes through glycolysis and cellular respiration, where its energy is captured in NADH and ATP.

- *Lipids* are broken down into their constituents, glycerol and fatty acids. Glycerol is converted into dihydroxyacetone phosphate (DAP), an intermediate in glycolysis, and fatty acids are converted into acetyl CoA in the mitochondria. In both cases, further oxidation to CO_2 and release of energy occur.

- *Proteins* are hydrolyzed to their amino acid building blocks. The 20 different amino acids feed into glycolysis or the citric acid cycle at different points. A specific example is shown in **Figure 7.17**, in which the amino acid glutamate is converted into α-ketoglutarate, an intermediate in the citric acid cycle.

ANABOLIC INTERCONVERSIONS Many catabolic pathways can operate in reverse. Glycolytic and citric acid cycle intermediates, instead of being oxidized to form CO_2, can be reduced and used to form glucose in a process called **gluconeogenesis** (which means "new formation of glucose"). Likewise, acetyl CoA can be used to form fatty acids. The most common fatty acids have an even number of carbons: 14, 16, or 18. These molecules are formed by adding two-carbon acetyl CoA "units" one at a time until the ap-

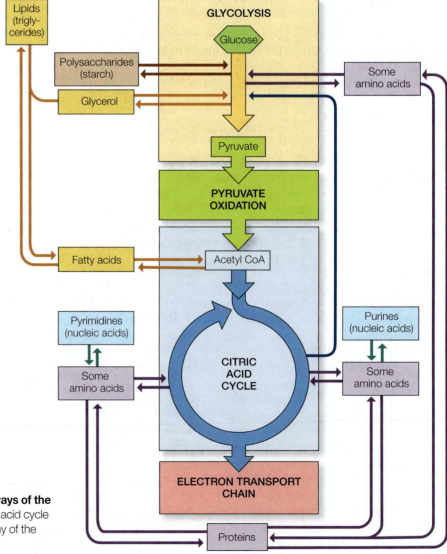

7.16 Relationships among the Major Metabolic Pathways of the Cell Note the central position of glycolysis and the citric acid cycle in this network of metabolic pathways. Also note that many of the pathways can operate in reverse.

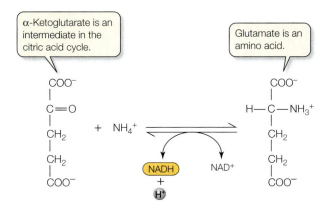

7.17 Coupling Metabolic Pathways This reaction, in which glutamate and α-ketoglutarate are interconverted, is catalyzed by the enzyme glutamate dehydrogenase.

propriate carbon chain length is reached. Amino acids can be formed by reversible reactions such as the one shown in Figure 7.17, and can then be polymerized into proteins.

Some intermediates in the citric acid cycle are used in the synthesis of various cellular constituents. For example, α-ketoglutarate is a starting point for purines and oxaloacetate for pyrimidines, both constituents of the nucleic acids DNA and RNA. α-Ketoglutarate is also a starting point for chlorophyll synthesis. Acetyl CoA is a building block for various pigments, plant growth substances, rubber, and the steroid hormones of animals, among other molecules.

Catabolism and anabolism are integrated

A carbon atom from a protein in your burger can end up in DNA or fat or CO_2, among other fates. How does the cell "decide" which metabolic pathway to follow? With all of these possible interconversions, you might expect that cellular concentrations of various biochemical molecules would vary widely. For example, you might think that the level of oxaloacetate in your cells would depend on what you eat (some food molecules form oxaloacetate) and on whether oxaloacetate is used up (in the citric acid cycle or in forming the amino acid aspartate). Remarkably, the levels of these substances in what is called the "metabolic pool"—the sum total of all the biochemical molecules in a cell—are quite constant. The cell regulates the enzymes of catabolism and anabolism so as to maintain a balance. This *metabolic homeostasis* gets upset only in unusual circumstances. Let's look one such unusual circumstance: undernutrition.

Glucose is an excellent source of energy. From Figure 7.16, you can see that lipids and proteins can also serve as energy sources. Any one, or all three, could be used to provide the energy your body needs. In reality, things are not so simple. Proteins, for example, have essential roles in your body as enzymes and structural elements; they are not stored for energy, and using them for energy might deprive you of a catalyst for a vital reaction.

Polysaccharides and fats (triglycerides) have no such catalytic roles. But polysaccharides, because they are somewhat polar, can bind a lot of water. Because they are nonpolar, fats do not bind as much water as polysaccharides do. So, in water, fats weigh less

than polysaccharides. In addition, fats are more reduced than carbohydrates (have more C—H bonds as opposed to C—OH bonds) and have more energy stored in their bonds. For these two reasons, fats are a better means of storing energy than polysaccharides. It is not surprising, then, that a typical person has about one day's worth of food energy stored as glycogen (a polysaccharide), a week's food energy as usable proteins in blood, and over a month's food energy stored as fats.

What happens if a person does not eat enough to produce sufficient ATP and NADH for anabolism and biological activities? This situation can be the result of a deliberate decision to lose weight, but for too many people, it is forced upon them because not enough food is available. In either case, the first energy stores in the body to be used are the glycogen stores in muscle and liver cells. These stores do not last long, and next come the fats.

The level of acetyl CoA rises as fatty acids are broken down. However, a problem remains: because fatty acids cannot cross from the blood to the brain, the brain can use only glucose as its energy source. With glucose already depleted, the body must convert something else to make glucose for the brain. This gluconeogenesis uses mostly amino acids, largely from the breakdown of proteins. Without sufficient food intake, both proteins (for glucose) and fats (for energy) are used up. After several weeks of starvation, fat stores become depleted, and the only energy source left is proteins. At this point, essential proteins, such as muscle proteins and the antibodies used to fight off infections, get broken down. The loss of such proteins can lead to severe illness and eventual death.

> Seasonal migrations between Canada and South America made by the tiny blackpoll warbler are fueled by 10–12 grams of fat, which is about equal to their lean body mass. The glycogen required to supply the same amount of energy would weigh 10 times more. The birds would not be able to fly (or even walk).

Metabolic pathways are regulated systems

We have described the relationships between metabolic pathways and noted that these pathways work together to provide homeostasis in the cell and organism. But how does the cell regulate interconversions between these pathways to maintain constant metabolic pools? In its most simplified sense, this is a problem of systems biology (see Figure 6.15).

Consider what happens to the starch in your burger bun. In the digestive system, starch is hydrolyzed to glucose, which enters the blood for distribution to the rest of the body. Before this happens, however, a regulatory check must be made: is there already enough glucose in the blood to supply the body's needs? If there is, the excess glucose is converted into glycogen and stored in the liver. If not enough glucose is supplied by food, that glycogen is broken down, or other molecules are used to make glucose by gluconeogenesis.

The end result is that the level of glucose in the blood is remarkably constant. We will describe the details of how this happens in Part Eight of this book. For now, it is important to realize

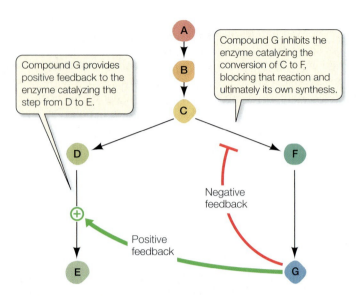

Compound G provides positive feedback to the enzyme catalyzing the step from D to E.

Compound G inhibits the enzyme catalyzing the conversion of C to F, blocking that reaction and ultimately its own synthesis.

Negative feedback

Positive feedback

7.18 Regulation by Negative and Positive Feedback
Allosteric regulation plays an important role in metabolic pathways. Excess accumulation of some products can shut down their synthesis or stimulate the synthesis of other products that require the same raw materials.

that the interconversions of glucose involve many steps, each catalyzed by an enzyme, and it is here that controls often reside.

Glycolysis, the citric acid cycle, and the electron transport chain are regulated by *allosteric* control of the enzymes involved. As described in Section 6.5, in metabolic pathways, a high concentration of the products of a later reaction can suppress the action of enzymes that catalyze an earlier reaction. On the other hand, an excess of the product of one pathway can speed up reactions in another pathway, diverting raw materials away from synthesis of the first product (**Figure 7.18**). These *negative and positive feedback control mechanisms* are used at many points in the energy-harvesting pathways, which are summarized in **Figure 7.19**.

■ *The main control point in glycolysis is the enzyme phosphofructokinase* (reaction 3 in Figure 7.5). This enzyme is allosterically inhibited by ATP and activated by ADP or AMP. As long as fermentation proceeds, yielding a relatively small amount of ATP, phosphofructokinase operates at full efficiency. But when cellular respiration begins producing 16 times more ATP than fermentation does, the abundant ATP allosterically inhibits the enzyme, and the conversion of fructose 6-phosphate into fructose 1,6-bisphosphate declines, as does the rate of glucose utilization.

■ *The main control point in the citric acid cycle is the enzyme isocitrate dehydrogenase, which converts isocitrate into α-ketoglutarate* (reaction 3 in Figure 7.8). NADH + H⁺ and ATP are feedback inhibitors of this reaction; ADP and NAD⁺ are activators. If too much ATP is accumulating, or if NADH + H⁺ is being produced faster than it can be used by the electron transport chain, the conversion of isocitrate is slowed, and the citric acid cycle is essentially shut down. A shutdown of the citric acid cycle would cause large amounts of isocitrate and citrate to accumulate if the conversion of acetyl CoA to citrate were not also slowed by abundant ATP and NADH + H⁺. An excess of citrate acts as an additional feedback inhibitor to slow the fructose 6-phosphate reaction early in glycolysis. Consequently, if the citric acid cycle has been slowed down because of abundant ATP (and not because of a lack of oxygen), glycolysis is shut down as well. Both processes resume when the

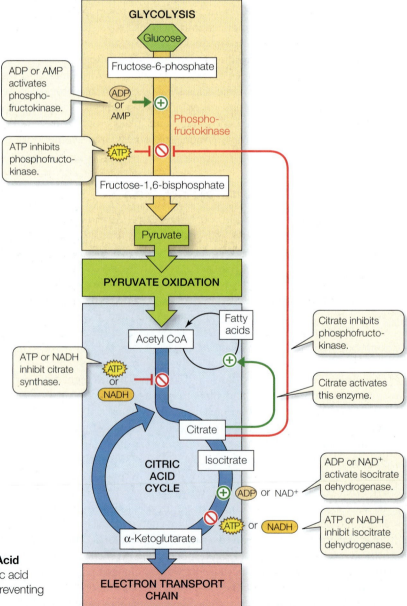

7.19 Allosteric Regulation of Glycolysis and the Citric Acid Cycle Allosteric feedback controls glycolysis and the citric acid cycle at crucial early steps, increasing their efficiency and preventing the excessive buildup of intermediates.

ATP level falls and they are needed again. Allosteric control keeps these processes in balance.

- *Another control point involves acetyl CoA.* If too much ATP is being made and the citric acid cycle shuts down, the accumulation of citrate diverts acetyl CoA to the synthesis of fatty acids for storage. That is one reason why people who eat too much accumulate fat. These fatty acids may be metabolized later to produce more acetyl CoA.

- *A final control point comes with cell differentiation.* For example, as illustrated in the opening of this chapter, a single protein, PPARδ, appears to control the proliferation of slow-twitch muscle fibers. These cells, which are full of mitochondria, catabolize fats and carbohydrates aerobically. The result is a steady release of ATP, whose hydrolysis can be used to power muscles in sustained activity. Long-distance running is a good way to combat obesity.

7.6 RECAP

Glucose can be made from intermediates in glycolysis and the citric acid cycle by a process called gluconeogenesis. The metabolic pathways of lipids and amino acids are tied to those of glucose metabolism. Key enzymes link and regulate the various pathways.

- Can you give an example of a catabolic interconversion of a lipid? Of an anabolic interconversion of a protein? See p. 154 and Figure 7.16

- Can you explain how phosphofructokinase serves as a control point for glycolysis? See p. 156 and Figure 7.19

- What would happen if there were no allosteric mechanism for sensing the level of acetyl CoA?

CHAPTER SUMMARY

7.1 How does glucose oxidation release chemical energy?

As a material is **oxidized**, the electrons it loses are transferred to another material, which is thereby **reduced**. Such **redox reactions** transfer large amounts of energy. Review Figure 7.2, Web/CD Activities 7.1, 7.2

The coenzyme **NAD** is a key electron carrier in biological redox reactions. It exists in two forms, one oxidized (NAD$^+$) and the other reduced (NADH + H$^+$).

Glycolysis operates in the presence or absence of O$_2$. Under **aerobic** conditions, **cellular respiration** continues the process of breaking down glucose. Under **anaerobic** conditions, **fermentation** occurs. Review Figure 7.4

The three pathways of cellular respiration are **pyruvate oxidation**, the **citric acid cycle**, and the **electron transport chain**.

7.2 What are the aerobic pathways of glucose metabolism?

Glycolysis consists of 10 enzyme-catalyzed reactions that occur in the cell cytosol. Two **pyruvate** molecules are produced for each partially oxidized molecule of glucose, providing the starting material for both cellular respiration and fermentation. Review Figure 7.5

The first five reactions of glycolysis require an investment of energy; the last five produce energy. The net gain is two molecules of ATP.

The enzyme-catalyzed transfer of phosphate groups to ADP is called **substrate-level phosphorylation**.

Pyruvate oxidation follows glycolysis and links glycolysis to the citric acid cycle. In this pathway, pyruvate is converted into **acetyl CoA**.

Acetyl CoA is the starting point of the citric acid cycle. It reacts with oxaloacetate to produce citrate. A series of eight enzyme-catalyzed reactions oxidize citrate and regenerate oxaloacetate, continuing the cycle. Review Figure 7.8, Web/CD Activity 7.3

7.3 How is energy harvested from glucose in the absence of oxygen?

In the absence of O$_2$, glycolysis is followed by fermentation. Together, these pathways partially oxidize glucose and generate end products such as **lactic acid** or **ethanol** along with a small amount of ATP. Review Figures 7.9 and 7.10

7.4 How does the oxidation of glucose form ATP?

Oxidation of electron carriers in the presence of O$_2$ releases energy that can be used to form ATP in a process called **oxidative phosphorylation**.

NADH and FADH$_2$ from glycolysis, pyruvate oxidation, and the citric acid cycle are oxidized by the electron transport chain, regenerating NAD$^+$ and FAD. Oxygen (O$_2$) is the final acceptor of electrons and protons, forming water (H$_2$O). Review Figure 7.11, Web/CD Activity 7.4, Web/CD Tutorial 7.1

The electron transport chain not only moves electrons across its carriers, but also pumps protons across the inner mitochondrial membrane, creating the **proton-motive force**.

The protons constituting the proton-motive force can return to the mitochondrial matrix via **ATP synthase**, which couples this movement of protons to the synthesis of ATP. This process is called **chemiosmosis**. Review Figure 7.13, Web/CD Tutorial 7.2

7.5 Why does cellular respiration yield so much more energy than fermentation?

For each molecule of glucose used, fermentation yields 2 molecules of ATP. In contrast, glycolysis operating with pyruvate oxidation, the citric acid cycle, and the electron transport chain yields up to 32 molecules of ATP per molecule of glucose. Review Figure 7.15, Web/CD Activity 7.5

7.6 How are metabolic pathways interrelated and controlled?

The **catabolic pathways** of carbohydrates, fats, and proteins feed into the energy-harvesting metabolic pathways. Review Figure 7.16

Anabolic pathways use intermediate components of the energy-harvesting pathways to synthesize fats, amino acids, and other essential building blocks.

The formation of glucose from glycolytic and citric acid cycle intermediates is called **gluconeogenesis**.

The rates of glycolysis and the citric acid cycle are regulated by allosteric feedback, by storage of excess acetyl CoA, and by cell differentiation. See Figure 7.19, Web/CD Activity 7.6

SELF-QUIZ

1. The role of oxygen gas in our cells is to
   a. catalyze reactions in glycolysis.
   b. produce CO_2.
   c. form ATP.
   d. accept electrons from the electron transport chain.
   e. react with glucose to split water.

2. Oxidation and reduction
   a. entail the gain or loss of proteins.
   b. are defined as the loss of electrons.
   c. are both endergonic reactions.
   d. always occur together.
   e. proceed only under aerobic conditions.

3. NAD^+ is
   a. a type of organelle.
   b. a protein.
   c. present only in mitochondria.
   d. a part of ATP.
   e. formed in the reaction that produces ethanol.

4. Glycolysis
   a. takes place in the mitochondrion.
   b. produces no ATP.
   c. has no connection with the electron transport chain.
   d. is the same thing as fermentation.
   e. reduces two molecules of NAD^+ for every glucose molecule processed.

5. Fermentation
   a. takes place in the mitochondrion.
   b. takes place in all animal cells.
   c. does not require O_2.
   d. requires lactic acid.
   e. prevents glycolysis.

6. Which statement about pyruvate is *not* true?
   a. It is the end product of glycolysis.
   b. It becomes reduced during fermentation.
   c. It is a precursor of acetyl CoA.
   d. It is a protein.
   e. It contains three carbon atoms.

7. The citric acid cycle
   a. takes place in the mitochondrion.
   b. produces no ATP.
   c. has no connection with the electron transport chain.
   d. is the same thing as fermentation.
   e. reduces two NAD^+ for every glucose processed.

8. The electron transport chain
   a. is located in the mitochondrial matrix.
   b. includes integral membrane proteins.
   c. always produces ATP.
   d. reoxidizes reduced coenzymes.
   e. operates simultaneously with fermentation.

9. Compared with fermentation, the aerobic pathways of glucose metabolism produce
   a. more ATP.
   b. pyruvate.
   c. fewer protons for pumping in mitochondria.
   d. less CO_2.
   e. more oxidized coenzymes.

10. Which statement about oxidative phosphorylation is *not* true?
    a. It is the formation of ATP by the electron transport chain.
    b. It is brought about by chemiosmosis.
    c. It requires aerobic conditions.
    d. It takes place in mitochondria.
    e. Its functions can be served equally well by fermentation.

FOR DISCUSSION

1. Trace the sequence of chemical changes that occurs in mammalian tissue when the oxygen supply is cut off. The first change is that the cytochrome *c* oxidase system becomes totally reduced, because electrons can still flow from cytochrome *c*, but there is no oxygen to accept electrons from cytochrome *c* oxidase. What are the remaining steps?

2. Some cells that use the aerobic pathways of glucose metabolism can also thrive by using fermentation under anaerobic conditions. Given the lower yield of ATP (per molecule of glucose) in fermentation, how can these cells function so efficiently under anaerobic conditions?

3. The drug antimycin A blocks electron transport in mitochondria. Explain what would happen if experiment 1 in Figure 7.14 were repeated in the presence of this drug.

4. You eat a burger that contains polysaccharides, proteins, and lipids. Using your knowledge of the integration of biochemical pathways, explain how the amino acids in the proteins and the glucose in the polysaccharides can end up as fats.

FOR INVESTIGATION

A protein present in the fat of newborn babies uncouples the synthesis of ATP from electron transport and instead generates heat. How would you investigate the hypothesis that this uncoupling protein adds a second proton channel to the mitochondrial membrane?

8 Photosynthesis: Energy from Sunlight

The ultimate energy source

If all the carbohydrates produced by photosynthesis in a year were in the form of sugar cubes, there would be 300 quadrillion of them. Lined up, these cubes would extend from Earth to the planet Pluto. That's a lot of photosynthesis!

Imagine what would happen if photosynthesis on Earth were reduced. That catastrophe actually happened about 65 million years ago, when a large meteorite slammed into Earth in what is now southern Mexico. Geological evidence suggests that the impact raised a huge cloud of dust, blocking the sun and curtailing photosynthesis—and hence plant growth, and thus the survival of species that depended on plants. The impact may have led to the extinction of the dinosaurs (among other species). Their extinction, however, benefited the early mammals, who thrived in the absence of competition from the great reptiles. (You might not even be here if photosynthesis had not been so drastically scaled back those many million years ago.)

Powered by sunlight, green plants use the reactions of photosynthesis to convert simple chemicals in the environment—carbon dioxide and water—into carbohydrates. The emergence of photosynthesis was a key event in the evolution of life, integrating an external energy source—the energy of sunlight—into the living world. Photosynthesizing organisms using solar energy to make their own food provide an entry point to the biosphere for chemical energy. Most other organisms depend on photosynthesizers for the raw materials of metabolism, such as glucose, as well as for atmospheric oxygen.

Globally, more than 10 billion tons of carbon is *fixed* by plants every year. By this we mean carbon molecules are converted from being part of a simple gas (CO_2) into more complex, reduced molecules (carbohydrates), making carbon available for use as food. As you might guess, Earth's "food chain" requires a lot of photosynthesis. On the African plain, it takes 6 acres of grassland to convert enough CO_2 into plant matter that supports the growth of a single grazing gazelle.

Humans consume a huge amount of Earth's photosynthetic output. To determine how much, Mark Imhoff and colleagues at the NASA Center in Maryland estimated the net productivity of Earth's photosynthetic organisms—that is, the amount of carbon dioxide they fix minus the amount they use for their own growth and reproduction. Having estimated available fixed carbon, the scientists then estimated

Primary Producers Covering less than 2 percent of Earth's surface, rainforests are photosynthetic dynamos, producing about one-fifth of the oxygen gas in the atmosphere.

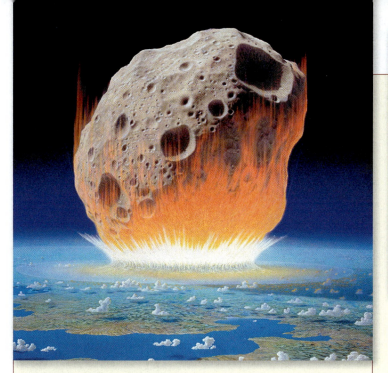

A Photosynthetic Disaster An artist's conception of the presumed meteorite impact of 65 million years ago, which raised a dust cloud that choked the atmosphere and severely reduced photosynthetic output across the planet.

human carbohydrate consumption. Direct consumption included all carbohydrates consumed as food, fuel, fiber, or timber. Indirect consumption included disposal of crops we do not use, fires set to clear the land, and land cleared for cities.

The astounding conclusion: humans appropriate one-third of all the carbon fixed each year, leaving two-thirds for the entire remainder of the biosphere. This is by far the greatest consumption ratio for any single species in known evolutionary history. Can we keep consuming this much photosynthetic output indefinitely? The 2002 United Nations Conference on Sustainability concluded that we need to take steps to protect our photosynthetic future. An important first step in examining ecological sustainability is to understand photosynthesis.

IN THIS CHAPTER we first explore how photosynthesis converts light energy into chemical energy in the form of reduced electron carriers and ATP. Then we'll see how these two sources of chemical energy are used to drive the synthesis of carbohydrates from carbon dioxide. We'll show how the relationships between these two processes are imperative for plant growth.

8.1 What Is Photosynthesis?

Photosynthesis (literally, "synthesis from light") is a metabolic process by which the energy of sunlight is captured and used to convert carbon dioxide (CO_2) and water (H_2O) into carbohydrate sugars ($C_6H_{12}O_6$) and oxygen gas (O_2) (**Figure 8.1**). By early in the nineteenth century, scientists had grasped these broad outlines of photosynthesis and had established several things about the way the process worked:

- The water for photosynthesis in land plants comes primarily from the soil and must travel from the roots to the leaves.
- Plants take in carbon dioxide, and release water and O_2, through tiny openings in leaves called *stomata* (singular *stoma*) (see Figure 8.1).
- Light is absolutely necessary for the production of oxygen and sugars.

By 1804, scientists had summarized photosynthesis as follows:

carbon dioxide + water + light energy → sugar + oxygen

In molecular terms, this equation seems to be the *reverse* of the overall equation for cellular respiration (see Section 7.1). More precisely, it can be written as:

$$6\ CO_2 + 6\ H_2O \rightarrow C_6H_{12}O_6 + 6\ O_2$$

While they have proved to be essentially correct, these equations left much more to know about the process of photosynthesis, which, in fact, is *not* the reverse of cellular respiration. What are the reactions of photosynthesis? What role does light play in these reactions? How do carbons become linked to form sugars? And where does the oxygen gas come from: from CO_2 or H_2O?

It took almost another century to determine the source of the O_2 released during photosynthesis. One of the first biological experiments to utilize radioisotopes traced the flow of oxygen in plants. Samuel Ruben and Martin Kamen allowed two groups of plants to carry on photosynthesis (**Figure 8.2**). Plants in the first group were supplied with water containing the oxygen isotope ^{18}O and with CO_2 containing only the common

8.1 The Ingredients for Photosynthesis A typical terrestrial plant uses light from the sun, water from the soil, and carbon dioxide from the atmosphere to form organic compounds by photosynthesis.

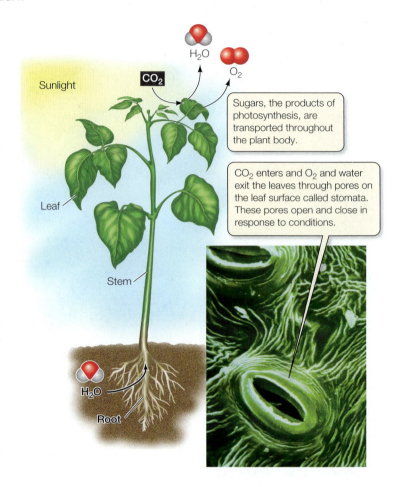

oxygen isotope ^{16}O; plants in the second group were supplied with CO_2 labeled with ^{18}O and water containing only ^{16}O. Oxygen gas from each group of plants was collected and analyzed. Oxygen gas containing ^{18}O was produced in abundance by the plants that had been given ^{18}O-labeled water, but *not* by those given ^{18}O-labeled CO_2.

These results showed that all the oxygen gas produced during photosynthesis comes from water, as reflected in the revised balanced equation:

$$6\ CO_2 + 12\ H_2O \rightarrow C_6H_{12}O_6 + 6\ O_2 + 6\ H_2O$$

Water appears on both sides of the equation because water is both used as a reactant (the twelve molecules on the left) and released as a product (the six new ones on the right). This revised equation accounts for all water molecules needed for all the oxygen gas produced.

Photosynthesis involves two pathways

Our equation summarizes the overall process of photosynthesis, but not the stages by which it is completed. Like glycolysis and the other metabolic pathways that harvest energy in cells, photosynthesis consists not of a single reaction, but of many reactions. The reactions of photosynthesis are commonly divided into two main pathways:

- The **light reactions** are driven by light energy. This pathway converts light energy into chemical energy in the form of ATP and a reduced electron carrier (NADPH + H$^+$).

- The **light-independent reactions** do not use light directly, but instead use ATP, NADPH + H$^+$ (made by the light reactions), and CO_2 to produce sugars. There are three different forms of the light-independent pathway that reduces CO_2: the *Calvin cycle*, C_4 *photosynthesis*, and *crassulacean acid metabolism*.

The light-independent reactions are sometimes called the *dark reactions* because they do not directly require light energy. However, both the light-dependent and light-independent reactions stop in the dark because ATP synthesis and NADP$^+$ reduction require light. The reactions of both pathways proceed within the chloroplast, but they reside in different parts of that organelle (**Figure 8.3**). The two pathways are linked by the exchange of ATP and ADP and of NADP$^+$ and NADPH, and the rate of each set of reactions depends on the rate of the other.

8.2 Water Is the Source of the Oxygen Produced by Photosynthesis Because only plants given isotope-labeled water released isotope-labeled O_2, this experiment showed that water is the source of the oxygen released during photosynthesis.

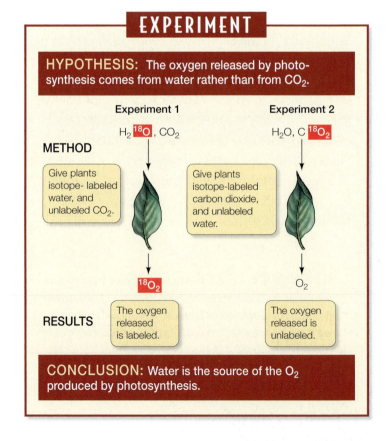

EXPERIMENT

HYPOTHESIS: The oxygen released by photosynthesis comes from water rather than from CO_2.

CONCLUSION: Water is the source of the O_2 produced by photosynthesis.

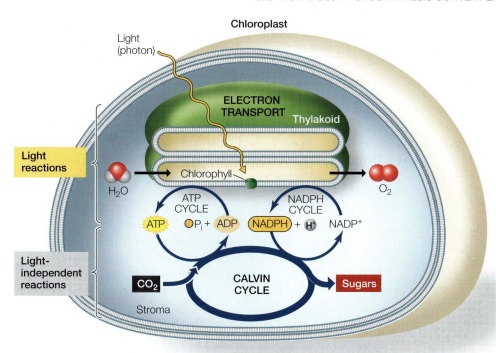

8.3 An Overview of Photosynthesis Photosynthesis comprises two pathways: the light reactions and the light-independent reactions. These reactions take place in the thylakoids and the stroma of chloroplasts, respectively.

We will describe the light reactions and the light-independent reactions separately and in detail. But since these two photosynthetic pathways are powered by the energy of sunlight, let's begin by discussing the physical nature of light and the specific photosynthetic molecules that capture its energy.

8.2 How Does Photosynthesis Convert Light Energy into Chemical Energy?

Light is a source of both energy and information. Later chapters will explore the many roles of light in the transmission of information. Here our focus is on light as a source of energy.

Light behaves as both a particle and a wave

Light is a form of **electromagnetic radiation**. It comes in discrete packets called **photons**. Light also behaves as if it were propagated in waves. The amount of energy contained in a single photon is inversely proportional to its **wavelength**: the shorter the wavelength, the greater the energy of the photons. To be active in a biological process, a photon must be absorbed by a receptive molecule, and it must have sufficient energy to perform the chemical work required.

Absorbing a photon excites a pigment molecule

When a photon meets a molecule, one of three things happens:

- The photon may bounce off the molecule—it may be *scattered* or *reflected*.

- The photon may pass through the molecule—it may be *transmitted*.

- The photon may be *absorbed* by the molecule.

Neither of the first two outcomes causes any change in the molecule. In the case of **absorption**, however, the photon disappears. Its energy, however, cannot disappear, because, according to the first law of thermodynamics, energy is neither created nor destroyed. Instead, when a molecule absorbs a photon, that molecule acquires the energy of the photon. It is thereby raised from a **ground state** (with lower energy) to an **excited state** (with higher energy) (**Figure 8.4A**).

The difference in free energy between the molecule's excited state and its ground state is approximately equal to the free energy of the absorbed photon (a small amount is lost to entropy). The increase in energy boosts one of the electrons within the molecule into a shell farther from its nucleus; this electron is now held less firmly (**Figure 8.4B**), making the molecule more chemically reactive.

Absorbed wavelengths correlate with biological activity

The electromagnetic spectrum (**Figure 8.5**) encompasses the wide range of wavelengths (and hence energy levels) that photons can have. The specific wavelengths absorbed by a particular molecule are characteristic of that type of molecule. Molecules that absorb wavelengths in the *visible spectrum*—that region of the spectrum that is visible to humans as light—are called **pigments**.

When a beam of white light (light containing visible light of *all* wavelengths) falls on a pigment, certain wavelengths of the light are absorbed. The remaining wavelengths, which are scattered or transmitted, make the pigment appear to us to be colored. For example, if a pigment absorbs both blue and red light—as chlorophyll does—what we see is the remaining light, which is primarily green.

If we plot the wavelengths of the light absorbed by a purified pigment, the result is an **absorption spectrum** for that pigment. If we plot the biological activity of a photosynthetic organism as a function of the wavelengths of light to which the organism is ex-

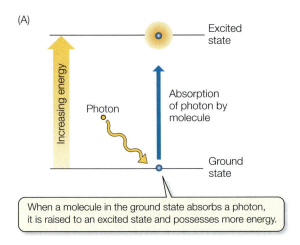

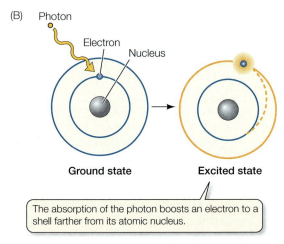

When a molecule in the ground state absorbs a photon, it is raised to an excited state and possesses more energy.

The absorption of the photon boosts an electron to a shell farther from its atomic nucleus.

8.4 Exciting a Molecule (A) When a molecule absorbs the energy of a photon, it is raised from a ground state to an excited state. (B) In the excited state, an electron is boosted to a more distant shell, where it is held less firmly.

posed, the result is an **action spectrum**. **Figure 8.6** shows the absorption spectrum for a pigment, chlorophyll *a*, isolated from the leaves of a plant and the action spectrum for photosynthetic activity for the same plant. A comparison of the two spectra shows that the wavelengths at which photosynthesis is maximal are the same wavelengths at which chlorophyll *a* absorbs light.

Photosynthesis uses energy absorbed by several pigments

The light energy used for photosynthesis is not absorbed by just one type of pigment. Instead, several different pigments with different absorption spectra absorb the energy that is eventually used for photosynthesis. In photosynthetic organisms of all kinds (plants, protists, and bacteria), these pigments include *chlorophylls*, *carotenoids*, and *phycobilins*.

8.5 The Electromagnetic Spectrum The portion of the electromagnetic spectrum that is visible to humans as light is shown in detail at the right.

CHLOROPHYLLS In plants, two **chlorophylls** predominate: chlorophyll *a* and chlorophyll *b*. These two molecules differ only slightly in their molecular structure. Both have a complex ring structure similar to that of the heme group of hemoglobin. In the center of each chlorophyll ring is a magnesium atom, and attached at a peripheral location on the ring is a long hydrocarbon "tail," which can anchor the chlorophyll molecule to integral proteins in the thylakoid membrane of a chloroplast (**Figure 8.7**; review the anatomy of a chloroplast in Figure 4.14).

An ancient Roman cookbook advised *omne holus smaragdinum fit, si cum nitro coquatur*—which means, "All green vegetables will be emerald-colored if cooked with nitrum." Nitrum is a natural form of sodium bicarbonate, which buffers chlorophyll and keeps it bright green.

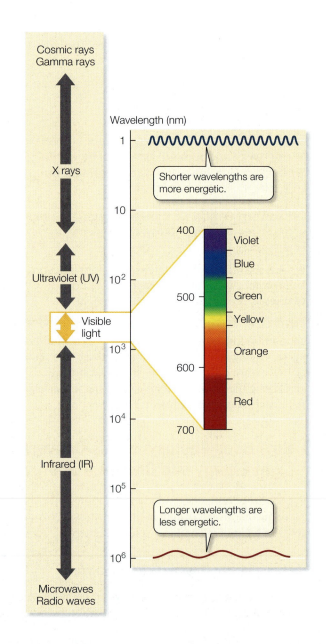

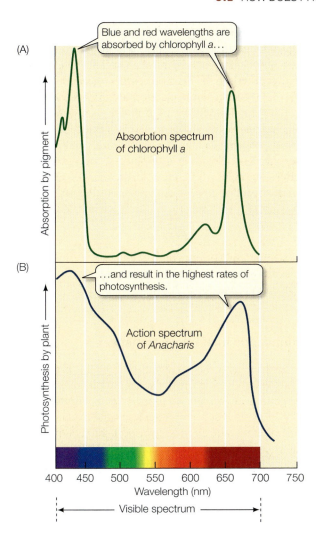

8.6 Absorption and Action Spectra The absorption spectrum (A) of the purified pigment chlorophyll *a* from the aquatic plant *Anacharis* is similar to the action spectrum (B) obtained when different wavelengths of light are shone on the intact plant and the rate of photosynthesis is measured.

ACCESSORY PIGMENTS We see in Figure 8.6 that chlorophyll absorbs blue and red wavelengths, which are near the two ends of the visible spectrum. Thus, if only chlorophyll were active in photosynthesis, much of the visible spectrum would go unused. However, all photosynthetic organisms possess **accessory pigments**, which absorb photons intermediate in energy between the red and the blue wavelengths and then transfer a portion of that energy to the chlorophylls. Among these accessory pigments are **carotenoids**, such as β-carotene, which absorb photons in the blue and blue-green wavelengths and appear deep yellow. The **phycobilins**, which are found in red algae and in cyanobacteria, absorb various yellow-green, yellow, and orange wavelengths.

Light absorption results in photochemical change

Any pigment molecule can become excited when its absorption spectrum matches the energies of incoming photons. After a pigment molecule absorbs a photon and enters an excited state (see Figure 8.4), that molecule returns to the ground state. When this

happens, some of the absorbed energy may be given off as heat, and the rest may be given off as light energy, or *fluorescence*. Because some of the absorbed light energy is lost as heat, the fluorescence has less energy and longer wavelengths than the absorbed light. When there is fluorescence, there are no permanent chemical changes or biological functions—no chemical work is done. If fluorescence does not occur, the pigment molecule may pass the absorbed energy along to another molecule, provided that the target molecule is very near, has the right orientation, and has the appropriate structure to receive the energy.

The pigments in photosynthetic organisms are arranged into energy-absorbing **antenna systems**. In these systems, the pigments are packed together and attached to thylakoid membrane proteins in such a way that the excitation energy from an absorbed photon can be passed along from one pigment molecule in the sys-

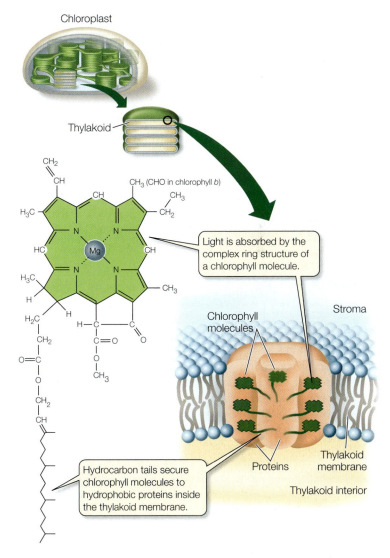

8.7 The Molecular Structure of Chlorophyll Chlorophyll consists of a complex ring structure (green area) with a magnesium atom at the center, plus a hydrocarbon "tail." The "tail" anchors the chlorophyll molecule to an integral protein in the thylakoid membrane. Chlorophyll *a* and chlorophyll *b* are identical except for the replacement of a methyl group ($-CH_3$) with an aldehyde group ($-CHO$) at the upper right.

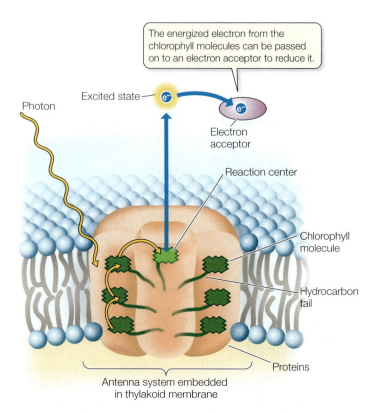

The energized electron from the chlorophyll molecules can be passed on to an electron acceptor to reduce it.

8.8 Energy Transfer and Electron Transport Rather than being lost as fluorescence, energy from a photon may be transferred from one pigment molecule to another. In an antenna system, an excited pigment molecule can transfer energy through a series of other pigment molecules to a pigment molecule in the reaction center. That molecule may become sufficiently excited that it gives up its excited electron, which can then be passed on to an electron acceptor.

tem to another (**Figure 8.8**). Excitation energy moves from pigments that absorb shorter wavelengths (higher energy) to pigments that absorb longer wavelengths (lower energy). Thus the excitation ends up in the one pigment molecule in the antenna system that absorbs the longest wavelengths; this molecule is in the **reaction center** of the antenna system.

It is the reaction center that converts the absorbed light energy into chemical energy. It is in the reaction center that a pigment molecule absorbs sufficient energy that it actually gives up its excited electron (is chemically oxidized) and becomes positively charged. In plants, the pigment molecule in the reaction center is always a molecule of chlorophyll *a*. There are many other chlorophyll *a* molecules in the antenna system, but all of them absorb light at shorter wavelengths than does the molecule in the reaction center.

Excited chlorophyll in the reaction center acts as a reducing agent

Chlorophyll has two vital roles in photosynthesis:

- It absorbs light energy and transforms it into chemical energy in the form of electrons.
- It transfers those electrons to other molecules.

We have dealt with the first role; now we turn to the second.

Photosynthesis harvests chemical energy by using the excited chlorophyll molecule in the reaction center as a *reducing agent* (*electron donor*) to reduce a stable electron acceptor (see Figure 8.8). Ground-state chlorophyll (symbolized Chl) is not much of a reducing agent, but excited chlorophyll (Chl*) is a good one. To understand the reducing capability of Chl*, recall that in an excited molecule, one of the electrons is zipping around in a shell farther away from its nucleus. Less tightly held, this electron can be passed on in a redox reaction to an oxidizing agent. Thus Chl* (but not Chl) can react with an oxidizing agent A in a reaction like this:

$$Chl^* + A \rightarrow Chl^+ + A^-$$

This, then, is the first consequence of light absorption by chlorophyll: the chlorophyll becomes a reducing agent (Chl*) and participates in a redox reaction. (The resulting Chl⁺ is a strong oxidizing agent, as described below.)

Reduction leads to electron transport

The oxidizing agent A that gets reduced by Chl* is the first in a chain of electron carriers in the thylakoid membrane of the chloroplast that participate in a process termed *electron transport*. This energetically "downhill" series of reductions and oxidations is similar to what occurs in the electron transport chain of mitochondria (see Section 7.4). The final electron acceptor is **NADP⁺ (nicotinamide adenine dinucleotide phosphate)**, which gets reduced:

$$NADP^+ + e^- \rightarrow NADPH + H^+$$

The energy-rich NADPH + H⁺ is a stable, reduced coenzyme. Its oxidized form is NADP⁺. Just as NAD⁺ couples the metabolic pathways of cellular respiration, NADP⁺ couples the two photosynthetic pathways. NADP⁺ is identical to NAD⁺ except that the former has an additional phosphate group attached to each ribose (see Figure 7.4). Whereas NAD⁺ participates in catabolism, NADP⁺ is used in anabolic (synthetic) reactions, such as carbohydrate synthesis from CO_2, that require energy from reducing power.

There are two different systems of electron transport in photosynthesis:

- *Noncyclic electron transport* produces NADPH + H⁺ and ATP.
- *Cyclic electron transport* produces only ATP.

We'll consider these two systems before considering the role of chemiosmosis in photophosphorylation—a process that is very similar to oxidative phosphorylation in mitochondria.

Noncyclic electron transport produces ATP and NADPH

In **noncyclic electron transport**, light energy is used to oxidize water, forming O_2, H⁺, and electrons.

When chlorophyll loses electrons upon excitation by light, it has an "electron hole" and thus a strong tendency to "grab" electrons from another molecule to replenish those it lost. In chemical terms, then, Chl⁺ (see above) is a strong oxidizing agent. The replenishing electrons come from water, splitting the H–O–H bonds. As the electrons are passed from water to chlorophyll, and ultimately to

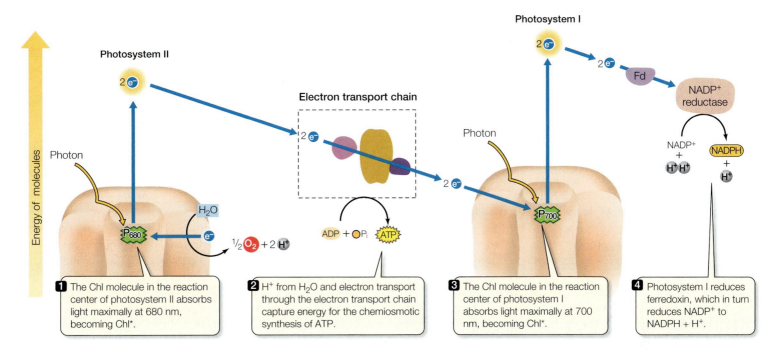

1 The Chl molecule in the reaction center of photosystem II absorbs light maximally at 680 nm, becoming Chl*.

2 H+ from H_2O and electron transport through the electron transport chain capture energy for the chemiosmotic synthesis of ATP.

3 The Chl molecule in the reaction center of photosystem I absorbs light maximally at 700 nm, becoming Chl*.

4 Photosystem I reduces ferredoxin, which in turn reduces $NADP^+$ to $NADPH + H^+$.

8.9 Noncyclic Electron Transport Uses Two Photosystems The energy of excited chlorophyll molecules in the reaction centers of photosystems I and II allows energized electrons to reduce carriers, setting up electron transport. The term "Z scheme" describes the path (blue arrows) of electrons as they travel through the two photosystems, using energy level as the *y*-axis of the "graph."

$NADP^+$, they pass through a chain of electron carriers in the thylakoid membrane. These redox reactions are exergonic, and some of the free energy released is ultimately used to form ATP by chemiosmosis.

TWO PHOTOSYSTEMS ARE REQUIRED Noncyclic electron transport requires the participation of two different **photosystems**—light-driven molecular units in the thylakoid membrane, each of which consists of many chlorophyll molecules and accessory pigments bound to proteins in *separate* energy-absorbing antenna systems.

■ **Photosystem I** uses light energy to reduce $NADP^+$ to $NADPH + H^+$.

■ **Photosystem II** uses light energy to oxidize water molecules, producing electrons, protons (H^+), and O_2.

The reaction center for photosystem I contains a chlorophyll *a* molecule called P_{700} because it can best absorb light with a wavelength of 700 nm. The reaction center for photosystem II contains a chlorophyll *a* molecule called P_{680} because it absorbs light maximally at 680 nm. Thus photosystem II requires photons that are somewhat more energetic (i.e., have shorter wavelengths) than those required by photosystem I. To keep noncyclic electron transport going, both photosystems must be constantly absorbing light, thereby boosting electrons to higher shells from which they may be captured by specific oxidizing agents. Photosystems I and II complement each other, interacting in a way that has been described in a model called the **Z scheme** (because the path of the electrons, when placed along an axis of rising energy level, resembles a sideways letter Z; **Figure 8.9**).

ELECTRON TRANSPORT: THE Z SCHEME In the Z scheme model describing the reactions of noncyclic electron transport from water to $NADP^+$, photosystem II comes before photosystem I. When photosystem II absorbs photons, electrons pass from P_{680} to the primary electron acceptor—the first carrier in the electron transport chain—and P_{680} is oxidized to P_{680}^+. Electrons from the oxidation of water are passed to P_{680}^+, reducing it once again to P_{680}, which can then absorb more photons. The electrons from photosystem II pass through a series of exergonic reactions in the electron transport chain that are indirectly coupled to proton pumping across the thylakoid membrane (described in Figure 8.11). This *chemiosmotic* pumping creates a proton gradient that produces energy for ATP synthesis.

In photosystem I, the reaction center containing P_{700} becomes excited to P_{700}^*, which leads to the reduction of an oxidizing agent called **ferredoxin** (Fd) and the production of P_{700}^+. P_{700}^+ returns to the ground state by accepting electrons passed through the electron transport chain from photosystem II.

With this accounting for the source of the electrons entering photosystem II, we can now consider the fate of the electrons from photosystem I. These electrons are used in the last step of noncyclic electron transport, in which two electrons and two protons are used to reduce a molecule of $NADP^+$ to $NADPH + H^+$.

In summary:

■ Noncyclic electron transport extracts electrons from water and passes them ultimately to $NADPH + H^+$, utilizing photons absorbed by photosystems I and II and resulting in ATP synthesis.

■ Noncyclic electron transport yields $NADPH + H^+$, ATP, and O_2.

Cyclic electron transport produces ATP but no NADPH

Noncyclic electron transport produces ATP and NADPH + H$^+$. However, as we will see, the light-independent reactions of photosynthesis use more ATP than NADPH + H$^+$. **Cyclic electron transport** occurs in some organisms when the ratio of NADPH + H$^+$ to NADP$^+$ in the chloroplast is high. This process, which produces only ATP, is called *cyclic* because an electron passed from an excited chlorophyll molecule at the outset cycles *back to the same chlorophyll molecule* at the end of the chain of reactions (**Figure 8.10**).

Before cyclic electron transport begins, P$_{700}$, the reaction center chlorophyll molecule of photosystem I, is in the ground state. It absorbs a photon and becomes P$_{700}$*. The P$_{700}$* then reacts with oxidized ferredoxin (Fd$_{ox}$) to produce reduced ferredoxin (Fd$_{red}$). The reaction is exergonic, releasing free energy. Reduced ferredoxin (Fd$_{red}$) passes its added electron to a different oxidizing agent, **plastoquinone** (**PQ**, a small organic molecule), which pumps two H$^+$ back across the thylakoid membrane. Thus Fd$_{red}$ reduces PQ, and PQ$_{red}$ passes the electron to the electron transport chain by way of **plastocyanin** (**PC**) until it completes its cycle by returning to P$_{700}$$^+$, resulting in a restoration of its uncharged form, P$_{700}$. By the time the electron from P$_{700}$* travels through the electron transport chain and comes back to reduce P$_{700}$$^+$, all the energy from the original photon has been released. This cycle is a series of redox reactions, each exergonic, and the released energy is stored in the form of a proton gradient that can be used to produce ATP.

Chemiosmosis is the source of the ATP produced in photophosphorylation

Section 7.4 considers the chemiosmotic mechanism for ATP formation in the mitochondrion. A similar chemiosmotic mechanism operates in **photophosphorylation**, the light-driven production of ATP from ADP and P$_i$ in the chloroplast. In chloroplasts, electron transport through the electron transport chain is coupled to the transport of protons (H$^+$) across the thylakoid membrane, which results in a proton gradient across the membrane (**Figure 8.11**).

> About 60 percent of the amino acid sequence in chloroplast ATP synthase is the same as that in human mitochondrial ATP synthase—a remarkable similarity, given that plants and animals had their most recent common ancestor more than a billion years ago.

The electron carriers in the thylakoid membrane are oriented so that protons move from the stroma—the interior matrix of the chloroplast—into the lumen of the thylakoid. Thus the lumen becomes acidic with respect to the stroma. This difference leads to the diffusion of H$^+$ back out of the thylakoid lumen through specific protein channels in the thylakoid membrane. These channels are enzymes—ATP synthases—that couple the diffusion of protons to the formation of ATP, just as in mitochondria (see Figure 7.14). The mechanisms of the two enzymes are also similar. What is different is their orientation: in plants, protons flow through the ATP synthase out of the thylakoid lumen, but in animals they flow *into* the mitochondrial matrix.

8.2 RECAP

Conversion of light energy into chemical energy occurs when pigments absorb photons. Light energy is transferred to electrons, which act to reduce a series of molecules in the chloroplast.

- How does chlorophyll absorb and transfer light energy? See pp. 165–166 and Figure 8.8

- Do you understand how electrons are produced in photosystem II and then flow to photosystem I? See pp. 166–167 and Figure 8.9

- How does cyclic electron transport in photosystem I result in the production of ATP? See p. 168 and Figure 8.10

We have seen how light energy drives the synthesis of ATP and NADPH + H$^+$. We now turn to the light-independent reactions of photosynthesis, which use these two energy-rich coenzymes to reduce CO$_2$ and form carbohydrates.

1 The Chl* molecule in the reaction center of Photosystem I passes electrons to an oxidizing agent, ferredoxin, leaving positively charged chlorophyll (Chl$^+$).

2 The carriers of the electron transport chain are in turn reduced.

3 Energy from electron flow is captured for chemiosmotic synthesis of ATP.

4 The last reduced electron carrier passes electrons to electron-deficient chlorophyll, allowing the reactions to start again.

Photosystem I

Electron transport chain

Energy of molecules

Photon

2 e$^-$

Fd

2 e$^-$

ADP + P$_i$

ATP

P$_{700}$

8.10 Cyclic Electron Transport Traps Light Energy as ATP Cyclic electron transport produces ATP, but no NADPH + H$^+$. The same chlorophyll molecule passes on the electrons that start the reactions and receives the electrons at the end of the reactions to start the process over again. Photosystem I and the electron transport molecules are the same as in noncyclic electron transport (see Figure 8.9). Photosystem II is not involved in cyclic electron transport.

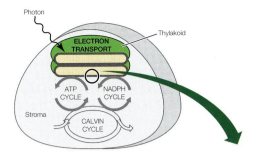

8.11 Chloroplasts Form ATP Chemiosmotically Protons (H$^+$) pumped across the thylakoid membrane from the stroma during electron transport make the lumen of the thylakoid more acidic than the stroma. Driven by this pH difference, the protons diffuse back to the stroma through ATP synthase channels, which couple the energy of proton diffusion to the formation of ATP from ADP + P$_i$.

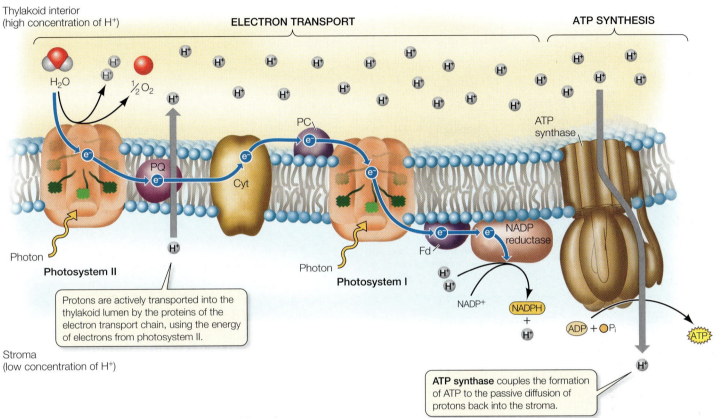

Protons are actively transported into the thylakoid lumen by the proteins of the electron transport chain, using the energy of electrons from photosystem II.

ATP synthase couples the formation of ATP to the passive diffusion of protons back into the stroma.

8.3 How Is Chemical Energy Used to Synthesize Carbohydrates?

Most of the enzymes that catalyze the reactions of CO$_2$ fixation are dissolved in the stroma of the chloroplast, and that is where those reactions take place. However, these enzymes use the energy in ATP and NADPH, produced in the thylakoids by the light reactions, to reduce CO$_2$ to carbohydrates. Because there is no stockpiling of these energy-rich coenzymes, those "light-independent" reactions take place *only in the light*, when these coenzymes are being generated.

Radioisotope labeling experiments revealed the steps of the Calvin cycle

To identify the sequence of reactions by which the carbon from CO$_2$ ends up in carbohydrates, scientists found a way to label CO$_2$ so that it could be followed after being taken up by a photosynthetic cell. In the 1950s, an experiment performed by Melvin Calvin, Andrew Benson, and their colleagues used radioactively labeled CO$_2$ in which some of the carbon atoms were not the normal ^{12}C, but its radioisotope ^{14}C. Although ^{14}C is distinguished by its emission of radiation, chemically it behaves virtually identically to nonradioactive ^{12}C. In general, enzymes do not distinguish be-

tween isotopes of an element in their substrates, so photosynthesizing cells utilize ^{14}CO$_2$ no differently than ^{12}CO$_2$.

Calvin and his colleagues exposed cultures of the unicellular green alga *Chlorella* to ^{14}CO$_2$ for 30 seconds. They then rapidly killed the cells and extracted their organic compounds. They separated the different compounds from one another by *paper chromatography* (**Figure 8.12**), a technique that had been invented just a few years previously by two British scientists. The algal extracts were dissolved in alcohol, and then applied to a sheet of filter paper, where they formed hydrogen bonds with the cellulose of the paper. The paper was put into a solvent called phenol-water, which crept up the paper by capillary action, much like water being absorbed by a paper towel. The various molecules in the algal extract were dissolved in the phenol-water and carried along. But as the solvent moved, some became less and less attracted to the solvent molecules compared with their attraction with the paper. Finally, they came out of solution and stayed where they were. Different molecules had different properties in this regard, and the use of a second solvent moving in a different direction provided even more separation. An X-ray film was exposed to the filter paper to reveal the positions of radioactive compounds.

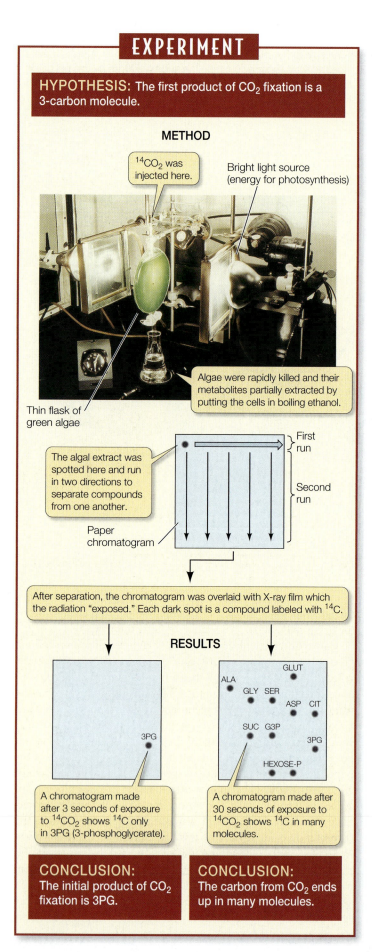

EXPERIMENT

HYPOTHESIS: The first product of CO_2 fixation is a 3-carbon molecule.

METHOD

$^{14}CO_2$ was injected here.

Bright light source (energy for photosynthesis)

Algae were rapidly killed and their metabolites partially extracted by putting the cells in boiling ethanol.

Thin flask of green algae

The algal extract was spotted here and run in two directions to separate compounds from one another.

First run

Second run

Paper chromatogram

After separation, the chromatogram was overlaid with X-ray film which the radiation "exposed." Each dark spot is a compound labeled with ^{14}C.

RESULTS

3PG

GLUT
ALA
GLY SER
ASP CIT
SUC G3P
3PG
HEXOSE-P

A chromatogram made after 3 seconds of exposure to $^{14}CO_2$ shows ^{14}C only in 3PG (3-phosphoglycerate).

A chromatogram made after 30 seconds of exposure to $^{14}CO_2$ shows ^{14}C in many molecules.

CONCLUSION:
The initial product of CO_2 fixation is 3PG.

CONCLUSION:
The carbon from CO_2 ends up in many molecules.

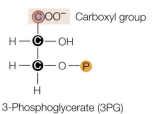

8.12 Tracing the Pathway of CO_2 The historical photograph at the top shows the apparatus Calvin and his colleagues used to follow radiolabeled carbon dioxide molecules ($^{14}CO_2$) as they were transformed by photosynthesis.

But many compounds in the algal extract, including monosaccharides and amino acids, contained ^{14}C. To discover the compound in which the labeled carbon first appears (suggesting the first step in the pathway of CO_2 fixation), Calvin and his team exposed the algae to $^{14}CO_2$ for just 3 seconds. This 3-second exposure revealed that only one compound was labeled—a 3-carbon sugar phosphate called 3-phosphoglycerate (3PG) (the ^{14}C is shown in red):

COO⁻ Carboxyl group

H — C — OH

H — C — O — P

H

3-Phosphoglycerate (3PG)

By tracing the steps with successive, increasingly long exposures, Calvin and his colleagues discovered the series of compounds through which the carbon taken up in CO_2 flows. Its pathway was discovered to be a cycle that "fixes" CO_2 in a larger molecule, produces a carbohydrate, and regenerates the initial CO_2 acceptor. This cycle was appropriately named the **Calvin cycle** (**Figure 8.13**).

The initial reaction in the Calvin cycle adds the 1-carbon CO_2 to an acceptor molecule, the 5-carbon compound **ribulose 1,5-bisphosphate** (**RuBP**). The product is an intermediate 6-carbon compound, which quickly breaks down and forms two 3-carbon molecules of 3PG (as Calvin and colleagues observed; **Figure 8.14**). The enzyme that catalyzes this fixation reaction, **ribulose bisphosphate carboxylase/oxygenase** (**rubisco**), is the most abundant protein in the world, constituting up to 50 percent of all the protein in every plant leaf.

The Calvin cycle is made up of three processes

The Calvin cycle uses the high-energy coenzymes made in the thylakoids during the light reactions (ATP and NADPH) to reduce CO_2 to a carbohydrate in the stroma. Three processes make up the cycle:

- *Fixation of CO_2.* As we have seen, this reaction is catalyzed by rubisco, and its product is 3PG.

- *Reduction of 3PG to form glyceraldehyde 3-phosphate (G3P).* This series of reactions involves a phosphorylation (using the ATP made in the light reactions) and a reduction (using the NADPH made in the light reactions).

- *Regeneration of the CO_2 acceptor, RuBP.* Most of the G3P ends up as RuMP (ribulose monophosphate), and ATP is used to convert this compound into RuBP. So for every "turn" of the cycle, with one CO_2 fixed, the CO_2 acceptor is regenerated.

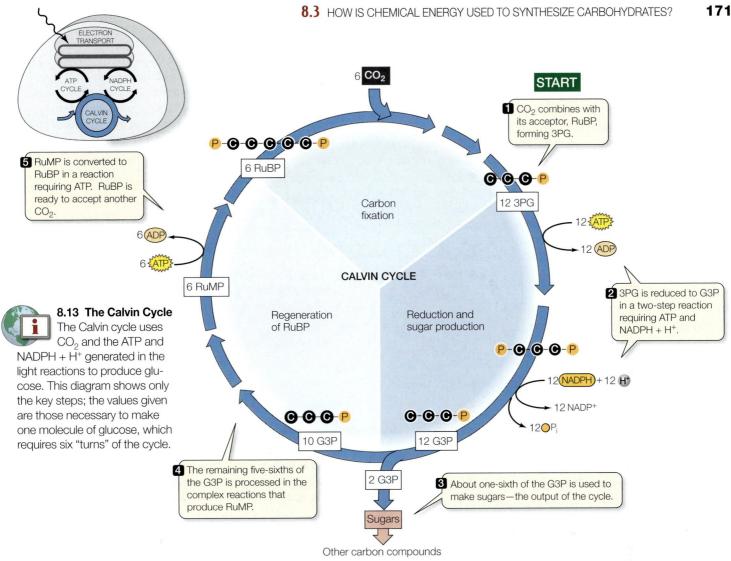

5 RuMP is converted to RuBP in a reaction requiring ATP. RuBP is ready to accept another CO_2.

1 CO_2 combines with its acceptor, RuBP, forming 3PG.

2 3PG is reduced to G3P in a two-step reaction requiring ATP and NADPH + H^+.

3 About one-sixth of the G3P is used to make sugars—the output of the cycle.

4 The remaining five-sixths of the G3P is processed in the complex reactions that produce RuMP.

8.13 The Calvin Cycle The Calvin cycle uses CO_2 and the ATP and NADPH + H^+ generated in the light reactions to produce glucose. This diagram shows only the key steps; the values given are those necessary to make one molecule of glucose, which requires six "turns" of the cycle.

The product of this cycle is glyceraldehyde 3-phosphate (G3P), which is a 3-carbon sugar phosphate, also called triose phosphate:

Glyceraldehyde 3-phosphate (G3P)

In a typical leaf, five-sixths of the G3P is recycled into RuBP. There are two fates for the remaining G3P:

- One-third of it ends up in the polysaccharide *starch*, which is stored in the chloroplast.

- Two-thirds of it is converted in the cytosol into the disaccharide *sucrose*, which is transported out of the leaf to other organs in the plant, where it is hydrolyzed to its constituent monosaccharides: *glucose* and *fructose*.

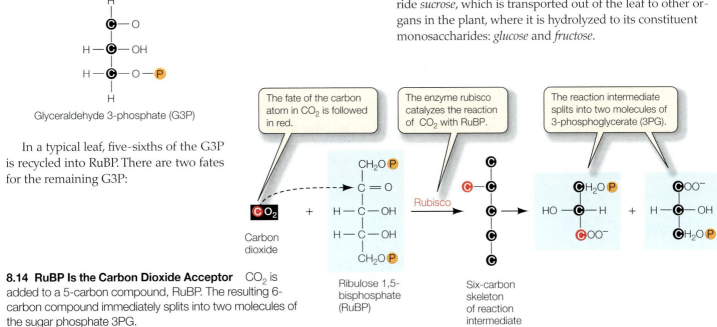

The fate of the carbon atom in CO_2 is followed in red.

The enzyme rubisco catalyzes the reaction of CO_2 with RuBP.

The reaction intermediate splits into two molecules of 3-phosphoglycerate (3PG).

Carbon dioxide

Ribulose 1,5-bisphosphate (RuBP)

Six-carbon skeleton of reaction intermediate

8.14 RuBP Is the Carbon Dioxide Acceptor CO_2 is added to a 5-carbon compound, RuBP. The resulting 6-carbon compound immediately splits into two molecules of the sugar phosphate 3PG.

These carbohydrates are subsequently used by the plant to make other compounds. Their carbon molecules are incorporated into amino acids, lipids, and the building blocks of nucleic acids.

The products of the Calvin cycle are of crucial importance to the entire biosphere, for the covalent bonds of the carbohydrates generated by the cycle represent the total energy yield from the harvesting of light by photosynthetic organisms. These organisms, which are also called *autotrophs* ("self-feeders"), release most of this energy by glycolysis and cellular respiration and use it to support their own growth, development, and reproduction. Much plant matter ends up being consumed by *heterotrophs* ("other-feeders"), such as animals, which cannot photosynthesize and depend on autotrophs for both raw materials and energy sources. Glycolysis and cellular respiration in heterotroph cells release free energy from food for use by the heterotrophs.

Light stimulates the Calvin cycle

The Calvin cycle uses NADPH and ATP, which, as we have seen, are made through photophosphorylation. Two other processes connect the light reactions with this CO_2 fixation pathway. Both connections are indirect, but significant:

- *Light-induced pH changes in the stroma activate some enzymes in the Calvin cycle.* Proton pumping from the stroma into the thylakoids increases the pH of the stroma from 7 to 8 (a tenfold decrease in H^+ concentration). This favors the activation of rubisco.

- *Light-induced electron flow reduces disulfide bonds to activate four Calvin cycle enzymes* (**Figure 8.15**). When ferredoxin is reduced in photosystem I (see Figure 8.9), it passes some electrons to a small, soluble protein called *thioredoxin*. This

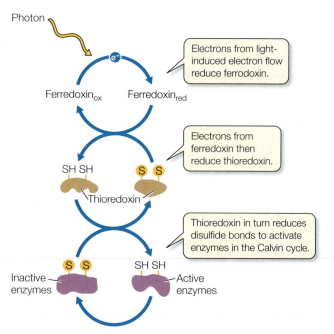

8.15 The Photochemical Reactions Stimulate the Calvin Cycle
By reducing (breaking) disulfide bridges, electrons from the light reactions activate enzymes in CO_2 fixation.

protein in turn passes electrons to four enzymes in the CO_2 fixation pathway. These enzymes all have disulfide bridges (see Figure 3.5) near their active sites, and reduction of the sulfurs breaks the bridges. The resulting change in their three-dimensional shape activates these four enzymes.

Although all green plants carry out the Calvin cycle, in some plant groups variations on (or additional steps in) the light-independent reactions have evolved in response to certain environmental conditions. Let's look at these environmental limitations and the metabolic bypasses that have evolved to circumvent them.

8.4 How Do Plants Adapt to the Inefficiencies of Photosynthesis?

One major limitation of rubisco is its tendency to react with O_2 instead of CO_2. This reaction leads to a process called photorespiration, which lowers the overall rate of CO_2 fixation. After examining this problem, we'll look at some biochemical pathways and features of plant anatomy that compensate for the limitations of rubisco.

Rubisco catalyzes RuBP reaction with O_2 as well as with CO_2

As its full name indicates, rubisco is an **oxygenase** as well as a **carboxylase**; that is, it can add O_2 to the acceptor molecule RuBP instead of CO_2. These two reactions compete with each other, so if RuBP reacts with O_2, it cannot react with CO_2. This reaction reduces the overall amount of CO_2 that is converted into carbohydrates, and therefore limits plant growth.

When O_2 is added to RuBP, one of the products is a 2-carbon compound, phosphoglycolate:

$$RuBP + O_2 \rightarrow phosphoglycolate + 3PG$$

Plants have evolved a metabolic pathway that can partially recover the carbon that has been channeled away from the Calvin cycle into phosphoglycolate. The phosphoglycolate forms glycolate, which diffuses into membrane-enclosed organelles called *peroxi-*

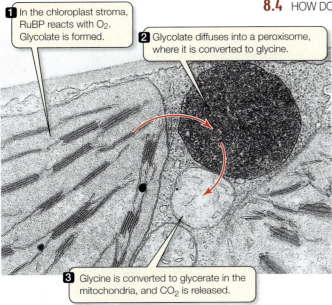

1 In the chloroplast stroma, RuBP reacts with O_2. Glycolate is formed.

2 Glycolate diffuses into a peroxisome, where it is converted to glycine.

3 Glycine is converted to glycerate in the mitochondria, and CO_2 is released.

8.16 Organelles of Photorespiration The reactions of photorespiration take place in the chloroplasts, peroxisomes, and finally, in the mitochondria.

somes (**Figure 8.16**). There, a series of reactions converts it into the amino acid glycine:

$$glycolate \rightarrow glycine$$

The glycine then diffuses into a mitochondrion, where two glycine molecules are converted into glycerate (a 3-carbon molecule) and CO_2:

$$2 \ glycine \rightarrow glycerate + CO_2$$

This pathway is called **photorespiration** because it consumes O_2 and releases CO_2. It uses ATP and NADPH produced in the light reactions, just like the Calvin cycle. The net effect is to take two 2-carbon molecules and make one 3-carbon molecule. So one carbon of the four is released as CO_2, and three of the carbons (75 percent) are recovered as fixed carbon. In other words, photorespiration reduces the net carbon fixed by the Calvin cycle by 25 percent.

How does rubisco "decide" whether to act as an oxygenase or a carboxylase? There are three factors involved:

- Rubisco has ten times more affinity for CO_2 than for O_2, and so favors CO_2 fixation.

- In the leaf, the relative concentrations of CO_2 and O_2 vary. If O_2 is relatively abundant, rubisco acts as an oxygenase, and photorespiration ensues. If CO_2 predominates, rubisco fixes it, and the Calvin cycle occurs.

- Photorespiration is more likely at high temperatures. On a hot, dry day, the stomata that allow water to evaporate from the leaf close to prevent water loss (see Figure 8.1). But this also prevents gases from entering and leaving the leaf. The CO_2 concentration in the leaf falls because CO_2 is being used up by photosynthetic reactions, and the O_2 concentration rises because of these same reactions. As the ratio of CO_2 to O_2 in the leaf falls, the oxygenase activity of rubisco is favored, and photorespiration proceeds.

C_4 plants can bypass photorespiration

In plants such as roses, wheat, and rice, the palisade **mesophyll** cells, which lie just below the surface of the leaf, are full of chloroplasts that contain abundant rubisco (**Figure 8.17A**). On a hot day, these leaves close their stomata to conserve water. The level of CO_2 in the air spaces of the leaves falls, and that of O_2 rises, as photosynthesis goes on. Under these conditions, rubisco acts as an oxygenase, and photorespiration occurs. Because the first product of CO_2 fixation in these plants is the 3-carbon molecule 3PG, they are called **C_3 plants**.

Corn, sugarcane, and other tropical grasses (**Figure 8.17B**) also close their stomata on a hot day, but their rate of photosynthesis does not fall, nor does photorespiration occur. They keep the ratio of CO_2 to O_2 around rubisco high so that rubisco continues to act as a carboxylase. They do this in part by making a 4-carbon compound, *oxaloacetate*, as the first product of CO_2 fixation, and so are called **C_4 plants**.

C_4 plants perform the normal Calvin cycle, but they have an additional early reaction that fixes CO_2 without losing carbon to photorespiration. Because this initial CO_2 fixation step can function

(A) Arrangement of cells in a C_3 leaf

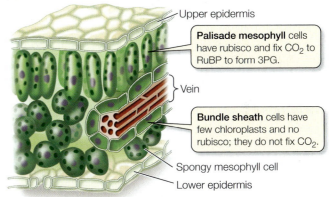

- Upper epidermis
- **Palisade mesophyll** cells have rubisco and fix CO_2 to RuBP to form 3PG.
- Vein
- **Bundle sheath** cells have few chloroplasts and no rubisco; they do not fix CO_2.
- Spongy mesophyll cell
- Lower epidermis

(B) Arrangement of cells in a C_4 leaf

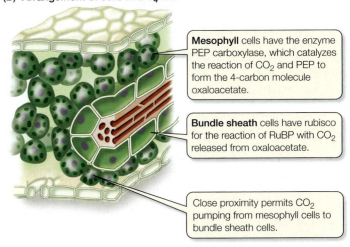

- **Mesophyll** cells have the enzyme PEP carboxylase, which catalyzes the reaction of CO_2 and PEP to form the 4-carbon molecule oxaloacetate.
- **Bundle sheath** cells have rubisco for the reaction of RuBP with CO_2 released from oxaloacetate.
- Close proximity permits CO_2 pumping from mesophyll cells to bundle sheath cells.

8.17 Leaf Anatomy of C_3 and C_4 Plants Carbon dioxide fixation occurs in different organelles and cells of the leaves in (A) C_3 plants and (B) C_4 plants.

8.18 The Anatomy and Biochemistry of C₄ Carbon Fixation (A) Carbon dioxide is fixed initially in the mesophyll cells, but enters the Calvin cycle in the bundle sheath cells. (B) The two cell types share an interconnected biochemical pathway for CO_2 assimilation.

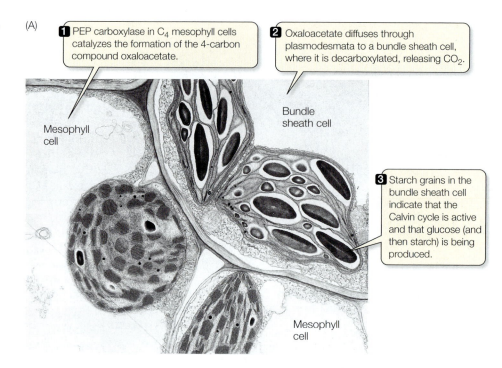

(A)

1 PEP carboxylase in C₄ mesophyll cells catalyzes the formation of the 4-carbon compound oxaloacetate.

2 Oxaloacetate diffuses through plasmodesmata to a bundle sheath cell, where it is decarboxylated, releasing CO_2.

Mesophyll cell

Bundle sheath cell

3 Starch grains in the bundle sheath cell indicate that the Calvin cycle is active and that glucose (and then starch) is being produced.

Mesophyll cell

even at low levels of CO_2 and high temperatures, C₄ plants very effectively optimize photosynthesis under conditions that inhibit it in C₃ plants. C₄ plants have two separate enzymes for CO_2 fixation, located in two different parts of the leaf (**Figure 8.18**; see also Figure 8.17B). The first enzyme, which is present in the cytosol of mesophyll cells near the surface of the leaf, fixes CO_2 to a 3-carbon acceptor compound, **phospho-enolpyruvate** (**PEP**), to produce the 4-carbon fixation product, oxaloacetate. This enzyme, **PEP carboxylase**, has two advantages over rubisco:

- It does not have oxygenase activity.
- It fixes CO_2 even at very low CO_2 levels.

So even on a hot day when the stomata are closed, the CO_2 concentration in the leaf is low, and the O_2 concentration is high, PEP carboxylase just keeps on fixing CO_2.

Oxaloacetate diffuses out of the mesophyll cells and into the **bundle sheath cells** (see Figure 8.17B), located in the interior of the leaf. The chloroplasts in the bundle sheath cells contain abundant rubisco. There, the 4-carbon oxaloacetate loses one carbon (is *decarboxylated*), forming CO_2 and regenerating the 3-carbon acceptor compound, PEP, in the mesophyll cells. Thus the role of PEP

TABLE 8.1

Comparison of Photosynthesis in C₃ and C₄ Plants

VARIABLE	C₃ PLANTS	C₄ PLANTS
Photorespiration	Extensive	Minimal
Perform Calvin cycle?	Yes	Yes
Primary CO_2 acceptor	RuBP	PEP
CO_2-fixing enzyme	Rubisco (RuBP carboxylase/ oxygenase)	PEP carboxylase and rubisco
First product of CO_2 fixation	3PG (3-carbon compound)	Oxaloacetate (4-carbon compound)
Affinity of carboxylase for CO_2	Moderate	High
Photosynthetic cells of leaf	Mesophyll	Mesophyll + bundle sheath
Classes of chloroplasts	One	Two

(B)

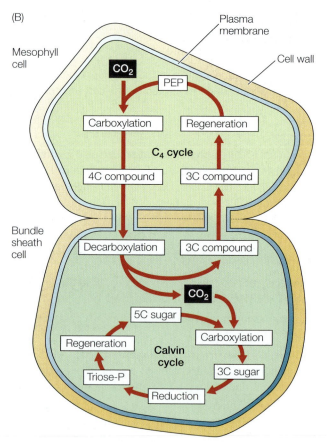

Plasma membrane

Mesophyll cell

Cell wall

CO_2

PEP

Carboxylation

Regeneration

C₄ cycle

4C compound

3C compound

Bundle sheath cell

Decarboxylation

3C compound

CO_2

5C sugar

Regeneration

Calvin cycle

Carboxylation

Triose-P

3C sugar

Reduction

is to bind CO_2 from the air in the leaf and carry it to the bundle sheath cells, where it is "dropped off" at rubisco. This process essentially pumps up the CO_2 concentration around rubisco, so that it acts as a carboxylase and begins the Calvin cycle.

Kentucky bluegrass, a C₃ plant, thrives on lawns in April and May. But in the heat of summer, it does not do as well, and

Bermuda grass (crabgrass), a C_4 plant, takes over the lawn. The same is true on a global scale for crops: C_3 plants, such as soybeans, rice, wheat, and barley, have been adapted for human food production in temperate climates, while C_4 plants, such as corn and sugarcane, originated and are grown in the tropics. **Table 8.1** compares C_3 and C_4 photosynthesis.

C_3 plants are certainly more ancient than C_4 plants. While C_3 photosynthesis appears to have begun about 3.5 billion years ago, C_4 plants appeared about 12 million years ago. A possible factor in the emergence of the C_4 pathway is the decline in atmospheric CO_2. When dinosaurs dominated Earth 100 million years ago, the concentration of CO_2 in the atmosphere was four times what it is now. As CO_2 levels declined thereafter, the more efficient C_4 plants would have had an advantage over their C_3 counterparts. In the last two hundred years, however, CO_2 levels began increasing. Measurement of atmospheric CO_2 at sites far away from industrial sources of the gas, such as the top of a volcano in Hawaii and bubbles in Antarctic ice, show that the level of this gas has risen significantly, from 250 parts per million in 1800 to 370 ppm now. This rise may be affecting photosynthesis and plant growth. Currently, the level of CO_2 is not enough for maximal activity of CO_2 fixation by rubisco, so photorespiration occurs and reduces the growth of C_3 plants, favoring C_4 plants. If CO_2 in the atmosphere increases even more, the reverse will occur, and C_3 plants will have a comparative advantage.

Climate scientists predict that atmospheric CO_2 could rise to 600 ppm by 2100. If this happens, the overall growth of crops such as rice and wheat should increase. This may or may not translate into more food, given that other effects of the human-spurred CO_2 increase (such as global warming) will also alter Earth's ecosystem.

CAM plants also use PEP carboxylase

Other plants besides the C_4 plants use PEP carboxylase to fix and accumulate CO_2. Such plants include some water-storing plants (called *succulents*) of the family Crassulaceae, many cacti, pineapples, and several other kinds of flowering plants. The CO_2 metabolism of these plants is called **crassulacean acid metabolism**, or **CAM**, after the family of succulents in which it was discovered. CAM is much like the metabolism of C_4 plants in that CO_2 is initially fixed into a 4-carbon compound. In CAM plants, however, the processes of initial CO_2 fixation and the Calvin cycle are separated in time, rather than in space.

- *At night*, when it is cooler and water loss is minimized, the stomata open. CO_2 is fixed in mesophyll cells to form the 4-carbon compound oxaloacetate, which is converted into malic acid.
- *During the day*, when the stomata close to reduce water loss, the accumulated malic acid is shipped to the chloroplasts, where its decarboxylation supplies the CO_2 for operation of the Calvin cycle, and the light reactions supply the necessary ATP and NADPH + H⁺.

8.4 RECAP

Rubisco catalyzes the fixation of CO_2 to RuBP but can also fix O_2 to that molecule. This diversion of rubisco decreases net CO_2 fixation. C_4 photosynthesis and crassulacean acid metabolism allow plants to get around this problem.

- Can you describe how photorespiration recovers some of the carbon that is channeled away from the Calvin cycle? See p. 173
- What do C_4 plants do to keep the concentration of CO_2 around rubisco high, and why? See pp. 173–174
- Can you describe CO_2 fixation in CAM plants? See p. 175

Now that we understand how photosynthesis produces carbohydrates, let's see how those carbohydrates fuel the metabolism of plants, and how the pathways of photosynthesis are connected to other metabolic pathways.

8.5 How Is Photosynthesis Connected to Other Metabolic Pathways in Plants?

Green plants are autotrophs and can synthesize all the molecules they need from simple starting materials: CO_2, H_2O, and phosphate, sulfate, and ammonium ions (NH_4^+). The NH_4^+ is needed for amino acids and comes either from the conversion of nitrogen-containing molecules in soil water taken up by the plant's roots, or from the conversion of N_2 gas from the atmosphere by bacteria, as we'll see in Chapter 26.

Plants use the carbohydrates generated by photosynthesis to provide energy for processes such as active transport and anabolism. Both cellular respiration and fermentation can occur in plants, although the former is far more common. Plant cellular respiration, unlike photosynthesis, takes place both in the light and in the dark. Because glycolysis occurs in the cytosol, respiration in the mitochondria, and photosynthesis in the chloroplasts, all these processes can proceed simultaneously.

Photosynthesis and respiration are closely linked through the Calvin cycle (**Figure 8.19**). The partitioning of G3P is particularly important:

- Some G3P from the Calvin cycle is part of the glycolysis pathway and can be converted into pyruvate. This pyruvate can be used in cellular respiration for energy, or its carbon skeletons can be used anabolically to make lipids, proteins, and other carbohydrates (see Figure 7.17).
- Some G3P can enter a pathway that is the reverse of glycolysis (*gluconeogenesis*; see Section 7.6). In this case, hexose-phosphates and then sucrose are formed and transported to the nonphotosynthetic tissues of the plant (such as the root).

Energy flows from sunlight to reduced carbon in photosynthesis to ATP in respiration. Energy can also be stored in the bonds of macromolecules such as polysaccharides, lipids, and proteins. For a plant to grow, energy storage (as body structures) must exceed

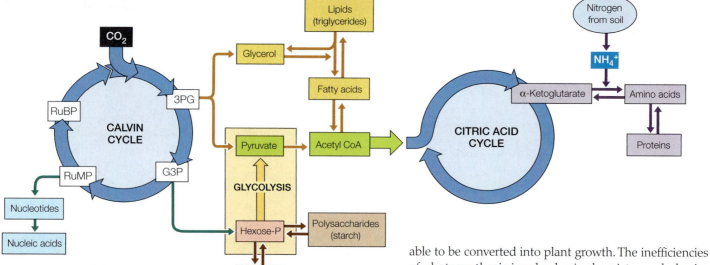

8.19 Metabolic Interactions in a Plant Cell The products of the Calvin cycle are used in the reactions of cellular respiration (glycolysis and the citric acid cycle).

energy release; that is, overall carbon fixation by photosynthesis must exceed respiration. This principle is the basis of the ecological food chain, as we will see in later chapters.

The opening of this chapter makes clear the extent to which people are dependent on photosynthesis. Given the uncertainties of the photosynthetic future (such as climate change), it would be wise to seek ways to reduce our dependence on photosynthesis or to improve photosynthetic efficiency. **Figure 8.20** shows the various ways in which solar energy is utilized and lost. In essence, only 5 percent of the sunlight that reaches Earth is avail-

able to be converted into plant growth. The inefficiencies of photosynthesis involve basic chemistry and physics (some light energy is not absorbed by photosynthetic pigments) as well as biology (plant anatomy and leaf exposure, photorespiration, inefficiencies in the metabolic pathways). While it is hard to change chemistry and physics, biologists can use their knowledge of plants to improve on the basic biology of photosynthesis. This in turn should result in more efficient use of resources and better food production.

8.5 RECAP

The products of photosynthesis are utilized in glycolysis and the citric acid cycle as well as in synthesizing lipids, proteins, and other carbohydrates.

■ Do you understand how the pathways of glycolysis and the citric acid cycle are linked to photosynthesis in plant cells? See p. 175 and Figure 8.19

ENERGY LOSS

SUNLIGHT 100% 50% → Wavelengths of light not part of absorption spectrum of photosynthetic pigments (e.g., green light)

Wavelengths of light that can be absorbed by photosynthetic pigments

30% → Light energy not absorbed due to plant structure (e.g., leaves not properly oriented to sun)

Light energy absorbed by photosynthetic pigments

10% → Inefficiency of light reactions converting light to chemical energy

Chemical energy available for CO_2 fixation

5% → Inefficiency of CO_2 fixation pathways

Chemical energy stored in carbohydrates 5%

8.20 Energy Losses During Photosynthesis As we face an increasingly uncertain photosynthetic future, understanding the inefficiencies of photosynthesis becomes increasingly important. The pathways of photosynthesis preserve only about 5 percent of the sun's energy output as chemical energy in carbohydrates.

CHAPTER SUMMARY

8.1 What is photosynthesis?

In the process of **photosynthesis**, plants and other organisms take in CO_2, water, and light energy, producing O_2 and carbohydrates. See Web/CD Tutorial 8.1

The **light reactions** of photosynthesis convert light energy into chemical energy. They produce ATP and reduce $NADP^+$ to $NADPH + H^+$. Review Figure 8.3

The **light-independent reactions** do not use light directly but instead use ATP and $NADPH + H^+$ to reduce CO_2, forming carbohydrates.

8.2 How does photosynthesis convert light energy into chemical energy?

Light is a form of **electromagnetic radiation**. It is emitted in particle-like packets called **photons** but has wavelike properties.

Molecules that absorb light in the visible spectrum are called **pigments**. Photosynthetic organisms have several pigments, most notably **chlorophylls**, but also **accessory pigments** such as **carotenoids** and **phycobilins**.

Absorption of a photon puts a pigment molecule in an **excited state** that has more energy than its **ground state**. Review Figure 8.4

Each compound has a characteristic **absorption spectrum**. An **action spectrum** reflects the biological activity of a photosynthetic organism for a given wavelength of light. Review Figure 8.6

The pigments in photosynthetic organisms are arranged into **antenna systems** that absorb energy from light and funnel this energy to a single chlorophyll *a* molecule in the **reaction center**. Chlorophyll acts as a reducing agent, transferring excited electrons to other molecules. Review Figure 8.8

Noncyclic electron transport uses **photosystems I** and **II** to produces ATP, $NADPH + H^+$, and O_2. **Cyclic electron transport** uses only photosystem I and produces only ATP. Review Figures 8.9 and 8.10

Chemiosmosis is the mechanism of ATP production in **photophosphorylation**. Review Figure 8.11, Web/CD Tutorial 8.2

8.3 How is chemical energy used to synthesize carbohydrates?

The **Calvin cycle** makes carbohydrates from CO_2. The cycle consists of three processes: fixation of CO_2, reduction and carbohydrate production, and regeneration of RuBP. See Web/CD Tutorial 8.3

RuBP is the initial CO_2 acceptor, and **3PG** is the first stable product of CO_2 fixation. The enzyme **rubisco** catalyzes the reaction of CO_2 and RuBP to form 3PG. Review Figure 8.13, Web/CD Activity 8.1

ATP and NADPH formed by the light reactions are used in the reduction of 3PG to form G3P. Light stimulates enzymes in the Calvin cycle, further integrating the two pathways.

8.4 How do plants adapt to the inefficiencies of photosynthesis?

Rubisco can catalyze a reaction between O_2 and RuBP in addition to the reaction between CO_2 and RuBP. At high temperatures and low CO_2 concentrations, the **oxygenase** function of rubisco is favored over its **carboxylase** function.

When rubisco functions as an oxygenase, the result is **photorespiration**, which significantly reduces the efficiency of photosynthesis.

In C_4 **plants**, CO_2 is fixed to PEP in mesophyll cells. The 4-carbon product releases its CO_2 to rubisco in the interior of the leaf, at the bundle sheath cells. Review Figure 8.17, Web/CD Activity 8.2

CAM plants operate much like C_4 plants, but their initial CO_2 fixation by **PEP carboxylase** is temporally separated from the Calvin cycle, rather than spatially separated as in C_4 plants.

8.5 How is photosynthesis connected to other metabolic pathways in plants?

Photosynthesis and cellular respiration are linked through the Calvin cycle, the citric acid cycle, and glycolysis. Review Figure 8.19

To survive, a plant must photosynthesize more than it respires.

Photosynthesis utilizes only a small portion of the energy of sunlight. Review Figure 8.20

SELF-QUIZ

1. In noncyclic photosynthetic electron transport, water is used to
   a. excite chlorophyll.
   b. hydrolyze ATP.
   c. reduce chlorophyll.
   d. oxidize NADPH.
   e. synthesize chlorophyll.

2. Which statement about light is true?
   a. An absorption spectrum is a plot of biological effectiveness versus wavelength.
   b. An absorption spectrum may be a good means of identifying a pigment.
   c. Light need not be absorbed to produce a biological effect.
   d. A given kind of molecule can occupy any energy level.
   e. A pigment loses energy as it absorbs a photon.

3. Which statement about chlorophylls is *not* true?
   a. Chlorophylls absorb light near both ends of the visible spectrum.

   b. Chlorophylls can accept energy from other pigments, such as carotenoids.
   c. Excited chlorophyll can either reduce another substance or fluoresce.
   d. Excited chlorophyll may be an oxidizing agent.
   e. Chlorophylls contain magnesium.

4. In cyclic electron transport,
   a. oxygen gas is released.
   b. ATP is formed.
   c. water donates electrons and protons.
   d. $NADPH + H^+$ forms.
   e. CO_2 reacts with RuBP.

5. Which of the following does *not* happen in noncyclic electron transport?
   a. Oxygen gas is released.
   b. ATP forms.
   c. Water donates electrons and protons.
   d. $NADPH + H^+$ forms.
   e. CO_2 reacts with RuBP.

6. In the chloroplasts,
   a. light leads to the pumping of protons out of the thylakoids.
   b. ATP forms when protons are pumped into the thylakoids.
   c. light causes the stroma to become more acidic than the thylakoids.
   d. protons return passively to the stroma through protein channels.
   e. proton pumping requires ATP.

7. Which statement about the Calvin cycle is *not* true?
   a. CO_2 reacts with RuBP to form 3PG.
   b. RuBP forms by the metabolism of 3PG.
   c. ATP and NADPH + H$^+$ form when 3PG is reduced.
   d. The concentration of 3PG rises if the light is switched off.
   e. Rubisco catalyzes the reaction of CO_2 and RuBP.

8. In C$_4$ photosynthesis,
   a. 3PG is the first product of CO_2 fixation.
   b. rubisco catalyzes the first step in the pathway.
   c. 4-carbon acids are formed by PEP carboxylase in bundle sheath cells.
   d. photosynthesis continues at lower CO_2 levels than in C$_3$ plants.
   e. CO_2 released from RuBP is transferred to PEP.

9. Photosynthesis in green plants occurs only during the day. Respiration in plants occurs
   a. only at night.
   b. only when there is enough ATP.
   c. only during the day.
   d. all the time.
   e. in the chloroplast after photosynthesis.

10. Photorespiration
   a. takes place only in C$_4$ plants.
   b. includes reactions carried out in peroxisomes.
   c. increases the yield of photosynthesis.
   d. is catalyzed by PEP carboxylase.
   e. is independent of light intensity.

FOR DISCUSSION

1. Both photosynthetic electron transport and the Calvin cycle stop in the dark. Which specific reaction stops first? Which stops next? Continue answering the question "Which stops next?" until you have explained why both pathways have stopped.

2. In what principal ways are the reactions of electron transport in photosynthesis similar to the reactions of oxidative phosphorylation discussed in Section 7.4?

3. Differentiate between cyclic and noncyclic electron transport in terms of (1) the products and (2) the source of electrons for the reduction of oxidized chlorophyll.

4. What two experimental techniques made it possible to elucidate the Calvin cycle? How were these techniques used in the investigation?

5. If water labeled with ^{18}O is added to a suspension of photosynthesizing chloroplasts, which of the following compounds will first become labeled with ^{18}O: ATP, NADPH, O$_2$, or 3PG? If water labeled with ^3H is added to a suspension of photosynthesizing chloroplasts, which of the same compounds will first become radioactive? If CO_2 labeled with ^{14}C is added to a suspension of photosynthesizing chloroplasts, which of those compounds will first become radioactive?

6. The Viking lander was sent to Mars in 1976 to detect signs of life. Explain the rationale behind the following experiments this unmanned probe performed:

   a. A scoop of dirt was inserted into a container and $^{14}CO_2$ was added. After a while during the Martian day, the $^{14}CO_2$ was removed and the dirt was heated to a high temperature. Scientists monitoring the experiment back on Earth looked for the release of $^{14}CO_2$ as a sign of life.

   b. The same experiment was performed, except that the dirt was heated to a high temperature for 30 minutes and then allowed to cool to Martian temperature right after scooping and *before* the $^{14}CO_2$ was added. If experiment *a* released $^{14}CO_2$, then this experiment should not release it, if living things were present.

FOR INVESTIGATION

Calvin's experiment (see Figure 8.12) laid the foundations for a full description of the pathway for fixation of CO_2. Given the metabolic interrelationships between pathways in plants, how would you do an experiment to follow the fate of fixed carbon through photosynthesis to proteins?

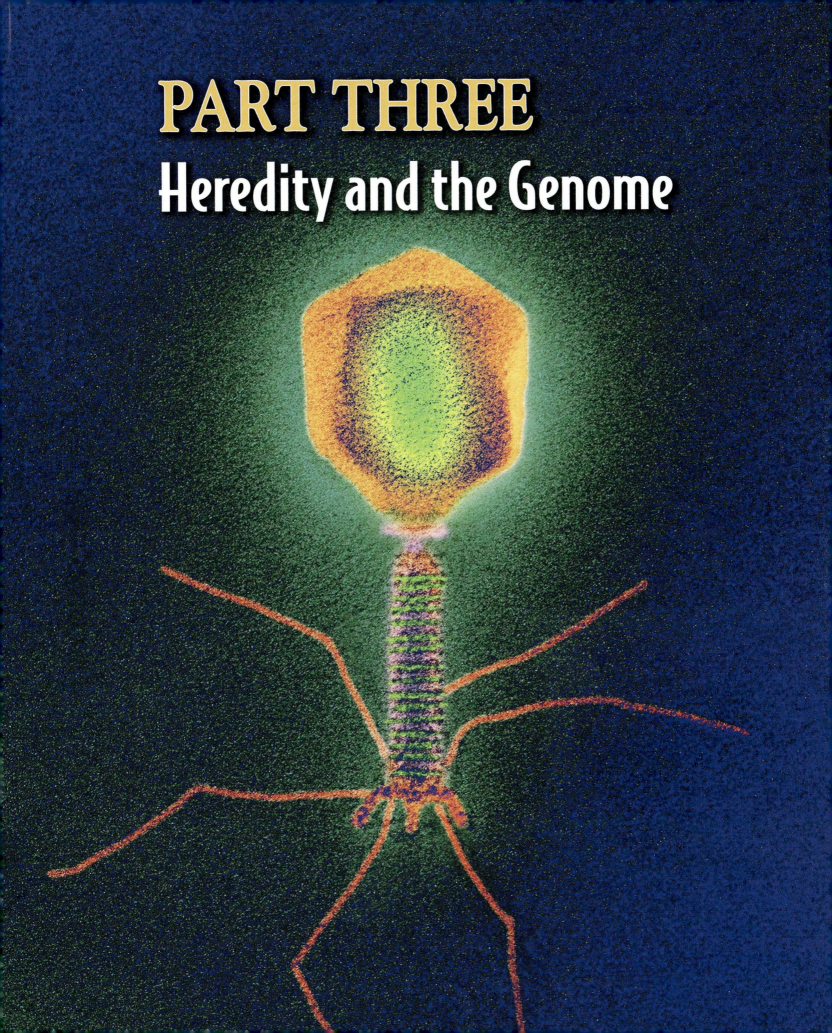

PART THREE
Heredity and the Genome

CHAPTER 9

Chromosomes, the Cell Cycle, and Cell Division

The immortal cells of Henrietta Lacks

On January 28, 1951, 31-year-old Henrietta Lacks found blood spotting her underwear. Sensing that something was wrong, the mother of five children convinced her husband to take her to nearby Johns Hopkins Hospital in Baltimore, Maryland. An examination of her cervix revealed the reason for the blood spots: a tumor the size of a quarter. Her doctor sent a piece of the tumor to a pathologist in the clinical laboratory, who confirmed that the tumor was malignant.

A week later, Henrietta was back in the hospital, where physicians treated her tumor with radium to try to kill it. Before the treatment began, however, they removed a small sample of cells from the tumor and sent them to the research laboratory of George and Margaret Gey, two scientists at Johns Hopkins who had been trying for 20 years to coax human cells to live and multiply outside the body, or in vitro. They

were attempting this in the belief that if they could get human cells to thrive in vitro, they could use those cells to find a cure for cancer. They hit paydirt with Henrietta's tumor cells, which grew more vigorously than any cells the Geys had previously cultured.

Unfortunately, the tumor cells also grew rapidly in Henrietta Lacks's body. Within a few months, cancerous cells had spread to almost all of her organs. She died on October 4, 1951. On that same day, George Gey appeared on national television displaying a test tube containing her cells—which he named HeLa cells—and saying that a cure for cancer was near.

Because of their robust ability to reproduce themselves, HeLa cells became a staple of much important basic and applied biomedical research. In controlled settings, the cells could be infected with viruses, and they were instrumental in developing the supply of poliovirus that led to the first vaccine against that disease. Although Henrietta herself had never been outside of Virginia and Maryland, her cells have traveled all over the world. HeLa cells even went into space aboard the space shuttle. Over the past half-century, tens of thousands of research articles have been published using information obtained from Henrietta's cells. But the hope that HeLa cells would lead to a quick cure for cancer has proved to be unfounded.

Cancer remains the second leading cause of death (after heart disease) in most of the world's developed nations. However, Henrietta Lacks would

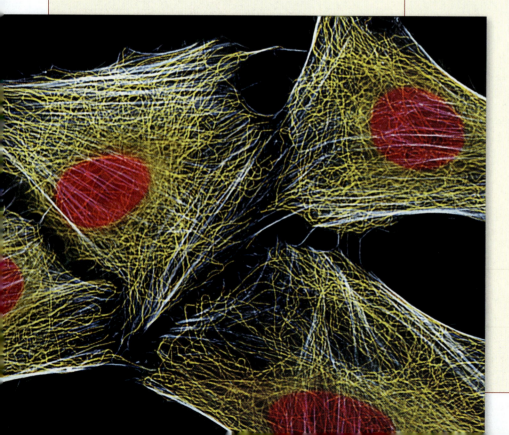

HeLa Cells These rapidly reproducing cancer cells have been cultured in many laboratories and have contributed greatly to biomedical research.

Henrietta Lacks Mrs. Lacks, shown here in front of her home in Baltimore, Maryland, died of cancer in 1951. She left a legacy in the form of cultured cells from the tumor that killed her.

probably have lived much longer if she had had access to a simple medical test that was first used in 1941. This test, called the Pap test, can detect precancerous cells in a woman's cervix, usually allowing them to be removed before they become cancerous. In the United States, the Pap test has prevented an estimated 90 percent of deaths from cervical cancer. Had such testing been performed in time, HeLa cells would never have developed.

In normal tissues, cell division (cell "birth") is offset by cell loss (cell "death"). Unlike most normal cells, most cancer cells, including HeLa cells, keep growing because they have a genetic imbalance that heavily favors cell division over cell death. Cancer treatments using radiation or drugs aim to alter this balance in favor of cell death.

IN THIS CHAPTER we will see how cells give rise to more cells. We will describe how prokaryotic cells produce two new organisms from an original single-celled organism. Then we will describe the two types of eukaryotic cell and nuclear division—mitosis and meiosis—and relate them to asexual and sexual reproduction in eukaryotic organisms. Finally, to balance our discussion of cell proliferation through division, we will describe the important process of programmed cell death, also known as apoptosis.

9.1 How Do Prokaryotic and Eukaryotic Cells Divide?

Unicellular organisms use cell division primarily to reproduce themselves, whereas in multicellular organisms it also plays important roles in growth and in the repair of tissues (**Figure 9.1**).

In order for any cell to divide, four events must occur:

- There must be a *reproductive signal*. This signal, which may come from either inside or outside the cell, initiates cell division.

- **Replication** of DNA (the genetic material) and other vital cell components must occur so that each of the two new cells will have identical genes and complete cell functions.

- The cell must distribute the replicated DNA to each of the two new cells. This process is called **segregation**.

- New material must be added to the cell membrane (and the cell wall, in organisms that have one) in order to separate the two new cells by a process called **cytokinesis**.

These four events proceed somewhat differently in prokaryotes and eukaryotes.

Prokaryotes divide by binary fission

In prokaryotes, cell division results in the reproduction of the entire single-celled organism. The cell grows in size, replicates its DNA, and then essentially divides into two new cells, a process called **binary fission**.

REPRODUCTIVE SIGNALS The reproductive rates of many prokaryotes respond to conditions in the environment. The bacterium *Escherichia coli*, a species commonly used in genetic studies, is a "cell division machine"; essentially, it divides continuously. Typically, cell division in *E. coli* takes 40 minutes at 37°C. But if abundant sources of carbohydrates and mineral nutrients are available, the division cycle speeds up, and the cells may divide in as little as 20 minutes. Another bacterium, *Bacillus subtilis*, stops dividing when food supplies are low, then resumes dividing when conditions improve. These observations

(A) Reproduction

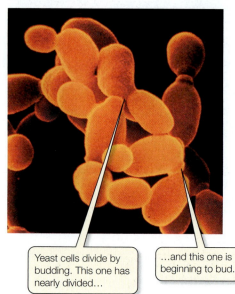

Yeast cells divide by budding. This one has nearly divided...

...and this one is beginning to bud.

(B) Growth

Cell division contributes to the growth of this root tissue.

9.1 Important Consequences of Cell Division
Cell division is the basis for (A) reproduction, (B) growth, and (C) repair and regeneration of tissues.

(C) Regeneration

Cell division contributes to the regeneration of a lizard's tail.

suggest that external factors, such as environmental conditions and nutrient concentrations, are signals for the initiation of cell division in prokaryotes.

REPLICATION OF DNA   As we saw in Section 4.3, a **chromosome** is a DNA molecule containing genetic information. When a cell divides, all of its chromosomes must be replicated, and each of the two resulting copies must find its way into one of the two new cells.

Most prokaryotes have only one chromosome, a single long DNA molecule with proteins bound to it. In the bacterium *E. coli*, the DNA is a continuous molecule, often called a *circular chromosome*. Although often drawn as circles, circular chromosomes are not perfectly round. If the bacterial DNA were arranged in an actual circle, it would be about 1.6 million nm (1.6 mm) in circumference. The bacterium itself is only about 1μm (1,000 nm) in diameter and about 4 μm long. Thus if the bacterial DNA were fully extended, it would form a circle over 100 times larger than the cell! To fit into the cell, the DNA must be compacted. The DNA molecule accomplishes some of this packaging by folding in on itself. Positively charged (basic) proteins bound to negatively charged (acidic) DNA contribute to this folding. Circular chromosomes are characteristic of almost all prokaryotes, as well as some viruses, and are also found in the chloroplasts and mitochondria of eukaryotic cells.

Two regions of the prokaryotic chromosome play a functional role in cell reproduction:

- *ori*: the site where replication of the circle starts (the *ori*gin of replication)

- *ter*: the site where replication ends (the *ter*minus of replication)

Chromosome replication takes place as the DNA is threaded through a "replication complex" of proteins near the center of the cell. (These proteins include the enzyme DNA polymerase, whose important role in replication will be discussed further in Section 11.3.) During the process of prokaryotic DNA replication, the cell grows, and the two daughter DNA's are segregated from one another at opposite ends of the cell.

SEGREGATION OF DNA   The two DNA's are replicated at the center of the cell and as they do so, the *ori* regions move towartd opposite ends of the cell. DNA adjacent to the *ori* region binds proteins that are essential for this segregation. This is an active process, since these binding proteins hydrolyze ATP (**Figure 9.2**). The prokaryotic cytoskeleton (see Section 4.2) may be involved in DNA segregation, either actively moving the DNA along, or passively acting as a "railroad track" along which DNA moves.

CYTOKINESIS   Cell separation, or cytokinesis, begins after chromosome replication is finished. The first event of cytokinesis is a pinching in of the plasma membrane to form a ring similar to a purse string. Fibers composed of a protein similar to eukaryotic tubulin (which makes up microtubules) are major components of this ring. As the membrane pinches in, new cell wall materials are synthesized, which finally separate the two cells.

Under optimal environmental conditions, a population of *E. coli* cells doubles in size every 20 minutes. Theoretically, in about one week a single *E. coli* cell could produce a ball of bacteria the size of the Earth! Thankfully for other organisms, the *E. coli* would run out of nutrients long before that happened.

Eukaryotic cells divide by mitosis or meiosis

Many complex eukaryotes, such as humans and flowering plants, originate from a single cell, the fertilized egg. This cell derives from

9.2 Prokaryotic Cell Division (A) The process of cell division in a bacterium. (B) These two cells of the bacterium *Pseudomonas aeruginosa* have almost completed cytokinesis.

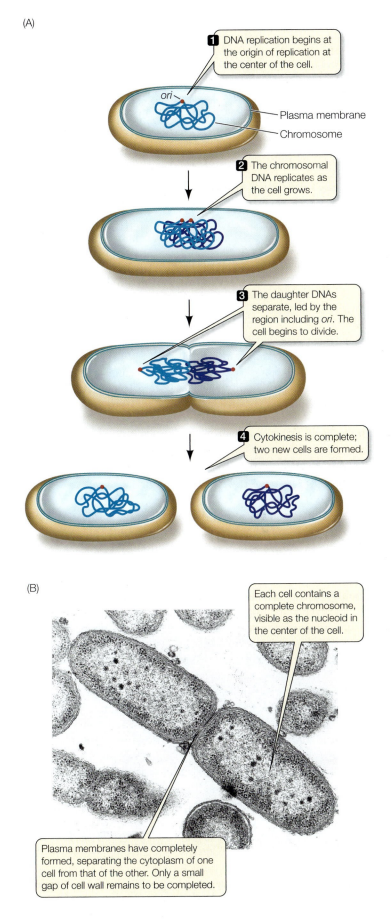

(A)

1 DNA replication begins at the origin of replication at the center of the cell.

ori
Plasma membrane
Chromosome

2 The chromosomal DNA replicates as the cell grows.

3 The daughter DNAs separate, led by the region including *ori*. The cell begins to divide.

4 Cytokinesis is complete; two new cells are formed.

(B)

Each cell contains a complete chromosome, visible as the nucleoid in the center of the cell.

Plasma membranes have completely formed, separating the cytoplasm of one cell from that of the other. Only a small gap of cell wall remains to be completed.

the union of two sex cells, called **gametes**, from the organism's parents—that is, a sperm and an egg—and thus contains genetic material from both parents. Specifically, the fertilized egg contains one set of chromosomes from the male parent and one set from the female parent.

The formation of a multicellular organism from a fertilized egg is called *development*. Development involves both cell reproduction and cell specialization. For example, an adult human has several trillion cells, all ultimately derived from a fertilized egg, yet many of those cells have specialized roles. How these cells become specialized for different functions is the subject of Chapter 43; here our focus is on cell reproduction.

As in prokaryotes, cell reproduction in eukaryotes entails reproductive signals, DNA replication, segregation, and cytokinesis. The details, however, are quite different:

■ Unlike prokaryotes, eukaryotic cells do not constantly divide whenever environmental conditions are adequate. In fact, eukaryotic cells that are part of a multicellular organism and have become specialized seldom divide. In a eukaryotic organism, the signals for cell division are related not to the environment of a single cell, but to the needs of the entire organism.

■ While most prokaryotes have a single main chromosome, eukaryotes usually have many (humans have 46), so the processes of replication and segregation, while basically the same as in prokaryotes, are more intricate. In eukaryotes, the newly replicated chromosomes are closely associated with each other (they are thus known as *sister chromatids*), and a different mechanism, called **mitosis**, is used to segregate them into two new nuclei.

■ Eukaryotic cells have a distinct nucleus, which has to be divided into two new nuclei, each containing an identical set of chromosomes. Thus, in eukaryotes, cytokinesis is distinct from segregation of the genetic material and can happen only after duplication of the entire nucleus.

■ Cytokinesis proceeds differently in plant cells (which have a cell wall) than in animal cells (which do not).

A second mechanism of nuclear division, **meiosis**, occurs only in cells that produce the gametes involved in sexual reproduction. That is, meiosis occurs only in cells that produce the sperm and eggs that will be contributed to a new organism. While the two products of mitosis are genetically identical to the cell that produced them—they both have the same DNA—the products of meiosis are not. As we will see in Section 9.5, meiosis generates diversity by shuffling the genetic material, resulting in new gene combinations. Meiosis plays a key role in sexual life cycles.

What determines whether a cell will divide? How does mitosis lead to identical cells, and meiosis to diversity? Why do most eukaryotic organisms reproduce sexually? In the sections that follow, we will describe the details of the two eukaryotic cell division processes, mitosis, and meiosis, and their roles in heredity, development, and evolution.

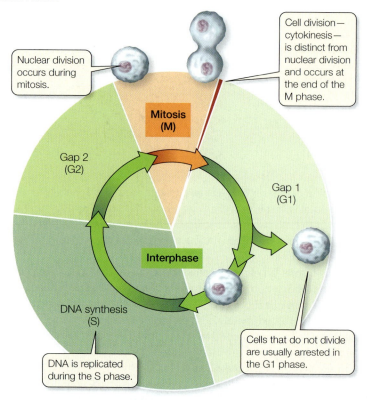

9.3 The Eukaryotic Cell Cycle The cell cycle consists of a mitotic (M) phase, during which mitosis and cytokinesis take place, and a long period of growth known as interphase. Interphase has three subphases (G1, S, and G2) in cells that divide.

9.2 How Is Eukaryotic Cell Division Controlled?

A cell lives and functions until it divides or dies. Or, if it is a gamete (egg or sperm), it lives until it fuses with another gamete. Some types of cells, such as red blood cells, lose the capacity to divide as they mature. Other cell types, such as cortical cells in plant stems, divide only rarely. Some cells, like the cells in a developing embryo, are specialized for rapid division.

The events that occur to produce two eukaryotic cells from one are referred to as the **cell cycle**. Between divisions—that is, for most of its life—a eukaryotic cell is in a condition called **interphase**. For most types of eukaryotic cells, the cell cycle has two phases: mitosis and interphase. In this section we will describe the events of interphase, especially those that trigger mitosis.

A given cell lives for one turn of the cell cycle and then becomes two cells. The cell cycle, when repeated again and again, is a constant source of new cells. However, even in tissues engaged in rapid growth, cells spend most of their time in interphase. Examination of any collection of dividing cells, such as the tip of a root or a slice of liver, will reveal that most of the cells are in interphase most of the time; only a small percentage of the cells will be in mitosis at any given moment.

Interphase has three subphases, called G1, S, and G2. The cell's DNA replicates during **S phase** (the S stands for synthesis). The period between the end of mitosis and the onset of S phase is called **G1**, or Gap 1. Another gap phase—**G2**—separates the end of S phase and the beginning of mitosis. Mitosis and cytokinesis are referred to as the **M phase** of the cell cycle (**Figure 9.3**).

Let's look at the events of interphase in more detail:

■ *G1 phase.* During G1, a cell is preparing for S phase, so at this stage each chromosome is a single, unreplicated structure. G1 is quite variable in length in different cell types. Some rapidly dividing embryonic cells dispense with it entirely, while other cells may remain in G1 for weeks or even years. In many cases, these cells enter a resting phase, called G0. Special internal and external signals are needed to prompt a cell to leave G0 and reenter the cell cycle at G1.

■ *The G1-to-S transition.* It is at the G1-to-S transition that the commitment to cell division (and thus another cell cycle) is made.

■ *S phase.* During S phase, the process of DNA replication, which we will describe in detail in Section 11.3, is completed. Where there was formerly one chromosome, there are now two sister chromatids joined together and awaiting segregation into two new cells by mitosis or meiosis.

■ *G2 phase.* During G2, the cell makes preparations for mitosis—for example, by synthesizing components of the microtubules that will move the chromatids to opposite ends of the dividing cell.

Cyclins and other proteins trigger events in the cell cycle

How are appropriate decisions to enter the S or M phases made? A first indication that there were substances that control these tran-

By catalyzing the phosphorylation of certain target proteins, Cdk's play important roles in initiating the steps of the cell cycle. The discovery that Cdk's induce cell division is a beautiful example of how research on different organisms and different cell types can converge on a single mechanism. One group of scientists, led by James Maller at the University of Colorado, was studying immature sea urchin eggs, trying to find out how they are stimulated to divide and form the precursor cells of the mature egg. A protein called *maturation promoting factor* was purified from maturing eggs, which by itself prodded immature eggs into division. Meanwhile, at the University of Washington, Leland Hartwell, who was studying the cell cycle in yeast (a single-celled eukaryote), found a strain that was stalled at the G1–S boundary because it lacked a Cdk. This yeast Cdk was discovered to be very similar to the sea urchin's maturation promoting factor. Similar Cdk's were soon found to control the G1-to-S transition in many other organisms, including humans.

Cdk's are not active by themselves. Binding to a second type of protein, called **cyclin**, activates a Cdk. This binding—an example of allosteric regulation (see Section 6.5)—activates the Cdk by altering its shape and exposing its active site (**Figure 9.5**). It is the cyclin–Cdk complex that acts as a protein kinase and triggers

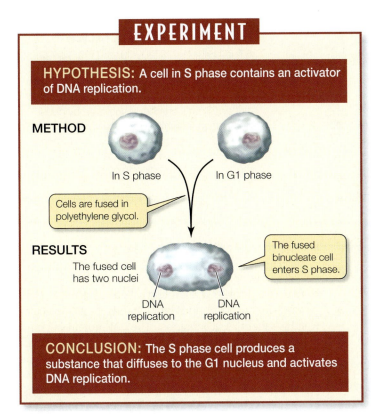

EXPERIMENT

HYPOTHESIS: A cell in S phase contains an activator of DNA replication.

METHOD

In S phase   In G1 phase

Cells are fused in polyethylene glycol.

RESULTS

The fused cell has two nuclei

The fused binucleate cell enters S phase.

DNA replication   DNA replication

CONCLUSION: The S phase cell produces a substance that diffuses to the G1 nucleus and activates DNA replication.

9.4 Regulation of the Cell Cycle   Cells can be induced to fuse by certain substances (such as the sugar alcohol polyethylene glycol) that disaggregate plasma membranes. Such fusion initially produces a binucleate cell. If an S phase cell is fused with an early G1 cell, the latter is stimulated by the former to enter S phase. FURTHER RESEARCH: How would you use this method to show that a cell in M phase produces an activator of mitosis?

sitions came from experiments using *cell fusion.* Fusing mammalian cells at different phases of the cell cycle showed that a cell in S phase produces a substance that activates DNA replication (**Figure 9.4**). Similar experiments point to a molecular activator for entry into M phase.

These transitions—from G1 to S and from G2 to M—depend on the activation of a type of protein called **cyclin-dependent kinase**, or **Cdk**. Recall from Section 7.2 that a *kinase* is an enzyme that catalyzes the transfer of a phosphate group from ATP to another molecule; this phosphate transfer is called *phosphorylation*.

$$\text{protein} + \text{ATP} \xrightarrow{\text{kinase}} \text{protein—P} + \text{ADP}$$

What does phosphorylation do to a protein? As discussed in Section 3.2, proteins have both hydrophilic regions (which tend to interact with water on the outside of the protein macromolecule) and hydrophobic regions (which tend to interact with one another on the inside of the macromolecule). These regions are important in giving a protein its three-dimensional shape. Phosphate groups are charged, so an amino acid with such a group tends to be on the outside of the protein. Thus phosphorylation changes the shape and function of a protein by changing its charges.

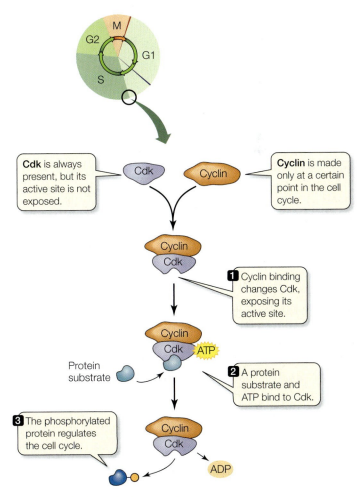

Cdk is always present, but its active site is not exposed.

Cyclin is made only at a certain point in the cell cycle.

Cyclin Cdk

❶ Cyclin binding changes Cdk, exposing its active site.

Cyclin Cdk ATP

Protein substrate

❷ A protein substrate and ATP bind to Cdk.

❸ The phosphorylated protein regulates the cell cycle.

Cyclin Cdk

ADP

9.5 Cyclin Binding Activates Cdk   Binding of a cyclin changes the three-dimensional structure of an inactive Cdk, making it a protein kinase.

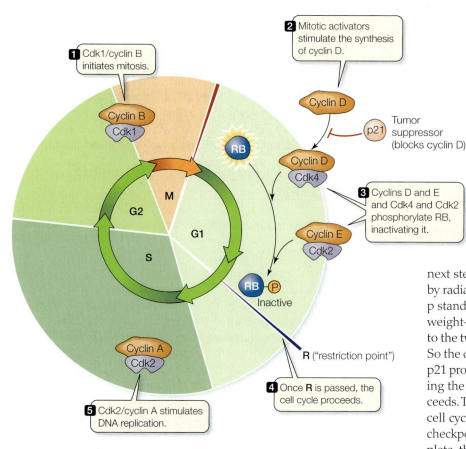

2 Mitotic activators stimulate the synthesis of cyclin D.

1 Cdk1/cyclin B initiates mitosis.

3 Cyclins D and E and Cdk4 and Cdk2 phosphorylate RB, inactivating it.

Tumor suppressor (blocks cyclin D)

Inactive

R ("restriction point")

4 Once **R** is passed, the cell cycle proceeds.

5 Cdk2/cyclin A stimulates DNA replication.

9.6 Cyclin-Dependent Kinases and Cyclins Trigger Transitions in the Cell Cycle Cdk's are activated upon binding with the appropriate cyclin. There are four cyclin/Cdk controls during the typical cell cycle in humans. RB is a tumor suppressor that has the potential to inhibit the cell cycle but it can be deactivated by phosphorylation, at which point the cell cycle proceeds past the restriction point. The tumor suppressor protein p21 can temporarily stop the cell cycle by binding to cyclin D.

next step can be taken. For example, if DNA is damaged by radiation during G1, a protein called p21 is made. (The p stands for "protein" and the 21 stands for its molecular weight—about 21,000 daltons.) The p21 protein can bind to the two G1 Cdk's, preventing their activation by cyclins. So the cell cycle stops while repairs are made to DNA. The p21 protein breaks down after the DNA is repaired, allowing the cyclins to bind to Cdk's so that the cell cycle proceeds. There are checkpoints at several other points in the cell cycle. For example, at the end of S phase, there is a checkpoint for complete DNA replication; if it is not complete, the cell cycle stops before mitosis.

Because cancer results from inappropriate cell division, it is not surprising that these cyclin–Cdk controls are disrupted in cancer cells. For example, some fast-growing breast cancers have too much cyclin D, which overstimulates Cdk4 and thus cell division. For example, a protein called p53 prevents normal cells from dividing by stimulating the synthesis of p21 and thereby inhibiting Cdk's. More than half of all human cancers contain defective p53, resulting in the absence of cell cycle controls. Proteins such as p53, p21, and RB that normally block the cell cycle are known as *tumor suppressors*.

The RB protein was named for retinoblastoma, a tumor that arises from uncontrolled cell division in the embryonic cells that produce the retina of the eye. This cancer affects about one infant in 20,000. Fortunately, the tumor can be removed, although the individual can lose sight in the affected eye.

Growth factors can stimulate cells to divide

Cyclin–Cdk complexes provide cells with an internal control on their progress through the cell cycle. Not all cells in an organism go through the cell cycle on a regular basis. Some cells either no longer go through the cell cycle, or go through it slowly and divide infrequently. If such cells are to divide, they must be stimulated by external chemical signals called **growth factors**. For example, when you cut yourself and bleed, specialized cell fragments called *platelets* gather at the wound to initiate blood clotting. The platelets pro-

the transition from G1 to S phase. Then cyclin breaks down and the Cdk becomes inactive.

Several different cyclin–Cdk combinations act at various stages of the mammalian cell cycle (**Figure 9.6**):

■ Cyclin D–Cdk4 acts during the middle of G1. It moves the cell past the **restriction point** (**R**), a key decision point beyond which the rest of the cell cycle is normally inevitable.

■ Cyclin E–Cdk2 also acts in the middle of G1; it works in concert with Cyclin D–Cdk4 to move the cell cycle past the restriction point.

■ Cyclin A–Cdk2 acts during the S phase to stimulate DNA replication.

■ Cyclin B–Cdk1 acts at the G2–M boundary, initiating the transition to mitosis.

The key to progress past the restriction point is a protein called **RB** (retinoblastoma protein). RB normally inhibits the cell cycle. But when RB is phosphorylated by a protein kinase, it becomes inactive and no longer blocks the restriction point, and the cell progresses past G1 into S phase. (Note the double negative here—a cell function happens because an inhibitor is inhibited. This phenomenon is common in the control of cellular metabolism.) The enzymes that catalyze RB phosphorylation are Cdk4 and Cdk2. So what is needed for a cell to pass the restriction point is the synthesis of cyclins D and E, which activate Cdk4 and Cdk2, which phosphorylate RB, which becomes inactivated.

The cyclin–Cdk complexes act as *checkpoints*, points at which a cell cycle's progress can be monitored to determine whether the duce and release a protein called *platelet-derived growth factor* that diffuses to the adjacent cells in the skin and stimulates them to divide and heal the wound.

Other growth factors include *interleukins*, which are made by one type of white blood cell and promote cell division in other cells that are essential for the body's immune system defenses. *Erythropoietin*, made by the kidney, stimulates the division of bone marrow cells and the production of red blood cells. In addition, many hormones promote division in specific cell types.

We will describe the physiological roles of growth factors in later chapters, but all of them act in a similar way. They bind to their target cells via specialized receptor proteins on the target cell surface. This specific binding triggers events within the target cell that initiate the cell cycle. Cancer cells often divide inappropriately because they make their own growth factors, or because they no longer require growth factors to start cycling.

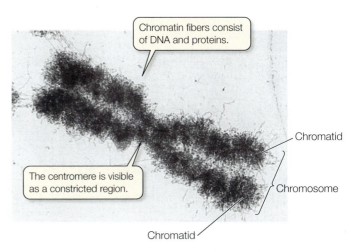

9.7 Chromosomes, Chromatids, and Chromatin A human chromosome in metaphase, shown as the cell prepares to divide.

9.2 RECAP

The eukaryotic cell cycle is under both external and internal controls. A series of kinases, themselves under control by cyclins, controls the eukaryotic cell cycle. External signals such as growth factors can initiate the cell cycle.

- Can you draw a cell cycle diagram showing the various phases of interphase? See p. 184 and Figure 9.3

- Do you understand how cyclins and RB protein control the progress of the cell cycle? See pp. 185–186 and Figure 9.6

- Do you understand the differences between external and internal controls of the cell cycle? See pp. 186–187

Having outlined the events of the eukaryotic cell cycle, now let's take a closer look at the events of the M phase—mitosis.

 # 9.3 What Happens during Mitosis?

The third essential step in the process of cell division— segregation of the replicated DNA—occurs during mitosis. Segregation is accomplished by packaging the huge DNA molecules and their associated proteins into very compact chromosomes. After segregation by mitosis, cytokinesis separates the two cells. Let's now look at these steps more closely.

Eukaryotic DNA is packed into very compact chromosomes

The eukaryotic chromosome shown in **Figure 9.7** consists of two gigantic, linear, double-stranded molecules of DNA complexed with many proteins to form a dense material called **chromatin**. Before S phase, each chromosome contains only one such double-stranded DNA molecule. After the DNA molecule replicates during S phase, however, there are two double-stranded DNA molecules, known as **sister chromatids**. The sister chromatids are

held together along most of their length by a protein complex called **cohesin**. They stay this way until mitosis, when most of the cohesin is removed, except in a region called the **centromere** at which the chromatids remain held together (see Figures 9.7 and 9.11). After replication, a second group of proteins called **condensins** coats the DNA molecules and makes them more compact.

If all the DNA in a typical human cell were put end to end, it would be nearly 2 meters long. Yet the nucleus is only 5 μm (0.000005 meters) in diameter. So, although the DNA in an interphase nucleus is "unwound," it is still impressively packed, as illustrated in **Figure 9.8**. This packing is achieved largely by proteins associated closely with the chromosomal DNA.

Chromosomes contain large quantities of proteins called **histones** (*histos*, "web" or "loom"). There are five classes of histones. All of them have a positive charge at cellular pH levels because of their high content of the basic amino acids lysine and arginine. These positive charges attract the negative phosphate groups on DNA. These DNA-histone interactions, as well as histone-histone interactions, result in the formation of beadlike units called **nucleosomes**. Each nucleosome contains the following components:

- Eight histone molecules, two each of four of the histone classes, united to form a core or spool

- 146 base pairs of DNA, 1.65 turns of it wound around the histone core

- Histone H1 (the remaining histone class) on the outside of the DNA, which may clamp it to the histone core

During interphase, a chromosome consists of a single DNA molecule running around vast numbers of nucleosomes like beads on a string. Between the nucleosomes stretches a variable amount of non-nucleosomal "linker" DNA. During this time, the DNA is exposed to the nuclear environment, so it is accessible to proteins involved in its replication and in the regulation of its expression, as we will see in Chapter 14.

During both mitosis and meiosis, the chromatin becomes ever more tightly coiled and condensed as the nucleosomes pack together. Further coiling of the chromatin continues up to the time at which the chromatids begin to move apart.

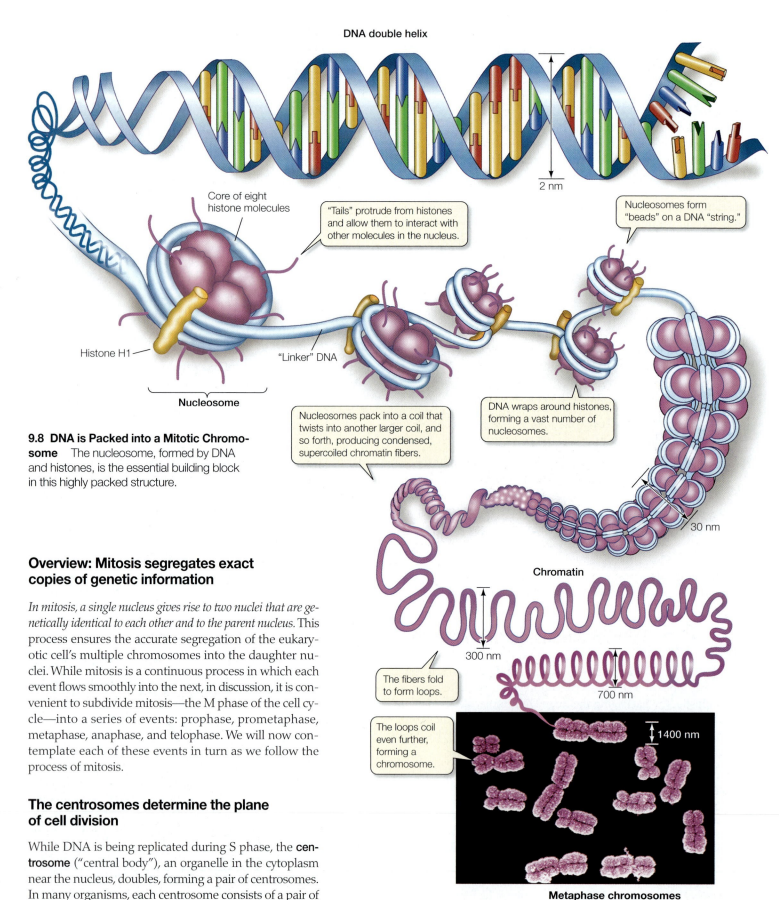

DNA double helix

2 nm

Core of eight histone molecules

"Tails" protrude from histones and allow them to interact with other molecules in the nucleus.

Nucleosomes form "beads" on a DNA "string."

Histone H1

"Linker" DNA

Nucleosome

9.8 DNA is Packed into a Mitotic Chromosome The nucleosome, formed by DNA and histones, is the essential building block in this highly packed structure.

Nucleosomes pack into a coil that twists into another larger coil, and so forth, producing condensed, supercoiled chromatin fibers.

DNA wraps around histones, forming a vast number of nucleosomes.

30 nm

Chromatin

300 nm

The fibers fold to form loops.

700 nm

The loops coil even further, forming a chromosome.

1400 nm

Metaphase chromosomes

Overview: Mitosis segregates exact copies of genetic information

In mitosis, a single nucleus gives rise to two nuclei that are genetically identical to each other and to the parent nucleus. This process ensures the accurate segregation of the eukaryotic cell's multiple chromosomes into the daughter nuclei. While mitosis is a continuous process in which each event flows smoothly into the next, in discussion, it is convenient to subdivide mitosis—the M phase of the cell cycle—into a series of events: prophase, prometaphase, metaphase, anaphase, and telophase. We will now contemplate each of these events in turn as we follow the process of mitosis.

The centrosomes determine the plane of cell division

While DNA is being replicated during S phase, the **centrosome** ("central body"), an organelle in the cytoplasm near the nucleus, doubles, forming a pair of centrosomes. In many organisms, each centrosome consists of a pair of **centrioles**, each one a hollow tube lined with nine microtubules. The two tubes are at right angles to each other.

At the G2-to-M transition, the two centrosomes separate from each other, moving to opposite ends of the nuclear envelope. The orientation of the centrosomes determines the plane at which the cell will divide, and therefore the spatial relationship of the two new cells to the parent cell. This relationship may be of little consequence to single free-living cells such as yeasts, but it is important for cells that make up part of a body tissue.

Surrounding the centrioles is a high concentration of tubulin dimers, and these proteins initiate the formation of microtubules, which will orchestrate chromosomal movement. (In plant cells, which lack centrosomes, a distinct microtubule organizing center at either end of the cell plays the same role.) The formation of microtubules will lead to the formation of the spindle structure that is required for the orderly segregation of the chromosomes.

Chromatids become visible and the spindle forms during prophase

During interphase, only the nuclear envelope, the nucleoli, and a barely discernible tangle of chromatin are visible under the light microscope. The appearance of the nucleus changes as the cell enters **prophase**—the beginning of mitosis. Most of the cohesin that has held the two products of DNA replication together since S phase is removed, so the individual chromatids become visible. They are still held together by a small amount of cohesin at the centromere. Late in prophase, specialized three-layered structures called **kinetochores** develop in the centromere region, one on each chromatid. These structures will be important in chromosome movements.

Each of the two centrosomes serves as a *mitotic center*, or *pole*, toward which the chromosomes will move (**Figure 9.9A**). Microtubules form between each pole and the chromosomes to make up a **spindle**, which serves both as a structure to which the chromosomes will attach and as a framework keeping the two poles apart. The spindle is actually two half-spindles: each microtubule runs from one pole to the middle of the spindle, where it overlaps with microtubules extending from the other half-spindle. The

microtubules are initially unstable, constantly forming and falling apart, until they contact microtubules from the other half-spindle and become more stable.

There are two types of microtubules in the spindle:

- *Polar microtubules* are the microtubules we have just described; they form the framework of the spindle. They have abundant tubulin around the centrioles. Tubulin dimers aggregate to form long fibers that extend into the middle region of the cell.

- *Kinetochore microtubules*, which form later, attach to the kinetochores on the chromosomes. The two sister chromatids in each chromosome pair will become attached by their kinetochore to kinetochore microtubules in opposite halves of the spindle (**Figure 9.9B**). This ensures that one chromatid of the pair will eventually move to one pole while the other chromatid moves to the opposite pole. Movement of the chromatids is the central feature and accomplishment of mitosis.

Chromosome movements are highly organized

It is during the next three phases of mitosis—prometaphase, metaphase, and anaphase—that the chromosomes move (**Figure 9.10**). During these phases, the centromeres holding the two chromatids together separate, and the former sister chromatids move away from each other in opposite directions.

PROMETAPHASE **Prometaphase** is marked by the disappearance of the nuclear envelope and the nucleoli. The material that composes them remains in the cytoplasm, however, to be reassembled when the daughter nuclei form. In prometaphase, the chromosomes begin to move toward the poles, but this movement is counteracted by two factors:

- A repulsive force from the poles pushes the chromosomes toward the middle region, or **equatorial plate** (metaphase plate), of the cell.

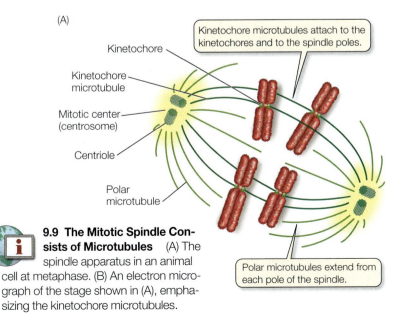

9.9 The Mitotic Spindle Consists of Microtubules (A) The spindle apparatus in an animal cell at metaphase. (B) An electron micrograph of the stage shown in (A), emphasizing the kinetochore microtubules.

Kinetochore

Kinetochore microtubule

Mitotic center (centrosome)

Centriole

Polar microtubule

Kinetochore microtubules attach to the kinetochores and to the spindle poles.

Polar microtubules extend from each pole of the spindle.

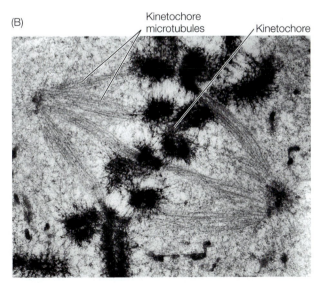

Kinetochore microtubules

Kinetochore

Interphase

Prophase

Prometaphase

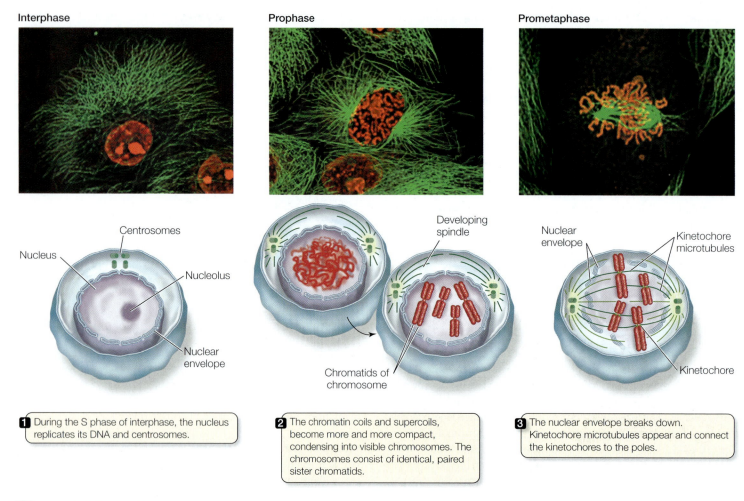

Centrosomes

Nucleus

Nucleolus

Nuclear envelope

Developing spindle

Chromatids of chromosome

Nuclear envelope

Kinetochore microtubules

Kinetochore

1 During the S phase of interphase, the nucleus replicates its DNA and centrosomes.

2 The chromatin coils and supercoils, become more and more compact, condensing into visible chromosomes. The chromosomes consist of identical, paired sister chromatids.

3 The nuclear envelope breaks down. Kinetochore microtubules appear and connect the kinetochores to the poles.

9.10 Mitosis   Mitosis results in two new nuclei that are genetically identical to each other and to the nucleus from which they were formed. In the micrographs, the green dye stains microtubules (and thus the spindle); the red dye stains the chromosomes. The chromosomes in the diagrams are stylized to emphasize the fates of the individual chromatids.

■ The two chromatids are still held together at the centromere by cohesin.

Thus, during prometaphase, the chromosomes appear to move aimlessly back and forth between the poles and the middle of the spindle. Gradually, the centromeres approach the equatorial plate.

METAPHASE   The cell is said to be in **metaphase** when all the centromeres arrive at the equatorial plate. Metaphase is the best time to see the sizes and shapes of chromosomes because they are maximally condensed. The chromatids are now clearly connected to one pole or the other by microtubules. At the end of metaphase, all of the chromatid pairs separate simultaneously.

ANAPHASE   Separation of the chromatids marks the beginning of **anaphase**, during which the two sister chromatids move to opposite ends of the spindle. Each chromatid contains one double-stranded DNA molecule and is now referred to as a **daughter chromosome**.

This separation occurs because the cohesin holding the sister chromatids together is hydrolyzed by a specific protease, appropriately called *separase*. Until this point, separase has been present but inactive, because it has been bound to an inhibitory subunit called *securin*. Once all the chromatids are connected to the spindle, securin is hydrolyzed, allowing separase to catalyze cohesin breakdown (**Figure 9.11**). In this way, chromosome alignment is connected to chromatid separation. This process, called the *spindle checkpoint*, apparently senses whether there are any kinetochores still unattached to the spindle. If there are, securin breakdown is blocked, and the sister chromatids stay together.

What propels this highly organized mass migration? Two things seem to move the chromosomes along. First, at the kinetochores are proteins that act as "molecular motors." These proteins, called *cytoplasmic dynein*, have the ability to hydrolyze ATP to ADP and phosphate, thus releasing energy to move the chromosomes along the microtubules toward the poles. These motor proteins account for about 75 percent of the force of motion. Second, the kinetochore microtubules shorten from the poles, drawing the chromosomes toward them, accounting for about 25 percent of the motions.

During anaphase the poles of the spindle are pushed farther apart, doubling the distance between them. The distance between poles increases because the overlapping polar microtubules extending from opposite ends of the spindle contain motor proteins that cause them to slide past each other, in much the same

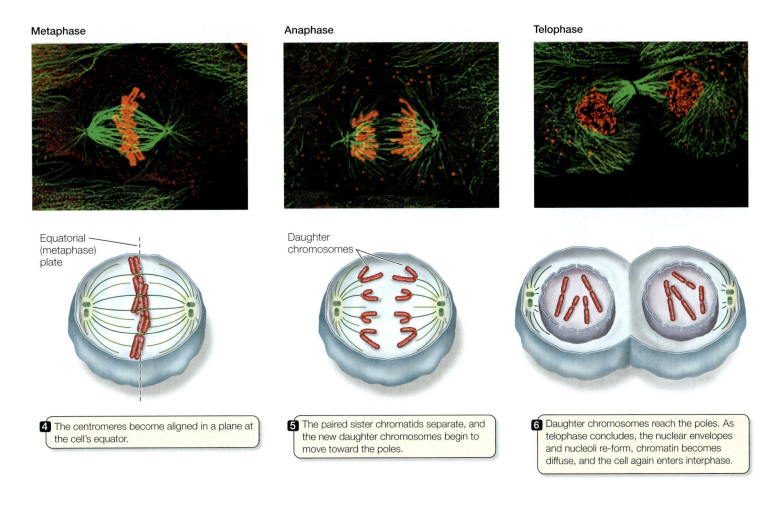

Metaphase

Equatorial (metaphase) plate

4 The centromeres become aligned in a plane at the cell's equator.

Anaphase

Daughter chromosomes

5 The paired sister chromatids separate, and the new daughter chromosomes begin to move toward the poles.

Telophase

6 Daughter chromosomes reach the poles. As telophase concludes, the nuclear envelopes and nucleoli re-form, chromatin becomes diffuse, and the cell again enters interphase.

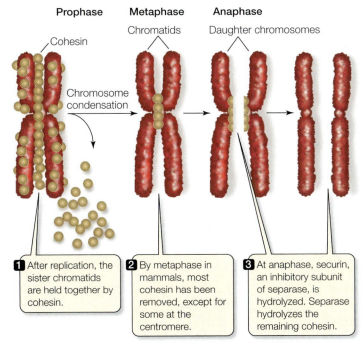

Prophase

Cohesin

Chromosome condensation

Metaphase

Chromatids

Anaphase

Daughter chromosomes

1 After replication, the sister chromatids are held together by cohesin.

2 By metaphase in mammals, most cohesin has been removed, except for some at the centromere.

3 At anaphase, securin, an inhibitory subunit of separase, is hydrolyzed. Separase hydrolyzes the remaining cohesin.

9.11 Chromatid Attachment and Separation Cohesin protein holds sister chromatids together. The enzyme separase hydrolyzes cohesin at the onset of anaphase, allowing the chromatids to separate into daughter chromosomes.

way that microtubules slide in cilia and flagella (see Figure 4.23A). This polar separation further separates one set of daughter chromosomes from the other.

Chromosomes move slowly, even in cellular terms. At about 1 μm per minute, it takes them 10–60 minutes to complete their journey to the poles—roughly equivalent to a person taking 9 million years to travel across the United States. This slow speed may ensure that the chromosomes segregate accurately.

Nuclei re-form during telophase

When the chromosomes stop moving at the end of anaphase, the cell enters **telophase**. Two sets of chromosomes (formerly referred to as daughter chromosomes), containing identical DNA and thus carrying identical sets of hereditary instructions, are now at the opposite ends of the spindle, which begins to break down. The chromosomes uncoil until they are once again the diffuse tangle of chromatin that is characteristic of interphase. The nuclear envelopes and nucleoli, which were disaggregated during prophase, coalesce and re-form their respective structures. When these and other changes are complete, telophase—and mitosis—is at an end, and each of the daughter nuclei enters another interphase.

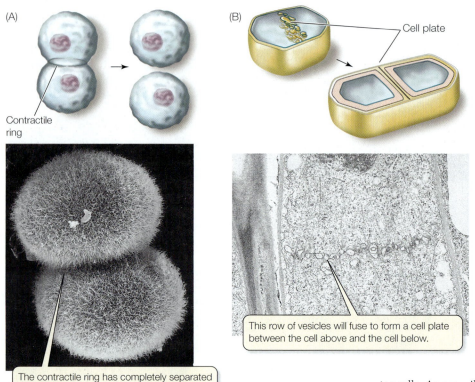

(A)

Contractile ring

The contractile ring has completely separated the cytoplasms of these two daughter cells, although their surfaces remain in contact.

(B)

Cell plate

This row of vesicles will fuse to form a cell plate between the cell above and the cell below.

9.12 Cytokinesis Differs in Animal and Plant Cells Plant cells must divide differently from animal cells because they have cell walls. (A) A sea urchin zygote (fertilized egg) that has just completed cytokinesis at the end of the first cell division of its development into an embryo. (B) A dividing plant cell in late telophase. The resulting two cells will be separated not by space, but by a solid cell wall.

contents to a *cell plate*, which is the beginning of a new cell wall (**Figure 9.12B**).

Following cytokinesis, both daughter cells contain all the components of a complete cell. A precise distribution of chromosomes is ensured by mitosis. In contrast, organelles such as ribosomes, mitochondria, and chloroplasts need not be distributed equally between daughter cells as long as some of each are present in both cells; accordingly, there is no mechanism with a precision comparable to that of mitosis to provide for their equal allocation to daughter cells. As we will see in Chapter 43, the unequal distribution of cytoplasmic components during development can have functional significance for the two new cells.

Mitosis is beautifully precise. Its result is two nuclei that are identical to each other and to the parent nucleus in chromosomal makeup, and hence in genetic constitution. Next, the two nuclei must be isolated in separate cells, which requires the division of the cytoplasm.

Cytokinesis is the division of the cytoplasm

Mitosis refers only to the division of the nucleus. The division of the cell's cytoplasm, which follows mitosis, is accomplished by cytokinesis. In different organisms, cytokinesis may be accomplished in different ways. The differences between the process in plants and in animals are substantial.

Animal cells usually divide by a furrowing of the plasma membrane, as if an invisible thread were cinching the cytoplasm between the two poles (**Figure 9.12A**). The "invisible thread" is actually microfilaments of actin and myosin (see Figure 4.20A) in what is called a *contractile ring* just beneath the plasma membrane. These two proteins interact to produce a contraction, just as they do in muscles, thus pinching the cell in two. These microfilaments assemble rapidly from actin monomers that are present in the interphase cytoskeleton. Their assembly appears to be under the control of calcium ions released from storage sites in the center of the cell.

Plant cell cytoplasm divides differently because plants have cell walls. In plant cells, as the spindle breaks down after mitosis, membranous vesicles derived from the Golgi apparatus appear at the equatorial plate, roughly midway between the two daughter nuclei. Propelled along microtubules by the motor protein kinesin, these vesicles fuse to form new plasma membrane and contribute their

The intricate process of mitosis results in two cells that are genetically identical. But, as mentioned earlier, there is another eukaryotic cell division process, called meiosis, that results in genetic diversity. What is the role of that process?

9.4 What Is the Role of Cell Division in Sexual Life Cycles?

The mitotic cell cycle repeats itself over and over. By this process, a single cell can give rise to a vast number of other cells. Meiosis, on the other hand, produces just four daughter cells, which may not undergo further duplications. Mitosis and meiosis are both involved in reproduction, but they have different reproductive roles.

Reproduction by mitosis results in genetic constancy

Asexual reproduction, sometimes called *vegetative reproduction*, is based on mitotic division of the nucleus. A cell undergoing mitosis may be an entire single-celled organism reproducing itself with each cell cycle, or it may be a cell in a multicellular organism that breaks off a piece to produce a new multicellular organism. Some multicellular organisms can reproduce themselves by releasing cells derived from mitosis and cytokinesis or by having a multicellular piece break away and grow on its own (**Figure 9.13**). In asexual reproduction, offspring are **clones** of the parent organism; that is, the offspring are *genetically identical* to the parent. If there is any variation among the offspring, it is likely to be due to *mutations*, or changes, in the genetic material. Asexual reproduction is a rapid and effective means of making new individuals, and it is common in nature.

Reproduction by meiosis results in genetic diversity

Unlike asexual reproduction, **sexual reproduction** results in an organism that is not identical to the parent organism. Sexual reproduction requires gametes created by meiosis; two parents each contribute one gamete to each of their offspring. Meiosis produces gametes—and thus offspring—that differ genetically not only from each parent, but from one another as well. Because of this genetic variation, some offspring may be better adapted than others to survive and reproduce in a particular environment. Meiosis thus generates the genetic diversity that is the raw material for natural selection and evolution.

In most multicellular organisms, **somatic cells**—those body cells that are *not* specialized for reproduction—each contain two sets of chromosomes, which are found in pairs. One chromosome of each pair comes from each of the organism's two parents. The members of such a **homologous pair** are similar in size and appearance (except for the sex chromosomes found in some species, as we will see in Section 10.4). The two chromosomes (the **homologs**) of a homologous pair bear corresponding, though generally not identical, genetic information.

Gametes, on the other hand, contain only a single set of chromosomes—that is, one homolog from each pair. The number of chromosomes in a gamete is denoted by n, and the cell is said to be **haploid**. Two haploid gametes fuse to form a new organism, the **zygote**, in a process called **fertilization**. The zygote thus has two sets of chromosomes, just as somatic cells do. Its chromosome number is denoted by $2n$, and the zygote is said to be **diploid**.

All sexual life cycles have certain hallmarks:

- There are two parents, each of which provides chromosomes to the offspring in the form of a gamete produced by meiosis.
- Each gamete is haploid; that is, it contains a single set of chromosomes.
- The two gametes—often identifiable as a female egg and a male sperm—fuse to produce a single cell, the zygote, or fertilized egg. The zygote thus contains two sets of chromosomes (is diploid).

(A)

9.13 Asexual Reproduction (A) Some cactus, such as this cholla, have brittle stems that break off easily. Fragments on the ground set down roots and mitotically grow an entire new plant that is genetically identical to the plant it came from. (B) These spool-shaped cells are asexual spores formed by a fungus. Each spore contains a nucleus produced by a mitotic division. A spore is genetically identical to the parent that fragmented to produce it.

(B)

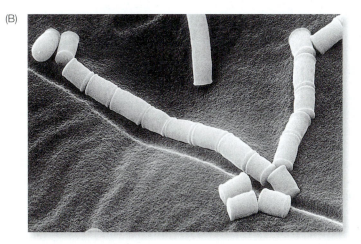

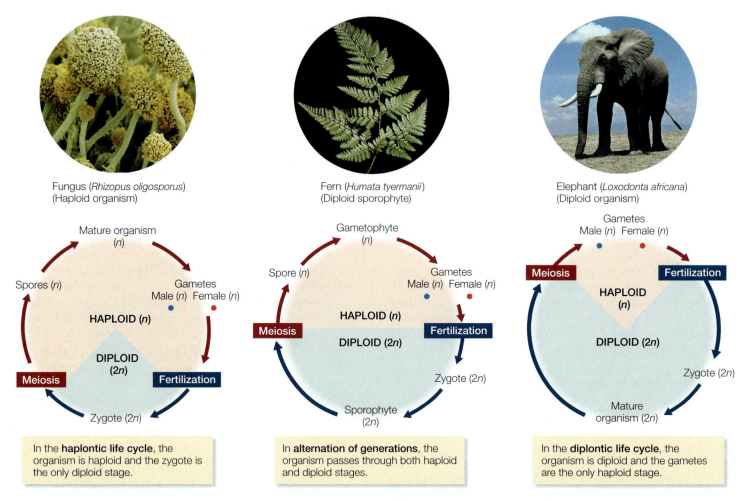

Fungus (*Rhizopus oligosporus*)
(Haploid organism)

Fern (*Humata tyermanii*)
(Diploid sporophyte)

Elephant (*Loxodonta africana*)
(Diploid organism)

In the **haplontic life cycle**, the organism is haploid and the zygote is the only diploid stage.

In **alternation of generations**, the organism passes through both haploid and diploid stages.

In the **diplontic life cycle**, the organism is diploid and the gametes are the only haploid stage.

As shown in **Figure 9.14**, after zygote formation, different kinds of sexual life cycles are observed:

- In **haplontic** organisms, such as most protists and many fungi, the tiny zygote is the only diploid cell in the life cycle; the mature organism is haploid. The zygote undergoes meiosis to produce haploid cells, or **spores**. These spores form the new organism, which may be single-celled or multicellular, by mitosis. The mature haploid organism produces gametes by mitosis, which fuse to form the diploid zygote.

- Most plants and some protists display **alternation of generations**, in which meiosis does not give rise to gametes, but to haploid spores. The spores divide by mitosis to form an alternate, haploid life stage (the *gametophyte*). It is this haploid stage that forms gametes by mitosis. The gametes fuse to form a diploid zygote, which divides by mitosis to become the diploid *sporophyte*.

- In **diplontic** organisms, which include animals and some plants, the gametes are the only haploid cells in the life cycle, and the mature organism is diploid. Gametes are formed by meiosis, which fuse to form a diploid zygote. The zygote divides by mitosis to form the mature organism.

These life cycles are described in greater detail in Part Six. In this chapter we will focus on the role of sexual reproduction in generating diversity among individual organisms.

 9.14 Fertilization and Meiosis Alternate in Sexual Reproduction In sexual reproduction, haploid (*n*) cells or organisms alternate with diploid (2*n*) cells or organisms.

TABLE 9.1

Numbers of Pairs of Chromosomes in Some Plant and Animal Species

COMMON NAME	SPECIES	NUMBER OF CHROMOSOME PAIRS
Mosquito	*Culex pipiens*	3
Housefly	*Musca domestica*	6
Toad	*Bufo americanus*	11
Rice	*Oryza sativa*	12
Frog	*Rana pipiens*	13
Alligator	*Alligator mississippiensis*	16
Rhesus monkey	*Macaca mulatta*	21
Wheat	*Triticum aestivum*	21
Human	*Homo sapiens*	23
Potato	*Solanum tuberosum*	24
Donkey	*Equus asinus*	31
Horse	*Equus caballus*	32
Dog	*Canis familiaris*	39
Carp	*Cyprinus carpio*	52

(A)

Centromeres (arrows) occupy characteristic positions on homologous chromosomes.

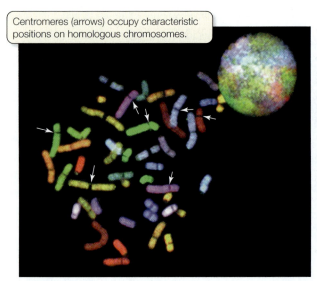

(B)

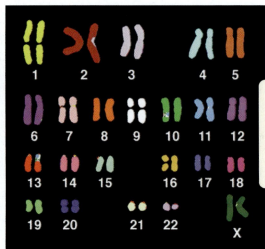

Humans have 23 pairs of chromosomes, including the sex chromosomes. This female's sex chromosomes are X and X; a male would have X and Y chromosomes.

9.15 The Human Karyotype (A) Chromosomes from a human cell in metaphase of mitosis. The DNA of each chromosome has a specific nucleotide sequence that is stained by a specific colored dye, so that each homologous pair shares a distinctive color. Each chromosome at this stage is composed of two chromatids, but they cannot be distinguished by this "chromosome painting" technique. The multicolored globe is an interphase nucleus. (B) This image, produced by computerized analysis of the image on the left, with homologous pairs lined up together and numbered, clearly reveals the human karyotype.

The essence of sexual reproduction is the random selection of half of a parent's diploid chromosome set to make a haploid gamete, followed by the fusion of two such haploid gametes to produce a diploid cell containing genetic information from both gametes. Both of these steps contribute to a shuffling of genetic information in the population, so that no two individuals have exactly the same genetic constitution. The diversity provided by sexual reproduction opens up enormous opportunities for evolution.

The number, shapes, and sizes of the metaphase chromosomes constitute the karyotype

When cells are in metaphase of mitosis, it is often possible to count and characterize their individual chromosomes. This is a relatively simple process in some organisms, thanks to techniques that can capture cells in metaphase and spread out their chromosomes. A photomicrograph of the entire set of chromosomes can then be made, and the images of the individual chromosomes can be placed in an orderly arrangement. Such a rearranged photomicrograph reveals the number, shapes, and sizes of the chromosomes in a cell, which together constitute its **karyotype** (**Figure 9.15**).

Individual chromosomes can be recognized by their lengths, the positions of their centromeres, and characteristic banding that is visible when they are stained and observed at high magnification. When the cell is diploid, the karyotype consists of homologous pairs of chromosomes—23 pairs for a total of 46 chromosomes in humans, and greater or smaller numbers of pairs in other diploid species. There is no simple relationship between the size of an organism and its chromosome number (**Table 9.1**).

9.4 RECAP

Meiosis is necessary for sexual reproduction, in which haploid gametes fuse to produce a diploid zygote. Sexual reproduction results in genetic diversity, the foundation of evolution.

- What is the difference, in terms of genetics, between asexual and sexual reproduction? See p. 193

- Can you describe how fertilization produces a diploid organism? See p. 193

- What general features do all sexual life cycles have in common? See p. 194 and Figure 9.14

Meiosis, unlike mitosis, results in daughter cells that are genetically different from, and have only half as many chromosomes as, the parent cell. How are these changes accomplished?

9.5 What Happens When a Cell Undergoes Meiosis?

Meiosis consists of *two* nuclear divisions that reduce the number of chromosomes to the haploid number in preparation for sexual reproduction. Although the *nucleus divides twice* during meiosis, the *DNA is replicated only once*. Unlike the products of mitosis, the products of meiosis are different both from one another and from the parent cell. To understand the process of meiosis and its specific details, it is useful to keep in mind the overall functions of meiosis:

- To reduce the chromosome number from diploid to haploid

- To ensure that each of the haploid products has a complete set of chromosomes

- To promote genetic diversity among the products

Throughout this discussion, refer to **Figure 9.16** to help you visualize each step.

Early prophase I

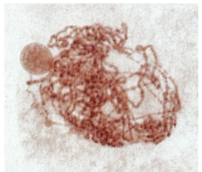

Mid-prophase I

Late prophase I–prometaphase

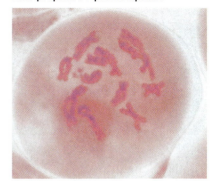

Centrosomes

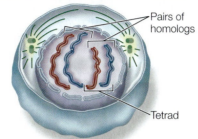

Pairs of homologs

Tetrad

1 The chromatin begins to condense following interphase.

2 Synapsis aligns homologs, and chromosomes condense further.

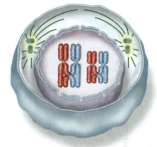

3 The chromosomes continue to coil and shorten. Crossing over results in an exchange of genetic material. In prometaphase the nuclear envelope breaks down.

Prophase II

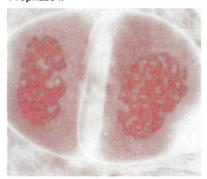

Metaphase II

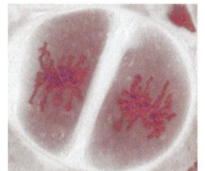

Anaphase II

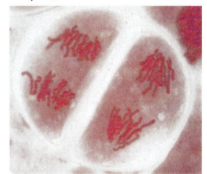

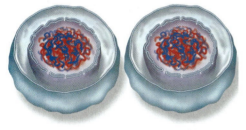

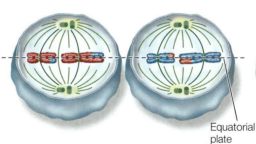

Equatorial plate

7 The chromosomes condense again, following a brief interphase (interkinesis) in which DNA does not replicate.

8 The centrosomes of the paired chromatids line up at the equatorial plates of each cell.

9 The chromatids finally separate, becoming chromosomes in their own right, and are pulled to opposite poles. Because of crossing over in prophase I, each new cell will have a different genetic makeup.

9.16 Meiosis In meiosis, two sets of chromosomes are divided among four nuclei, each of which then has half as many chromosomes as the original cell. Four haploid cells are the result of two successive nuclear divisions. The micrographs show meiosis in the male reproductive organ of a lily; the diagrams show corresponding phases in an animal. (For instructional purposes, the chromosomes from one parent are colored blue and those from the other parent are red.)

Metaphase I

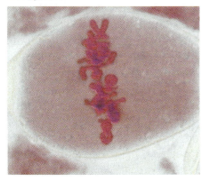

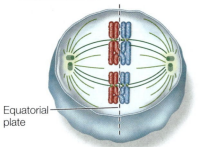

Equatorial plate

4 The homologous pairs line up on the equatorial (metaphase) plate.

Anaphase I

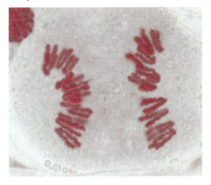

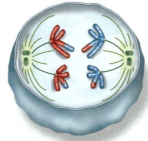

5 The homologous chromosomes (each with two chromatids) move to opposite poles of the cell.

Telophase I

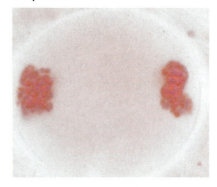

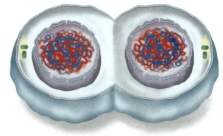

6 The chromosomes gather into nuclei, and the original cell divides.

Telophase II

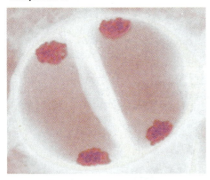

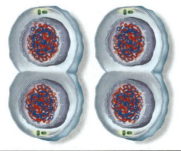

10 The chromosomes gather into nuclei, and the cells divide.

Products

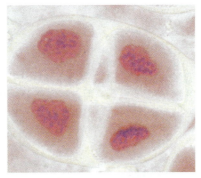

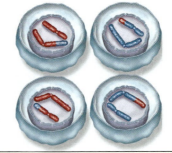

11 Each of the four cells has a nucleus with a haploid number of chromosomes.

The first meiotic division reduces the chromosome number

Two unique features characterize the first of the two meiotic divisions, **meiosis I**. The first is that homologous chromosomes come together to pair along their entire lengths. No such pairing occurs in mitosis. The second is that after metaphase I, the homologous chromosomes separate. The individual chromosomes, each consisting of two sister chromatids, remain intact until the end of metaphase II in the second meiotic division.

Like mitosis, meiosis I is preceded by an interphase with an S phase, during which each chromosome is replicated. As a result, each chromosome consists of two sister chromatids, held together by cohesin proteins.

Meiosis I begins with a long prophase I (the first three frames of Figure 9.16), during which the chromosomes change markedly. The homologous chromosomes pair by adhering along their lengths, a process called **synapsis**. This pairing process lasts from prophase I to the end of metaphase I.

By the time chromosomes can be clearly seen under the light microscope, the two homologs are already tightly joined. This joining begins at the telomeres and is mediated by the recognition of homologous DNA sequences on homologous chromosomes. In addition, a special group of proteins may form a scaffold called the *synaptonemal complex*, which runs lengthwise along the homologous chromosomes and appears to join them together.

The four chromatids of each pair of homologous chromosomes form what is called a **tetrad**, or *bivalent*. In other words, a tetrad consists of four chromatids, two from each of two homologous chromosomes. For example, there are 46 chromosomes in a hu-

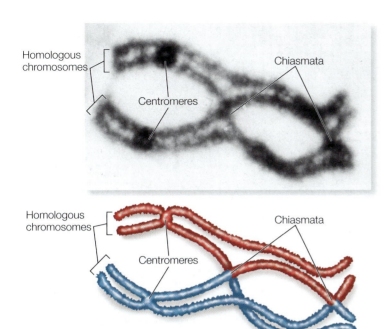

9.17 Chiasmata: Evidence of Exchange between Chromatids
This micrograph shows a pair of homologous chromosomes, each with two chromatids, during prophase I of meiosis in a salamander. Two chiasmata are visible.

Prophase I is followed by prometaphase I, during which the nuclear envelope and the nucleoli disaggregate. A spindle forms, and microtubules become attached to the kinetochores of the chromosomes. In meiosis I, the kinetochores of both chromatids in each chromosome become attached to the same half-spindle. Thus the entire chromosome, consisting of two chromatids, will migrate to one pole. Which member of a homologous chromosome pair becomes attached to each half-spindle, and thus which member will go to which pole, is random. By metaphase I, all the chromosomes have moved to the equatorial plate. Up to this point, homologous pairs are held together by chiasmata.

The homologous chromosomes separate in anaphase I, when the individual chromosomes, each still consisting of two chromatids, are pulled to the poles, with one homolog of a pair going to one pole and the other homolog going to the opposite pole. (Note that this process differs from the separation of *chromatids* during mitotic anaphase.) Each of the two daughter nuclei from this division thus contains only one set of chromosomes, not the two sets that were present in the original diploid nucleus. However, because they consist of two chromatids rather than just one, each of these chromosomes has twice the mass as that of a chromosome at the end of a mitotic division.

In some organisms, there is a telophase I, with the reaggregation of nuclear envelopes. When there is a telophase I, it is followed

man diploid cell at the beginning of meiosis, so there are 23 homologous pairs of chromosomes, each with two chromatids (that is, 23 tetrads), for a total of 92 chromatids during prophase I.

Throughout prophase I and metaphase I, the chromatin continues to coil and compact, so that the chromosomes appear ever thicker. At a certain point, the homologous chromosomes seem to repel each other, especially near the centromeres, but they are held together by physical attachments mediated by cohesins. These cohesins are different from the ones holding the two sister chromatids together. Regions having these attachments take on an X-shaped appearance (**Figure 9.17**) and are called **chiasmata** (singular *chiasma*, "cross").

A chiasma reflects an exchange of genetic material between nonsister chromatids on homologous chromosomes—what geneticists call **crossing over** (**Figure 9.18**). The chromosomes usually begin exchanging material shortly after synapsis begins, but chiasmata do not become visible until later, when the homologs are repelling each other. Crossing over increases genetic variation among the products of meiosis by reshuffling genetic information among the homologous pairs. We will have a great deal to say about crossing over and its genetic consequences in the coming chapters.

There seems to be plenty of time for the complicated events of prophase I to occur. Whereas mitotic prophase is usually measured in minutes, and all of mitosis seldom takes more than an hour or two, meiosis can take much longer. In human males, the cells in the testis that undergo meiosis take about a week for prophase I and about a month for the entire meiotic cycle. In the cells that will become eggs, prophase I begins long before a woman's birth, during her early fetal development, and ends as much as decades later, during the monthly ovarian cycle (see Figure 42.13).

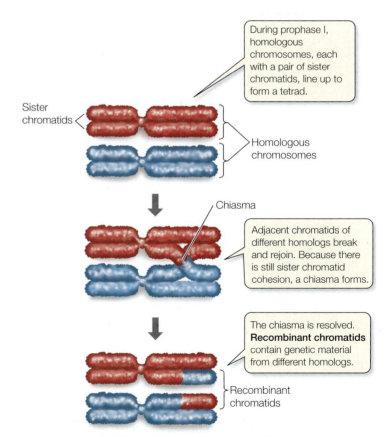

9.18 Crossing Over Forms Genetically Diverse Chromosomes
The exchange of genetic material by crossing over may result in new combinations of genetic information on the recombinant chromosomes. The two different colors distinguish the chromosomes contributed by the male and female parent.

by an interphase, called **interkinesis**, similar to the mitotic interphase. During interkinesis the chromatin is partially uncoiled; however, there is no replication of the genetic material, because each chromosome already consists of two chromatids. Furthermore, the sister chromatids at interkinesis are generally not genetically identical, because crossing over in prophase I has reshuffled genetic material between the maternal and paternal chromosomes. In other organisms, the chromosomes move directly into the second meiotic division.

The second meiotic division separates the chromatids

Meiosis II is similar to mitosis in many ways. In each of the two nuclei produced by meiosis I, the chromosomes line up at the equatorial plate at metaphase II. The centromeres of the sister chromatids separate because of cohesin breakdown, and the daughter chromosomes move to the poles in anaphase II.

There are three major differences between meiosis II and mitosis:

- DNA replicates before mitosis, but not before meiosis II.

- In mitosis, the sister chromatids that make up a given chromosome are identical. In meiosis II, they may differ over part of their length if they participated in crossing over during prophase I.

- The number of chromosomes on the equatorial plate in meiosis II is half the number in the mitotic nucleus.

The result of meiosis is four nuclei; each nucleus is haploid and has a single set of unreplicated chromosomes that differs from those of the other nuclei in its exact genetic composition. The differences among the four haploid nuclei result from crossing over during prophase I and from the random segregation of homologous chromosomes during anaphase I.

The activities and movements of chromosomes during meiosis result in genetic diversity

What are the consequences of the synapsis and segregation of homologous chromosomes during meiosis? In mitosis, each chromosome behaves independently of its homolog; its two chromatids are sent to opposite poles at anaphase. If a mitotic division begins with *x* chromosomes, we end up with *x* chromosomes in each daughter nucleus, and each chromosome consists of one chromatid. Each of the two sets of chromosomes (one of paternal and one of maternal origin) is divided equally and distributed equally to each daughter cell. In meiosis, things are very different. **Figure 9.19** compares the two processes.

In meiosis, chromosomes of maternal origin pair with their paternal homologs during synapsis. Segregation of the homologs during meiotic anaphase I ensures that each pole receives one member of each homologous pair. For example, at the end of meiosis I in humans, each daughter nucleus contains 23 of the original 46 chromosomes. In this way, the chromosome number is decreased from diploid to haploid. Furthermore, meiosis I guarantees that each daughter nucleus gets one full set of chromosomes.

The products of meiosis I are genetically diverse for two reasons:

- Synapsis during prophase I allows the maternal chromosome in each homologous pair to exchange segments with the paternal homolog by crossing over. The resulting **recombinant** chromatids contain some genetic material from each parent.

- It is a matter of chance which member of a homologous pair goes to which daughter cell at anaphase I. For example, if there are two homologous pairs of chromosomes in the diploid parent nucleus, a particular daughter nucleus could get paternal chromosome 1 and maternal chromosome 2, or paternal 2 and maternal 1, or both maternal, or both paternal. It all depends on the way in which the homologous pairs line up at metaphase I. This phenomenon is termed **independent assortment**.

Note that of the four possible chromosome combinations just described, only two produce daughter nuclei that are the same as one of the parental types (except for any material exchanged by crossing over). The greater the number of chromosomes, the less probable that the original parental combinations will be reestablished, and the greater the potential for genetic diversity. Most species of diploid organisms do indeed have more than two pairs of chromosomes. In humans, with 23 chromosome pairs, 2^{23} (8,388,608) different combinations can be produced just by the mechanism of independent assortment. Taking the extra genetic shuffling afforded by crossing over into account, the number of possible combinations is virtually infinite.

Meiotic errors lead to abnormal chromosome structures and numbers

In the complex process of cell division, things occasionally go wrong. A pair of homologous chromosomes may fail to separate during meiosis I, or sister chromatids may fail to separate during meiosis II or during mitosis. This phenomenon is called **nondisjunction**. Conversely, homologous chromosomes may fail to remain together. Either problem can result in the production of *aneuploid* cells. **Aneuploidy** is a condition in which one or more chromosomes are either lacking or present in excess (**Figure 9.20**).

One cause of aneuploidy may be a lack of cohesins. Recall that in meiosis, these proteins, formed after DNA replication, hold the two homologous chromosomes together into metaphase I (see Figure 9.11). They ensure that when the chromosomes line up at the equatorial plate, one homolog will face one pole and the other homolog will face the other pole. Without this "glue," the two homologs may line up randomly at metaphase I, just like chromosomes during mitosis, and there is a 50 percent chance that both will go to the same pole. If, for example, during the formation of a human egg, both members of the chromosome 21 pair go to the same pole during anaphase I, the resulting eggs will contain either two of chromosome 21 or none at all. If an egg with two of these chromosomes is fertilized by a normal sperm, the resulting zygote will have three copies of the chromosome: it will be **trisomic** for chromosome 21. A child with an extra chromosome 21 has the symptoms of *Down syndrome*: impaired intelligence; characteristic abnormalities of the hands, tongue, and eyelids; and an increased susceptibility to cardiac abnormalities and diseases such as leukemia. If an egg that did not receive chromosome 21 is fertilized by a normal sperm, the zygote will have only one copy: it will be **monosomic** for chromosome 21.

Other abnormal chromosomal events can also occur. In a process called **translocation**, a piece of a chromosome may break away and become attached to another chromosome. For example, a particular large part of one chromosome 21 may be translocated to another chromosome. Individuals who inherit this translocated piece along with two normal chromosomes 21 will have Down syndrome.

Trisomies (and the corresponding monosomies) are surprisingly common in human zygotes, with 10–30 percent of all conceptions showing aneuploidy. But most of the embryos that develop from such zygotes do not survive to birth, and those that do often die before the age of 1 year. Trisomies and monosomies for most chro-

mosomes other than chromosome 21 are lethal to the embryo. At least one-fifth of all recognized pregnancies are spontaneously terminated (miscarried) during the first 2 months, largely because of such trisomies and monosomies. (The actual proportion of spontaneously terminated pregnancies is certainly higher, because the earliest ones often go unrecognized.)

Polyploids can have difficulty in cell division

As mentioned in Section 9.4, both diploid and haploid nuclei can divide by mitosis. Multicellular diploid and multicellular haploid individuals both develop from single-celled beginnings by mitotic divisions. Likewise, mitosis may proceed in diploid organisms even when a chromosome is missing from one of the haploid sets or when there is an extra copy of one of the chromosomes (as in people with Down syndrome).

Organisms with complete extra sets of chromosomes may sometimes be produced by artificial breeding or by natural acci-

9.19 Mitosis and Meiosis: A Comparison Meiosis differs from mitosis chiefly by the synapsis of homologs and by the failure of the centromeres to separate at the end of metaphase I.

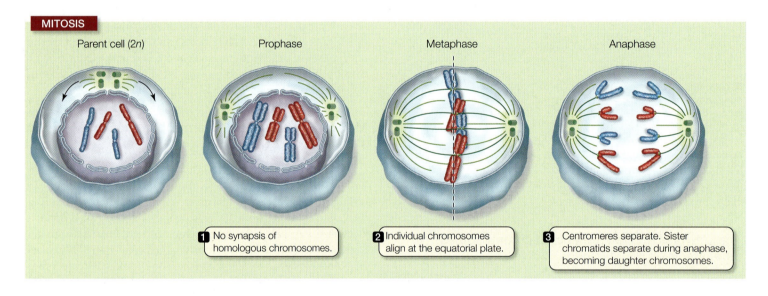

MITOSIS

Parent cell (2n) | Prophase | Metaphase | Anaphase

1 No synapsis of homologous chromosomes.

2 Individual chromosomes align at the equatorial plate.

3 Centromeres separate. Sister chromatids separate during anaphase, becoming daughter chromosomes.

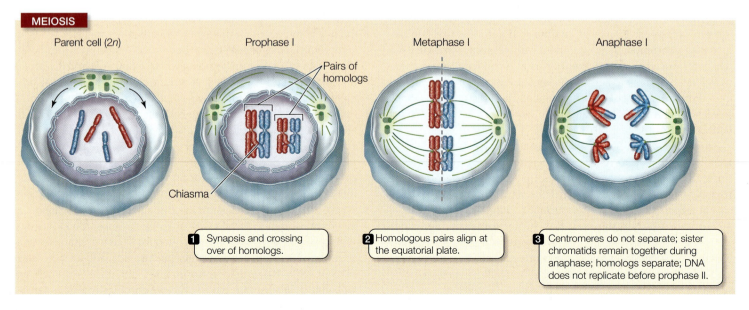

MEIOSIS

Parent cell (2n) | Prophase I | Metaphase I | Anaphase I

Pairs of homologs

Chiasma

1 Synapsis and crossing over of homologs.

2 Homologous pairs align at the equatorial plate.

3 Centromeres do not separate; sister chromatids remain together during anaphase; homologs separate; DNA does not replicate before prophase II.

9.20 Nondisjunction Leads to Aneuploidy Nondisjunction occurs if homologous chromosomes fail to separate during meiosis I. The result is aneuploidy: one or more chromosomes are either lacking or present in excess.

dents. Under some circumstances, triploid (3*n*), tetraploid (4*n*), and higher-order **polyploid** nuclei may form. Each of these *ploidy levels* represents an increase in the number of complete sets of chromosomes present.

If a nucleus has one or more extra full sets of chromosomes, its abnormally high ploidy in itself does not prevent mitosis, because in mitosis each chromosome behaves independently of the others. In meiosis, however, homologous chromosomes must synapse to begin division. If even one chromosome has no homolog, anaphase I cannot send representatives of that chromosome to both poles. A diploid nu-

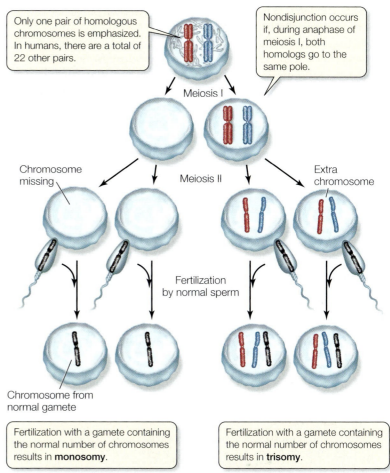

Only one pair of homologous chromosomes is emphasized. In humans, there are a total of 22 other pairs.

Nondisjunction occurs if, during anaphase of meiosis I, both homologs go to the same pole.

Meiosis I

Chromosome missing

Meiosis II

Extra chromosome

Fertilization by normal sperm

Chromosome from normal gamete

Fertilization with a gamete containing the normal number of chromosomes results in **monosomy**.

Fertilization with a gamete containing the normal number of chromosomes results in **trisomy**.

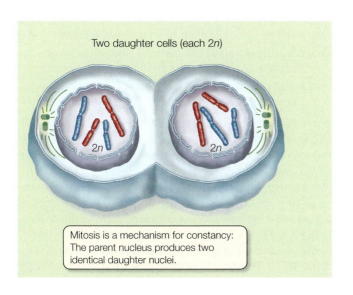

Two daughter cells (each 2*n*)

2*n* 2*n*

Mitosis is a mechanism for constancy: The parent nucleus produces two identical daughter nuclei.

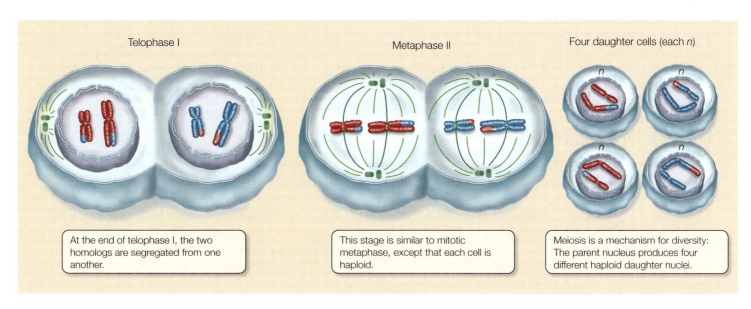

Telophase I

Metaphase II

Four daughter cells (each *n*)

n *n* *n* *n*

At the end of telophase I, the two homologs are segregated from one another.

This stage is similar to mitotic metaphase, except that each cell is haploid.

Meiosis is a mechanism for diversity: The parent nucleus produces four different haploid daughter nuclei.

cleus can undergo normal meiosis; a haploid one cannot. Similarly, a tetraploid nucleus has an even number of each kind of chromosome, so each chromosome can pair with its homolog. But a triploid nucleus cannot undergo normal meiosis because one-third of the chromosomes would lack partners. This limitation has important consequences for the fertility of triploid, tetraploid, and other chromosomally unusual organisms. Such organisms are common in modern agriculture. Modern bread wheat plants, for example, are hexaploids, the result of naturally occurring crosses between three different grasses, each having its own diploid set of 14 chromosomes.

9.5 RECAP

Meiosis produces four daughter cells in which the chromosome number is reduced from diploid to haploid. Because of the independent assortment of chromosomes and the crossing over of homologous chromatids, the four products of meiosis are not genetically identical. Meiotic errors, such as failure of a homologous chromosome pair to separate, can lead to abnormal numbers of chromosomes.

- Can you describe how crossing over and independent assortment result in unique daughter nuclei? See p. 198 and Figure 9.18

- Do you understand the differences between meiosis and mitosis? See p. 199 and Figure 9.19

- What is aneuploidy, and how can it arise from nondisjunction during meiosis? See p. 199 and Figure 9.20

As mentioned at the start of this chapter, an essential role of cell division in complex eukaryotes is to replace cells that die. What happens to those cells?

9.6 How Do Cells Die?

Cells die in one of two ways. The first type of cell death, **necrosis**, occurs when cells either are damaged by toxins or starved of oxygen or essential nutrients. These cells usually swell up and burst, releasing their contents into the extracellular environment. This process often results in inflammation (see Section 18.2). The scab that forms around a wound is a familiar example of necrotic tissue.

More typically, cell death is due to **apoptosis** (Greek, "falling apart"). Apoptosis is a genetically programmed series of events that result in cell death. These two modes of cell death are compared in **Table 9.2**.

Why would a cell initiate apoptosis, which is essentially "cell suicide"? There are two possible reasons:

- *The cell is no longer needed by the organism.* For example, before birth, a human fetus has weblike hands, with connective tissue between the fingers. As development proceeds, this unneeded tissue disappears as its cells undergo apoptosis (see Figure 19.14).

- *The longer cells live, the more prone they are to genetic damage that could lead to cancer.* This is especially true of cells in the blood and in the epithelia lining organs such as the intestine, which are exposed to high levels of toxic substances. Such cells normally die after only days or weeks.

In the developing human brain, there are many possible connections among nerve cells, yet only a few of these connections survive to birth. The rest of the brain cells—as many as half of them—die.

The events of apoptosis are very similar in most organisms. The cell becomes isolated from its neighbors, cuts up its chromatin into nucleosome-sized pieces, and forms membranous lobes, or "blebs," that break up into cell fragments (**Figure 9.21A**). In a remarkable example of the economy of nature, the surrounding living cells usually ingest the remains of the dead cell. The genetic signals that lead to apoptosis are also shared by many organisms.

Like the cell division cycle, the cell death cycle is controlled by signals, which may come from either inside or outside the cell (**Figure 9.21B**). These signals include the lack of a mitotic signal (such as a growth factor) and the recognition of damaged DNA. External signals (or a lack of them) can cause a receptor protein in the plasma membrane to change its shape and in turn activate a class of enzymes called **caspases**. Internal signals can cause mitochondria to release molecules that in turn activate caspases. In either case, the result is that the cell turns into a killing machine as the

---

TABLE 9.2

Two Different Ways for Cells to Die

	NECROSIS	APOPTOSIS
Stimuli	Low O_2, toxins, ATP depletion, damage	Specific, genetically programmed physiological signals
ATP required	No	Yes
Cellular pattern	Swelling, organelle disruption, tissue death	Chromatin condensation, membrane blebbing, single-cell death
DNA breakdown	Random fragments	Nucleosome-sized fragments
Plasma membrane	Bursts	Blebbed (see Figure 9.21A)
Fate of dead cells	Ingested by white blood cells	Ingested by neighboring cells
Reaction in tissue	Inflammation	No inflammation

(A)

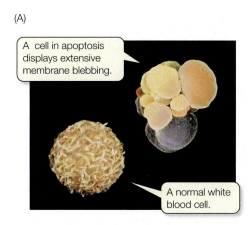

A cell in apoptosis displays extensive membrane blebbing.

A normal white blood cell.

(B)

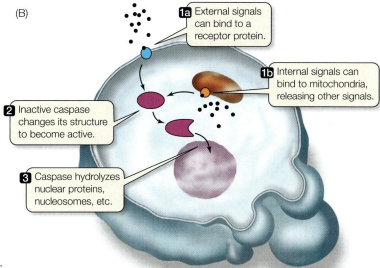

1a External signals can bind to a receptor protein.

1b Internal signals can bind to mitochondria, releasing other signals.

2 Inactive caspase changes its structure to become active.

3 Caspase hydrolyzes nuclear proteins, nucleosomes, etc.

caspases hydrolyze proteins of the nuclear envelope, nucleosomes, and plasma membrane. As we will see in Section 17.5, many of the drugs used to treat diseases of excess cell proliferation, such as cancer, work through these signals.

9.21 Apoptosis: Programmed Cell Death (A) Many cells are genetically programmed to "self-destruct" when they are no longer needed, or when they have lived long enough to accumulate a burden of DNA damage that might harm the organism. (B) Both external and internal signals stimulate caspases, the enzymes that break down specific cell constituents, resulting in apoptosis.

9.6 RECAP

The growth of a population of cells depends on the relative rates of cell reproduction and cell death. Cell death can occur either by necrosis or by apoptosis. Apoptosis is governed by precise molecular controls.

- What are some differences between apoptosis and necrosis? See Table 9.2

- Can you think of some situations in which apoptosis is necessary? See p. 202

- How is apoptosis regulated? See Figure 9.21

We have now looked at the cell cycle and how cells divide by fission, by mitosis, and by meiosis. We have seen how meiosis produces the genetic complement that will be passed on to the next generation. We will spend the next five chapters in the "realm of the genome," looking at heredity as it was explicated by Gregor Mendel in the nineteenth century and at the exponential amount of knowledge that has been acquired since the elucidation of the genetic code. Virtually no area of modern life is untouched by the field of genetics.

CHAPTER SUMMARY

9.1 How do prokaryotic and eukaryotic cells divide?

Cell division is necessary for the reproduction, growth, and repair of organisms.

Cell division must be initiated by a reproductive signal. Before a cell can divide, the genetic material (DNA) must be **replicated** and **segregated** to separate portions of the cell. **Cytokinesis** then divides the cytoplasm into two cells.

In prokaryotes, most cellular DNA is a single molecule, usually in the form of a circular **chromosome**. Prokaryotes reproduce by **binary fission**. Review Figure 9.2

In eukaryotes, cells divide by either **mitosis** or **meiosis**. Eukaryotic cell division follows the same general pattern as binary fission, but with significant differences in detail. For example, eukaryotic cells have a distinct nucleus (not just a chromosome) that must be replicated. Replicated chromosomes, called **sister chromatids**, must be separated by mitosis.

Cells that produce gametes undergo a special kind of nuclear division called meiosis; the two nuclei produced by meiosis are not genetically identical.

9.2 How is eukaryotic cell division controlled?

The eukaryotic **cell cycle** has two main phases: **interphase** (during which cells are not dividing) and mitosis or **M phase** (when cells are dividing).

During most of the cell cycle, the cell is in interphase, which is divided into three subphases: **S, G1**, and **G2**. DNA is replicated during the S phase. Mitosis and cytokinesis take place during the M phase. Review Figure 9.3

Cyclin/Cdk complexes regulate the passage of cells through checkpoints in the cell cycle. The suppressor protein **RB** inhibits the cell cycle. Cdk4 and Cdk2 act in concert to inactivate RB and allow the cell cycle to progress beyond the **restriction point**. Review Figure 9.6

CHAPTER SUMMARY

Controls external to the cell, such as **growth factors** and hormones, can also stimulate the cell to begin a division cycle.

9.3 What happens during mitosis?

See Web/CD Tutorial 9.1

In mitosis, a single nucleus gives rise to two nuclei that are genetically identical to each other and to the parent nucleus.

DNA is wrapped around proteins called **histones**, forming beadlike units called **nucleosomes**. A eukaryotic chromosome contains strings of nucleosomes bound to proteins in a complex called **chromatin**. Review Figure 9.8

At mitosis, the replicated chromatids are held together at the **centromere**. Each chromatid consists of one double-stranded DNA molecule. Review Figure 9.9, Web/CD Activity 9.1

Mitosis can be divided into several phases, called **prophase**, **prometaphase**, **metaphase**, **anaphase**, and **telophase**.

During mitosis, sister chromatids, attached by **cohesin** protein, line up at the **equatorial plate**, and attach to the **spindle**. The chromatids separate (becoming **daughter chromosomes**) and migrate to opposite ends of the cell. Review Figure 9.10, Web/CD Activity 9.2

Nuclear division is usually followed by cytokinesis. Animal cell cytoplasm usually divides by a contractile ring made up of actin microfilaments. In plant cells, cytokinesis is accomplished by vesicles that fuse to form a cell plate. Review Figure 9.12

9.4 What is the role of cell division in sexual life cycles?

Asexual reproduction produces a **clone**, a new organism that is genetically identical to the parent. Any genetic variation is the result of mutations.

In **sexual reproduction**, two **haploid** gametes—one from each parent—unite in **fertilization** to form a genetically unique, **diploid zygote**. There are several patterns of sexual life cycles: **haplontic** life cycles, **alternation of generations**, and **diplontic** life cycles. Review Figure 9.14, Web/CD Activity 9.3

In sexually reproducing organisms, certain cells in the adult undergo meiosis, a process by which a diploid cell produces haploid gametes. Each gamete contains a random selection of one of each **pair of homologous chromosomes** from the parent.

The numbers, shapes, and sizes of the chromosomes constitute the **karyotype** of an organism.

9.5 What happens when a cell undergoes meiosis?

See Web/CD Tutorial 9.2

Meiosis consists of two nuclear divisions, **meiosis I** and **meiosis II**, that collectively reduce the chromosome number from diploid to haploid. It ensures that each haploid cell contains one member of each chromosome pair, and results in four genetically diverse haploid cells, usually gametes. Review Figure 9.16, Web/CD Activity 9.4

In anaphase I, entire chromosomes, each with two chromatids, migrate to the poles. By the end of meiosis I, there are two nuclei, each with the haploid number of chromosomes.

In meiosis II, the sister chromatids separate. No DNA replication precedes this division, which in other aspects is similar to mitosis.

During prophase I of the first meiotic division, homologous chromosomes undergo **synapsis** to form pairs in a **tetrad**. Chromatids can form junctions called **chiasmata** and genetic material may be exchanged between the two homologs by **crossing over**. Review Figure 9.18

Both crossing over during prophase I and the **independent assortment** of the homologs that migrate to each pole during anaphase I ensure that the genetic composition of each haploid gamete is different from that of the parent cell and from that of the other gametes.

In **nondisjunction**, one member of a homologous pair of chromosomes fails to separate from the other, and both go to the same pole. Pairs of homologous chromosomes may also fail to stick together when they should. Both problems can lead to one gamete having an extra chromosome and another lacking that chromosome. Review Figure 9.20

The union of a gamete with an abnormal chromosome number with a normal haploid gamete at fertilization results in **aneuploidy**. Such genetic abnormalities are invariably harmful or lethal to the organism.

9.6 How do cells die?

Cells may die by **necrosis**, or they may self-destruct by **apoptosis**, a genetically programmed series of events that includes the detachment of the cell from its neighbors and the fragmentation of its nuclear DNA. Review Table 9.2

Apoptosis is regulated by external and internal signals. These signals result in activation of a class of enzymes called **caspases** that hydrolyze proteins of the cell, leading to its destruction. Review Figure 9.21

SELF-QUIZ

1. Which statement about eukaryotic chromosomes is *not* true?
   a. They sometimes consist of two chromatids.
   b. They sometimes consist only of a single chromatid.
   c. They normally possess a single centromere.
   d. They consist only of proteins.
   e. They are clearly visible as defined bodies under the light microscope.

2. Nucleosomes
   a. are made of chromosomes.
   b. consist entirely of DNA.
   c. consist of DNA wound around a histone core.
   d. are present only during mitosis.
   e. are present only during prophase.

3. Which statement about the cell cycle is *not* true?
   a. It consists of mitosis and interphase.
   b. The cell's DNA replicates during G1.
   c. A cell can remain in G1 for weeks or much longer.
   d. DNA is not replicated during G2.
   e. Cells enter the cell cycle as a result of internal or external signals.

4. Which statement about mitosis is *not* true?
   a. A single nucleus gives rise to two identical daughter nuclei.
   b. The daughter nuclei are genetically identical to the parent nucleus.
   c. The centromeres separate at the onset of anaphase.
   d. Homologous chromosomes synapse in prophase.
   e. The centrosomes organize the microtubules of the spindle fibers.

5. Which statement about cytokinesis is true?
   a. In animals, a cell plate forms.
   b. In plants, it is initiated by furrowing of the membrane.
   c. It follows mitosis.
   d. In plant cells, actin and myosin play an important part.
   e. It is the division of the nucleus.

6. Apoptosis
   a. occurs in all cells.
   b. involves the formation of the plasma membrane.
   c. does not occur in an embryo.
   d. is a series of programmed events resulting in cell death.
   e. is the same as necrosis.

7. In meiosis,
   a. meiosis II reduces the chromosome number from diploid to haploid.
   b. DNA replicates between meiosis I and meiosis II.
   c. the chromatids that make up a chromosome in meiosis II are identical.

   d. each chromosome in prophase I consists of four chromatids.
   e. homologous chromosomes separate from one another in anaphase I.

8. In meiosis,
   a. a single nucleus gives rise to two daughter nuclei.
   b. the daughter nuclei are genetically identical to the parent nucleus.
   c. the centromeres separate at the onset of anaphase I.
   d. homologous chromosomes synapse in prophase I.
   e. no spindle forms.

9. A plant has a diploid chromosome number of 12. An egg cell of that plant has 5 chromosomes. The most probable explanation is
   a. normal mitosis.
   b. normal meiosis.
   c. nondisjunction in meiosis I.
   d. nondisjunction in meiosis I and II.
   e. nondisjunction in mitosis.

10. The number of daughter chromosomes in a human cell in anaphase II of meiosis is
    a. 2.
    b. 23.
    c. 46.
    d. 69.
    e. 92.

FOR DISCUSSION

1. The story of HeLa cells raises some important issues in the ethics of biomedical research. Henrietta Lacks was not asked for permission to use her cells, and neither she nor her surviving family benefited in any direct way from the many commercial uses of her cells. What types of regulations are needed to ensure ethical handling of opportunities such as that presented by Mrs. Lacks's cells?

2. Compare the roles of cohesins in mitosis, meiosis I, and meiosis II.

3. Compare and contrast mitosis (and subsequent cytokinesis) in animals and plants.

4. Contrast mitotic prophase and prophase I of meiosis. Contrast mitotic anaphase and anaphase I of meiosis.

5. Compare the sequence of events in the mitotic cell cycle with the sequence of events in apoptosis.

FOR INVESTIGATION

1. Suggest two ways in which, with the help of a microscope, one might determine the relative duration of the various phases of mitosis.

2. Describing the events and controls of the cell cycle is much easier if the cells under investigation are synchronous; that is, if a population of cells are all in the same stage of the cell cycle. This can be accomplished with various chemicals. But some populations of cells are naturally synchronous. The anther (male sex organ) of a

lily plant contains cells that become pollen grains (male gametes). As anthers develop in the flower, their length correlates precisely with the stage of the meiotic cycle in those cells. These stages each take many days, so that an anther that is 1.5 millimeters long, for example, contains cells in early prophase I. How would you use lily anthers to investigate the roles of cyclins and Cdk's in the meiotic cell cycle?

10 Genetics: Mendel and Beyond

The wisdom of the rabbis

In the Middle Eastern desert of 1,800 years ago, the rabbi faced a dilemma. A Jewish woman had given birth to a son. As required by the laws set down by God's commandment to Abraham almost 2,000 years previously and later reiterated by Moses, the mother brought her 8-day-old son to the rabbi for ritual penile circumcision. The rabbi knew that the woman's two previous sons had bled to death when their foreskins were cut. Yet the biblical commandment remained: unless he was circumcised, the boy could not be counted among those with whom God had made His solemn covenant. After consultation with other rabbis, it was decided to exempt this, the third son.

Almost a thousand years later, in the twelfth century, the physician and biblical commentator Moses Maimonides reviewed this and numerous other cases in the rabbinical literature and stated that in such instances the third son should not be circumcised. Furthermore, the exemption should apply whether the mother's son was "from her first husband or from her second husband." The bleeding disorder, he reasoned, was clearly carried by the mother and passed on to her sons.

Without any knowledge of our modern concepts of genes and genetics, the rabbis had linked a human disease (which we now know as hemophilia A) to a pattern of inheritance (which we know as sex linkage). Only in the past few decades have the precise biochemical nature of hemophilia A and its genetic determination been worked out.

Humans normally have two copies each of 22 of the 23 chromosomes in the human karyotype, as described in Chapter 9. Thus, even if a given gene on one of the chromosomes is mutant, the normal gene on the second copy of that chromosome can usually produce a functional protein. But one pair of chromosomes is different. In the case of X and Y chromosomes, males receive only one copy of each; females receive two copies of the X chromosome (but no Y chromosome). The genetic mutation that causes the blood clotting malfunction of hemophilia is located on the X chromosome, and males carrying the mutation have no "back-up" normal gene. Color blindness, a physical condition with only minor ramifications for most individuals who suffer from it, has a similar pattern of transmission.

How do we account for and predict such patterns of inheritance? Much about inheritance was intuited even before scientists and

An Ancient Ritual A male infant undergoes ritual circumcision in accordance with Jewish laws. Sons of Jewish mothers who carry the gene for hemophilia may be exempted from this ritual.

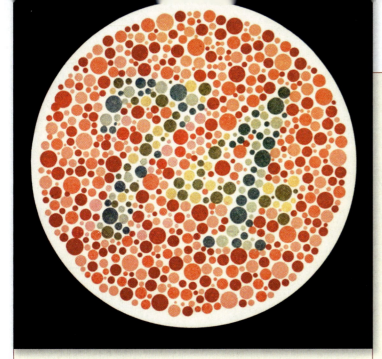

Test for a Sex-Linked Trait   Like hemophilia, the mutant allele for red-green color blindness is carried on the X chromosome. Unlike hemophilia, however, this condition is not usually deleterious. In the simple test shown here, a person with normal color vision sees the number 74; people with the most typical type of color blindness see 21; and severely color blind people cannot distinguish any numeral.

scholars knew that genes and chromosomes existed—as proven by the ruling of that wise rabbi almost two thousand years ago. Indeed, the foundations of the science of inheritance and genetic transmission were laid in the 1860s by some of the most amazing experiments and feats of data analysis in the history of biological science. It was almost 50 years before the significance of these experiments and their analyses by Gregor Mendel was recognized by the scientific community. Once that recognition was finally achieved, however, natural science and medicine began to move forward at an unprecedented pace.

IN THIS CHAPTER we will discuss how the units of inheritance—genes—are transmitted from generation to generation. We will show that many of the rules that govern inheritance can be explained by the behavior of chromosomes during meiosis. We will describe the interactions of genes with one another and with the environment, and we will see how the specific positions of genes on chromosomes affect diversity.

10.1 What Are the Mendelian Laws of Inheritance?

Much of the early study of biological inheritance was done with plants and animals of economic importance. Records show that people were deliberately cross-breeding date palm trees and horses as early as 5,000 years ago. By the early nineteenth century plant breeding was widespread, especially for ornamental flowers such as tulips. Plant breeders of that time were operating under two key assumptions about how inheritance worked. Only one of those assumptions turned out to be correct.

- *Each parent contributes equally to offspring (correct).* In the 1770s, the German botanist Josef Gottlieb Kölreuter studied the offspring of **reciprocal crosses**, in which plants are *crossed* (mated with each other) in opposite directions. For example, in one cross, males that have white flowers are mated with females that have red flowers, while in a complementary cross, red-flowered males and white-flowered females are mated. In Kölreuter's studies, such reciprocal crosses always gave identical results, showing that both parents contributed equally to the offspring.

- *Hereditary determinants blend in offspring (incorrect).* Kölreuter and others proposed that there were hereditary determinants in the egg and sperm cells. When these determinants came together in a single cell after mating, they were believed to blend together. If a plant that had one form of a characteristic (say, red flowers) was crossed with one that had a different form of that characteristic (blue flowers), the offspring would have a blended combination of the two parents' characteristics (purple flowers). According to the blending theory, it was thought that once heritable elements were combined, they could not be separated again (like inks of different colors mixed together). The red and blue hereditary determinants were thought to be forever blended into the new purple one.

In his experiments in the 1860s, Gregor Mendel confirmed the first of these two assumptions, but refuted the second.

Mendel brought new methods to experiments on inheritance

Gregor Mendel was an Austrian monk, not an academic scientist (**Figure 10.1**). He was well qualified, however, to undertake scientific investigations. After his 1850 failure in an examination for a teaching certificate in natural science, he undertook intensive studies in physics, chemistry, mathematics, and various aspects of biology at the University of Vienna. His studies in physics and mathematics strongly influenced his use of experimental and quantitative methods in his studies of heredity, and it was those quantitative experiments that were key to his successful deductions.

Over the seven years he spent working out the principles of inheritance in plants, Mendel made crosses between and noted the resulting characteristics of 24,034 plants. Analysis of his meticulously gathered data suggested to him a new theory of how inheritance might work. His work culminated in a public lecture in 1865 and a detailed written publication in 1866. Mendel's paper appeared in a journal that was received by 120 libraries, and he sent reprinted copies (of which he had obtained 40) to several distinguished scholars. However, his theory was not readily accepted. In fact, it was mostly ignored.

One reason Mendel's paper received so little attention was that most prominent biologists of his time were not in the habit of thinking in mathematical terms, even the simple terms Mendel used. Even Charles Darwin, whose theory of evolution by natural selection was predicated on heritable variation among individuals, failed to understand the significance of Mendel's findings. In fact, Darwin performed breeding experiments on snapdragons similar to Mendel's work on peas and got data similar to Mendel's, but he failed to question the assumption that parental contributions blend in offspring.

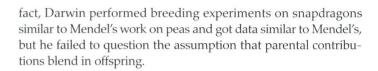

Mendel's work may have gone unnoticed in part because he had little credibility as a biologist. In fact his lowest test scores were in biology. Whatever the reasons, Mendel's pioneering paper had no discernible influence on the scientific community for more than 30 years.

By 1900, the events of meiosis had been observed and described, and Mendel's discoveries burst into sudden prominence as a result of independent experiments by three plant geneticists: Hugo DeVries, Carl Correns, and Erich von Tschermak. Each carried out crossing experiments, each published his principal findings in 1900, and each cited Mendel's 1866 paper. These three men realized that chromosomes and meiosis provided a physical explanation for the theory that Mendel had proposed to explain the data from his crosses.

That Mendel was able to achieve his remarkable insights before the discovery of genes and meiosis was largely due to his experimental methods. His work is a definitive example of extensive preparation, fortunate choice of experimental subject, meticulous execution, and imaginative yet logical interpretation. Let's take a closer look at these experiments and the conclusions and hypotheses that emerged.

Mendel devised a careful research plan

Mendel chose to study the common garden pea because of its ease of cultivation, the feasibility of controlled pollination, and the availability of varieties with contrasting traits. He controlled pollination, and thus fertilization, of his parent plants by manually moving pollen from one plant to another (**Figure 10.2**). Thus he knew the parentage of the offspring in his experiments. The pea plants Mendel studied produce male and female sex organs and gametes in the same flower. If untouched, they naturally *self-pollinate*—that is, the female organ of each flower receives pollen from the male organs of the same flower. Mendel made use of this natural phenomenon in some of his experiments.

Mendel began by examining different varieties of peas in a search for heritable characters and traits suitable for study:

- A **character** is an observable physical feature, such as flower color.

- A **trait** is a particular form of a character, such as purple flowers or white flowers.

- A **heritable trait** is one that is passed from parent to offspring.

10.1 Gregor Mendel and His Garden The Austrian monk Gregor Mendel (left) did his pathbreaking experiments in genetics in a garden of the monastery at Brno, in what is now the Czech Republic.

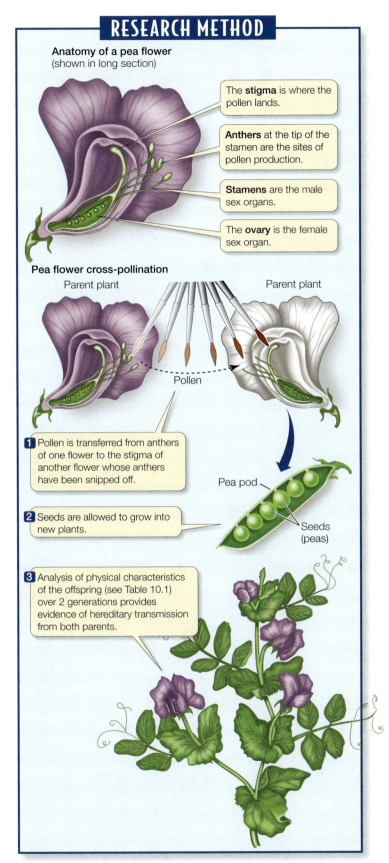

RESEARCH METHOD

Anatomy of a pea flower
(shown in long section)

The **stigma** is where the pollen lands.

Anthers at the tip of the stamen are the sites of pollen production.

Stamens are the male sex organs.

The **ovary** is the female sex organ.

Pea flower cross-pollination

Parent plant   Parent plant

Pollen

1 Pollen is transferred from anthers of one flower to the stigma of another flower whose anthers have been snipped off.

Pea pod

2 Seeds are allowed to grow into new plants.

Seeds (peas)

3 Analysis of physical characteristics of the offspring (see Table 10.1) over 2 generations provides evidence of hereditary transmission from both parents.

10.2 A Controlled Cross between Two Plants Plants were widely used in early genetic studies because it is easy to control which individuals mate with which. Mendel used the garden pea (*Pisum sativum*) in many of his experiments.

Mendel looked for characters with well-defined, contrasting alternative traits, such as purple flowers versus white flowers. Furthermore, these traits had to be **true-breeding**, meaning that the observed trait was the only form present for many generations. In other words, if they were true-breeding, peas with white flowers, when crossed with one another, would give rise only to progeny with white flowers for many generations; tall plants bred to tall plants would produce only tall progeny.

Mendel isolated each of his true-breeding strains by repeated inbreeding (done by crossing of sibling plants that were seemingly identical or by allowing individuals to self-pollinate) and selection. In most of his work, Mendel concentrated on the seven pairs of contrasting traits shown in **Table 10.1**. Before performing any experimental cross, he made sure that each potential parent was from a true-breeding strain—an essential point in his analysis of his experimental results.

Mendel then performed his crosses in the following manner:

■ He collected pollen from one parental strain and placed it on the stigma (female organ) of flowers of the other strain whose anthers (male organs) had been removed (so that the recipient plant could not fertilize itself). The plants providing and receiving the pollen were the **parental generation**, designated **P**.

■ In due course, seeds formed and were planted. The seeds and the resulting new plants constituted the **first filial generation**, or F_1. Mendel and his assistants examined each F_1 plant to see which traits it bore and then recorded the number of F_1 plants expressing each trait.

■ In some experiments the F_1 plants were allowed to self-pollinate and produce a **second filial generation**, F_2. Again, each F_2 plant was characterized and counted.

Mendel's first experiments involved monohybrid crosses

The term *hybrid* refers to offspring of crosses between organisms differing in one or more traits. In Mendel's first experiment, he crossed two true-breeding parental (P) lineages differing in just *one* trait, producing *monohybrids* (the F_1 generation). He subsequently planted the F_1 seeds and allowed the resulting plants to self-pollinate to produce the F_2 generation. This technique is referred to as a **monohybrid cross**, even though in this case, the monohybrid plants were not literally crossed, but self-pollinated.

Mendel performed the same experiment for all seven pea-plant traits. His method is illustrated in **Figure 10.3**, using the seed shape trait as an example. He took pollen from pea plants of a true-breeding strain with wrinkled seeds and placed it on the stigmas of flowers of a true-breeding strain with spherical seeds. He also performed the complementary cross, in which the parental source of each trait is reversed: he placed pollen from the spherical-seeded strain on the stigmas of flowers of the wrinkled-seeded strain. In all cases, all F_1 seeds produced by crosses of the P plants were spherical—it was as if the wrinkled seed trait had disappeared completely.

The following spring, Mendel grew 253 F_1 plants from these spherical seeds. Each of these plants was allowed to self-polli-

TABLE 10.1

Mendel's Results from Monohybrid Crosses

PARENTAL GENERATION PHENOTYPES				F₂ GENERATION PHENOTYPES			
	DOMINANT	RECESSIVE		DOMINANT	RECESSIVE	TOTAL	RATIO
	Spherical seeds × Wrinkled seeds			5,474	1,850	7,324	2.96:1
	Yellow seeds × Green seeds			6,022	2,001	8,023	3.01:1
	Purple flowers × White flowers			705	224	929	3.15:1
	Inflated pods × Constricted pods			882	299	1,181	2.95:1
	Green pods × Yellow pods			428	152	580	2.82:1
	Axial flowers × Terminal flowers			651	207	858	3.14:1
	Tall stems × Dwarf stems (1 m) (0.3 m)			787	277	1,064	2.84:1

nate to produce F_2 seeds. In all, 7,324 F_2 seeds were produced, of which 5,474 were spherical and 1,850 wrinkled (Figure 10.3 and Table 10.1).

Mendel observed that the wrinkled seed trait was never expressed in the F_1 generation, even though it reappeared in the F_2 generation. This lead him to conclude that the spherical seed trait was **dominant** to the wrinkled seed trait, which he called **recessive**. In each of the other six pairs of traits Mendel studied, one trait proved to be dominant over the other trait. The trait that disappears in the F_1 generation of true-breeding crosses is always the recessive trait.

Mendel also observed that the ratio of the two traits in the F_2 generation was always the same—approximately 3:1—for each of the seven pea-plant traits he studied. That is, *three-fourths of the F_2 generation showed the dominant trait and one-fourth showed the recessive trait* (see Table 10.1). For example, Mendel's monohybrid cross for seed shape produced a ratio of 5,474:1,850 = 2.96:1. The two reciprocal crosses in the parental generation yielded similar outcomes in the F_2; it did not matter which parent contributed the pollen, just as Kölreuter had shown.

REJECTION OF THE BLENDING THEORY Mendel's monohybrid cross experiments showed that inheritance cannot be the result of a blending phenomenon. According to the blending theory, Mendel's F_1 seeds should have had an appearance intermediate between those of the two parents—in other words, they should have been slightly wrinkled. Furthermore, the blending theory offered no explanation for the reappearance of the wrinkled trait in the F_2 seeds after its apparent absence in the F_1 seeds.

SUPPORT FOR THE PARTICULATE THEORY Given the absence of blending and the reappearance of the wrinkled seed trait in the F_2 generation of his monohybrid cross experiments, Mendel proposed that the units responsible for the inheritance of specific traits are present as *discrete particles* that occur in pairs and segregate (separate) from one another during the formation of gametes. According to his **particulate theory**, the units of inheritance retain their integrity in the presence of other units. Mendel concluded that each pea plant has two units (particles) of inheritance for each character, one from each parent. He proposed that during the production of gametes, only one of these paired units is given to a gamete. He concluded that while each gamete contains one unit, the resulting zygote contains two, because it is produced by the fusion of two gametes. This conclusion is the core of Mendel's model of inheritance. Mendel's unit of inheritance is now called a **gene**. The totality of all the genes of an organism is that organism's **genome**.

Mendel reasoned that in his experiments, the two true-breeding parent plants had different forms of the gene affecting seed shape (although he did not use the term "gene"). The true-breeding spherical-seeded parent had two genes of the same form, which we will call S, and the parent with wrinkled seeds had two copies of an alternative form of the gene, which we will call s. The SS parent would produce gametes having a single S gene, and the ss parent would produce gametes having a single s gene. The cross producing the F_1 generation would donate an S from one parent and an s from the other to each seed; the F_1 offspring would thus be Ss. We say that S is dominant over s because the trait specified by s is not evident—is not *expressed*—when both forms of the gene are present.

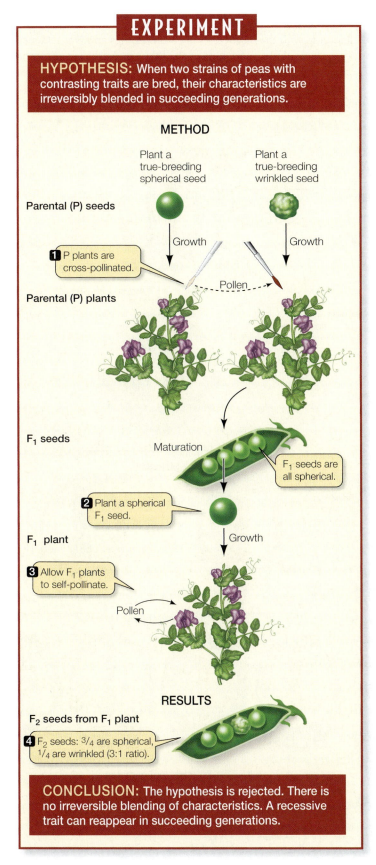

EXPERIMENT

HYPOTHESIS: When two strains of peas with contrasting traits are bred, their characteristics are irreversibly blended in succeeding generations.

METHOD

Plant a true-breeding spherical seed

Plant a true-breeding wrinkled seed

Parental (P) seeds

Growth

Growth

1 P plants are cross-pollinated.

Pollen

Parental (P) plants

F₁ seeds

Maturation

F₁ seeds are all spherical.

2 Plant a spherical F₁ seed.

F₁ plant

Growth

3 Allow F₁ plants to self-pollinate.

Pollen

RESULTS

F₂ seeds from F₁ plant

4 F₂ seeds: ³⁄₄ are spherical, ¹⁄₄ are wrinkled (3:1 ratio).

CONCLUSION: The hypothesis is rejected. There is no irreversible blending of characteristics. A recessive trait can reappear in succeeding generations.

10.3 Mendel's Monohybrid Experiments The pattern Mendel observed in the F_2 generation—³⁄₄ of the seeds spherical, ¹⁄₄ wrinkled—was the same no matter which strain contributed the pollen in the parental generation.

Alleles are different forms of a gene

The different forms of a gene (*S* and *s* in this case) are called **alleles**. Individuals that are true-breeding for a trait contain two copies of the same allele. For example, all the individuals in a population of a strain of true-breeding peas with wrinkled seeds must have the allele pair *ss*; if the dominant *S* allele were present, the plants would produce spherical seeds.

We say that the individuals that produce wrinkled seeds are **homozygous** for the allele *s*, meaning that they have two copies of the same allele (*ss*). Some peas with spherical seeds—the ones with the genotype *SS*—are also homozygous. However, not all plants with spherical seeds have the *SS* genotype. Some spherical-seeded plants, like Mendel's F_1, are **heterozygous**: they have two different alleles of the gene in question (in this case, *Ss*). An individual that is homozygous for a character is sometimes called a *homozygote*; a *heterozygote* is heterozygous for the character in question.

As a somewhat more complex example of inheritance, let's consider three gene pairs. An individual with the alleles *AABbcc* is homozygous for the *A* and *C* genes, because it has two *A* alleles and two *c* alleles, but heterozygous for the *B* gene, because it contains the *B* and *b* alleles.

The physical appearance of an organism is its **phenotype**. Mendel correctly supposed the phenotype to be the result of the **genotype**, or genetic constitution, of the organism showing the phenotype. Spherical seeds and wrinkled seeds are two phenotypes, which are the result of *three* genotypes: the wrinkled seed phenotype is produced by the genotype *ss*, whereas the spherical seed phenotype is produced by two genotypes, *SS* and *Ss*.

What's in a name—of a gene? Names such as *tall*, *short*, *spherical*, and *wrinkled* describe a trait. So what do *Drosophila* geneticists name a fruit fly that is impaired in learning experiments? Why, *dunce*, of course! A mutation that prevents heart formation? Call it *tinman*. A gene that produces extra bristles on the face is *groucho*; flies with no external genitalia are *ken* and *barbie*.

Mendel's first law says that the two copies of a gene segregate

How does Mendel's model of inheritance explain the ratios of traits seen in the F_1 and F_2 generations? Consider first the F_1, in which all progeny have the spherical seed phenotype. According to Mendel's model, when any individual produces gametes, the two copies of a gene separate, so that each gamete receives only one copy. This is *Mendel's first law*, the **law of segregation**. Thus from each parent of the P generation, every individual in the F_1 inherits one gene copy, and has the genotype *Ss* (**Figure 10.4**).

Now let's consider the composition of the F_2 generation. Half of the gametes produced by the F_1 generation have the *S* allele and the other half the *s* allele. Since both *SS* and *Ss* plants produce spherical seeds while *ss* produces wrinkled seeds, in the F_2 generation there are three ways to get a spherical-seeded plant, but

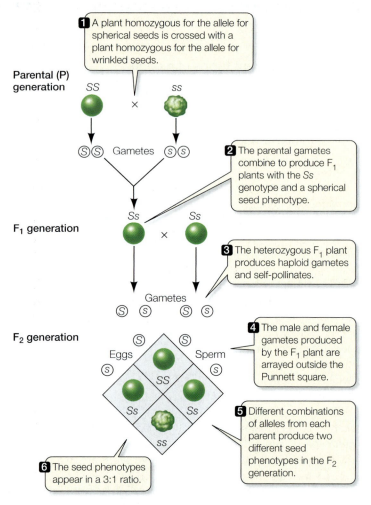

1 A plant homozygous for the allele for spherical seeds is crossed with a plant homozygous for the allele for wrinkled seeds.

Parental (P) generation

SS × ss

Gametes

2 The parental gametes combine to produce F₁ plants with the Ss genotype and a spherical seed phenotype.

F₁ generation

Ss × Ss

3 The heterozygous F₁ plant produces haploid gametes and self-pollinates.

Gametes

F₂ generation

Eggs   Sperm

4 The male and female gametes produced by the F₁ plant are arrayed outside the Punnett square.

5 Different combinations of alleles from each parent produce two different seed phenotypes in the F₂ generation.

6 The seed phenotypes appear in a 3:1 ratio.

10.4 Mendel's Explanation of Inheritance Mendel concluded that inheritance depends on discrete factors from each parent that do not blend in the offspring.

only one way to get a wrinkled-seeded plant (s from both parents)—predicting a 3:1 ratio remarkably close to the values Mendel found experimentally for all seven of the traits he compared (see Table 10.1).

The allele combinations that will result from a cross can be predicted using a **Punnett square**, a method devised in 1905 by the British geneticist Reginald Crundall Punnett. This device ensures that we consider all possible combinations of gametes when calculating expected genotype frequencies. A Punnett square looks like this:

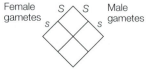

Female gametes   S S   Male gametes
             s   s

It is a simple grid with all possible male gamete (haploid sperm) genotypes shown along one side and all possible female gamete (haploid egg) genotypes along another side. The grid is completed by filling in each square with the diploid genotype that can be generated from each combination of gametes (see Figure 10.4). In this example, to fill in, say, the rightmost square, we put in the S from the egg (female gamete) and the s from the pollen (male gamete), yielding Ss.

Mendel did not live to see his theory placed on a sound physical footing with the discoveries of chromosomes and DNA. Genes are now known to be regions of the DNA molecules in chromosomes. More specifically, a gene is a sequence of DNA that resides at a particular site on a chromosome, called a **locus** (plural **loci**), and encodes a particular character. Genes are expressed in the pheno-

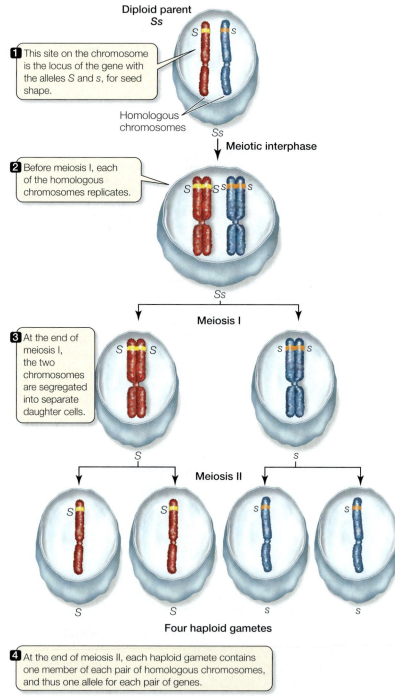

Diploid parent
Ss

1 This site on the chromosome is the locus of the gene with the alleles S and s, for seed shape.

Homologous chromosomes

Ss

Meiotic interphase

2 Before meiosis I, each of the homologous chromosomes replicates.

Ss

Meiosis I

3 At the end of meiosis I, the two chromosomes are segregated into separate daughter cells.

Meiosis II

Four haploid gametes

4 At the end of meiosis II, each haploid gamete contains one member of each pair of homologous chromosomes, and thus one allele for each pair of genes.

10.5 Meiosis Accounts for the Segregation of Alleles Although Mendel had no knowledge of chromosomes or meiosis, we now know that a pair of alleles resides on homologous chromosomes, and that meiosis segregates those alleles.

type mostly as proteins with particular functions, such as enzymes. So a dominant gene can be thought of as a region of DNA that is expressed as a functional enzyme, while a recessive gene typically expresses a nonfunctional enzyme. Mendel arrived at his law of segregation with no knowledge of chromosomes or meiosis, but today we can picture the different alleles of a gene segregating as chromosomes separate in meiosis I (**Figure 10.5**).

Mendel verified his hypothesis by performing a test cross

Mendel set out to test his hypothesis that there were two possible allele combinations (*SS* and *Ss*) in the spherical-seeded F₁ generation. He did so by performing a **test cross**, which is a way of finding out whether an individual showing a dominant trait is homozygous or heterozygous. In a test cross, the individual in question is crossed with an individual known to be homozygous for the recessive trait—an easy individual to identify, because in order to have the recessive phenotype, it must be homozygous for the recessive trait.

For the seed shape gene that we have been considering, the recessive homozygote used for the test cross is *ss*. The individual being tested may be described initially as *S_* because we do not yet know the identity of the second allele. We can predict two possible results:

- If the individual being tested is homozygous dominant (*SS*), all offspring of the test cross will be *Ss* and show the dominant trait (spherical seeds) (**Figure 10.6, left**).

- If the individual being tested is heterozygous (*Ss*), then approximately half of the offspring of the test cross will be heterozygous and show the dominant trait (*Ss*), but the other half will be homozygous for, and will show, the recessive trait (*ss*) (**Figure 10.6, right**).

The second prediction matches the results that Mendel obtained; thus Mendel's hypothesis accurately predicted the results of his test cross.

With his first hypothesis confirmed, Mendel went on to ask another question: How do different pairs of genes behave in crosses when considered together?

Mendel's second law says that copies of different genes assort independently

Consider an organism that is heterozygous for two genes (*SsYy*), in which the *S* and *Y* alleles came from its mother and *s* and *y* came from its father. When this organism produces gametes, do the alleles of maternal origin (*S* and *Y*) go together to one gamete and those of paternal origin (*s* and *y*) to another gamete? Or can a single gamete receive one maternal and one paternal allele, *S* and *y* (or *s* and *Y*)?

To answer these questions, Mendel performed another series of experiments. He began with peas that differed in two seed characters: seed shape and seed color. One true-breeding parental strain produced only spherical, yellow seeds (*SSYY*), and the other produced only wrinkled, green ones (*ssyy*). A cross between these two strains produced an F₁ generation in which all the plants were

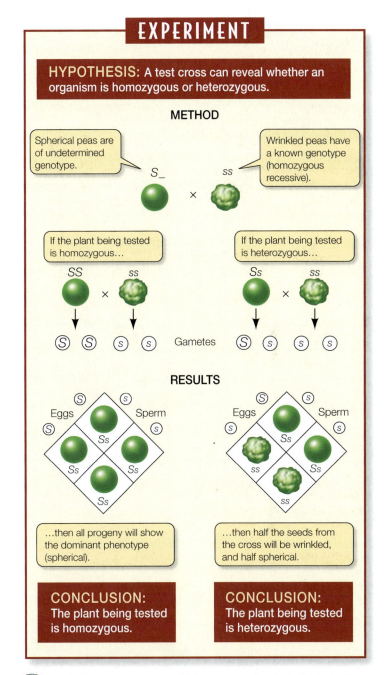

10.6 Homozygous or Heterozygous? An individual with a dominant phenotype may be homozygous or heterozygous. Its genotype can be determined by crossing it with a homozygous recessive individual and observing the phenotypes of the progeny produced. This procedure is known as a test cross. FURTHER RESEARCH: What would be the result if the "tester" plant was homozygous for spherical instead of wrinkled seeds?

SsYy. Because the *S* and *Y* alleles are dominant, the F₁ seeds were all spherical and yellow.

Mendel continued this experiment to the F₂ generation by performing a **dihybrid cross** (a cross between individuals that are identical double heterozygotes) with F₁ plants (although again, in this case, this was done by allowing the F₁ plants to self-pollinate). There are two possible ways in which such doubly heterozygous plants might produce gametes, as Mendel saw it (remember that he had never heard of chromosomes or meiosis):

1. The alleles could maintain the associations they had in the parental generation (that is, they could be *linked*).

In this case, the F₁ plants should produce two types of gametes (*SY* and *sy*), and the F₂ progeny resulting from self-pollination of the F₁ plants should consist of three times as many plants bearing spherical, yellow seeds as ones with wrinkled, green seeds. Were such results to be obtained, there might be no reason to suppose that seed shape and seed color were regulated by two different genes, because spherical seeds would always be yellow and wrinkled ones always green.

2. The segregation of *S* from *s* could be independent of the segregation of *Y* from *y* (that is, that the two genes could be *unlinked*).

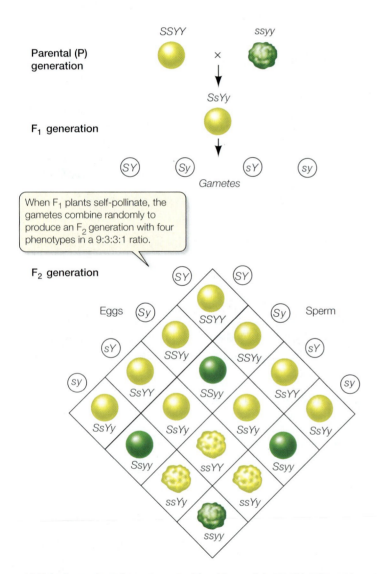

When F₁ plants self-pollinate, the gametes combine randomly to produce an F₂ generation with four phenotypes in a 9:3:3:1 ratio.

10.7 Independent Assortment The 16 possible combinations of gametes in this dihybrid cross result in 9 different genotypes. Because *S* and *Y* are dominant over *s* and *y*, respectively, the 9 genotypes result in four phenotypes in a ratio of 9:3:3:1. These results show that the two genes segregate independently.

In this case, four kinds of gametes should be produced by the F₁ in equal numbers: *SY*, *Sy*, *sY*, and *sy*. When these gametes combine at random, they should produce an F₂ having nine different genotypes. The F₂ progeny could have any of three possible genotypes for shape (*SS*, *Ss*, or *ss*) and any of three possible genotypes for color (*YY*, *Yy*, or *yy*). The combined nine genotypes should produce four phenotypes (spherical yellow, spherical green, wrinkled yellow, wrinkled green). Putting these data into a Punnett square, we can predict that these four phenotypes will occur in a ratio of 9:3:3:1 (**Figure 10.7**).

Mendel's dihybrid crosses supported the second prediction: four different phenotypes appeared in the F₂ in a ratio of about 9:3:3:1. The parental traits appeared in new combinations (spherical green and wrinkled yellow) in some progeny. Such new combinations are called **recombinant** phenotypes.

These results led Mendel to the formulation of what is now known as *Mendel's second law*: Alleles of different genes assort independently of one another during gamete formation. That is, the segregation of the alleles of gene A is independent of the segregation of the alleles of gene B. We now know that this **law of independent assortment** is not as universal as the law of segregation, because it applies to genes located on separate chromosomes, but not always to those located on the same chromosome, as we will see in Section 10.4. However, it is correct to say that *chromosomes* segregate independently during the formation of gametes, and so do any two genes on separate homologous chromosome pairs (**Figure 10.8**).

One of Mendel's major contributions to the science of genetics was his use of the rules of statistics and probability to analyze his masses of data from hundreds of crosses producing thousands of plants. His mathematical analyses revealed clear patterns in the data that allowed him to formulate his hypotheses. Ever since Mendel, geneticists have used simple mathematics in the same ways that Mendel did.

Punnett squares or probability calculations: A choice of methods

Punnett squares provide one way of solving problems in genetics, and probability calculations provide another. Many people find it easiest to use the principles of probability, some of which are intuitive and familiar. For example, when we flip a coin, the law of probability states that it has an equal probability of landing "heads" or "tails." For any given toss of a fair coin, the probability of heads is independent of what happened in all the previous tosses. A run of ten straight heads implies nothing about the next toss. No "law of averages" increases the likelihood that the next toss will come up tails, and no "momentum" makes an eleventh occurrence of heads any more likely. On the eleventh toss, the odds of getting heads are still 50-50.

The basic conventions of probability are simple:

- If an event is absolutely certain to happen, its probability is 1.
- If it cannot possibly happen, its probability is 0.
- All other events have a probability between 0 and 1.

A coin toss results in heads approximately half the time, so the probability of heads is ½—as is the probability of tails.

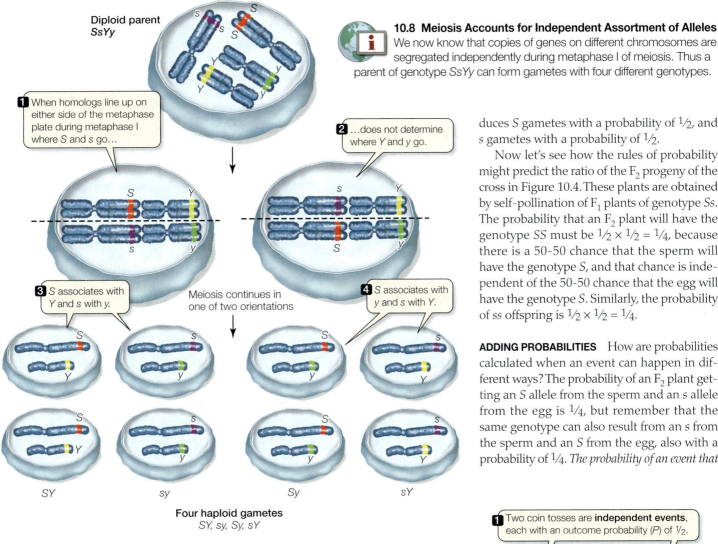

Diploid parent
SsYy

1 When homologs line up on either side of the metaphase plate during metaphase I where *S* and *s* go…

2 …does not determine where *Y* and *y* go.

3 *S* associates with *Y* and *s* with *y*.

Meiosis continues in one of two orientations

4 *S* associates with *y* and *s* with *Y*.

SY   *sy*   *Sy*   *sY*

Four haploid gametes
SY, sy, Sy, sY

10.8 Meiosis Accounts for Independent Assortment of Alleles We now know that copies of genes on different chromosomes are segregated independently during metaphase I of meiosis. Thus a parent of genotype *SsYy* can form gametes with four different genotypes.

duces *S* gametes with a probability of $1/2$, and *s* gametes with a probability of $1/2$.

Now let's see how the rules of probability might predict the ratio of the F_2 progeny of the cross in Figure 10.4. These plants are obtained by self-pollination of F_1 plants of genotype *Ss*. The probability that an F_2 plant will have the genotype *SS* must be $1/2 \times 1/2 = 1/4$, because there is a 50-50 chance that the sperm will have the genotype *S*, and that chance is independent of the 50-50 chance that the egg will have the genotype *S*. Similarly, the probability of *ss* offspring is $1/2 \times 1/2 = 1/4$.

ADDING PROBABILITIES How are probabilities calculated when an event can happen in different ways? The probability of an F_2 plant getting an *S* allele from the sperm and an *s* allele from the egg is $1/4$, but remember that the same genotype can also result from an *s* from the sperm and an *S* from the egg, also with a probability of $1/4$. *The probability of an event that*

MULTIPLYING PROBABILITIES How can we determine the probability of two independent events happening together? If two coins (a penny and a dime, say) are tossed, each acts independently of the other. What, then, is the probability of both coins coming up heads? Half the time, the penny comes up heads; of that fraction, half the time the dime also comes up heads. Therefore, the *joint probability* of both coins coming up heads is half of one-half, or $1/2 \times 1/2 = 1/4$. To find the joint probability of independent events, then, we multiply the probabilities of the individual events (**Figure 10.9**). How does this method apply to genetics?

To see how joint probability is calculated in genetics problems, let's consider the monohybrid cross. The probabilities of two events are involved: gamete formation and random fertilization.

Calculating the probabilities involved in gamete formation is straightforward. A homozygote can produce only one type of gamete, so, for example, the probability of an *SS* individual producing gametes with the genotype *S* is 1. The heterozygote *Ss* pro-

1 Two coin tosses are **independent events**, each with an outcome probability (*P*) of $1/2$.

2 This outcome is the result of two independent events. The joint probability is $1/2 + 1/2 = 1/4$ (**multiplication rule**).

$P = 1/2$

$P = 1/2$

$P = 1/2$

$P = 1/2$

$1/2 \times 1/2 = 1/4$

$1/2 \times 1/2 = 1/4$

$1/2 \times 1/2 = 1/4$

$1/2 \times 1/2 = 1/4$

Because there are two ways to arrive at a heterozygote, we add the probabilities of the two individual outcomes: $1/4 + 1/4 = 1/2$ (**addition rule**).

10.9 Using Probability Calculations in Genetics Like the results of a coin toss, the probability of any given combination of alleles from a sperm and an egg appearing in the offspring of a cross can be obtained by multiplying the probabilities of each event. Since a heterozygote can be formed in two ways, these two probabilities are added together.

can occur in two or more different ways is the sum of the individual probabilities of those ways. Thus the probability that an F$_2$ plant will be a heterozygote is equal to the sum of the probabilities of the two ways of forming a heterozygote: $1/4 + 1/4 = 1/2$ (see Figure 10.9). The three genotypes are therefore expected in the ratio $1/4$ *SS*:$1/2$ *Ss*:$1/4$ *ss*—hence the 1:2:1 ratio of genotypes and the 3:1 ratio of phenotypes seen in Figure 10.4.

PROBABILITY AND THE DIHYBRID CROSS If F$_1$ plants heterozygous for two independent characters self-pollinate, the resulting F$_2$ plants express four different phenotypes. The proportions of these phenotypes are easily determined by probability calculations. Let's see how this works for the experiment shown in Figure 10.7.

Using the principles described above, we can calculate that the probability that an F$_2$ seed will be spherical is $3/4$: the probability of an *Ss* heterozygote ($1/2$) plus the probability of an *SS* homozygote ($1/4$) = $3/4$. By the same reasoning, the probability that a seed will be yellow is also $3/4$. The two characters are determined by separate genes and are independent of each other, so the joint probability that a seed will be both spherical and yellow is $3/4 \times 3/4 = 9/16$. What is the probability of F$_2$ seeds being both wrinkled and yellow? The probability of being yellow is again $3/4$; the probability of being wrinkled is $1/2 \times 1/2 = 1/4$. The joint probability that a seed will be both wrinkled and yellow, then, is $1/4 \times 3/4 = 3/16$. The

same probability applies, for similar reasons, to spherical, green F$_2$ seeds. Finally, the probability that F$_2$ seeds will be both wrinkled and green is $1/4 \times 1/4 = 1/16$. Looking at all four phenotypes, we see they are expected in the ratio of 9:3:3:1.

Probability calculations and Punnett squares give the same results. Learn to do genetics problems both ways, and then decide which method you prefer.

Mendel's laws can be observed in human pedigrees

How are Mendel's laws of inheritance applied to humans? Mendel worked out his laws by performing many planned crosses and counting many offspring. Neither of these approaches is possible with humans, so human geneticists rely on **pedigrees**: family trees

10.10 Pedigree Analysis and Inheritance (A) This pedigree represents a family affected by Huntington's disease, which results from a rare dominant allele. Everyone who inherits this allele is affected. (B) The family in this pedigree carries the allele for albinism, a recessive trait. Because the trait is recessive, heterozygotes do not have the albino phenotype, but they can pass the allele on to their offspring. Affected persons must inherit the allele from two heterozygous parents or (rarely) from one homozygous and one heterozygous parent. In this family, the heterozygous parents are cousins, but the same result could occur if the parents were unrelated but heterozygous.

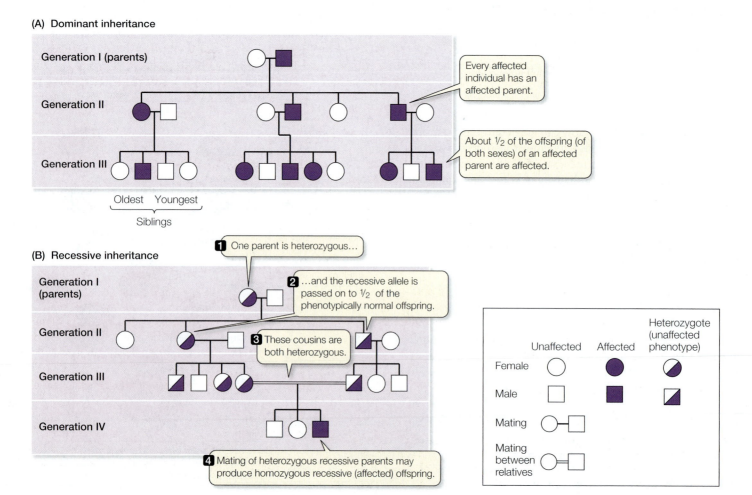

(A) Dominant inheritance

Generation I (parents)

Generation II

Generation III

Oldest   Youngest
Siblings

Every affected individual has an affected parent.

About $1/2$ of the offspring (of both sexes) of an affected parent are affected.

(B) Recessive inheritance

Generation I (parents)

Generation II

Generation III

Generation IV

1 One parent is heterozygous…

2 …and the recessive allele is passed on to $1/2$ of the phenotypically normal offspring.

3 These cousins are both heterozygous.

4 Mating of heterozygous recessive parents may produce homozygous recessive (affected) offspring.

	Unaffected	Affected	Heterozygote (unaffected phenotype)
Female	○	●	◐
Male	□	■	◪
Mating	○—□		
Mating between relatives	○=□		

that show the occurrence of phenotypes (and alleles) in several generations of related individuals.

Because humans have such small numbers of offspring, human pedigrees do not show the clear proportions of offspring phenotypes that Mendel saw in his pea plants. For example, when a man and a woman who are both heterozygous for a recessive allele (say, *Aa*) have children together, each child has a 25 percent probability of being a recessive homozygote (*aa*). Thus if this couple were to have dozens of children, one-fourth of them would be recessive homozygotes (*aa*). But the offspring of a single couple are likely to be too few to show the exact one-fourth proportion. In a family with only two children, for example, both could easily be *aa* (or *Aa*, or *AA*).

What if we want to know whether a recessive allele is carried by both the mother and the father? Human geneticists assume that any allele that causes an abnormal phenotype (such as a genetic disease) is rare in the human population. This means that if some members of a given family have a rare allele, it is highly unlikely that an outsider marrying into that family will have that same rare allele.

Human geneticists may wish to know whether a particular rare allele that causes an abnormal phenotype is dominant or recessive. **Figure 10.10A** is a pedigree showing the pattern of inheritance of a rare *dominant allele*. The following are the key features to look for in such a pedigree:

- Every affected person has an affected parent.
- About half of the offspring of an affected parent are also affected.
- The phenotype occurs equally in both sexes.

Compare this pattern with **Figure 10.10B**, which shows the pattern of inheritance of a rare *recessive* allele:

- Affected people usually have two parents who are not affected.
- In affected families, about one-fourth of the children of unaffected parents are affected.
- The phenotype occurs equally in both sexes.

In pedigrees showing inheritance of a recessive phenotype, it is not uncommon to find a marriage of two relatives. This observation is a result of the rarity of recessive alleles that give rise to abnormal phenotypes. For two phenotypically normal parents to have an affected child (*aa*), the parents must both be heterozygous (*Aa*). If a particular recessive allele is rare in the general population, the chance of two people marrying who are both carrying that allele is quite low. On the other hand, if that allele is present in a family, two cousins might share it (see Figure 10.10B). This is why studies on populations isolated either culturally (by religion, as with the Amish in the United States) or geographically (as on islands) have been so valuable to human geneticists. People in these groups tend to have large families, or to marry among themselves, or both.

Because the major use of pedigree analysis is in the clinical evaluation and counseling of patients with inherited abnormalities, a single pair of alleles is usually followed. However, just as pedigree analysis shows the segregation of alleles, it also can show independent assortment if two different allele pairs are considered.

10.1 RECAP

Mendel showed that genetic determinants are particulate and do not "blend" or disappear when the genes from two gametes combine. Mendel's first law states that the two copies of a gene segregate during gamete formation. His second law states that genes assort independently during gamete formation. The frequencies with which different allele combinations will be expressed in offspring can be calculated with a Punnett square or using probability theory.

- What results seen in the F_1 and F_2 generations of Mendel's monohybrid cross experiments refuted the blending theory of inheritance? See p. 210 and Figures 10.3 and 10.4

- Can you explain Mendel's experiment on segregation of alleles in terms of meiosis? See pp. 211–212 and Figure 10.5

- Can you explain, in terms of meiosis, how Mendel's dihybrid cross experiments suggested independent assortment of alleles? See pp. 213–214 and Figures 10.7 and 10.8

- Draw human pedigrees for dominant and recessive inheritance? See pp. 216–217 and Figure 10.10

The laws of inheritance as articulated by Mendel remain valid today; his discoveries laid the groundwork for all future studies of genetics. Inevitably, however, we have learned that things are more complicated. Let's take a look at some of these complications, beginning with the interactions between alleles at different loci.

10.2 How Do Alleles Interact?

In many cases, alleles do not show the simple relationships between dominance and recessiveness that we have described. Existing alleles are subject to mutation, and thus may give rise to new alleles, so there can be many alleles for a single character. Furthermore, a single allele may have multiple phenotypic effects.

New alleles arise by mutation

Different alleles of a gene exist because genes are subject to **mutations**, which are rare, stable, and inherited changes in the genetic material. In other words, an allele can mutate to become a different allele. Mutation, which will be discussed in detail in Section 12.6, is a random process; different copies of the same allele may be changed in different ways.

Geneticists usually define one particular allele of a gene as the **wild type**; this allele is the one that is present in most individuals in nature ("the wild") and gives rise to an expected trait or phenotype. Other alleles of that gene, often called *mutant alleles*, may pro-

Possible genotypes	CC, Cc^{ch}, Cc^h, Cc	c^{ch}c^{ch}	c^{ch}c^h, c^{ch}c	c^hc^h, c^hc	cc
Phenotype	Dark gray	Chinchilla	Light gray	Point restricted	Albino

duce a different phenotype. The wild-type and mutant alleles reside at the same locus and are inherited according to the rules set forth by Mendel. A genetic locus with a wild-type allele that is present less than 99 percent of the time (the rest of the alleles being mutant) is said to be **polymorphic** (Greek *poly*, "many," and *morph*, "form").

Many genes have multiple alleles

Because of random mutations, more than two alleles of a given gene may exist in a group of individuals. (Any one individual has only two alleles—one from its mother and one from its father.) In fact, there are many examples of such multiple alleles.

Coat color in rabbits, for example, is determined by one gene with four alleles. Any rabbit with the *C* allele (paired with any of the

10.11 Inheritance of Coat Color in Rabbits   There are four alleles of the gene for coat color in these Netherlands dwarf rabbits. Different combinations of two alleles give different coat colors. The dominance hierarchy is $C > c^{ch} > c^h > c$.

four) is dark gray, and a rabbit with *cc* is albino. The intermediate colors result from the different allele combinations shown in **Figure 10.11**.

Multiple alleles increase the number of possible phenotypes. In Mendel's monohybrid cross, there was just one pair of alleles (*Ss*) and two possible phenotypes (resulting from *SS* or *Ss* and *ss*). The four alleles of the rabbit coat color gene produce five different phenotypes.

Dominance is not always complete

In the single pairs of alleles studied by Mendel, dominance is *complete* when an individual is heterozygous. That is, an *Ss* individual always expresses the *S* phenotype. However, many genes have alleles that are not dominant or recessive to one another. Instead, the heterozygotes show an intermediate phenotype—at first glance, like that predicted by the old blending theory of inheritance. For example, if a true-breeding red snapdragon is crossed with a true-breeding white one, all the F₁ flowers are pink. That this phenomenon can still be explained in terms of Mendelian genetics, rather than blending, is readily demonstrated by a further cross.

The blending theory predicts that if one of the pink F₁ snapdragons is crossed with a true-breeding white one, all the offspring should be a still lighter pink. In fact, approximately half of the offspring are white, and half are the same shade of pink as the F₁ parent. When the F₁ pink snapdragons are allowed to self-pollinate, the resulting F₂ plants show a ratio of 1 red:2 pink:1 white (**Figure 10.12**). Clearly the hereditary particles—

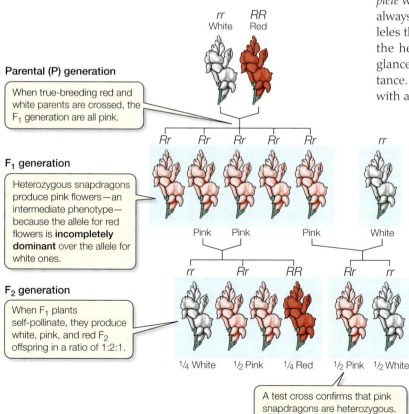

Parental (P) generation

When true-breeding red and white parents are crossed, the F₁ generation are all pink.

F₁ generation

Heterozygous snapdragons produce pink flowers—an intermediate phenotype—because the allele for red flowers is **incompletely dominant** over the allele for white ones.

F₂ generation

When F₁ plants self-pollinate, they produce white, pink, and red F₂ offspring in a ratio of 1:2:1.

¼ White   ½ Pink   ¼ Red   ½ Pink   ½ White

A test cross confirms that pink snapdragons are heterozygous.

10.12 Incomplete Dominance Follows Mendel's Laws An intermediate phenotype can occur in heterozygotes when neither allele is dominant. The heterozygous phenotype (here, pink flowers) may give the appearance of a blended trait, but the traits of the parental generation reappear in their original forms in succeeding generations, as predicted by Mendel's laws of inheritance.

10.13 ABO Blood Reactions Are Important in Transfusions This graph shows the results of mixing red blood cells of types A, B, AB, and O with serum containing anti-A or anti-B antibodies. As you look down the columns, note that each of the types, when mixed separately with anti-A and with anti-B, gives a unique pair of results; this is the basic method by which blood is typed. People with type O blood are good blood donors because O cells do not react with either anti-A or anti-B antibodies. People with type AB blood are good recipients, since they make neither type of antibody.

Blood type of cells	Genotype	Antibodies made by body	Reaction to added antibodies	
			Anti-A	Anti-B
A	I^AI^A or I^Ai^O	Anti-B		
B	I^BI^B or I^Bi^O	Anti-A		
AB	I^AI^B	Neither anti-A nor anti-B		
O	i^Oi^O	Both anti-A and anti-B		

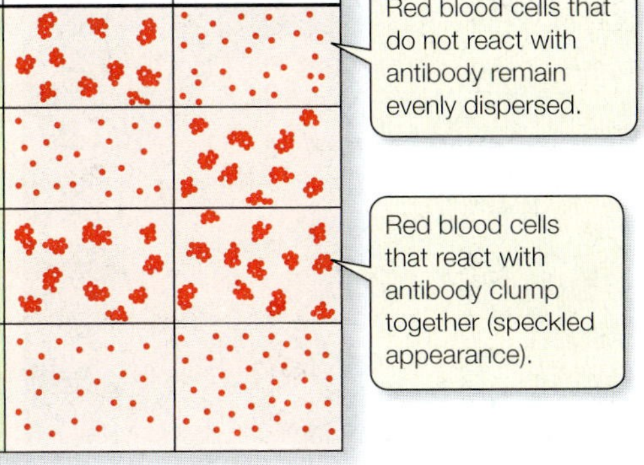

Red blood cells that do not react with antibody remain evenly dispersed.

Red blood cells that react with antibody clump together (speckled appearance).

the genes—have not blended; they are readily sorted out in the F$_2$ generation.

We can understand these results in terms of the Mendelian laws of inheritance. When heterozygotes show a phenotype intermediate between those of the two homozygotes, the gene is said to be governed by **incomplete dominance**. In other words, neither of the two alleles is dominant. Incomplete dominance is common in nature. In fact, Mendel's study of seven pea-plant traits is unusual in that all seven traits happened to be characterized by complete dominance.

In codominance, both alleles at a locus are expressed

Sometimes the two alleles at a locus produce two different phenotypes that *both* appear in heterozygotes, a phenomenon called **codominance**. A good example of codominance is seen in the ABO blood group system in humans.

Early attempts at blood transfusion frequently killed the patient. Around 1900, the Austrian scientist Karl Landsteiner mixed blood cells and *serum* (blood from which cells have been removed) from different individuals. He found that only certain combinations of blood are compatible. In other combinations, the red blood cells from one individual form clumps in the presence of serum from the other individual. This discovery led to our ability to administer compatible blood transfusions that do not kill the recipient.

Clumps form in incompatible transfusions because specific proteins in the serum, called *antibodies*, react with foreign, or "nonself," cells. The antibodies react with proteins on the surface of nonself cells, called *antigens*. (We will learn much more about the function of antibodies and antigens in Chapter 18.) Blood compatibility is determined by a set of three alleles (I^A, I^B, and i^O) at one locus, which determine the antigens on the surface of red blood cells. Different combinations of these alleles in different people produce four different blood types, or phenotypes: A, B, AB, and O (**Figure 10.13**). The AB phenotype found in individuals of genotype I^AI^B is an example of codominance—these individuals produce cell surface antigens of both the A and B types.

Some alleles have multiple phenotypic effects

Mendel's principles were further extended when it was discovered that a single allele can influence more than one phenotype. When a single allele has more than one distinguishable phenotypic ef-

fect, we say that the allele is **pleiotropic**. A familiar example of pleiotropy involves the allele responsible for the coloration pattern (light body, darker extremities) of Siamese cats. The same allele is also responsible for the characteristic crossed eyes of Siamese cats. Although these effects appear to be unrelated, both result from the same protein produced under the influence of the allele.

10.2 RECAP

Genes are subject to random mutations that give rise to new alleles; thus many genes have more than two alleles. Dominance is not necessarily an all-or-nothing phenomenon.

- Can you explain how the experiment in Figure 10.12 demonstrates incomplete dominance? See pp. 218–219

- How does the AB blood type result from codominance? See p. 219 and Figure 10.13

Thus far we have treated the phenotype of an organism, with respect to a given character, as a simple result of the alleles of a single gene. In many cases, however, *several genes interact* to determine a phenotype. To complicate things further, the physical environment may interact with the genetic constitution of an individual in determining the phenotype.

10.3 How Do Genes Interact?

Epistasis occurs when the phenotypic expression of one gene is affected by another gene. For example, two genes determine coat color in Labrador retrievers:

- Allele *B* (black pigment) is dominant to *b* (brown)

- Allele *E* (pigment deposition in hair) is dominant to *e* (no deposition, so hair is yellow)

So a dog with *BB* or *Bb* is black; one with *bb* is brown; and one with *ee* is yellow regardless of the *B/b* alleles present. Clearly, gene *E* de-

A dog with alleles *B* and *E* is black.

A dog with alleles *bb* and *E* is brown.

A dog with *ee* is yellow, regardless of its *B/b* alleles.

(A) Black labrador (*B_E_*)   (B) Chocolate labrador (*bbE_*)   (C) Yellow labrador (*_ _ee*)

10.14 Genes May Interact Epistatically Epistasis occurs when one gene alters the phenotypic effect of another gene. In Labrador retrievers, the *E/e* gene determines the expression of the *B/b* gene.

termines the expression of *B/b* (**Figure 10.14**). If two dogs that are *BbEe* are mated, the phenotypic ratio among the puppies will be ⁹⁄₁₆ black: ³⁄₁₆ brown: ⁴⁄₁₆ yellow. Can you show why?

The song *Camptown Races* by Stephen Foster laments that "Somebody bet on the bay." A bay horse is dark brown. Palominos are blonde, chestnuts are reddish brown, and so on. Horses come in a wide array of colors and patterns, the result of epistasis involving multiple alleles of at least seven genes. Skin color in humans is likewise determined by multiple alleles and genes.

Hybrid vigor results from new gene combinations and interactions

Early in the twentieth century a paper called "The composition of a field of maize" by G. H. Shull had a lasting impact on the field of applied genetics. Farmers have known for centuries that matings among close relatives (known as **inbreeding**) can result in offspring of lower quality than matings between unrelated individuals. The problems with inbreeding arise because close relatives tend to have the same recessive alleles, some of which may be harmful, as we saw in our discussion of human pedigrees in Section 10.1. In fact, it has long been known that if one crosses two different true-breeding, homozygous genetic strains of a plant or animal, the offspring are phenotypically much stronger, larger, and in general more "vigorous" than either of the parents (**Figure 10.15**).

Shull began his experiment with two of the thousands of existing varieties of corn (maize). Both varieties produced about 20 bushels of corn per acre. But when he crossed them, the yield of their offspring was an astonishing 80 bushels per acre. This phenomenon is known as **heterosis** (short for *heterozygosis*), or *hybrid vigor*. The cultivation of hybrid corn spread rapidly in the United States and all over the world, quadrupling grain production. The

practice of hybridization has spread to many other crops and animals used in agriculture. For example, beef cattle that are cross-bred are larger and live longer than cattle bred within their own genetic strain.

The mechanism by which heterosis works is not known. A widely accepted hypothesis is *overdominance*, in which the heterozygous condition in certain important genes is superior to either homozygote. Another hypothesis is that the homozygotes have alleles that inhibit growth, and these are less active or absent in the heterozygote.

The environment affects gene action

The phenotype of an individual does not result from its genotype alone. Genotype and environment interact to determine the phenotype of an organism. Environmental variables such as light, temperature, and nutrition can affect the expression of a genotype as a phenotype.

Parent   Parent   Hybrid offspring

10.15 Hybrid Vigor in Corn The heterozygous F₁ offspring is larger and more vigorous than either of its homozygous parents.

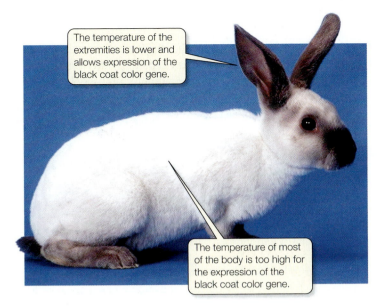

The temperature of the extremities is lower and allows expression of the black coat color gene.

The temperature of most of the body is too high for the expression of the black coat color gene.

10.16 The Environment Influences Gene Expression This rabbit expresses a coat pattern known as "chocolate point." Its genotype specifies dark fur, but the enzyme for dark fur is inactive at normal body temperature, so only the rabbit's extremities—the coolest regions of the body—express this phenotype.

A familiar example of this phenomenon involves "point restriction" coat patterns found in Siamese cats and certain rabbit breeds (**Figure 10.16**). These animals have a genotype that should result in dark fur all over the body. However, an enzyme that produces the dark fur has a mutation that renders it inactive at temperatures above a certain point (usually around 35°C). The animals maintain a body temperature above this point, and so their fur is mostly light. However, the extremities—feet, ears, nose, and tail—are cooler, about 25°C, so the fur on these regions is dark.

A simple experiment shows that the dark fur is temperature-dependent. If a patch of white fur on a point-restricted rabbit's back is removed and an ice pack is placed on the skin where the patch was, the fur that grows back will be dark. This indicates that the gene for dark fur was there all along; it's the environment that inhibited its expression.

Two parameters describe the effects of genes and environment on the phenotype:

- **Penetrance** is the proportion of individuals in a group with a given genotype that actually show the expected phenotype.

- **Expressivity** is the degree to which a genotype is expressed in an individual.

For an example of environmental effects on expressivity, consider how Siamese cats kept indoors or outdoors in different climates might look.

Most complex phenotypes are determined by multiple genes and the environment

The differences between individual organisms in simple characters, such as those that Mendel studied in pea plants, are discrete and **qualitative**. For example, the individuals in a population of pea plants are either short or tall. For most complex characters, however, such as height in humans, the phenotype varies more or less continuously over a range. Some people are short, others are tall, and many are in between the two extremes. Such variation within a population is called **quantitative**, or *continuous*, variation (**Figure 10.17**).

Sometimes this variation is largely genetic. For instance, much of human eye color is the result of a number of genes controlling the synthesis and distribution of dark melanin pigment. Dark eyes have a lot of it, brown eyes less, and green, gray, and blue eyes even less. In the latter cases, the distribution of other pigments in the eye is what determines light reflection and color.

In most cases, however, quantitative variation is due to *both genes and environment*. Height in humans certainly falls into this category. If you look at families, you often see that parents and their offspring all tend to be tall or short. However, nutrition also plays a role in height: American 18-year-olds today are about 20 percent

10.17 Quantitative Variation Quantitative variation is produced by the interaction of genes and environment. These students (women in white on the left; men in blue on the right) show continuous variation in height that is the result of interactions between many alleles and the environment.

taller than their great-grandparents were at the same age, a difference that is certainly not genetic.

Geneticists call the genes that together determine such complex characters **quantitative trait loci**. Identifying these loci is a major challenge, and an important one. For example, the amount of grain that a variety of rice produces in a growing season is determined by many interacting genetic factors. Crop plant breeders have worked hard to decipher these factors in order to breed higher-yielding rice strains. In a similar way, human characteristics such as disease susceptibility and behavior are caused in part by quantitative trait loci.

10.3 RECAP

In epistasis, one gene affects the expression of another. Perhaps the most challenging problem for genetics is the explanation of complex phenotypes that are caused by many interacting genes and the environment.

- Can you explain the difference between penetrance and expressivity? See p. 221

- How is quantitative variation different from qualitative variation? See p. 221

In the next section we'll see how the discovery that genes occupy specific positions on chromosomes enabled Mendel's successors not only to provide a physical explanation for his model of inheritance, but also to provide an explanation for those cases where Mendel's second law does not apply.

10.4 What Is the Relationship between Genes and Chromosomes?

The observation that genes located on the same chromosome do not always follow Mendel's law of independent assortment raised questions that led to new insights: What is the pattern of inheritance of such genes? How do we determine where genes are located on a chromosome, and the distances between them?

The answers to these and many other genetic questions were worked out in studies of the fruit fly *Drosophila melanogaster*. Its small size, the ease with which it could be bred, and its short generation time made this animal an attractive experimental subject. Beginning in 1909, Thomas Hunt Morgan and his students pioneered the study of *Drosophila* in Columbia University's famous "fly room," where they discovered the phenomena described in this section. *Drosophila* remains extremely important in studies of

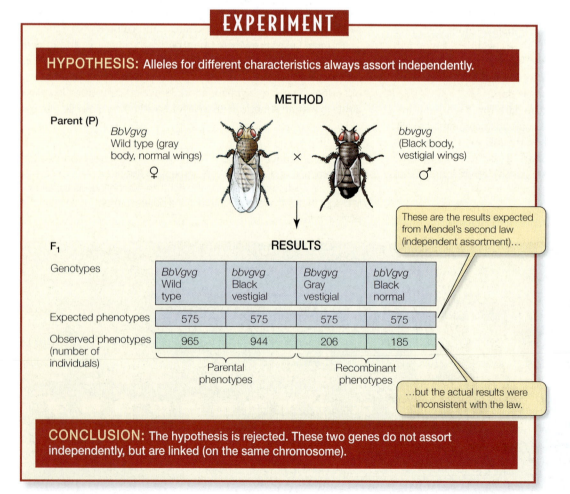

EXPERIMENT

HYPOTHESIS: Alleles for different characteristics always assort independently.

METHOD

Parent (P)

BbVgvg
Wild type (gray body, normal wings)
♀

×

bbvgvg
(Black body, vestigial wings)
♂

These are the results expected from Mendel's second law (independent assortment)…

RESULTS

F₁

Genotypes	*BbVgvg* Wild type	*bbvgvg* Black vestigial	*Bbvgvg* Gray vestigial	*bbVgvg* Black normal
Expected phenotypes	575	575	575	575
Observed phenotypes (number of individuals)	965	944	206	185

Parental phenotypes · Recombinant phenotypes

…but the actual results were inconsistent with the law.

CONCLUSION: The hypothesis is rejected. These two genes do not assort independently, but are linked (on the same chromosome).

10.18 Some Alleles Do Not Assort Independently Morgan's studies showed that the genes for body color and wing size in *Drosophila* are linked, so that their alleles do not assort independently. Linkage accounts for the departure of the phenotype ratios Morgan observed from those predicted by Mendel's law of independent assortment. FURTHER RESEARCH: Look again at Mendel's dihybrid cross (Figure 10.7). If the genes for seed shape and seed color were linked, what would these results be?

chromosome structure, population genetics, the genetics of development, and the genetics of behavior.

Genes on the same chromosome are linked

Some of the crosses Morgan performed with fruit flies yielded phenotypic ratios that were not in accord with those predicted by Mendel's law of independent assortment. Morgan crossed *Drosophila* with two known genotypes, *BbVgvg* × *bbvgvg*,* for two different characters, body color and wing shape:

- *B* (wild-type gray body), is dominant over *b* (black body)
- *Vg* (wild-type wing) is dominant over *vg* (vestigial, a very small wing)

Morgan expected to see four phenotypes in a ratio of 1:1:1:1, but that is not what he observed. The body color gene and the wing size gene were not assorting independently; rather, they were, for the most part, inherited together (**Figure 10.18**).

These results became understandable to Morgan when he considered the possibility that the two loci are on the same chromosome—that is, that they might be linked. After all, since the number of genes in a cell far exceeds the number of chromosomes, each chromosome must contain many genes. We now say that the full set of loci on a given chromosome constitutes a **linkage group**. The number of linkage groups in a species equals its number of homologous chromosome pairs.

Suppose, now, that the *Bb* and *Vgvg* loci are indeed located on the same chromosome. Why, then, didn't *all* of Morgan's F₁ flies have the parental phenotypes—that is, why did his cross result in anything other than gray flies with normal wings (wild-type) and black flies with vestigial wings? If linkage were *absolute*—that is, if chromosomes always remained intact and unchanged—we would expect to see just those two types of progeny. However, this is not always what happens.

Genes can be exchanged between chromatids

Absolute linkage is extremely rare. If linkage were absolute, Mendel's law of independent assortment would apply only to loci on different chromosomes. What actually happens is more complex, and therefore more interesting. Because chromosomes can break, recombination of genes can occur. That is, genes at different loci on the same chromosome do sometimes separate from one another during meiosis.

Genes may recombine when two homologous chromosomes physically exchange corresponding segments during prophase I of meiosis—that is, by crossing over (**Figure 10.19**; see also Figure 9.18). As described in Section 9.5, DNA is replicated during the S phase, so that by prophase I, when homologous chromosome pairs

---

*Do you recognize this type of cross? It is a *test cross* for the two gene pairs; see Figure 10.6.

10.19 Crossing Over Results in Genetic Recombination Genes at different loci on the same chromosome can be separated from one another and recombined by crossing over. Such recombination occurs during prophase I of meiosis.

come together to form tetrads, each chromosome consists of two chromatids. The exchange event involves only two of the four chromatids in a tetrad, one from each member of the homologous pair, and can occur at any point along the length of the chromosome. The chromosome segments involved are exchanged reciprocally, so both chromatids involved in crossing over become recombinant (that is, each chromatid ends up with genes from both of the organism's parents). Usually several exchange events occur along the length of each homologous pair.

When crossing over takes place between two linked genes, not all the progeny of a cross have the parental phenotypes. Instead, recombinant offspring appear as well, as they did in Morgan's cross. They appear in proportions called **recombinant frequencies**, which are calculated by dividing the number of recombinant prog-

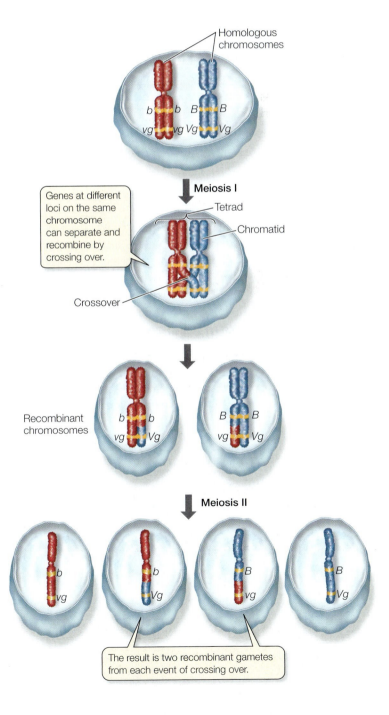

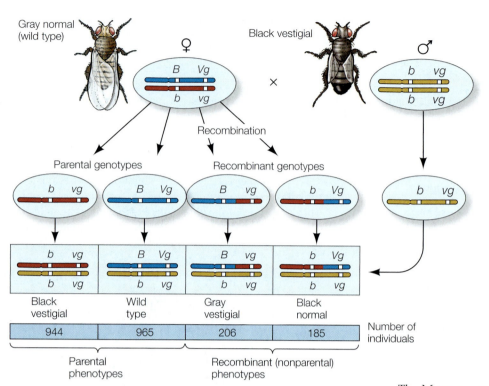

10.20 Recombinant Frequencies The frequency of recombinant offspring (those with a phenotype different from either parent) can be calculated.

Gray normal (wild type)

Black vestigial

♀ × ♂

Recombination

Parental genotypes   Recombinant genotypes

Black vestigial	Wild type	Gray vestigial	Black normal
944	965	206	185

Number of individuals

Parental phenotypes

Recombinant (nonparental) phenotypes

$$\text{Recombinant frequency} = \frac{391 \text{ recombinants}}{2{,}300 \text{ total offspring}} = 0.17$$

eny by the total number of progeny (**Figure 10.20**). Recombinant frequencies will be greater for loci that are farther apart on the chromosome than for loci that are closer together because an exchange event is more likely to occur between genes that are far apart than between genes that are close together.

Geneticists can make maps of chromosomes

If two loci are very close together on a chromosome, the odds of crossing over between them are small. In contrast, if two loci are far apart, crossing over could occur between them at many points. This pattern is a consequence of the mechanism of crossing over: the

farther apart two genes are, the more places there are in the chromosome for breakage and reunion of chromatids to occur. In a population of cells undergoing meiosis, a greater proportion of the cells will undergo recombination between two loci that are far apart than between two loci that are close together. In 1911, Alfred Sturtevant, then an undergraduate student in T. H. Morgan's fly room, realized how this simple insight could be used to show where different genes lie on a chromosome in relation to one another.

The Morgan group had determined recombinant frequencies for many pairs of linked *Drosophila* genes. Sturtevant used those recombinant frequencies to create **genetic maps** that showed the arrangement of genes along the chromosome (**Figure 10.21**). Ever since Sturtevant demonstrated this method, geneticists have mapped the chromosomes of eukaryotes, prokaryotes, and viruses, assigning distances between genes in **map units**. A map unit corresponds to a recombinant frequency of 0.01; it is also referred to as a **centimorgan (cM)**, in honor of the founder of the fly room. You, too, can work out a genetic map (**Figure 10.22**).

10.21 Steps toward a Genetic Map Because the chance of a recombinant genotype occurring increases with the distance between two loci on a chromosome, Sturtevant was able to derive this partial map of a *Drosophila* chromosome from the Morgan group's data on the recombinant frequencies of five recessive traits. He used an arbitrary unit of distance—the map unit, or centimorgan (cM)—equivalent to a recombinant frequency of 0.01.

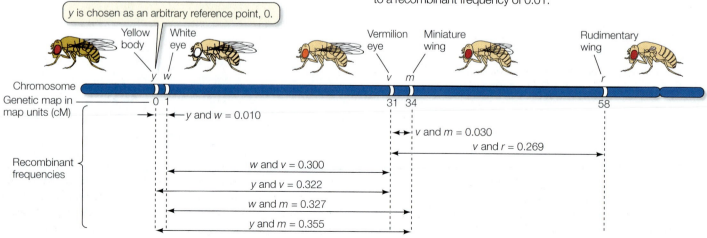

y is chosen as an arbitrary reference point, 0.

Yellow body | White eye | Vermilion eye | Miniature wing | Rudimentary wing

Chromosome

Genetic map in map units (cM)

0 1 — *y* and *w* = 0.010

31 34 — *v* and *m* = 0.030

58 — *v* and *r* = 0.269

Recombinant frequencies

w and *v* = 0.300

y and *v* = 0.322

w and *m* = 0.327

y and *m* = 0.355

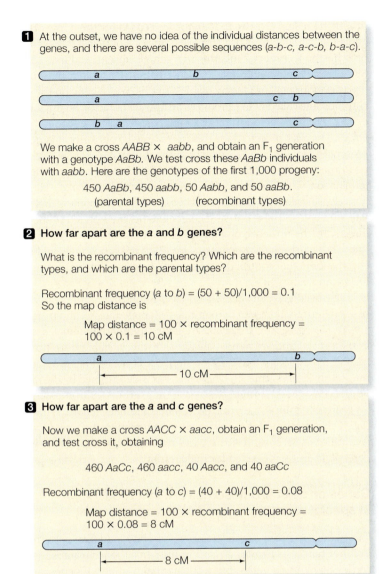

1 At the outset, we have no idea of the individual distances between the genes, and there are several possible sequences (*a-b-c*, *a-c-b*, *b-a-c*).

We make a cross *AABB* × *aabb*, and obtain an F₁ generation with a genotype *AaBb*. We test cross these *AaBb* individuals with *aabb*. Here are the genotypes of the first 1,000 progeny:

450 *AaBb*, 450 *aabb*, 50 *Aabb*, and 50 *aaBb*.
(parental types)   (recombinant types)

2 How far apart are the *a* and *b* genes?

What is the recombinant frequency? Which are the recombinant types, and which are the parental types?

Recombinant frequency (*a* to *b*) = (50 + 50)/1,000 = 0.1
So the map distance is

Map distance = 100 × recombinant frequency =
100 × 0.1 = 10 cM

3 How far apart are the *a* and *c* genes?

Now we make a cross *AACC* × *aacc*, obtain an F₁ generation, and test cross it, obtaining

460 *AaCc*, 460 *aacc*, 40 *Aacc*, and 40 *aaCc*

Recombinant frequency (*a* to *c*) = (40 + 40)/1,000 = 0.08

Map distance = 100 × recombinant frequency =
100 × 0.08 = 8 cM

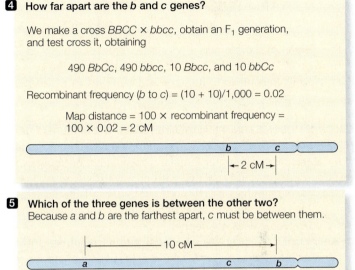

10.22 Map These Genes The object of this exercise is to determine the order of three loci (*a*, *b*, and *c*) on a chromosome, as well as the map distances (in cM) between them.

4 How far apart are the *b* and *c* genes?

We make a cross *BBCC* × *bbcc*, obtain an F₁ generation, and test cross it, obtaining

490 *BbCc*, 490 *bbcc*, 10 *Bbcc*, and 10 *bbCc*

Recombinant frequency (*b* to *c*) = (10 + 10)/1,000 = 0.02

Map distance = 100 × recombinant frequency =
100 × 0.02 = 2 cM

5 Which of the three genes is between the other two?
Because *a* and *b* are the farthest apart, *c* must be between them.

These numbers add up perfectly. In most real cases, they will not add up perfectly because of multiple crossovers.

Linkage is revealed by studies of the sex chromosomes

In Mendel's work, reciprocal crosses always gave identical results; it did not matter, in general, whether a dominant allele was contributed by the mother or by the father. But in some cases, the parental origin of a chromosome does matter. For example, human males inherit a bleeding disorder called hemophilia from their mother, not from their father. To understand the types of inheritance in which the parental origin of an allele is important, we must consider the ways in which sex is determined in different species.

SEX DETERMINATION BY CHROMOSOMES In corn, every diploid adult has both male and female reproductive structures. The tissues in these two types of structures are genetically identical, just as roots and leaves are genetically identical. Plants such as corn, in which the same individual produces both male and female gametes, are said to be *monoecious* (Greek, "one house"). Other plants, such as date palms and oak trees, and most animals are *dioecious* ("two houses"), meaning that some individuals can produce only male gametes and the others can produce only female gametes. In other words, dioecious organisms have two sexes.

In most dioecious organisms, sex is determined by differences *in the chromosomes*, but such determination operates in different ways in different groups of organisms. For example, in many animals, including humans, sex is determined by a single **sex chromosome**, or by a pair of them. Both males and females have two copies of each of the rest of the chromosomes, which are called **autosomes**.

The sex chromosomes of female mammals consist of a pair of X chromosomes. Male mammals, on the other hand, have one X chromosome and a sex chromosome that is not found in females: the Y chromosome. Females may be represented as XX and males as XY:

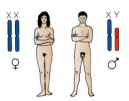

Male mammals produce *two* kinds of gametes. Each gamete has a complete set of autosomes, but half the gametes carry an X chromosome and the other half carry a Y. When an X-bearing sperm fertilizes an egg, the resulting XX zygote is female; when a Y-bearing sperm fertilizes an egg, the resulting XY zygote is male.

The situation is different in birds, in which males are XX and females are XY (to avoid confusion, these chromosomes are called ZZ and ZW):

In these organisms, the female produces two types of gametes, carrying Z or W. Whether the egg is Z or W determines the sex of the offspring, in contrast to humans and fruit flies, in which the sperm, carrying either X or Y, determines the sex.

SEX CHROMOSOME ABNORMALITIES REVEALED THE GENE THAT DETERMINES SEX   There must be genes on the Y (or W) chromosome that determine sex (male or female, respectively). But how can we be sure? As previously pointed out, one way to determine cause (in the case of mammals, a Y chromosome gene) and effect (in this case, maleness) is to look at cases of biological error, in which the expected outcome does not happen.

Abnormal sex chromosome constitutions resulting from nondisjunction in meiosis (see Section 9.5) tell us something about the functions of the X and Y chromosomes. As you will recall, nondisjunction occurs when a pair of sister chromosomes (in meiosis I) or sister chromatids (in meiosis II) fail to separate. As a result, a gamete may have one too few or one too many chromosomes. Assuming fertilization by another gamete with the full haploid chromosome set, the resulting offspring are aneuploid, with fewer or more chromosomes than normal.

In humans, XO individuals sometimes appear. (The O implies that a chromosome is missing—that is, individuals that are XO have only one sex chromosome.) Human XO individuals are females who are physically moderately abnormal but mentally normal; usually they are also sterile. The XO condition in humans is called *Turner syndrome*. It is the only known case in which a person can survive with only one member of a chromosome pair (here, the XY pair), although most XO conceptions are spontaneously terminated early in development. XXY individuals also occur; this condition, which affects males, is called *Klinefelter syndrome*, and results in overlong limbs and sterility.

These observations suggested that the gene that determines maleness is located on the Y chromosome. Observations of people with other types of chromosomal abnormalities helped researchers to pinpoint the location of that gene:

■ Some XY individuals are phenotypically women but lack a small portion of the Y chromosome.

■ Some men are genetically XX but have a small piece of the Y chromosome attached to another chromosome.

It was clear that the Y fragments that are respectively missing and present in these two cases contained the maleness-determining gene, which was named *SRY* (*sex-determining region on the Y* chromosome).

The *SRY* gene encodes a protein involved in **primary sex determination**—that is, the determination of the kinds of gametes that an individual will produce and the organs that will make them. In the presence of functional SRY protein, an embryo develops sperm-producing testes. (Notice that *italic type* is used for the name of a gene, but roman type is used for the name of a protein.) If the embryo has no Y chromosome, the *SRY* gene is absent, and thus the SRY protein is not made. In the absence of the SRY protein, the embryo develops egg-producing ovaries. In this case, a gene on the X chromosome called *DAX1* produces an anti-testis factor. So the role of *SRY* in a male is to inhibit the maleness inhibitor encoded by *DAX1*. The SRY protein does this in male cells, but since it is not present in females, *DAX1* can act to inhibit maleness.

Primary sex determination is not the same as **secondary sex determination**, which results in the outward manifestations of maleness and femaleness (such as body type, breast development, body hair, and voice). These outward characteristics are not determined directly by the presence or absence of the Y chromosome. Rather, they are determined by genes scattered on the autosomes and X chromosome that control the actions of hormones, such as testosterone and estrogen.

Superficially, *Drosophila melanogaster* follows the same pattern of sex determination seen in mammals—females are XX and males are XY. However, XO *Drosophila* are males (rather than females, as in mammals) and are almost always indistinguishable from normal XY males except that they are sterile. XXY individuals are normal, fertile *females*. Thus, in *Drosophila*, sex is determined by the *ratio of X chromosomes to autosome sets*. If there is one X chromosome for each set of autosomes, the individual is a female; if there is only one X chromosome for the two sets of autosomes, the individual is a male. The Y chromosome plays no sex-determining role in *Drosophila*, but it is needed for male fertility.

Genes on sex chromosomes are inherited in special ways

Genes on sex chromosomes do not show the Mendelian patterns of inheritance. In *Drosophila* and in humans, the Y chromosome carries few known genes, but a substantial number of genes, affecting a great variety of characters, are carried on the X chromosome. Any such gene is present in two copies in females but in only one copy in males. Therefore, males will always be **hemizygous** for genes on the X chromosome—they will have only one copy of each, and it will be expressed. Thus reciprocal crosses do not give identical results for characters whose genes are carried on the sex chromosomes, and these characters do not show the usual Mendelian ratios for the inheritance of genes located on autosomes.

The first, and still one of the best, examples of inheritance of characters governed by loci on the sex chromosomes (**sex-linked** inheritance) is that of eye color in *Drosophila*. The wild-type eye color of these flies is red. In 1910, Morgan discovered a mutation that causes white eyes. He experimented by crossing flies of the wild-type and mutant phenotypes. His results demonstrated that the eye color locus is on the X chromosome.

■ When a homozygous red-eyed female was crossed with a (hemizygous) white-eyed male, all the sons and daughters had red eyes, because red is dominant over white and all the progeny had inherited a wild-type X chromosome from their mothers (**Figure 10.23A**).

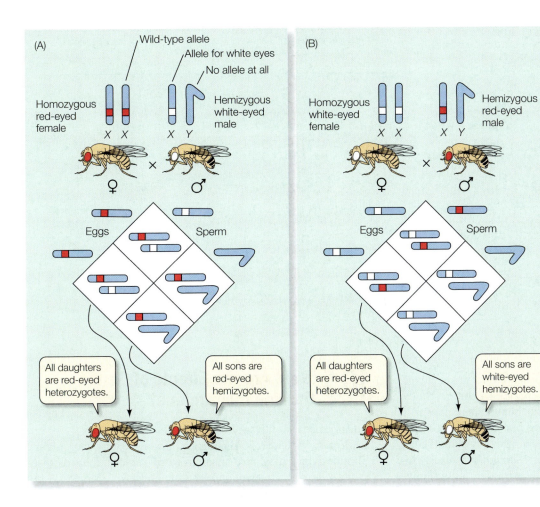

10.23 Eye Color Is a Sex-Linked Trait in *Drosophila* Morgan demonstrated that a mutant allele that causes white eyes in *Drosophila* is carried on the X chromosome. Note that in this case, the reciprocal crosses do not have the same results.

■ In the reciprocal cross, in which a white-eyed female was mated with a red-eyed male, all the sons were white-eyed and all the daughters were red-eyed (**Figure 10.23B**).

The sons from the reciprocal cross inherited their only X chromosome from their white-eyed mother; the Y chromosome they inherited from their father does not carry the eye color locus. The daughters, on the other hand, got an X chromosome bearing the white allele from their mother and an X chromosome bearing the red allele from their father; they were therefore red-eyed heterozygotes.

■ When heterozygous females were mated with red-eyed males, half their sons had white eyes, but all their daughters had red eyes.

Together, these three results showed that eye color was carried on the X chromosome and not on the Y.

Humans display many sex-linked characters

The human X chromosome carries about 2,000 known genes. The alleles at these loci follow the same pattern of inheritance as those for white eyes in *Drosophila*. One gene on the human X chromosome, for example, has a mutant recessive allele that leads to red-green color blindness, a hereditary disorder, as described at the start of this chapter. Red-green color blindness appears in individuals who are homozygous or hemizygous for the recessive mutant allele.

Pedigree analysis of X-linked recessive phenotypes (**Figure 10.24**) reveals the following patterns:

■ The phenotype appears much more often in males than in females, because only one copy of the rare allele is needed for its expression in males, while two copies must be present in females.

■ A male with the mutation can pass it on only to his daughters; all his sons get his Y chromosome.

■ Daughters who receive one mutant X chromosome are heterozygous **carriers**. They are phenotypically normal, but they can pass the mutant X to both sons and daughters (but do so only half of the time, on average, since half of their X chromosomes carry the normal allele).

■ The mutant phenotype can skip a generation if the mutation passes from a male to his daughter (who will be phenotypically normal) and thus to her son.

Red-green color blindness (see p. 207) is an X-linked recessive phenotype, as are several important human diseases. Human mutations inherited as X-linked dominant phenotypes are rarer than X-linked recessives because dominant phenotypes appear in every generation, and because people carrying the harmful mutation, even as heterozygotes, often fail to survive and reproduce. (Look at the four points above and try to determine what would happen if the mutation were dominant.)

10.24 Red-Green Color Blindness Is a Sex-Linked Trait in Humans   The mutant allele for red-green color blindness is inherited as an X-linked recessive.

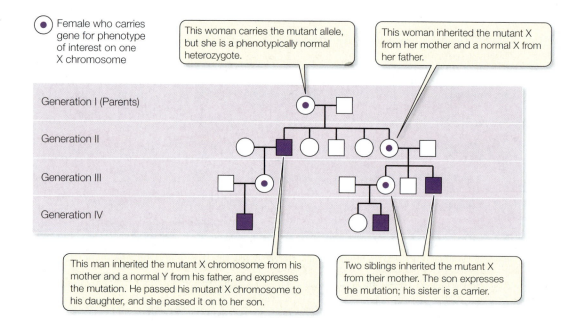

○ Female who carries gene for phenotype of interest on one X chromosome

This woman carries the mutant allele, but she is a phenotypically normal heterozygote.

This woman inherited the mutant X from her mother and a normal X from her father.

Generation I (Parents)

Generation II

Generation III

Generation IV

This man inherited the mutant X chromosome from his mother and a normal Y from his father, and expresses the mutation. He passed his mutant X chromosome to his daughter, and she passed it on to her son.

Two siblings inherited the mutant X from their mother. The son expresses the mutation; his sister is a carrier.

The small human Y chromosome carries several dozen genes. Among them is the maleness determinant, *SRY*. Interestingly, for some genes on the Y, there are similar, but not identical, genes on the X. For example, one of the proteins that make up ribosomes is encoded by a gene on the Y that is expressed only in male cells, while the X-linked counterpart is expressed in both sexes. This means that there are "male" and "female" ribosomes; the significance of this phenomenon is unknown. Y-linked alleles are passed only from father to son. (You can verify this with a Punnett square.)

10.4 RECAP

Simple Mendelian ratios are not observed when genes are linked on the same chromosome. Linkage is indicated by atypical frequencies of gametes evidenced by offspring in a test cross. Sex linkage in humans refers to genes on the X chromosome that have no counterpart on the Y.

- Can you explain the concept of linkage and its implications for the results of genetic crosses? See p. 223 and Figures 10.19 and 10.20

- How does a sex-linked gene behave differently in genetic crosses than a gene on an autosome? See pp. 226–227 and Figure 10.23

The genes we've discussed so far in this chapter are all in the cell nucleus. But other organelles, including mitochondria and plastids, also carry genes. What are they, and how are they inherited?

10.5 What Are the Effects of Genes Outside the Nucleus?

The nucleus is not the only organelle in a eukaryotic cell that carries genetic material. As described in Section 4.3, mitochondria and plastids, which may have arisen from prokaryotes that colonized other cells, contain small numbers of genes. For example, in humans, there are about 24,000 genes in the nuclear genome and 37 in the mitochondrial genome. Plastid genomes are about five times larger than those of mitochondria. In any case, several of the genes of cytoplasmic organelles are important for organelle assembly and function, so it is not surprising that mutations of these genes can have profound effects on the organism.

The inheritance of organelle genes differs from that of nuclear genes for several reasons:

- In most organisms, mitochondria and plastids are inherited only from the mother. As you will learn in Chapter 43, eggs contain abundant cytoplasm and organelles, but the only part of the sperm that survives to take part in the union of haploid gametes is the nucleus. So you have inherited your mother's mitochondria (with their genes), but not your father's.

- There may be hundreds of mitochondria or plastids in a cell. So a cell is not diploid for organelle genes; rather, it is highly polyploid.

- Organelle genes tend to mutate at much faster rates than nuclear genes, so there are multiple alleles of organelle genes.

The phenotypes of mutations in the DNA of organelles reflect the organelles' roles. For example, in plants and some photosynthetic protists, certain plastid gene mutations affect the proteins that assemble chlorophyll molecules into photosystems and result in a phenotype that is essentially white instead of green. Mitochondrial gene mutations that affect one of the complexes in the electron transport chain result in less ATP production. These mutations have especially noticeable effects in tissues with a high energy requirement, such as the nervous system, muscles, and kidneys. In 1995, Greg LeMond, a professional cyclist who had won the famous Tour de France three times, was forced to retire because of muscle weakness caused by a mitochondrial mutation.

10.5 RECAP

Genes in the genomes of organelles, specifically plastids and mitochondria, do not behave in a Mendelian fashion.

■ Can you explain why genes carried in the organelle genomes are usually inherited only from the mother?

CHAPTER SUMMARY

10.1 What are the Mendelian laws of inheritance?

Physical features of organisms, or **characters**, can exist in different forms, or **traits**. A **heritable trait** is one that can be passed from parent to offspring. A **phenotype** is the physical appearance of an organism; a **genotype** is the genetic constitution of the organism.

The different forms of a **gene** are called **alleles**. Organisms that have two identical alleles for a trait are called **homozygous**; organisms that have two different alleles for a trait are called **heterozygous**. A gene resides at a particular site on a chromosomes called a **locus**.

Mendel's experiments included **reciprocal crosses** and **monohybrid** crosses of **true-breeding** pea plants. Analysis of his meticulously tabulated data led Mendel to propose a **particulate theory** of inheritance stating that discrete units (now called genes) are responsible for the inheritance of specific traits, to which both parents contribute equally.

Mendel's first law, the **law of segregation**, states that when any individual produces gametes, the two copies of a gene separate, so that each gamete receives only one member of the pair. Thus every individual in the F_1 inherits one copy from each parent. Review Figures 10.4 and 10.5

Mendel used a **test cross** to find out whether an individual showing a dominant phenotype is homozygous or heterozygous. Review Figure 10.6, Web/CD Activity 10.1

Mendel's use of **dihybrid crosses** to study the inheritance of two characters led to his second law: the **law of independent assortment**. The independent assortment of genes in meiosis leads to **recombinant** phenotypes. Review Figures 10.7 and 10.8, Web/CD Tutorial 10.1

Probability calculations and **pedigrees** help geneticists trace Mendelian inheritance patterns. Review Figures 10.9 and 10.10

10.2 How do alleles interact?

New alleles arise by random **mutation**. Many genes have multiple alleles. A **wild-type** allele gives rise to the predominant form of a trait. When the wild-type allele is present at a locus less than 99 percent of the time, the locus is said to be **polymorphic**. Review Figure 10.11

In **incomplete dominance**, neither of two alleles is dominant. The heterozygous phenotype is intermediate between the homozygous phenotypes. Review Figure 10.12

Codominance exists when two alleles at a locus produce two different phenotypes that both appear in heterozygotes.

An allele that affects more than one trait is said to be **pleiotropic**.

10.3 How do genes interact?

In **epistasis**, one gene affects the expression of another. Review Figure 10.14

Environmental conditions can affect the expression of a genotype.

Penetrance is the proportion of individuals in a group with a given genotype that show the expected phenotype. **Expressivity** is the degree to which a genotype is expressed in an individual.

Variation in phenotype can be **qualitative** (discrete) or **quantitative** (graduated, continuous). Most quantitative traits are the result of the effects of several genes and the environment. Genes that together determine quantitative characters are called **quantitative trait loci**.

10.4 What is the relationship between genes and chromosomes?

See Web/CD Tutorial 10.2

Each chromosome carries many genes. Genes on the same chromosome are referred to as a **linkage group**.

Genes on the same chromosome can recombine by crossing over. The resulting recombinant chromosomes have new combinations of alleles. Review Figures 10.19 and 10.20

Sex chromosomes carry genes that determine whether the organism will produce male or female gametes. All other chromosomes are called **autosomes**. The specific functions of X and Y chromosomes differ among groups of organisms.

Primary sex determination in mammals is usually a function of the presence or absence of the *SRY* gene. **Secondary sex determination** results in the outward manifestations of maleness or femaleness.

In fruit flies and mammals, the X chromosome carries many genes, but the Y chromosome has only a few. Males have only one allele (are **hemizygous**) for X-linked genes, so rare recessive alleles show up phenotypically more often in males than in females. Females may be unaffected **carriers** of such alleles.

10.5 What are the effects of genes outside the nucleus?

Cytoplasmic organelles such as plastids and mitochondria contain small numbers of genes. In many organsisms, cytoplasmic genes are inherited only from the mother because male gametes contribute only their nucleus (i.e., no cytoplasm) to the zygote at fertilization.

See Web/CD Activities 10.2 and 10.3 for a concept review of this chapter.

SELF-QUIZ

1. In a simple Mendelian monohybrid cross, tall plants are crossed with short plants, and the F_1 plants are allowed to self-pollinate. What fraction of the F_2 generation are both tall *and* heterozygous?
   a. 1/8
   b. 1/4
   c. 1/3
   d. 2/3
   e. 1/2

2. The phenotype of an individual
   a. depends at least in part on the genotype.
   b. is either homozygous or heterozygous.
   c. determines the genotype.
   d. is the genetic constitution of the organism.
   e. is either monohybrid or dihybrid.

3. The ABO blood groups in humans are determined by a multiple-allele system in which I^A and I^B are codominant and dominant to i^O. A newborn infant is type A. The mother is type O. Possible genotypes of the father are
   a. A, B, or AB
   b. A, B, or O
   c. O only
   d. A or AB
   e. A or O

4. Which statement about an individual that is homozygous for
   an allele is *not* true?
   a. Each of its cells possesses two copies of that allele.
   b. Each of its gametes contains one copy of that allele.
   c. It is true-breeding with respect to that allele.
   d. Its parents were necessarily homozygous for that allele.
   e. It can pass that allele to its offspring.

5. Which statement about a test cross is *not* true?
   a. It tests whether an unknown individual is homozygous or heterozygous.
   b. The test individual is crossed with a homozygous recessive individual.
   c. If the test individual is *heterozygous*, the progeny will have a 1:1 ratio.
   d. If the test individual is *homozygous*, the progeny will have a 3:1 ratio.

e. Test cross results are consistent with Mendel's model of inheritance.

6. Linked genes
   a. must be immediately adjacent to one another on a chromosome.
   b. have alleles that assort independently of one another.
   c. never show crossing over.
   d. are on the same chromosome.
   e. always have multiple alleles.

7. In the F_2 generation of a dihybrid cross
   a. four phenotypes appear in the ratio 9:3:3:1 if the loci are linked.
   b. four phenotypes appear in the ratio 9:3:3:1 if the loci are unlinked.
   c. two phenotypes appear in the ratio 3:1 if the loci are unlinked.
   d. three phenotypes appear in the ratio 1:2:1 if the loci are unlinked.
   e. two phenotypes appear in the ratio 1:1 whether or not the loci are linked.

8. The genetic sex of a human is determined by
   a. ploidy, with the male being haploid.
   b. the Y chromosome.
   c. X and Y chromosomes, the male being XX.
   d. the number of X chromosomes, the male being XO.
   e. Z and W chromosomes, the male being ZZ.

9. In epistasis
   a. nothing changes from generation to generation.
   b. one gene alters the effect of another.
   c. a portion of a chromosome is deleted.
   d. a portion of a chromosome is inverted.
   e. the behavior of two genes is entirely independent.

10. In humans, spotted teeth are caused by a dominant sex-linked gene. A man with spotted teeth whose mother had normal teeth marries a woman with normal teeth. Therefore,
    a. all of their daughters will have normal teeth.
    b. all of their daughters will have spotted teeth.
    c. all of their children will have spotted teeth.
    d. half of their sons will have spotted teeth.
    e. all of their sons will have spotted teeth.

GENETIC PROBLEMS

1. Using the Punnett squares below, show that for typical dominant and recessive autosomal traits, it does not matter which parent contributes the dominant allele and which the recessive allele. Cross true-breeding tall plants (*TT*) with true-breeding dwarf plants (*tt*).

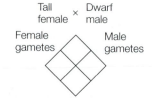

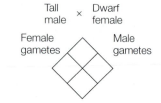

2. Show diagrammatically what occurs when the F_1 offspring of the cross in Question 1 self-pollinate.

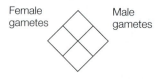

3. A new student of genetics suspects that a particular recessive trait in fruit flies (dumpy wings, which are somewhat smaller and more bell-shaped than the wild type) is sex-linked. A single mating between a fly with dumpy wings (*dp*; female) and a fly with wild-type wings (*Dp*; male) produces three dumpy-winged females and two wild-type males in the F_1 generation. On the basis of these data, is the trait sex-linked or autosomal? What were the genotypes of the parents? Explain how these conclusions can be reached on the basis of so few data.

4. The photograph shows the shells of 15 bay scallops, *Argopecten irradians*. These scallops are hermaphroditic; that is, a single individual can reproduce sexually, as did the pea plants of the F₁ generation in Mendel's experiments. Three color schemes are evident: yellow, orange, and black and white. The color-determining gene has three alleles. The top row shows a yellow scallop and a representative sample of its offspring, the middle row shows a black-and-white scallop and its offspring, and the bottom row shows an orange scallop and its offspring. Assign a suitable symbol to each of the three alleles participating in color control; then determine the genotype of each of the three parent individuals and tell what you can about the genotypes of the different offspring. Explain your results carefully.

5. The sex of some fishes is determined by the same XY system as in humans. An allele of one locus on the Y chromosome of the fish *Lebistes* causes a pigmented spot to appear on the dorsal fin. A male fish that has a spotted dorsal fin is mated with a female fish that has an unspotted fin. Describe the phenotypes of the F₁ and the F₂ generations from this cross.

6. In *Drosophila melanogaster*, the recessive allele p, when homozygous, determines pink eyes. Pp or PP results in wild-type eye color. Another gene, on another chromosome, has a recessive allele, sw, that produces short wings when homozygous. Consider a cross between females of genotype $PPSwSw$ and males of genotype $ppswsw$. Describe the phenotypes and genotypes of the F₁ generation and of the F₂ generation produced by allowing the F₁ progeny to mate with one another.

7. On the same chromosome of *Drosophila melanogaster* that carries the p (pink eyes) locus, there is another locus that affects the wings. Homozygous recessives, $byby$, have blistery wings, while the dominant allele By produces wild-type wings. The P and By loci are very close together on the chromosome; that is, the two loci are tightly linked. In answering these questions, assume that no crossing over occurs.

   a. For the cross $PPByBy \times ppbyby$, give the phenotypes and genotypes of the F₁ and of the F₂ generations produced by interbreeding of the F₁ progeny.
   b. For the cross $PPbyby \times ppByBy$, give the phenotypes and genotypes of the F₁ and of the F₂ generations.
   c. For the cross of Question 7b, what further phenotype(s) would appear in the F₂ generation if crossing over occurred?
   d. Draw a nucleus undergoing meiosis at the stage in which the crossing over (Question 7c) occurred. In which generation (P, F₁, or F₂) did this crossing over take place?

8. Consider the following cross of *Drosophila melanogaster* (alleles are as described in Question 6): Males with genotype $Ppswsw$ are crossed with females of genotype $ppSwsw$. Describe the phenotypes and genotypes of the F₁ generation.

9. In the Andalusian fowl, a single pair of alleles for one gene controls the color of the feathers. Three colors are observed: blue, black, and splashed white. Crosses among these three types yield the following results:

PARENTS	PROGENY
Black × blue	Blue and black (1:1)
Black × splashed white	Blue
Blue × splashed white	Blue and splashed white (1:1)
Black × black	Black
Splashed white × splashed white	Splashed white

   a. What progeny would result from the cross blue × blue?
   b. If you wanted to sell eggs, all of which would yield blue fowl, how should you proceed?

10. In *Drosophila melanogaster*, white (w), eosin (w^e), and wild-type red (w^+) are multiple alleles at a single locus for eye color. This locus is on the X chromosome. A female that has eosin (pale orange) eyes is crossed with a male that has wild-type eyes. All the female progeny are red-eyed; half the male offspring have eosin eyes, and half have white eyes.
   a. What is the order of dominance of these alleles?
   b. What are the genotypes of the parents and progeny?

11. Red-green color blindness is a recessive trait. Two people with normal vision have two sons, one color-blind and one with normal vision. If the couple also has daughters, what proportion of them will have normal vision? Explain.

12. A mouse with an agouti coat is mated with an albino mouse of genotype $aabb$. Half of the offspring are albino, one-fourth are black, and one-fourth are agouti. What are the genotypes of the agouti parents and of the various kinds of offspring? (*Hint*: See the section on epistasis.)

13. The disease Leber's optic neuropathy is caused by a mutation in a gene carried on mitochondrial DNA. What would be the phenotype of their first child if a man with this disease married a woman who did not have the disease? What would be the result if the wife had the disease and the husband did not?

FOR INVESTIGATION

Sometimes scientists get lucky. Consider Mendel's dihybrid cross (see Figure 10.7). Peas have a haploid number of 7 chromosomes, so many of their genes are linked. What would Mendel's results have been if the genes for seed color and seed shape were *linked* with a map distance of 10 units? Now, consider Morgan's fruit flies (see Figure 10.21). Suppose that the genes for body color and wing shape were *not* linked? What results would Morgan have obtained?

11 DNA and Its Role in Heredity

A structure for our times

In Michael Crichton's novel *Jurassic Park* and its film counterpart, fictional scientists were depicted using biotechnology to produce living dinosaurs for display in a theme park. In the story, the scientists isolated the DNA of dinosaurs from fossilized insects that had sucked the reptiles' blood. The insects, which had been preserved intact in amber (fossilized tree resin), yielded DNA that could be used to produce living individuals of long-extinct organisms such as *Tyrannosaurus rex*.

The premise of Crichton's novel was based on an actual scientific paper that claimed to show reptilian DNA sequences in a fossil insect. Unfortunately, the scientific report was not upheld; the "preserved" DNA turned out to be a contaminant from modern organisms.

Despite the fact that the preservation of intact DNA over millions of years is highly improbable, the popular success of *Jurassic Park* did bring the idea of DNA as the genetic material to the attention of millions of readers and viewers. Indeed, even before the novel and movie, the image of the DNA double helix was a familiar secular icon.

The double helix was first proposed by James Watson and Francis Crick in a short paper in the scientific journal *Nature*. A drawing of the structure made by Crick's wife, Odile, accompanied the article, and its simplicity and elegance made it an instant hit, not just with scientists, but with the general public. As Watson put it later, "A structure this pretty just had to exist."

Deoxyribonucleic acid—DNA—and its double-helical structure has become one of the great symbols of science of our era. It is not just trumpeted on the covers of newsmagazines as the "secret of life," but has moved from academic obscurity to common speech. One sees advertisements about a company whose customers get "into the DNA of business." A perfume is named "DNA" and is advertised as "The Essence of Life." A digital media software system is called the "DNA Server."

This is not the first time such a powerful symbol has emerged from science. Think of the mushroom cloud of a nuclear explosion and the Bohr model of the atom, with its electrons whizzing around the nucleus. Salvador Dali was the first well-known artist to use the DNA double helix in his whimsical creations. A portrait of Sir John Sulston, a Nobel prize-

Resurrecting the Rex Scientists and artists have been creating inanimate reconstructions of dinosaurs for more than 100 years. Michael Crichton's novel *Jurassic Park* was based on the fictional premise that DNA retrieved from fossils could produce living dinosaurs such as *Tyrannosaurus rex*.

11.1 What Is the Evidence that the Gene is DNA?

By the early twentieth century, geneticists had associated the presence of genes with chromosomes. Research began to focus on exactly which chemical component of chromosomes comprised this genetic material.

By the 1920s, scientists knew that chromosomes were made up of DNA and proteins. At this time a new dye was developed that could bind specifically to DNA and turned red in direct proportion to the amount of DNA present in a cell. This technique provided circumstantial evidence that DNA was the genetic material:

- *It was in the right place.* DNA was confirmed to be an important component of the nucleus and the chromosomes, which were known to carry genes.

- *It varied among species.* When cells from different species were stained with the dye and their color intensity measured, each species appeared to have its own specific amount of nuclear DNA.

- *It was present in the right amounts.* The amount of DNA in somatic cells (body cells not specialized for reproduction) was twice that in reproductive cells (eggs or sperm)—as might be expected for diploid and haploid cells, respectively.

But circumstantial evidence is *not* a scientific demonstration of cause and effect. After all, proteins are also present in cell nuclei. Science relies on experiments to test hypotheses. The convincing demonstration that DNA is the genetic material came from two sets of experiments, one on bacteria and the other on viruses.

DNA from one type of bacterium genetically transforms another type

The history of biology is filled with incidents in which research on one specific topic has—with or without answering the question originally asked—contributed richly to another, apparently unrelated area. Such a case of serendipity is seen in the work of Frederick Griffith, an English physician.

Bejeweled with DNA The double helix of DNA has become an iconic symbol of modern science and culture. Artists and designers make use of the widely recognized shape in many ways.

winning geneticist, is made of tiny bacterial colonies, each containing a piece of Sulston's DNA. The Brazilian artist Eduardo Kac translated a sentence from the Bible into a DNA nucleotide base sequence and incorporated this DNA into bacteria. The viewer can turn on an ultraviolet lamp to change the DNA sequence and the biblical verse it represents. DNA sculptures abound, and jewelry made with the double helix motif is called the "strands of life" collection.

But it is not only DNA's structure that stirs our society. It is what that structure symbolizes, which is nothing less than the promise and perils of our rapidly expanding knowledge of genetics.

IN THIS CHAPTER we will first describe the key experiments that led to the determination that the genetic material is DNA. We will then describe the structure of the DNA molecule and how this structure determines its function. We will describe the processes by which DNA is replicated, repaired, and maintained. Finally, we present two practical applications arising from our knowledge of DNA replication: the polymerase chain reaction and DNA sequencing.

In the 1920s, Griffith was studying the bacterium *Streptococcus pneumoniae*, or pneumococcus, one of the agents that cause pneumonia in humans. He was trying to develop a vaccine against this devastating illness (antibiotics had not yet been discovered). Griffith was working with two strains of pneumococcus:

■ Cells of the S strain produced colonies that looked smooth (S). Covered by a polysaccharide capsule, these cells were protected from attack by a host's immune system. When S cells were injected into mice, they reproduced and caused pneumonia (the strain was *virulent*).

■ Cells of the R strain produced colonies that looked rough (R), lacked the protective capsule, and were not virulent.

Griffith inoculated some mice with heat-killed S pneumococci. These heat-killed bacteria did not produce infection. However, when Griffith inoculated other mice with a mixture of living R bacteria and heat-killed S bacteria, to his astonishment, the mice died of pneumonia (**Figure 11.1**). When he examined blood from the hearts of these mice, he found it full of living bacteria—many of them with characteristics of the virulent S strain! Griffith concluded that, in the presence of the dead S pneumococci, some of the living R pneumococci had been transformed into virulent S strain organisms.

Did this transformation of the bacteria depend on something that happened in the mouse's body? No. It was shown that simply incubating living R and heat-killed S bacteria together in a test tube yielded the same transformation. Years later, another group of scientists discovered that a cell-free extract of heat-killed S cells also transformed R cells. (A *cell-free extract* contains all the contents of ruptured cells, but no intact cells.) This result demonstrated that some substance—called at the time a chemical **transforming principle**—from the dead S pneumococci could cause a heritable change in the affected R cells. This was an extraordinary discovery: treatment with a chemical substance permanently changed an inherited characteristic. Now it remained to identify the chemical structure of this substance.

The transforming principle is DNA

Identifying the transforming principle was a crucial step in the history of biology. It was accomplished over a period of years by Oswald Avery and his colleagues at what is now Rockefeller University. They treated samples known to contain the pneumococcal transforming principle in a variety of ways to destroy different types of molecules—proteins, nucleic acids, carbohydrates, and lipids—and tested the treated samples to see if they had retained transforming activity.

11.1 Genetic Transformation of Nonvirulent Pneumococci
Frederick Griffith's experiments demonstrated that something in the virulent S strain could transform nonvirulent R strain bacteria into a lethal form, even when the S strain bacteria had been killed by high temperatures. FURTHER RESEARCH: How would you show that heat-killed R strain bacteria can transform living S strain bacteria?

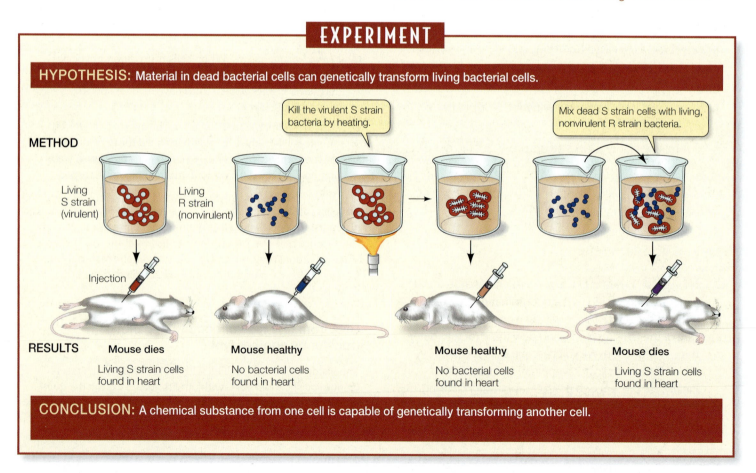

EXPERIMENT

HYPOTHESIS: Material in dead bacterial cells can genetically transform living bacterial cells.

METHOD

Living S strain (virulent)

Living R strain (nonvirulent)

Kill the virulent S strain bacteria by heating.

Mix dead S strain cells with living, nonvirulent R strain bacteria.

Injection

RESULTS

Mouse dies
Living S strain cells found in heart

Mouse healthy
No bacterial cells found in heart

Mouse healthy
No bacterial cells found in heart

Mouse dies
Living S strain cells found in heart

CONCLUSION: A chemical substance from one cell is capable of genetically transforming another cell.

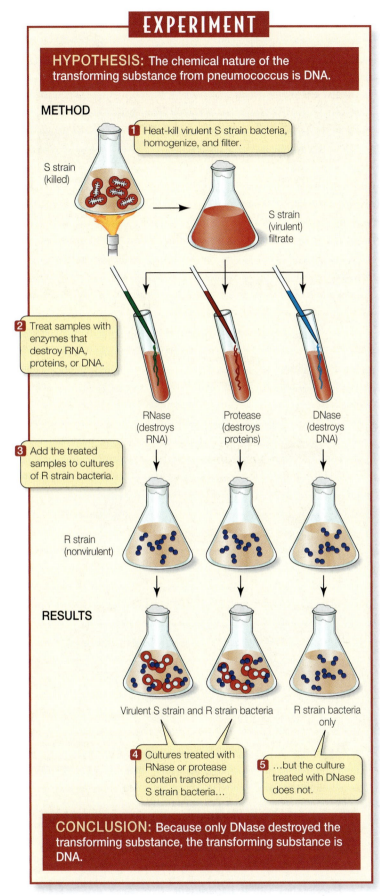

EXPERIMENT

HYPOTHESIS: The chemical nature of the transforming substance from pneumococcus is DNA.

METHOD

1 Heat-kill virulent S strain bacteria, homogenize, and filter.

S strain (killed)

S strain (virulent) filtrate

2 Treat samples with enzymes that destroy RNA, proteins, or DNA.

RNase (destroys RNA)

Protease (destroys proteins)

DNase (destroys DNA)

3 Add the treated samples to cultures of R strain bacteria.

R strain (nonvirulent)

RESULTS

Virulent S strain and R strain bacteria

R strain bacteria only

4 Cultures treated with RNase or protease contain transformed S strain bacteria...

5 ...but the culture treated with DNase does not.

CONCLUSION: Because only DNase destroyed the transforming substance, the transforming substance is DNA.

11.2 Genetic Transformation by DNA Experiments by Avery, MacLeod, and McCarty showed that DNA from virulent S strain pneumococci was responsible for transformation in Griffith's experiments (see Figure 11.1).

The answer was always the same: if the DNA in the sample was destroyed, transforming activity was lost, but there was no loss of activity when proteins, carbohydrates, or lipids were destroyed (**Figure 11.2**). As a final step, Avery, with Colin MacLeod and Maclyn McCarty, isolated virtually pure DNA from a sample containing pneumococcal transforming principle and showed that it caused bacterial transformation. (We now know that the gene encoding the enzyme that catalyzes the synthesis of the pneumococcal polysaccharide capsule was transferred during transformation.)

The work of Avery, MacLeod, and McCarty was a milestone in establishing that DNA is the genetic material in bacterial cells. However, when it was first published (in 1944), it had little impact, for two reasons. First, most scientists did not believe that DNA was chemically complex enough to be the genetic material, especially given the much greater chemical complexity of proteins. Second, and perhaps more important, bacterial genetics was a new field of study—it was not yet clear that bacteria even *had* genes.

Viral replication experiments confirmed that DNA is the genetic material

The questions about bacteria were soon resolved as researchers identified genes and mutations. Bacteria and viruses seemed to undergo genetic processes similar to those in fruit flies and pea plants. Experiments with these relatively simple systems were designed to discover the nature of the genetic material.

In 1952, Alfred Hershey and Martha Chase of the Carnegie Laboratory of Genetics published a paper that had a much greater immediate impact than Avery's 1944 paper. The Hershey–Chase experiment, which sought to determine whether DNA or protein was the genetic material, was carried out with a virus that infects bacteria. This virus, called bacteriophage T2, consists of little more than a DNA core packed inside a protein coat (**Figure 11.3**). The virus is thus made of the two materials that were, at the time, the leading candidates for the genetic material.

When bacteriophage T2 attacks a bacterium, part (but not all) of the virus enters the bacterial cell. About 20 minutes later, the cell bursts, releasing dozens of viruses. Clearly the virus is somehow able to replicate itself inside the bacterium. Hershey and Chase deduced that the entry of some viral component affects the genetic program of the host bacterial cell, transforming it into a bacteriophage factory. They set out to determine which part of the virus—protein or DNA—enters the bacterial cell. To trace the two components of the virus over its life cycle, Hershey and Chase labeled each component with a specific radioisotope:

■ Proteins contain some *sulfur* (in the amino acids cysteine and methionine), an element not present in DNA. Sulfur has a radioactive isotope, ^{35}S. Hershey and Chase grew bacteriophage T2 in a bacterial culture in the presence of ^{35}S, so the proteins of the resulting viruses were labeled with the radioisotope.

■ The deoxyribose–phosphate "backbone" of DNA is rich in *phosphorus* (see Figure 3.24), an element that is not present in most proteins. Phosphorus also has a radioisotope, ^{32}P. The researchers grew another batch of T2 in a bacterial culture in the presence of ^{32}P, thus labeling the viral DNA with ^{32}P.

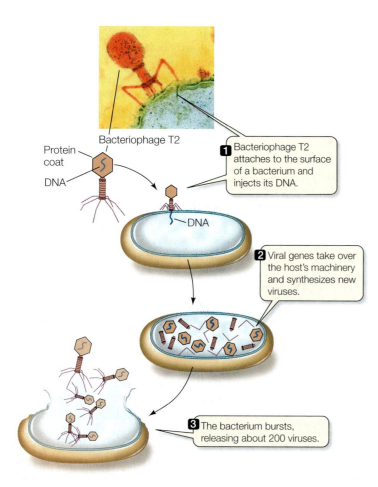

11.3 Bacteriophage T2: Reproduction Cycle Bacteriophage T2 is parasitic on *E. coli*, depending on the bacterium to produce new viruses. The external structures of bacteriophage T2 consist entirely of protein; its DNA is injected into the host bacterium.

Using these radioactively labeled viruses, Hershey and Chase performed their revealing experiments (**Figure 11.4**). In one experiment, they allowed ^{32}P-labeled bacteriophage to infect bacteria; in the other, the bacteria were infected by ^{35}S-labeled bacteriophage. After a few minutes, they agitated each mixture of infected bacteria vigorously in a kitchen blender, which (without bursting the bacteria) stripped away the parts of the virus that had not penetrated the bacteria. Then they separated the bacteria from the rest of the material in a *centrifuge*. Spinning solutions or suspensions at high speed in a centrifuge causes solutes or particles to separate and form a gradient according to their density. The lighter remains of the viruses (those parts that had not penetrated the bacteria) were captured in the supernatant fluid, while the heavier bacterial cells segregated in a "pellet" in the bottom of the container. The scientists found that the supernatant fluid contained most of the ^{35}S (and thus the viral protein), while most of the ^{32}P (and thus the viral DNA) had stayed with the bacteria. These results suggested that it was DNA that had been transferred to the bacteria, and that DNA therefore was the compound responsible for redirecting the genetic program of the bacterial cell.

Hershey and Chase performed other similar but longer-range experiments, allowing a progeny (offspring) generation of viruses

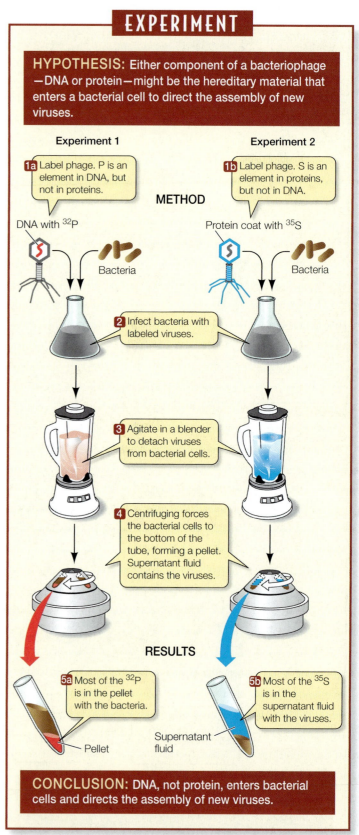

11.4 The Hershey–Chase Experiment This classic experiment demonstrated that DNA, not protein, is the genetic material. When bacterial cells were infected by radiolabeled T2 bacteriophage, only labeled DNA was found in the bacteria; labeled protein remained in the viral matter.

to be collected. The resulting viruses contained almost none of the original ^{35}S and none of the parental viral protein. They did, however, contain about one-third of the original ^{32}P—and thus, presumably, one-third of the original DNA. Because DNA was carried over in the virus from generation to generation but protein was not, the logical conclusion was that the hereditary information of the virus is contained in the DNA.

Eukaryotic cells can also be genetically transformed by DNA

With the publication of the evidence for DNA as the genetic material in bacteria and viruses, the question arose as to whether DNA could be similarly shown to be the genetic material in complex eukaryotes. Some simple experiments, such as injecting a white duck with DNA from a brown duck and reporting the recipient turning brown, or feeding flatworms DNA from worms that had learned a simple task and then seeing the recipients immediately get smarter, were reported, but the results were dubious because no one could duplicate them. It would be impossible for a large molecule such as DNA to get into all the cells of the body, let alone avoid hydrolysis into nucleotides in the digestive system.

However, genetic transformation of eukaryotic cells by DNA (called **transfection**) *can* be demonstrated. The key is to use a genetic **marker**, a gene whose presence in the recipient cell confers an observable phenotype. In the experiments with pneumococcus, these phenotypes were the smooth polysaccharide capsule and virulence. In eukaryotes, researchers usually use a nutritional or antibiotic resistance marker gene that permits the growth of transformed recipient cells but not of nontransformed cells. For example, in the absence of the gene that codes for thymidine kinase, an enzyme needed to make use of thymidine, mammalian cells do not grow. When DNA containing the thymidine kinase marker gene is added to a culture of mammalian cells lacking this gene, however, some cells will grow, demonstrating that they have been transfected with the gene (**Figure 11.5**). Any cell can be transfected in this way, even an egg cell. In this case, a whole new genetically transformed organism can result; such an organism is referred to as *transgenic*. Transformation in eukaryotes is the final line of evidence for DNA as the genetic material.

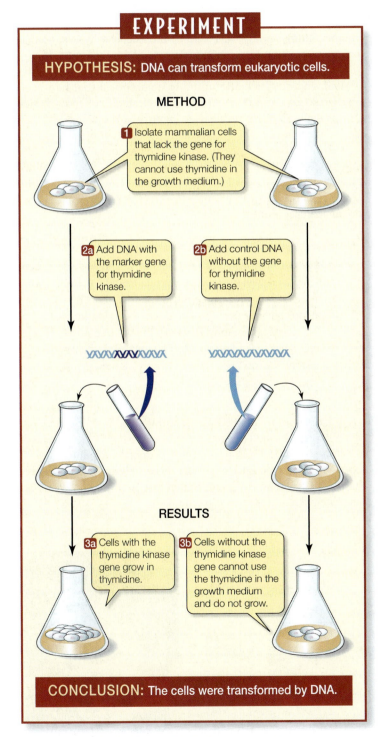

EXPERIMENT

HYPOTHESIS: DNA can transform eukaryotic cells.

METHOD

1 Isolate mammalian cells that lack the gene for thymidine kinase. (They cannot use thymidine in the growth medium.)

2a Add DNA with the marker gene for thymidine kinase.

2b Add control DNA without the gene for thymidine kinase.

RESULTS

3a Cells with the thymidine kinase gene grow in thymidine.

3b Cells without the thymidine kinase gene cannot use the thymidine in the growth medium and do not grow.

CONCLUSION: The cells were transformed by DNA.

11.5 Transfection in Eukaryotic Cells The use of a marker gene demonstrates that mammalian cells can be genetically transformed by DNA.

11.1 RECAP

Experiments on bacteria and on viruses demonstrated that DNA is the genetic material.

- At the time of Griffith's experiments in the 1920s, what circumstantial evidence suggested to scientists that DNA might be the genetic material? See p. 233

- Why were the experiments of Avery, MacLeod, and McCarty definitive evidence that DNA was the genetic material? See pp. 234–235 and Figure 11.2

- What attributes of bacteriophage T2 were key to the Hershey–Chase experiments demonstrating that DNA is the genetic material? See pp. 235–236 and Figure 11.4

As soon as scientists became convinced that the genetic material was DNA, they began efforts to learn its precise three-dimensional chemical structure. In determining the structure of DNA, scientists hoped to find the answers to two questions: How is DNA replicated between nuclear divisions, and how does it direct the synthesis of specific proteins? They were eventually able to answer both questions.

11.2 What Is the Structure of DNA?

The structure of DNA was deciphered only after many types of experimental evidence and theoretical considerations were considered together. The crucial evidence was obtained by *X-ray crystallography*. Some chemical substances, when they are isolated and purified, can be made to form crystals. The positions of atoms in a crystallized substance can be inferred from the pattern of diffraction of X-rays passed through it (**Figure 11.6A**). The attempt to characterize DNA would have been impossible without the crystallographs prepared in the early 1950s by the English chemist Rosalind Franklin (**Figure 11.6B**). Franklin's work, in turn, depended on the success of the English biophysicist Maurice Wilkins, who prepared a sample containing very uniformly oriented DNA fibers. These DNA preparations provided samples for diffraction that were far better than previous ones, and the crystallographs Franklin prepared from them suggested a spiral or helical molecule.

The chemical composition of DNA was known

The chemical composition of DNA also provided important clues to its structure. Biochemists knew that DNA was a polymer of *nucleotides*. Each nucleotide of DNA consists of a molecule of the sugar deoxyribose, a phosphate group, and a nitrogen-containing base (see Figures 3.23 and 3.24). The only differences among the four nucleotides of DNA are their nitrogenous bases: the purines **adenine (A)** and **guanine (G)**, and the pyrimidines **cytosine (C)** and **thymine (T)**.

In 1950, Erwin Chargaff at Columbia University reported some observations of major importance. He and his colleagues found that DNA from many different species—and from different sources within a single organism—exhibits certain regularities. In almost all DNA, the following rule holds: The amount of adenine equals the amount of thymine (A = T), and the amount of guanine equals the amount of cytosine (G = C) (**Figure 11.7**). As a result, the total abundance of purines (A + G) equals the total abundance of pyrimidines (T + C). The structure of DNA could not have been worked out without this observation, now known as *Chargaff's rule*, yet its significance was overlooked for at least three years.

Watson and Crick described the double helix

The solution to the puzzle of the structure of DNA was accelerated by *model building*: the assembly of three-dimensional representations of possible molecular structures using known relative molecular dimensions and known bond angles. This technique, originally exploited in structural studies by the American biochemist Linus Pauling, was used by the English physicist Francis Crick and the American geneticist James D. Watson (**Figure 11.8A**), then both at the Cavendish Laboratory of Cambridge University.

Watson and Crick attempted to combine all that had been learned so far about DNA structure into a single coherent model. The crystallographers' results (see Figure 11.6) convinced Watson and Crick that the DNA molecule is **helical** (cylindrically spiral). The results of density measurements and previous model building suggested that there are two polynucleotide chains in the molecule. Modeling studies had also led to the conclusion that the two chains in DNA run in opposite directions—that is, that they are **antiparallel** (**Figure 11.8B**).

In 1952, the American biochemist Linus Pauling was also working to unravel the molecular structure of DNA. He attempted to travel to a meeting in London, where he hoped to convince the British researchers to allow him to view their crystallography. However, the U.S. Department of State revoked Pauling's passport based on his stance against the Korean War.

In late February of 1953, Crick and Watson built a model out of tin that established the general structure of DNA. This structure explained all the known chemical properties of DNA, and it opened the door to understanding its biological functions. There have been minor amendments to that first published structure, but its principal features remain unchanged.

11.6 X-Ray Crystallography Helped Reveal the Structure of DNA (A) The positions of atoms in a crystallized chemical substance can be inferred by the pattern of diffraction of X-rays passed through it. The pattern of DNA is both highly regular and repetitive. (B) Rosalind Franklin's crystallography helped scientists to visualize the helical structure of the DNA molecule.

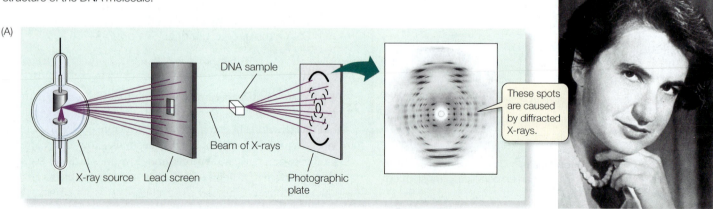

(A) X-ray source   Lead screen   DNA sample   Beam of X-rays   Photographic plate

These spots are caused by diffracted X-rays.

(B)

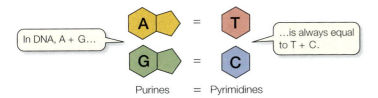

11.7 Chargaff's Rule In DNA, the total abundance of purines is equal to the total abundance of pyrimidines.

Four key features define DNA structure

Four features summarize the molecular architecture of the DNA molecule:

- It is a *double-stranded helix* of uniform diameter.
- It is *right-handed*. [Hold your right hand with the thumb pointing up (the axis of the helix) and the fingers curled around (the sugar–phosphate backbone) and you have the idea.]
- It is *antiparallel* (the two strands run in opposite directions).
- The outer edges of the nitrogenous bases are *exposed* in the major and minor grooves.

THE HELIX The sugar–phosphate "backbones" of the polynucleotide chains coil around the outside of the helix, and the nitrogenous bases point toward the center. The two chains are held together by hydrogen bonding between specifically paired bases (**Figure 11.9**). Consistent with Chargaff's rule,

- Adenine (A) pairs with thymine (T) by forming two hydrogen bonds.
- Guanine (G) pairs with cytosine (C) by forming three hydrogen bonds.

Every *base pair* consists of one purine (A or G) and one pyrimidine (T or C). This pattern is known as **complementary base pairing**.

Because the AT and GC pairs are of equal length, they fit into a fixed distance between the two chains (like rungs on a ladder), and the diameter of the helix is thus uniform. The base pairs are flat, and their stacking in the center of the molecule is stabilized by hydrophobic interactions (see Section 2.2), contributing to the overall stability of the double helix.

ANTIPARALLEL STRANDS What does it mean to say that the two DNA strands are *antiparallel*? The direction of each strand is determined by examining the bonds between the alternating phosphate groups and sugars that make up the backbones of each strand. Look closely at the five-carbon sugar (deoxyribose) mole-

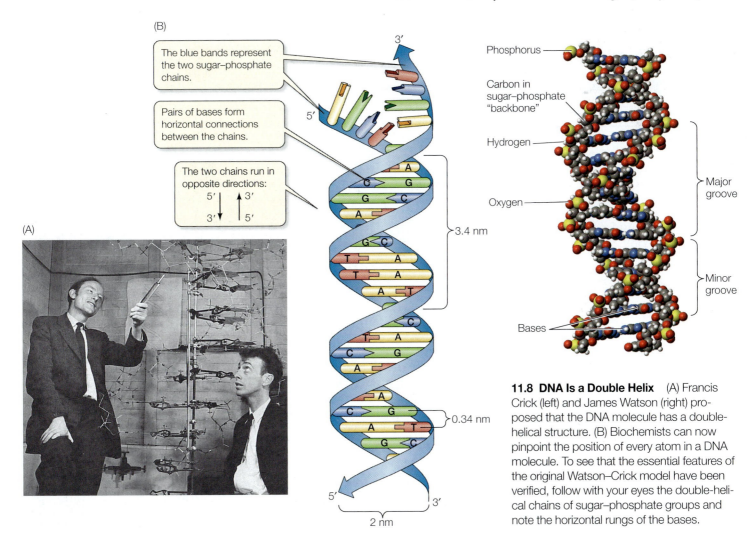

11.8 DNA Is a Double Helix (A) Francis Crick (left) and James Watson (right) proposed that the DNA molecule has a double-helical structure. (B) Biochemists can now pinpoint the position of every atom in a DNA molecule. To see that the essential features of the original Watson–Crick model have been verified, follow with your eyes the double-helical chains of sugar–phosphate groups and note the horizontal rungs of the bases.

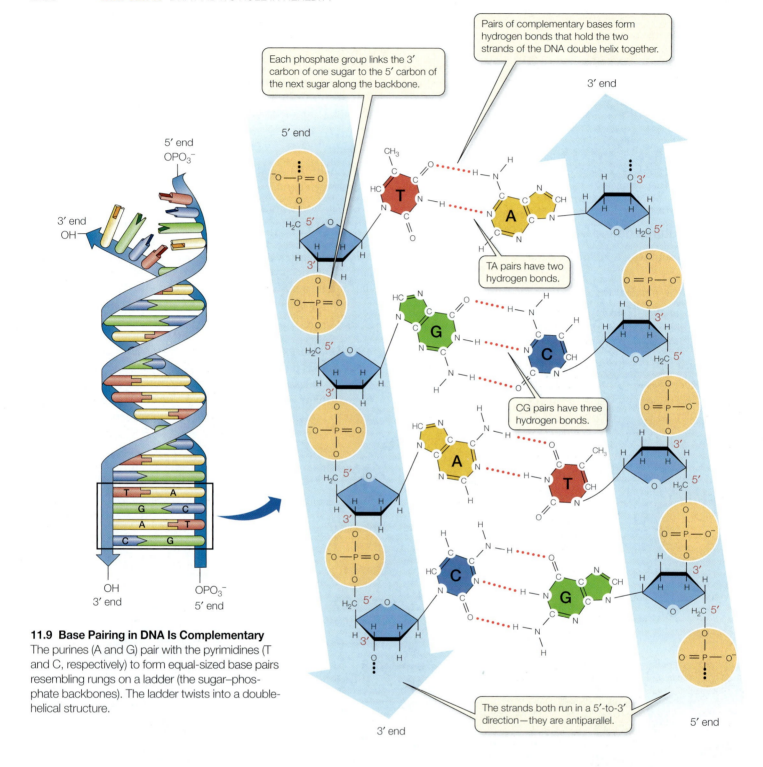

11.9 Base Pairing in DNA Is Complementary
The purines (A and G) pair with the pyrimidines (T and C, respectively) to form equal-sized base pairs resembling rungs on a ladder (the sugar–phosphate backbones). The ladder twists into a double-helical structure.

Each phosphate group links the 3′ carbon of one sugar to the 5′ carbon of the next sugar along the backbone.

Pairs of complementary bases form hydrogen bonds that hold the two strands of the DNA double helix together.

TA pairs have two hydrogen bonds.

CG pairs have three hydrogen bonds.

The strands both run in a 5′-to-3′ direction—they are antiparallel.

cules in Figure 11.9. The number followed by a prime (′) designates the position of a carbon atom in the sugar. In the sugar–phosphate backbone of DNA, the phosphate groups connect to the 3′ carbon of one deoxyribose molecule and the 5′ carbon of the next, linking successive sugars together.

Thus the two ends of a polynucleotide chain differ. At one end of a chain is a free (not connected to another nucleotide) 5′ phosphate group (—OPO$_3^-$); this end is called the 5′ end. At the other end is a free 3′ hydroxyl group (—OH); this end is called the 3′ end. In a DNA double helix, the 5′ end of one strand is paired with the 3′ end of the other strand, and vice versa. In other words, were you

to draw an arrow for each strand running from 5′ to 3′, the arrows would point in opposite directions.

BASE EXPOSURE AT THE GROOVES Looking back at Figure 11.8B, note the major and minor grooves in the helix. From these grooves, the exposed outer edges of the flat, hydrogen-bonded base pairs are accessible for potential hydrogen bonding. As seen in Figure 11.8, two hydrogen bonds join each AT base pair, while three hydrogen bonds join each GC base pair. Hydrogen-bonding opportunities also exist at the C=O group in T and the "N" group in A; the GC base pair offers additional hydrogen bonding possibilities.

Thus the surfaces of the AT and GC base pairs offer slightly different, chemically distinct surfaces that another molecule, such as a protein, could recognize and bind to. Access to the exposed base-pair sequences in the major and minor grooves is the key to protein–DNA interactions in the replication and expression of the genetic information in DNA.

The double-helical structure of DNA is essential to its function

The genetic material performs four important functions, and the DNA structure proposed by Watson and Crick was elegantly suited to three of them.

- *The genetic material stores an organism's genetic information.* With its millions of nucleotides, the base sequence of a DNA molecule could encode and store an enormous amount of information and could account for species and individual differences. DNA fits this role nicely.

- *The genetic material is susceptible to mutation,* or permanent changes in the information it encodes. For DNA, mutations might be simple changes in the linear sequence of base pairs.

- *The genetic material is precisely replicated* in the cell division cycle. Replication could be accomplished by complementary base pairing, A with T and G with C. In the original publication of their findings in the journal *Nature* in 1953, Watson and Crick coyly pointed out, "It has not escaped our notice that the specific pairing we have postulated immediately suggests a possible copying mechanism for the genetic material."

- *The genetic material is expressed as the phenotype.* This function is not obvious in the structure of DNA. However, as we will see in the next chapter, the nucleotide sequence of DNA is copied into RNA, which is in turn converted into a linear sequence of amino acids—a protein. The folded forms of proteins provide much of the phenotype of an organism.

11.2 RECAP

DNA is a double helix made up of two antiparallel polynucleotide chains. The two chains are joined by hydrogen bonds between the nucleotide bases, which pair specifically, A with T and G with C. Chemical groups of the bases that are exposed in the grooves of the helix can be recognized by other molecules.

- Can you describe some of the evidence that Watson and Crick used to come up with the double helix model for DNA? See pp. 238–239

- Do you understand how the double-helical structure of DNA relates to its function?

Once the structure of DNA was understood, it was possible to discover how DNA replicates itself. Let's examine the experiments that taught us how this elegant process works.

11.3 How Is DNA Replicated?

The mechanism of DNA replication that had suggested itself to Watson and Crick was soon confirmed. First, experiments showed that single strands of DNA could be replicated in a test tube containing simple substrates and an enzyme. Then a truly classic experiment showed that each of the two strands of the double helix can serve as a template for a new strand of DNA.

Three modes of DNA replication appeared possible

The prediction that the DNA molecule contains the information needed for its own replication was confirmed by the work of Arthur Kornberg, then at Washington University in St. Louis. He showed that DNA with the same base composition as parental DNA can be synthesized in a test tube containing three substances:

- The substrates, deoxyribonucleoside triphosphates dATP, dCTP, dGTP, and dTTP

- A **DNA polymerase** enzyme

- DNA, which serves as a **template** to guide the incoming nucleotides

Recall that a nucleoside is a nitrogen base attached to a sugar. The four deoxyribonucleoside triphosphates each consist of a nitrogen base attached to deoxyribose, which in turn is attached to three phosphate groups.

The next question was which of three possible replication patterns was occurring:

- *Semiconservative replication,* in which each parent strand serves as a template for a new strand, and the two new DNA molecules each have one old and one new strand (**Figure 11.10A**)

- *Conservative replication,* in which the original double helix serves as a template for, but does not contribute to, a new double helix (**Figure 11.10B**)

- *Dispersive replication,* in which fragments of the original DNA molecule serve as templates for assembling two new molecules, each containing old and new parts, perhaps at random (**Figure 11.10C**)

Watson and Crick's original paper suggested that DNA replication was semiconservative, but Kornberg's experiment did not provide a basis for choosing among these three models.

Meselson and Stahl demonstrated that DNA replication is semiconservative

The work of Matthew Meselson and Franklin Stahl convinced the scientific community that the pattern seen in DNA is **semiconservative replication**. Working at the California Institute of Technology, Meselson and Stahl devised a simple way to distinguish old parent strands of DNA from newly copied ones: *density labeling.*

The key to their experiment was the use of a "heavy" isotope of nitrogen. Heavy nitrogen (^{15}N) is a rare, nonradioactive isotope that makes molecules containing it more dense than chemically identical molecules containing the common isotope, ^{14}N. Meselson, Stahl, and Jerome Vinograd grew two cultures of the bacterium *Escherichia coli* for many generations:

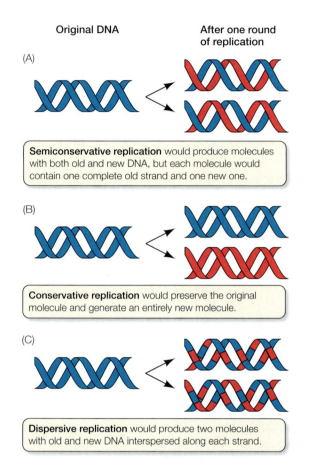

Original DNA   After one round of replication

(A)

Semiconservative replication would produce molecules with both old and new DNA, but each molecule would contain one complete old strand and one new one.

(B)

Conservative replication would preserve the original molecule and generate an entirely new molecule.

(C)

Dispersive replication would produce two molecules with old and new DNA interspersed along each strand.

11.10 Three Models for DNA Replication   In each model, original DNA is shown in blue and newly synthesized DNA in red.

- One culture was grown in a medium whose nitrogen source (ammonium chloride, NH_4Cl) was made with ^{15}N instead of ^{14}N. As a result, all the DNA in the bacteria was "heavy."

- Another culture was grown in a medium containing ^{14}N, and all the DNA in these bacteria was "light."

When extracts from the two cultures were combined and centrifuged, two separate DNA bands formed, showing that this method could distinguish DNA samples of slightly different densities.

Next, the researchers grew another *E. coli* culture on ^{15}N medium, then transferred it to normal ^{14}N medium and allowed the bacteria to continue growing (**Figure 11.11**). Under the conditions they used, *E. coli* cells divide, replicating their DNA every 20 minutes. Meselson and Stahl collected some of the bacteria after each division and extracted DNA from the samples. They found that the density gradient was different in each bacterial generation:

- At the time of the transfer to the ^{14}N medium, the DNA was uniformly labeled with ^{15}N, and hence was relatively dense.

- After one generation in the ^{14}N medium, when the DNA had been duplicated once, all the DNA was of an intermediate density.

- After two generations, there were two equally large DNA bands: one of low density and one of intermediate density.

- In samples from subsequent generations, the proportion of low-density DNA increased steadily.

The results of this experiment can be explained only by the semiconservative model of DNA replication. In the first round of DNA replication in the ^{14}N medium, the strands of the double helix—both heavy with ^{15}N—separated. Each strand then acted as the template for a second strand, which contained only ^{14}N and hence was less dense. Each double helix then consisted of one ^{15}N strand and one ^{14}N strand, and was of intermediate density. In the second replication, the ^{14}N-containing strands directed the synthesis of partners with ^{14}N, creating low-density DNA, and the ^{15}N strands formed new ^{14}N partners (see Figure 11.10).

The crucial observation demonstrating the semiconservative model was that intermediate-density DNA (^{15}N–^{14}N) appeared in the first generation and continued to appear in subsequent generations. With the other models, the results would have been quite different (see Figure 11.10):

- In conservative replication, the first generation would have had both high-density DNA (^{15}N–^{15}N) and low-density DNA (^{14}N–^{14}N), but no intermediate-density DNA.

- In dispersive replication, the density of the new DNA would have been half that of the parent DNA, but DNA of this density would not continue to appear in subsequent generations.

The Meselson–Stahl experiment, called by some scientists among the most elegant ever done by biologists, was an excellent example of the scientific method. It began with three hypotheses—the three models of DNA replication—and was designed so that the results could differentiate between them.

We all began life as a fertilized egg, with only one set of double-stranded DNA molecules from our parents. Given semiconservative replication, do we still have those original parental strands? There is some evidence that we may. During mitosis, stem cells in the adult body preferentially retain DNA with the "old" strands. A similar mechanism may operate during early development.

There are two steps in DNA replication

Semiconservative DNA replication in the cell involves a number of different enzymes and other proteins. It takes place in two steps:

- The DNA double helix is unwound to separate the two template strands and make them available for new base pairing.

- New nucleotides are joined by phosphodiester linkages to each growing new strand in a sequence determined by complementary base pairing with the bases on the template strand.

A key observation is that *nucleotides are added to the growing new strand at the 3' end*—the end at which the DNA strand has a free

EXPERIMENT

HYPOTHESIS: DNA replicates semiconservatively.

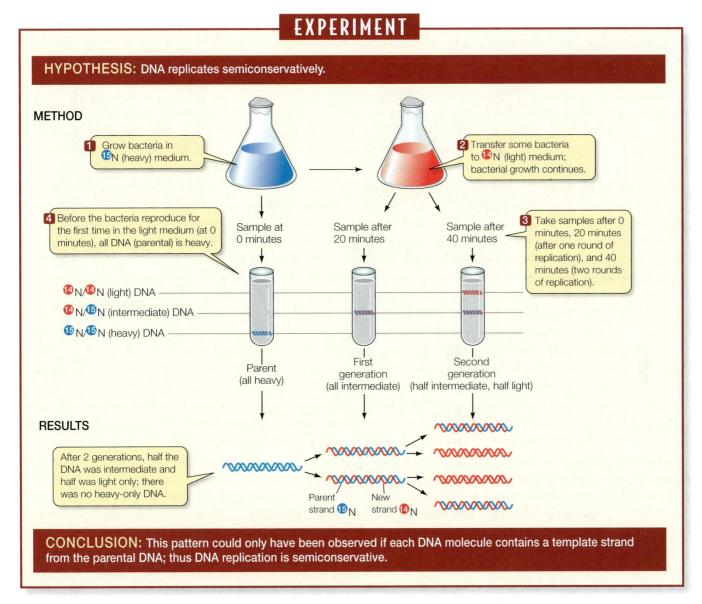

METHOD

1 Grow bacteria in ¹⁵N (heavy) medium.

2 Transfer some bacteria to ¹⁴N (light) medium; bacterial growth continues.

4 Before the bacteria reproduce for the first time in the light medium (at 0 minutes), all DNA (parental) is heavy.

Sample at 0 minutes

Sample after 20 minutes

Sample after 40 minutes

3 Take samples after 0 minutes, 20 minutes (after one round of replication), and 40 minutes (two rounds of replication).

¹⁴N/¹⁴N (light) DNA

¹⁴N/¹⁵N (intermediate) DNA

¹⁵N/¹⁵N (heavy) DNA

Parent (all heavy)

First generation (all intermediate)

Second generation (half intermediate, half light)

RESULTS

After 2 generations, half the DNA was intermediate and half was light only; there was no heavy-only DNA.

Parent strand ¹⁵N

New strand ¹⁴N

CONCLUSION: This pattern could only have been observed if each DNA molecule contains a template strand from the parental DNA; thus DNA replication is semiconservative.

11.11 The Meselson–Stahl Experiment A centrifuge was used to separate DNA molecules labeled with isotopes of different densities. This experiment revealed a pattern that supports the semiconservative model of DNA replication. FURTHER RESEARCH: If you continued this experiment for two more generations (as Meselson and Stahl actually did), what would be the composition (in terms of low- and intermediate-density) of the fourth generation DNA?

hydroxyl (—OH) group on the 3′ carbon of its terminal deoxyribose (**Figure 11.12**). One of the three phosphate groups in a deoxyribonucleoside triphosphate is attached to the 5′ position of the sugar. The bonds linking the other two phosphate groups to the nucleotide are broken, releasing energy for the reaction.

DNA is threaded through a replication complex

DNA is replicated through the interaction of the template strand with a huge protein complex called the **replication complex**, which catalyzes the reactions involved. All chromosomes have at least one base sequence, called the **origin of replication** (*ori*), to which this replication complex initially binds. As noted above, this binding is based on the recognition of the different nucleotide bases by proteins. DNA replicates *in both directions* from the origin of replication, forming two **replication forks**. Both of the separated strands of the parent molecule act as templates simultaneously, and the formation of the new strands is guided by complementary base pairing.

Until recently, DNA replication was depicted as a locomotive (the replication complex) moving along a railroad track (the DNA) (**Figure 11.13A**). The current view is that this model may not be correct. Instead, the replication complex seems to be stationary, attached to nuclear structures, and it is the DNA that moves, essentially threading through the complex as single strands and emerging as double strands (**Figure 11.13B**). All replication complexes contain several proteins with different roles in DNA replication; we will describe these proteins as we examine the steps of the process.

11.12 Each New DNA Strand Grows from Its 5′ End to Its 3′ End

The DNA strand at the right (blue) is the template for the synthesis of the complementary strand that is growing at the left (pink).

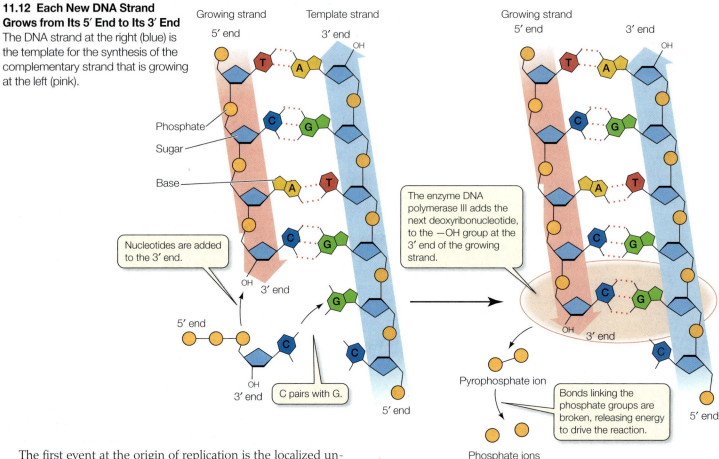

Nucleotides are added to the 3′ end.

C pairs with G.

The enzyme DNA polymerase III adds the next deoxyribonucleotide, to the —OH group at the 3′ end of the growing strand.

Bonds linking the phosphate groups are broken, releasing energy to drive the reaction.

Pyrophosphate ion

Phosphate ions

The first event at the origin of replication is the localized unwinding (denaturation) of DNA. There are several forces that hold the two strands together, including hydrogen bonding and the hydrophobic interactions of the bases. An enzyme called **DNA helicase** uses energy from ATP hydrolysis to unwind the DNA, and special proteins called **single-strand binding proteins** bind to the unwound strands to keep them from reassociating into a double helix. This process makes each of the two template strands available for complementary base pairing.

SMALL CIRCULAR CHROMOSOMES REPLICATE FROM A SINGLE ORIGIN Small circular chromosomes, such as the 1–4-million-base-pair DNA of bacteria, have a single origin of replication. As the DNA moves through the replication complex, the replication forks grow around the circle (**Figure 11.14A**). Two interlocking circular DNAs are formed, and they are separated by an enzyme called **DNA topoisomerase**.

DNA polymerases are very fast. In the bacterium *E. coli*, replication can be as fast as 1,000 bases per second, and it takes 20–40 minutes to replicate the bacterium's 4.7 million base pairs. Human polymerases are slower (50 bases per second), and human chromosomes are much larger (about 80 million base pairs). In this case, to get the job done in an hour, it takes many polymerases working at many replication forks.

LARGE LINEAR CHROMOSOMES HAVE MANY ORIGINS In large linear chromosomes, such as a human chromosome, there are hundreds of origins of replication. Origins of replication that are adjacent to one another along the linear chromosome can be bound by replication complexes at the same time and replicated simultaneously. So there are many replication forks in eukaryotic DNA (**Figure 11.14B**).

(A) Replication complex moves

Two replication complexes move apart as the DNA replicates in two directions away from the origin.

(B) DNA moves

The replication complexes are stationary and the DNA threads through.

11.13 Two Views of DNA Replication
(A) It was once thought that the replication complex moved along DNA like a locomotive moving along a railroad track. (B) More recent evidence suggests that the DNA is threaded through the stationary replication complex.

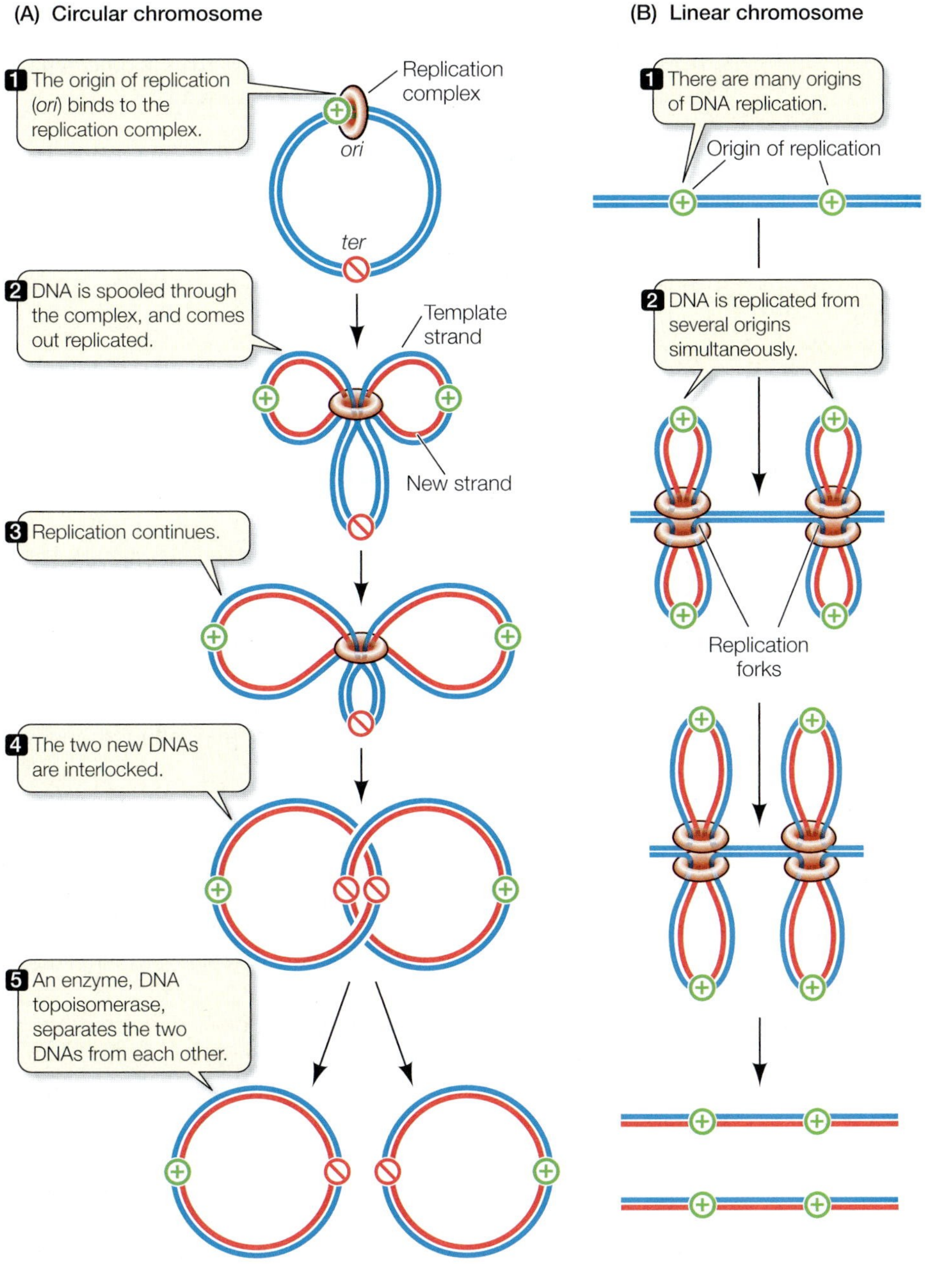

(A) Circular chromosome

1 The origin of replication (*ori*) binds to the replication complex.

Replication complex

ori

ter

2 DNA is spooled through the complex, and comes out replicated.

Template strand

New strand

3 Replication continues.

4 The two new DNAs are interlocked.

5 An enzyme, DNA topoisomerase, separates the two DNAs from each other.

(B) Linear chromosome

1 There are many origins of DNA replication.

Origin of replication

2 DNA is replicated from several origins simultaneously.

Replication forks

11.14 Replication in Small Circular and Large Linear Chromosomes (A) Small circular chromosomes have a single origin (*ori*) and terminus (*ter*) of replication. (B) Larger linear chromosomes have many origins of replication.

DNA polymerases add nucleotides to the growing chain

DNA polymerases are much larger than their substrates, the deoxyribonucleoside triphosphates, and the template DNA, which is very thin (**Figure 11.15A**). Molecular models of the enzyme–substrate–template complex from bacteria show that the enzyme is shaped like an open right hand with a palm, a thumb, and fingers (**Figure 11.15B**). The palm holds the active site of the enzyme and brings together the substrate and the template. The finger regions rotate inward and have precise shapes that can recognize the different shapes of the four nucleotide bases.

NO DNA BEGINS WITHOUT A PRIMER DNA polymerases can elongate a polynucleotide strand by covalently linking new nucleotides to a previously existing strand, but they cannot start a strand from scratch. Therefore, a "starter" strand, called a **primer**, is required. In DNA replication, the primer is a short single strand of RNA (**Figure 11.16**). This RNA strand, complementary to the DNA template strand, is synthesized one nucleotide at a time by an enzyme called **primase**. DNA polymerase then adds nucleotides to the 3' end of the primer and continues until the replication of that section of DNA has been completed. Then the RNA

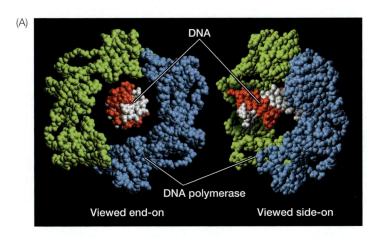

(A) DNA

DNA polymerase

Viewed end-on   Viewed side-on

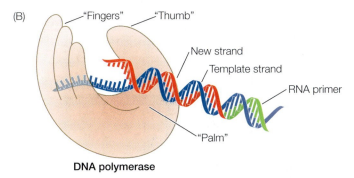

(B) "Fingers"   "Thumb"

New strand

Template strand

RNA primer

"Palm"

DNA polymerase

11.15 DNA Polymerase Binds to the Template Strand (A) The DNA polymerase enzyme (blue and green) is much larger than the DNA molecule (red and white). (B) DNA polymerase is shaped like a hand, and in this side-on view, its "fingers" can be seen curling around the DNA. These "fingers" can recognize the distinctive shapes of the four bases.

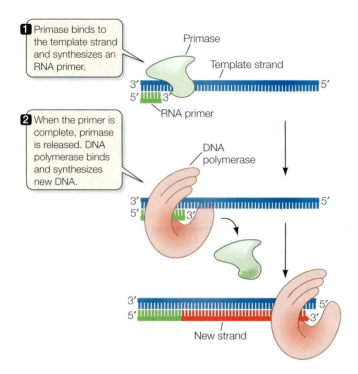

① Primase binds to the template strand and synthesizes an RNA primer.

② When the primer is complete, primase is released. DNA polymerase binds and synthesizes new DNA.

Primase

Template strand

RNA primer

DNA polymerase

New strand

11.16 No DNA Forms without a Primer DNA polymerases require a primer—a "starter" strand of DNA or RNA to which they can add new nucleotides.

primer is degraded, DNA is added in its place, and the resulting DNA fragments are connected by the action of other enzymes. When DNA replication is complete, each new strand consists only of DNA.

CELLS CONTAIN SEVERAL DIFFERENT DNA POLYMERASES Most cells contain more than one DNA polymerase, but only one of them is responsible for chromosomal DNA replication. The others are involved in primer removal and DNA repair. Fourteen DNA polymerases have been identified in humans; the one catalyzing most replication is DNA polymerase δ. In the bacterium *E. coli* there are five DNA polymerases; the one responsible for replication is

DNA polymerase III. Various other proteins play roles in other replication tasks; some of these are shown in **Figure 11.17**.

THE TWO DNA STRANDS GROW DIFFERENTLY As Figure 11.17 shows, the DNA at the replication fork opens up like a zipper in one direction. Study **Figure 11.18** and try to imagine what is happening over a short period of time. Remember that the two DNA strands are antiparallel; that is, the 3' end of one strand is paired with the 5' end of the other.

■ One newly replicating strand (the **leading strand**) is pointing in the "right" direction to grow continuously at its 3' end as the fork opens up.

■ The other new strand (the **lagging strand**) is pointing in the "wrong" direction: as the fork opens up further, its exposed 3' end gets farther and farther away from the fork, and an unreplicated gap is formed, which would get bigger and bigger if there were not a special mechanism to overcome this problem.

Synthesis of the lagging strand requires working in relatively small, discontinuous stretches (100 to 200 nucleotides at a time in eukaryotes; 1,000 to 2,000 at a time in prokaryotes). These discontinuous stretches are synthesized just as the leading strand is, by the addition of new nucleotides one at a time to the 3' end of the new strand, but the synthesis of this new strand moves in the direction opposite to that in which the replication fork is moving. These stretches of new DNA for the lagging strand are called **Okazaki fragments**, after their discoverer, the Japanese biochemist Reiji Okazaki. While the leading strand grows continuously "forward," the lagging strand grows in shorter, "backward" stretches with gaps between them.

A single primer suffices for synthesis of the leading strand, but each Okazaki fragment requires its own primer. In bacteria, DNA polymerase III synthesizes Okazaki fragments by adding nucleotides to a primer until it reaches the primer of the previous fragment. At this point, DNA polymerase I (discovered by Arthur Kornberg) removes the old primer and replaces it with DNA. Left behind is a tiny nick—the final phosphodiester linkage between the adjacent Okazaki fragments is missing. The enzyme **DNA ligase** catalyzes the formation of that bond, linking the fragments and making the lagging strand whole (**Figure 11.19**).

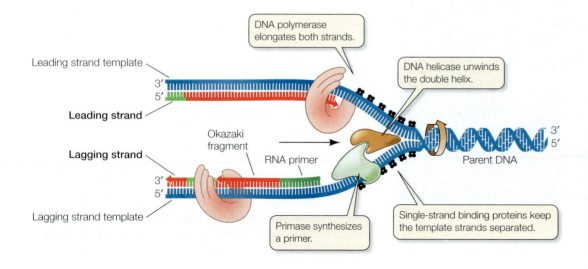

Leading strand template

Leading strand

Lagging strand

Lagging strand template

DNA polymerase elongates both strands.

Okazaki fragment

RNA primer

Primase synthesizes a primer.

DNA helicase unwinds the double helix.

Parent DNA

Single-strand binding proteins keep the template strands separated.

11.17 Many Proteins Collaborate in the Replication Complex Several proteins in addition to DNA polymerase are involved in DNA replication. The two molecules of DNA polymerase shown here are actually part of the same complex.

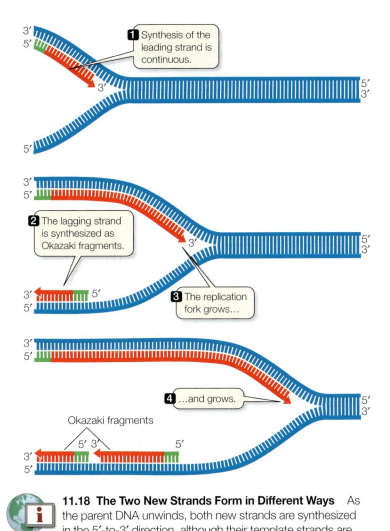

11.18 The Two New Strands Form in Different Ways As the parent DNA unwinds, both new strands are synthesized in the 5'-to-3' direction, although their template strands are antiparallel. The leading strand grows continuously forward, but the lagging strand grows in short discontinuous stretches called Okazaki fragments. Eukaryotic Okazaki fragments are hundreds of nucleotides long, with gaps between them.

Working together, DNA helicase, the two DNA polymerases, primase, DNA ligase, and the other proteins of the replication complex do the job of DNA synthesis with a speed and accuracy that are almost unimaginable. In *E. coli*, the replication complex makes new DNA at a rate in excess of 1,000 base pairs per second, committing errors in fewer than one base in a million.

A SLIDING DNA CLAMP MAKES DNA POLYMERASE PROCESSIVE
How do DNA polymerases work so fast? We saw in Section 6.4 that an enzyme catalyzes a chemical reaction:

substrate binds to enzyme → one product is formed →
enzyme is released → cycle repeats

It is hard to envision a reaction so fast that it would be possible to go through such a cycle for each nucleotide added to DNA. Instead, DNA polymerases are **processive**—that is, they catalyze many polymerizations each time they bind to a DNA molecule:

substrates bind to one enzyme → many products are formed →
enzyme is released → cycle repeats

The newly replicated strand is stabilized by a **sliding DNA clamp** (**Figure 11.20**). This protein has multiple identical subunits assem-

bled into a doughnut shape. The doughnut's "hole" is just large enough to encircle the DNA double helix, along with a single layer of water molecules for lubrication. The clamp binds to DNA just behind DNA polymerase, keeping it associated tightly with the newly replicated DNA. If the clamp is absent, DNA polymerase dissociates from DNA after 20–100 polymerizations. With the clamp, it can polymerize up to 50,000 nucleotides before it detaches.

Telomeres are not fully replicated

As we have just seen, replication of the lagging strand occurs by the addition of Okazaki fragments to RNA primers. When the ter-

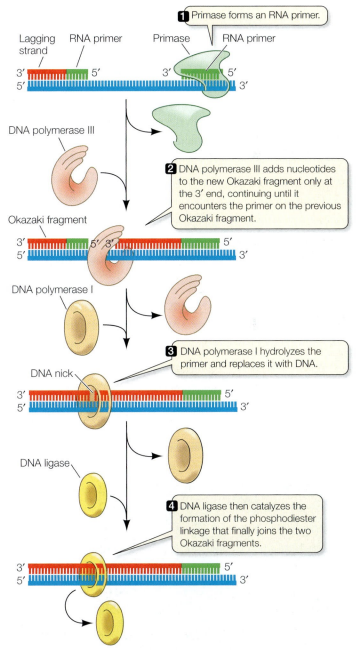

11.19 The Lagging Strand Story In bacteria, DNA polymerase I and DNA ligase cooperate with DNA polymerase III to complete the complex task of synthesizing the lagging strand.

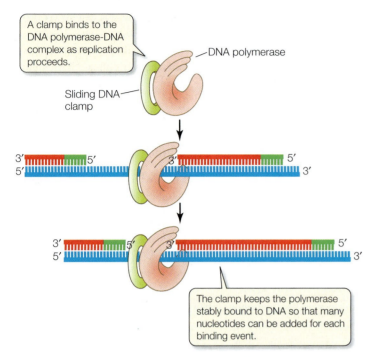

A clamp binds to the DNA polymerase-DNA complex as replication proceeds.

DNA polymerase

Sliding DNA clamp

The clamp keeps the polymerase stably bound to DNA so that many nucleotides can be added for each binding event.

11.20 A Sliding DNA Clamp Increases the Efficiency of DNA Polymerization   The clamp keeps DNA polymerase bound to DNA, so that thousands of nucleotides can be polymerized every time the enzyme binds to a template strand.

cut off the single-stranded region, along with some of the intact double-stranded end. Thus the chromosome becomes slightly shorter with each cell division.

In many eukaryotes, there are repetitive sequences at the ends of chromosomes called **telomeres**. In humans, the telomere sequence is TTAGGG, and it is repeated about 2,500 times. These repeats bind special proteins that maintain the stability of chromosome ends. Each human chromosome can lose 50–200 base pairs of telomeric DNA after each round of DNA replication and cell division. After 20–30 divisions, the chromosomes are unable to take part in cell division, and the cell dies.

This phenomenon explains in part why cells do not last the entire lifetime of the organism: their telomeres are lost. Yet constantly dividing cells, such as bone marrow stem cells and gamete-producing cells, maintain their telomeric DNA. An enzyme, appropriately called **telomerase**, catalyzes the addition of any lost telomeric sequences (**Figure 11.21B**). Telomerase contains an RNA sequence that acts as a template for the telomeric repeat sequence.

Telomerase is expressed in more than 90 percent of human cancers and may be an important factor in the ability of cancer cells to divide continuously. Since most normal cells do not have this ability, telomerase is an attractive target for drugs designed to attack tumors specifically.

There is also interest in telomerase and aging. When a gene expressing high levels of telomerase is added to human cells in culture, their telomeres do not shorten. Instead of dying after 20–30

minal RNA primer is removed, no DNA can be synthesized to replace it because there is no DNA 3′ end to extend (that is, there is no complementary DNA strand). So the new chromosome formed by DNA replication has a bit of single-stranded DNA at each end (**Figure 11.21A**). This situation activates mechanisms that

11.21 Telomeres and Telomerase   (A) Removal of the RNA primer at the 3′ end of the lagging strand leaves a region of DNA—the telomere—unreplicated. (B) The enzyme telomerase binds to the 3′ end and extends the lagging strand of DNA. An RNA sequence embedded in telomerase provides a template so that, overall, the DNA does not get shorter. (C) Bright fluorescent staining marks the telomeric regions on these blue-stained human chromosomes.

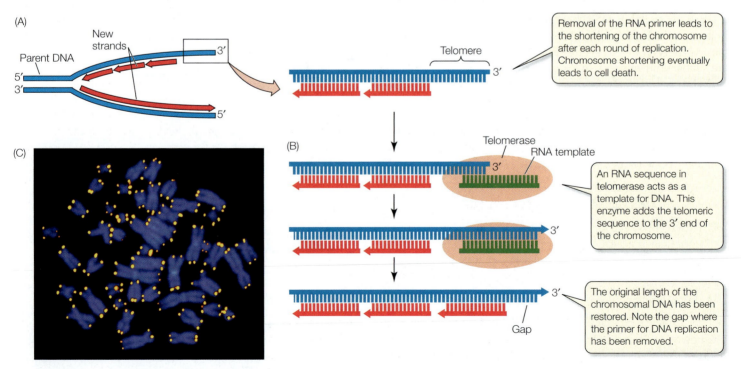

(A)

Parent DNA
New strands

Telomere

Removal of the RNA primer leads to the shortening of the chromosome after each round of replication. Chromosome shortening eventually leads to cell death.

(B)

Telomerase
RNA template

An RNA sequence in telomerase acts as a template for DNA. This enzyme adds the telomeric sequence to the 3′ end of the chromosome.

The original length of the chromosomal DNA has been restored. Note the gap where the primer for DNA replication has been removed.

Gap

(C)

cell generations, the cells become immortal. It remains to be seen how this finding relates to the aging of a large organism.

11.3 RECAP

Meselson and Stahl showed that DNA replication is semiconservative: each parent strand serves as a template for a new strand. A complex of proteins, most notably DNA polymerases, is involved in replication. New DNA is polymerized in one direction only, but since the two strands are antiparallel, one strand is made continuously and the other in Okazaki fragments that are eventually joined.

- Do you see how the Meselson–Stahl experiment differentiated between the three models for DNA replication? See pp. 242–243 and Figures 11.10 and 11.11
- Can you recognize five enzymes needed for DNA replication and explain the role of each? See pp. 244–245 and Figures 11.15, 11.16, 11.17, and 11.21
- Do you undersand how the leading strand of DNA be replicated continuously while the lagging strand must be replicated in fragments? See p. 246 and Figure 11.18

The complex process of DNA replication is amazingly accurate, but it is not perfect. What happens when things go wrong?

11.4 How Are Errors in DNA Repaired?

DNA is accurately replicated and faithfully maintained. The price of failure can be great: the transmission of genetic information is at stake, as is the functioning and even the life of a cell or multicellular organism. Yet the replication of DNA is not perfectly accurate, and the DNA of nondividing cells is subject to damage by natural chemical alterations of the bases as well as by environmental agents. In the face of these threats, how has life gone on so long?

The preservers of life are DNA repair mechanisms. DNA polymerases initially make a significant number of mistakes in assembling polynucleotide strands. The observed error rate of one for every 10^5 bases replicated would result in about 60,000 mutations every time a human cell divided. Fortunately, our cells have at least three DNA repair mechanisms at their disposal:

- A **proofreading** mechanism corrects errors in replication as DNA polymerase makes them.
- A **mismatch repair** mechanism scans DNA immediately after it has been replicated and corrects any base-pairing mismatches.
- An **excision repair** mechanism removes abnormal bases that have formed because of chemical damage and replaces them with functional bases.

Every time it introduces a new nucleotide into a growing polynucleotide strand, DNA polymerase performs a **proofreading** function (**Figure 11.22A**). When a DNA polymerase recognizes a mispairing of bases, it removes the improperly introduced nucleotide and tries again. (Other proteins of the replication complex also play roles in proofreading.) The error rate for this process is only about 1 in 10,000 base pairs, and it lowers the overall error rate for replication to about one base in every 10^{10} bases replicated.

After DNA has been replicated, a second set of proteins surveys the newly replicated molecule and looks for mismatched base pairs that were missed in proofreading (**Figure 11.22B**). For example, this **mismatch repair** mechanism might detect an AC base pair instead of an AT pair. But how does the repair mechanism "know" whether the AC pair should be repaired by removing the C and replacing it with T or by removing the A and replacing it with G?

> Individuals who suffer from a condition known as xeroderma pigmentosum lack an excision repair mechanism that normally corrects the damage caused by ultraviolet radiation. They can develop skin cancers after even a brief exposure to sunlight.

The mismatch repair mechanism can detect the "wrong" base because a DNA strand is chemically modified some time after replication. In prokaryotes, methyl groups ($—CH_3$) are added to some adenines. Immediately after replication, methylation has not yet occurred on the newly replicated strand, so the new strand is "marked" (distinguished by being unmethylated) as the one in which errors should be corrected.

When mismatch repair fails, DNA sequences are altered. One form of colon cancer arises in part from a failure of mismatch repair.

DNA molecules can also be damaged during the life of a cell (for example, when it is in G1). High-energy radiation, chemicals from the environment, and random spontaneous chemical reactions can all damage DNA. **Excision repair** mechanisms deal with these kinds of damage.

Certain enzymes constantly "inspect" the cell's DNA (**Figure 11.22C**). When they find mispaired bases, chemically modified bases, or points at which one strand has more bases than the other (with the result that one or more bases of one strand form an unpaired loop), these enzymes cut the defective strand. Another enzyme cuts away the bases adjacent to and including the offending base, and DNA polymerase and DNA ligase synthesize and seal up a new (usually correct) base sequence to replace the excised one.

11.4 RECAP

DNA replication is not perfect; in addition, DNA may be naturally altered or damaged. Repair mechanisms exist that detect and repair mismatched or damaged DNA.

- Do you understand the roles of DNA proofreading, mismatch repair, and excision repair? See Figure 11.22

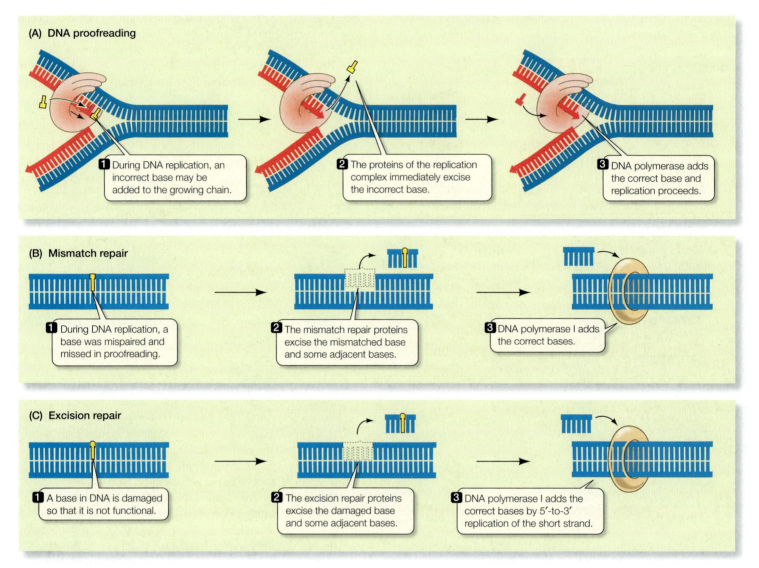

11.22 DNA Repair Mechanisms The proteins of the replication complex also play roles in the life-preserving DNA repair mechanisms, helping to ensure the exact replication of template DNA and repair any damage that occurs.

Understanding how DNA is replicated and repaired has allowed scientists to develop techniques for studying genes. We'll look at just two of those techniques next.

11.5 What Are Some Applications of Our Knowledge of DNA Structure and Replication?

The principles underlying DNA replication in cells have been used to develop two laboratory techniques that have been vital in analyzing genes and genomes. The first technique allows researchers to make multiple copies of short DNA sequences, and the second allows them to determine the base sequence of a DNA molecule.

The polymerase chain reaction makes multiple copies of DNA

Since DNA can be replicated in the laboratory, it is possible to make multiple copies of a DNA sequence. The **polymerase chain reaction (PCR)** technique essentially automates this process by copying a short region of DNA many times in a test tube.

PCR is a cyclic process in which a sequence of steps is repeated over and over again (**Figure 11.23**):

- Double-stranded fragments of DNA are separated into single strands by heating (*denatured*).

- A short, artificially synthesized primer is added to the mixture, along with the four deoxyribonucleoside triphosphates (dATP, dGTP, dCTP, and dTTP) and DNA polymerase.

- DNA polymerase catalyzes the production of complementary new strands.

A single cycle takes a few minutes to double the amount of DNA, leaving the new DNA in the double-stranded state. Repeating the cycle many times leads to an exponential increase in the number

RESEARCH METHOD

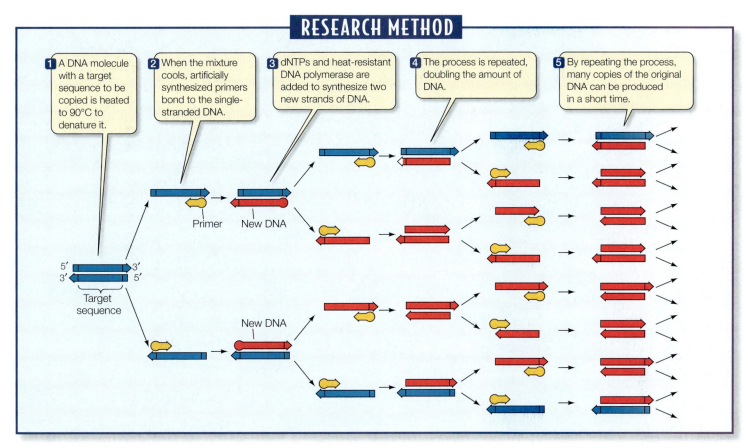

1 A DNA molecule with a target sequence to be copied is heated to 90°C to denature it.

2 When the mixture cools, artificially synthesized primers bond to the single-stranded DNA.

3 dNTPs and heat-resistant DNA polymerase are added to synthesize two new strands of DNA.

4 The process is repeated, doubling the amount of DNA.

5 By repeating the process, many copies of the original DNA can be produced in a short time.

Primer   New DNA

5'   3'
3'   5'

Target
sequence

New DNA

11.23 The Polymerase Chain Reaction The steps in this cyclic process are repeated many times to produce multiple copies of a DNA fragment.

of copies of the DNA sequence; this process is referred to as *amplifying* the sequence.

The PCR technique requires that the base sequences at the 3' end of each strand of the target DNA sequence be known so that complementary primers, usually 15–20 bases long, can be made in the laboratory. Because of the uniqueness of DNA sequences, usually only two primers of this length will bind to only one region of DNA in an organism's genome. This specificity in the face of the incredible diversity of target DNA is a key to the power of PCR.

One initial problem with PCR was its temperature requirements. To denature the DNA, it must be heated to more than 90°C—a temperature that destroys most DNA polymerases. The PCR technique would not be practical if new polymerase had to be added after denaturation in each cycle.

This problem was solved by nature: in the hot springs at Yellowstone National Park, as well as in other high-temperature locations, lives a bacterium called, appropriately, *Thermus aquaticus*. The means by which this organism survives temperatures up to 95°C was investigated by Thomas Brock and his colleagues. They discovered that *T. aquaticus* has an entire metabolic machinery that is heat-resistant, including DNA polymerase that does not denature at these high temperatures.

Scientists pondering the problem of copying DNA by PCR read Brock's basic research articles and got a clever idea: Why not use *T. aquaticus* DNA polymerase in the PCR technique? It could with-

stand the 90°C denaturation temperature and would not have to be added during each cycle. The idea worked, and it earned biochemist Kerry Mullis a Nobel prize. PCR has had an enormous impact on genetic research. Some of its most striking applications will be described in Chapters 13 through 17.

The nucleotide sequence of DNA can be determined

Another important technique allows researchers to determine the base sequence of a DNA molecule. This **DNA sequencing** technique relies on the use of artificially altered nucleosides. As we saw earlier in this chapter, the deoxyribonucleoside triphosphates (dNTPs) that are the normal substrates for DNA replication contain the sugar deoxyribose. If that sugar is replaced with 2,3-dideoxyribose, the resulting *di*deoxyribonucleoside triphosphate (ddNTP) will still be added by DNA polymerase to a growing polynucleotide chain. However, because ddNTPs lack a hydroxyl group (—OH) at the 3' position, the next nucleotide cannot be added (**Figure 11.24A**). Thus synthesis stops at the position where ddNTP has been incorporated into the growing end of a DNA strand.

To determine the sequence of DNA, a fragment of DNA (usually no more than 700 base pairs long) is denatured. The resulting single strands of DNA are placed in a test tube and mixed with

- DNA polymerase, to synthesize the complementary strand

- Short, artificially synthesized primers appropriate for the DNA sequence

- The four dNTPs (dATP, dGTP, dCTP, and dTTP)

RESEARCH METHOD

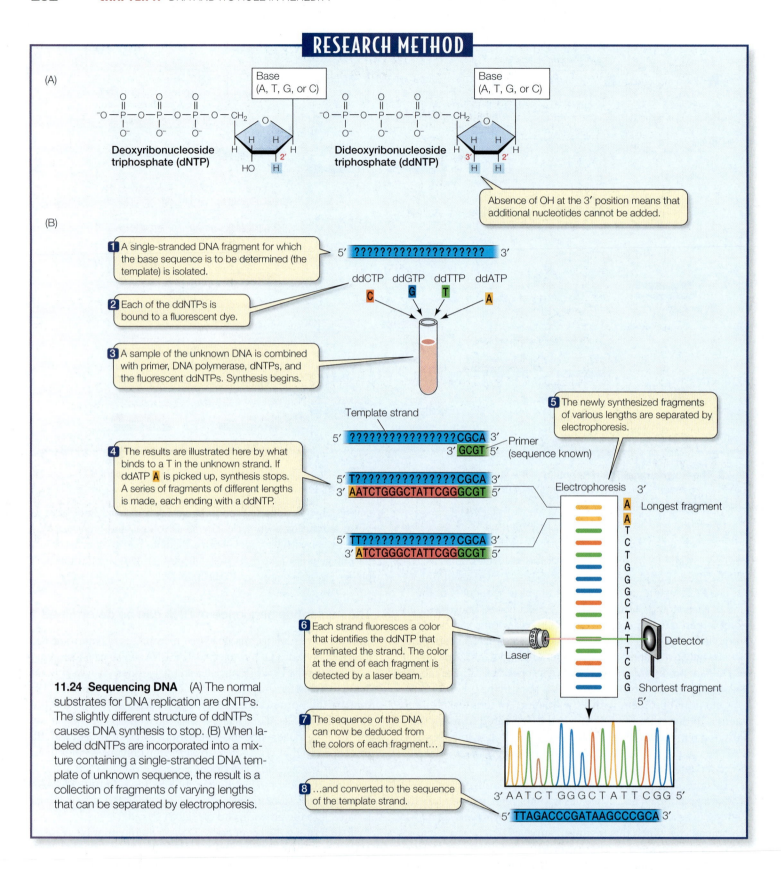

11.24 Sequencing DNA (A) The normal substrates for DNA replication are dNTPs. The slightly different structure of ddNTPs causes DNA synthesis to stop. (B) When labeled ddNTPs are incorporated into a mixture containing a single-stranded DNA template of unknown sequence, the result is a collection of fragments of varying lengths that can be separated by electrophoresis.

(A)

Base (A, T, G, or C)

Deoxyribonucleoside triphosphate (dNTP)

Base (A, T, G, or C)

Dideoxyribonucleoside triphosphate (ddNTP)

Absence of OH at the 3' position means that additional nucleotides cannot be added.

(B)

1 A single-stranded DNA fragment for which the base sequence is to be determined (the template) is isolated.

5' ????????????????????? 3'

ddCTP ddGTP ddTTP ddATP
C   G   T   A

2 Each of the ddNTPs is bound to a fluorescent dye.

3 A sample of the unknown DNA is combined with primer, DNA polymerase, dNTPs, and the fluorescent ddNTPs. Synthesis begins.

Template strand

5' ??????????????CGCA 3'
                3' GCGT 5' Primer (sequence known)

4 The results are illustrated here by what binds to a T in the unknown strand. If ddATP A is picked up, synthesis stops. A series of fragments of different lengths is made, each ending with a ddNTP.

5' T?????????????????CGCA 3'
3' AATCTGGGCTATTCGGGCGT 5'

5' TT????????????????CGCA 3'
3' ATCTGGGCTATTCGGGCGT 5'

5 The newly synthesized fragments of various lengths are separated by electrophoresis.

Electrophoresis 3'

A Longest fragment
A
T
C
T
G
G
G
C
T
A
T
T
C
G
G Shortest fragment
5'

6 Each strand fluoresces a color that identifies the ddNTP that terminated the strand. The color at the end of each fragment is detected by a laser beam.

Laser   Detector

7 The sequence of the DNA can now be deduced from the colors of each fragment...

8 ...and converted to the sequence of the template strand.

3' A A T C T G G G C T A T T C G G 5'

5' TTAGACCCGATAAGCCCGCA 3'

■ Small amounts of the four ddNTPs, each bonded to a fluorescent "tag" that emits a different color of light

DNA replication proceeds, and the test tube soon contains a mixture of the template DNA strands and shorter, new complementary strands. The new strands, each ending with a fluorescent ddNTP, are of varying lengths. For example, each time a T is reached on the template strand, DNA polymerase adds either a dATP or a ddATP to the growing complementary strand. If dATP is added, the strand continues to grow. If ddATP is added, growth stops.

After DNA replication has been allowed to proceed for a while, the new DNA fragments are denatured from their templates. The fragments are then subjected to *electrophoresis* (see Figure 16.2). This technique sorts the DNA fragments by length and can detect differences in fragment length as short as one base. During the electrophoresis run, the fragments pass in order of increasing length through a laser beam that excites the fluorescent tags. The light emitted is then detected, and the resulting information—that is, which color of fluorescence, and therefore which ddNTP, is at the end of a strand of which length—is fed into a computer. The computer processes this information and prints out the DNA sequence of the fragment (**Figure 11.24B**). DNA sequencing has formed the basis of the new science of genomics.

11.5 RECAP

Knowledge of the mechanisms of DNA replication led to the development of techniques for making multiple copies of DNA sequences and for determining the nucleotide sequence of DNA molecules.

- Do you understand the role of primers in PCR? See pp. 250–251 and Figure 11.23

- Can you explain why dideoxyribonucleosides are used in DNA sequencing? See pp. 251–252 and Figure 11.24

CHAPTER SUMMARY

11.1 What is the evidence that the gene is DNA?

Griffith's experiments in the 1920s demonstrated that some substance in cells—then called a **transforming principle**—can cause heritable change in other cells. Review Figure 11.1

The location and quantity of DNA in the cell suggested that DNA might be the genetic material. Experiments by Avery, MacLeod, and McCarty isolated the transforming principle from bacteria and identified it as DNA. Review Figure 11.2

The Hershey–Chase experiment established conclusively that DNA (and not protein) is the genetic material by tracing the DNA of radiolabeled viruses with which they infected bacterial cells. Review Figure 11.4

Genetic transformation of eukaryotic cells is called **transfection**. Transformation and transfection can be studied with the aid of a **marker** gene that confers a known and observable phenotype. Review Figure 11.5

11.2 What is the structure of DNA?

Chargaff's rule states that the amount of **adenine** in DNA is equal to the amount of **thymine**, and that the amount of **guanine** is equal to the amount of **cytosine**; thus the total abundance of purines (A + G) equals the total abundance of pyrimidines (T + C).

X-ray crystallography showed that the DNA molecule is **helical**. Watson and Crick proposed that DNA is a double-stranded helix in which the strands are **antiparallel**. Review Figure 11.8

Complementary base pairing between A and T and between G and C accounts for Chargaff's rule. The bases are held together by hydrogen bonding. Review Figure 11.9

11.3 How is DNA replicated?

See Web/CD Tutorial 11.1

Meselson and Stahl showed the replication of DNA to be **semiconservative**. Each parent strand acts as a **template** for the synthesis of a new strand; thus the two replicated DNA molecules each contain one parent strand and one newly synthesized strand. Review Figure 11.11, Web/CD Tutorial 11.2

In DNA replication, the enzyme **DNA polymerase** catalyzes the addition of nucleotides to the 3′ end of each strand. Which nucleotides are added is determined by complementary base pairing with the template strand. Review Figure 11.12

The **replication complex** is a huge protein complex that attaches to the chromosome at the **origin of replication** (*ori*).

Replication proceeds from the origin of replication on both strands in the 5′-to-3′ direction, forming two **replication forks**.

Many proteins assist in DNA replication. **DNA helicase** separates the strands, and **single-strand binding proteins** keep the strands from reassociating.

In prokaryotes, two interlocking circular DNAs are formed; they are separated by an enzyme called **DNA topoisomerase**. Review Figure 11.14

Primase catalyzes the synthesis of a short RNA **primer** to which nucleotides are added by DNA polymerase. Review Figure 11.16

The **leading strand** is synthesized continuously and the **lagging stand** in pieces called **Okazaki fragments**. The fragments are joined together by **DNA ligase**. Review Figures 11.18 and 11.19, Web/CD Tutorial 11.3

The speed with which DNA polymerization proceeds is attributed to the **processive** nature of DNA polymerases, which can catalyze many polymerizations at a time. A **sliding DNA clamp** helps ensure the stability of this process. Review Figure 11.20

DNA replication leaves a short, unreplicated sequence, the **telomere**, at the 3′ end of the chromosome. Unless the enzyme **telomerase** is present, the sequence is removed. After mutiple cell cycles, the telomeres shorten, leading to chromosome instability and cell death. Review Figure 11.21

11.4 How are errors in DNA repaired?

DNA polymerases make about one error in 10^5 bases replicated. DNA is also subject to natural alteration and chemical damage. DNA can be repaired by three different mechanisms: **proofreading**, **mismatch repair**, and **excision repair**. Review Figure 11.22

11.5 What are some applications of our knowledge of DNA structure and replication?

The **polymerase chain reaction** technique uses DNA polymerase to make multiple copies of DNA in the laboratory. Review Figure 11.23

DNA sequencing techniques use the principles of DNA replication to determine the nucleotide sequence of DNA. Review Figure 11.24

SELF-QUIZ

1. Griffith's studies of *Streptococcus pneumoniae*
   a. showed that DNA is the genetic material of bacteria.
   b. showed that DNA is the genetic material of bacteriophages.
   c. demonstrated the phenomenon of bacterial transformation.
   d. proved that prokaryotes reproduce sexually.
   e. proved that protein is not the genetic material.

2. In the Hershey–Chase experiment,
   a. DNA from parent bacteriophages appeared in progeny bacteriophages.
   b. most of the phage DNA never entered the bacteria.
   c. more than three-fourths of the phage protein appeared in progeny phages.
   d. DNA was labeled with radioactive sulfur.
   e. DNA formed the coat of the bacteriophages.

3. Which statement about complementary base pairing is *not* true?
   a. It plays a role in DNA replication.
   b. In DNA, T pairs with A.
   c. Purines pair with purines, and pyrimidines pair with pyrimidines.
   d. In DNA, C pairs with G.
   e. The base pairs are of equal length.

4. In semiconservative replication of DNA,
   a. the original double helix remains intact and a new double helix forms.
   b. the strands of the double helix separate and act as templates for new strands.
   c. polymerization is catalyzed by RNA polymerase.
   d. polymerization is catalyzed by a double-helical enzyme.
   e. DNA is synthesized from amino acids.

5. Which of the following does *not* occur during DNA replication?
   a. Unwinding of the parent double helix
   b. Formation of short pieces that are connected by DNA ligase
   c. Complementary base pairing
   d. Use of a primer
   e. Polymerization in the 3′-to-5′ direction

6. The primer used for DNA replication
   a. is a short strand of RNA added to the 3′ end.
   b. is present only once on the leading strand.
   c. remains on the DNA after replication.
   d. ensures that there will be a free 5′ end to which nucleotides can be added.
   e. is added to only one of the two template strands.

7. One strand of DNA has the sequence 5′-ATTCCG-3′. The complementary strand for this is
   a. 5′-TAAGGC-3′
   b. 5′-ATTCCG-3′
   c. 5′-ACCTTA-3′
   d. 5′-CGGAAT-3′
   e. 5′-GCCTTA-3′

8. The role of DNA ligase in DNA replication is to
   a. add more nucleotides to the growing strand one at a time.
   b. open up the two DNA strands to expose template strands.
   c. ligate base to sugar to phosphate in a nucleotide.
   d. bond Okazaki fragments to one another.
   e. remove incorrectly paired bases.

9. The polymerase chain reaction
   a. is a method for sequencing DNA.
   b. is used to transcribe specific genes.
   c. amplifies specific DNA sequences.
   d. does not require DNA replication primers.
   e. uses a DNA polymerase that denatures at 55°C.

10. What is the correct order for the following events in excision repair of DNA?
    (1) Base-paired DNA is made complementary to the template
    (2) Damaged bases are recognized
    (3) DNA ligase seals the new strand to existing DNA
    (4) Part of a single strand is excised
    a. 1234
    b. 2134
    c. 2413
    d. 3421
    e. 4231

FOR DISCUSSION

1. Suppose that Meselson and Stahl had continued their experiment on DNA replication for another ten bacterial generations. Would there still have been any ^{14}N–^{15}N hybrid DNA present? Would it still have appeared in the centrifuge tube? Explain.

2. If DNA replication were conservative rather than semiconservative, what results would Meselson and Stahl have observed? Diagram the results using the conventions of Figure 11.10.

3. Using the following information, calculate the number of origins of DNA replication on a human chromosome: DNA polymerase adds nucleotides at 3,000 base pairs per minute in one direction; replication is bidirectional; S phase lasts 300 minutes; there are 120 million base pairs per chromosome. With a typical chromosome 3 μm long, how many origins are there per μm?

4. The drug dideoxycytidine, used to treat certain viral infections, is a nucleotide made with 2′,3′-dideoxyribose. This sugar lacks —OH groups at both the 2′ and the 3′ positions. Explain why this drug stops the growth of a DNA chain when added to DNA.

FOR INVESTIGATION

Outline a series of experiments using radioactive isotopes to show that bacterial DNA and not protein enters the host cell and is responsible for bacterial transformation.

12 From DNA to Protein: Genotype to Phenotype

Toxic avenger at the ribosome

In 1978, Georgi Markov, a Bulgarian journalist who had written articles critical of the then-Communist government of Bulgaria, was living in exile in London. As he stood one evening at a bus stop near Waterloo Station, a man—possibly a Bulgarian secret agent—brushed up against him and, seemingly by accident, poked him with an umbrella. Markov felt a sharp pain. Within a few hours, he started to feel weak. A high temperature, vomiting, and more severe symptoms soon followed. Two days later he was dead.

Police investigators found a tiny perforated pellet embedded in Markov's leg, and in that pellet was a small amount of ricin, a highly toxic molecule isolated from the seeds of the tropical castor bean plant, *Ricinus communis*. The seeds of *Ricinus* have been used for centuries as a source of castor oil, a natural product once frequently administered to children to "clean out" the digestive tract. Castor oil is used today in the plastics industry. The toxin ricin is a protein that is not present in the seed oil, and people found out the hard way that it is one of the most poisonous substances made by any organism. About 1 milligram (an amount the size of the head of a pin) can kill a human.

Markov's murder is not the only case of deliberate use of ricin. Small amounts of ricin were found in caves in Afghanistan occupied by the terrorist group Al-Qaeda, and the poison may have been used in the war between Iran and Iraq that raged throughout much of the 1980s. In the 1990s, four members of an anti-tax group were arrested for plotting to use home-grown ricin to kill a U.S. government official. And on February 3, 2004, the U.S. Senate offices were closed when ricin was found in a mailroom.

Much has been written about the possible ways ricin might be used in a terrorist attack. That is unlikely, however, because relatively large amounts of it would be needed to harm a significant number of people. Unlike bacteria such as those that cause anthrax, ricin molecules are proteins; they do not reproduce.

Ricin enters cells by binding to membrane glycoproteins and glycolipids that contain the sugar galactose. Since many cell surface molecules contain this sugar, ricin can bind to most cells. After being endocytosed and released into the cell cytoplasm, ricin kills the cell by blocking protein synthesis. More specifically, it catalyzes the modifi-

Ricinus communis, the Castor Bean Plant This brightly colored tropical plant produces ricin, a lethal toxin that inhibits protein synthesis at the ribosome.

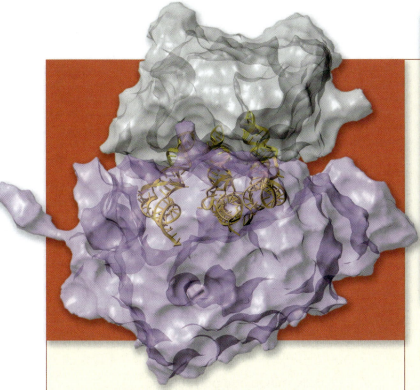

Ricin's Target Ricin inactivates the ribosome, which is the site of protein synthesis. Ribosomes are large aggregates of macromolecules, containing dozens of proteins and ribosomal RNA in two subunits (violet and gray) and three molecules of transfer RNA (gold).

cation and cleavage of one of the large RNA molecules that make up the eukaryotic ribosome—the "workbench" of protein synthesis. A single ricin molecule in the cytoplasm can modify 1500 ribosomes, killing the cell in minutes.

Proteins are the major phenotypic expression of the genotype—the genetic information encoded in a cell's DNA. Ricin inhibits the cell's ability to express the genotype as phenotype through protein synthesis, and therefore ricin-poisoned cells cannot survive.

IN THIS CHAPTER we will see how genes are expressed as proteins. We begin with evidence for the relationship between genes and proteins, and then fill in some of the details of transcription (the copying of the gene sequence of DNA into a sequence of RNA) and translation (the use of a sequence of RNA to make a polypeptide with a defined sequence of amino acids). Finally, we will define mutations and their phenotypes in specific molecular terms.

12.1 What Is the Evidence that Genes Code for Proteins?

In Chapter 11, we defined genes as sequences of DNA and learned that genes are expressed as physical characteristics known as the *phenotype*. Here, we define the phenotype as proteins. What is the evidence for this definition?

The molecular basis of phenotypes was actually discovered before it was known that DNA was the genetic material. Scientists had studied the chemical differences between individuals carrying wild-type and mutant alleles in organisms as diverse as humans and bread molds. They found that the major phenotypic differences were the result of differences in specific proteins.

Experiments on bread mold established that genes determine enzymes

Because of life's basis in the cell theory, scientists investigating a biological phenomenon can reasonably assume that what is found in one organism applies to others. Thus they often search for a *model organism*, one that is easy to grow in the laboratory or observe in the field and which shows the phenomenon to be studied. In previous chapters we have seen several examples of model organisms, including:

- Pea plants (*Pisum sativum*) used by Mendel in his genetics experiments
- Fruit flies (*Drosophila*) used by Morgan in his genetics experiments
- *E. coli* used by Meselson and Stahl to study DNA replication

To this list we now add the common bread mold, *Neurospora crassa*. *Neurospora* is a type of fungus known as an ascomycete (see Chapter 30). This mold is haploid for most of its life, making its genetics straightforward (since there are no dominant–recessive relationships). It is simple to culture and grows well in the laboratory. In the 1940s, George W. Beadle and Edward L. Tatum at Stanford University undertook studies to chemically define the phenotype in *Neurospora*.

The roles of enzymes in biochemistry were being described at the time Beadle and Tatum began their work. They hypothesized that the expression of a gene as a phenotype could occur through an enzyme. They grew *Neurospora* on a minimal nutritional medium containing sucrose, minerals, and a vitamin. Using this medium, the enzymes of wild-type *Neurospora* could catalyze the metabolic reactions needed to make all the chemical constituents of their cells, including proteins. These wild-type strains are called *prototrophs* ("original eaters").

Mutations provide a powerful way to determine cause and effect in biology. Nowhere has this been so evident as in the elucidation of biochemical pathways. Beadle and Tatum treated wild-type *Neurospora* with X rays, which act as a *mutagen* (something known to cause mutations). When they examined the treated molds, they found that some mutant strains could no longer grow on the minimal medium, but grew only if they were supplied with additional nutrients. The scientists hypothesized that these *auxotrophs* ("increased eaters") must have suffered mutations in genes that coded for the enzymes used to synthesize the nutrients they now needed to obtain from their environment. For each auxotrophic strain, Beadle and Tatum were able to find a single compound that, when added to the minimal medium, supported the growth of that strain. This result suggested that mutations have simple effects, and that each mutation causes a defect in only one enzyme in a metabolic pathway. These conclusions became known as the **one-gene, one-enzyme hypothesis** (**Figure 12.1**).

One group of auxotrophs, for example, could grow only if the minimal medium was supplemented with the amino acid arginine. (Wild-type *Neurospora* makes its own arginine.) These mutant strains were designated *arg* mutants. Beadle and Tatum found several different *arg* mutant strains. They proposed two alternative hypotheses to explain why these different genetic strains had the same phenotype:

- The different *arg* mutants could have mutations in *the same gene*, as in the case of the different eye color alleles of fruit flies. In this case, the gene might code for an enzyme involved in arginine synthesis.

- The different *arg* mutants could have mutations in *different genes*, each coding for a separate function that leads to arginine production. These independent functions might be different enzymes along the same biochemical pathway.

Some of the *arg* mutant strains fell into each of the two categories. Genetic crosses showed that some of the mutations were at the same chromosomal locus, and were different alleles of the same gene. Other mutations were at different loci, or on different chromosomes, and so were not alleles of the same gene. Beadle and Tatum concluded that these different genes participated in governing a single biosynthetic pathway—in this case, the pathway leading to arginine synthesis (see the Interpretation in Figure 12.1).

By growing different *arg* mutants in the presence of various compounds suspected to be intermediates in the biosynthetic pathway for arginine, Beadle and Tatum were able to classify each mutation as affecting one enzyme or another and to order the compounds along the pathway. Then they broke open the wild-type and mutant cells and examined them for enzyme activities.

12.1 One Gene, One Enzyme Beadle and Tatum studied several ▶ *arg* mutants of *Neurospora*. The different *arg* mutant strains required the addition of different compounds to their growth medium in order to synthesize the arginine required for their growth. Step through the figure to follow the reasoning that upheld the "one-gene, one-enzyme" hypothesis. FURTHER RESEARCH: If a diploid *Neurospora* spore were made from two haploid cells, one with mutant 3 and the other with mutant 2, what would be its phenotype?

The results confirmed their hypothesis: each mutant strain was indeed missing a single active enzyme in the pathway.

The gene–enzyme connection had been proposed 40 years earlier by the Scottish physician Archibald Garrod, who studied the inherited human disease alkaptonuria. In 1908, Garrod linked the biochemical phenotype of alkaptonuria to a missing enzyme, and thus to an abnormal gene. Today we know of hundreds of examples of such hereditary diseases.

One gene determines one polypeptide

The gene–enzyme relationship has undergone several modifications in light of our current knowledge of molecular biology. Many proteins, including many enzymes, are composed of more than one polypeptide chain, or subunit (that is, they have a quaternary structure; see Section 3.2). Look at the illustration of hemoglobin in Figure 3.9: this protein has four polypeptides, two of each of type of chain. In the case of hemoglobin, each polypeptide chain is specified by its own separate gene. Thus it is more correct to speak of a **one-gene, one-polypeptide relationship**.

In other words, *the function of a gene is to control the production of a single, specific polypeptide.* This statement remains true despite the fact that we know now of genes that code for forms of RNA that are not translated into polypeptides, and have discovered still other gene sequences that do not themselves produce physical polypeptides but instead are involved in controlling which *other* DNA sequences are expressed (i.e., produce polypeptides).

12.1 RECAP

Beadle and Tatum's studies of mutations in bread molds led to our understanding of the one-gene, one-polypeptide relationship: the function of a gene is to code for a specific polypeptide.

- What is a model organism, and why was *Neurospora* a good model for studying biochemical genetics? See p. 257

- How were Beadle and Tatum's experiments on *Neurospora* set up to determine, on the basis of phenotypes of mutant strains, the order of a biochemical pathway? See p. 258 and Figure 12.1

- Do you understand the distinction between the phrases "one gene, one protein" and "one gene, one polypeptide"?

EXPERIMENT

HYPOTHESIS: Genes determine enzymes in a biochemical pathway.

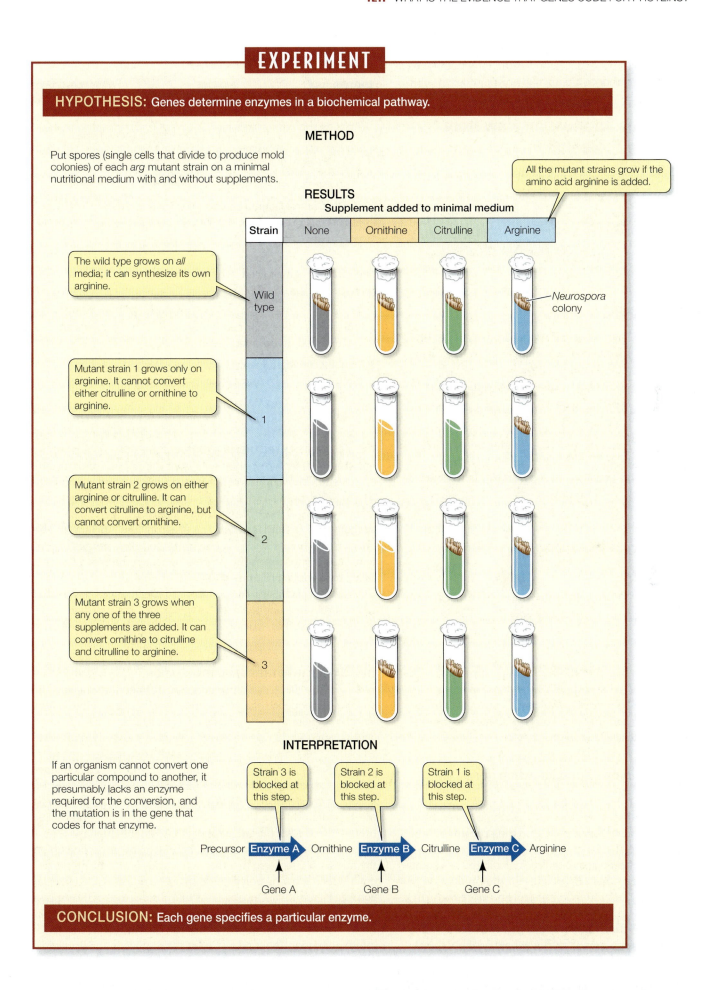

METHOD

Put spores (single cells that divide to produce mold colonies) of each *arg* mutant strain on a minimal nutritional medium with and without supplements.

All the mutant strains grow if the amino acid arginine is added.

RESULTS

Supplement added to minimal medium

Strain	None	Ornithine	Citrulline	Arginine

The wild type grows on *all* media; it can synthesize its own arginine.

Wild type — *Neurospora* colony

Mutant strain 1 grows only on arginine. It cannot convert either citrulline or ornithine to arginine.

1

Mutant strain 2 grows on either arginine or citrulline. It can convert citrulline to arginine, but cannot convert ornithine.

2

Mutant strain 3 grows when any one of the three supplements are added. It can convert ornithine to citrulline and citrulline to arginine.

3

INTERPRETATION

If an organism cannot convert one particular compound to another, it presumably lacks an enzyme required for the conversion, and the mutation is in the gene that codes for that enzyme.

Strain 3 is blocked at this step.

Strain 2 is blocked at this step.

Strain 1 is blocked at this step.

Precursor → **Enzyme A** → Ornithine → **Enzyme B** → Citrulline → **Enzyme C** → Arginine

Gene A   Gene B   Gene C

CONCLUSION: Each gene specifies a particular enzyme.

Now that we have established the one-gene, one-polypeptide relationship, how does it work? That is, how is the information encoded in DNA used to produce a particular polypeptide?

12.2 How Does Information Flow from Genes to Proteins?

The expression of a gene to form a polypeptide occurs in two major steps:

■ *Transcription* copies the information of a DNA sequence (a gene) into corresponding information in an RNA sequence.

■ *Translation* converts this RNA sequence into the amino acid sequence of a polypeptide.

RNA differs from DNA

RNA is a key intermediary between DNA and polypeptide. **RNA** (**ribonucleic acid**) is a polynucleotide similar to DNA (see Figure 3.24), but it differs from DNA in three ways:

■ RNA generally consists of only one polynucleotide strand.

■ The sugar molecule found in RNA is ribose, rather than the deoxyribose found in DNA.

■ Although three of the nitrogenous bases (adenine, guanine, and cytosine) in RNA are identical to those in DNA, the fourth base in RNA is **uracil** (**U**), which is similar to thymine but lacks the methyl (—CH$_3$) group.

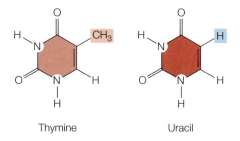

Thymine   Uracil

The bases in RNA can pair with those in a single strand of DNA. This pairing obeys the same complementary base-pairing rules as in DNA, except that *adenine pairs with uracil* instead of thymine. Single-stranded RNA can fold into complex shapes by internal base pairing, as we will see later in this chapter.

Information flows in one direction when genes are expressed

Soon after he and James Watson proposed their three-dimensional structure for DNA, Francis Crick pondered the problem of how DNA is functionally related to proteins. This led him to propose what he called the **central dogma** of molecular biology. The central dogma, simply stated, is that DNA codes for the production of RNA, RNA codes for the production of protein (more correctly, polypeptide), and protein does not code for the production of protein, RNA, or DNA (**Figure 12.2**). In Crick's words, "once 'information' has passed into protein it cannot get out again."

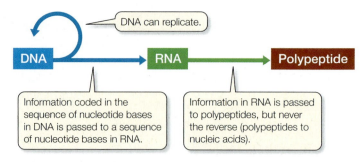

12.2 The Central Dogma   Information flows from DNA to RNA to polypeptide, as indicated by the arrows.

The central dogma raised two questions:

■ How does genetic information get from the nucleus to the cytoplasm? (As Section 4.3 explains, most of the DNA of a eukaryotic cell is confined to the nucleus, but proteins are synthesized in the cytoplasm.)

■ What is the relationship between a specific nucleotide sequence in DNA and a specific amino acid sequence in a protein?

To answer these questions, Crick proposed two hypotheses.

THE MESSENGER HYPOTHESIS AND TRANSCRIPTION   To answer the question of how information gets from the nucleus and into the cytoplasm, Crick and his colleagues proposed the *messenger hypothesis*. They proposed that an RNA molecule forms as a complementary copy of one DNA strand of a particular gene. This **messenger RNA**, or **mRNA**, then travels from the nucleus to the cytoplasm, where it serves as a template for the synthesis of proteins at the ribosomes. The process by which this RNA forms is called **transcription** (**Figure 12.3**).

Crick's hypothesis has been tested repeatedly for genes that code for proteins, and the result is always the same: each gene sequence in DNA that codes for a protein is expressed as a sequence in mRNA.

THE ADAPTER HYPOTHESIS AND TRANSLATION   To answer the question of how a DNA sequence gets transformed into the specific amino acid sequence of a polypeptide, Crick proposed the *adapter hypothesis*: there must be an *adapter molecule* that can both bind a specific amino acid and recognize a sequence of nucleotides. He envisioned such adapters as molecules with two regions, one serving the binding function and the other serving the recognition function. In due course, such adapter molecules were found: they are known as **transfer RNA**, or **tRNA**. Because they recognize the genetic message of mRNA and simultaneously carry specific amino acids, tRNAs can *translate* the language of DNA into the language of proteins. The tRNA adapters, carrying bound amino acids, line up on the mRNA sequence so that the amino acids are in the proper sequence for a growing polypeptide chain—a process called **translation** (see Figure 12.3). Once again, actual observations of the expression of thousands of genes have confirmed the hypothesis that tRNA acts as the intermediary between the nucleotide sequence information in mRNA and the amino acid sequence in a protein.

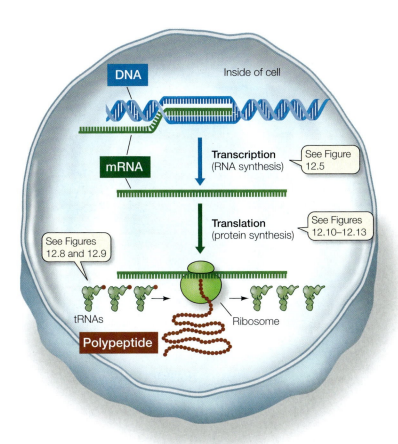

RNA-to-RNA. Instead, after infecting a host cell, they make a DNA copy of their genome and use it to make more RNA. This RNA is then used both as a template for making more copies of the viral genome and as mRNA to produce viral proteins.

Synthesis of DNA from RNA is called **reverse transcription**, and not surprisingly, such viruses are called **retroviruses**.

12.2 RECAP

The central dogma of molecular biology states that DNA codes for the production of RNA and RNA codes for the production of protein (polypeptides). Proteins do *not* code for the production of protein, RNA, or DNA. Transcription is the process that copies a DNA sequence into mRNA. Translation is the process by which this information is converted into protein. Transfer RNA recognizes the genetic information in messenger RNA and brings the appropriate amino acid into position in a growing polypeptide chain.

- Do you understand the central dogma of molecular biology? See p. 260 and Figure 12.2

- Can you describe the roles of mRNA and tRNA in gene expression? See p. 260 and Figure 12.3

12.3 From Gene to Protein This diagram summarizes the processes of gene expression in prokaryotes. In eukaryotes, the processes are somewhat more complex.

Summarizing the main features of the central dogma, the messenger hypothesis, and the adapter hypothesis, we may say that a given gene is transcribed to produce a messenger RNA (mRNA) molecule complementary to one of the DNA strands, and that transfer RNA (tRNA) molecules translate the sequence of bases in the mRNA into the appropriate sequence of linked amino acids during protein synthesis.

RNA viruses are exceptions to the central dogma

Certain viruses present exceptions to the central dogma. As we saw in Section 11.1, *viruses* are acellular infectious particles that reproduce inside cells. Many viruses, such as the tobacco mosaic virus, influenza viruses, and poliovirus, have RNA rather than DNA as their genetic material. With its nucleotide sequence, RNA could potentially act as an information carrier and be expressed as protein. But if RNA is usually single-stranded, how does it replicate? The viruses generally solve this problem by transcribing from RNA to RNA, making an RNA strand that is complementary to their genome. This "opposite" strand is then used to make multiple copies of the viral genome by transcription:

Human immunodeficiency viruses (HIV) and certain rare tumor viruses also have RNA as their genome, but do not replicate it as

Let's look at the physical processes underlying the central dogma in more detail. We'll begin by describing how the information in DNA is transcribed to produce RNA.

12.3 How Is the Information Content in DNA Transcribed to Produce RNA?

In normal prokaryotic and eukaryotic cells, RNA synthesis is directed by DNA. Transcription—the formation of a specific RNA from a specific DNA—requires several components:

- A DNA template for complementary base pairing
- The appropriate ribonucleoside triphosphates (ATP, GTP, CTP, and UTP) to act as substrates
- An *RNA polymerase* enzyme

Within each gene, only one of the two strands of DNA—the **template strand**—is transcribed. The other, complementary DNA strand, referred to as the *non-template strand*, remains untranscribed. For different genes in the same DNA molecule, different strands may be transcribed. That is, the strand that is the non-template strand in one gene may be the template strand in another.

Not only mRNA is produced by transcription. The same process is responsible for the synthesis of tRNA and ribosomal RNA (rRNA), whose important roles in protein synthesis will be described below. Like polypeptides, these RNAs are encoded by specific genes.

RNA polymerases share common features

RNA polymerases from both prokaryotes and eukaryotes catalyze the synthesis of RNA from template DNA. There is only one RNA polymerase in bacteria, while there are three in eukaryotes. Yet all share a common structure that resembles a crab claw (**Figure 12.4**). Catalysis occurs in several steps:

- The enzyme recognizes certain bases within the DNA double helix and binds to them.

- Once the template DNA has bound to the enzyme, the "pincers" close, keeping DNA in a double-stranded form called a *closed complex*.

- A conformational change in the RNA polymerase occurs, denaturing a short (10 base pairs) stretch of DNA and forming an *open complex*.

- The open complex makes the unpaired bases within DNA available to pair with ribonucleotides, and RNA synthesis begins.

Like DNA polymerases, RNA polymerases are *processive*; that is, a single enzyme–template binding event results in the polymerization of hundreds of RNA bases. Unlike DNA polymerases, RNA polymerases do not require a primer.

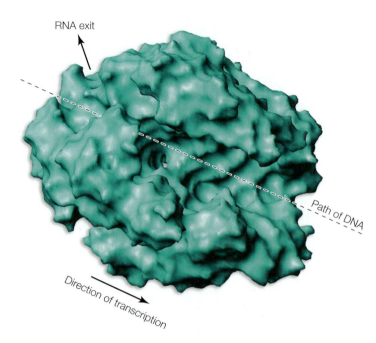

RNA exit

Path of DNA

Direction of transcription

12.4 RNA Polymerase   This enzyme from yeast is similar to most other RNA polymerases.

Transcription occurs in three steps

Transcription can be divided into three distinct processes: *initiation*, *elongation*, and *termination*.

INITIATION   Initiation, which begins transcription, requires a **promoter**, a special sequence of DNA to which RNA polymerase binds very tightly (**Figure 12.5A**). There is at least one promoter for each gene (or, in prokaryotes, each set of genes). Promoters are important control sequences that "tell" the RNA polymerase three things:

- Where to start transcription

- Which strand of DNA to transcribe

- The direction to take from the start

A promoter, which is a specific sequence in the DNA that reads in a particular direction, orients the RNA polymerase and thus "aims" it at the correct strand to use as a template. Promoters function somewhat like the punctuation marks that determine how a sequence of words is to be read as a sentence. Part of each promoter is the **initiation site**, where transcription begins. Groups of nucleotides lying "upstream" from the initiation site (5′ on the non-template strand and 3′ on the template strand) help the RNA polymerase bind.

Although every gene has a promoter, not all promoters are identical. Some promoters are more effective at transcription initiation than others. Furthermore, there are differences between transcription initiation in prokaryotes and in eukaryotes (which will be explored in Chapters 13 and 14).

ELONGATION   Once RNA polymerase has bound to the promoter, it begins the process of **elongation** (**Figure 12.5B**). RNA polymerase unwinds the DNA about 10 base pairs at a time and reads the tem-plate strand in the 3′-to-5′ direction. Like DNA polymerase, RNA polymerase adds new nucleotides to the 3′ end of the growing strand, but does not require a primer to get this process started. The new RNA elongates from the first base, which forms its 5′ end, to its 3′ end. The RNA transcript is thus antiparallel to the DNA template strand.

Unlike DNA polymerases, RNA polymerases do not proofread and correct their work (see Section 11.4). Transcription errors occur at a rate of one for every 10^4 to 10^5 bases. Because many copies of RNA are made, however, and because they often have only a relatively short life span, these errors are not as potentially harmful as mutations in DNA.

TERMINATION   What tells RNA polymerase to stop adding nucleotides to a growing RNA transcript? Just as initiation sites in the DNA template strand specify the starting point for transcription, particular base sequences specify its **termination** (**Figure 12.5C**). The mechanisms of termination are complex and of more than one kind. For some genes, the newly formed transcript falls away from the DNA template and the RNA polymerase. For others, a helper protein pulls the transcript away.

In prokaryotes, which have no nuclear envelope and can have ribosomes located near the chromosome, translation often begins near the 5′ end of the mRNA before transcription of the mRNA molecule is complete. In eukaryotes, the situation is more complicated. First, there is a spatial separation of transcription (in the nucleus) and translation (in the cytoplasm). Second, the first product of transcription is a pre-mRNA that is longer than the final mRNA and must undergo considerable processing before it can be translated. The advantages of this processing and its mechanisms will be discussed in Section 14.3.

The information for protein synthesis lies in the genetic code

The **genetic code** relates genes (DNA) to mRNA and mRNA to the amino acids that make up proteins. The genetic code specifies which amino acids will be used to build a protein. You can think of the genetic information in an mRNA molecule as a series of sequential, nonoverlapping three-letter "words." Each sequence of

12.5 DNA Is Transcribed to Form RNA DNA is partially unwound by RNA polymerase to serve as a template for RNA synthesis. The RNA transcript is formed and then peels away, allowing the DNA that has already been transcribed to rewind into a double helix. Three distinct processes—initiation, elongation, and termination—constitute DNA transcription. RNA polymerase is much larger in reality than indicated here, covering about 50 base pairs.

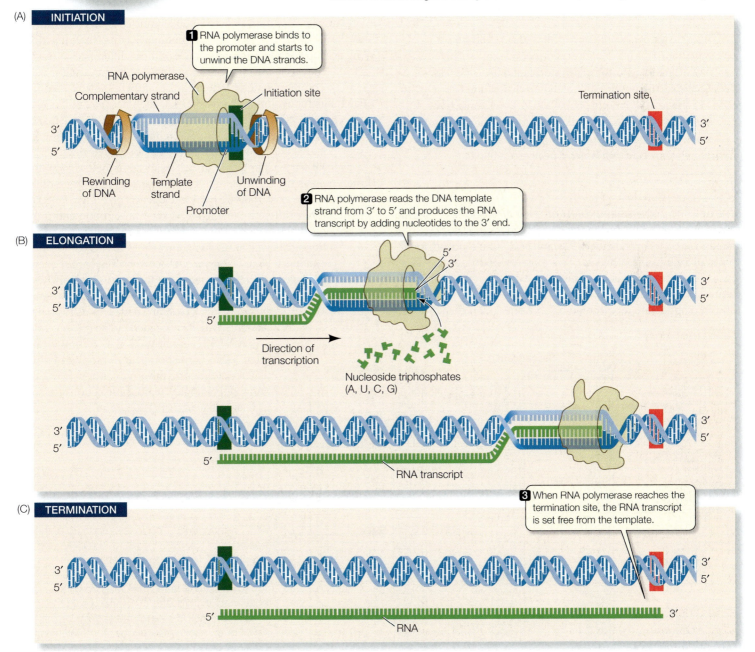

(A) **INITIATION**

1 RNA polymerase binds to the promoter and starts to unwind the DNA strands.

RNA polymerase

Complementary strand

Initiation site

Termination site

3′
5′

3′
5′

Rewinding of DNA

Template strand

Unwinding of DNA

Promoter

2 RNA polymerase reads the DNA template strand from 3′ to 5′ and produces the RNA transcript by adding nucleotides to the 3′ end.

(B) **ELONGATION**

3′
5′

5′
3′

5′

Direction of transcription

Nucleoside triphosphates (A, U, C, G)

3′
5′

5′

RNA transcript

3′
5′

3 When RNA polymerase reaches the termination site, the RNA transcript is set free from the template.

(C) **TERMINATION**

3′
5′

3′
5′

5′

3′

RNA

three nucleotide bases (the three "letters") along the mRNA polynucleotide chain specifies a particular amino acid. Each three-letter "word" is called a **codon**. Each codon is complementary to the corresponding triplet of bases in the DNA molecule from which it was transcribed. Thus the genetic code is the means of relating codons to their specific amino acids.

The complete genetic code is shown in **Figure 12.6**. Notice that there are many more codons than there are different amino acids in proteins. Combinations of the four available "letters" (the bases) give 64 (4^3) different three-letter codons, yet these codons determine only 20 amino acids. AUG, which codes for methionine, is also the **start codon**, the initiation signal for translation. Three of the codons (UAA, UAG, UGA) are **stop codons**, or termination signals for translation; when the translation machinery reaches one of these codons, translation stops, and the polypeptide is released from the translation complex.

A severe anemic condition, α-thalassemia, results from a mutation in the gene for the α-polypeptide chain of hemoglobin. In this gene, the mRNA stop codon UAA normally occurs at base position 142, and the polypeptide chain is 141 amino acids long. In people with α-thalassemia, however, position 142 is mutated to GAA (glutamine). The next stop codon doesn't occur until position 173, resulting in a protein molecule with larger, defective α subunits.

THE GENETIC CODE IS REDUNDANT BUT NOT AMBIGUOUS After the start and stop codons, the remaining 60 codons are far more than enough to code for the other 19 amino acids—and indeed, for almost all amino acids, there is more than one codon. Thus we say that the genetic code is *redundant*. For example, leucine is represented by six different codons (see Figure 12.6). Only methionine and tryptophan are represented by only one codon each.

The term *redundant* should not be confused with *ambiguous*. If the code were ambiguous, a single codon could specify either of two (or more) different amino acids, and there would be doubt about which amino acid should be incorporated into a growing polypeptide chain. Redundancy in the code simply means that there is more than one clear way to say, "Put leucine here." The genetic code is *not* ambiguous: a given amino acid may be encoded by more than one codon, but a codon can code for only one amino acid.

THE GENETIC CODE IS (NEARLY) UNIVERSAL Over 40 years of experiments on thousands of organisms from all three domains reveal that the genetic code is nearly *universal*, applying to all the species on our planet. That is, for almost every species, the codons that specify the amino acids are the same. Thus the code must be an ancient one that has been maintained intact throughout the evolution of living organisms. Exceptions are known: within mitochondria and chloroplasts, the code differs slightly from that in prokaryotes and in the nuclei of eukaryotic cells; in one group of protists, UAA and UAG code for glutamine rather than functioning as stop codons. The significance of these differences is not yet clear. What is clear is that the exceptions are few and slight.

The common genetic code means that there is also a common language for evolution. As natural selection has resulted in gradual changes in the genomes of different types of organisms, the raw material of genetic variation has remained the same. The common code also has profound implications for genetic engineering, as we will see in Chapter 16, since it means that the code for a human gene is the same as that for a bacterial gene. Where differences in code exist, they are more like dialects of a single language than entirely different languages. So the transcription and translation machinery of a bacterium could theoretically utilize genes from a human as well as its own genes.

The codons in Figure 12.6 are *mRNA codons*. The base sequence of the DNA strand that is transcribed to produce the mRNA is complementary and antiparallel to these codons. Thus, for example, 3'-AAA-5' in the template DNA strand corresponds to phenylalanine (which is encoded by the mRNA codon 5'-UUU-3'); similarly, 3'-ACC-5' in the template DNA corresponds to tryptophan (which is encoded by the mRNA codon 5'-UGG-3'). How did biologists learn these codons for specific amino acids?

12.6 The Genetic Code Genetic information is encoded in mRNA in three-letter units—codons—made up of the bases uracil (U), cytosine (C), adenine (A), and guanine (G). To decode a codon, find its first letter in the left column, then read across the top to its second letter, then read down the right column to its third letter. The amino acid the codon specifies is given in the corresponding row. For example, AUG codes for methionine, and GUA codes for valine.

EXPERIMENT

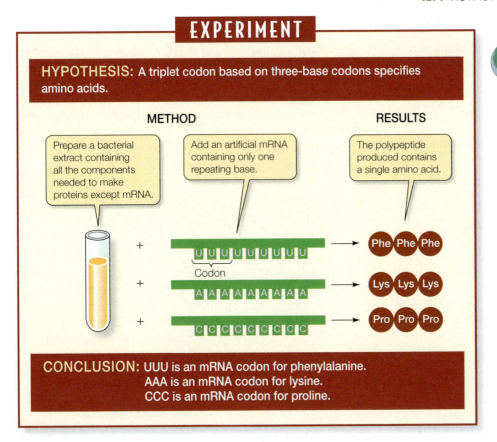

HYPOTHESIS: A triplet codon based on three-base codons specifies amino acids.

METHOD

Prepare a bacterial extract containing all the components needed to make proteins except mRNA.

Add an artificial mRNA containing only one repeating base.

RESULTS

The polypeptide produced contains a single amino acid.

U U U U U U U U U U
Codon

→ Phe Phe Phe

A A A A A A A A A A

→ Lys Lys Lys

C C C C C C C C C C

→ Pro Pro Pro

CONCLUSION: UUU is an mRNA codon for phenylalanine.
AAA is an mRNA codon for lysine.
CCC is an mRNA codon for proline.

12.7 Deciphering the Genetic Code Nirenberg and Matthaei used a test-tube protein synthesis system to determine the amino acids specified by synthetic mRNAs of known codon composition. FURTHER RESEARCH: What would be the result if the artificial mRNA were poly G?

Biologists used artificial messengers to decipher the genetic code

Molecular biologists "broke" the genetic code in the early 1960s. The problem they addressed was perplexing: how could more than 20 "code words" be written with an "alphabet" consisting of only four "letters"? In other words, how could four bases (A, U, G, and C) code for 20 different amino acids?

A triplet code, based on three-letter codons, was considered likely. Since there are only four letters (A, G, C, U), a one-letter code clearly could not unambiguously encode 20 amino acids; it could encode only four of them. A two-letter code could unambiguously code for only $4 \times 4 = 16$ codons—still not enough. But a triplet code could account for up to $4 \times 4 \times 4 = 64$ codons, more than enough to encode the 20 amino acids.

Marshall W. Nirenberg and J. H. Matthaei, at the U.S. National Institutes of Health, made the first decoding breakthrough in 1961 when they realized that they could use a simple artificial polynucleotide instead of a complex natural mRNA as a messenger. They could then identify the polypeptide that the artificial messenger encoded. They prepared an artificial mRNA in which all the bases were uracil (this artificial mRNA was aptly named *poly U*). When poly U was added to a test tube containing all the ingredients necessary for bacterial protein synthesis (ribosomes, all the amino acids, activating enzymes, tRNAs, and other factors), a polypeptide formed. This polypeptide chain was composed of only one kind of amino acid: phenylalanine (Phe). Poly U coded for poly Phe! Accordingly, UUU appeared to be the mRNA code word—the codon—for phenylalanine. Following up on this success, Nirenberg and Matthaei soon showed that CCC codes for proline and AAA for lysine (**Figure 12.7**). (Poly G presented some chemical problems and

was not tested initially.) UUU, CCC, and AAA were three of the easiest codons; different approaches were required to work out the rest.

Other scientists later found that simple artificial mRNAs only three nucleotides long—each amounting to a codon—could bind to a ribosome, and that the resulting complex could then cause the binding of the corresponding tRNA with its specific amino acid. Thus, for example, simple UUU caused the tRNA carrying phenylalanine to bind to the ribosome. After this discovery, complete deciphering of the genetic code was relatively simple. To discover which amino acid a codon represents, Nirenberg repeated his experiment using a sample of artificial mRNA for that codon and observed which amino acid became bound to it.

12.3 RECAP

Transcription, which is catalyzed by an RNA polymerase, proceeds in three steps: initiation, elongation, and termination. The genetic code relates the information in mRNA (as a linear sequence of codons) to protein (a linear sequence of amino acids).

- Describe the steps of gene transcription that produce mRNA. See p. 262 and Figure 12.5

- How do RNA polymerases work? See p. 262

- Can you explain how the genetic code was elucidated? See pp. 264–265 and Figure 12.7

We now turn to the second step in gene expression, the translation of the nucleotide sequence in mRNA into an amino acid sequence in a polypeptide chain.

12.4 How Is RNA Translated into Proteins?

As Crick's adapter hypothesis proposed, the translation of mRNA into proteins requires a molecule that links the information contained in mRNA codons with specific amino acids in proteins. That function is performed by tRNA. Two key events must take place to ensure that the protein made is the one specified by mRNA:

■ tRNA must read mRNA codons correctly.

■ tRNA must deliver the amino acids that correspond to the mRNA codons it has read.

Once the tRNAs "decode" the mRNA and deliver the appropriate amino acids, components of the ribosome catalyze the formation of peptide bonds between amino acids. We now turn to these two steps.

Transfer RNAs carry specific amino acids and bind to specific codons

The codon in mRNA and the amino acid in a protein are related by way of an *adapter*—a specific tRNA with an attached amino acid. For each of the 20 amino acids, there is at least one specific type ("species") of tRNA molecule. The tRNA molecule has three functions:

■ It carries (is "charged with") an amino acid.

■ It associates with mRNA molecules.

■ It interacts with ribosomes.

Its molecular structure relates clearly to all of these functions. A tRNA molecule has about 75 to 80 nucleotides. It has a *conformation* (a three-dimensional shape) that is maintained by complementary base pairing (hydrogen bonding) within its own sequence (**Figure 12.8**).

The conformation of a tRNA molecule is exquisitely suited for its interaction with specific binding sites on ribosomes. At the 3′ end of every tRNA molecule is its *amino acid attachment site*: a site to which its specific amino acid binds covalently. At about the midpoint of the tRNA sequence is a group of three bases, called the

anticodon, that constitutes the site of complementary base pairing (via hydrogen bonding) with mRNA. Each tRNA species has a unique anticodon, which is complementary to the mRNA codon for that tRNA's amino acid. At contact, the codon and the anticodon are antiparallel to each other. As an example of this process, consider the amino acid arginine:

■ The DNA sequence that codes for arginine is 3′-GCC-5′, which is transcribed, by complementary base pairing, to produce the mRNA codon 5′-CGG-3′.

■ That mRNA codon binds by complementary base pairing to a tRNA with the anticodon 3′-GCC-5′, which is charged with arginine.

Recall that 61 different codons encode the 20 amino acids in proteins (see Figure 12.6). Does this mean that the cell must produce 61 different tRNA species, each with a different anticodon? No. The cell gets by with about two-thirds that number of tRNA species because the specificity for the base at the 3′ end of the codon (and the 5′ end of the anticodon) is not always strictly observed. This phenomenon, called *wobble*, allows the alanine codons GCA, GCC, and GCU, for example, all to be recognized by the same tRNA. Wobble is allowed in some matches, but not in others; of most importance, it does not allow the genetic code to be ambiguous.

It took 3 years for Cornell University chemist Robert Holley and his team to sequence the 80 nucleotides of alanine-tRNA from yeast, an effort for which Holley won the 1968 Nobel prize in medicine. With today's technology, sequencing a nucleic acid with 80 nucleotides takes a few seconds.

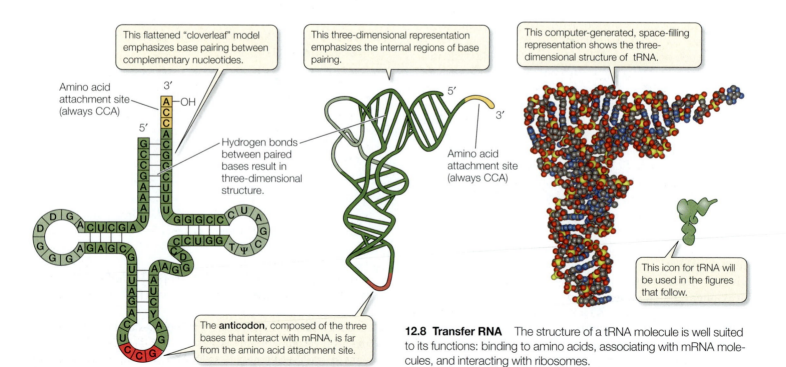

12.8 Transfer RNA The structure of a tRNA molecule is well suited to its functions: binding to amino acids, associating with mRNA molecules, and interacting with ribosomes.

(A)

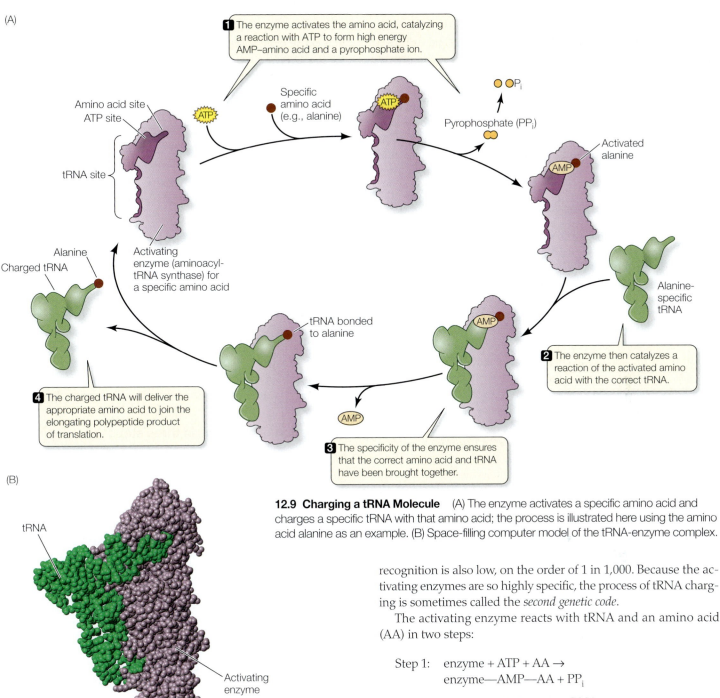

1 The enzyme activates the amino acid, catalyzing a reaction with ATP to form high energy AMP–amino acid and a pyrophosphate ion.

Amino acid site
ATP site
tRNA site

Specific amino acid (e.g., alanine)

Pyrophosphate (PP$_i$)

Activated alanine

Alanine-specific tRNA

Alanine
Charged tRNA

Activating enzyme (aminoacyl-tRNA synthase) for a specific amino acid

tRNA bonded to alanine

AMP

2 The enzyme then catalyzes a reaction of the activated amino acid with the correct tRNA.

4 The charged tRNA will deliver the appropriate amino acid to join the elongating polypeptide product of translation.

3 The specificity of the enzyme ensures that the correct amino acid and tRNA have been brought together.

(B)

tRNA

Activating enzyme

12.9 Charging a tRNA Molecule (A) The enzyme activates a specific amino acid and charges a specific tRNA with that amino acid; the process is illustrated here using the amino acid alanine as an example. (B) Space-filling computer model of the tRNA-enzyme complex.

Activating enzymes link the right tRNAs and amino acids

The charging of each tRNA with its correct amino acid is achieved by a family of activating enzymes, known more formally as *amino-acyl-tRNA synthetases* (**Figure 12.9**). Each activating enzyme is specific for one amino acid and for its corresponding tRNA. The enzyme has a three-part active site that recognizes three smaller molecules: a specific amino acid, ATP, and a specific tRNA. Since tRNA is large and has a complex three-dimensional structure, the activating enzyme's recognition of tRNA is quite specific and has a very low error rate. Remarkably, the error rate for amino acid

recognition is also low, on the order of 1 in 1,000. Because the activating enzymes are so highly specific, the process of tRNA charging is sometimes called the *second genetic code*.

The activating enzyme reacts with tRNA and an amino acid (AA) in two steps:

Step 1:   enzyme + ATP + AA →
           enzyme—AMP—AA + PP$_i$

Step 2:   enzyme—AMP—AA + tRNA →
           enzyme + AMP + tRNA—AA

The amino acid is attached to the 3′ end of the tRNA (to a free OH group on the ribose) with an energy-rich bond, forming charged tRNA. This bond will provide the energy for the synthesis of the peptide bond that will join adjacent amino acids.

A clever experiment by Seymour Benzer and his colleagues at Purdue University demonstrated the importance of the specificity of the attachment of tRNA to its amino acid. In their laboratory, the amino acid cysteine, already properly attached to its tRNA, was chemically modified to become a different amino acid, alanine. Which component—the amino acid or the tRNA—would be recognized when this hybrid charged tRNA was put into a protein-synthesizing system? The answer was: the tRNA. Everywhere in the synthesized protein where cysteine was supposed to be, alanine appeared instead. The cysteine-specific tRNA had delivered its cargo (alanine) to every mRNA "address" where cysteine was called for. This experiment showed that the protein synthesis machinery recognizes the anticodon of the charged tRNA, not the amino acid attached to it. If activating enzymes in nature did what Benzer did in the laboratory and charged tRNAs with the wrong amino acids, those amino acids would be inserted into proteins at inappropriate places, leading to alterations in protein shape and function.

The ribosome is the workbench for translation

The **ribosome** is the molecular workbench where the task of translation is accomplished. Its structure enables it to hold mRNA and charged tRNAs in the right positions, thus allowing a polypeptide chain to be assembled efficiently. A given ribosome does not specifically produce just one kind of protein. A ribosome can use any mRNA and all species of charged tRNAs, and thus can be used to make many different polypeptide products. The mRNA, as a linear sequence of codons, specifies the polypeptide sequence to be made.

Although ribosomes are small in contrast to other cellular organelles, their mass of several million daltons makes them large in comparison with charged tRNAs. Each ribosome consists of two subunits, a large one and a small one (**Figure 12.10**). In eukaryotes, the large subunit consists of three different molecules of **ribosomal RNA (rRNA)** and about 45 different protein molecules, arranged in a precise pattern. The small subunit consists of one rRNA molecule and 33 different protein molecules. When not active in the translation of mRNA, the ribosomes exist as separated subunits.

The ribosomes of prokaryotes are somewhat smaller than those of eukaryotes, and their ribosomal proteins and RNAs are different. Mitochondria and chloroplasts also contain ribosomes, some of which are similar to those of prokaryotes.

The different proteins and rRNAs in a ribosomal subunit are held together by ionic and hydrophobic forces, not covalent bonds. If these forces are disrupted by a detergent, for example, the proteins and rRNAs separate from one another. When the detergent is removed, the entire complex structure self-assembles. This is like separating the pieces of a jigsaw puzzle and having them fit together again without human hands to guide them!

On the large subunit of the ribosome are three sites to which tRNA can bind (see Figure 12.10). A charged tRNA traverses these three sites in order:

- The *A* (amino acid) *site* is where the charged tRNA anticodon binds to the mRNA codon, thus lining up the correct amino acid to be added to the growing polypeptide chain.

- The *P* (polypeptide) *site* is where the tRNA adds its amino acid to the growing polypeptide chain.

- The *E* (exit) *site* is where the tRNA, having given up its amino acid, resides before being released from the ribosome and going back to the cytosol to pick up another amino acid and begin the process again.

An important role of the ribosome is to make sure that the mRNA–tRNA interactions are accurate; that is, that a charged tRNA with the correct anticodon (e.g., 3′-UAC-5′) binds to the appropriate codon in mRNA (e.g., 5′-AUG-3′). When this occurs, hydrogen bonds form between the base pairs. But these hydrogen bonds are not enough to hold the tRNA in place. The rRNA of the small ribosomal subunit plays a role in validating the three-base-pair match. If hydrogen bonds have not formed between all three base pairs, the tRNA must be the wrong one for that mRNA codon, and that tRNA is ejected from the ribosome.

Translation takes place in three steps

Like transcription, translation occurs in three steps: initiation, elongation, and termination.

INITIATION The translation of mRNA begins with the formation of an **initiation complex**, which consists of a charged tRNA bearing what will be the first amino acid of the polypeptide chain and a small ribosomal subunit, both bound to the mRNA (**Figure 12.11**). The rRNA of the small ribosomal subunit first binds to a complementary ribosome binding site (known as the *Shine–Dalgarno sequence*) on the mRNA. This sequence is "upstream" (toward the 5′ end) of the actual start codon that begins translation.

Recall that the mRNA start codon in the genetic code is AUG (see Figure 12.6). The anticodon of a methionine-charged tRNA binds to this start codon by complementary base pairing to complete the initiation complex. Thus the first amino acid in a polypeptide chain is always methionine. Not all mature proteins have methionine as their N-terminal amino acid, however. In many cases, the initiator methionine is removed by an enzyme after translation.

After the methionine-charged tRNA has bound to the mRNA, the large subunit of the ribosome joins the complex. The methionine-charged tRNA now lies in the P

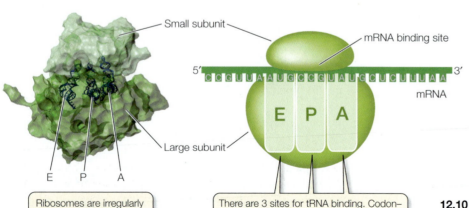

12.10 Ribosome Structure Each ribosome consists of a large and a small subunit. The subunits remain separate when they are not in use for protein synthesis.

Small subunit

mRNA binding site

5′ CCGUUAAUGCCGUAUGCUCUUUAA 3′

mRNA

E P A

Large subunit

E P A

Ribosomes are irregularly shaped and composed of two subunits.

There are 3 sites for tRNA binding. Codon–anticodon interactions between tRNA and mRNA occur only at the P and A sites.

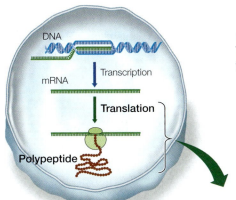

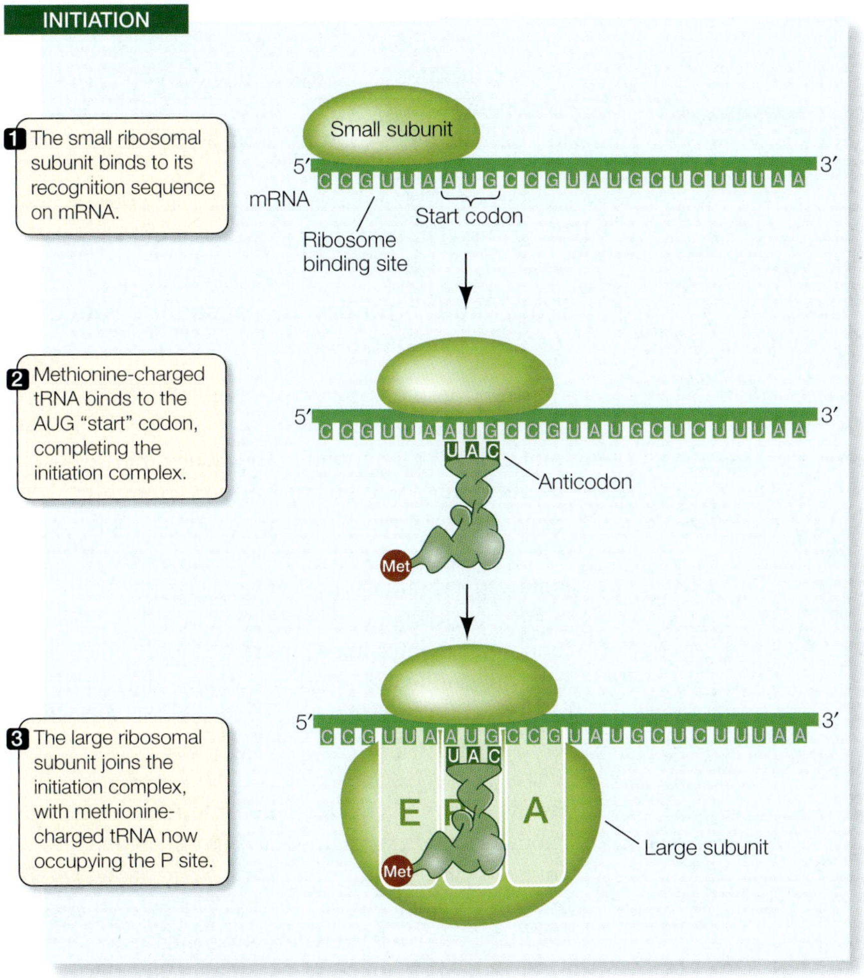

12.11 The Initiation of Translation Translation begins with the formation of an initiation complex.

INITIATION

1 The small ribosomal subunit binds to its recognition sequence on mRNA.

Small subunit

5′ CCGUUAAUGCCGUAUGCUCUUUAA 3′

mRNA

Start codon

Ribosome binding site

2 Methionine-charged tRNA binds to the AUG "start" codon, completing the initiation complex.

5′ CCGUUAAUGCCGUAUGCUCUUUAA 3′

UAC

Anticodon

Met

3 The large ribosomal subunit joins the initiation complex, with methionine-charged tRNA now occupying the P site.

5′ CCGUUAAUGCCGUAUGCUCUUUAA 3′

UAC

E P A

Met

Large subunit

ELONGATION   A charged tRNA whose anticodon is complementary to the second codon of the mRNA now enters the open A site of the large ribosomal subunit (**Figure 12.12**). The large subunit then catalyzes two reactions:

■ It breaks the bond between the tRNA in the P site and its amino acid.

■ It catalyzes the formation of a peptide bond between that amino acid and the one attached to the tRNA in the A site.

Because the large subunit performs these two actions, it is said to have *peptidyl transferase activity*. In this way, methionine (the amino acid in the P site) becomes the N terminus of the new protein. The second amino acid is now bound to methionine, but remains attached to its tRNA at the A site.

How does the large ribosomal subunit catalyze this binding? Harry Noller and his colleagues at the University of California at Santa Cruz found that if they removed almost all the proteins from the large subunit, it still catalyzed peptide bond formation. But if the RNA was destroyed, so was peptidyl transferase activity. Part of the rRNA in the large subunit interacts with the end of the charged tRNA where the amino acid is attached. Thus RNA appears to be the catalyst, probably tRNA. This situation is unusual, because proteins are the usual catalysts in biological systems. The recent purification and crystallization of ribosomes has allowed scientists to examine their structure in detail, and the catalytic role of rRNA in peptidyl transferase activity has been confirmed, supporting the hypothesis that RNA, and catalytic RNA in particular, evolved before DNA (see Section 3.6).

After the first tRNA releases its methionine, it moves to the E site and is then dissociated from the ribosome, returning to the cytosol to become charged with another methionine. The second tRNA, now bearing a *dipeptide* (a two-amino-acid chain), is shifted to the P site as the ribosome moves one codon along the mRNA in the 5′-to-3′ direction.

The elongation process continues, and the polypeptide chain grows, as the steps are repeated:

■ The next charged tRNA enters the open A site, where its anticodon binds with the mRNA codon.

■ Its amino acid forms a peptide bond with the amino acid chain in the P site, so that it picks up the growing polypeptide chain from the tRNA in the P site.

■ The tRNA in the P site is transferred to the E site and then released. The ribosome shifts one codon, so that the entire tRNA–polypeptide complex moves to the newly vacated P site.

All these steps are assisted by proteins called *elongation factors*.

site of the ribosome, and the A site is aligned with the second mRNA codon. These ingredients—mRNA, two ribosomal subunits, and methionine-charged tRNA—are put together properly by a group of proteins called *initiation factors*.

The prokaryotic ribosome is smaller and has a different collection of proteins than the eukaryotic ribosome. Some antibacterial antibiotics work by binding and inhibiting specific ribosomal proteins that are essential to the bacterium but do not exist in the eukaryotic ribosome.

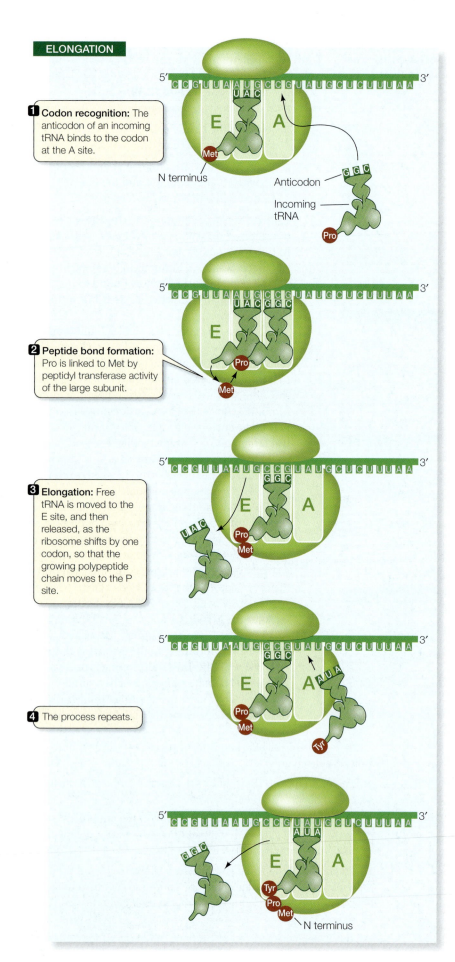

ELONGATION

1 Codon recognition: The anticodon of an incoming tRNA binds to the codon at the A site.

N terminus

Anticodon

Incoming tRNA

2 Peptide bond formation: Pro is linked to Met by peptidyl transferase activity of the large subunit.

3 Elongation: Free tRNA is moved to the E site, and then released, as the ribosome shifts by one codon, so that the growing polypeptide chain moves to the P site.

4 The process repeats.

N terminus

12.12 The Elongation of Translation   The polypeptide chain elongates as the mRNA is translated.

TERMINATION   The elongation cycle ends, and translation is terminated, when a stop codon—UAA, UAG, or UGA—enters the A site (**Figure 12.13**). These codons encode no amino acids, nor do they bind tRNA. Rather, they bind a protein *release factor*, which allows hydrolysis of the bond between the polypeptide chain and the tRNA in the P site.

The newly completed polypeptide thereupon separates from the ribosome. Its C terminus is the last amino acid to join the chain. Its N terminus, at least initially, is methionine, as a consequence of the AUG start codon. In its amino acid sequence, it contains information specifying its conformation, as well as its ultimate cellular destination.

Table 12.1 summarizes the nucleic acid signals for initiation and termination of transcription and translation.

Polysome formation increases the rate of protein synthesis

Several ribosomes can work simultaneously at translating a single mRNA molecule, producing multiple molecules of the protein at the same time. As soon as the first ribosome has moved far enough from the Shine–Dalgarno sequence, a second initiation complex can form, then a third, and so on. An assemblage consisting of a strand of mRNA with its beadlike ribosomes and their growing polypeptide chains is called a **polyribosome**, or **polysome** (**Figure 12.14**). Cells that are actively synthesizing proteins contain large numbers of polysomes and few free ribosomes or ribosomal subunits.

A polysome is like a cafeteria line, in which patrons follow one another, adding items to their trays. At any moment, the person at the start has a little food (a newly initiated protein); the person at the end has a complete meal (a completed protein). However, in the polysome cafeteria, everyone gets the same meal: many copies of the same protein are made from a single mRNA.

TABLE 12.1

Signals that Start and Stop Transcription and Translation

	TRANSCRIPTION	TRANSLATION
Initiation	Promoter sequence in DNA	AUG start codon in mRNA
Termination	Terminator sequence in DNA	UAA, UAG, or UGA stop codon in mRNA

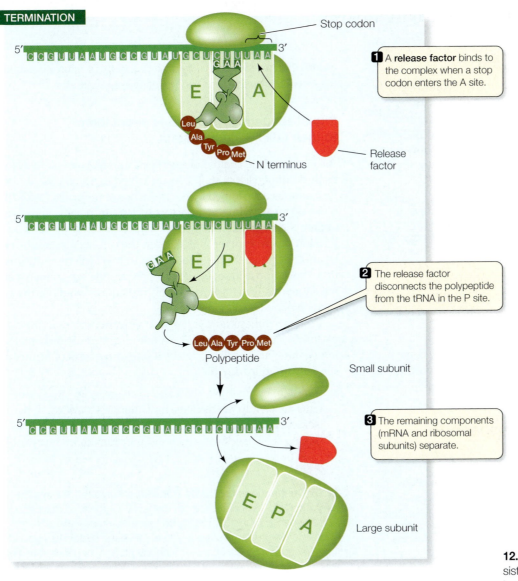

Stop codon

1 A **release factor** binds to the complex when a stop codon enters the A site.

N terminus

Release factor

12.13 The Termination of Translation Translation terminates when the A site of the ribosome encounters a stop codon on the mRNA.

2 The release factor disconnects the polypeptide from the tRNA in the P site.

Polypeptide

Small subunit

3 The remaining components (mRNA and ribosomal subunits) separate.

Large subunit

12.14 A Polysome (A) A polysome consists of multiple ribosomes and their growing polypeptide chains moving in single file along an mRNA molecule. (B) An electron microscopic view of a polysome.

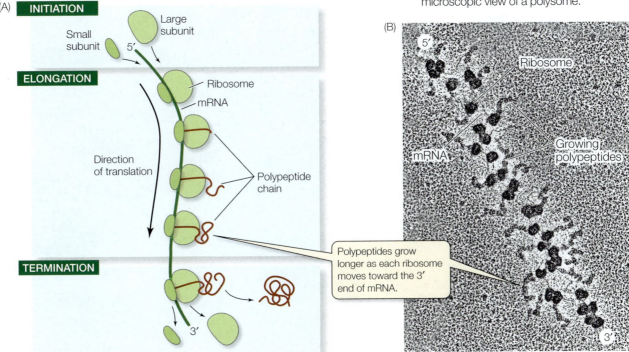

(A) INITIATION

Large subunit

Small subunit

5′

ELONGATION

Ribosome

mRNA

Direction of translation

Polypeptide chain

TERMINATION

3′

Polypeptides grow longer as each ribosome moves toward the 3′ end of mRNA.

(B)

5′

Ribosome

mRNA

Growing polypeptides

3′

A key step in protein synthesis is the attachment of an amino acid to its proper tRNA, which is carried out by an activating enzyme. Translation of the genetic information from mRNA into protein occurs at the ribosome. Multiple ribosomes may act on a single mRNA to make multiple copies of the protein for which it codes.

- Do you understand how an amino acid is attached to a specific tRNA, and why the term "second genetic code" is associated with this process? See p. 267 and Figure 12.9

- Describe the events of initiation, elongation, and termination of translation. See pp. 268-270 and Figures 12.11, 12.12, and 12.13

The polypeptide chain that is released from the ribosome is not necessarily a functional protein. Let's look at some of the posttranslational changes that affect the fate and function of polypeptides.

12.5 What Happens to Polypeptides after Translation?

Especially in eukaryotic cells, the site of a polypeptide's function may be far away from its point of synthesis in the cytoplasm; it may need to be moved into an organelle, or even out of the cell. In addition, polypeptides are often modified by the addition of new chemical groups that have functional significance. In this section we examine these two *posttranslational* aspects of protein synthesis.

Signal sequences in proteins direct them to their cellular destinations

As a polypeptide chain emerges from the ribosome, it folds into its three-dimensional shape. As described in Section 3.2, its conformation is determined by the sequence of the amino acids that make up the protein, and by factors such as the polarity and charge of their R groups. Ultimately, a polypeptide's conformation allows it to interact with other molecules in the cell, such as a substrate or another polypeptide. In addition to this structural information, the amino acid sequence of a polypeptide can contain a **signal sequence**—an "address label" indicating where in the cell the polypeptide belongs.

Protein synthesis always begins on free ribosomes in the cytoplasm. As a polypeptide chain is made, the information contained in its amino acid sequence gives it one of two sets of further instructions (**Figure 12.15**):

- *"Finish translation and be released to an organelle."* Such proteins are sent to the nucleus, mitochondria, plastids, or peroxisomes, depending on the address in their instructions; or, lacking such specific instructions, they remain in the cytosol.

12.15 Destinations for Newly Translated Polypeptides in a Eukaryotic Cell Signal sequences on newly synthesized polypeptides bind to specific receptor proteins on the outer membrane of the organelle to which they are "addressed." Once the protein has bound to it, the receptor forms a channel in the membrane, and the protein enters the organelle.

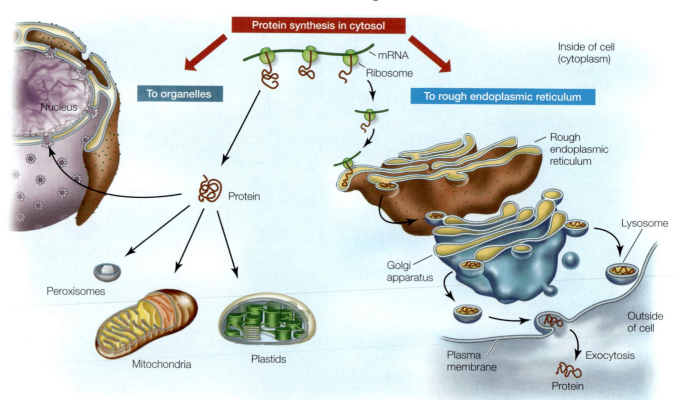

■ *"Stop translation, go to the endoplasmic reticulum, and finish synthesis there."* After protein synthesis is completed, such proteins may be retained in the ER, and sent to the Golgi apparatus. From there, they may be sent to the lysosomes, to the plasma membrane, or, lacking such specific instructions, they may be secreted from the cell via vesicles that fuse with the plasma membrane.

Proteins destined to remain in the cytoplasm by default have no signal.

DESTINATION: NUCLEUS, MITOCHONDRION, OR CHLOROPLAST After translation, some folded polypeptides have a short exposed sequence of amino acids that acts like a postal "zip code," directing them to an organelle. These signal (or *localization*) sequences are either at the N terminus or in the interior of the amino acid chain. For example, the following sequence directs a protein to the nucleus:

—Pro—Pro—Lys—Lys—Lys—Arg—Lys—Val—

This amino acid sequence would occur in the histone proteins associated with nuclear DNA, but not in citric acid cycle enzymes, which are addressed to the mitochondria.

Signal sequences have a conformation that allows them to bind to a specific receptor protein, appropriately called a **docking protein**, on the outer membrane of the appropriate organelle. Once the protein has bound to it, the receptor forms a channel in the membrane, allowing the protein to pass through to its organelle destination. In this process, the protein is usually unfolded by a chaperonin (see Figure 3.12) so that it can pass through the channel, then refolds into its normal conformation.

DESTINATION: ENDOPLASMIC RETICULUM If a specific hydrophobic sequence of 15–30 amino acids occurs at the N terminus of a polypeptide chain, the polypeptide is sent initially to the ER and then to the Golgi, where it can be modified for eventual transport to the lysosomes, the plasma membrane, or out of the cell. In the cytoplasm, before translation is finished and while the polypeptide is still attached to a ribosome, the signal sequence binds to a **signal recognition particle** composed of protein and RNA (**Figure 12.16**). This binding blocks further protein synthesis until the ribosome becomes attached to a specific receptor protein in the membrane of the rough ER. Once again, the receptor protein is converted into a channel, through which the growing polypeptide passes. The elongating polypeptide may be retained in the ER membrane itself, or it may enter the interior space—the lumen—of the ER. In either case, an enzyme in the lumen of the ER removes the signal sequence from the polypeptide chain.

At this point, protein synthesis resumes, and the chain grows longer until its sequence is completed. If the finished protein enters the ER lumen, it can be transported to its appropriate location—to other cellular compartments or to the outside of the cell—via the ER and the Golgi apparatus without mixing with other molecules in the cytoplasm.

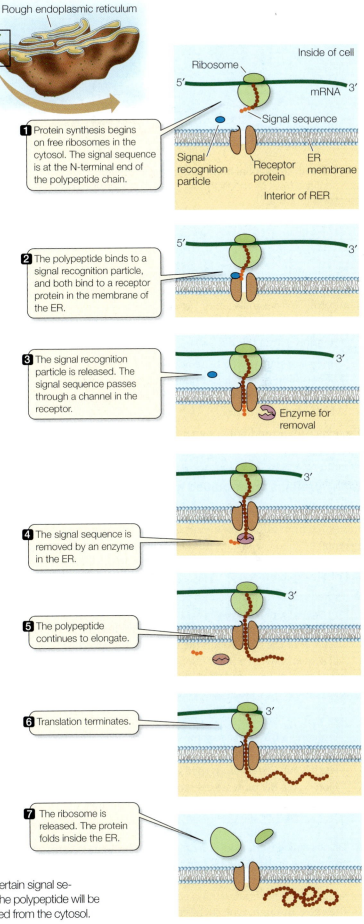

12.16 A Signal Sequence Moves a Polypeptide into the ER When a certain signal sequence of amino acids is present at the beginning of a polypeptide chain, the polypeptide will be taken into the endoplasmic reticulum. The finished protein is thus segregated from the cytosol.

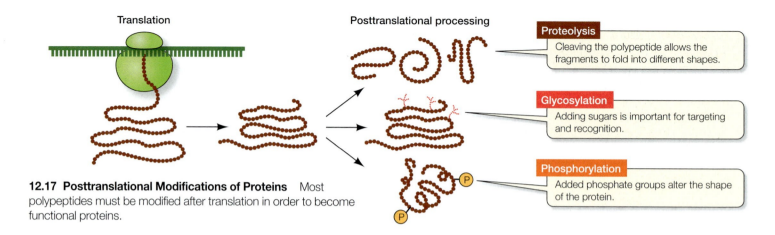

12.17 Posttranslational Modifications of Proteins Most polypeptides must be modified after translation in order to become functional proteins.

Additional signals are needed to further direct the protein (remember that the signal sequence that sent it to the ER has been removed). These signals are of two kinds:

- Some are sequences of amino acids that allow the protein's retention within the ER.

- Others are sugars, which are added in the Golgi apparatus. The resulting *glycoproteins* end up either at the plasma membrane or in a lysosome (or plant vacuole), depending on which sugars are added.

Proteins with no additional signals pass from the ER through the Golgi apparatus and are secreted from the cell.

Many proteins are modified after translation

Most finished proteins are not identical to the polypeptide chains translated from mRNA on the ribosomes. Instead, most polypeptides are modified in any of a number of ways after translation (**Figure 12.17**). These modifications are essential to the final functioning of the protein.

- **Proteolysis** is the cutting of a polypeptide chain. Cleavage of the signal sequence from the growing polypeptide chain in the ER is an example of proteolysis; the protein might move back out of the ER through the membrane channel if the signal sequence were not cut off. Some proteins are actually made from *polyproteins* (long polypeptides) that are cut into final products by enzymes called *proteases*.

Proteases are essential to some viruses, including HIV, because the large viral polyprotein cannot fold properly unless it is cut. Certain drugs used to treat AIDS work by inhibiting the HIV protease, thereby preventing the formation of proteins needed for viral reproduction.

- **Glycosylation** is the addition of sugars to proteins to form glycoproteins. In both the ER and the Golgi apparatus, resident enzymes catalyze the addition of various sugars or short sugar chains to certain amino acid R groups on proteins as they pass through. One such type of "sugar coating" is essen-

tial for addressing proteins to lysosomes, as mentioned above. Other types are important in the conformation and the recognition functions of proteins at the cell surface. Still other attached sugars help to stabilize proteins stored in vacuoles in plant seeds.

- **Phosphorylation**, the addition of phosphate groups to proteins, is catalyzed by protein kinases. The charged phosphate groups change the conformation of a protein, often exposing the active site of an enzyme or a binding site for another protein.

12.5 RECAP

Signal sequences in polypeptides "address" them to their appropriate destinations inside or outside the cell. Many polypeptides are modified after translation.

- Do you understand how signal sequences determine where a protein will go after it is made?
  See pp. 272–273 and Figure 12.16

- Can you explain some ways in which posttranslational modifications alter protein structure and function?
  See p. 274 and Figure 12.17

All of the processes we have just described result in a functional protein only if the amino acid sequence of that protein is correct. If the sequence is not correct, cellular dysfunction may result. Changes in the DNA—mutations—are a major source of errors in amino acid sequences.

12.6 What Are Mutations?

In Chapter 10, we described mutations as inherited changes in genes, and we saw that the new alleles that result may produce altered phenotypes (short pea plants instead of tall, for example). Now that we understand the chemical nature of genes and how they are expressed as phenotypes, we will return to the concept of mutations for a more specific definition.

Errors in DNA replication can occur in any cell undergoing the cell cycle, and these errors are passed on to the daughter cells.

Mutations in multicellular organisms can be divided into two types:

- **Somatic mutations** are those that occur in *somatic* (body) cells. These mutations are passed on to the daughter cells after mitosis, and to the offspring of those cells in turn, but are not passed on to sexually produced offspring. A mutation in a single human skin cell, for example, could result in a patch of skin cells, all with the same mutation, but would not be passed on to a person's children.

- **Germ line mutations** are those that occur in the cells of the *germ line*—the specialized cells that give rise to gametes. A gamete with the mutation passes it on to a new organism at fertilization.

Some mutations cause their phenotypes only under certain *restrictive* conditions. They are not detectable under other, *permissive* conditions. These phenotypes are known as **conditional mutants**. Many conditional mutants are temperature-sensitive; that is, they show the altered phenotype only at a certain temperature (recall the rabbit in Figure 10.16). The mutant allele in such an organism may code for an enzyme with an unstable tertiary structure that is altered at the restrictive temperature.

All mutations are alterations in the nucleotide sequence of DNA. At the molecular level, we can divide mutations into two categories:

- **Point mutations** are mutations of single base pairs and so are limited to single genes: one allele (usually dominant) becomes another allele (usually recessive) because of an alteration (gain, loss, or substitution) of a single nucleotide (which, after DNA replication, becomes a mutant base pair).

- **Chromosomal mutations** are more extensive alterations than point mutations. They may change the position or orientation of a DNA segment without actually removing any genetic information, or they may cause a segment of DNA to be irretrievably lost or duplicated.

Point mutations change single nucleotides

Point mutations result from the addition or subtraction of a nucleotide base, or the substitution of one base for another, in the DNA. Point mutations can be the result of errors in DNA replication that are not corrected in proofreading, or they can be caused by environmental mutagens such as chemicals and radiation.

Point mutations in DNA usually result in changes in mRNA, but changes in mRNA may or may not result in changes in the protein. *Silent mutations* have no effect on the protein. *Missense* and *nonsense* mutations result in changes in the protein, some of them drastic.

SILENT MUTATIONS Because of the redundancy of the genetic code, some base substitutions result in no change in amino acids when the altered mRNA is translated; for this reason, they are called **silent mutations**. For example, there are four mRNA codons that code for proline: CCA, CCC, CCU, and CCG (see Figure 12.6). If the template strand of DNA has the sequence 5'-CGG-3', it will be transcribed as 5'-CCG-3' in mRNA, and proline-charged tRNA will bind to it at the ribosome. If there is a mutation such that

the codon in the template DNA now reads AGG, the mRNA codon will be CCU—and the tRNA that binds it will still carry proline:

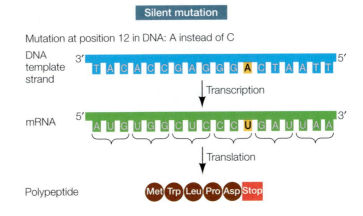

Silent mutations are quite common, and they result in genetic diversity that is not expressed as phenotypic differences.

MISSENSE MUTATIONS In contrast to silent mutations, some base substitutions change the genetic message such that one amino acid substitutes for another in the protein. These changes are called **missense mutations**:

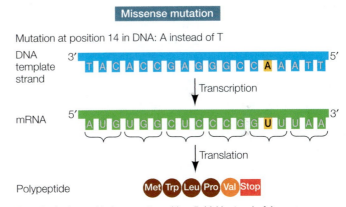

Result: Amino acid change at position 5: Val instead of Asp

A specific example of a missense mutation is the sickle allele for human β-globin. Sickle-cell disease results from a defect in hemoglobin, a protein in human red blood cells that carries oxygen. The sickle allele of the gene that codes for β-globin subunits of hemoglobin differs from the normal allele by one base, and thus codes for a polypeptide that differs by one amino acid from the normal protein. Individuals who are homozygous for this recessive allele have defective, sickle-shaped red blood cells (**Figure 12.18**) that cause abnormalities in blood circulation and lead to serious illness.

A missense mutation may cause a protein not to function, but often its effect is only to reduce the functional efficiency of the protein. Therefore, individuals carrying missense mutations may survive, even though the affected protein is essential to life. Through evolution, some missense mutations can even improve functional efficiency.

NONSENSE MUTATIONS Nonsense mutations, another type of mutation in which one base is substituted for another, are more often disruptive than missense mutations. In a nonsense mutation, the base substitution causes a stop codon, such as UAG, to form in the mRNA product:

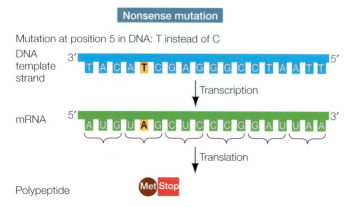

Nonsense mutation

Mutation at position 5 in DNA: T instead of C

DNA template strand

mRNA

Translation

Polypeptide Met Stop

Result: Only one amino acid translated; no protein made

A nonsense mutation results in a shortened protein, since translation does not proceed beyond the point where the mutation occurred. Such short proteins are usually not functional.

FRAME-SHIFT MUTATIONS Not all point mutations are base substitutions. Single base pairs may be inserted into or deleted from DNA. Such mutations are known as **frame-shift mutations** because they interfere with the decoding of the genetic message by throwing it out of register:

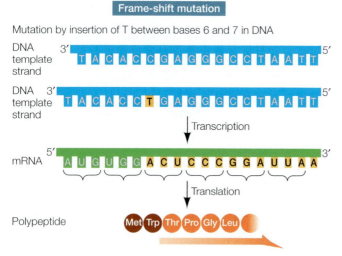

Frame-shift mutation

Mutation by insertion of T between bases 6 and 7 in DNA

DNA template strand

DNA template strand

Transcription

mRNA

Translation

Polypeptide Met Trp Thr Pro Gly Leu

Result: All amino acids changed beyond the insertion

Think again of codons as three-letter words, each corresponding to a particular amino acid. Translation proceeds codon by codon; if a base is added to the mRNA or subtracted from it, translation proceeds perfectly until it comes to the one-base insertion or deletion. From that point on, the three-letter words in the genetic message are one letter out of register. In other words, such mutations shift the "reading frame" of the message. Frame-shift mutations almost always lead to the production of nonfunctional proteins.

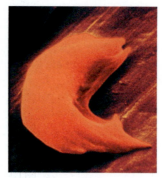

Sickle-cell phenotype

Normal phenotype

12.18 Sickled and Normal Red Blood Cells The misshapen red blood cell on the left is caused by a missense mutation and an incorrect amino acid in one of the two polypeptides of hemoglobin.

Chromosomal mutations are extensive changes in the genetic material

Changes in single nucleotides are not the most dramatic changes that can occur in the genetic material. Whole DNA molecules can break and rejoin, grossly disrupting the sequence of genetic information. There are four types of such chromosomal mutations: *deletions, duplications, inversions,* and *translocations.* These mutations can be caused by severe damage to chromosomes resulting from mutagens or by drastic errors in chromosome replication.

- **Deletions** remove part of the genetic material (**Figure 12.19A**). Like frame-shift point mutations, their consequences can be severe unless they affect unnecessary genes or are masked by the presence, in the same cell, of normal alleles of the deleted genes. It is easy to imagine one mechanism that could produce deletions: a DNA molecule might break at two points, and the two end pieces might rejoin, leaving out the DNA between the breaks.

- **Duplications** can be produced at the same time as deletions (**Figure 12.19B**). Duplication would arise if homologous chromosomes broke at different positions and then reconnected to the wrong partners. One of the two chromosomes produced by this mechanism would lack a segment of DNA (it would have a deletion), and the other would have two copies (a duplication) of the segment that was deleted from the first chromosome.

- **Inversions** can also result from breaking and rejoining of chromosomes. A segment of DNA may be removed and reinserted into the same location in the chromosome, but "flipped" end over end so that it runs in the opposite direction (**Figure 12.19C**). If the break site includes part of a DNA segment that codes for a protein, the resulting protein will be drastically altered and almost certainly nonfunctional.

- **Translocations** result when a segment of DNA breaks off, moves from its chromosome, and is inserted into a different chromosome. Translocations may be reciprocal, as in **Figure 12.19D**, or nonreciprocal, as in the mutation involving duplication and deletion in Figure 12.19B illustrates. Translocations often lead to duplications and deletions and may result in sterility if normal chromosome pairing in meiosis cannot occur.

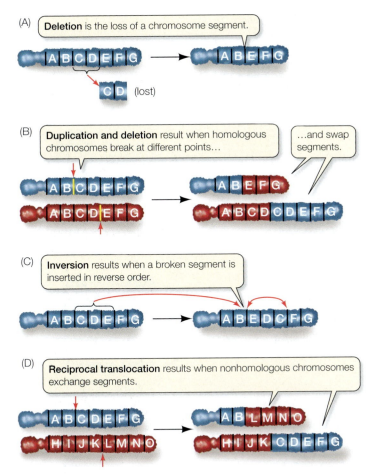

(A) **Deletion** is the loss of a chromosome segment.

C D (lost)

(B) **Duplication and deletion** result when homologous chromosomes break at different points... ...and swap segments.

(C) **Inversion** results when a broken segment is inserted in reverse order.

(D) **Reciprocal translocation** results when nonhomologous chromosomes exchange segments.

12.19 Chromosomal Mutations Chromosomes may break during replication, and parts of chromosomes may then rejoin incorrectly

Mutations can be spontaneous or induced

It is useful to distinguish two types of mutations in terms of their causes:

- **Spontaneous mutations** are permanent changes in the genetic material that occur without any outside influence. In other words, they occur simply because the machinery of the cell is imperfect.
- **Induced mutations** occur when some agent outside the cell— a **mutagen**—causes a permanent change in DNA.

Spontaneous mutations may occur by several mechanisms:

- *The four nucleotide bases of DNA are somewhat unstable.* They can exist in two different forms (called *tautomers*), one of which is common and one rare. When a base temporarily forms its rare tautomer, it can pair with the wrong base. For example, C normally pairs with G, but if C is in its rare tautomer at the time of DNA replication, it pairs with (and DNA polymerase will insert) A. The result is a point mutation: G → A (**Figure 12.20A,C**).
- *Bases may change because of a chemical reaction.* For example, loss of an amino group in cytosine (a reaction called *deamination*) forms uracil. When DNA replicates, instead of a G

opposite what was C, DNA polymerase adds an A (base-pairs with U).

- *DNA polymerase can make errors in replication* (see Section 11.4)—for example, inserting a T opposite a G. Most of these errors are repaired by the proofreading function of the replication complex, but some errors escape detection and become permanent.
- *Meiosis is not perfect.* Nondisjunction—failure of homologous chromosomes to separate during meiosis—can occur, leading to one too many or one too few chromosomes (aneuploidy; see Figure 9.20). Random chromosome breakage and rejoining can produce deletions, duplications, and inversions, or, when involving nonhomologous chromosomes, translocations.

Mutagens can alter DNA by several mechanisms, thus inducing mutations.

- *Some chemicals can alter the nucleotide bases.* For example, nitrous acid (HNO_2) and its relatives can turn cytosine in DNA into uracil by deamination: they convert an amino group on cytosine ($-NH_2$) into a keto group ($-C=O$). This alteration has the same result as a spontaneous deamination: instead of a G, DNA polymerase inserts an A (**Figure 12.20B,C**).
- *Some chemicals add groups to the bases.* For instance, benzpyrene, a component of cigarette smoke, adds a large chemical group to guanine, making it unavailable for base pairing. When DNA polymerase reaches such a modified guanine, it inserts any of the four bases; of course, three-fourths of the time the inserted base will not be cytosine, and a mutation results.
- *Radiation damages the genetic material.* Radiation can damage DNA in two ways. First, ionizing radiation (X-rays) produces highly reactive chemical species called *free radicals*, which can change bases in DNA to unrecognizable (by DNA polymerase) forms. It can also break the sugar–phosphate backbone of DNA, causing chromosomal abnormalities. Second, ultraviolet radiation from the sun (or a tanning lamp) is absorbed by thymine in DNA, causing it to form interbase covalent bonds with adjacent nucleotides. This, too, plays havoc with DNA replication.

Mutations have both costs and benefits. The costs are obvious, since mutations often produce an organism that is less fit for its current environment. Somatic mutations can lead to cancer—an effect we will return to in Chapter 17. But germ line mutations are also essential to life, because they provide the genetic diversity on which the forces of evolution act.

Mutations are the raw material of evolution

Without mutation, there would be no evolution. As we will see in Part Five of this book, mutation does not drive evolution, but it provides the genetic diversity on which natural selection and other agents of evolution act.

All mutations are rare events, but mutation frequencies vary from organism to organism and from gene to gene within a given organism. The frequency of mutation is usually much lower than one mutation per 10^4 base pairs per DNA replication, and some-

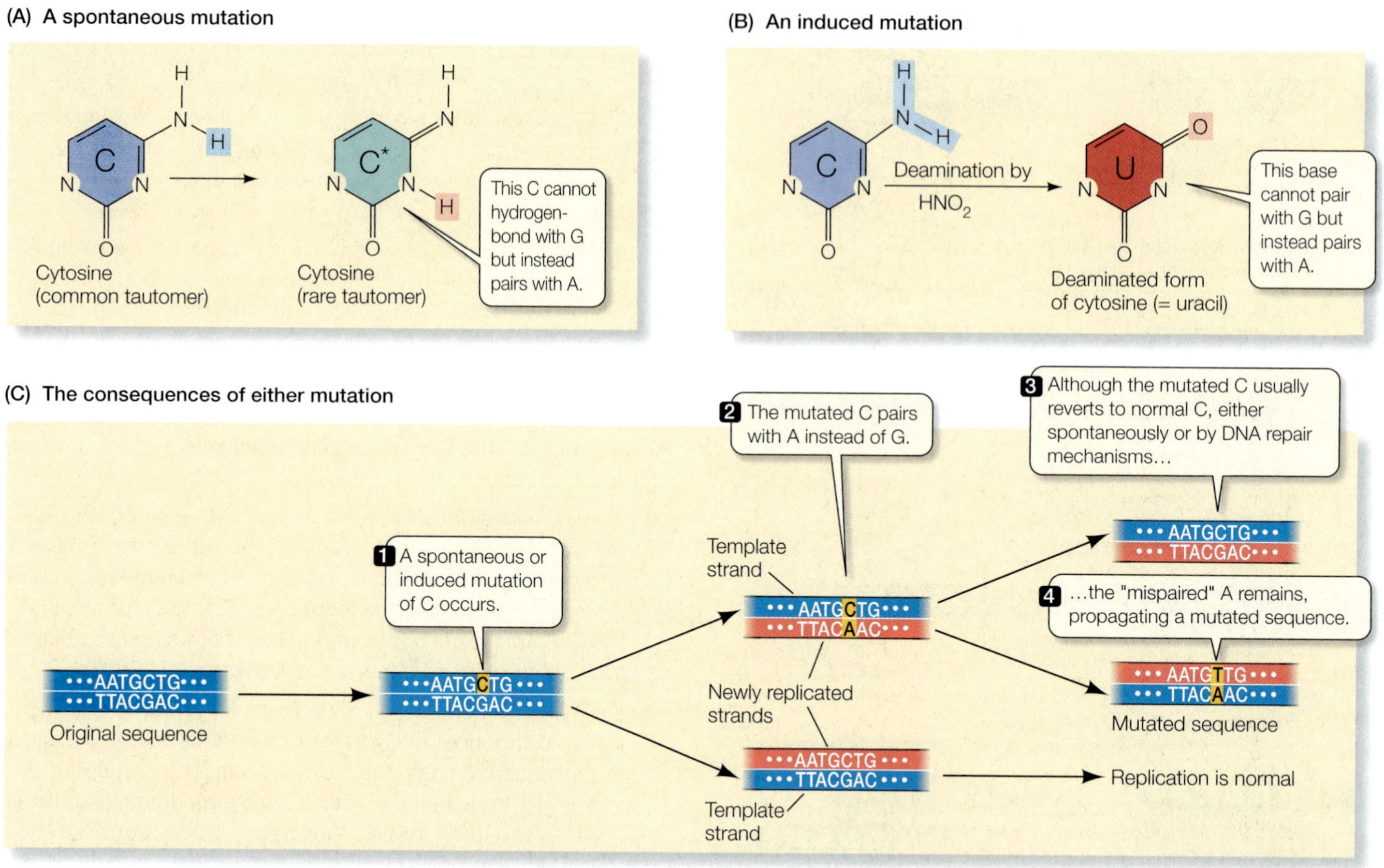

(A) A spontaneous mutation

Cytosine
(common tautomer)

Cytosine
(rare tautomer)

This C cannot hydrogen-bond with G but instead pairs with A.

(B) An induced mutation

Deamination by
HNO_2

Deaminated form
of cytosine (= uracil)

This base cannot pair with G but instead pairs with A.

(C) The consequences of either mutation

2 The mutated C pairs with A instead of G.

3 Although the mutated C usually reverts to normal C, either spontaneously or by DNA repair mechanisms…

1 A spontaneous or induced mutation of C occurs.

Template strand

•••AATGCTG•••
•••TTACGAC•••

4 …the "mispaired" A remains, propagating a mutated sequence.

•••AATGCTG•••
•••TTACAAC•••

Newly replicated strands

•••AATGTTG•••
•••TTACAAC•••

Mutated sequence

•••AATGCTG•••
•••TTACGAC•••

Original sequence

•••AATGCTG•••
•••TTACGAC•••

•••AATGCTG•••
•••TTACGAC•••

Template strand

Replication is normal

12.20 Spontaneous and Induced Mutations   (A) All four nitrogenous bases in DNA exist in both a prevalent (common) form and a rare form. When a base spontaneously forms its rare tautomer, it can pair with a different base. (B) Mutagenic chemicals such as nitrous acid can induce changes in the bases. (C) In both spontaneous and induced mutations, the result is a permanent change in the DNA sequence following replication.

times as low as one mutation per 10^9 base pairs per replication. Most mutations are point mutations in which one nucleotide is substituted for another during the synthesis of a new DNA strand.

Mutations can harm the organism that carries them, or they can be neutral (have no effect on the organism's ability to survive or produce offspring). Once in a while, a mutation improves an organism's adaptation to its environment, or it becomes favorable when environmental conditions change.

Most of the complex creatures living on Earth have more genes than the simpler creatures do. Humans, for example, have 20 times more genes than prokaryotes have. How did these new genes arise? Whole genes might be duplicated, and the bearer of the duplication would have a surplus of genetic information that might be turned to good use. Subsequent mutations in one of the two copies of the gene might not have an adverse effect on survival because the other copy of the gene would continue to produce functional protein. The extra gene might mutate over and over again without ill effect because its original function would be fulfilled by the original copy.

If the random accumulation of mutations in the extra gene led to the production of a useful protein (for example, an enzyme with an altered specificity for the substrates it binds, allowing it to catalyze different—but related—reactions), natural selection would tend to perpetuate the existence of this new gene. New copies of genes may also arise through the activity of *transposable elements*, which are discussed in Chapters 13 and 14.

12.6 RECAP

Mutations are alterations in the nucleotide sequence of DNA. They may take the form of changes in single nucleotides or extensive rearrangements of the genetic material in chromosomes. If they occur in somatic cells, they will be passed on to daughter cells; if they occur in germ line cells, they will be passed on to offspring.

■ Can you distinguish the various kinds of chromosomal mutations: deletions, duplications, inversions, and translocations? See p. 276 and Figure 12.19

■ Do you understand the difference between spontaneous and induced mutations? Can you give an example of each? See p. 277 and Figure 12.20

CHAPTER SUMMARY

12.1 What is the evidence that genes code for proteins?

Beadle and Tatum's experiments on metabolic enzymes in the bread mold *Neurospora* led to the **one-gene, one-enzyme hypothesis**. We now know that there is a **one-gene, one-polypeptide relationship**. Review Figure 12.1

12.2 How does information flow from genes to proteins?

The **central dogma** of molecular biology states that DNA codes for RNA and RNA codes for protein. Protein does not code for protein, RNA, or DNA. Review Figure 12.2

The process by which the information in DNA is copied to RNA is called **transcription**. The process by which a protein is built from the information in RNA is called **translation**. Review Figure 12.3

Certain RNA viruses are exceptions to the central dogma. These **retroviruses** synthesize DNA from RNA in **reverse transcription**.

The product of transcription, **messenger RNA (mRNA)**, travels from the nucleus to the cytoplasm. **Transfer RNA (tRNA)** molecules translate the genetic information in mRNA into a corresponding sequence of amino acids to produce a polypeptide.

12.3 How is the information content in DNA transcribed to produce RNA?

In a given gene, only one of the two strands of DNA (the **template strand**) acts as a template for transcription. **RNA polymerase** is the catalyst for transcription.

RNA transcription from DNA proceeds in three steps: **initiation**, **elongation**, and **termination**. Review Figure 12.5, Web/CD Tutorial 12.1

Initiation requires a **promoter**, to which DNA polymerase binds. Part of each promoter is the **initiation site**, where transcription begins.

Elongation of the RNA molecule proceeds from the 5′ to 3′ end.

Particular base sequences specify termination, at which point transcription ends and the RNA transcript separates from the DNA template.

The **genetic code** is a "language" of triplets of mRNA nucleotide bases (**codons**) corresponding to 20 specific amino acids; there are **start** and **stop codons** as well. The code is redundant (an amino acid may be represented by more than one codon), but not ambiguous (no single codon represents more than one amino acid). Review Figure 12.6, Web/CD Activity 12.1, Web/CD Tutorial 12.2

12.4 How is RNA translated into proteins?

See Web/CD Tutorial 12.3

In translation, amino acids are linked in an order specified by the codons in mRNA. This task is achieved by tRNAs, which bind to (are charged with) specific amino acids.

Each tRNA species has an amino acid attachment site as well as an **anticodon** complementary to a specific mRNA codon. A specific activating enzyme charges each tRNA with its specific amino acid. Review Figures 12.8 and 12.9

The **ribosome** is the molecular workbench where translation takes place. It has one large and one small subunit, both made of **ribosomal RNA** and proteins.

Three sites on the large subunit of the ribosome interact with tRNA anticodons. The A site is where the charged tRNA anticodon binds to the mRNA codon; the P site is where the tRNA adds its amino acid to the growing polypeptide chain; and the E site is where the tRNA is released. Review Figure 12.10

Translation occurs in three steps: **initiation**, **elongation**, and **termination**.

The **initiation complex** consists of tRNA bearing the first amino acid, the small ribosomal subunit, and mRNA. The mRNA binds to a specific complementary sequence on rRNA. Review Figure 12.11

The growing polypeptide chain is elongated by the formation of peptide bonds between amino acids, catalyzed by RNA. Review Figure 12.12

When a stop codon reaches the A site, it terminates translation by binding a release factor. Review Figure 12.13

In a **polysome**, more than one ribosome moves along a strand of mRNA at one time. Review Figure 12.14

12.5 What happens to polypeptides after translation?

Signal sequences of amino acids direct polypeptides to their cellular destinations. Review Figure 12.15

Destinations in the cytoplasm include organelles, which proteins enter upon recognizing surface receptors called **docking proteins**.

Proteins "addressed" to the ER bind to a **signal recognition particle**. Review Figure 12.16

Posttranslational modifications of polypeptides include **proteolysis**, in which a polypeptide is cut into smaller fragments; **glycosylation**, in which sugars are added; and **phosphorylation**, in which phosphate groups are added. Review Figure 12.17

12.6 What are mutations?

Mutations are heritable changes in DNA. **Somatic mutations** are passed on to daughter cells, but only **germ line mutations** are passed on to sexually produced offspring.

Point mutations result from alterations in single base pairs of DNA. **Silent mutations** result in no change in amino acids when the altered mRNA is translated into a polypeptide. **Missense**, **nonsense**, and **frame-shift** mutations do cause changes in the amino acids produced. Review pp. 275–276

Chromosomal mutations (**deletions**, **duplications**, **inversions**, or **translocations**) involve large regions of a chromosome. Review Figure 12.19

Spontaneous mutations occur because of instabilities in DNA or chromosomes. **Induced** mutations occur when a **mutagen** damages DNA. Review Figure 12.20

Mutations, although often detrimental to an individual organism, are the raw material of evolution.

SELF-QUIZ

1. Which of the following is *not* a difference between RNA and DNA?
   a. RNA has uracil; DNA has thymine.
   b. RNA has ribose; DNA has deoxyribose.
   c. RNA has five bases; DNA has four.
   d. RNA is a single polynucleotide strand; DNA is a double strand.
   e. RNA is relatively smaller than human chromosomal DNA.

2. Normally, *Neurospora* can synthesize all 20 amino acids. A certain strain of this mold cannot grow in minimal nutritional medium, but grows only when the amino acid leucine is added to the medium. This strain
   a. is dependent on leucine for energy.
   b. has a mutation affecting the biochemical pathway leading to the synthesis of proteins.
   c. has a mutation affecting the biochemical pathway leading to the synthesis of all 20 amino acids.
   d. has a mutation affecting the biochemical pathway leading to the synthesis of leucine.
   e. has a mutation affecting the biochemical pathways leading to the syntheses of 19 of the 20 amino acids.

3. An mRNA has the sequence 5′-AUGAAAUCCUAG-3′. What is the template DNA strand for this sequence?
   a. 5′-TACTTTAGGATC-3′
   b. 5′-ATGAAATCCTAG-3′
   c. 5′-GATCCTAAAGTA-3′
   d. 5′-TACAAATCCTAG-3′
   e. 5′-CTAGGATTTCAT-3′

4. The adapters that allow translation of the four-letter nucleic acid language into the 20-letter protein language are called
   a. aminoacyl-tRNA synthetases.
   b. transfer RNAs.
   c. ribosomal RNAs.
   d. messenger RNAs.
   e. ribosomes.

5. At a certain location in a gene, the non-template strand of DNA has the sequence GAA. A mutation alters the triplet to GAG. This type of mutation is called

   a. silent.
   b. missense.
   c. nonsense.
   d. frame-shift.
   e. translocation.

6. Transcription
   a. produces only mRNA.
   b. requires ribosomes.
   c. requires tRNAs.
   d. produces RNA growing from the 5′ end to the 3′ end.
   e. takes place only in eukaryotes.

7. Which statement about translation is *not* true?
   a. It is RNA-directed polypeptide synthesis.
   b. An mRNA molecule can be translated by only one ribosome at a time.
   c. The same genetic code operates in almost all organisms and organelles.
   d. Any ribosome can be used in the translation of any mRNA.
   e. There are both start and stop codons.

8. Which statement about RNA is *not* true?
   a. Transfer RNA functions in translation.
   b. Ribosomal RNA functions in translation.
   c. RNAs are produced by transcription.
   d. Messenger RNAs are produced on ribosomes.
   e. DNA codes for mRNA, tRNA, and rRNA.

9. The genetic code
   a. is different for prokaryotes and eukaryotes.
   b. has changed during the course of recent evolution.
   c. has 64 codons that code for amino acids.
   d. has more than one codon for many amino acids
   e. is ambiguous.

10. A mutation that results in the codon UAG where there had been UGG is
    a. a nonsense mutation.
    b. a missense mutation.
    c. a frame-shift mutation.
    d. a large-scale mutation.
    e. unlikely to have a significant effect.

FOR DISCUSSION

1. How is it possible that a point mutation, consisting of the replacement of a single nucleotide base in DNA by a different base, might not result in an error in protein production?

2. Har Gobind Khorana at the University of Wisconsin synthesized artificial mRNAs such as poly CA (CACA …) and poly CAA (CAACAACAA …). He found that poly CA codes for a polypeptide consisting of threonine (Thr) and histidine (His), in alternation (His–Thr–His–Thr …). There are two possible codons in poly CA, CAC and ACA. One of these must code for histidine and the other for threonine—but which is which? The answer comes from results with poly CAA, which produces three different polypeptides: poly Thr, poly Gln (glutamine), and poly Asn (asparagine). (An artificial mRNA can be read, inefficiently, beginning at any point in the chain; there is no specific initiation signal. Thus poly CAA can be read as a polymer of CAA, of ACA, or of AAC.) Compare the results of the poly CA and poly CAA experi-

ments, and determine which codon codes for threonine and which for histidine.

3. Look back at Question 2. Using the genetic code in Figure 12.6 as a guide, deduce what results Khorana would have obtained had he used poly UG and poly UGG as artificial messengers. In fact, very few such artificial messengers would have given useful results. For an example of what could happen, consider poly CG and poly CGG. If poly CG were the messenger, a mixed polypeptide of arginine and alanine (Arg–Ala–Ala–Arg …) would be obtained; poly CGG would give three polypeptides: poly Arg, poly Ala, and poly Gly (glycine). Can any codons be determined from only these data? Explain.

4. Errors in transcription occur about 100,000 times as often as do errors in DNA replication. Why can this high rate be tolerated in RNA synthesis but not in DNA synthesis?

FOR INVESTIGATION

Beadle and Tatum's experiments showed that a biochemical pathway could be deduced from mutant strains. In bacteria, the biosynthesis of the amino acid tryptophan (T) from the precursor chorismate (C) involves four intermediate chemical compounds, which we will call D, E, F, and G. Here are the phenotypes of various mutant strains, where each strain has a mutation in a gene for a different enzyme, + means growth with the stated addition to the medium, and 0 means no growth. Based on these data, order the compounds (C, D, E, F, G, and T) and enzymes (1, 2, 3, 4, and 5) in a biochemical pathway.

MUTANT STRAIN	ADDITION TO MEDIUM					
	C	D	E	F	G	T
1	0	0	0	0	+	+
2	0	+	+	0	+	+
3	0	+	0	0	+	+
4	0	+	+	+	+	+
5	0	0	0	0	0	+

13 The Genetics of Viruses and Prokaryotes

Mutation of a bird virus results in human infection

On May 9, 1997, a 3-year-old boy in Hong Kong developed a cough and fever. His physician treated the boy with antibiotics and aspirin, but the fever got worse. The boy was hospitalized on May 15. Unfortunately, his fever progressed to lung failure, and he died 6 days later.

In an attempt to find the cause of the boy's death, researchers took fluid drawn from his lungs prior to his death and added it to mammalian kidney cells in a laboratory culture. Two days later, the cells were dying, with each cell releasing hundreds of influenza virus particles. Public health professionals at Hong Kong Hospital searched for signs that the boy's flu was one of the strains that had infected people earlier that winter. They tested for glycoproteins on the virus surface that would allow it to attach to human cells. None of the tests worked, indicating that the virus was not typical human flu. What was this deadly new virus?

By August they were able to determine that the little boy had been infected with H5N1, a flu virus previously known to infect only chickens. The boy's day care provider had kept chicks for the children to play with, and several of the chicks died. Genetic tests showed that the nucleotide sequences of the viruses in the birds and in the boy matched. Comparison of this sequence with that of flu virus confined to birds revealed that a mutation in the gene for the viral surface glycoprotein had made the avian virus capable of binding to and infecting human cells.

By December, more human cases of "bird flu" appeared in Hong Kong; of 18 people infected, 6 died. Although there was no clear connection between the infected individuals and diseased birds, all of the victims had visited live poultry markets in the week before developing symptoms. Tests of live chickens from these markets revealed a high rate of H5N1 infection. Hong Kong health officials immediately closed the border to mainland China and directed the slaughter of every chicken on the island. Within days, over 1.5 million birds were killed and, in all likelihood, a major epidemic was avoided.

But the bird flu virus was not wiped out by this action. H5N1 has been found in birds other than chickens, and on continents other than Asia. There have been more cases of human infection, but so far prompt action to kill infected birds has prevented widespread occurrence of the disease.

Chicken Farming   With the spread of bird flu throughout the Far East, raising chickens there has become a hazardous occupation.

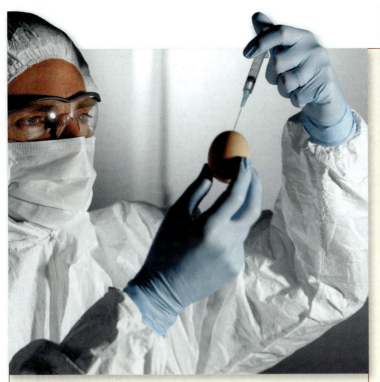

Searching for a Vaccine   Avian flu virus is injected into an egg in part of the procedure to develop a vaccine. The development of flu vaccines is always a race against time. Because viruses evolve so rapidly, new vaccines are constantly needed.

We haven't always been so lucky. The "Spanish flu" epidemic of 1918, which may have started with a single soldier, spread to Europe with U.S. troops fighting in World War I. The resulting pandemic led to 40 million deaths worldwide. Flu pandemics in 1957 and 1968 killed a million people each. In all three pandemics, a single gene mutation made it possible for an animal influenza virus to infect people.

When will the next flu pandemic occur? The answers lie in the molecular genetics of viruses and their evolution.

IN THIS CHAPTER we will describe the growth and reproduction of viruses and bacteria, and we will see how these organisms transmit their genes. First we'll examine the nature of viruses and see how they infect, reproduce in, and express their genes in host cells. Next we'll see how prokaryotes can exchange genes. We will also describe how the expression of prokaryotic genes is regulated and what DNA sequencing has revealed about the prokaryotic genome.

13.1 How Do Viruses Reproduce and Transmit Genes?

Many prokaryotes and viruses have served as model organisms for studying the structure, function, and transmission of genes (**Figure 13.1**). They are excellent model organisms compared with more complex eukaryotes for several reasons:

- *Their genomes are small.* A typical bacterium contains about a thousandth as much DNA as a single human cell, and a typical bacteriophage contains about a hundredth as much DNA as a bacterium.

- *They reproduce quickly.* A single milliliter of growth medium can contain more than 10^9 cells of the bacterium *Escherichia coli*, and its numbers can double every 25 minutes.

- *They are usually haploid,* which makes genetic analyses easier.

Viruses are not cells

Unlike the organisms that make up the three domains of the living world, **viruses** are *acellular*; that is, they are not cells and do not consist of cells. Most viruses are composed of only nucleic acid and a few proteins. Viruses do not carry out two of the basic functions of cellular life: they do not regulate the transport of substances into and out of themselves by membranes, and they perform no metabolic functions—that is, they do not take in nutrients or expel wastes. But they can reproduce in systems that do perform these functions: living cells.

Most viruses are much smaller than even the smallest bacteria (**Table 13.1**). Viruses have become well understood only within the last half century, but the first step on this path of discovery was taken by the Russian botanist Dmitri Ivanovsky in 1892. He was trying to find the cause of tobacco mosaic disease, which results in the destruction of photosynthetic tissues in plants and can devastate a tobacco crop. Ivanovsky passed an extract of diseased tobacco leaves through a fine porcelain filter, a technique that had been used previously by physicians and veterinarians to isolate disease-causing bacteria.

To Ivanovsky's surprise, the disease agent in this case was not retained on the filter. It passed through, and the liquid fil-

13.1 Model Organisms   Viruses and bacteria are valuable model organisms for studying genetics and molecular biology. (A) Bacteriophage T4, a commonly studied virus, is about 10 times smaller than (B) *Escherichia coli*, a commonly studied bacterium.

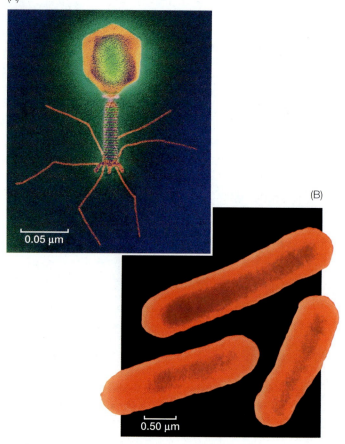

(B)

trate still caused tobacco mosaic disease. But instead of concluding that the agent was smaller than a bacterium, he assumed that his filter was faulty. Louis Pasteur's recent demonstration that bacteria could cause disease was the dominant idea at the time, and Ivanovsky chose not to challenge it. But, as often happens in science, someone soon came along who did. In 1898, the Dutch microbiologist Martinus Beijerinck repeated Ivanovsky's experiment and also showed that the tobacco mosaic disease agent could diffuse through an agar gel. He called the tiny agent *contagium vivum fluidum*, which later became shortened to *virus*.

Almost 40 years later, the disease agent was crystallized by Wendell Stanley (who won the Nobel prize for his efforts). The crystalline viral preparation became infectious again when it was dissolved. It was soon shown that crystallized viral preparations consist of proteins and nucleic acids. Still, it was not until the electron microscope allowed direct observation of viruses in the 1950s that it became clear how different they are from bacteria and other organisms.

Viruses reproduce only with the help of living cells

Whole viruses never arise directly from preexisting viruses. Viruses are *obligate intracellular parasites*; that is, they develop and reproduce only within the cells of specific hosts. The cells of animals, plants, fungi, protists, and prokaryotes (both bacteria and archaea) can serve as hosts to viruses. Viruses use the host's DNA replication and protein synthesis machinery to reproduce themselves, usually destroying the host cell in the process. The host cell releases progeny viruses, which then infect new hosts.

Viruses outside of host cells exist as individual particles called **virions**. The virion, the basic unit of a virus, consists of a central core of either DNA or RNA (but not both) surrounded by a **capsid**, or coat, composed of one or more proteins. Because they lack the distinctive cell wall and ribosomal biochemistry of bacteria, *viruses are not affected by antibiotics* that target these structures.

TABLE 13.1

Relative Sizes of Microorganisms

MICROORGANISM	TYPE	TYPICAL SIZE RANGE (μm^3)
Protists	Eukaryote	5,000–50,000
Photosynthetic bacteria	Prokaryote	5–50
Spirochetes	Prokaryote	0.1–2.0
Mycoplasmas	Prokaryote	0.01–0.1
Poxviruses	Virus	0.01
Influenza virus	Virus	0.0005
Poliovirus	Virus	0.00001

Viruses are classified according to several characteristics:

- Whether the genome is DNA or RNA
- Whether the nucleic acid is single-stranded or double-stranded
- Whether the shape of the virion is simple or complex
- Whether the virion is surrounded by a membrane

Some of these variations are shown in **Figure 13.2**.

Another important characteristic of a virus is the type of organism it infects and the manner in which the infection occurs. Most viruses simply infect and immediately replicate in their host cells. Others can infect a host cell but postpone reproduction, remaining inactive in the host cell until conditions for replication are favorable.

Bacteriophage reproduce by a lytic cycle or a lysogenic cycle

Viruses that infect bacteria are known as **bacteriophage** or **phage** (Greek *phagos*, "one that eats"; note that "phage" is both singular and plural). They recognize their prospective hosts by means of proteins in the capsid, which bind to specific receptor proteins or carbohydrates in the host's cell wall. The virions, whose nucleic acids must penetrate the cell wall for successful infection, are often equipped with tail assemblies that inject the phage's nucleic acid through the cell wall into the host bacterium. After the nucleic acid has entered the host, one of two things happens, depending on the type of phage:

- The virus reproduces immediately and kills the host cell.
- The virus postpones reproduction by integrating its nucleic acid into the host cell's genome.

13.2 Virions Come in Various Shapes (A) The tobacco mosaic virus (a plant virus) consists of an inner helix of RNA covered with a helical array of protein molecules. (B) Many animal viruses, such as this adenovirus, have a capsid as an outer shell. Inside the capsid is a spherical mass of proteins and DNA. (C) In some viruses, such as this herpes virus, a membrane envelope surrounds the capsid.

The Hershey–Chase experiment (see Figure 11.4) involved the first type of viral reproductive cycle, called the **lytic cycle**, so named because the infected bacterium *lyses* (bursts), releasing progeny phage. The alternative fate is the **lysogenic cycle**, in which the infected bacterium does not lyse, but instead harbors the viral nucleic acid in its genome and passes it along over many generations until conditions trigger a lytic cycle. Most viruses reproduce only by the lytic cycle; others undergo both types of reproductive cycles (**Figure 13.3**).

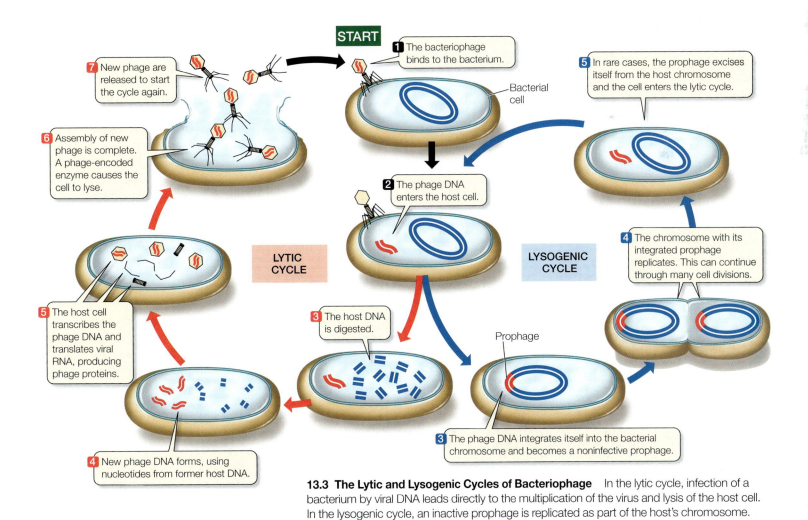

START

1 The bacteriophage binds to the bacterium.

Bacterial cell

2 The phage DNA enters the host cell.

LYTIC CYCLE

3 The host DNA is digested.

4 New phage DNA forms, using nucleotides from former host DNA.

5 The host cell transcribes the phage DNA and translates viral RNA, producing phage proteins.

6 Assembly of new phage is complete. A phage-encoded enzyme causes the cell to lyse.

7 New phage are released to start the cycle again.

3 The phage DNA integrates itself into the bacterial chromosome and becomes a noninfective prophage.

Prophage

LYSOGENIC CYCLE

4 The chromosome with its integrated prophage replicates. This can continue through many cell divisions.

5 In rare cases, the prophage excises itself from the host chromosome and the cell enters the lytic cycle.

13.3 The Lytic and Lysogenic Cycles of Bacteriophage In the lytic cycle, infection of a bacterium by viral DNA leads directly to the multiplication of the virus and lysis of the host cell. In the lysogenic cycle, an inactive prophage is replicated as part of the host's chromosome.

If bacteria are everywhere, then bacteriophage are everywhere, and in much larger numbers. There are up to 100 million viruses in a milliliter of seawater; similar numbers are found in soil water. Scientists estimate that in nature, there are some 10 phage per every bacterium.

THE LYTIC CYCLE   A virus that reproduces only by the lytic cycle is called a **virulent** virus. Once a virulent phage has bound to and injected its nucleic acid into a bacterium, that nucleic acid takes over the host's synthetic machinery. It does so in two stages (**Figure 13.4**):

- The viral genome contains promoter sequences that attract host RNA polymerase. In the *early stage*, viral genes that lie adjacent to this promoter are transcribed. These *early genes* often code for proteins that shut down host transcription, stimulate viral genome replication, and stimulate viral gene transcription. Viral nuclease enzymes digest the host's chromosome, providing nucleotides for the synthesis of viral genomes.

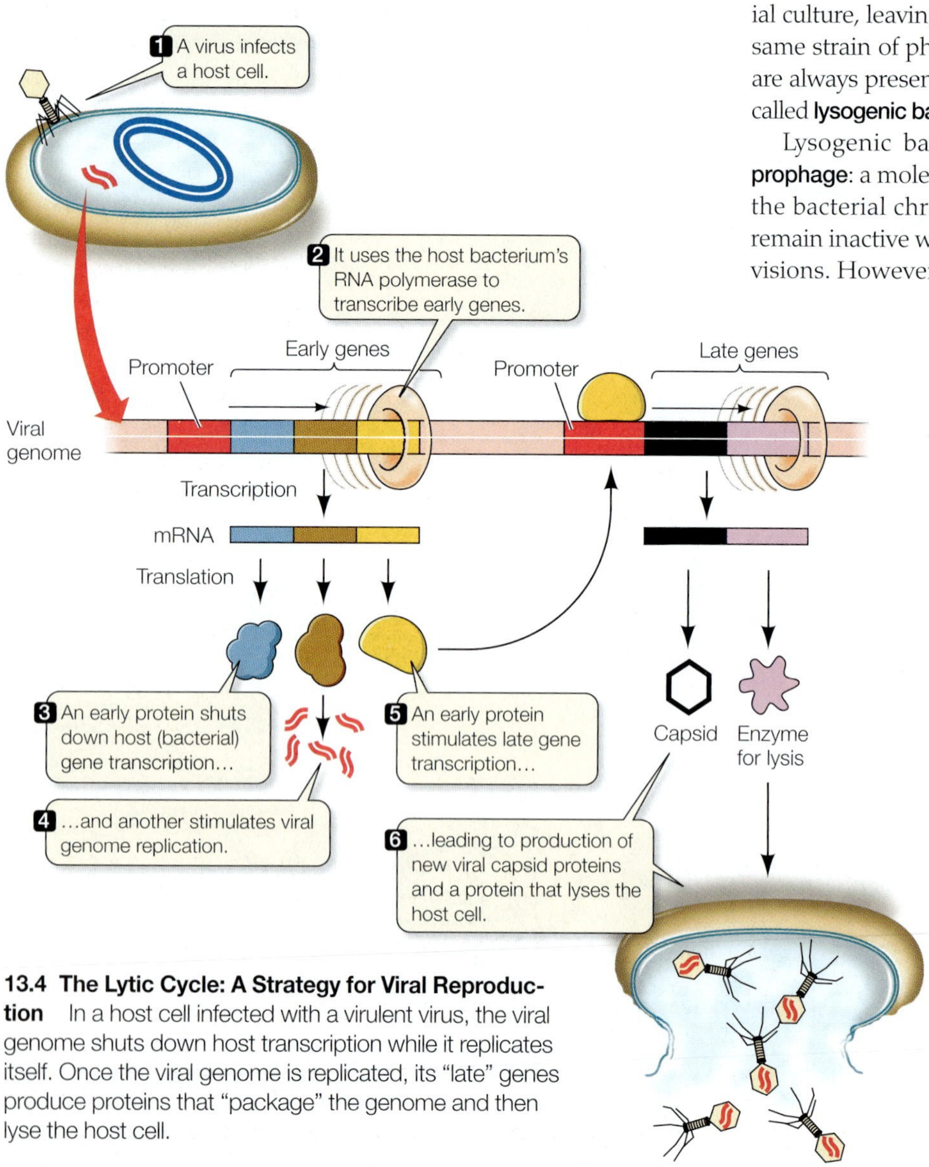

- In the *late stage*, viral *late genes*, which code for the proteins of the viral capsid and those that lyse the host cell to release the new virions, are transcribed.

The whole process—from binding and infection to release of phage by lysis of the host cell—takes about half an hour. However, this sequence of transcriptional events is carefully controlled: premature lysis of the host cell before virions are assembled and ready for release would stop the infection.

Two viruses can infect a cell at the same time. This is an unusual event because once a lytic cycle is under way, there is usually not enough time for an additional infection. In addition, an early viral protein may prevent further infections. When two different viral genomes are in the same host cell, however, there exists the possibility of genetic recombination by crossing over (as in prophase I of meiosis in eukaryotes; see Figure 9.18). This phenomenon enables genetically different but related viruses to swap genes and create new strains.

THE LYSOGENIC CYCLE   Viral infection does not always result in lysis of the host cell. Some phage seem to disappear from a bacterial culture, leaving the bacteria "immune" to further attack by the same strain of phage. In such cultures, however, a few free phage are always present. Bacteria harboring viruses that are not lytic are called **lysogenic bacteria**, and the viruses are called **temperate** viruses.

Lysogenic bacteria contain a noninfective entity called a **prophage**: a molecule of phage DNA that has been integrated into the bacterial chromosome (see Figure 13.3). The prophage can remain inactive within the bacterial genome through many cell divisions. However, an occasional lysogenic bacterium can be induced to activate its prophage. This activation results in a lytic cycle, in which the prophage excises itself from the host chromosome and reproduces.

The capacity to switch between the lysogenic and the lytic cycle is very useful to the phage because it enhances opportunities to produce the maximum number of progeny viruses. When its host cell is growing and reproducing rapidly, the phage is lysogenic. When the host is stressed or damaged by mutagens, the prophage is released from its inactive state, and the lytic cycle proceeds.

USING LYTIC BACTERIOPHAGE COULD BE USEFUL IN TREATING BACTERIAL INFECTIONS   Because lytic bacteriophage destroy their bacterial hosts, scientists perceived that they might be useful in treating infectious diseases caused by bacteria. Indeed, one of the early discoverers of phage, the French-Canadian microbiologist Felix D'Herelle, noted in 1917 (before antibiotics were discovered) that when some patients with bacterial dysentery were recovering from the disease, the quantity of phage near the bacteria was much higher than when the disease was at its peak.

13.4 The Lytic Cycle: A Strategy for Viral Reproduction   In a host cell infected with a virulent virus, the viral genome shuts down host transcription while it replicates itself. Once the viral genome is replicated, its "late" genes produce proteins that "package" the genome and then lyse the host cell.

D'Herelle tried using phage to control infections of chickens by the bacterium *Salmonella gallinarum*. To do this, he divided chickens into two groups, one that was given phage and another that was not. Then he exposed both groups to the infectious bacteria. The phage-protected group did not get the bacterial disease. Later, he used phage successfully to treat people in Egypt infected with plague-causing bacteria and people in India with infectious cholera.

The emergence of antibiotics and of phage-resistant bacteria reduced interest in phage therapy. However, interest has revived now that bacterial resistance to antibiotics is becoming common. Bacteriophage are even being investigated as a means of treating edible fruits and vegetables to prevent bacterial contamination. In addition to advancing our understanding of fundamental biological processes, the study of bacteriophage has opened the door to investigations of viruses that infect eukaryotes.

Animal viruses have diverse reproductive cycles

Almost all vertebrates are susceptible to viral infections, but among invertebrates, such infections are common only in arthropods (the group that includes insects and crustaceans). One group of viruses, called *arboviruses* (short for "arthropod-borne viruses"), is transmitted to vertebrates through insect bites. Although they are carried within the arthropod host's cells, arboviruses apparently do not harm that host; they affect only the bitten and infected vertebrate. The arthropod acts as a **vector**—an intermediate carrier—by transmitting the virus from one vertebrate host to another.

Animal viruses are very diverse. Some are just particles consisting of proteins surrounding a nucleic acid. Others have a membrane derived from the host cell's plasma membrane and are called *enveloped* viruses. Some animal viruses have DNA as their genetic material; others have RNA. In most cases, the viral genome is small, coding for only a few proteins.

Like that of bacteriophage, the lytic cycle of animal viruses can be divided into early and late stages (see Figure 13.4). Animal viruses enter cells in one of three ways:

■ A naked virion (without a membrane envelope) is taken up by endocytosis, which traps it within a membranous vesicle inside the host cell. The membrane of the vesicle breaks down, releasing the virion into the cytoplasm, and the host cell digests the protein capsid, liberating the viral nucleic acid, which takes charge of the host cell.

■ Enveloped viruses may also be taken up by endocytosis (see Figure 13.5) and released from a vesicle. In these viruses,

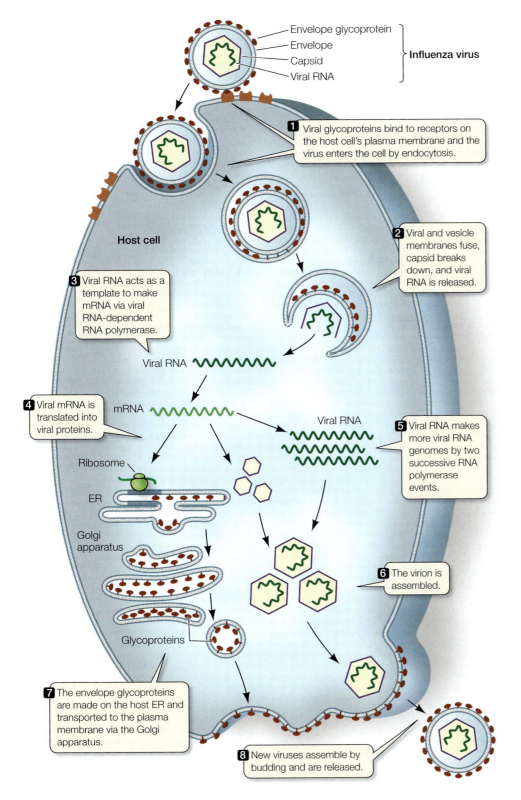

1 Viral glycoproteins bind to receptors on the host cell's plasma membrane and the virus enters the cell by endocytosis.

2 Viral and vesicle membranes fuse, capsid breaks down, and viral RNA is released.

3 Viral RNA acts as a template to make mRNA via viral RNA-dependent RNA polymerase.

4 Viral mRNA is translated into viral proteins.

5 Viral RNA makes more viral RNA genomes by two successive RNA polymerase events.

6 The virion is assembled.

7 The envelope glycoproteins are made on the host ER and transported to the plasma membrane via the Golgi apparatus.

8 New viruses assemble by budding and are released.

Envelope glycoprotein
Envelope
Capsid
Viral RNA
} Influenza virus

Host cell

Viral RNA

mRNA

Viral RNA

Ribosome

ER

Golgi apparatus

Glycoproteins

13.5 The Reproductive Cycle of the Influenza Virus The enveloped influenza virus is taken into the host cell by endocytosis. Once inside, fusion of the vesicle and viral membranes releases the viral genome, which replicates and assembles new virions.

the viral membrane is studded with glycoproteins that bind to receptors on the host cell's plasma membrane.

■ More commonly, the membranes of the host and the enveloped virus fuse, releasing the rest of the virion into the cell (see Figure 13.6).

Following viral reproduction, enveloped viruses usually escape from the host cell by a budding process in which they acquire a membrane envelope from the host cell's plasma membrane.

Both influenza virus and human immunodeficiency virus (HIV) are single-stranded RNA viruses, yet their life cycles illustrate two very different strategies of infection and genome replication. Influenza virus is taken up into a membrane vesicle by endocytosis (**Figure 13.5**). Fusion of the viral and vesicle membranes releases the virion into the cell. The virus carries its own enzyme to replicate its RNA genome. This enzyme is an RNA-dependent RNA polymerase using RNA as a template (as opposed to the DNA-dependent RNA polymerases that use a DNA template; see Section 12.3). This newly synthesized viral RNA strand is then used as mRNA to make, by complementary base pairing, more copies of the viral genome.

Retroviruses such as HIV have a more complex reproductive cycle (**Figure 13.6**). The virus enters a host cell by direct fusion of the viral envelope and the host plasma membrane. A distinctive feature of the retroviral life cycle is RNA-directed DNA synthesis. This process, driven by the viral enzyme **reverse transcriptase**, produces a DNA **provirus** consisting of cDNA (complementary DNA transcribed from the RNA genome), which is the form of the viral genome that gets integrated into the host's DNA. The provirus resides in the host chromosome permanently and is occasionally activated to produce new virions. When this happens, the provirus is transcribed as mRNA, which is then translated into viral proteins using the host cell's protein-synthesizing machinery. Viral glycoproteins are inserted into the host

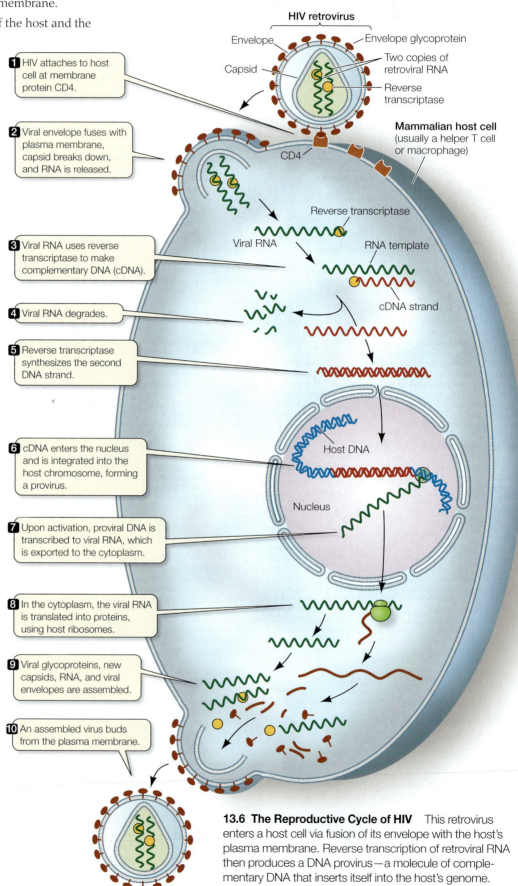

1 HIV attaches to host cell at membrane protein CD4.

2 Viral envelope fuses with plasma membrane, capsid breaks down, and RNA is released.

3 Viral RNA uses reverse transcriptase to make complementary DNA (cDNA).

4 Viral RNA degrades.

5 Reverse transcriptase synthesizes the second DNA strand.

6 cDNA enters the nucleus and is integrated into the host chromosome, forming a provirus.

7 Upon activation, proviral DNA is transcribed to viral RNA, which is exported to the cytoplasm.

8 In the cytoplasm, the viral RNA is translated into proteins, using host ribosomes.

9 Viral glycoproteins, new capsids, RNA, and viral envelopes are assembled.

10 An assembled virus buds from the plasma membrane.

HIV retrovirus

Envelope — Envelope glycoprotein

Capsid — Two copies of retroviral RNA

Reverse transcriptase

Mammalian host cell (usually a helper T cell or macrophage)

CD4

Reverse transcriptase

Viral RNA — RNA template

cDNA strand

Host DNA

Nucleus

13.6 The Reproductive Cycle of HIV This retrovirus enters a host cell via fusion of its envelope with the host's plasma membrane. Reverse transcription of retroviral RNA then produces a DNA provirus—a molecule of complementary DNA that inserts itself into the host's genome.

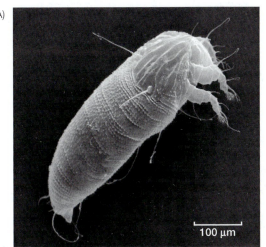

13.7 Wheat Streak Mosaic Virus (A) A tiny mite is the vector for wheat streak mosaic virus. (B) Photosynthetic tissues damaged by the virus are visible on these wheat plants as yellow streaks.

cell's plasma membrane, which will become the viral envelope. Other viral proteins form capsids, which enclose viral RNA molecules. Release of virions from the cell is by a process of budding very similar to exocytosis. Almost every step in this complex cycle can, in principle, be attacked by therapeutic drugs; this fact is used by researchers in their quest to conquer AIDS, the deadly condition caused by HIV infection in humans, as is discussed further in Section 18.7.

Animal viruses, including human viruses, take a severe toll on human and animal health. But our well-being is also challenged by plant viruses and the diseases they cause.

Many plant viruses spread with the help of vectors

Viral diseases of flowering plants are common. Plant viruses can be transmitted *horizontally*, from one plant to another, or *vertically*, from parent to offspring. To infect a plant cell, viruses must pass through a cell wall as well as a plasma membrane. Most plant viruses accomplish this through their association with vectors,

which are often insects. When an insect vector penetrates a cell wall with its proboscis (snout), virions can move from the insect into the plant. Once inside a plant cell, the virus reproduces and spreads to other cells in the plant. Within a structure such as a leaf, the virus spreads through the plasmodesmata, the cytoplasmic connections between cells (see Figure 15.20).

An example of a virus that causes an economically important plant disease is the wheat streak mosaic virus (**Figure 13.7**). It enters the leaf of a wheat plant via a tiny (1 mm long) insect, the mite *Aceria tosichella*. Yellow streaks are seen in the leaves as the infection spreads and destroys photosynthetic tissues. Without the chemical energy from photosynthesis, the plant's production of wheat grains can be severely reduced. This disease was first detected in Canada in 1960 and in the Unites States in 1964. The only way to control it is to eliminate the host insects (or plants) to break the viral life cycle. There is serious concern about the use of this virus as a bioterrorism agent, as its spread could wreak havoc on food supplies.

13.1 RECAP

Viruses are not cells. They consist of nucleic acid and a few proteins, and require a host cell to reproduce. In the lytic cycle, the viral genome directs the host cell to generate new virions along with proteins that cause the host cell to lyse and release them. In the lysogenic cycle, viruses may remain inactive in host cells for long periods.

- How can bacteria and viruses make good model organisms? See p. 283

- Do you understand the lytic and lysogenic cycles of bacteriophage? See pp. 284–285 and Figure 13.3

- Describe the HIV life cycle. See p. 288 and Figure 13.6

How does a phage manage to switch between the lysogenic and the lytic cycle? How does it know when it is to its advantage to make this switch? Let's examine these questions and other aspects of the regulation of viral gene expression.

13.2 How Is Gene Expression Regulated in Viruses?

The mechanisms used by viruses within a host cell for the regulation of gene expression are similar to those used by their hosts. Even a "simple" biological agent such as a virus is faced with complicated molecular decisions when its genome enters a cell. For example, the viral genome must direct the shutdown of host transcription and translation, then redirect the host's protein synthesis machinery to virus production and host cell lysis. All the genes involved in this process must be activated in the right order. In temperate viruses, which can insert their genome (or a DNA copy) into the host chromosome, an additional issue arises: When should the provirus leave the host chromosome and undergo a lytic cycle?

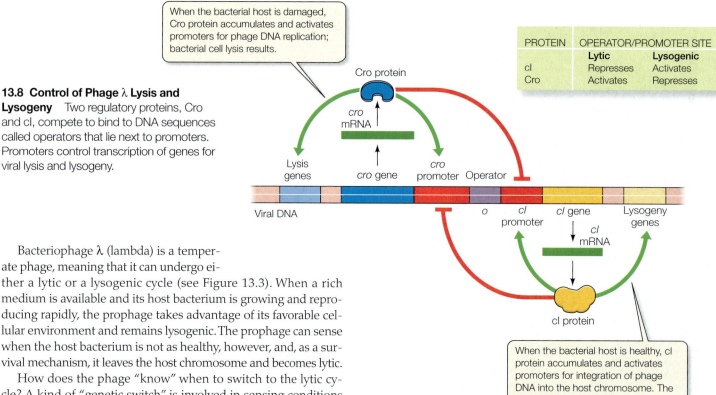

13.8 Control of Phage λ Lysis and Lysogeny Two regulatory proteins, Cro and cI, compete to bind to DNA sequences called operators that lie next to promoters. Promoters control transcription of genes for viral lysis and lysogeny.

[Figure annotations:]

When the bacterial host is damaged, Cro protein accumulates and activates promoters for phage DNA replication; bacterial cell lysis results.

Cro protein

cro mRNA

Lysis genes — cro gene — cro promoter — Operator

Viral DNA

o — cl promoter — cl gene — cl mRNA — Lysogeny genes

cI protein

When the bacterial host is healthy, cI protein accumulates and activates promoters for integration of phage DNA into the host chromosome. The phage enters the lysogenic cycle.

PROTEIN	OPERATOR/PROMOTER SITE	
	Lytic	**Lysogenic**
cI	Represses	Activates
Cro	Activates	Represses

Bacteriophage λ (lambda) is a temperate phage, meaning that it can undergo either a lytic or a lysogenic cycle (see Figure 13.3). When a rich medium is available and its host bacterium is growing and reproducing rapidly, the prophage takes advantage of its favorable cellular environment and remains lysogenic. The prophage can sense when the host bacterium is not as healthy, however, and, as a survival mechanism, it leaves the host chromosome and becomes lytic.

How does the phage "know" when to switch to the lytic cycle? A kind of "genetic switch" is involved in sensing conditions within the host. Two viral regulatory proteins, cI and Cro, compete for two promoters on phage DNA. The two promoters control the transcription of the viral genes involved in the lytic and the lysogenic cycles, respectively, and the two regulatory proteins have opposite effects on the two promoters (**Figure 13.8**). Additional DNA sequences called operators can bind proteins that affect RNA polymerase binding to adjacent promoters.

Phage infection is essentially a "race" between these two regulatory proteins. In a healthy *E. coli* host cell, Cro synthesis is low, so cI "wins," and the phage enters a lysogenic cycle. If the host cell is damaged by mutagens or other stress, Cro synthesis is high, promoters for phage DNA and viral capsid proteins are activated, and bacterial lysis ensues. The two regulatory proteins are made very early in phage infection, and each has a binding site for a specific DNA sequence.

The life cycle of phage λ, which has been greatly simplified here, is a paradigm for viral infections throughout the biological world. The lessons learned from transcriptional controls in this system have been applied again and again to other viruses, including HIV.

13.2 RECAP

Special viral proteins that interact with host and viral DNA sequences are the keys to the regulation of viral gene expression.

■ Describe the genetic switch that regulates the bacteriophage λ life cycle between lytic and lysogenic. See Figure 13.8

Now that we've seen how viruses reproduce, transmit, and regulate the expression of their genes, let's see how these processes work in their hosts—the prokaryotes.

13.3 How Do Prokaryotes Exchange Genes?

In contrast to viruses, prokaryotes (organisms in the domains Bacteria and Archaea) are living cells that carry out all the basic functions of life. Prokaryotes usually reproduce asexually, but nonetheless have several ways of recombining their genes. Whereas in eukaryotes, genetic recombination occurs between the genomes of two parents, recombination in prokaryotes results from the interaction of the genome of one cell with a much smaller sample of genes—a DNA fragment—from another cell.

The reproduction of prokaryotes gives rise to clones

Most prokaryotes reproduce by the division of a single cell into two identical offspring (see Figure 9.2). In this way, a single cell gives rise to a **clone**—a population of genetically identical individuals. Prokaryotes can reproduce very rapidly. A population of *E. coli*, as we saw above, can double every 20 minutes as long as conditions remain favorable. That is one of the reasons that this bacterium is used so widely in research as a model organism.

Simple, reliable methods exist for isolating single bacterial cells and rapidly growing them into clones for identification and study. Pure cultures of *E. coli* or other bacteria can be grown on the surface of a solid *minimal nutrient medium* containing a sugar, minerals, a nitrogen source such as ammonium chloride (NH_4Cl), and a solidifying agent such as agar (**Figure 13.9**). If the number of cells spread on the medium is small, each cell will give rise to a small, rapidly growing *bacterial colony*. If a large number of cells is spread on the medium, their growth will produce one continuous layer—

13.9 Growing Bacteria in the Laboratory A population of *Escherichia coli* doubles every 20 minutes in laboratory culture. The three techniques of culture shown here are used for different applications.

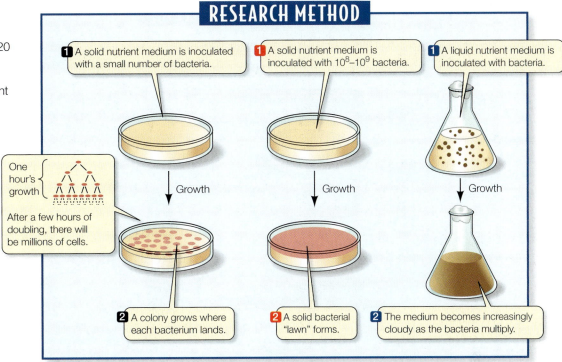

RESEARCH METHOD

1 A solid nutrient medium is inoculated with a small number of bacteria.

1 A solid nutrient medium is inoculated with 10^8–10^9 bacteria.

1 A liquid nutrient medium is inoculated with bacteria.

One hour's growth

After a few hours of doubling, there will be millions of cells.

Growth

Growth

Growth

2 A colony grows where each bacterium lands.

2 A solid bacterial "lawn" forms.

2 The medium becomes increasingly cloudy as the bacteria multiply.

a *bacterial lawn*. Bacteria can also be grown in a liquid medium. We'll see examples of all these techniques in this chapter.

Bacteria have several ways of recombining their genes

The existence and heritability of mutations in bacteria has attracted the attention of geneticists. If there were no form of exchange of genetic information between individuals, bacteria would not be useful for genetic analysis. But how can these asexually reproducing organisms exchange genetic information? In eukaryotes, genetic recombination occurs between homologous chromosomes from two parents during meiosis. While prokaryotes usually reproduce asexually, they still have several ways of recombining their genes.

CONJUGATION The most important way in which bacteria recombine their genes is through the interaction of the genome of one cell with a sample of genes—a DNA fragment—from another cell. In 1946, Joshua Lederberg and Edward Tatum demonstrated that such exchanges do occur, although they are rare events.

Initially, Lederberg and Tatum grew two auxotrophic (nutrient-requiring) mutant strains of *E. coli*. Like the *Neurospora* studied by Beadle and Tatum (see Figure 12.1), these strains could not grow on a minimal nutritional medium, but required supplementation with nutrients that they could not synthesize for themselves because of an enzyme defect.

- Strain 1 required the amino acid methionine and the vitamin biotin for growth; it could make its own threonine and leucine. So its phenotype (and genotype) is given as $met^-bio^-thr^+leu^+$.

- Strain 2 required neither methionine nor biotin, but could not grow without the amino acids threonine and leucine. Its phenotype is given as $met^+bio^+thr^-leu^-$.

Lederberg and Tatum mixed these two mutant strains and cultured them together for several hours in a liquid medium supplemented with methionine, biotin, threonine, and leucine, so that both strains could grow. The bacteria were then removed from the medium by centrifugation, washed, and transferred to minimal medium, which lacked all four supplements. Neither strain 1 nor strain 2 had been able to grow on this medium before because of their nutritional requirements. Nevertheless, after the two strains had been cultured together, a few bacterial colonies did appear on the minimal medium (**Figure 13.10**). Because they were growing in the minimal medium, these colonies must have consisted of bacteria that were $met^+bio^+thr^+leu^+$; that is, they must have been prototrophic. These colonies appeared at a rate of approximately one for every 10 million cells originally placed on the plates (10^{-7}).

Where did these prototrophic colonies come from? Lederberg and Tatum were able to rule out mutation, and other investigators ruled out transformation (a process we discussed in Section 11.1 and which we'll look at in more detail shortly). A third possibility is that the two strains of *E. coli* had exchanged genetic material, producing some cells containing met^+ and bio^+ alleles from strain 2 and thr^+ and leu^+ alleles from strain 1. Later experiments showed that such an exchange, during **conjugation**, had indeed occurred. One bacterial cell—the recipient—had received DNA from another cell—the donor—that included the two wild-type ($^+$) alleles that were missing in the recipient. Recombination had then created a genotype with four wild-type alleles.

The physical contact between two bacteria required for conjugation can be observed under the electron microscope (**Figure 13.11**). It is initiated by a thin projection called a **sex pilus** (plural *pili*). Once the sex pili bring the two cells into proximity, the actual transfer of DNA occurs through a thin cytoplasmic bridge called a **conjugation tube** that forms between the cells. Since the bacter-

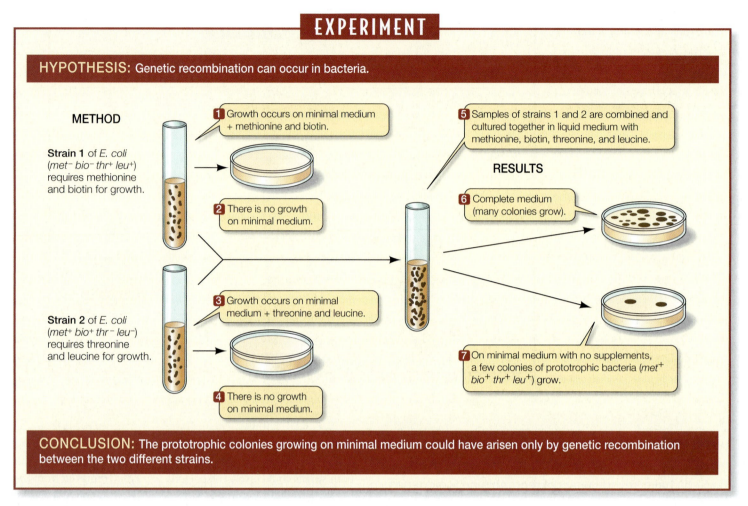

EXPERIMENT

HYPOTHESIS: Genetic recombination can occur in bacteria.

METHOD

Strain 1 of *E. coli* (*met⁻ bio⁻ thr⁺ leu⁺*) requires methionine and biotin for growth.

1 Growth occurs on minimal medium + methionine and biotin.

2 There is no growth on minimal medium.

Strain 2 of *E. coli* (*met⁺ bio⁺ thr⁻ leu⁻*) requires threonine and leucine for growth.

3 Growth occurs on minimal medium + threonine and leucine.

4 There is no growth on minimal medium.

5 Samples of strains 1 and 2 are combined and cultured together in liquid medium with methionine, biotin, threonine, and leucine.

RESULTS

6 Complete medium (many colonies grow).

7 On minimal medium with no supplements, a few colonies of prototrophic bacteria (*met⁺ bio⁺ thr⁺ leu⁺*) grow.

CONCLUSION: The prototrophic colonies growing on minimal medium could have arisen only by genetic recombination between the two different strains.

13.10 Lederberg and Tatum's Experiment After being grown together, a mixture of two complementary auxotrophic strains of *E. coli* contained a few cells that gave rise to new prototrophic colonies. This experiment proved that genetic recombination takes place in prokaryotes. FURTHER RESEARCH: How would you set up experiments to distinguish reversion mutations (back to wild type; for example, *met⁻ bio⁻ thr⁺ leu⁺* to *met⁺ bio⁺ thr⁺ leu⁺*) from recombination in these experiments?

ial chromosome is circular, it must be made linear (cut) before it can pass through the tube. Contact between the cells is brief—only rarely long enough for the entire donor genome to enter the recipient cell. Therefore, the recipient cell usually receives only a portion of the donor DNA.

Once the donor DNA fragment is inside the recipient cell, it can recombine with the recipient cell's genome. In much the same way that chromosomes pair up, gene for gene, in prophase I of meiosis, the donor DNA can line up beside its homologous genes in the recipient, and crossing over can occur. Enzymes that can cut and rejoin DNA molecules are active in bacteria, so gene(s) from the donor can become integrated into the genome of the recipient,

thus changing the recipient's genetic constitution (**Figure 13.12**), even though only about half the transferred genes become integrated in this way.

TRANSFORMATION Frederick Griffith obtained the first evidence for the transfer of prokaryotic genes more than 75 years ago when he discovered the transforming principle (see Figure 11.1). We can now explain Griffith's results: DNA had leaked from dead cells of

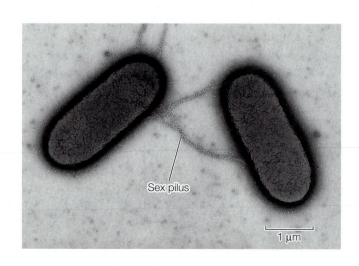

Sex pilus

1 μm

13.11 Bacterial Conjugation Sex pili draw two bacteria into close contact, so that a cytoplasmic conjugation tube can form. DNA is transferred from one cell to the other via the conjugation tube.

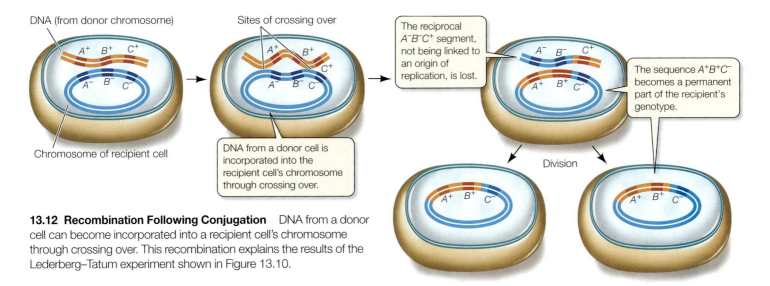

13.12 Recombination Following Conjugation   DNA from a donor cell can become incorporated into a recipient cell's chromosome through crossing over. This recombination explains the results of the Lederberg–Tatum experiment shown in Figure 13.10.

pathogenic pneumococci and was taken up as free DNA by living nonvirulent pneumococci, which became virulent as a result. This phenomenon, called **transformation**, occurs in nature in some species of bacteria when cells die and their DNA leaks out (**Figure 13.13A**). Once transforming DNA is inside a host cell, an event very similar to recombination occurs, and new genes can be incorporated into the host chromosome.

TRANSDUCTION   When bacteriophage undergo a lytic cycle, they package their DNA in capsids, as we saw in Section 13.1. These capsids generally form before the viral DNA is inserted into them. Sometimes bacterial DNA fragments are inserted into an empty phage capsid instead of, or along with, the phage DNA (**Figure 13.13B**). So, when the new virion infects another bacterium, the bacterial DNA is injected into the new host cell. This mechanism of DNA transfer is called **transduction**. Needless to say, it does not result in a productive viral infection. Instead, the incoming DNA fragment can recombine with the host chromosome, resulting in the replacement of host cell genes with bacterial genes from the virus's former host.

Plasmids are extra chromosomes in bacteria

In addition to their main chromosome, many bacteria harbor additional smaller, circular chromosomes called **plasmids**. They contain at most a few dozen genes and, most important, an origin of replication (*ori*, the sequence where DNA replication starts), which defines them as chromo-

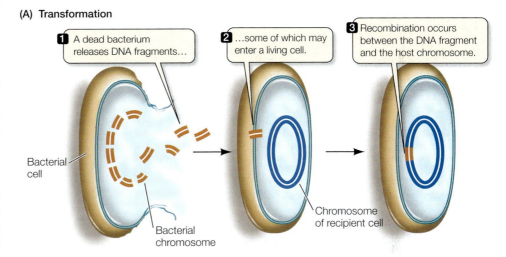

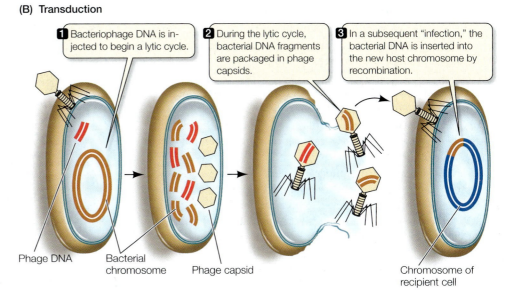

13.13 Transformation and Transduction   When a new DNA fragment enters a bacterial cell, recombination can occur. (A) Transforming DNA can leak from dead bacterial cells and be taken up by a living bacterium, which may incorporate the new genes into its chromosome. (B) In transduction, some viruses carry bacterial DNA fragments from one cell to another.

somes. Plasmids usually replicate at the same time as the main bacterial chromosome, but not always.

Plasmids are *not* viruses. They do not take over the cell's molecular machinery or make a capsid to help them move from cell to cell. Instead, they can move between cells during conjugation, thereby adding some new genes to the recipient bacterium (**Figure 13.14**). Because plasmids exist independently of the main chromosome, they do not need to recombine with the main chromosome to add their genes to the recipient cell's genome.

There are several types of plasmids, classified according to the kinds of genes they carry. Some code for catabolic enzymes, others enable conjugation, and still others code for genes that circumvent antibiotic attack.

SOME PLASMIDS CARRY SPECIAL GENES Some plasmids, called **metabolic factors**, carry genes that allow their recipients to carry out unusual metabolic functions. For example, some bacteria can actually thrive on unusual hydrocarbons found in oil spills, and can use them as a carbon source. Plasmids carry the genes for the enzymes that break down such hydrocarbons.

Bacteria such as *Phenylobacterium immobile* can break down oil from oil spills to less harmful CO_2. In a process called bioremediation, nutrients that promote bacterial growth are added to the polluted area, along with the bacteria. Bioremediation has been used to help clean up major oil spills, as well as smaller releases in urban areas.

Plasmids called **fertility factors**, or **F factors**, encode the genes needed for conjugation. F factors have approximately 25 genes, including the ones that make both the sex pilus and the conjugation tube. A cell harboring an F factor is referred to as F$^+$. It can transfer a copy of the F factor to an F$^-$ cell, making the recipient F$^+$. Sometimes the F factor integrates itself into the main chromosome (at which point it is no longer a plasmid); when it does, it can bring along other genes from that chromosome when it moves through the conjugation tube from one cell to another.

SOME PLASMIDS ARE RESISTANCE FACTORS Resistance factors, also called *R factors* or *R plasmids*, may carry genes coding for proteins that destroy or modify antibiotics. Other R factors provide resistance to heavy metals that bacteria encounter in their environment. R factors first came to the attention of biologists in 1957 during an epidemic of dysentery in Japan, when it was discovered that some strains of the *Shigella* bacterium, which causes dysentery, were resistant to several antibiotics. Researchers found that resistance to the entire spectrum of antibiotics in use could be transferred by conjugation even when no genes on the main chromosome were transferred. Eventually it was shown that the genes for antibiotic resistance are carried on plasmids. Each R factor carries one or more genes conferring resistance to particular antibiotics, as well as genes that code for proteins involved in the conjugation with a recipient bacterium. As far as biologists can determine, R factors providing resistance to naturally occurring antibiotics existed long before antibiotics were discovered and used

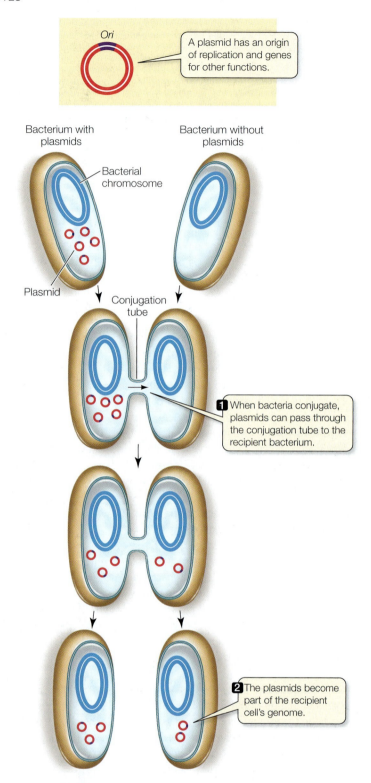

A plasmid has an origin of replication and genes for other functions.

Bacterium with plasmids

Bacterium without plasmids

Bacterial chromosome

Plasmid

Conjugation tube

1 When bacteria conjugate, plasmids can pass through the conjugation tube to the recipient bacterium.

2 The plasmids become part of the recipient cell's genome.

13.14 Gene Transfer by Plasmids When plasmids enter a cell via conjugation, their genes can be expressed in the recipient cell.

by humans. However, R factors seem to have become more abundant in modern times, possibly because the heavy use of antibiotics in hospitals selects for bacterial strains bearing them.

Antibiotic resistance poses a serious threat to human health, and the inappropriate use of antibiotics contributes to this prob-

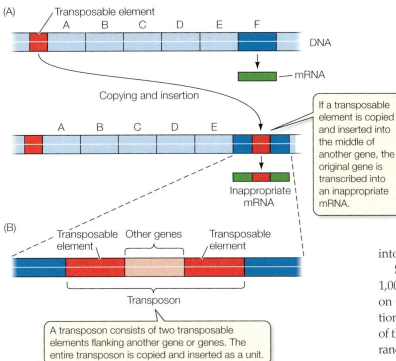

(A)
Transposable element

Copying and insertion

If a transposable element is copied and inserted into the middle of another gene, the original gene is transcribed into an inappropriate mRNA.

Inappropriate mRNA

(B)
Transposable element / Other genes / Transposable element

Transposon

A transposon consists of two transposable elements flanking another gene or genes. The entire transposon is copied and inserted as a unit.

13.15 Transposable Elements and Transposons
(A) Transposable elements are segments of DNA that can be inserted at new locations, either on the same chromosome or on a different chromosome. (B) Transposons consist of transposable elements combined with other genes.

inserted either at a new location on the same chromosome or into another chromosome. These DNA sequences are called **transposable elements**. Their insertion often produces phenotypic effects by disrupting the genes into which they are inserted (**Figure 13.15A**).

Some transposable elements are relatively short sequences, 1,000–2,000 base pairs long. Such sequences are found at many sites on the *E. coli* main chromosome. In one mechanism of transposition, the transposable element replicates independently of the rest of the chromosome. The copy then inserts itself at other, seemingly random sites on the chromosome. The genes encoding the enzymes necessary for this insertion are found within the transposable element itself. Other transposable elements are cut from their original sites and inserted elsewhere without replication. Longer transposable elements (about 5,000 base pairs) carry one or more additional genes and are called **transposons** (**Figure 13.15B**).

Transposable elements have contributed to the evolution of plasmids. R factors probably originally gained their genes for antibiotic resistance through the activity of transposable elements. One piece of evidence for this conclusion is that each resistance gene in an R factor was originally part of a transposon.

As we have seen, rapid asexual reproduction can produce huge clones of prokaryotes. These genetically identical cells are all equally vulnerable to any change in the environment. Recombination by means of conjugation, transformation, and transduction, or the acquisition of new genes by means of plasmids and transposable elements, introduces genetic diversity into bacterial populations, and this diversity may allow at least some cells to survive under changing conditions.

lem. You have probably gone to a physician because of a sore throat, which can have either a viral or a bacterial cause. The best way to determine the causative agent is for the doctor to take a small sample from your inflamed throat, culture it, and identify any bacteria that are present. But perhaps you feel you cannot wait another day for the results. Impatient, you ask the doctor to give you something to make you feel better. The doctor prescribes an antibiotic, which you take. The sore throat gradually gets better, and you think that the antibiotic did the job.

But suppose the infection is viral. In that case, the antibiotic has done nothing to combat the infection, which just runs its normal course. However, it may do something harmful: by killing many normal bacteria in your body, the antibiotic may select for bacteria harboring R factors. These bacteria may survive and reproduce in the presence of the antibiotic, and may soon become quite numerous. You may continue to harbor these bacteria or pass them on to other people. The next time you get a bacterial infection, there may be a ready supply of resistant bacteria in your body, and antibiotics may be ineffective.

Acquisition of antibiotic resistance in pathogenic bacteria provides an example of evolution in action. In the years after they were first discovered in the twentieth century, antibiotics were very successful in combating diseases that had plagued humans for millennia, such as cholera, tuberculosis, and leprosy. But over time, resistant bacteria have appeared. This is classic natural selection: genetic variation exists among bacteria, and those that have survived the onslaught of antibiotics must have a genetic constitution that allows them to do so.

Transposable elements move genes among plasmids and chromosomes

As we have seen, plasmids, viruses, and even phage capsids (in the case of transduction) can transport genes from one bacterial cell to another. There is another type of "gene transport" that occurs within the individual cell. It relies on segments of DNA that can be

13.3 RECAP

Bacteria can exchange genetic information through conjugation, transformation, and transduction. Plasmids are small chromosomes that can be transferred between bacteria. Transposable elements and transposons are small pieces of DNA that can move from one place to another in the genome.

- Describe the experiment that demonstrated that genetic recombination occurs in bacteria. See p. 291 and Figure 13.10

- How is DNA exchanged between bacteria during transformation? Transduction? See pp. 292–293 and Figure 13.13

- Do you understand the importance of plasmids in spreading antibiotic resistance? See pp. 294–295

Antibiotics are not the only threats to the survival of prokaryotes. Such threats are numerous, and include changes in environmental temperature and scarcity of resources. Prokaryotes may respond to changes in their environment not only by exchanging genes, but also by regulating the expression of their genes.

13.4 How Is Gene Expression Regulated in Prokaryotes?

Prokaryotes conserve energy and resources by making proteins only when they are needed. The protein content of a bacterium can change rapidly when conditions warrant. There are several ways in which a prokaryotic cell could shut off the supply of an unneeded protein. The cell can:

- downregulate the transcription of mRNA for that protein
- hydrolyze the mRNA after it is made and prior to translation
- prevent translation of the mRNA at the ribosome
- hydrolyze the protein after it is made
- inhibit the function of the protein

Whichever mechanism is used, it must do two things:

- It must respond to environmental signals. (If you find yourself feeling hungry, you might go to the kitchen.)
- It must be efficient. (Why wander through every room in the house if all you want is a midnight snack in the kitchen?)

The earlier the cell intervenes in the process of protein synthesis, the less energy it has to expend. Selective blocking of transcription is far more efficient than transcribing the gene, translating the message, and then degrading or inhibiting the protein. While examples of all five mechanisms for regulating protein levels are found in nature, prokaryotes generally use the most efficient one: transcriptional regulation.

Regulating gene transcription conserves energy

As a normal inhabitant of the human intestine, *E. coli* must be able to adjust to sudden changes in its chemical environment. Its host may present it with one foodstuff one hour and another the next. This variation presents the bacterium with a metabolic challenge. Glucose is its preferred energy source, and is the easiest sugar to metabolize, but not all of its host's foods contain an abundant supply of glucose. For example, the bacterium may suddenly be deluged with milk, whose predominant sugar is lactose. Lactose is a β-galactoside—a disaccharide containing galactose β-linked to glucose (see Section 3.3). To be taken up and metabolized by *E. coli*, lactose must be acted on by three proteins:

- *β-galactoside permease* is a carrier protein in the bacterial plasma membrane that moves the sugar into the cell.
- *β-galactosidase* is an enzyme that catalyzes the hydrolysis of lactose to glucose and galactose.
- *β-galactoside transacetylase* is an enzyme that transfers acetyl groups from acetyl CoA to certain β-galactosides. Its role in the metabolism of lactose is not clear.

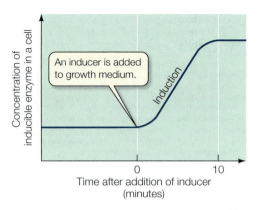

13.16 An Inducer Stimulates the Synthesis of an Enzyme It is most efficient for a cell to produce an enzyme only when it is needed. Some enzymes are induced by the presence of the substance they act upon (for example, β-galactosidase is induced by the presence of lactose).

When *E. coli* is grown on a medium that contains glucose, but no lactose or other β-galactosides, the levels of these three proteins are extremely low—the cell does not waste energy and materials making the unneeded enzymes. If, however, the environment changes such that lactose is the predominant sugar available and very little glucose is present, the bacterium promptly begins making all three enzymes, and they increase rapidly in abundance. For example, there are only two molecules of β-galactosidase present in an *E. coli* cell when glucose is present in the medium. But when glucose is absent, the presence of lactose can induce the synthesis of 3,000 molecules of β-galactosidase per cell!

If lactose is removed from *E. coli*'s environment, synthesis of the three enzymes that process it stops almost immediately. The enzyme molecules that have already formed do not disappear; they are merely diluted during subsequent cell divisions until their concentration falls to the original low level within each bacterium.

Compounds that stimulate the synthesis of a protein (such as lactose in our example) are called **inducers** (**Figure 13.16**). The proteins that are produced are called **inducible** proteins, whereas proteins that are made all the time at a constant rate are called **constitutive** proteins.

We have now seen two basic ways of regulating the rate of a metabolic pathway. Section 6.5 described allosteric regulation of enzyme activity (the rate of enzyme-catalyzed reactions); this mechanism allows rapid fine-tuning of metabolism. Regulation of protein synthesis—that is, regulation of the concentration of enzymes—is slower, but produces a greater savings of energy. **Figure 13.17** compares these two modes of regulation.

A single promoter can control the transcription of adjacent genes

The genes that encode the synthesis of the three enzymes that process lactose in *E. coli* are called **structural genes**, indicating that they specify the primary structure (the amino acid sequence) of a protein molecule. In other words, structural genes are genes that can be transcribed into mRNA.

13.17 Two Ways to Regulate a Metabolic Pathway Feedback from the end product of a metabolic pathway can block enzyme activity (allosteric regulation), or it can stop the transcription of genes that code for the enzymes in the pathway (transcriptional regulation).

Regulation of enzyme activity

The end product feeds back, inhibiting the activity of enzyme 1 only, and quickly stopping the pathway.

Precursor → Enzyme 1 → A → Enzyme 2 → B → Enzyme 3 → C → Enzyme 4 → D → Enzyme 5 → End product

Gene 1   Gene 2   Gene 3   Gene 4   Gene 5

Regulation of enzyme concentration

The end product blocks the transcription of all five genes. No enzymes are produced.

The three structural genes involved in the metabolism of lactose lie adjacent to one another on the *E. coli* chromosome. This arrangement is no coincidence: their DNA is transcribed into a single, continuous molecule of mRNA. Because this particular mRNA governs the synthesis of all three lactose-metabolizing enzymes, either all or none of these enzymes are made, depending on whether their common message—their mRNA—is present in the cell.

The three genes share a single promoter. As we saw in Section 12.3, a *promoter* is a DNA sequence to which RNA polymerase binds to initiate transcription. We also noted that some promoters are more effective at transcription initiation than others. The promoter for these three *E. coli* structural genes can be very effective, so the maximum rate of mRNA synthesis can be high. There is, however, a mechanism to shut down mRNA synthesis when the enzymes are not needed. That mechanism is called an *operon*, and was elegantly worked out by François Jacob and Jacques Monod.

Operons are units of transcription in prokaryotes

Prokaryotes shut down transcription by placing an obstacle between the promoter and the structural genes it regulates. A short stretch of DNA called the **operator** lies in this position. It can bind very tightly with a special type of protein molecule, called a **repressor**, to create such an obstacle:

■ When the repressor protein is bound to the operator, it blocks the transcription of mRNA.

■ When the repressor is not attached to the operator, mRNA synthesis proceeds constitutively.

The whole unit, consisting of the closely linked structural genes and the DNA sequences that control their transcription, is called an **operon**. An operon always consists of a promoter, an operator, and two or more structural genes (**Figure 13.18**). The promoter and operator are binding sites on DNA and are not transcribed.

E. coli has numerous mechanisms to control the transcription of operons; we will focus on three of them here. Two of these con-

trol mechanisms depend on interactions of a repressor protein with the operator, and the third depends on interactions of other proteins with the promoter.

Operator–repressor control induces transcription in the *lac* operon

The operon containing the genes for the three lactose-metabolizing proteins of *E. coli* is called the *lac operon* (see Figure 13.18). As we have just seen, RNA polymerase can bind to the promoter, and a repressor protein can bind to the operator.

The repressor protein has two binding sites: one for the operator and the other for inducers. The inducers of the *lac* operon, as we know, are molecules of lactose. Binding with an inducer changes the shape of the repressor protein (by allosteric modification; see Section 6.3). This change in shape prevents the repressor from binding to the operator (**Figure 13.19**). As a result, RNA polymerase can bind to the promoter and start transcribing the structural genes of the *lac* operon. The mRNA transcribed from these genes is translated on ribosomes to synthesize the three proteins required for metabolizing lactose.

What happens if the concentration of lactose drops? As the lactose concentration decreases, the inducer (lactose) molecules separate from the repressor. Free of lactose molecules, the repressor

13.18 The *lac* Operon of *E. coli* The *lac* operon of *E. coli* is a segment of DNA that includes a promoter, an operator, and the three structural genes that code for lactose-metabolizing enzymes.

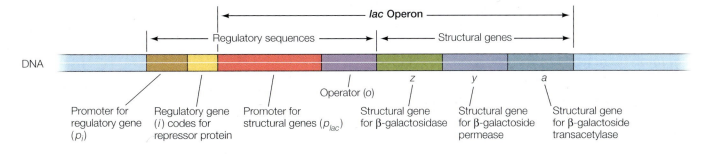

lac Operon

Regulatory sequences — Structural genes

DNA

Operator (*o*)

z   *y*   *a*

Promoter for regulatory gene (*p_i*)

Regulatory gene (*i*) codes for repressor protein

Promoter for structural genes (*p_lac*)

Structural gene for β-galactosidase

Structural gene for β-galactoside permease

Structural gene for β-galactoside transacetylase

13.19 The *lac* Operon: An Inducible System Lactose (the inducer) leads to synthesis of the enzymes in the lactose-metabolizing pathway by preventing the repressor protein (which would have stopped transcription) from binding to the operator.

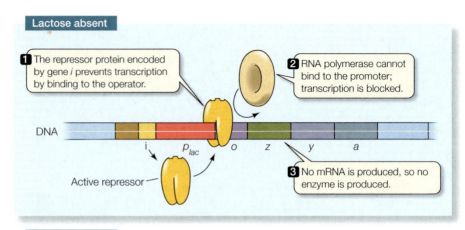

Lactose absent

1 The repressor protein encoded by gene *i* prevents transcription by binding to the operator.

2 RNA polymerase cannot bind to the promoter; transcription is blocked.

DNA

i p_{lac} *o*   *z*   *y*   *a*

Active repressor

3 No mRNA is produced, so no enzyme is produced.

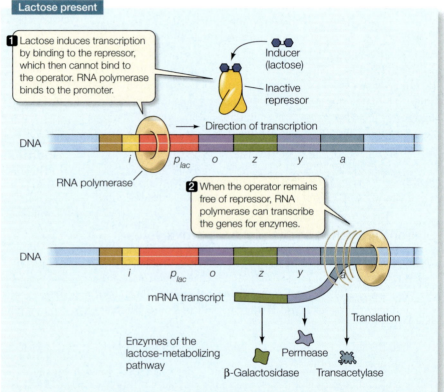

Lactose present

1 Lactose induces transcription by binding to the repressor, which then cannot bind to the operator. RNA polymerase binds to the promoter.

Inducer (lactose)

Inactive repressor

Direction of transcription

DNA

i p_{lac} *o*   *z*   *y*   *a*

RNA polymerase

2 When the operator remains free of repressor, RNA polymerase can transcribe the genes for enzymes.

DNA

i p_{lac} *o*   *z*   *y*   *a*

mRNA transcript

Translation

Enzymes of the lactose-metabolizing pathway

β-Galactosidase   Permease   Transacetylase

returns to its original shape and binds to the operator, and transcription of the *lac* operon stops. Translation stops soon thereafter because the mRNA that is already present breaks down quickly. Thus it is the presence or absence of lactose—the inducer—that regulates the binding of the repressor to the operator, and therefore the synthesis of the proteins needed to metabolize it. In other words, the *lac* operon is an *inducible system*.

Repressor proteins are encoded by **regulatory genes**. The regulatory gene that codes for the repressor of the *lac* operon is called the *i* (inducibility) gene. The *i* gene happens to lie close to the operon that it regulates (see Figure 13.18), but some regulatory genes are distant from their operons. Like all other genes, the *i* gene itself has a promoter, which can be designated p_i. Because this promoter does not bind RNA polymerase very effectively, only enough mRNA to synthesize about ten molecules of repressor protein per cell per generation is produced. This quantity of the repressor is enough to regulate the operon effectively—producing more would be a waste of energy. There is no operator between p_i and the *i* gene. Therefore, the repressor of the *lac* operon is a constitutive protein; that is, it is made at a constant rate that is not subject to environmental control.

Let's review the important features of inducible systems such as the *lac* operon:

- In the absence of inducer, the operon is turned off.

- Control is exerted by a regulatory protein—the repressor—that turns the operon off.

- Adding inducer changes the repressor and turns the operon on.

- Regulatory genes produce proteins whose sole function is to regulate the expression of other genes.

- Certain other DNA sequences (operators and promoters) do not code for proteins, but are binding sites for regulatory or other proteins.

Operator–repressor control represses transcription in the *trp* operon

We have seen how *E. coli* benefits from having an inducible system for lactose metabolism: only when lactose is present does the system switch on. Equally valuable to a bacterium is the ability to switch off the synthesis of certain enzymes in response to the excessive accumulation of their end products. For example, when the amino acid tryptophan, an essential constituent of proteins, is present in ample concentrations, it is advantageous to stop making the enzymes for tryptophan synthesis. When the synthesis of a protein can be turned off in response to such a biochemical cue, that protein is said to be **repressible**.

In *repressible systems*, such as the *trp* operon that controls the synthesis of tryptophan, the repressor protein cannot shut off its operon unless it first binds to a **corepressor**, which may be either the metabolic end product itself (tryptophan in this case) or an analog of it (**Figure 13.20**). If the end product is absent, the repressor protein cannot bind to the operator, and the operon is transcribed at the maximal rate. If the end product is present, the repressor binds to the operator, and the operon is turned off.

The difference between inducible and repressible systems is small, but significant:

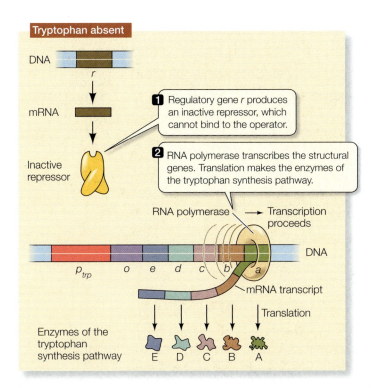

Tryptophan absent

DNA

1 Regulatory gene *r* produces an inactive repressor, which cannot bind to the operator.

mRNA

Inactive repressor

2 RNA polymerase transcribes the structural genes. Translation makes the enzymes of the tryptophan synthesis pathway.

RNA polymerase → Transcription proceeds

DNA

p_{trp} o e d c b a

mRNA transcript

Translation

Enzymes of the tryptophan synthesis pathway   E D C B A

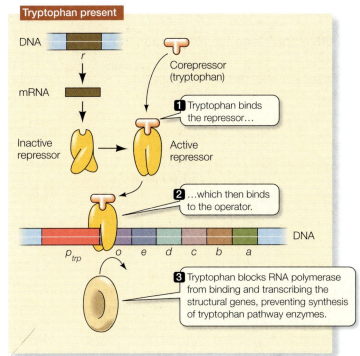

Tryptophan present

DNA

Corepressor (tryptophan)

mRNA

1 Tryptophan binds the repressor…

Inactive repressor → Active repressor

2 …which then binds to the operator.

DNA

p_{trp} o e d c b a

3 Tryptophan blocks RNA polymerase from binding and transcribing the structural genes, preventing synthesis of tryptophan pathway enzymes.

13.20 The *trp* Operon: A Repressible System Because tryptophan activates an otherwise inactive repressor, it is called a corepressor.

- In inducible systems, the *substrate* of a metabolic pathway (the inducer) interacts with a regulatory protein (the repressor), rendering the repressor incapable of binding to the operator and thus *allowing* transcription.

- In repressible systems, the *product* of a metabolic pathway (the corepressor) interacts with a regulatory protein to make it capable of binding to the operator, thus *blocking* transcription.

In general, inducible systems control catabolic pathways (which are turned on only when the substrate is available), whereas repressible systems control anabolic pathways (which are turned off until the product becomes unavailable). In both systems, the regulatory molecule functions by binding to the operator. Next we will consider an example of control by binding to the promoter.

Protein synthesis can be controlled by increasing promoter efficiency

Suppose an *E. coli* cell lacks a sufficient supply of glucose, its preferred energy source, but has access to lactose, another sugar. The *lac* operon encodes enzymes that catabolize lactose, and can increase transcription of those enzymes by increasing the efficiency of the promoter (**Figure 13.21**).

In the *lac* operon and other similar operons, the promoter binds RNA polymerase in a series of steps. First,

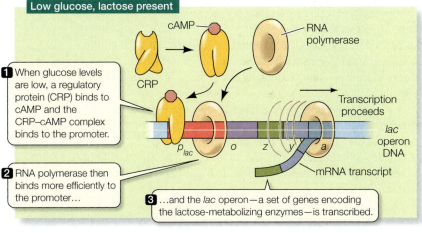

Low glucose, lactose present

cAMP

RNA polymerase

CRP

1 When glucose levels are low, a regulatory protein (CRP) binds to cAMP and the CRP–cAMP complex binds to the promoter.

Transcription proceeds

p_{lac} o z y a

lac operon DNA

2 RNA polymerase then binds more efficiently to the promoter…

mRNA transcript

3 …and the *lac* operon—a set of genes encoding the lactose-metabolizing enzymes—is transcribed.

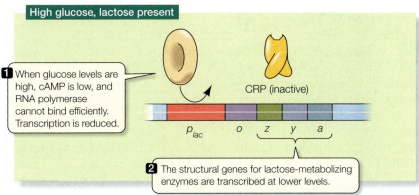

High glucose, lactose present

1 When glucose levels are high, cAMP is low, and RNA polymerase cannot bind efficiently. Transcription is reduced.

CRP (inactive)

p_{lac} o z y a

2 The structural genes for lactose-metabolizing enzymes are transcribed at lower levels.

13.21 Catabolite Repression Regulates the *lac* Operon The promoter for the *lac* operon does not function efficiently in the absence of cAMP, as occurs when glucose levels are high. High glucose levels thus repress the catabolites (metabolic enzymes) for lactose.

TABLE 13.2

Positive and Negative Regulation in the *lac* Operon[a]

GLUCOSE	cAMP LEVELS	RNA POLYMERASE BINDING TO PROMOTER	LACTOSE	*lac* REPRESSOR	TRANSCRIPTION OF *lac* GENES?	LACTOSE USED BY CELLS?
Present	Low	Absent	Absent	Active and bound to operator	No	No
Present	Low	Present, not efficient	Present	Inactive and not bound to operator	Low level	No
Absent	High	Present, very efficient	Present	Inactive and not bound to operator	High level	Yes
Absent	High	Absent	Absent	Active and bound to operator	No	No

[a]Negative regulators are in red type.

a regulatory protein called CRP (short for *c*AMP *r*eceptor *p*rotein) binds the nucleotide adenosine 3′,5′-cyclic monophosphate, better known as cyclic AMP, or cAMP. Next, the CRP–cAMP complex binds to DNA just upstream (5′) of the promoter. This binding results in more efficient binding of RNA polymerase to the promoter, and thus an elevated level of transcription of the structural genes.

When glucose becomes abundant in the medium, the bacterium does not need to break down alternative food molecules, so synthesis of the enzymes that catabolize these molecules diminishes or ceases. The presence of glucose decreases the synthesis of these enzymes by lowering the cellular concentration of cAMP. The lower cAMP concentration leads to less CRP binding to the promoter, less efficient binding of RNA polymerase, and reduced transcription of the structural genes. This mechanism is called **catabolite repression**.

CRP controls the efficiency of RNA polymerase binding in many other operons, such as those that metabolize the sugars arabinose and galactose. In general, when an adequate level of glucose is present, *E. coli* uses the CRP system to turn off pathways that use other energy sources. As you will see in later chapters of this book, cAMP is a widely used signaling molecule in eukaryotes as well as in prokaryotes. The use of cAMP in such widely diverse situations as a bacterium sensing glucose levels and a human sensing hunger demonstrates the prevalence of common themes in biochemistry and natural selection.

The inducible *lac* and repressible *trp* systems of *E. coli*—the two operator–repressor systems—are examples of **negative control** of transcription because the regulatory protein (the repressor) in each case prevents transcription. The catabolite repression mechanism is an example of **positive control** of transcription because the regulatory molecule (the CRP–cAMP complex) is a transcriptional activator. The relationships between these positive and negative regulators in the *lac* operon are summarized in **Table 13.2**.

The control of gene expression by regulatory proteins is not unique to prokaryotes. It occurs in viruses, as we saw earlier, and as we will see in the next chapter, it is important in eukaryotes as well. Throughout nature, genomes contain not only sequences that are transcribed, but nontranscribed sequences that, by binding proteins, can determine whether a particular gene will be transcribed.

13.4 RECAP

Gene expression in prokaryotes is most commonly regulated through control of transcription. Operons consist of a set of closely linked genes and DNA sequences that control their transcription. Operons can be regulated by both negative and positive controls.

- What are the key differences between an inducible system and a repressible system? See pp. 298–299 and Figures 13.19 and 13.20

- Describe the difference between positive and negative control of transcription. See pp. 299–300 and Table 13.2

Clearly, an amazing amount of knowledge about prokaryotic genomes has been recently acquired, and advanced sequencing and genomic techniques promise to deliver more information about specific prokaryotic genes and their functions. To what uses might we hope to put this information?

13.5 What Have We Learned from the Sequencing of Prokaryotic Genomes?

When DNA sequencing first became possible in the late 1970s, the first organisms to be sequenced were the simplest viruses. Soon, over 150 viral genomes, including those of important animal and plant pathogens, had been sequenced. Information on how these viruses infect their hosts and reproduce came quickly as a result. But the manual sequencing techniques used on viruses were not up to the task of elucidating the genomes of prokaryotes and eukaryotes, the smallest of which are a hundred times larger than those of a bacteriophage. In the past decade, however, the automated sequencing techniques described in Section 11.5 have rapidly added many prokaryotic sequences to biologists' store of knowledge.

In 1995, a team led by Craig Venter and Hamilton Smith determined the first sequence of a free-living cellular organism, the bacterium *Haemophilus influenzae*. Many more prokaryotic sequences have followed. These sequences have revealed not only how prokaryotes apportion their genes to perform different cellular functions, but also how their specialized functions are carried out. A beginning has even been made on the provocative question of what the minimal requirements for a living cell might be.

Three types of information can be obtained from a genomic sequence, including:

■ *Open reading frames,* which are the coding regions of genes. For protein-coding genes, these regions can be recognized by the start and stop codons for translation.

■ *Amino acid sequences of proteins.* These sequences can be deduced from the DNA sequences of open reading frames by applying the genetic code (see Figure 12.6).

■ *Regulatory sequences,* such as promoters and terminators for transcription.

Perhaps the record for speedy genome sequencing was that of the virus that causes SARS (Severe Adult Respiratory Syndrome). It took less than two weeks for a team led by Steven Jones at the British Columbia Cancer Center in Vancouver to sequence the viral genome.

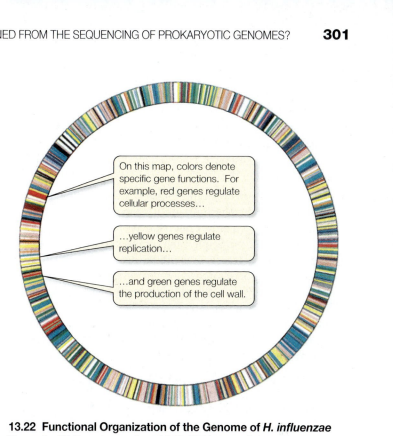

13.22 Functional Organization of the Genome of *H. influenzae*
The entire DNA sequence has 1,830,137 base pairs.

FUNCTIONAL GENOMICS   Functional genomics is the assignment of functions to the products of genes described by genomic sequencing. This field, less than a decade old, is now a major occupation of biologists. Let's see how its methods were applied to the bacterium *H. influenzae* once its sequence was known.

The only host for *H. influenzae* is humans. It lives in the upper respiratory tract and can cause ear infections or, more seriously, meningitis in children. Its single circular chromosome has 1,830,137 base pairs (**Figure 13.22**). In addition to its origin of replication and the genes coding for rRNAs and tRNAs, this bacterial chromosome has 1,743 regions containing amino acid codons as well as the transcriptional (promoter) and translational (start and stop codons) information needed for protein synthesis—that is, regions that are likely to be genes that code for proteins.

When this sequence was first announced, only 1,007 (58 percent) of the bacterium's genes had amino acid sequences that corresponded to proteins with known functions—in other words, only 58 percent were genes that the researchers, based on their knowledge of the functions of bacteria, expected to find. The remaining 42 percent of its genes coded for proteins that were unknown to researchers. The roles of most of the unknown proteins have been identified since that time by a process known as **annotation**.

Most annotation of genes and proteins with known functions confirmed a century of biochemical description of bacterial enzymatic pathways. For example, genes for enzymes making up entire pathways of glycolysis, fermentation, and electron transport were found. Some of the remaining gene sequences for unknown proteins may code for membrane proteins, including those involved in active transport. Another important finding was that highly infective strains of *H. influenzae* have genes coding for sur-

face proteins that attach the bacterium to the human respiratory tract, while noninfective strains lack those genes.

COMPARATIVE GENOMICS   Soon after the sequence of *H. influenzae* was announced, smaller (*Mycoplasma genitalium*, 580,070 base pairs) and larger (*E. coli*, 4,639,221 base pairs) prokaryotic sequences were completed. Thus began a new era in biology, the era of **comparative genomics**, in which the genome sequences of different organisms are compared to see what genes one organism has that are missing in another, in order to relate the results to physiology.

M. genitalium, for example, lacks the enzymes needed to synthesize amino acids, which the other two prokaryotes possess. This finding reveals that *M. genitalium* is a parasite, which must obtain all its amino acids from its environment, the human urogenital tract. *E. coli* has 55 regulatory genes coding for transcriptional activators and 58 for repressors; *M. genitalium* has only 3 genes for activators. Comparisons such as these have led to the formulation of specific questions about how an organism lives the way it does. We'll see many more applications of comparative genomics in the next chapter.

The sequencing of prokaryotic genomes has many potential benefits

Prokaryotic genome sequencing promises to provide insights into microorganisms that cause human diseases as well as those involved in global ecological cycles (see Chapter 56).

■ *Chlamydia trachomatis* causes the most common sexually transmitted disease in the United States. Because it is an intracellular parasite, it has been very hard to study. Among its

900 genes are several for ATP synthesis—something scientists used to think this bacterium could not do.

- *Rickettsia prowazekii* causes typhus; it infects people bitten by louse vectors. Of its 634 genes, 6 code for proteins that are essential for its virulence. The sequences of these genes are being used to develop vaccines.

- Not a lot is known about the Sargasso Sea, a floating mass of warm water in the North Atlantic Ocean where a large mat of *Sargassum* seaweed grows, but relatively few other life forms have been observed. Sequencing of cells in this water has shown that it harbors its own unique ecosystem, with over 1,000 bacterial species, most of them never described before, that are specially adapted to live among the *Sargassum* mats. Earth's oceans may contain up to 2 million prokaryotic species, and the soils another 4 million. The roles of 99 percent of them in Earth's ecology remain obscure. Sequencing their genomes may provide some answers.

- *Mycobacterium tuberculosis* causes tuberculosis. It has a large genome for a prokaryote, coding for 4,000 proteins. Over 250 of these proteins are used to metabolize lipids, so this may be the main way that the bacterium gets its energy. Some of its genes code for previously unidentified cell surface proteins; these genes are targets for potential vaccines.

- *Streptomyces coelicolor* and its close relatives produce two-thirds of all the antibiotics currently in clinical use, including streptomycin, tetracycline, and erythromycin. The genome sequence of this bacterium reveals 22 clusters of genes responsible for antibiotic production, of which only 4 were previously known. This finding may lead to more and better antibiotics to combat resistant pathogens.

- In addition to the well-known carbon dioxide, another important gas contributing to the atmospheric "greenhouse effect" and global warming is methane (CH_4; see Figure 2.8). Some bacteria, such as *Methanococcus*, produce methane in the stomachs of cows. Others, such as *Methylococcus*, remove methane from the air and use it as a carbon–energy source. The genomes of both of these bacteria have been sequenced. Understanding the genes involved in methane production and oxidation may help us meet the challenge of global warming.

- *E. coli* strain O157:H7 in hamburger can cause severe illness when in-

gested, as happens to at least 70,000 people a year in the United States. Its genome has 5,416 genes, of which 1,387 are different from those in the familiar (and harmless) laboratory strains of this bacterium. Remarkably, many of these unique genes are also present in other pathogenic bacteria, such as *Salmonella* and *Shigella*. This finding suggests that there is extensive genetic exchange between these species, and that "superbugs" may be on the horizon.

Will defining the genes required for cellular life lead to artificial life?

When the genomes of prokaryotes and eukaryotes are compared, a striking conclusion arises: certain genes are present in all organisms (*universal genes*). There are also some (nearly) universal gene segments—one that codes for an ATP binding site, for example—

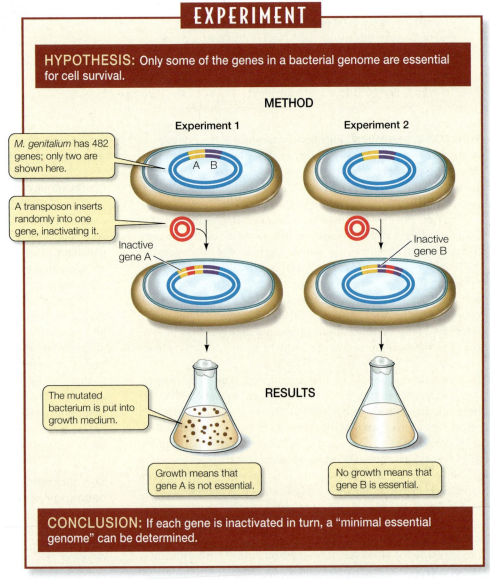

EXPERIMENT

HYPOTHESIS: Only some of the genes in a bacterial genome are essential for cell survival.

METHOD

Experiment 1 — Experiment 2

M. genitalium has 482 genes; only two are shown here.

A transposon inserts randomly into one gene, inactivating it.

Inactive gene A — Inactive gene B

RESULTS

The mutated bacterium is put into growth medium.

Growth means that gene A is not essential. — No growth means that gene B is essential.

CONCLUSION: If each gene is inactivated in turn, a "minimal essential genome" can be determined.

13.23 Using Transposon Mutagenesis to Determine the Minimal Genome By inactivating the genes in a bacterial genome one by one, scientists can determine which genes are essential for the cell's survival.

that are present in many genes in many organisms. These findings suggest that there is some ancient, minimal set of DNA sequences common to all cells. One way to identify these sequences is to look for them (or, more realistically, to have a computer look for them).

Another way to define the minimal genome is to take the organism with the simplest genome, deliberately mutate one gene at a time, and see what happens. *Mycoplasma genitalium* has the smallest known genome—only 482 genes. Even so, some of its genes are dispensable under some circumstances. It has genes for metabolizing both glucose and fructose. In the laboratory, the organism can survive on a medium supplying only one of those sugars, making the genes for metabolizing the other sugar unnecessary. But what about other genes? Experiments using transposons as mutagens have addressed this question. When the bacterium is exposed to transposons, they insert themselves into a gene at random, mutating and inactivating it (**Figure 13.23**). The mutated bacteria are sequenced to determine which gene was mutated, and then tested for growth and survival.

The astonishing result of these studies is that *M. genitalium* can survive in the laboratory without the services of 100 of its genes, leaving a minimal genome of 382 genes! But is this really what it takes to make a viable organism? The only way to find out would be to build the genome, gene by gene. A team led by Craig Venter at a private company is trying to do this. Since it is possible to synthesize artificial DNA molecules, and since the DNA sequences of all of the genes of *M. genitalium* are known, making and then assembling the genome in the laboratory should be straightforward. The genome will then be inserted into an empty bacterial cell. If it starts transcribing mRNA and making proteins, it may result in a viable cell. If so, human-created life will be a reality.

In addition to the obvious technical feat of creating artificial life, this technique could have many important applications. New microbes could be made with entirely new abilities, such as degrading oil spills, making synthetic fibers, reducing tooth decay, or converting cellulose to ethanol for use as fuel. On the other hand, fears of the misuse or mishandling of this knowledge are not unfounded. It would be equally possible to develop synthetic bacteria with detrimental properties, such as species toxic to people, plants, or animals, and use them as agents of biological warfare or bioterrorism. As described at the beginning of this chapter, the mutation of avian flu into a lethal human pathogen suggests the danger that new species, however small, can sometimes pose. The "genomics genie" is, for better or worse, already out of the bottle.

13.5 RECAP

DNA sequencing is used to study the genomes of prokaryotes of importance to humans and ecosystems. Functional genomics uses these sequences to determine the functions of gene products; comparative genomics uses sequence similarities with genes with known roles in other organisms to help identify functions.

- Can you describe how open reading frames are recognized in a genomic sequence? What kind of information can be derived from open reading frames? See p. 301

- Can you give some examples of prokaryotic genomes that have been sequenced and what the sequence has shown? See p. 302

- How are selective inactivation studies being used to determine the minimal genome? See pp. 302–303 and Figure 13.23

CHAPTER SUMMARY

13.1 How do viruses reproduce and transmit genes?

Prokaryotes and viruses are useful model organisms for the study of genetics and molecular biology because they contain much less DNA than eukaryotes, grow and reproduce rapidly, and are usually haploid.

Viruses are not cells, and rely on host cells to reproduce.

The basic unit of a virus is a **virion**, which consists of a nucleic acid genome (DNA or RNA) and a protein coat, called a **capsid**.

Bacteriophage are viruses that infect bacteria.

Virulent viruses undergo a **lytic cycle**, which causes the host cell to burst, releasing new virions.

Temperate viruses can also undergo a **lysogenic cycle**, in which a molecule of their DNA, called a **prophage**, is inserted into the host chromosome, where it replicates for generations. Such hosts are referred to as **lysogenic bacteria**. Review Figure 13.3

Some viruses have promoters that bind host RNA polymerase, which they use to transcribe their own genes and proteins. Review Figure 13.4

Many animal and plant viruses are spread by **vectors**, such as insects.

The viruses that infect animals include both RNA and DNA viruses. Enveloped viruses have a membrane derived from the host's plasma membrane. Review Figure 13.5

A **retrovirus** uses reverse transcriptase to generate a cDNA **provirus** from its RNA genome. The provirus is incorporated into the animal host's DNA and can be activated to produce new virions. Review Figure 13.6

13.2 How is gene expression regulated in viruses?

Whether a temperate phage undergoes a lytic or a lysogenic cycle depends on the state of cellular environment offered by the host, which is assessed by regulatory proteins that compete for promoters on phage DNA. Review Figure 13.8

13.3 How do prokaryotes exchange genes?

Most prokaryotes reproduce by the division of single cells, giving rise to **clones** of genetically identical cells. Allowed to multiply on a

CHAPTER SUMMARY

solidified nutrient medium, they will produce bacterial colonies or a bacterial lawn. Review Figure 13.9

An experiment by Lederberg and Tatum demonstrated that bacteria can exchange genetic information. Review Figure 13.10

In **conjugation**, two bacterial cells are brought into proximity by **sex pili**. A portion of the donor cell's chromosome moves into the recipient cell through a **conjugation tube** and may become incorporated into the recipient cell's chromosome.

In **transformation**, bacteria take up free DNA released by dead bacterial cells.

In **transduction**, phage capsids carry bacterial DNA from one bacterium to another. Review Figure 13.13

Plasmids are small bacterial chromosomes that are independent of the main chromosome and may be transferred between cells. Plasmids may act as **metabolic factors**, **fertility factors**, or **resistance factors**. Review Figure 13.14

Transposable elements are short DNA sequences that can move from one place to another in the bacterial genome, often interrupting gene sequences. Transposable elements that carry other genes are called **transposons**. Review Figure 13.15

13.4 How is gene expression regulated in prokaryotes?

In prokaryotes, the synthesis of some proteins is regulated so that they are made only when they are needed. Proteins that are made only in the presence of a particular compound—an **inducer**—are **inducible** proteins. Proteins that are made at a constant rate regardless of conditions are **constitutive** proteins.

An **operon** consists of a promoter, an **operator**, and two or more **structural genes**. Promoters and operators do not code for pro-

teins, but serve as binding sites for regulatory proteins. Review Figure 13.18

Regulatory genes code for regulatory proteins, such as **repressors**. When a repressor binds to an operator, transcription of the structural gene is inhibited. Review Figure 13.19, Web/CD Tutorial 13.1

The *lac* operon is an example of an inducible system, in which the presence of an inducer keeps the repressor from binding the operator and allows transcription.

A **repressible** protein is one whose synthesis is turned off by the presence of a particular compound—a **corepressor**. The *trp* operon is an example of a repressible system, in which the presence of a corepressor causes the repressor to bind to the operator and stops transcription. Review Figure 13.20, Web/CD Tutorial 13.2

In catabolite repression transcription is enhanced by the binding of a regulatory protein to the promoter. Review Figure 13.21

13.5 What have we learned from the sequencing of prokaryotic genomes?

Functional genomics relates gene sequences to protein functions.

In **comparative genomics**, the genome sequences of different organisms are compared.

Sequencing studies have revealed universal genes that are present in all organisms and unique genes associated with organism functions.

By mutating individual genes in a small genome, scientists can determine the minimal genome required for cellular life. These studies could lead to the creation of artificial cells. Review Figure 13.23

See Web/CD Activity 13.1 for a concept review of this chapter.

SELF-QUIZ

1. Which of the following statements about the *lac* operon is *not* true?
   a. When lactose binds to the repressor, the repressor can no longer bind to the operator.
   b. When lactose binds to the operator, transcription is stimulated.
   c. When the repressor binds to the operator, transcription is inhibited.
   d. When lactose binds to the repressor, the shape of the repressor is changed.
   e. When the repressor is mutated, one possibility is that it does not bind to the operator.

2. Which of the following is *not* a type of viral reproduction?
   a. DNA virus in a lytic cycle
   b. DNA virus in a lysogenic cycle
   c. RNA virus by a double-stranded RNA intermediate
   d. RNA virus by reverse transcription to make cDNA
   e. RNA virus by acting as tRNA

3. In the lysogenic cycle of bacteriophage λ,
   a. a repressor, cI, blocks the lytic cycle.
   b. a bacteriophage carries DNA between bacterial cells.
   c. both early and late phage genes are transcribed.
   d. the viral genome is made into RNA, which stays in the host cell.

   e. many new viruses are made immediately, regardless of host health.

4. An operon is
   a. a molecule that can turn genes on and off.
   b. an inducer bound to a repressor.
   c. a series of regulatory sequences controlling transcription of protein-coding genes.
   d. any long sequence of DNA.
   e. a promoter, an operator, and a group of linked structural genes.

5. Which statement is true of both transformation and transduction?
   a. DNA is transferred between viruses and bacteria.
   b. Neither occurs in nature.
   c. Small fragments of DNA move from one cell to another.
   d. Recombination between the incoming DNA and host cell DNA does not occur.
   e. A conjugation tube is used to transfer DNA between cells.

6. Plasmids
   a. are circular protein molecules.
   b. are required by bacteria.
   c. are tiny bacteria.
   d. may confer resistance to antibiotics.
   e. are a form of transposable element.

7. The minimal genome can be estimated for a prokaryote
   a. by counting the total number of genes.
   b. by comparative genomics.
   c. as about 5,000 genes.
   d. by transposon mutagenesis, one gene at a time.
   e. by leaving out genes coding for tRNA.

8. When tryptophan accumulates in a bacterial cell,
   a. it binds to the operator, preventing transcription of adjacent genes.
   b. it binds to the promoter, allowing transcription of adjacent genes.
   c. it binds to the repressor, causing it to bind to the operator.
   d. it binds to the genes that code for enzymes.
   e. it binds to RNA and initiates a negative feedback loop to reduce transcription.

9. The promoter in the *lac* operon is
   a. the region that binds the repressor.
   b. the region that binds RNA polymerase.
   c. the gene that codes for the repressor.
   d. a structural gene.
   e. an operon.

10. The CRP–cAMP system
    a. produces many catabolites.
    b. requires ribosomes.
    c. operates by an operator–repressor mechanism.
    d. is an example of positive control of transcription.
    e. relies on operators.

FOR DISCUSSION

1. Viruses sometimes carry DNA from one cell to another by transduction. Sometimes a segment of bacterial DNA is incorporated into a phage protein coat without any phage DNA. These particles can infect a new host. Would the new host become lysogenic if the phage originally came from a lysogenic host? Why or why not?

2. Compare the life cycles of the viruses that cause influenza and AIDS (Figures 13.5 and 13.6) with respect to:
   a. how the virus enters the cell.
   b. how the virion is released in the cell.
   c. how the viral genome is replicated.
   d. how new viruses are produced.

3. Compare promoters adjacent to early and late genes in the bacteriophage lytic cycle.

4. The repressor protein that acts on the *lac* operon of *E. coli* is encoded by a regulatory gene. The repressor protein is made in small quantities and at a constant rate per cell. Would you surmise that the promoter for this repressor protein is efficient or inefficient? Is synthesis of the repressor constitutive, or is it under environmental control?

5. A key characteristic of a repressible enzyme system is that the repressor protein must react with a corepressor (typically, the end product of a pathway) before it can bind to the operator of an operon to shut the operon off. How is this system different from an inducible enzyme system?

FOR INVESTIGATION

A patient was admitted to the hospital with a urinary tract infection, and antibiotic-sensitive *Klebsiella* bacteria were identified as the cause. He was treated with ampicillin and kanamycin. However, during his hospital stay, the bacteria became resistant to both antibiotics. The patient in the bed next to him had *E. coli* that were resistant to these two antibiotics. Three possible explanations for the development of resistance in *Klebsiella* are (A) spontaneous mutation, (B) bacterial conjugation and transfer of resistance genes on the main chromosome, and (C) transfer of a plasmid containing the resistance genes. How would you investigate these possibilities?

14 The Eukaryotic Genome and Its Expression

Endangered genomes

The derivations of the common and scientific names of the cheetah, *Acinonyx jubatus*, tell the story:

- The word "cheetah" is from the Hindi *chiita*, meaning "spotted." The small black spots on yellow fur are unmistakable in this animal.

- *Acinonyx* in Greek means "no-move claw." Alone among the cats, cheetahs cannot fully retract their claws—an advantage in running fast and in hunting.

- *Jubatus* in Latin means "maned"; manes are characteristic of cheetah cubs.

This sleek, muscular cat is a solitary hunter, preying on smaller mammals such as gazelles and hares. It stalks its victim to within about 10–30 meters, then chases the prey at speeds up to 110 kilometers (70 miles) per hour. The chase is usually over in a minute or less.

There are only about 12,000 cheetahs in the world today, almost all of them in Africa. Some of the decline in their numbers is due to human intervention. For instance, farmers have hunted them down because they were (mistakenly) blamed for killing cattle. But something else is involved. When the sequences of the two parental strands of DNA from an individual cheetah are compared, there is a high degree of homozygosity: both alleles for protein-coding genes are usually the same. In addition, the *same sequences are present in almost all cheetahs*. It's as if they all came from a single set of parents!

Fossil remains of cheetahs and of several related extinct cat species provide part of the explanation for this remarkable genetic homogeneity. The modern cheetah probably evolved in Africa about 15 million years ago, then migrated to Asia and North America. Until the end of the last glaciation (about 10,000 years ago), cheetahs were widespread. Then, something (we're not sure what) happened. Many related species—such as the saber-toothed tiger—died out. But a few cheetahs apparently survived, and we presume that the genomes of all the cheetahs alive today are descended from those few. As we will learn in Chapter 22, such an event is called a *bottleneck*.

With their homogeneous genomes, cheetahs lack the genetic variability that can help a species survive adverse conditions. For example, cheetahs are very susceptible to diseases because they lack the genetic diversity that might allow them to mount an attack on new pathogens.

The cheetah is not the only species with high genomic homogeneity. The few Florida panthers that remain are likewise homogeneous in their DNA sequences. Like the cheetah, they are susceptible to diseases and genetic defects. In the Florida panther's case, however, their decline in numbers is recent and

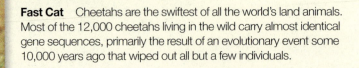

Fast Cat   Cheetahs are the swiftest of all the world's land animals. Most of the 12,000 cheetahs living in the wild carry almost identical gene sequences, primarily the result of an evolutionary event some 10,000 years ago that wiped out all but a few individuals.

Endangered Cat   The Florida panther, *Felis concolor coryi*, is a highly endangered North American cougar. The U.S. Fish and Wildlife Service estimates the number of living adult Florida panthers to be between 30 and 50.

is due solely to human intervention. Overhunted during the nineteenth century, their habitat in the southeastern United States was almost totally eliminated during the twentieth century, having been cleared to make way for human habitation.

When a variety of alleles is present in a population, it means there is a lot of room for new variations of the protein (or proteins) the gene transcribes. The complex process of turning a gene into a protein is the subject of this chapter.

IN THIS CHAPTER we will look at the organization and expression of eukaryotic genomes, which are significantly different from prokaryotic genomes. We'll begin by highlighting some of these differences and describing some of the model organisms that have made our understanding of eukaryotic genomes possible. In the rest of this chapter we look at the many points at which the complex process of eukaryotic gene expression can be regulated before, during, and after transcription and translation.

14.1 What Are the Characteristics of the Eukaryotic Genome?

As genomes have been sequenced and their expression studied, and as the roles of the proteins they encode are annotated, a number of major differences have emerged between eukaryotic genomes and their prokaryotic counterparts (**Table 14.1**). Key differences between eukaryotic and prokaryotic genomes include:

■ In terms of haploid DNA content, *eukaryotic genomes are larger than those of prokaryotes.* This difference is not surprising, given that in multicellular organisms there are many cell types and many jobs to do; many proteins—all encoded by DNA—are needed to do those jobs. A typical virus contains enough DNA to code for only a few proteins—about 10,000 base pairs (bp). The most thoroughly studied prokaryote, *E. coli*, has sufficient DNA (over 4.6 million bp) to make several thousand different proteins and regulate their synthesis. Humans have considerably more genes and regulatory sequences: some 6 billion bp (2 meters of DNA) are crammed into each diploid human cell. However, the amount of DNA in a organism does not always correlate with its complexity. For example, the lily (which produces beautiful flowers each spring, but produces fewer proteins than a human does) has 18 times more DNA than a human.

■ *Eukaryotic genomes have more regulatory sequences*—and many more regulatory proteins that bind to them—than prokaryotic genomes do. The great complexity of eukaryotes requires a great deal of regulation, which is evident in the many points of control associated with the expression of the eukaryotic genome.

■ *Much of eukaryotic DNA is noncoding.* Interspersed throughout many eukaryotic genomes are various kinds of DNA sequences that are not transcribed into mRNA. And coding regions of genes sometimes contain sequences that do not appear in the mRNA that is translated at the ribosome.

■ *Eukaryotes have multiple chromosomes.* The genomic "encyclopedia" of a eukaryote is separated into multiple

TABLE 14.1

A Comparison of Prokaryotic and Eukaryotic Genes and Genomes

CHARACTERISTIC	PROKARYOTES	EUKARYOTES
Genome size (base pairs)	10^4–10^7	10^8–10^{11}
Repeated sequences	Few	Many
Noncoding DNA within coding sequences	Rare	Common
Transcription and translation separated in cell	No	Yes
DNA segregated within a nucleus	No	Yes
DNA bound to proteins	Some	Extensive
Promoters	Yes	Yes
Enhancers/silencers	Rare	Common
Capping and tailing of mRNA	No	Yes
RNA splicing required (spliceosomes)	Rare	Common
Number of chromosomes in genome	One	Many

"volumes." This separation requires that each chromosome have, at a minimum, three defining DNA sequences that we have described in previous chapters: an origin of replication (*ori*) recognized by the DNA replication machinery; a centromere region that holds the replicated chromosomes together before mitosis; and a telomeric sequence at each end of the chromosome.

■ *In eukaryotes, transcription and translation are physically separated.* The nuclear envelope separates DNA and its transcription (inside the nucleus) from the sites where mRNA is translated into polypeptides (the ribosomes in the cytoplasm). This separation allows for many points of regulation before translation begins: in the synthesis of a primary (pre-mRNA) transcript, in its processing into mature mRNA, and in its transport to the cytoplasm for translation (**Figure 14.1**).

Model organisms reveal the characteristics of eukaryotic genomes

Most of the lessons learned from eukaryotic genomics have come from several simple model organisms that have been studied extensively: the yeast *Saccharomyces cerevisiae*, the nematode (roundworm) *Caenorhabditis elegans*, the fruit fly *Drosophila melanogaster*, and—representing plants—the thale cress, *Arabidopsis thaliana*.

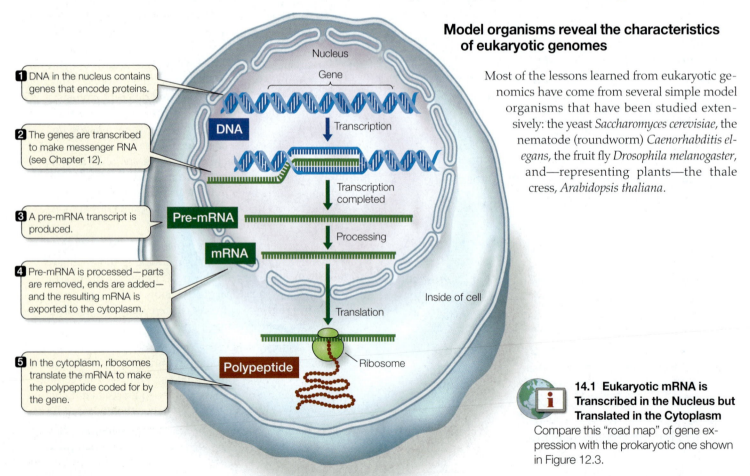

1 DNA in the nucleus contains genes that encode proteins.

2 The genes are transcribed to make messenger RNA (see Chapter 12).

3 A pre-mRNA transcript is produced.

4 Pre-mRNA is processed—parts are removed, ends are added—and the resulting mRNA is exported to the cytoplasm.

5 In the cytoplasm, ribosomes translate the mRNA to make the polypeptide coded for by the gene.

Nucleus
Gene
DNA
Transcription
Transcription completed
Pre-mRNA
Processing
mRNA
Inside of cell
Translation
Polypeptide
Ribosome

14.1 Eukaryotic mRNA is Transcribed in the Nucleus but Translated in the Cytoplasm Compare this "road map" of gene expression with the prokaryotic one shown in Figure 12.3.

The genome of the puffer fish *Fugu rubripes* is the most compact known among vertebrates. It has about the same number of genes as the human genome in about one-eighth as much DNA. The two genomes are so similar that, as Sydney Brenner put it, "the *Fugu* genome is the *Reader's Digest* version of the Book of Man."

YEAST: THE BASIC EUKARYOTIC MODEL Yeasts are single-celled eukaryotes. Like most eukaryotes, they have membrane-enclosed organelles, such as a nucleus and endoplasmic reticulum, as well as a life cycle with alternation of haploid and diploid generations (see Figure 9.14).

In comparison with the prokaryote *E. coli*, whose genome has about 4.6 million bp on a single circular chromosome, the genome of budding yeast (*Saccharomyces cerevisiae*) has 16 linear chromosomes and a haploid content of more than 12 million bp. More than 600 scientists around the world collaborated in mapping and sequencing the yeast genome. When they began, they knew of about 1,000 yeast genes coding for RNAs or proteins. The final sequence revealed 5,800 genes, and annotation has assigned probable roles to most of them. Some of these genes are homologous to genes found in prokaryotes, but many are not.

The most striking difference between the yeast genome and that of *E. coli* is in the number of genes for protein targeting (**Table 14.2**). Both of these single-celled organisms appear to use about the same numbers of genes to perform the basic functions of cell survival. It is the compartmentalization of the eukaryotic yeast cell into organelles that requires it to have so many more genes. This finding is direct, quantitative confirmation of something we have known for a century: the eukaryotic cell is structurally more complex than the prokaryotic cell.

Genes encoding several other types of proteins are present in the yeast and other eukaryotic genomes, but have no homologs in prokaryotes:

- Genes encoding histones that package DNA into nucleosomes
- Genes encoding cyclin-dependent kinases that control cell division
- Genes encoding proteins involved in the processing of mRNA

THE NEMATODE: UNDERSTANDING EUKARYOTIC DEVELOPMENT In 1965, Sydney Brenner, fresh from being part of the team that first isolated mRNA, looked for a simple organism in which to study multicellularity. He settled on *Caenorhabditis elegans*, a millimeter-long nematode (roundworm) that normally lives in the soil. But it also lives in the laboratory, where it has become a favorite model organism of developmental biologists (see Section 19.4). The nematode has a transparent body, which scientists can watch over 3 days as a fertilized egg divides and forms an adult worm made up of nearly 1,000 cells. In spite of its small number of cells, the nematode has a nervous system, digests food, reproduces sexually, and ages. So it is not surprising that an intense effort was made to sequence the genome of this model organism.

The *C. elegans* genome is eight times larger than that of yeast (97 million bp) and has four times as many protein-coding genes (19,099). Once again, sequencing revealed far more genes than expected: when the sequencing effort began, researchers estimated that the worm would have about 6,000 genes and about that many proteins. Clearly, it has far more. About 3,000 genes in the worm have direct homologs in yeast; these genes code for basic eukaryotic cell functions. What do the rest of the genes—the bulk of the worm genome—do?

In addition to surviving, growing, and dividing, as single-celled organisms do, multicellular organisms must have genes for holding cells together to form tissues, for cell differentiation to divide up tasks among those tissues, and for intercellular communication pathways to coordinate their activities (**Table 14.3**). Many of the genes so far identified in *C. elegans* that are not present in yeast perform these roles, which will be described in the remainder of this chapter and the next one.

DROSOPHILA MELANOGASTER: RELATING GENETICS TO GENOMICS The fruit fly *Drosophila melanogaster* is a famous model organism, as its study elucidated many principles of genetics (see Section 10.4). Over 2,500 mutations of *Drosophila melanogaster* had been described by the 1990s, when genome sequencing began, and this fact alone was a good reason for sequencing its DNA. The fruit fly is a much larger organism than *C. elegans*, both in size (it has 10 times more cells) and complexity, and it undergoes a complicated developmental transformation from egg to larva to pupa to adult.

Not surprisingly, the fly's genome is larger than that of *C. elegans* (about 180 million bp). But as we mentioned earlier, genome size does not necessarily correlate with the number of genes encoded. In this case, the larger fruit fly genome

TABLE 14.2

Comparison of the Genomes of *E. coli* and Yeast

	E. COLI	YEAST
Genome length (base pairs)	4,640,000	12,068,000
Number of protein-coding genes	4,300	5,800
Proteins with roles in:		
Metabolism	650	650
Energy production/storage	240	175
Membrane transport	280	250
DNA replication/repair/ recombination	120	175
Transcription	230	400
Translation	180	350
Protein targeting/secretion	35	430
Cell structure	180	250

TABLE 14.3

C. elegans Genes Essential to Multicellularity

FUNCTION	PROTEIN/DOMAIN	NUMBER OF GENES
Transcription control	Zinc finger; homeobox	540
RNA processing	RNA binding domains	100
Nerve impulse transmission	Gated ion channels	80
Tissue formation	Collagens	170
Cell interactions	Extracellular domains; glycotransferases	330
Cell–cell signaling	G protein-linked receptors; protein kinases; protein phosphatases	1,290

contains fewer genes (13,449) than does the smaller nematode genome. Annotation has given us a snapshot of the functions of the *Drosophila* genes so far characterized (**Figure 14.2**). The distribution of their functions is typical of complex eukaryotes.

Several additional findings from the fruit fly genome are notable:

■ 18,941 different mRNAs are transcribed from the 13,449 genes. This means that the fruit fly genome codes for more proteins than it has genes—a discovery of significance in understanding other genomes, including our own.

■ An additional 514 genes code for RNAs that are not translated into proteins. These include genes for tRNAs and rRNAs (see Chapter 12), but there are also 123 genes coding for small RNAs that remain in the nucleus. We will look at their possible roles later in this chapter.

■ Comparative genomics has revealed many protein-coding sequences that are similar in fruit flies and other organisms. Not surprisingly, the similarities are strongest with other *Drosophila* species. But a survey of genes involved in human diseases shows that about half have sequences similar to sequences found in fruit flies. Finding out the function of a gene in fruit flies may lead to finding out its function in humans.

***ARABIDOPSIS*: STUDYING THE GENOMES OF PLANTS**
About 250,000 species of flowering plants dominate land and fresh water. But in the history of life, the flowering plants are fairly young, having evolved only about 200 million years ago. Given the pace of mutation and other genetic changes, the differences among these plants are likely to be relatively small. So, although it is the genomes of the plants used by people as food and fiber that hold the greatest interest for us, it is not surprising that instead of sequencing the huge genomes of wheat (16 billion bp) or corn (3 billion bp), scientists first chose to sequence a simpler flowering plant.

Arabidopsis thaliana, the thale cress, is a member of the mustard family and has long been a favorite model organism for study by plant biologists. It is small (hundreds could grow and reproduce in the space occupied by this page), is easy to manipulate, and has a small (119 million bp) genome.

The *Arabidopsis* genome has about 26,000 protein-coding genes, but remarkably, many of these genes are duplicates and probably originated by chromosomal rearrangements. When these duplicate genes are subtracted from the total, about 15,000 unique genes are left—a number similar to the numbers of genes found in the fruit fly and nematode genomes. Indeed, many of the genes found in these animals have homologs in the plant, suggesting that plants and animals have a common ancestor.

But *Arabidopsis* has some genes that distinguish it as a plant (**Table 14.4**). These genes include those involved in photosynthesis, in the transport of water into the root and throughout the plant, in the assembly of the cell wall, in the uptake and metabolism of inorganic substances from the environment, and in the synthesis of specific molecules used for defense against herbivores (organisms that eat plants).

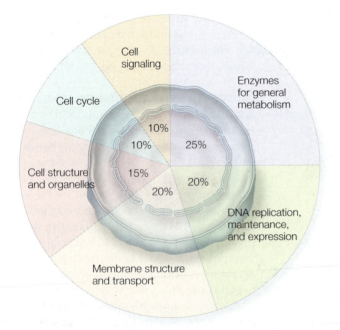

14.2 Functions of the Eukaryotic Genome The functions of the protein-coding genes in the genome of *Drosophila melanogaster* show a pattern that is typical of many complex organisms.

TABLE 14.4

Arabidopsis Genes Unique to Plants

FUNCTION		NUMBER OF GENES
Cell wall and growth		420
Water channels		300
Photosynthesis		139
Defense and metabolism		94

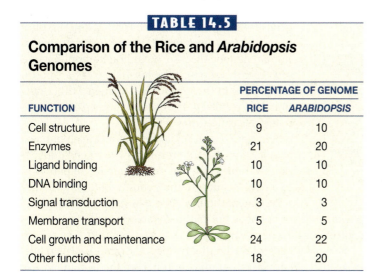

TABLE 14.5

Comparison of the Rice and *Arabidopsis* Genomes

FUNCTION	PERCENTAGE OF GENOME	
	RICE	ARABIDOPSIS
Cell structure	9	10
Enzymes	21	20
Ligand binding	10	10
DNA binding	10	10
Signal transduction	3	3
Membrane transport	5	5
Cell growth and maintenance	24	22
Other functions	18	20

Justifying its position as a model plant, these "plant" genes in *Arabidopsis* were also found in the genome of rice, the first major crop plant whose sequence has been determined. Rice (*Oryza sativa*) is the world's most important crop; it is the staple diet for 3 billion people. In fact, two *O. sativa* sequences of have been deciphered: that of *O. sativa indica*, the rice subspecies grown in China and most of tropical Asia, and that of the subspecies *japonica*, which is grown in Japan and other temperate climates (such as the United States). Both genomes are about the same size (430 million bp), yet in this much larger genome is a set of genes remarkably similar to that of *Arabidopsis* (**Table 14.5**). And many of the genes in rice are also present in the much larger genomes of corn and wheat.

Of course, rice as a whole, and each subspecies, has its own particular set of genes that make it unique. The *indica* subspecies is estimated to have 46,000–55,000 genes, while *japonica* has some 32,000–50,000—both higher numbers of genes than in *Arabidopsis*. These "extra" genes include genes for characters that are specific to rice, such as the physiology that allows rice to grow for part of the season submerged in water; the nutrient-packed seeds that sustain human lives; and resistance to certain plant diseases, such as viruses and fungi. Analyses of these and other rice genes will no doubt lead to significant improvements in this crop, and to the improvement of the other grain crops as well.

Eukaryotic genomes contain many repetitive sequences

Studies of the genomes of eukaryotic model organisms have revealed that they contain numerous repetitive DNA sequences that do not code for polypeptides.

HIGHLY REPETITIVE SEQUENCES Two types of *highly repetitive sequences* are found in eukaryotic genomes:

- *Minisatellites* are 10–40 base pairs long and are repeated up to several thousand times. Because DNA polymerase tends to make errors in copying these sequences, the number of

copies present varies among individuals. For example, one person might have 300 minisatellites at a particular locus, whereas another person will have 500 minisatellites at that same locus. This variation provides a set of molecular genetic markers that can be used to identify an individual, as we will see in Section 16.1.

- *Microsatellites* are very short (1–3 bp) sequences, present in small clusters of 15–100 copies. They are scattered all over the genome and are also known as simple sequence repeats.

Why call these noncoding repetitive sequences "satellites"? The term conveys that the character of these sequences is different from that of the rest of the genome. For example, the ratio of GC base pairs to AT base pairs may set a satellite apart from the rest of the genome. Recall that there are three hydrogen bonds between G and C, and only two between A and T. This makes the GC base pair somewhat more tightly held than the AT pair. Therefore GC-rich DNA is denser than AT-rich DNA. So, if a satellite has a different proportion of GC base pairs from the rest of DNA, it will have a different density. DNAs of different densities can be separated in a centrifuge, so it is easy to separate some satellites from the rest of the genome.

These highly repetitive sequences are not represented in mature mRNA transcripts. While laboratory scientists have made use of these sequences in genetic studies, their roles in eukaryotes remain unclear.

MODERATELY REPETITIVE SEQUENCES While highly repetitive sequences are not transcribed into mRNA, some *moderately repetitive DNA* sequences are transcribed. These sequences code for tRNAs and rRNAs, which are used in protein synthesis.

The cell makes tRNAs and rRNAs constantly, but even at the maximum rate of transcription, single copies of the DNA sequences coding for them would be inadequate to supply the large amounts of these molecules needed by most cells; hence the genome has multiple copies of these sequences. Since these moderately repetitive sequences are transcribed into RNA, they are properly termed "genes"; that is, there are genes that code for rRNA and tRNA.

In mammals, four different rRNA molecules make up the ribosome: the 18S, 5.8S, 28S, and 5S rRNAs. (The S stands for *Svedberg unit*, a measure of how a substance behaves in a centrifuge that is related to the size of a molecule.) The 18S, 5.8S, and 28S rRNAs are transcribed from a repeated DNA sequence as a single precursor RNA molecule, which is twice the size of the three ultimate products (**Figure 14.3**). Several posttranscriptional steps cut this precursor into the final three rRNA products and discard the noncoding "spacer" RNA. The sequence encoding these RNAs is moderately repetitive in humans: a total of 280 copies of the sequence are located in clusters on five different chromosomes.

TRANSPOSONS Outside of the moderately repetitive DNA sequences that code for rRNA, most moderately repetitive sequences are not stably integrated into the genome. Instead, these sequences can move from place to place in the genome, and thus are called *transposable elements* or **transposons**. Transposons make up over 40 percent of the human genome, far more than the 3–10 percent found in the other sequenced eukaryotes.

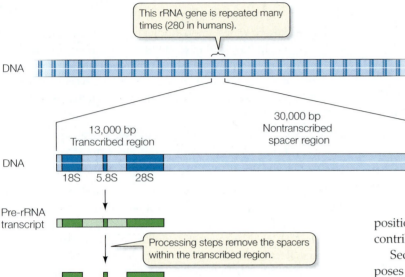

14.3 A Moderately Repetitive Sequence Codes for rRNA This rRNA gene, along with its nontranscribed spacer region, is repeated 280 times in the human genome, with clusters on five chromosomes. Once this gene has been transcribed, posttranscriptional processing removes the spacers within the transcribed region and separates the primary transcript into the three final rRNA products.

There are four main types of transposons in eukaryotes:

1. *SINEs* (short *i*nterspersed *e*lements) are up to 500 bp long and are transcribed, but not translated. There are about 1.5 million of them in the human genome, making up 15 percent of human DNA. A single type, the 300 bp *Alu* element, accounts for 11 percent of the human genome; it is present in a million copies scattered over all the chromosomes.

2. *LINEs* (long *i*nterspersed *e*lements) are up to 7,000 bp long, and some are transcribed and translated into proteins. They constitute about 17 percent of the human genome.

Both of these elements are present in more than 100,000 copies. They move about the genome in a distinctive way: they make an RNA copy of themselves, which acts as a template for new DNA, which then inserts itself at a new location in the genome. In this "copy and paste" mechanism, the original sequence stays where it is, and the copy inserts itself at a new location.

3. *Retrotransposons* also make an RNA copy of themselves when they move about the genome. They constitute about 8 percent of the human genome. Some of them code for some of the proteins necessary for their own transposition, and others do not.

4. *DNA transposons* do not use an RNA intermediate, but actually move to a new spot in the genome without replicating (**Figure 14.4**).

What role do these moving sequences play in the cell? There are few answers to this question. The best answer so far seems to be that transposons are cellular parasites that simply replicate themselves. These replications can lead to the insertion of a transposon at a new location, which can have important consequences. For example, the insertion of a transposon into the coding region of a

gene results in a mutation (see Figure 14.4). This phenomenon accounts for a few rare forms of several human genetic disease, including hemophilia and muscular dystrophy. If the insertion of a transposon takes place in the germ line, a gamete with a new mutation results. If the insertion takes place in a somatic cell, cancer may result.

If a transposon replicates not just itself but also an adjacent gene, the result may be a gene duplication. A transposon can carry a gene, or a part of it, to a new location in the genome, shuffling the genetic material and creating new genes. Clearly, transposition stirs the genetic pot in the eukaryotic genome and thus contributes to genetic variation.

Section 4.5 described the theory of endosymbiosis, which proposes that chloroplasts and mitochondria are the descendants of once free-living prokaryotes. Transposons may have played a role in endosymbiosis. In living eukaryotes, although chloroplasts and mitochondria contain some DNA, the nucleus contains most of the genes that encode the organelles' proteins. If the organelles were once independent, they must originally have contained all of those genes. How did the genes move to the nucleus? They may have done so by DNA transpositions between organelles and nucleus, which still occur today. The DNA that remains in the organelles may be the remnants of more complete prokaryotic genomes.

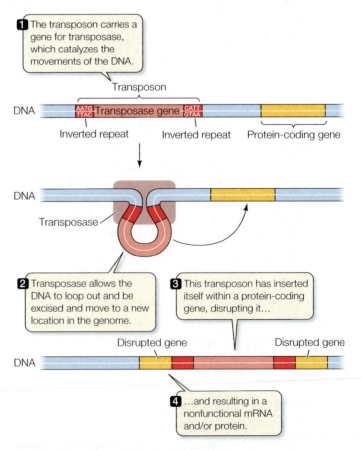

14.4 DNA Transposons and Transposition At the end of each DNA transposon is an inverted repeat sequence that helps in the transposition process.

The sequencing of the genomes of model organisms has demonstrated common features of the eukaryotic genome, which include the presence of repetitive sequences and transposons.

- Can you describe some of the major differences between prokaryotic and eukaryotic genomes? See pp. 307–308 and Table 14.1

- What is one function of genes found in *C. elegans* that has no counterpart in the genome of yeast? See p. 309

- What purpose is served by having multiple copies of sequences coding for rRNA in the mammalian genome? See p. 311

- What effects can transposons have on a genome? See p. 312 and Figure 14.4

Studies of eukaryotic genomes have revealed that they contain coding as well as noncoding sequences. We turn next to the subject at the heart of molecular genetics: genes that code for proteins, and the surprising finding that noncoding regions exist even within these genes.

14.2 What Are the Characteristics of Eukaryotic Genes?

Like their prokaryotic counterparts, many protein-coding genes in eukaryotes exist in only one copy per haploid genome. But eukaryotic genes have two distinctive characteristics that are uncommon among prokaryotes:

- They contain noncoding internal sequences.

- They form gene families—groups of structurally and functionally related "cousins" within the genome.

Protein-coding genes contain noncoding sequences

The structure and transcription of a typical eukaryotic gene is diagrammed in **Figure 14.5**. Preceding the coding region of a eukaryotic gene is a *promoter*, to which an RNA polymerase binds to begin the transcription process. Unlike the prokaryotic enzyme, however, a eukaryotic RNA polymerase does not recognize the promoter sequence by itself, but requires help from other molecules, as we'll see below. At the other end of the gene, after the coding region, is a DNA sequence appropriately called the **terminator**, which signals the end of transcription. It is important that you distinguish the terminator from the stop codon:

- The *terminator* sequence is usually after the stop codon and signals the end of transcription by RNA polymerase.

- The *stop codon* is within the coding region and, when transcribed into mRNA, signals the end of translation at the ribosome.

Eukaryotic protein-coding genes may also contain noncoding base sequences, called **introns**. One or more introns may be interspersed with the coding sequences, which are called **exons**. Transcripts of the introns appear in the primary mRNA transcript, called **pre-mRNA**, but they are removed by the time the mature mRNA—the final mRNA that will be translated—leaves the nucleus. Pre-mRNA processing involves cutting introns out of the pre-mRNA transcript and splicing together the remaining exon transcripts.

How can we locate introns within a eukaryotic gene? The easiest way is by **nucleic acid hybridization**, the method that originally

14.5 Transcription of a Eukaryotic Gene The β-globin gene diagrammed here is about 1,600 bp long. The three exons—the protein-coding sequences—contain 441 base pairs (codons for 146 amino acids plus a stop codon). The two introns—noncoding sequences of DNA containing almost 1,000 bp between them—are initially transcribed, but are spliced out of the pre-mRNA transcript.

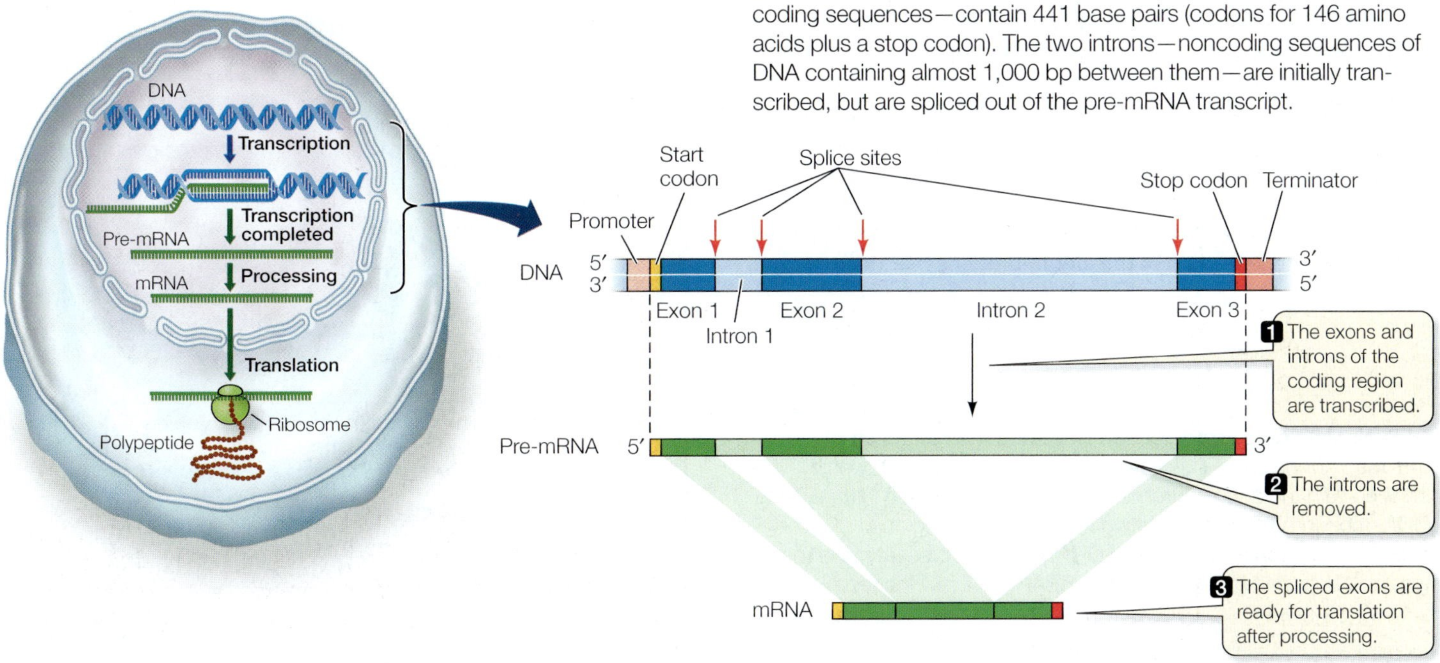

RESEARCH METHOD

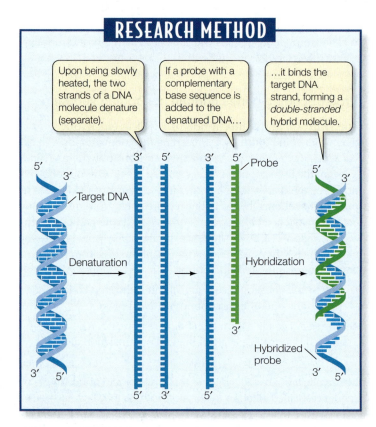

Upon being slowly heated, the two strands of a DNA molecule denature (separate).

If a probe with a complementary base sequence is added to the denatured DNA…

…it binds the target DNA strand, forming a *double-stranded* hybrid molecule.

14.6 Nucleic Acid Hybridization Base pairing permits the detection of a sequence complementary to the probe.

revealed the existence of introns. This method, outlined in **Figure 14.6**, has been crucial for studying the relationship between eukaryotic genes and their transcripts. It involves two steps:

- The target DNA is denatured to break the hydrogen bonds between the base pairs and separate the two strands.
- A single-stranded nucleic acid from another source (called a **probe**) is incubated with the denatured DNA. If the probe has a base sequence complementary to the target DNA, a probe–target double helix forms by hydrogen bonding between the bases. Because the two strands are from different sources, the resulting double-stranded molecule is called a *hybrid*.

Biologists used nucleic acid hybridization to examine the β-globin gene, which encodes one of the globin proteins that make up he-

14.7 Nucleic Acid Hybridization Revealed the Existence of Introns When an mRNA transcript of the β-globin gene was hybridized with the double-stranded DNA of that gene, the introns in the DNA "looped out," demonstrating that the coding region of a eukaryotic gene can contain noncoding DNA that is not present in the mature mRNA transcript. FURTHER RESEARCH: Draw the result assuming that there were three exons and two introns.

EXPERIMENT

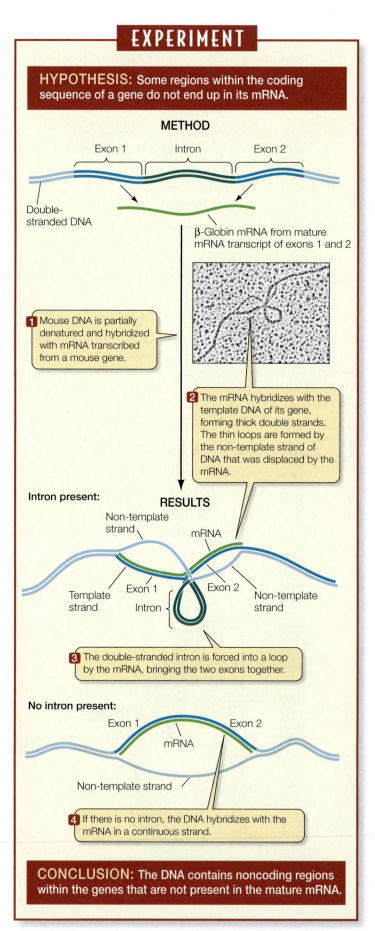

HYPOTHESIS: Some regions within the coding sequence of a gene do not end up in its mRNA.

METHOD

Exon 1   Intron   Exon 2

Double-stranded DNA

β-Globin mRNA from mature mRNA transcript of exons 1 and 2

1 Mouse DNA is partially denatured and hybridized with mRNA transcribed from a mouse gene.

2 The mRNA hybridizes with the template DNA of its gene, forming thick double strands. The thin loops are formed by the non-template strand of DNA that was displaced by the mRNA.

Intron present:

RESULTS

Non-template strand

mRNA

Template strand   Exon 1   Exon 2   Non-template strand

Intron

3 The double-stranded intron is forced into a loop by the mRNA, bringing the two exons together.

No intron present:

Exon 1   Exon 2

mRNA

Non-template strand

4 If there is no intron, the DNA hybridizes with the mRNA in a continuous strand.

CONCLUSION: The DNA contains noncoding regions within the genes that are not present in the mature mRNA.

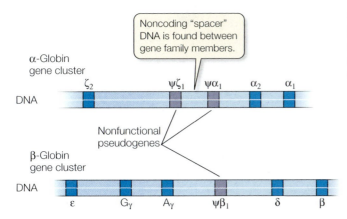

14.8 The Globin Gene Family The α-globin and β-globin clusters of the human globin gene family are located on different chromosomes. The genes of each cluster are separated by noncoding "spacer" DNA. The nonfunctional pseudogenes are indicated by the Greek letter psi (ψ). The γ gene has two variants, A_γ and G_γ.

moglobin. Follow the experiment in **Figure 14.7** carefully as we describe what they did and what happened.

The researchers first denatured DNA containing the β-globin gene by heating it slowly, then added previously isolated, mature β-globin mRNA. As expected, the mRNA bound to the DNA by complementary base pairing. The researchers expected to obtain a linear (1:1) matchup of the mRNA to the coding DNA. That expectation was only partly met: there were indeed stretches of RNA–DNA hybridization, but some looped structures were also visible. These loops were the introns, stretches of DNA that did not have complementary bases on the mature mRNA.

Later studies would show complete hybridization of *pre*-mRNA to DNA, revealing that the introns were indeed part of the pre-mRNA transcript. Somewhere on the path from primary transcript (pre-mRNA) to mature mRNA, the introns had been removed, and the exons had been spliced together. We will examine this splicing process in the next section.

Most (but not all) vertebrate genes contain introns, as do many other eukaryotic genes (and even a few prokaryotic ones). The largest human gene, for the muscle protein called titin, has 363 exons, which together code for 38,138 amino acids. Introns interrupt, but do not scramble, the DNA sequence that codes for a polypeptide chain. The base sequence of the exons, taken in order, is complementary to that of the mature mRNA product. In some cases, the separated exons code for different functional regions, or *domains*, of the protein. For example, the globin polypeptides that make up hemoglobin each have two domains: one for binding to a nonprotein pigment called heme, and another for binding to the other globin subunits. These two domains are encoded by different exons in the globin genes.

Gene families are important in evolution and cell specialization

About half of all eukaryotic protein-coding genes exist in only one copy in the haploid genome (two copies in somatic cells). The rest are present in multiple copies. Over evolutionary time, the copies of a gene may undergo separate mutations, giving rise to a group of closely related genes called a **gene family**. Some gene families, such as the genes encoding the globin proteins that make up

hemoglobin, contain only a few members; other families, such as the genes encoding the immunoglobulins that make up antibodies, have hundreds of members.

Like the members of any family, the DNA sequences in a gene family are usually different from one another. As long as one member retains the original DNA sequence and thus codes for the proper protein, the other members can mutate slightly, extensively, or not at all. The availability of such "extra" genes is important for "experiments" in evolution: If the mutated gene is useful, it may be selected for in succeeding generations. If the mutated gene is a total loss, the functional copy is still there to save the day.

The gene family encoding the globins is a good example of the gene families found in vertebrates. These proteins are found in hemoglobin as well as in myoglobin (an oxygen-binding protein present in muscle). The globin genes all arose from a single common ancestor gene long ago. In humans, there are three functional members of the alpha-globin (α-globin) cluster and five in the beta-globin (β-globin) cluster (**Figure 14.8**). In adults, each hemoglobin molecule is a tetramer containing two identical α-globin subunits, two identical β-globin subunits, and four heme pigments (each held inside a globin subunit) (see Figure 3.9).

During human development, different members of the globin gene cluster are expressed at different times and in different tissues (**Figure 14.9**). This *differential gene expression* has great physiological significance. For example, γ-globin, a subunit found in the hemoglobin of the human fetus, binds O_2 more tightly than adult hemoglobin does. This specialized form of hemoglobin ensures that in the placenta, where the maternal and fetal circulation come close to one another, O_2 will be transferred from the mother's blood to the developing fetus's blood. Just before birth, the synthesis of fetal hemoglobin proteins in the liver stops, and the bone marrow cells take over, making the adult forms. Thus hemoglobins with different binding affinities for O_2 are provided at different stages of human development.

In addition to genes that encode proteins, many gene families include nonfunctional **pseudogenes**, designated with the Greek letter psi (ψ) (see Figure 14.8). These pseudogenes are the "black sheep" of a gene family: they result from mutations that cause a loss of function, rather than an enhanced or new function. The DNA sequence of a pseudogene may not differ greatly from that of other family members. It may simply lack a promoter, for example, and thus fail to be transcribed. Or it may lack the recognition sites needed for the removal of introns (a process we will describe in the next section) and thus be transcribed into pre-mRNA, but not correctly processed into a useful mature mRNA. In some gene families, pseudogenes outnumber functional genes. Because some members of the family are functional, there appears to be little selection pressure for evolution to eliminate pseudogenes.

14.9 Differential Expression in the Globin Gene Family
During human development, different members of the globin gene family are expressed at different times and in different tissues.

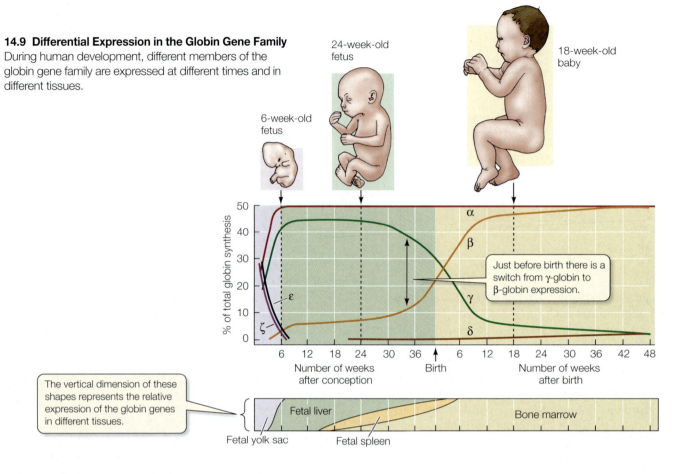

6-week-old fetus

24-week-old fetus

18-week-old baby

Just before birth there is a switch from γ-globin to β-globin expression.

% of total globin synthesis

Number of weeks after conception

Birth

Number of weeks after birth

The vertical dimension of these shapes represents the relative expression of the globin genes in different tissues.

Fetal liver

Bone marrow

Fetal yolk sac

Fetal spleen

14.2 RECAP

Most eukaryotic genes contain noncoding sequences called introns, which are removed from the pre-mRNA transcript. Many eukaryotic genes belong to groups of closely related genes called gene families.

- Can you describe the method of nucleic acid hybridization? See p. 314 and Figure 14.6

- Can you describe the experiment that showed that the β-globin gene contains introns? See p. 315 and Figure 14.7

- What are gene families, and how might they be important in evolution? See p. 315

We now know that eukaryotic protein-coding genes contain some sequences that do not appear in the mature mRNA that is translated into proteins. How are these noncoding sequences eliminated from the pre-mRNA transcript?

14.3 How Are Eukaryotic Gene Transcripts Processed?

The primary transcript of a eukaryotic gene is modified in two ways before it leaves the nucleus: both ends of the pre-mRNA are modified, and the introns are removed.

The primary transcript of a protein-coding gene is modified at both ends

Two steps in the processing of pre-mRNA take place in the nucleus, one at each end of the molecule (**Figure 14.10**):

- A **G cap** is added to the 5′ end of the pre-mRNA as it is transcribed. The G cap is a chemically modified molecule of guanosine triphosphate (GTP). It apparently facilitates the binding of mRNA to the ribosome for translation and protects the mRNA from being digested by *ribonucleases* that break down RNAs.

- A **poly A tail** is added to the 3′ end of pre-mRNA at the end of transcription. Near the 3′ end of pre-mRNA, and after the last codon, is the sequence AAUAAA. This sequence acts as a signal for an enzyme to cut the pre-mRNA. Immediately after this cleavage, another enzyme adds 100 to 300 adenine bases ("poly A") to the 3′ end of the pre-mRNA. This "tail" may assist in the export of the mRNA from the nucleus and is important for mRNA stability.

Splicing removes introns from the primary transcript

The next step in the processing of eukaryotic pre-mRNA within the nucleus is deletion of the introns. If these RNA sequences were not removed, an mRNA translating to a very different amino acid sequence, and possibly a nonfunctional protein, would result. A process called **RNA splicing** removes the introns and splices the exons together.

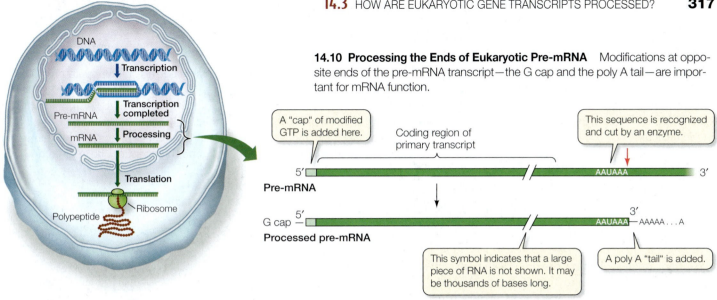

14.10 Processing the Ends of Eukaryotic Pre-mRNA Modifications at opposite ends of the pre-mRNA transcript—the G cap and the poly A tail—are important for mRNA function.

> A "cap" of modified GTP is added here.

Coding region of primary transcript

> This sequence is recognized and cut by an enzyme.

> This symbol indicates that a large piece of RNA is not shown. It may be thousands of bases long.

> A poly A "tail" is added.

As soon as the pre-mRNA is transcribed, it is bound by several **small nuclear ribonucleoprotein particles** (**snRNPs**, commonly pronounced "snurps"). There are several types of these RNA–protein particles in the nucleus.

At the boundaries between introns and exons are **consensus sequences**—short stretches of DNA that appear, with little variation ("consensus"), in many different genes. The RNA in one of the snRNPs has a stretch of bases complementary to the consensus sequence at the 5′ exon–intron boundary, and it binds to the pre-mRNA by complementary base pairing. Another snRNP binds to the pre-mRNA near the 3′ intron–exon boundary (**Figure 14.11**).

Next, using energy from ATP, proteins are added to form a large RNA–protein complex called a **spliceosome**. This complex cuts the pre-mRNA, releases the introns, and joins the ends of the exons together to produce mature mRNA.

Molecular studies of human genetic diseases have provided insights into consensus sequences and splicing machinery. People with a genetic disease called beta thalassemia, for example, make inadequate amounts of the β-globin subunit of hemoglobin. These people suffer from severe anemia because they have an inadequate supply of red blood cells. In some cases, the genetic mutation that causes the disease occurs at a consensus sequence in the β-globin gene. Consequently, β-globin pre-mRNA cannot be spliced correctly, and nonfunctional β-globin mRNA is made.

This finding offers an excellent example of how mutations can elucidate cause-and-effect relationships in biology. In the logic of science, merely linking two phenomena (for example, consensus sequences and splicing) does not prove that one is necessary for the other. In an experiment, the scientist alters one phenomenon (for example, the bases of the consensus sequence) to see whether the other (for example, splicing) occurs. In beta thalassemia, nature has done this experiment for us.

14.11 The Spliceosome: An RNA Splicing Machine The binding of snRNPs to consensus sequences bordering the introns on the pre-mRNA lines up the splicing machinery. After the snRNPs bind to the pre-mRNA, other proteins join the complex to form a spliceosome. This mechanism determines the exact position of each cut in the pre-mRNA with great precision.

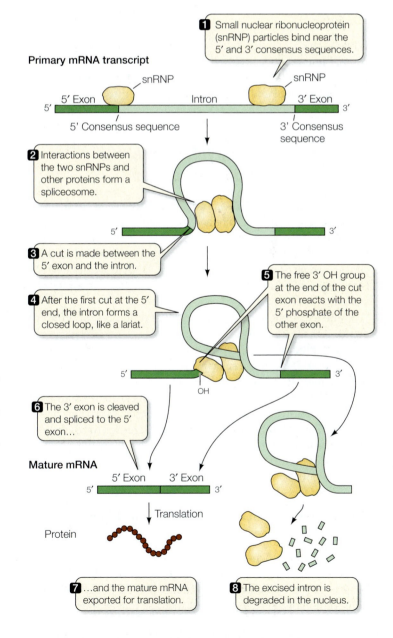

1 Small nuclear ribonucleoprotein (snRNP) particles bind near the 5′ and 3′ consensus sequences.

2 Interactions between the two snRNPs and other proteins form a spliceosome.

3 A cut is made between the 5′ exon and the intron.

4 After the first cut at the 5′ end, the intron forms a closed loop, like a lariat.

5 The free 3′ OH group at the end of the cut exon reacts with the 5′ phosphate of the other exon.

6 The 3′ exon is cleaved and spliced to the 5′ exon…

7 …and the mature mRNA exported for translation.

8 The excised intron is degraded in the nucleus.

After processing is completed in the nucleus, the mature mRNA leaves that organelle through the nuclear pores. A protein called TAP binds to the 5′ end of processed mRNA. This protein in turn binds to others, which are recognized by a receptor at the nuclear pore. Unprocessed or incompletely processed pre-mRNAs remain in the nucleus.

14.3 RECAP

The primary transcript of a eukaryotic gene is modified while still in the nucleus. First, its 5′ and 3′ ends are modified, and then its introns are spliced out.

- Can you describe how the pre-mRNA transcript is modified at the 5′ and 3′ ends? See p. 316 and Figure 14.10

- How does RNA splicing happen? What are the consequences if it does not happen correctly? See p. 317 and Figure 14.11

While each of a multicellular organism's somatic cells contains a complete set of genes, no cell expresses all of those genes. Each cell type usually expresses only those genes that it needs for its own development and function. How a cell controls gene expression is the subject of the next section.

14.4 How Is Eukaryotic Gene Transcription Regulated?

For development to proceed normally, and for each cell in a multicellular organism to acquire and maintain its proper specialized function, certain proteins must be synthesized at just the right times and in just the right cells. Thus the expression of eukaryotic genes must be precisely regulated. Unlike DNA replication, which is regulated in every cell on an all-or-none basis, gene expression is highly selective.

Gene expression can be regulated at a number of different points in the process of transcribing and translating the gene into a protein (**Figure 14.12**). In this section we will describe the mechanisms that result in the selective transcription of specific genes. Some of these mechanisms involve nuclear proteins that alter chromosome function or structure. In other cases, the regulation of transcription involves changes in the DNA itself: genes may be selectively replicated to provide more templates for transcription, or even rearranged on the chromosome. The following two sections will examine the regulation of eukaryotic gene expression after transcription.

Specific genes can be selectively transcribed

The brain cells and the liver cells of a mouse have some proteins in common and others that are characteristic of each cell type. Yet both cells have the same DNA sequences and, therefore, the same genes. Are the differences in protein content due to differential transcription of the genes? Or is it the case that all the genes are transcribed in both cell types, and some mechanism that acts after transcription is responsible for the differences in proteins?

These two alternatives— *transcriptional regulation* and *posttranscriptional regulation*—can be distinguished by examining the actual mRNA sequences made within the nucleus of each cell type. Such analyses indicate that for some proteins, the mechanism of regulation is differential gene transcription. Both brain and liver

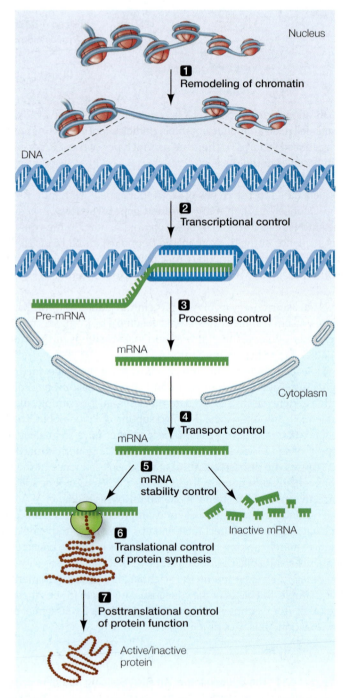

14.12 Potential Points for the Regulation of Gene Expression Gene expression can be regulated before transcription (1), during transcription (2, 3), after transcription but before translation (4, 5), at translation (6), or after translation (7).

cells, for example, transcribe "housekeeping" genes—those that encode proteins involved in the basic metabolic processes that occur in every living cell, such as glycolysis enzymes. But liver cells transcribe some genes for liver-specific proteins, and brain cells transcribe some genes for brain-specific proteins. And neither cell type transcribes the genes for proteins that are characteristic of muscle, blood, bone, or the other specialized cell types in the body.

CONTRASTING EUKARYOTES AND PROKARYOTES Unlike prokaryotes, in which functionally related genes are often grouped into operons that are transcribed as a unit, eukaryotes tend to have solitary genes. Thus the regulation of several genes at once requires common control elements in each of the genes that allow all of the genes to respond to the same signal.

In contrast to the single RNA polymerase in bacteria, eukaryotes have three different RNA polymerases. Each eukaryotic polymerase catalyzes the transcription of a specific type of gene. Only one (RNA polymerase II) transcribes protein-coding genes. The other two transcribe the DNA that codes for rRNA (polymerase I) and for tRNA and small nuclear RNAs (polymerase III).

The diversity of eukaryotic polymerases is reflected in the diversity of eukaryotic promoters, which tend to be much more varied in their sequences than prokaryotic promoters. Furthermore, most eukaryotic genes have additional sequences that can regulate the rate of their transcription. Whether a eukaryotic gene is transcribed depends on the sum total of the effects of all of these DNA and protein elements; thus there are many points of possible regulation.

Finally, the initiation of transcription in eukaryotes is very different from that in prokaryotes, in which RNA polymerase directly recognizes the promoter. In eukaryotes, many proteins are involved in initiating transcription. We will confine the following discussion to RNA polymerase II, which catalyzes the transcription of most protein-coding genes, but the mechanisms for the other two RNA polymerases are similar.

TRANSCRIPTION FACTORS As Section 13.4 described, the prokaryotic promoter is a sequence of DNA near the 5′ end of the coding region of a gene or operon where RNA polymerase begins transcription. A prokaryotic promoter has two essential sequences. One is the **recognition sequence**—the sequence recognized by RNA polymerase. The second, closer to the initiation site, is the **TATA box** (so called because it is rich in AT base pairs), where DNA begins to denature so that the template strand can be exposed.

Some plants can attract bacteria and fungi to their roots to provide them with important nutrients. A chemical signal generated by the plant causes the microbes to make oligosaccharides called "nod factors." Nod factors, in turn, cause the plant to activate two transcription factors that increase the expression of genes that result in a fruitful symbiosis.

Things are different in eukaryotes. Eukaryotic RNA polymerase II cannot simply bind to the promoter and initiate transcription. Rather, it does so only after various regulatory proteins, called **tran-**

scription factors, have assembled on the chromosome (**Figure 14.13**). First, the protein TFIID ("TF" stands for transcription factor) binds to the TATA box. Its binding changes both its own shape and that of the DNA, presenting a new surface that attracts the binding of other transcription factors to form a *transcription complex*. RNA polymerase II does not bind until several other proteins have bound to this complex.

Some DNA sequences, such as the TATA box, are common to the promoters of many eukaryotic genes and are recognized by transcription factors that are found in all the cells of an organism.

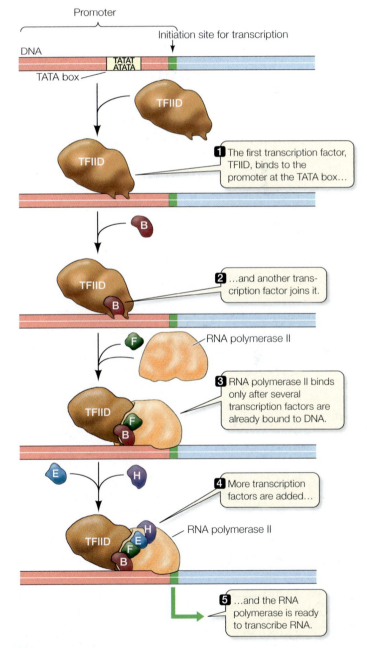

14.13 The Initiation of Transcription in Eukaryotes Except for TFIID, which binds to the TATA box, each transcription factor in this transcription complex has binding sites only for the other proteins in the complex, and does not bind directly to DNA. B, E, F, and H are transcription factors.

Other sequences found in promoters are specific to only a few genes and are recognized by transcription factors found only in certain tissues. These specific transcription factors play an important role in *differentiation*, the specialization of cells during development.

REGULATORS, ENHANCERS, AND SILENCERS IN DNA   In addition to the promoter, two other types of regulatory DNA sequences bind proteins that activate RNA polymerase. The recently discovered **regulator sequences** are clustered just upstream of the promoter. Various *regulator proteins* (seven for the β-globin gene) may bind to these regulator sequences (**Figure 14.14**). The resulting complexes bind to the adjacent transcription complex and activate it.

Much farther away—up to 20,000 bp away—from the promoter are **enhancer sequences**. Enhancer sequences bind *activator* proteins, and this binding strongly stimulates the transcription complex. How enhancers exert their influence is not clear. In one proposed model, the DNA bends (it is known to do so) so that the activator protein is in contact with the transcription complex.

In addition, there are *negative* regulatory sequences on DNA, called **silencer** sequences, that have the opposite effect from enhancers. Silencers turn off transcription by binding proteins appropriately called *repressor* proteins.

How do these proteins and DNA sequences—transcription factors, regulators, enhancers, activators, silencers, and repressors—regulate transcription? Apparently, in most tissues, a small amount of RNA is transcribed from all genes, but the combination of these factors determines the rate of transcription. In the immature red blood cells in bone marrow, for example, which make a large amount of β-globin, transcription of the β-globin gene is stimulated by the binding of seven regulator proteins and six activator proteins. But in white blood cells in the same bone marrow, these thirteen proteins are not made and do not bind to the regulator and enhancer sequences adjacent to the β-globin gene; consequently, the β-globin gene is hardly transcribed at all.

PROTEIN STRUCTURES INVOLVED IN PROTEIN–DNA INTERACTIONS
The regulation and coordination of gene expression requires the binding of many specialized proteins to DNA. Among DNA-binding proteins, there are four common structural themes in the protein domains that bind to DNA. These themes, called *motifs*, consist of different combinations of structural elements and special components: *helix-turn-helix, zinc finger, leucine zipper,* and *helix-loop-helix* (**Figure 14.15**). DNA-binding proteins with specific motifs in their binding domains are involved in the activation and inactivation of certain types of genes, both during development and in the adult organism.

Let's look at how just one of these motifs works. As pointed out in Section 11.2, the complementary bases in DNA not only form hydrogen bonds with each other, but can form additional hydrogen bonds with proteins, particularly at points exposed in the major and minor grooves (see Figure 11.8B). In this way, an intact DNA double helix can be recognized by a protein whose structure:

- fits into the major or minor groove
- has amino acids that can project into the interior of the double helix
- has amino acids that can form hydrogen bonds with the interior bases

Proteins with a helix-turn-helix motif, in which two α-helices are connected via a non-helical turn, fit these three criteria. The interior-facing "recognition" helix is the one whose amino acids interact with the bases inside the DNA. The exterior-facing helix sits on the sugar–phosphate backbone, ensuring that the interior helix is presented to the bases in the correct configuration. Many repressor proteins have this helix-turn-helix configuration in their structure. Binding of these repressors to DNA prevents other proteins from interacting with that DNA, and this is what inhibits transcription.

COORDINATING THE EXPRESSION OF SEVERAL GENES   How do eukaryotic cells coordinate the regulation of several genes whose transcription must be turned on at

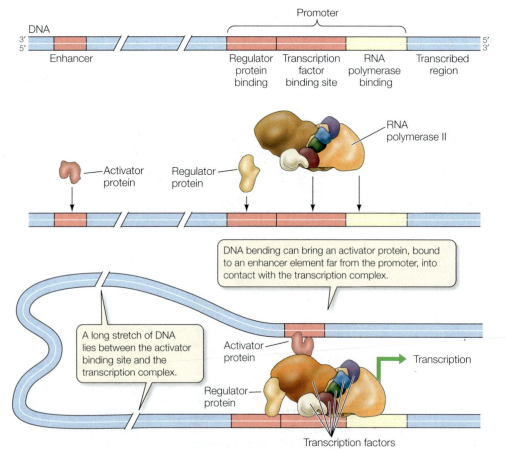

14.14 Transcription Factors, Regulators, and Activators   The actions of many proteins determine whether and where RNA polymerase II will transcribe DNA.

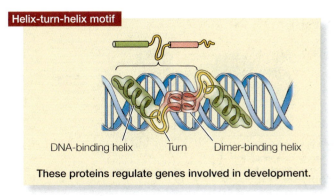

Helix-turn-helix motif

DNA-binding helix — Turn — Dimer-binding helix

These proteins regulate genes involved in development.

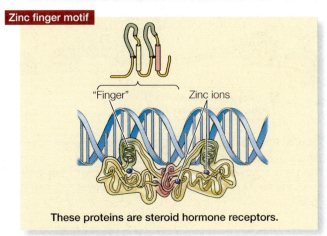

Leucine zipper motif

Leucine — Zipper

These proteins regulate cell division genes.

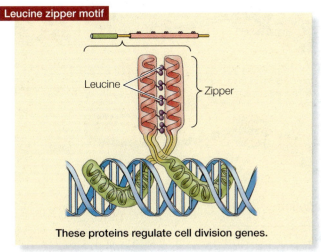

Zinc finger motif

"Finger" — Zinc ions

These proteins are steroid hormone receptors.

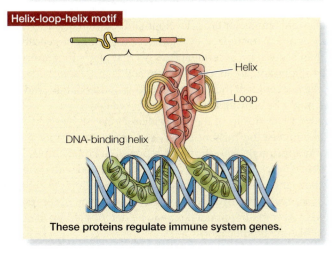

Helix-loop-helix motif

Helix — Loop

DNA-binding helix

These proteins regulate immune system genes.

14.15 Protein–DNA Interactions The DNA-binding domains of most regulatory proteins contain one of four structural motifs.

the same time? In prokaryotes, in which related genes are linked together in an operon, a single regulatory system can regulate several adjacent genes. But in eukaryotes, the several genes whose regulation requires coordination may be far apart on a chromosome, or even on different chromosomes.

In such a case, regulation can be achieved if the various genes all have the same regulator sequences, which bind the same regulator proteins. One of the many examples of this phenomenon is provided by the response of an organism to a stressor—for example, that of plants to drought. Under conditions of drought stress, a plant must synthesize a number of proteins, but the genes for those proteins are scattered throughout the genome. However, each of these genes has a specific regulator sequence near its promoter, called the *stress response element* (SRE). The binding of a regulator protein to this element stimulates RNA synthesis (**Figure 14.16**). The proteins made from these genes are involved not only in water conservation, but also in protecting the plant against ex-

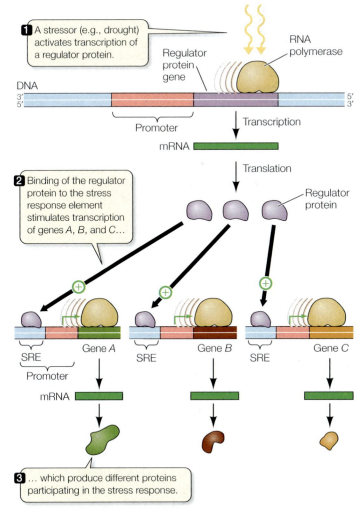

14.16 Coordinating Gene Expression A single environmental signal, such as drought stress, causes the synthesis of a transcriptional regulator protein that acts on many genes.

cess salt in the soil and against freezing. This finding has considerable importance for agriculture, in which crops are often grown under less than optimal conditions.

Gene expression can be regulated by changes in chromatin structure

Other mechanisms that regulate transcription act on the structure of chromatin and chromosomes. As Section 9.3 describes, chromatin contains a number of proteins as well as DNA. The packaging of DNA into nucleosomes by these nuclear proteins can make DNA physically inaccessible to RNA polymerase and the rest of the transcription apparatus, much as the binding of a repressor to the operator in the prokaryotic *lac* operon prevents transcription (see Section 13.4). Chromatin structure at both the local and whole-chromosome levels affects transcription.

CHROMATIN REMODELING Recall that DNA is wound around proteins called histones to form a structure called a nucleosome. Nucleosomes block both the initiation and elongation steps of transcription. In a process called **chromatin remodeling**, two types of remodeling proteins inactivate these two blocks (**Figure 14.17**). To allow initiation, the first remodeling protein binds upstream of the initiation site, disaggregating the nucleosomes so that the transcription complex can bind and RNA polymerase can begin transcription. To allow elongation, the second remodeling protein binds once transcription is under way, allowing the transcription complex to move through the nucleosomes.

THE HISTONE CODE How are nucleosomes disassembled to allow transcription (and then reassembled)? Each histone protein has an approximately 20–amino acid "tail" at its N terminus that sticks out of the compact structure. This tail has certain amino acids (notably lysine) that are positively charged. These amino acids are targets for enzymes that add acetyl groups to the positively charged amino acids, thus changing their charges:

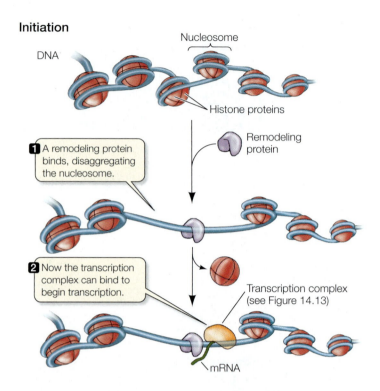

Initiation

1 A remodeling protein binds, disaggregating the nucleosome.

2 Now the transcription complex can bind to begin transcription.

Elongation

1 A second remodeling protein can bind to the nucleosome…

2 …allowing transcription without disaggregation.

14.17 Local Remodeling of Chromatin for Transcription Initiation of transcription requires that nucleosomes change their structure, becoming less compact. This makes DNA accessible to the transcription complex. During elongation of RNA, however, they can remain intact.

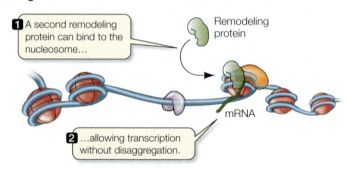

Lysine in histone   Acetyl-CoA   Acetyl-lysine

Reducing the positive charge of the histone tails reduces the affinity of the histones for DNA, opening up the compact nucleosome. Ordinarily, because histone proteins are positively charged and DNA is negatively charged (owing to its phosphate groups), the attachment of these two molecules is electrostatic. Without this electrostatic attraction, DNA is not so closely held to the nucleosome, and chromatin remodeling proteins can bind to the looser nucleosome–DNA complex.

While gene activation calls for enzymes (*histone acetyltransferases*) to add acetyl groups, gene repression requires other enzymes (*histone deacetylases*) to *remove* the acetyl groups. The latter enzymes are targets for drug development. Certain diseases such as cancer are characterized by a greater proportion of deacetylation than acetylation at certain genes, so that genes that normally block cell division are inactive. A drug acting as a *histone deacetylase inhibitor* could tilt the activity ratio toward acetylation, and the genes might be activated, turning off the cell cycle.

In what David Allis of the Rockefeller University in New York City has dubbed the "histone code," several types of histone modification affect gene activation and repression. For example, histones are subject to methylation, which is associated with gene inactivation, and phosphorylation, as well as acetylation. All of these effects are reversible. So whether a eukaryotic gene becomes activated by chromatin remodeling may be determined by the pattern of histone modification.

The Barr body is the condensed, inactive member of a pair of X chromosomes in the cell. The other X is not condensed and is active in transcription.

14.18 A Barr Body in the Nucleus of a Female Cell The number of Barr bodies per nucleus is equal to the number of X chromosomes minus one. Thus normal males (XY) have no Barr body, whereas normal females (XX) have one.

WHOLE-CHROMOSOME EFFECTS Some transcriptional regulation mechanisms act on entire chromosomes. Under a microscope, two kinds of chromatin can be distinguished in the stained interphase nucleus: *euchromatin* and *heterochromatin*. Euchromatin is diffuse and stains lightly; it contains the DNA that is transcribed into mRNA. Heterochromatin is condensed and stains darkly; any genes it contains are generally not transcribed.

Perhaps the most dramatic example of heterochromatin is seen in the inactive X chromosome of mammals. A normal female mammal has two X chromosomes; a normal male has an X and a Y. The X and Y chromosomes probably arose from a pair of autosomes about 300 million years ago. Over time, mutations in the Y chromosome resulted in maleness-determining genes (see Section 10.4), and the Y chromosome gradually lost most of the genes it once shared with its X homolog. As a result, there is a great difference between females and males in the "dosage" of X-linked genes. Each female cell has two copies of the genes on the X chromosome, and therefore has the potential to produce twice as much of the protein products of these genes as a male cell has. Nevertheless, for 75 percent of genes on the X, transcription is generally the same in males and in females. How does this happen?

Mary Lyon, Liane Russell, and Ernest Beutler independently suggested in 1961 that one of the X chromosomes in each cell of a female is, to a significant extent, transcriptionally inactivated early in embryonic development. That is, they proposed that a copy of the X remains inactive in each embryonic cell, and in all the cells arising from it. In a given embryonic cell, the "choice" of which X in the pair of Xs to inactivate is random. Recall that one X in a female comes from her father and one from her mother. Thus, in one embryonic cell, the paternal X might be the one remaining transcriptionally active, but in a neighboring cell, the maternal X might be active.

During interphase, a single, stainable nuclear body, called a *Barr body* (after its discoverer, Murray Barr), can be seen in cells of human females under the light microscope (**Figure 14.18**). This clump of heterochromatin, which is not present in males, is the inactivated X chromosome. The number of Barr bodies in a nucleus is equal to the number of X chromosomes minus one (the one represents the X chromosome that remains transcriptionally active). So a female with the normal two X chromosomes will have one Barr body, a rare female with three Xs will have two, an XXXX female will have three, and an XXY male will have one. These observations suggest that the interphase cells of each person, male or female, have a single active X chromosome, making the dosage of the expressed X chromosome genes constant across both sexes.

Condensation of the inactive X chromosome makes its DNA sequences physically unavailable to the transcriptional machinery. One mechanism of condensation is the addition of a methyl group ($-CH_3$) to the 5′ position of cytosine on DNA. Methylation of cytosines is associated with transcriptionally inactive genes. For example, many cytosines of the DNA of the inactive X chromosome are methylated, while few cytosines on the active X are methylated. Methylated DNA appears to bind certain chromosomal proteins that may be responsible for heterochromatin formation.

The otherwise inactive X chromosome has one gene that is only lightly methylated and is transcriptionally active, called *Xist* (for *X inactivation-specific transcript*). *Xist* is heavily methylated on, and not transcribed from, the other, "active" X chromosome. The RNA transcribed from *Xist* does not leave the nucleus and is not an mRNA. Instead, it appears to bind to the X chromosome from which it is transcribed, and this binding somehow leads to a spreading of inactivation along the chromosome. This RNA transcript is known as an **interference RNA** (**Figure 14.19**).

How does the transcriptionally active X overcome the effects of *Xist* RNA? Apparently, there is an anti-*Xist* gene, appropriately called *Tsix*. This gene codes for an RNA that binds by complementary base pairing to *Xist* RNA at the active X chromosome.

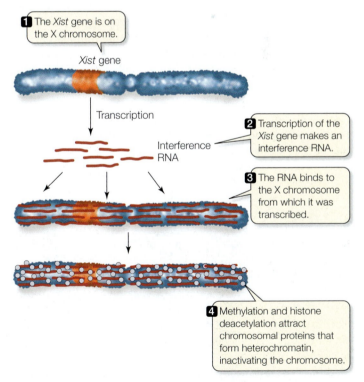

1 The *Xist* gene is on the X chromosome.

Xist gene

Transcription

2 Transcription of the *Xist* gene makes an interference RNA.

Interference RNA

3 The RNA binds to the X chromosome from which it was transcribed.

4 Methylation and histone deacetylation attract chromosomal proteins that form heterochromatin, inactivating the chromosome.

14.19 A Model for X Chromosome Inactivation Interference RNA and chromosomal proteins combine to inactivate the X chromosome.

Selective gene amplification results in more templates for transcription

Another way for one cell to make more of a certain gene product than another cell does is to make more copies of the appropriate gene and transcribe them all. The process of creating more copies of a gene in order to increase its transcription is called **gene amplification**.

As described earlier, the genes that code for three of the four human ribosomal RNAs are linked together in a unit, and this unit is repeated several hundred times in the genome to provide multiple templates for rRNA synthesis (rRNA is the most abundant kind of RNA in the cell). In some circumstances, however, even this moderate repetition is not enough to satisfy the demands of the cell.

The mature eggs of frogs and fishes, for example, have up to a trillion ribosomes. These ribosomes are used for the massive protein synthesis that follows fertilization. The precursor cell that will differentiate into the egg contains fewer than 1,000 copies of the rRNA gene cluster, and would take 50 years to make a trillion ribosomes if it transcribed those rRNA genes at peak efficiency. How does the egg end up with so many ribosomes (and so much rRNA)?

The cell solves this problem by selectively amplifying its rRNA gene clusters until there are more than a million copies (**Figure 14.20**). In fact, this gene complex goes from being 0.2 percent of the total genome to 68 percent. These million copies, transcribed at the maximum rate, are just enough to make the necessary trillion ribosomes in a few days.

The mechanism for selective amplification of a single gene is not clearly understood, but it has important medical implications. In some cancers, a cancer-causing gene called an *oncogene* becomes amplified (see Section 17.4). In addition, when some tumors are treated with a drug that targets a single protein, amplification of the gene for the target protein leads to an excess of that protein, and the cell becomes resistant to the prescribed dose of the drug.

There are many ways in which gene expression can be regulated even after the gene has been transcribed. The processing of pre-mRNA described in Section 14.3 is merely one opportunity for doing so.

14.5 How Is Eukaryotic Gene Expression Regulated After Transcription?

As we have seen, pre-mRNA is processed by cutting out the introns and splicing the exons together. If exons are selectively deleted from the pre-mRNA, different proteins can be synthesized. The longevity of mRNA in the cytoplasm can also be regulated: the longer an mRNA exists in the cytoplasm, the more of its protein can be made.

Different mRNAs can be made from the same gene by alternative splicing

Most primary mRNA transcripts contain several introns (see Figure 14.5). We have seen how the splicing mechanism recognizes the boundaries between exons and introns. What would happen if the β-globin pre-mRNA, which has two introns, were spliced from the start of the first intron to the end of the second? Not only

14.20 Transcription from Multiple Genes for rRNA Elongating strands of rRNA transcripts form arrowhead-shaped regions, each centered on a DNA sequence that codes for three of the four ribosomal subunits.

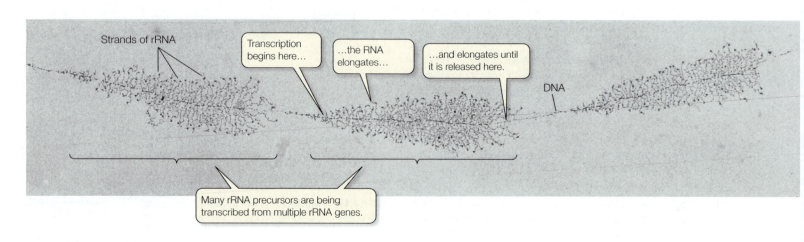

Strands of rRNA

Transcription begins here…

…the RNA elongates…

…and elongates until it is released here.

DNA

Many rRNA precursors are being transcribed from multiple rRNA genes.

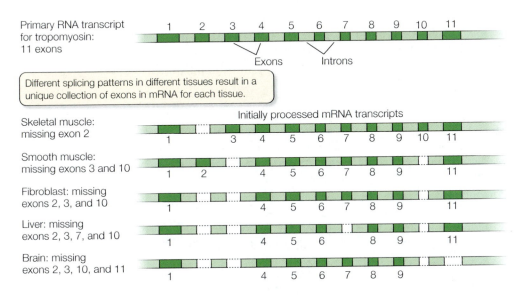

14.21 Alternative Splicing Results in Different Mature mRNAs and Proteins In mammals, the protein tropomyosin is encoded by a gene that has 11 exons. Tropomyosin pre-mRNA is spliced differently in different tissues, resulting in five different forms of the protein.

that half of all human genes are alternatively spliced. Alternative splicing may be a key to the differences in levels of complexity among organisms.

the two introns, but also the middle exon, would be spliced out. An entirely new protein (certainly not a β-globin) would be made, and the functions of normal β-globin would be lost.

Such **alternative splicing** can be a deliberate mechanism for generating a family of different proteins from a single gene. In mammals, for example, a single pre-mRNA for the structural protein tropomyosin is spliced differently in five different tissues to give five different mature mRNAs. These mRNAs are translated into the five different forms of tropomyosin found in skeletal muscle, smooth muscle, fibroblast, liver, and brain (**Figure 14.21**).

Before the sequencing of the human genome began, most scientists estimated that they would find between 100,000 and 150,000 genes. You can imagine their surprise when the actual sequence revealed only about 24,000 genes! In fact, there are many more human mRNAs than there are human genes, and most of this variation comes from alternative splicing. Indeed, recent surveys show

The stability of mRNA can be regulated

DNA, as the genetic material, must remain stable, and as we saw in Section 11.3, there are elaborate mechanisms for repairing it if it becomes damaged. RNA, however, has no such repair mechanisms. After it arrives in the cytoplasm, mRNA is subject to breakdown catalyzed by ribonucleases, which exist both in the cytoplasm and in lysosomes. The less time an mRNA spends in the cytoplasm, the less of its protein can be translated. But not all eukaryotic mRNAs have the same life span. Differences in the stabilities of mRNAs provide another mechanism for posttranscriptional regulation of protein synthesis.

Specific AU-rich nucleotide sequences within some mRNAs mark them for rapid breakdown by a ribonuclease complex called the **exosome**. Signaling molecules such as growth factors, for example, are made only when needed, and then break down rapidly. Their mRNAs are highly unstable because they contain an AU-rich sequence.

Small RNAs can break down mRNAs

Very small RNAs—about 20 bases long—that are complementary to a region in mRNA can bind by base pairing to that mRNA before it gets to the ribosome. This binding causes the target mRNA to break down, but even if it survives, its translation is inhibited because tRNA cannot bind to the already base-paired region. Although only recently discovered, these **micro RNAs** are a common mechanism of posttranscriptional regulation. There are about 250 genes in the human genome that code for micro RNAs.

Small RNAs begin as 70-nucleotide, double-stranded molecules. A protein complex appropriately named *dicer* cuts the RNAs down to "micro" size and directs them to their target mRNAs (**Figure 14.22**). They have been implicated in controlling genes with a wide range of functions, ranging from the development of the nervous system in the roundworm, to apoptosis in the fruit fly, to flower development in plants, to development of the blood system in humans. As we will see in Chapter 16, small RNAs are under development as drugs to block the expression of certain human genes in diseases.

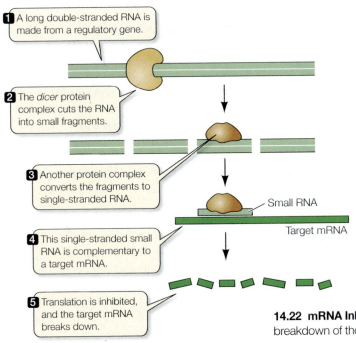

1 A long double-stranded RNA is made from a regulatory gene.

2 The *dicer* protein complex cuts the RNA into small fragments.

3 Another protein complex converts the fragments to single-stranded RNA.

4 This single-stranded small RNA is complementary to a target mRNA.

Small RNA

Target mRNA

5 Translation is inhibited, and the target mRNA breaks down.

14.22 mRNA Inhibition by Small RNAs Small RNAs result in inhibition of translation and breakdown of the target mRNA.

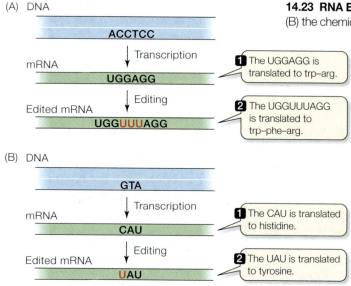

14.23 RNA Editing RNA can be edited by (A) the insertion of new nucleotides or by (B) the chemical alteration of existing nucleotides.

RNA can be edited to change the encoded protein

The sequence of mRNA can be changed after transcription and splicing by **RNA editing**. This editing can occur in two ways (**Figure 14.23**):

- *Insertion of nucleotides.* In the parasitic protist *Trypanosoma brucei*, certain mRNAs have been found that have a longer base sequence than predicted by the gene coding for them. Stretches of uracil are added after transcription, changing the protein that is made.

- *Alteration of nucleotides.* An enzyme can catalyze the deamination of cytosine, forming uracil. This process can affect a membrane channel protein in the mammalian nervous system that normally allows calcium and sodium to pass through. Editing of a certain cytosine in the mRNA for this protein to uracil changes the amino acid at that position in the polypeptide chain from histidine to tyrosine, and the channel protein no longer allows the passage of calcium.

14.5 RECAP

One of the most important means of posttranscriptional regulation is alternative RNA splicing, which allows more than one protein to be made from a gene. The stability of mRNA in the cytoplasm can also be regulated. Micro RNAs and RNA editing are two recently discovered mechanisms of regulation.

- Do you understand how a single pre-mRNA sequence can encode several different proteins?
  See pp. 324–325 and Figure 14.21

- How do micro RNAs regulate gene expression?
  See p. 325 and Figure 14.22

- What is RNA editing? See p. 326 and Figure 14.23

Even after mature mRNA is produced and transported to the cytoplasm, the presence of mRNA does not necessarily mean that it will be translated into a functional protein. Eukaryotes have several mechanisms for regulating gene expression during and after translation.

14.6 How Is Gene Expression Controlled During and After Translation?

Is the amount of a protein in a cell determined by the amount of its mRNA? Recently, the relationships between mRNAs and proteins in yeast cells were examined. For about a third of the dozens of genes surveyed, a clear relationship between mRNA and protein held: more of one led to more of the other. But for two-thirds of the proteins, no apparent relationship was observed. The concentrations of these proteins in the cell must therefore be determined by factors acting after the mRNA is made.

The initiation and extent of translation can be regulated

One way to regulate translation is through the G cap on mRNA. As we saw in Section 14.3, mRNA is capped at its 5′ end by a modified guanosine triphosphate molecule (see Figure 14.10). An mRNA that is capped with an unmodified GTP molecule is not translated. For example, stored mRNA in the egg cells of the tobacco hornworm moth has a G cap at its 5′ end, but the GTP molecule is not modified and the stored mRNA is not translated. After the egg is fertilized, however, the cap *is* modified, allowing the mRNA to be translated to produce the proteins needed for early embryonic development.

Conditions within a cell can influence translational processes. Within mammalian cells, for example, free iron ions (Fe^{2+}) are bound by a storage protein called *ferritin*. When iron is present in excess, ferritin synthesis rises dramatically. Yet the amount of ferritin mRNA remains constant. The increase in ferritin synthesis is due to an increased rate of mRNA translation. When the iron level in the cell is low, a translational repressor protein binds to ferritin mRNA and prevents its translation by blocking its attachment to a ribosome. When the iron level rises, the excess iron ions bind to the repressor and alter its three-dimensional structure, causing it to detach from the mRNA, and translation of ferritin proceeds.

Translational control can be used to keep a proper balance where several subunits associate to form a functional unit, as in hemoglobin molecules. As we saw in Section 14.2, a hemoglobin molecule consists of four globin subunits and four heme pigments. If globin synthesis does not equal heme synthesis, some heme stays free in the cell, waiting for a globin partner. Excess heme increases the rate of translation of globin mRNA by removing a block to the initiation of translation at the ribosome.

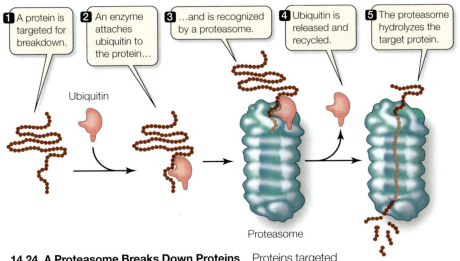

1 A protein is targeted for breakdown.

2 An enzyme attaches ubiquitin to the protein…

3 …and is recognized by a proteasome.

4 Ubiquitin is released and recycled.

5 The proteasome hydrolyzes the target protein.

Ubiquitin

Proteasome

14.24 A Proteasome Breaks Down Proteins Proteins targeted for breakdown are bound to ubiquitin, which "leads" the target protein to a proteasome (a complex composed of many polypeptides).

Posttranslational controls regulate the longevity of proteins

Most gene products—proteins—are modified after translation. Some of these are covalent changes, such as the addition of sugars (glycosylation), the addition of phosphate groups, or the removal of a signal sequence after a protein has crossed a membrane (see Figure 12.5).

One way to regulate the action of a protein in a cell is to regulate its lifetime in the cell. In many cases, an enzyme attaches a 76-amino acid protein called **ubiquitin** (so named because it is *ubiquitous*, or widespread) to a lysine in a protein targeted for breakdown. Other ubiquitin chains then attach to the primary one, forming a *polyubiquitin complex*. The protein–polyubiquitin complex then

binds to a huge protein complex called a **proteasome** (**Figure 14.24**). The entryway to this "molecular chamber of doom" is a hollow cylinder. This part of the complex uses energy from ATP to cut off the ubiquitin for recycling and unfold its targeted protein "victim." The protein then passes by three different proteases (thus the name of the complex), which digest it into small peptides and amino acids.

The cellular concentrations of many proteins are determined not by differential expression of their genes, but by their degradation in proteasomes. Cyclins, for example, are degraded at just the right time during the cell cycle (see Section 9.2), and transcriptional regulators are broken down after they are used, lest the affected genes be always "on." Viruses can hijack this system. Human papillomavirus, which causes cervical cancer, targets the cell division inhibitory protein p53 for proteasomal degradation, resulting in unregulated cell division—i.e., cancer.

14.6 RECAP

Proteins in the cell can be targeted for breakdown by ubiquitin and then hydrolyzed in proteasomes.

- Do you understand the role of translational repressors? See p. 326

- Can you describe how a proteasome works? See p. 327 and Figure 14.24

CHAPTER SUMMARY

14.1 What are the characteristics of the eukaryotic genome?

There are many differences between prokaryotic and eukaryotic genomes and their mechanisms of expression. Review Table 14.1

Unlike prokaryotic DNA, eukaryotic DNA is contained within a nucleus, so that transcription and translation are physically separated. Review Figure 14.1, Web/CD Activity 14.1

Various types of highly repetitive sequences, known as satellites, characterize eukaryotic DNA. Most are not transcribed.

Some moderately repetitive DNA sequences are transcribed; many of these code for tRNA and rRNA, of which many copies are needed. Review Figure 14.3

Transposons are moderately repetitive sequences that are able to move about the genome. Review Figure 14.4

14.2 What are the characteristics of eukaryotic genes?

A typical eukaryotic protein-coding gene is flanked by promoter and **terminator** sequences and contains noncoding internal sequences, called **introns**. The coding sequences of such a gene are called **exons**. Review Figure 14.5

Introns are transcribed to the primary mRNA transcript (**pre-mRNA**), but are later removed and do not appear in the mature mRNA transcript.

In **nucleic acid hybridization**, DNA is denatured and incubated with a single-stranded **probe** that binds to the DNA by complementary base pairing. Review Figure 14.6

Some eukaryotic genes exist as members of **gene families**. Proteins may be made from these closely related genes at different times and in different tissues. Some members of gene families may be nonfunctional **pseudogenes**. Review Figure 14.8

14.3 How are eukaryotic gene transcripts processed?

The transcribed pre-mRNA is altered by the addition of a **G cap** at the 5' end and a **poly A tail** at the 3' end. Review Figure 14.10

The introns are removed from pre-mRNA by **RNA splicing**. A complex of **snRNPs** and enzymes, called a **spliceosome**, forms at the **consensus sequences** that lie between introns and exons. The spliceosome cuts out the introns and splices the exons together. Review Figure 14.11, Web/CD Tutorial 14.1

CHAPTER SUMMARY

14.4 How is eukaryotic gene transcription regulated?

Eukaryotic gene expression can be regulated before or during transcription, during pre-mRNA processing, and during or after translation. Review Figure 14.12, Web/CD Activity 14.2

Eukaryotes have three different RNA polymerases. RNA polymerase II transcribes protein-coding genes.

For transcription to occur, the protein TFIID must bind to the **TATA box** on the promoter, and other **transcription factors** must assemble on that protein, before RNA polymerase can bind at the promoter. Review Figure 14.13, Web/CD Tutorial 14.2

Other regulatory DNA sequences include **regulator** sequences, which bind regulator proteins and activate transcription, **enhancer** sequences, which bind activator proteins and stimulate transcription, and **silencer** sequences, which bind repressor proteins and turn off transcription. Review Figure 14.14

The DNA-binding domains of most DNA-binding proteins have one of four structural motifs.

Chromatin remodeling allows the transcription complex to bind to DNA and to move through the nucleosomes. Review Figure 14.17

Interference RNA, transcribed from the *Xist* gene, is important in inhibiting transcription of the inactive X chromosome. Review Figure 14.19

Some genes are selectively **amplified** in some cells. The extra copies of these genes result in increased transcription of their protein product.

14.5 How is eukaryotic gene expression regulated after transcription?

Alternative splicing of pre-mRNA can produce different proteins. Review Figure 14.21

Not all RNAs have the same life span. Regulating the stability of mRNA in the cytoplasm is a posttranscriptional mechanism for regulating protein synthesis. Specific AU-rich sequences can, for example, be rapidly broken down by an **exosome**. **Small RNAs** can base-pair with target mRNA sequences, preventing their translation and breaking them down. Review Figure 14.22

mRNA can be **edited** by the addition of new nucleotides or by the chemical alteration of existing nucleotides. Review Figure 14.23

14.6 How is gene expression controlled during and after translation?

Translational repressors can inhibit the translation of mRNA.

Proteasomes can degrade proteins that have been targeted for breakdown by attachment of **ubiquitin**. Review Figure 14.24

SELF-QUIZ

1. Eukaryotic protein-coding genes differ from their prokaryotic counterparts in that eukaryotic genes
   a. are double-stranded.
   b. are present in only a single copy.
   c. contain introns.
   d. have a promoter.
   e. transcribe mRNA.

2. Comparison of the genomes of yeast and bacteria shows that only yeast has many genes for
   a. energy metabolism.
   b. cell wall synthesis.
   c. intracellular protein targeting.
   d. DNA-binding proteins.
   e. RNA polymerase.

3. The genomes of the fruit fly and the nematode are similar to that of yeast, except that the former organisms have many genes for
   a. intercellular signaling.
   b. synthesis of polysaccharides.
   c. cell cycle regulation.
   d. intracellular protein targeting.
   e. transposable elements.

4. Which of the following does *not* occur after mRNA is transcribed?
   a. Binding of RNA polymerase II to the promoter
   b. Capping of the 5′ end
   c. Addition of a poly A tail to the 3′ end
   d. Splicing out of the introns
   e. Transport to the cytosol

5. Which statement about RNA splicing is *not* true?
   a. It removes introns.
   b. It is performed by small nuclear ribonucleoprotein particles (snRNPs).
   c. It always removes the same introns.
   d. It is usually directed by consensus sequences.
   e. It shortens the RNA molecule.

6. Eukaryotic transposons
   a. always use RNA for replication.
   b. are approximately 50 bp long.
   c. are made up of either DNA or RNA.
   d. do not contain genes coding for transposition.
   e. make up about 40 percent of the human genome.

7. Which statement about selective gene transcription in eukaryotes is *not* true?
   a. Different classes of RNA polymerase transcribe different parts of the genome.
   b. Transcription requires transcription factors.
   c. Genes are transcribed in groups called operons.
   d. Both positive and negative regulation occur.
   e. Many proteins bind at the promoter.

8. Heterochromatin
   a. contains more DNA than does euchromatin.
   b. is transcriptionally inactive.
   c. is responsible for all negative transcriptional control.
   d. clumps the X chromosome in human males.
   e. occurs only during mitosis.

9. Translational control
   a. is not observed in eukaryotes.
   b. is a slower form of regulation than transcriptional control.
   c. can be achieved by only one mechanism.
   d. requires that mRNA be uncapped.
   e. ensures that heme synthesis equals globin synthesis.

10. Control of gene expression in eukaryotes includes all of the following *except*
    a. alternative splicing of RNA transcripts.
    b. binding of proteins to DNA.
    c. transcription factors.
    d. feedback inhibition of enzyme activity by allosteric control.
    e. DNA methylation.

FOR DISCUSSION

1. In rats, a gene 1,440 bp long codes for an enzyme made up of 192 amino acids. Discuss this apparent discrepancy. How long would the initial and final mRNA transcripts be?

2. The genomes of rice, wheat, and corn are similar to one other and to that of *Arabidopsis* in many ways. Discuss how these plants might nevertheless have very different proteins.

3. The activity of the enzyme dihydrofolate reductase (DHFR) is high in some tumor cells. This activity makes the cells resistant to the anticancer drug methotrexate, which targets DHFR. Assuming that you had the complementary DNA for the gene that encodes DHFR, how would you show whether this increased activity was due to increased transcription of the single-copy DHFR gene or to amplification of the gene?

4. Describe the steps in the production of a mature, translatable mRNA from a eukaryotic gene that contains introns. Compare this to the situation in prokaryotes (see Section 13.3).

5. A protein-coding gene has three introns. How many different proteins can be made from alternative splicing of the pre-mRNA transcribed from this gene?

FOR INVESTIGATION

Nucleic acid hybridization has had many uses in biological research and clinical medicine. It is possible to make a DNA probe that fluoresces under ultraviolet light. Suppose you want to determine whether the fetus a woman is carrying has an extra copy of chromosome 21 (and so has Down syndrome). You take a sample of cells from the fetus. These cells are mostly in interphase. Assuming you have isolated the DNA for several genes that map on chromosome 21, outline the experiments you would do to determine whether the fetus has Down syndrome.

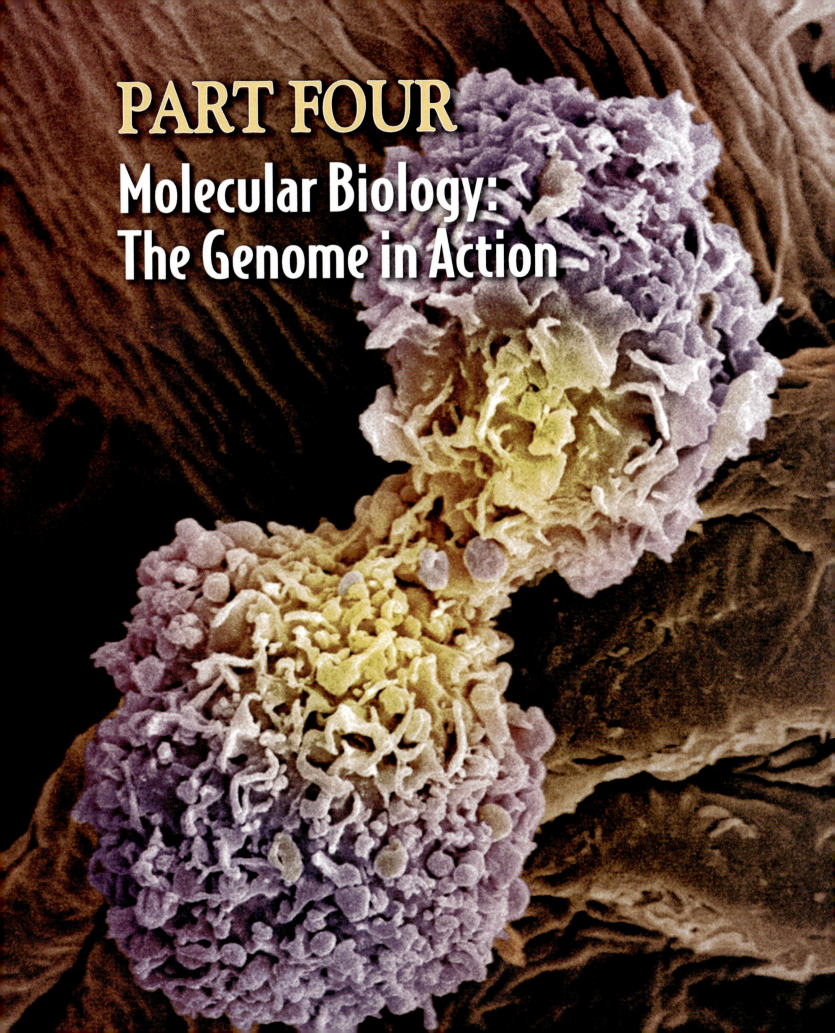

PART FOUR

Molecular Biology:
The Genome in Action

15 Cell Signaling and Communication

Have a cup of signals

It's probably happened to you: it's late, you have a paper due tomorrow, and you've put it off until the last minute. You're exhausted, but you need to stay awake and alert so that you can get your work done. What do you do? Well, for starters, you might have a cup (or several cups) of coffee. Many people turn to coffee when they need to wake themselves up or give themselves an energy boost.

Legend has it that the energy-inducing effects of coffee were first noticed a thousand years ago in what is now Ethiopia. A goat herder named Kaldi is said to have noticed that his goats became very frisky after they ate the berries of a certain plant. His curiosity aroused, Kaldi ate some of the berries himself, and thoroughly enjoyed the result. Word of his discovery spread throughout the region. Soon monks at a nearby monastery found that eating the berries kept them awake during their late-night prayers. The monks came up with the idea of drying the berries for

storage and transport, and subsequently learned that pulverizing the dried berries and heating the powder in water resulted in an enjoyable beverage. Coffee shops were not far behind.

On a typical day, at least 90 percent of North Americans and Europeans consume caffeine in some form—in tea (which may contain up to 90 mg of caffeine per 8-ounce cup), cola (50 mg), chocolate (20 mg in a dark bar), as well as coffee (up to 180 mg per cup). Caffeine is our most popular drug, and like many drugs it is a *signal molecule*. To understand the effects of caffeine, we must first understand the pathways by which the body's cells respond to signals in their environment.

A cell's response to any signal molecule takes place in three sequential steps. First, the signal binds to a receptor protein in the cell, often on the outside surface of the plasma membrane. Second, the binding of the signal causes a message to be conveyed to the inside of the cell and amplified. Third, the cell changes its activity in response to the signal.

Caffeine acts in different ways in different tissues. A tired person's brain produces adenosine molecules that bind to specific receptor proteins, resulting in decreased brain activity and increased drowsiness. Caffeine's molecular structure is similar to that of adenosine, so it occupies the adenosine receptors without inhibiting brain cell function, much like a competitive enzyme inhibitor, and alertness is restored. Adenosine also opens up, or *dilates*, the blood vessels supplying the brain, causing headaches; thus caffeine is a common ingredient in headache remedies.

A Signal to the Body The caffeine in coffee signals cells in the body of this coffee drinker. The effects of these signals help him stay alert.

Traditional Ethiopian Coffee Ceremony   Coffee plays an important ceremonial role in the traditional culture of Ethiopia, where (according to legend) people originally discovered its energizing effects.

In heart and liver cells, caffeine indirectly stimulates the same signaling pathway normally stimulated by epinephrine, the "fight-or-flight" hormone. The result is an increased heartbeat rate. Muscles tighten up and the liver is stimulated to convert glycogen into glucose and release it into the bloodstream.

How can a single molecule from outside the body have so many biological effects?

IN THIS CHAPTER we describe types of signals that affect cells, which include both chemicals produced by other cells and substances from outside the body, as well as physical and environmental factors such as light. We'll learn that whatever the signal, it affects only those cells that have the appropriate receptor protein to respond to the specific signal. We'll trace the steps of signal transduction by which the receptor communicates that a signal has been received, thus effecting a change in cell function.

CHAPTER OUTLINE

15.1 **What** Are Signals, and How Do Cells Respond to Them?

15.2 **How** Do Signal Receptors Initiate a Cellular Response?

15.3 **How** Is a Response to a Signal Transduced through the Cell?

15.4 **How** Do Cells Change in Response to Signals?

15.5 **How** Do Cells Communicate Directly?

15.1 What Are Signals, and How Do Cells Respond to Them?

Both prokaryotic and eukaryotic cells process information from their environment. This information can be in the form of a physical stimulus, such as the light reaching your eyes as you read this book, or chemicals that bathe a cell, such as lactose in the medium surrounding *E. coli*. It may come from outside the organism, such as the scent of a female moth seeking a mate in the dark, or from a neighboring cell within the organism, as in the heart, where thousands of muscle cells contract in unison by transmitting signals to one another.

Of course, the mere presence of a signal does not mean that a cell will respond to it, just as you do not pay close attention to every sound in your environment as you study. To respond to a signal, the cell must have a specific *receptor* that can detect it. This section provides examples of some types of cellular signals and one model *signal transduction pathway*—the series of steps that lead to a cell's response to a signal. After discussing signals in this section, we will consider their receptors in Section 15.2.

Cells receive signals from the physical environment and from other cells

The physical environment is full of signals. Our sense organs allow us to respond to light, odors and tastes (chemical signals), temperature, touch, and sound. Bacteria and protists respond to even minute chemical changes in their environment. Plants respond to light as a signal. For example, at sunset, at night, or in the shade, not only the amount but also the wavelengths of the light reaching Earth's surface differ from that of full sunlight in the daytime. These variations act as signals that affect plant growth and reproduction. Some plants also respond to temperature: when the weather gets cold, they respond either by becoming tolerant to cold or by accelerating flowering. Even magnetism can be a signal: some bacteria and birds orient themselves to Earth's magnetic poles, like a needle on a compass.

A cell deep inside a large multicellular organism is far away from the exterior environment. Instead, its environment consists of other cells and extracellular fluids. Cells receive their

(A)

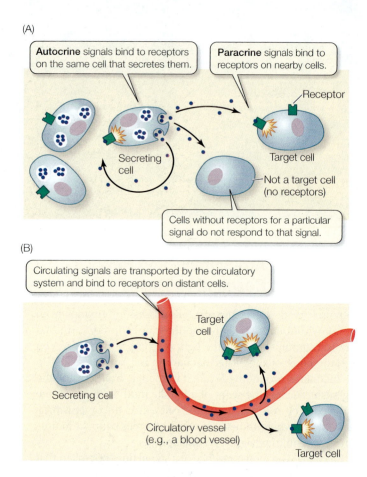

Autocrine signals bind to receptors on the same cell that secretes them.

Paracrine signals bind to receptors on nearby cells.

Receptor

Secreting cell

Target cell

Not a target cell (no receptors)

Cells without receptors for a particular signal do not respond to that signal.

(B)

Circulating signals are transported by the circulatory system and bind to receptors on distant cells.

Target cell

Secreting cell

Circulatory vessel (e.g., a blood vessel)

Target cell

15.1 Chemical Signaling Systems (A) A signal molecule can diffuse to act on the cell that produces it, or on a nearby cell. (B) Many signals act on distant cells, to which they must be transported by the organism's circulatory system.

nutrients from, and pass their wastes into, extracellular fluids. Cells also receive signals—mostly chemical signals—from their extracellular fluid environment. Most of these chemical signals come from other cells. Cells also respond to chemical signals coming from the environment via the digestive and respiratory systems. And cells can respond to concentrations of certain chemicals, such as CO_2 and H^+, whose presence in the extracellular fluids results from the metabolic activities of other cells.

Inside a large multicellular organism, chemical signals made by the body itself reach a target cell by local diffusion or by circulation within the blood. **Autocrine** signals diffuse to and affect the cells that make them, while **paracrine** signals diffuse to and affect nearby cells (**Figure 15.1A**). Signals to distant cells are called *hormones* and usually travel through the circulatory system (**Figure 15.1B**).

In all cases, in order for a signal to be transmitted, the target cell must be able to receive or sense the signal and respond to it, and the response must have some effect on the function of the cell. Depending on the target cell and the signal, these effects range from the cell's entering the cell cycle to heal a wound, to moving to a new location in the embryo to form a tissue, to releasing enzymes that digest food, to sending messages to the brain about the book you are reading.

A signal transduction pathway involves a signal, a receptor, transduction, and effects

The entire signaling process—from the signal's affecting a receptor, to conveying the message to the cytoplasm, to the cell's final response—is called a **signal transduction pathway**. Let's look at an example of such a pathway in *E. coli*. In Section 13.4, we saw that this bacterium responds to changes in the nutrient content of its environment by altering its transcription of certain genes, such as those in the *lac* operon. The bacterium must also be able to sense and respond to other kinds of changes in its environment, such as changes in solute concentration.

In the human intestine, where *E. coli* lives, the solute concentration around the bacterium often rises far above that inside the cell. The principle of diffusion tells us that when this happens, water will diffuse out of the cell, and solutes will move into the cell. But the bacterium must maintain homeostasis, so it must perceive and respond to this environmental signal (**Figure 15.2, step 1**). The pathway by which *E. coli* does so has much in common with signal transduction pathways in more complex multicellular eukaryotes. The pathway involves two major components: a receptor and a responder.

RECEPTOR A receptor is the first component of a signal transduction pathway. The *E. coli* receptor protein for changes in solute concentration is called EnvZ. EnvZ is a transmembrane protein that extends through the bacterium's plasma membrane into the space between the plasma membrane and the highly porous outer membrane, which forms a complex with the cell wall. When the solute concentration of the extracellular environment rises, so does the solute concentration in the space between the two membranes. This change in its aqueous medium causes the part of the receptor protein sticking into the intermembrane space to undergo a change in *conformation* (its three-dimensional shape).

As we saw in Section 6.5, changing the tertiary structure of one part of a protein often leads to changes in distant parts of the protein. In the case of the bacterial EnvZ receptor, the conformational change in the intermembrane domain of the protein is transmitted to the domain that lies in the cytoplasm, initiating the events of signal transduction. This conformational change exposes an active site, so that EnvZ becomes a *protein kinase*, an enzyme that catalyzes the addition of a phosphate group from ATP to one of EnvZ's own histidine molecules. In other words, EnvZ phosphorylates itself (**Figure 15.2, step 2**).

RESPONDER A responder is the second component of a signal transduction pathway. The charged phosphate group added to the histidine causes the cytoplasmic domain of the EnvZ protein to change its shape again. It now binds to a second protein, OmpR, which takes the phosphate group from EnvZ. This phosphorylation changes the shape of OmpR in turn (**Figure 15.2, step 3**). This change in a responder is a key event in signaling for three reasons:

- The signal on the outside of the cell has now been *transduced* to a protein that lies totally within the cell's cytoplasm.

- The altered responder can *do something*. In the case of the phosphorylated OmpR, that "something" is to bind to a promoter on *E. coli* DNA adjacent to the sequence that codes for

15.2 A Model Signal Transduction Pathway *E. coli* responds to the signal of an increase in solute concentration in its environment. The basic steps of such signal transduction pathways occur in all living organisms.

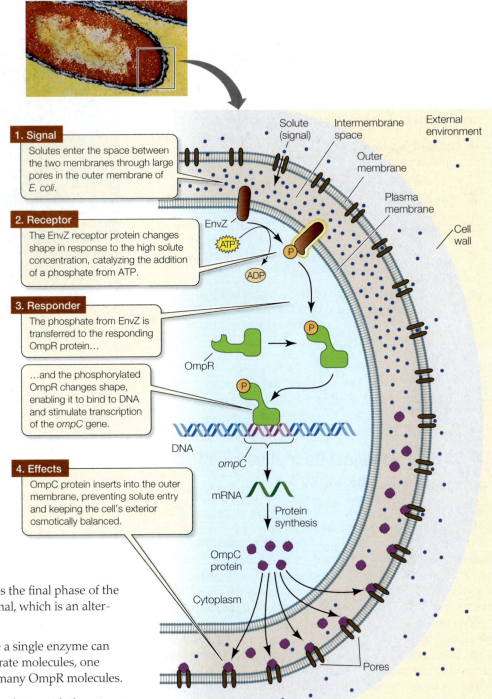

1. Signal
Solutes enter the space between the two membranes through large pores in the outer membrane of *E. coli*.

2. Receptor
The EnvZ receptor protein changes shape in response to the high solute concentration, catalyzing the addition of a phosphate from ATP.

3. Responder
The phosphate from EnvZ is transferred to the responding OmpR protein...

...and the phosphorylated OmpR changes shape, enabling it to bind to DNA and stimulate transcription of the *ompC* gene.

4. Effects
OmpC protein inserts into the outer membrane, preventing solute entry and keeping the cell's exterior osmotically balanced.

the protein OmpC. This binding begins the final phase of the signaling pathway: the *effect* of the signal, which is an alteration in cell function.

■ The signal has been *amplified*. Because a single enzyme can catalyze the conversion of many substrate molecules, one EnvZ molecule alters the structure of many OmpR molecules.

Phosphorylated OmpR is a transcription factor with the correct three-dimensional structure to bind to the promoter of the *ompC* gene, resulting in an increase in the transcription of that gene. Translation of *ompC* mRNA results in the production of OmpC protein, which leads to the response that regulates *E. coli*'s solute concentration (**Figure 15.2, step 4**). The OmpC protein is inserted into the outer membrane of the bacterial cell, where it blocks pores and prevents solutes from entering the intermembrane space. As a result, the solute concentration in the intermembrane space is lowered, and homeostasis is restored. Thus the *E. coli* cell can go on behaving just as if the external environment had a normal solute concentration.

Many of the same elements we have highlighted in this prokaryotic signal transduction system will reappear in many other signal transduction pathways in eukaryotic organisms:

■ A receptor protein *changes its conformation* upon interacting with a signal.

■ A conformational change in the receptor protein gives it *protein kinase* activity, resulting in the transfer of a phosphate group from ATP to a target protein.

■ This *phosphorylation* alters the function of a responder protein.

■ The signal is *amplified*.

■ A *transcription factor* is activated.

■ The *synthesis of a specific protein* is turned on.

■ The action of the protein *alters cell activity*.

The general features of signal transduction pathways described in this section will recur in more detail throughout the chapter. First let's consider more closely the nature of the receptors that bind signal molecules.

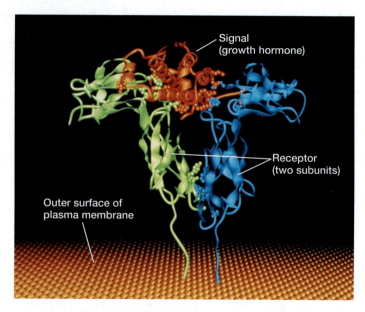

15.3 A Signal Bound to Its Receptor Human growth hormone is shown bound to its receptor, a transmembrane protein. Only the extracellular regions of the receptor are shown.

15.2 How Do Signal Receptors Initiate a Cellular Response?

Although any given cell in a multicellular organism is bombarded with many signals, it responds to only a few of them, because no cell makes receptors for all signals. Which cells make which receptors is determined by the regulatory processes described in Chapter 14; in short, if a cell transcribes the gene encoding a particular receptor and the resulting mRNA is translated, the cell will have that receptor.

A receptor protein that binds to a chemical signal does so very specifically, in much the same way as an enzyme binds to a substrate. Just as there are many types of enzymes with diverse specificities, there are many kinds of signal receptor proteins. This specificity of binding ensures that only those cells that make a specific receptor will respond to a given signal.

Receptors have specific binding sites for their signals

A specific chemical signal molecule fits into a three-dimensional site on its receptor (**Figure 15.3**). A molecule that binds to a receptor site in another molecule in this way is called a **ligand**. As you saw with the example in *E. coli*, binding of the signaling ligand causes the receptor protein to change its three-dimensional shape, and that conformational change initiates a cellular response. The ligand does not contribute further to this response. In fact, the ligand signal usually is not metabolized into useful products. Its role is purely to "knock on the door." (This is in sharp contrast to the enzyme–substrate interactions described in Chapter 6, whose whole purpose is to change the substrate into a useful product.)

Receptors bind to their ligands according to chemistry's law of mass action:

$$R + L \rightleftharpoons RL$$

This means that the binding is reversible, although for most ligand–receptor complexes, the equilibrium point is far to the right—that is, binding is favored. Reversibility is important, however, because if the ligand were never released, the receptor would be continuously stimulated.

As with enzymes, inhibitors can bind to the ligand binding site on a receptor protein. Both natural and artificial inhibitors of receptor binding are important in medicine. For example, over two-thirds of the drugs that alter human behavior bind to specific receptors in the brain.

Receptors can be classified by location

Receptors can be classified by their location in the cell, which largely depends on the nature of their ligands. The chemistry of signal molecules is quite variable, but they can be divided into two classes (**Figure 15.4**):

- *Ligands with cytoplasmic receptors:* Small or nonpolar ligands can diffuse across the phospholipid bilayer of the plasma membrane and enter the cell. Estrogen, for example, is a lipid-soluble steroid hormone that can easily diffuse across the plasma membrane and enter the cell; it binds to a receptor in the cytoplasm.

- *Ligands with plasma membrane receptors:* Large or polar ligands cannot cross the plasma membrane. Insulin, for example, is a protein hormone that cannot diffuse through the plasma membrane; instead, it binds to a receptor that is a transmembrane protein with an extracellular binding domain.

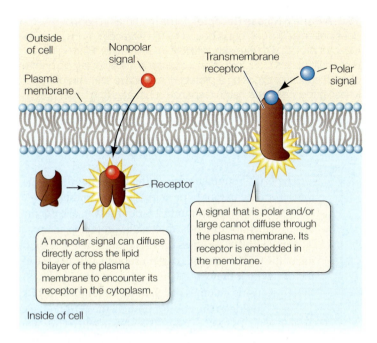

Outside of cell

Nonpolar signal

Transmembrane receptor

Polar signal

Plasma membrane

Receptor

A signal that is polar and/or large cannot diffuse through the plasma membrane. Its receptor is embedded in the membrane.

A nonpolar signal can diffuse directly across the lipid bilayer of the plasma membrane to encounter its receptor in the cytoplasm.

Inside of cell

15.4 Two Locations for Receptors Receptors can be located in the plasma membrane or in the cytoplasm of the cell.

In complex eukaryotes such as mammals, there are three well-studied types of plasma membrane receptors that bear examination: *ion channels*, *protein kinases*, and *G protein-linked receptors*.

ION CHANNEL RECEPTORS As described in Section 5.3, the plasma membranes of many types of cells contain channel proteins that can be open or closed. These **ion channels** act as "gates," allowing ions such as Na^+, K^+, Ca^{2+}, or Cl^- to enter or leave the cell (see Figure 5.10). The gate-opening mechanism is an alteration in the three-dimensional shape of the channel protein upon ligand binding. Some ion channels are plasma membrane receptors for signal

molecules; others act later in signal transduction pathways. Each type of ion channel receptor has its own signal. These signals include sensory stimuli such as light, sound, and electric charge differences across the plasma membrane, as well as chemical ligands such as hormones and neurotransmitters.

The acetylcholine receptor, which is located in the plasma membrane of vertebrate skeletal muscle cells, is an example of a gated ion channel. This receptor protein is a sodium channel that binds the ligand *acetylcholine*, which is a *neurotransmitter*, a chemical signal released from neurons (nerve cells) (**Figure 15.5**). When *two* molecules of acetylcholine bind to the receptor, it opens for about a thousandth of a second. That is enough time for Na^+, which is more concentrated outside the cell than inside, to rush into the cell. The change in Na^+ concentration in the cell initiates a series of events that result in muscle contraction.

> Acetylcholine receptors are present in brain cells as well as skeletal muscle. Nicotine binds strongly to these receptors and activates many pathways in the brain. The effects of nicotine may be transiently pleasurable, but ultimately result in addiction.

PROTEIN KINASE RECEPTORS Like the EnvZ receptor protein of *E. coli*, some eukaryotic receptor proteins become **protein kinases** when they are activated: that is, they catalyze the transfer of a phosphate group from ATP to a specific protein, referred to as the *target protein*. This phosphorylation can alter the conformation and activity of the target protein.

The receptor for insulin is an example of a protein kinase receptor. Insulin is a protein hormone made by the mammalian pancreas. Its receptor has two copies each of two different polypeptide subunits (**Figure 15.6**). As with acetylcholine, two molecules of insulin must bind to the receptor. When insulin binds to its extracellular subunits, the receptor protein changes its shape to expose a protein kinase active site in the cytoplasm. Like the *E. coli* receptor described above, the insulin receptor *autophosphorylates* (phosphorylates itself). Then, as a protein kinase, it catalyzes the phosphorylation of certain cytoplasmic proteins, appropriately called insulin response substrates. These proteins then initiate many cellular responses, including the insertion of glucose transporters (see Figure 5.12) into the plasma membrane.

G PROTEIN-LINKED RECEPTORS A third category of eukaryotic plasma membrane receptors is the *seven-transmembrane-spanning G protein-linked receptors*. This long name identifies a fascinating group of receptors, all of which are composed of a single protein with seven transmembrane regions. These seven regions pass through the phospholipid bilayer, separated by short loops that extend either outside or inside the

Outside of cell

Acetylcholine (ACh)

Na^+

1 Acetylcholine binds to two of the five AChR subunits, causing the channel to change shape and open.

Plasma membrane

2 The channel is lined with negatively charged amino acids, allowing Na^+ to flow into the cell.

Acetylcholine receptor (AChR)

Inside of cell

3 Na^+ buildup in cells leads to muscle contraction.

15.5 A Gated Ion Channel The acetylcholine receptor (AChR) is a gated ion channel for sodium ions. It is made up of five polypeptide subunits. When acetylcholine molecules (ACh) bind to two of the subunits, the gate opens and Na^+ flows into the cell.

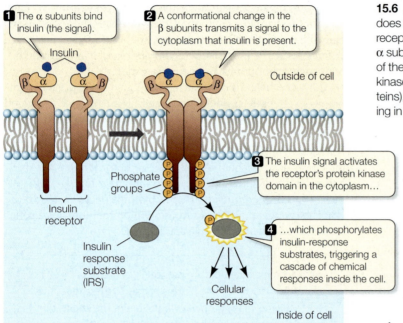

1 The α subunits bind insulin (the signal).

2 A conformational change in the β subunits transmits a signal to the cytoplasm that insulin is present.

Insulin

Outside of cell

Insulin receptor

Phosphate groups

3 The insulin signal activates the receptor's protein kinase domain in the cytoplasm…

Insulin response substrate (IRS)

4 …which phosphorylates insulin-response substrates, triggering a cascade of chemical responses inside the cell.

Cellular responses

Inside of cell

15.6 A Protein Kinase Receptor The mammalian hormone insulin does not enter the cell, but is bound by the extracellular domain of a receptor protein with four subunits (two α and two β). Binding to the α subunit causes a conformational change in the cytoplasmic domain of the β subunits, exposing a protein kinase active site. This protein kinase phosphorylates insulin response substrates (the target proteins), triggering further responses within the cell and eventually resulting in the transport of glucose across the membrane into the cell.

The GTP-bound subunit of the G protein then separates from the parent G protein, diffusing in the plane of the phospholipid bilayer until it encounters an **effector protein** to which it can bind. An effector protein is just what its name implies: it causes an effect in the cell. The binding of the GTP-bearing G protein subunit activates the effector—which may be an enzyme or an ion channel—thereby causing changes in cell function (**Figure 15.7C**).

After binding to the effector protein, the GTP on the G protein is hydrolyzed to GDP. The now inactive G protein subunit separates from the effector protein. The G protein subunit must form a complex with other subunits before binding to yet another activated receptor. When an activated receptor is bound, the G protein exchanges its GDP for GTP, and the cycle begins again.

By means of their diffusing subunits, G proteins can either activate or inhibit an effector protein. An example of an *activating* response involves the receptor for epinephrine (adrenaline), a hor-

cell. Ligand binding on the extracellular side of the receptor protein changes the shape of its cytoplasmic region, exposing a binding site for a mobile membrane **G protein**.

The term *G protein* describes a class of signaling proteins with three polypeptide subunits that are characterized by their ability to bind GDP and GTP (guanosine diphosphate and triphosphate, respectively; these are nucleoside phosphates like ADP and ATP). A G protein has two important binding sites: one for the G protein-linked receptor and the other for GDP/GTP. The inactive G protein binds GDP to one of its subunits (**Figure 15.7A**). When the G protein binds to an activated receptor protein, GDP is exchanged for GTP (**Figure 15.7B**). At the same time, the ligand is released from the extracellular side of the receptor.

15.7 A G Protein-Linked Receptor (A) The inactive G protein has binding sites for its linked receptor and for GDP/GTP. (B) Binding of an extracellular signal—in this case, a hormone—activates the G protein-linked receptor. The activated receptor binds the G protein, which is then activated by the exchange of GDP for GTP. (C) The GTP-bound subunit separates from the parent G protein and activates an effector protein—in this case, an enzyme that catalyzes a reaction in the cytoplasm, amplifying the signal. This figure is a generalized diagram that could apply to any member of the large family of G proteins and the signals they react to.

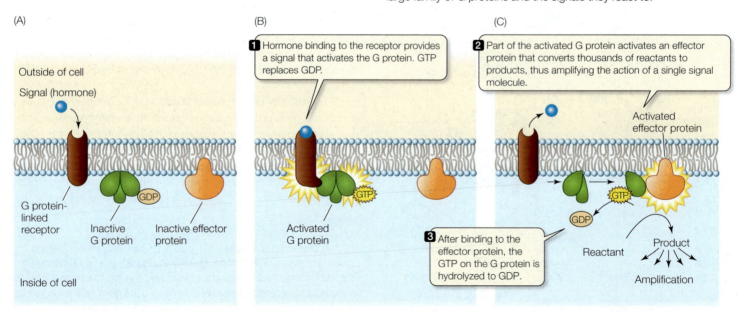

(A)

Outside of cell

Signal (hormone)

G protein-linked receptor

Inactive G protein

Inactive effector protein

Inside of cell

(B)

1 Hormone binding to the receptor provides a signal that activates the G protein. GTP replaces GDP.

Activated G protein

(C)

2 Part of the activated G protein activates an effector protein that converts thousands of reactants to products, thus amplifying the action of a single signal molecule.

Activated effector protein

3 After binding to the effector protein, the GTP on the G protein is hydrolyzed to GDP.

Reactant

Product

Amplification

mone made by the adrenal gland in response to stress or heavy exercise. In heart muscle, this hormone binds to its G protein-linked receptor, activating a G protein. The GTP-bound subunit then activates a membrane-bound enzyme to produce a small molecule, cyclic AMP that has many effects on the cell (as we will see below), including glucose mobilization for energy and muscle contraction.

G protein-mediated *inhibition* occurs when the same hormone, epinephrine, binds to its receptor in the smooth muscle cells surrounding blood vessels lining the digestive tract. Again, the epinephrine-bound receptor changes its shape and activates a G protein, and the GTP-bound subunit binds to a target enzyme. But in this case, the enzyme is inhibited instead of being activated. As a result, the muscles relax and the blood vessel diameter increases, allowing more nutrients to be carried away from the digestive system to the rest of the body. Thus the same signal and initial signaling mechanism can have different consequences in different cells, depending on the nature of the responding cell.

CYTOPLASMIC RECEPTORS **Cytoplasmic receptors** are located inside the cell and bind to signals that can diffuse across the plasma membrane. Binding to the signaling ligand causes the receptor to change its shape so that it can enter the cell nucleus, where it acts as a transcription factor affecting gene expression. But this general view is somewhat simplified. The receptor for the hormone cortisol, for example, is normally bound to a chaperone protein, which blocks it from entering the nucleus. Binding of the hormone causes the receptor to change its shape so that the chaperone is released (**Figure 15.8**). This allows the receptor to fold into an appropriate conformation for entering the nucleus and initiating transcription of mRNA.

15.8 A Cytoplasmic Receptor The receptor for cortisol is bound to a chaperone protein. Binding of the signal releases the chaperone and allows the receptor protein to enter the cell's nucleus, where it functions as a transcription factor.

15.2 RECAP

Receptors for signals are proteins that bind, or are changed by, specific signals and then initiate a response in the cell. These receptors may be at the plasma membrane or inside the cell.

- Do you understand the specificity of receptors for ligands? See p. 336
- What are the three different kinds of plasma membrane receptors seen in complex eukaryotes? See pp. 337–338

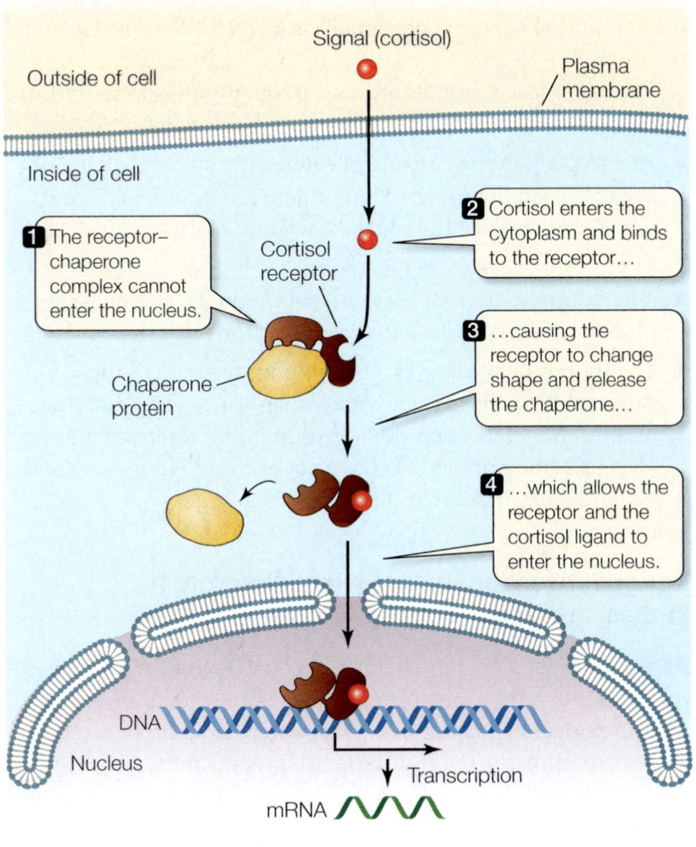

Now that we have discussed signals and receptors, let's examine the characteristics of the transducers that mediate between the receptor and the cellular response.

15.3 How Is a Response to a Signal Transduced through the Cell?

As we have just seen, the same signal may produce different responses in different tissues. When epinephrine, for example, binds to receptors on heart muscle cells, it stimulates muscle contraction, but when it binds to receptors on smooth muscle cells in the blood vessels of the digestive system, it slows muscle contraction. These different responses to the same signal–receptor complex are mediated by the events of signal transduction. These events, which are critical to the cell's response, may be either direct or indirect:

- **Direct transduction** is a function of the receptor itself and occurs at the plasma membrane (**Figure 15.9A**).
- In **indirect transduction**, which is more common, another molecule, termed a **second messenger**, mediates a further interaction between the receptor and the cell's response (**Figure 15.9B**).

In both cases, the signal initiates a **cascade** of events, in which proteins interact with other proteins until the final responses are achieved. Through such a cascade, a weak initial signal can be both *amplified* and *distributed* to cause several different responses in the target cell.

Protein kinase cascades amplify a response to ligand binding

We have seen that when a signal binds to a protein kinase receptor, the receptor's conformation changes, exposing a protein kinase active site, which catalyzes the phosphorylation of target proteins.

(A)

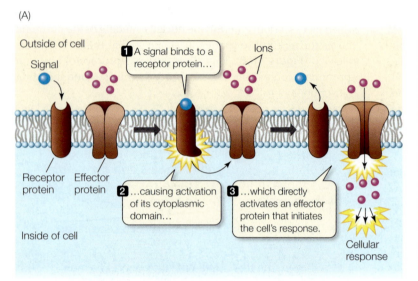

(B)

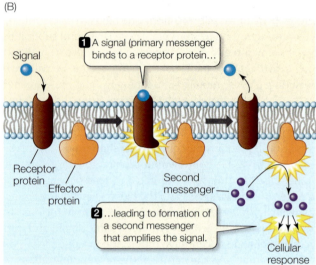

15.9 Direct and Indirect Signal Transduction (A) All the events of direct transduction occur at the plasma membrane at or near the receptor. (B) In indirect transduction, a second messenger mediates the events.

This process is an example of direct signal transduction. Protein kinase receptors are important in binding signals that stimulate cell division in both plants and animals. For example, the growth factors described in Section 9.2 that serve as external inducers of the cell cycle work by binding to protein kinase receptors.

The complete signal transduction pathway that occurs after a protein kinase receptor binds a growth factor was worked out through studies on a cell that went wrong. Many human bladder cancers contain an abnormal form of a protein called Ras (so named because it was first isolated from a *rat sarcoma* tumor). Investigations of these bladder cancers showed that this Ras protein was a G protein, but was always active because it was permanently bound to GTP, and thus caused continuous cell division. If the cancer cells' Ras protein was inhibited, they stopped dividing. This discovery has led to a major effort to develop specific Ras inhibitors for cancer treatment.

What does Ras do in normal, noncancerous cells? Researchers knew that cells must be stimulated by growth factors (signals) in order to enter the cell cycle and divide (see Section 9.2). One hypothesis was that Ras was an intermediary between the binding of a growth factor to its receptor and the ultimate response of cell division. To investigate this hypothesis, the researchers treated cells in a culture dish with both a Ras inhibitor and a growth factor. Cell division did not occur, confirming their hypothesis.

After this discovery, the next step was to work out what the activated growth factor receptor did to Ras, and what Ras did to stimulate further events in signal transduction. This signaling pathway has been worked out, and it is an example of a more general phenomenon, called a **protein kinase cascade** (**Figure 15.10**). Such cascades are key to the external regulation of many cellular activities. Indeed, the eukaryotic genome codes for hundreds, even thousands, of such kinases.

The unbound receptors for growth factors exist in the plasma membrane as separate polypeptide chains (subunits). When the growth factor signal binds to a subunit, it associates with another subunit to form a dimer, which changes its shape to expose a protein kinase active site. Its kinase activity sets off a series of events, activating several other protein kinases in turn. The final phosphorylated, activated protein kinase—MAP kinase—moves into the nucleus and phosphorylates target proteins that are necessary for cell division.

Protein kinase cascades are useful signal transducers for three reasons:

- At each step in the cascade of events, the signal is *amplified*, because each newly activated protein kinase is an enzyme, which can catalyze the phosphorylation of many target proteins.

- The information from a signal that originally arrived at the plasma membrane is *communicated* to the nucleus.

- The multitude of steps provides some *specificity* to the process. As we have seen with epinephrine, signal binding and receptor activation do not result in the same response in all cells. Different target proteins at each step in the cascade can provide variation in the response.

Second messengers can stimulate protein kinase cascades

As we have just seen, protein kinase receptors initiate a protein kinase cascade right at the plasma membrane. However, the stimulation of events in the cell is more often indirect. In a series of clever experiments, Earl Sutherland, Edwin Krebs, and Edmond Fischer showed that in many cases, a small, water-soluble chemical messenger mediates between the plasma membrane receptor and cytoplasmic events. These researchers were investigating the activation of the liver enzyme *glycogen phosphorylase* by the hormone epinephrine. Glycogen phosphorylase catalyzes the hydrolysis of glycogen stored in the liver so that the resulting glucose

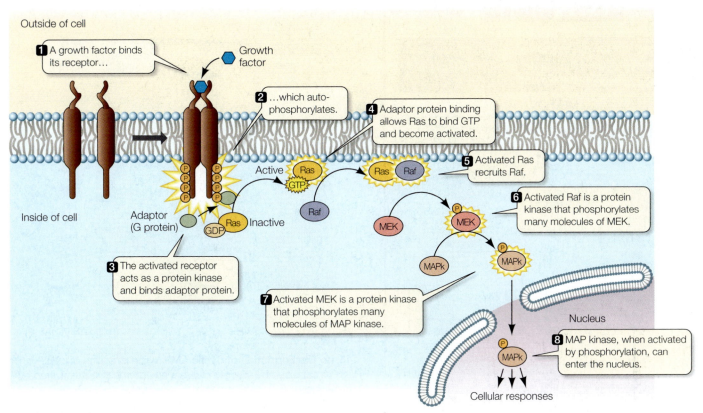

15.10 A Protein Kinase Cascade In a protein kinase cascade, a series of proteins are sequentially activated. In this example, the growth factor receptor protein stimulates the G protein Ras, which mediates a cascading series of reactions. The final product of the cascade, MAP kinase (MAPk), enters the nucleus and causes changes in transcription.

molecules can be released to the blood to fuel the fight-or-flight response (see Figure 41.4). It is normally present in liver cell cytoplasm, but is inactive except in the presence of epinephrine.

The researchers found that glycogen phosphorylase could be activated in liver cells that had been broken open, but only if the entire cell contents, including the plasma membrane fragments, were present. Under these circumstances epinephrine bound to the plasma membranes, but the active phosphorylase was present in the cytoplasm. The researchers hypothesized that there must be some chemical messenger that transmits the message of epinephrine binding (at the membrane) to the phosphorylase (in the cytoplasm). To investigate the production of this message, they separated the plasma membrane fragments from the cytoplasm of broken liver cells followed the sequence of steps described in **Figure 15.11**. This experiment confirmed their hypothesis that hormone binding to the membrane receptor had caused the production of a small, water-soluble molecule that then diffused into the cytoplasm, where it activated the enzyme. This small molecule was later determined to be *cyclic AMP* (*cAMP*), the same molecule involved in the *lac* operon regulatory system in *E. coli* (see Section 13.4). Here, cAMP was working as a **second messenger**.

Second messengers are substances that are released into the cytoplasm after the first messenger—the signal—binds to its re-

ceptor. In contrast to the specificity of receptor binding, second messengers affect many processes in the cell, and they allow a cell to respond to a single event at the plasma membrane with many events inside the cell. Like protein kinase cascades, second messengers amplify the signal: a single epinephrine molecule, for example, leads to the production of many molecules of cAMP, which then activate many enzyme targets. Second messengers themselves do not have enzymatic activity; rather, they act as *cofactors* or *allosteric regulators* of target enzymes (see Chapter 6).

Cyclic AMP is a common and widely used second messenger. *Adenylyl cyclase*, the effector enzyme that catalyzes the formation of cAMP from ATP, is located on the cytoplasmic surface of the plasma membrane of target cells (**Figure 15.12**). Usually it is activated by the binding of G proteins, which are themselves activated by receptors (see above).

Cyclic AMP has two major target types. In many kinds of sensory cells, cAMP binds to *ion channels* and thus opens them. Cyclic AMP may also bind to a *protein kinase* in the cytoplasm, whose active site is exposed as a result. A protein kinase cascade (see Figure 15.10) ensues, leading to the final effects in the cell.

Second messengers can be derived from lipids

Phospholipids, in addition to their roles as structural components of the plasma membrane, are involved in signal transduction. When certain phospholipids are hydrolyzed into their component parts by enzymes called *phospholipases*, second messengers are formed.

EXPERIMENT

HYPOTHESIS: A second messenger mediates between receptor activation at the plasma membrane and enzyme activation in the cytoplasm.

METHOD

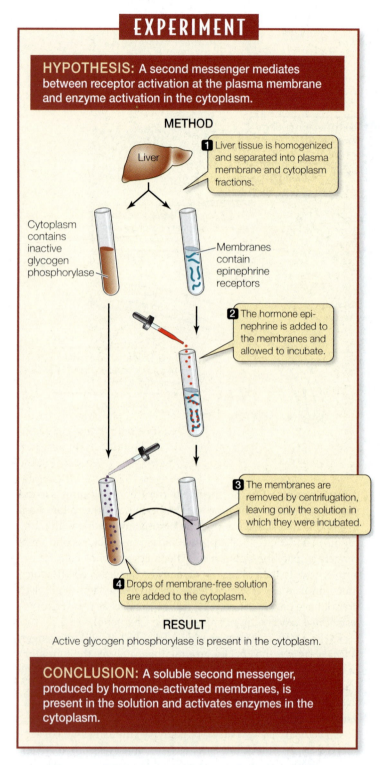

1 Liver tissue is homogenized and separated into plasma membrane and cytoplasm fractions.

Liver

Cytoplasm contains inactive glycogen phosphorylase

Membranes contain epinephrine receptors

2 The hormone epinephrine is added to the membranes and allowed to incubate.

3 The membranes are removed by centrifugation, leaving only the solution in which they were incubated.

4 Drops of membrane-free solution are added to the cytoplasm.

RESULT

Active glycogen phosphorylase is present in the cytoplasm.

CONCLUSION: A soluble second messenger, produced by hormone-activated membranes, is present in the solution and activates enzymes in the cytoplasm.

15.11 The Discovery of a Second Messenger Sutherland and colleagues separated the components of a cell signaling pathway. In this manner, they were able to show that a soluble second messenger was present. FURTHER RESEARCH: The soluble molecule produced in this experiment was later identified as cAMP. How would you show that cAMP, and not ATP, is the second messenger in this system?

The best-studied of the lipid-derived second messengers come from hydrolysis of the **phosphatidyl inositol-bisphosphate (PIP2)**, which, like all phospholipids, has a hydrophobic portion (two fatty

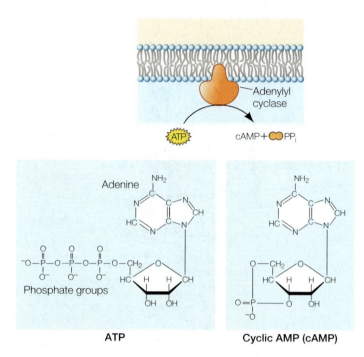

Adenylyl cyclase

ATP → cAMP + PP$_i$

ATP

Cyclic AMP (cAMP)

15.12 The Formation of Cyclic AMP The formation of cAMP from ATP is catalyzed by adenylyl cyclase, an enzyme that is activated by G proteins.

acid tails attached to a molecule of glycerol, which together form **diacylglycerol**, or **DAG**) embedded in the plasma membrane; and a hydrophilic portion (**inositol triphosphate**, or **IP$_3$**) projecting into the cytoplasm.

As with cAMP, the receptors involved in this second-messenger system are often G protein-linked receptors. The G protein subunit activated by the receptor diffuses within the plasma membrane and activates an enzyme, phospholipase C. This enzyme cleaves off the IP$_3$ from PIP2, leaving the glycerol and the two attached fatty acids (DAG) in the phospholipid bilayer:

$$\text{PIP2} \xrightarrow{\text{Phospholipase C}} \text{IP}_3 + \text{DAG}$$
in membrane   released to   in membrane
cytoplasm

IP$_3$ and DAG are both second messengers and have different modes of action that build on each other. DAG activates a membrane-bound enzyme, protein kinase C (PKC). PKC is dependent on Ca^{2+} (hence the "C"), and that is where IP$_3$ plays an essential role. IP$_3$ diffuses through the cytoplasm to the smooth endoplasmic reticulum, where it opens an ion channel, releasing Ca^{2+} into the cytoplasm. There, in combination with DAG, the Ca^{2+} causes PKC to become active. PKC can then phosphorylate a wide variety of proteins, leading to the ultimate response of the cell (**Figure 15.13**).

This pathway is apparently the target for lithium, used clinically for many years as a psychoactive drug to treat bipolar (manic-depressive) disorder. This serious illness occurs in about 1 person in 100, in whom an overactive IP$_3$/DAG signal transduction pathway in the brain leads to excessive brain activity in certain regions.

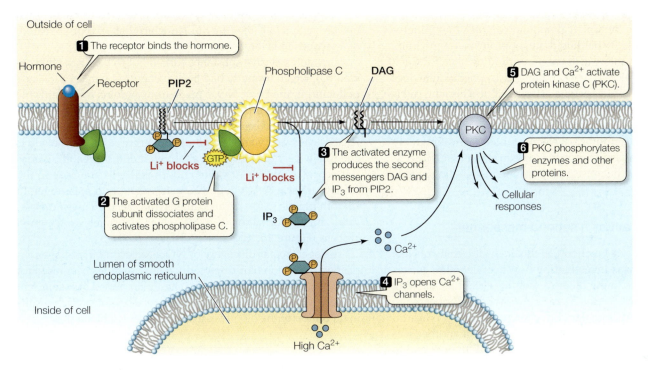

15.13 The IP$_3$/DAG Second-Messenger System Phospholipase C hydrolyzes the phospholipid PIP2 into its components, IP$_3$ and DAG, both of which are second messengers. Lithium ions (Li$^+$), used to treat bipolar disorder, block this pathway (red type).

Lithium ions (Li$^+$) "tone down" this pathway in two ways, as indicated by the red notations in Figure 15.13. Li$^+$ inhibits G protein activation of phospholipase C, and also inhibits the synthesis of IP$_3$. The overall result is that brain activity returns to normal.

Calcium ions are involved in many signal transduction pathways

Calcium ions are scarce inside most cells, which have a cytoplasmic Ca^{2+} concentration of only about 0.1 mM; Ca^{2+} concentrations outside the cell and within the endoplasmic reticulum are usually much higher. This concentration difference is maintained by active transport proteins at the plasma and ER membranes that pump Ca^{2+} out of the cytoplasm. In contrast to cAMP and the lipid-derived second messengers, intracellular Ca^{2+} concentration cannot be increased by making more Ca^{2+}. Instead, the opening and closing of ion channels and the action of membrane pumps regulate ion levels in a cellular compartment.

There are many signals that can cause calcium channels to open, including IP$_3$ (as we saw in the previous section). The entry of a sperm into an egg is a very important signal that causes a massive opening of Ca^{2+} channels; (**Figure 15.14**). Whatever the signal, the open calcium channels result in a dramatic increase in cytoplasmic Ca^{2+} concentration, up to a hundredfold within a fraction of a second. As we saw earlier, this increase activates protein kinase C. In addition, Ca^{2+} controls other ion channels and stimulates secretion by exocytosis.

A distinctive aspect of Ca^{2+} signaling is that the ion can stimulate its own release from intracellular stores. For example, in some plant leaf cells, the hormone abscisic acid binds to gated calcium channels in the plasma membrane and opens them, causing the ion to rush into the cells. This influx is not enough to trigger the cell's response, however. The ion binds to calcium channels in the endoplasmic reticulum and in the membranes of vacuoles, causing those organelles to release their Ca^{2+} stores as well.

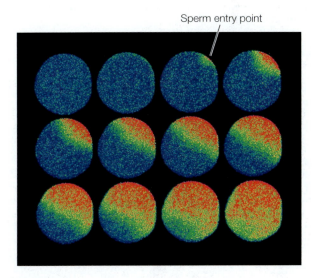

15.14 Calcium Ions as a Second Messenger The concentration of Ca^{2+} can be measured by a dye that fluoresces and turns red when it binds the ion. Here, fertilization causes a wave of Ca^{2+}, photographed at 5-second intervals, to pass through the egg of a starfish. Calcium signaling is used by virtually all animal groups to deliver the message that fertilization is complete and the mitotic divisions, or *cleavages*, that will result in a new organism can begin.

In some cases, Ca^{2+} acts via a calcium-binding protein called *calmodulin*, and it is the Ca^{2+}/calmodulin complex that initiates the cellular response by binding to target proteins. Calmodulin, which is present in many cells, has four binding sites for Ca^{2+}. When the cytoplasmic Ca^{2+} concentration is low, calmodulin does not bind enough Ca^{2+} to become activated. But when the cell is stimulated by a signal that causes a rise in the Ca^{2+} concentration, all four binding sites are filled. The calmodulin then changes its shape and binds to a number of cellular targets, activating them in turn. One such target is a protein kinase in smooth muscle cells that phosphorylates the muscle protein myosin, initiating muscle contraction.

Nitric oxide can act as a second messenger

Pharmacologist Robert Furchgott, at the State University of New York in Brooklyn, was investigating how acetylcholine causes the smooth muscles lining blood vessels to relax, thus allowing more blood to flow to certain organs. Acetylcholine appeared to stimulate the IP$_3$/DAG signal transduction pathway to produce an influx of Ca^{2+}, leading to an increase in the level of another second messenger, *cyclic GMP* (cGMP). Cyclic GMP then bound to a protein kinase, which in turn stimulated a protein kinase cascade leading to muscle relaxation. So far, the pathway seemed straightforward.

While the signal transduction pathway Furchgott was studying seemed to work in intact animals, it did not work on isolated strips of artery tissue. When Furchgott switched to tubular sections of artery, however, signal transduction did occur. There turned out to be a crucial difference between these two tissue preparations: in the strips, the delicate inner layer of cells that lines blood vessels had been lost. Furchgott hypothesized that this layer, the *endothelium*, was producing some chemical that diffused into the smooth muscle cells and was needed for their response to acetylcholine.

The substance was not easy to isolate. It seemed to break down quickly, with a half-life (the time in which half of it disappeared) of 5 seconds in living tissue.

Furchgott's elusive substance turned out to be a gas, **nitric oxide (NO)**, which formerly had been recognized only as a toxic air pollutant! In the body, NO is made from arginine by the enzyme *NO synthase*. This enzyme is activated by Ca^{2+}, which enters the endothelial cells through a channel opened by PIP2, which is released when acetylcholine binds to its receptor. Nitric oxide is chemically very unstable, and although it diffuses readily, it does not get far. Conveniently, the endothelial cells are close to the smooth muscle cells, where NO acts as a second messenger. In smooth muscle, NO activates an enzyme called guanylyl cyclase, catalyzing the formation of cGMP, which in turn relaxes the muscle cells (**Figure 15.15**).

The spectacular discovery of NO as a second messenger explained the action of nitroglycerin, a drug that has been used for over a century to treat angina, the chest pain caused by insufficient blood flow to the heart. Nitroglycerin releases NO, which results in relaxation of the blood vessels and increased blood flow.

Developed to treat angina by the NO pathway, the drug Viagra was only modestly useful for that purpose. However, men taking it reported more pronounced penile erections. During sexual stimulation, NO acts as a second messenger, causing an increase in cGMP and ensuing relaxation of the smooth muscles in the corpus cavernosum of the penis, which fills with blood, resulting in an erection.

15.15 Nitric Oxide as a Second Messenger Nitric oxide (NO) is an unstable gas, which nevertheless serves as a second messenger mediating between a signal (acetylcholine) and its effect (the relaxation of smooth muscles).

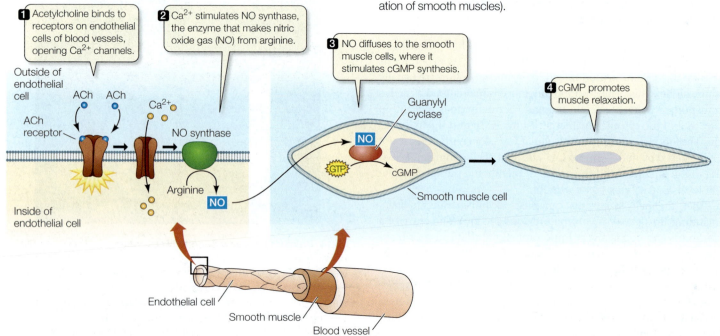

Signal transduction is highly regulated

There are several ways in which cells can regulate the activity of a transducer. The concentration of NO, which breaks down quickly, can be regulated only by how much of it is made. The concentration of Ca^{2+}, on the other hand, is determined by both membrane pumps and ion channels, as we have seen. To regulate protein kinase cascades, G proteins, and cAMP, there are enzymes that convert the activated transducer back to its inactive precursor (**Figure 15.16**).

The balance between these two enzyme activities (making and breaking down active molecules) is what determines the ultimate cellular response to a signal. Cells can alter this balance in several ways:

■ *Synthesis or breakdown of the enzymes involved.* For example, synthesis of adenylyl cyclase and breakdown of phosphodiesterase (which breaks down cAMP) would tilt the balance in favor of more cAMP in the cell.

■ *Activation or inhibition of the enzymes by other molecules.* Simple examples include signal binding to a receptor activating G proteins, and caffeine inhibiting phosphodiesterase in muscle cells. Viagra inhibits the phosphodiesterase that normally breaks down cGMP.

Because cell signaling is so important in diseases such as cancer, a search is under way for new drugs that can modulate the activities of the enzymes regulating the concentrations of molecules in signal transduction pathways.

15.16 Regulation of Signal Transduction Some signals lead to the production of active transducers such as (A) enzymes, (B) G proteins, and (C) cAMP. Other enzymes (red type) inactivate these transducers.

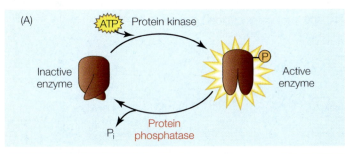

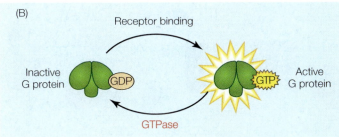

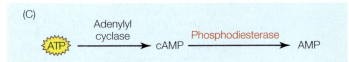

We have seen how the binding of a signal to its receptor initiates the response of a cell to the signal, and how the direct or indirect transduction of the signal to the inside of the cell amplifies the signal. In the next section we will consider the third and final step in the signal transduction process, the actual effects of the signal on cell function.

15.4 How Do Cells Change in Response to Signals?

The effects of a signal on cell function take three primary forms: the opening of ion channels, changes in the activities of enzymes, or differential gene transcription.

Ion channels open in response to signals

The opening of ion channels is a key step in the response of the nervous system to signals. Sensory neurons in the sense organs, for example, become stimulated through the opening of ion channels. We will focus here on one such signal transduction pathway, that for the sense of smell, which responds to gaseous molecules in the environment.

The sense of smell is well developed in mammals, some of which have more than 1,000 genes for odor signal receptors—making this the largest gene family known. Each of the thousands of neurons in the nose expresses one of these receptors. The identification of which chemical signal, or *odorant*, activates which receptor is just getting under way.

> The average human is born with about 950 genes for odor signal receptor proteins, but very few people express more than 400 of them; some express far fewer, which may explain why you are able to smell certain things your roommate cannot, or vice versa.

When an odorant molecule binds to its receptor, a G protein becomes activated, which in turn activates adenylyl cyclase, which catalyzes the formation of the second messenger cAMP. This molecule then binds to ion channels, causing them to open (**Figure 15.17**). The resulting influx of Na^+ and Ca^{2+} causes the neuron to become stimulated so that it sends a signal to the brain that a particular odor is present.

Enzyme activities change in response to signals

Proteins will change their shape, and thus their function, if they are modified either covalently or noncovalently. We have seen examples of both types of modification in our description of signal transduction. Protein kinases add phosphate groups to a target protein, and this covalent change alters the protein's conformation. Cyclic AMP binds to target proteins allosterically, and this noncovalent interaction changes the protein's conformation. In both cases, previously inaccessible active sites are exposed, and the target protein goes on to perform a new cellular role.

The G protein-mediated protein kinase cascade stimulated by epinephrine in liver cells results in the phosphorylation of two key enzymes in glycogen metabolism, with opposite effects:

- *Inhibition.* Glycogen synthase, which catalyzes the joining of glucose molecules to synthesize the energy-storing molecule glycogen, is inactivated by phosphorylation. Thus the epinephrine signal prevents glucose from being stored in glycogen (**Figure 15.18, step 1**).

- *Activation.* Phosphorylase kinase is activated when a phosphate group is added to it. It goes on to stimulate a protein kinase cascade that ultimately leads to the activation by phosphorylation of glycogen phosphorylase, the other key enzyme in glucose metabolism. This enzyme liberates glucose molecules from glycogen (**Figure 15.18, steps 2 and 3**).

Thus the same signaling pathway inhibits the storage of glucose as glycogen (by inhibiting glycogen synthase) and promotes the release of glucose through glycogen breakdown (by activating glycogen phosphorylase). As we mentioned earlier, the released glucose fuels the fight-or-flight response to epinephrine.

The *amplification* of the signal in this pathway is impressive; as detailed in Figure 15.18, each single molecule of epinephrine that arrives at the plasma membrane ultimately results in *10,000 molecules* of blood glucose:

1	molecule of epinephrine bound to the membrane activates
20	molecules of cAMP, which activate
20	molecules of protein kinase A, which activate
100	molecules of phosphorylase kinase, which activate
1,000	molecules of glycogen phosphorylase, which produce
10,000	molecules of glucose 1-phosphate, which produce
10,000	molecules of blood glucose

15.17 A Signal Transduction Pathway Leads to the Opening of Ion Channels In the signal transduction pathway for the sense of smell, the final effect is the opening of ion channels. The resulting influx of Na^+ and Ca^{2+} stimulates the transmission of a scent message to a specific region of the brain.

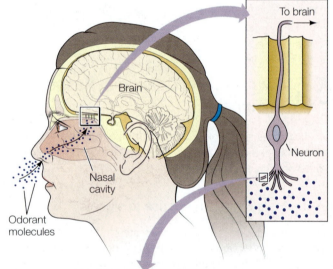

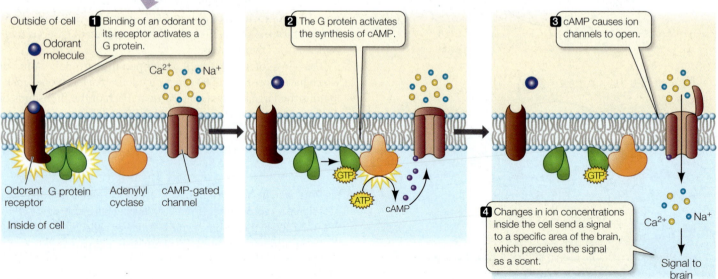

1 Binding of an odorant to its receptor activates a G protein.

2 The G protein activates the synthesis of cAMP.

3 cAMP causes ion channels to open.

4 Changes in ion concentrations inside the cell send a signal to a specific area of the brain, which perceives the signal as a scent.

Outside of cell
Odorant molecule
Ca^{2+} Na^+
Odorant receptor   G protein   Adenylyl cyclase   cAMP-gated channel
Inside of cell
GTP
ATP
cAMP
Ca^{2+} Na^+
Signal to brain

To brain
Brain
Neuron
Nasal cavity
Odorant molecules

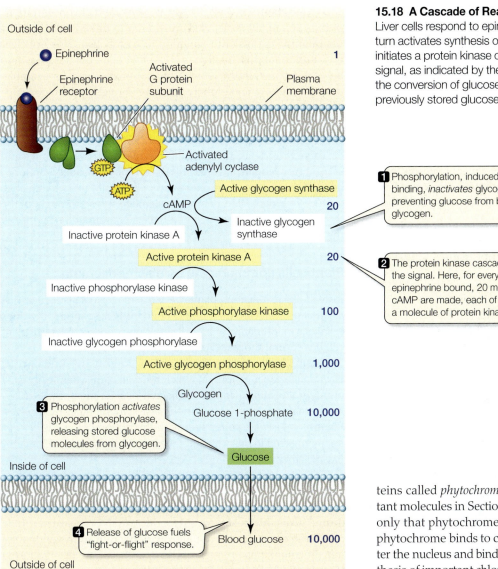

15.18 A Cascade of Reactions Leads to Altered Enzyme Activity
Liver cells respond to epinephrine by activating G proteins, which in turn activates synthesis of the second messenger cAMP. Cyclic AMP initiates a protein kinase cascade, greatly amplifying the epinephrine signal, as indicated by the blue numbers. The cascade both inhibits the conversion of glucose to glycogen and stimulates the release of previously stored glucose.

1 Phosphorylation, induced by epinephrine binding, *inactivates* glycogen synthase, preventing glucose from being stored as glycogen.

2 The protein kinase cascade amplifies the signal. Here, for every molecule of epinephrine bound, 20 molecules of cAMP are made, each of which activates a molecule of protein kinase A.

3 Phosphorylation *activates* glycogen phosphorylase, releasing stored glucose molecules from glycogen.

4 Release of glucose fuels "fight-or-flight" response.

teins called *phytochromes*. We will say more about these important molecules in Section 37.5, but for now it is important to know only that phytochromes are activated by red light. An activated phytochrome binds to cytoplasmic regulatory proteins, which enter the nucleus and bind to promoters of genes involved in the synthesis of important chloroplast proteins. Synthesis of these chloroplast proteins is the key to plant "greening."

Signals can initiate gene transcription

Plasma membrane receptors are involved in activating a broad range of gene expression responses. The Ras signaling pathway, for example, ends in the nucleus (see Figure 15.10). The final protein kinase in the Ras signaling cascade, MAPk, enters the nucleus and phosphorylates a transcription factor, which stimulates the transcription of a number of genes involved in cell proliferation.

As described in Section 15.2, lipid-soluble hormones can diffuse through the plasma membrane and meet their receptors in the cytoplasm. In this case, binding of the ligand allows the ligand–receptor complex to enter the nucleus, where it binds to hormone-responsive elements at the promoters of a number of genes (see Figure 15.8). In some cases, transcription is stimulated, and in others it is inhibited.

In plants, light acts as a signal to initiate the formation of chloroplasts. Between the light signal and its effect on the chloroplast is a transcription-mediated signal transduction pathway. In bright sunlight, red wavelengths of light are absorbed by receptor pro-

15.4 RECAP

Cells respond to signal transduction by activating enzymes, opening membrane channels, or initiating gene transcription.

- Describe the role that cAMP plays in the sense of smell. See pp. 345–346 and Figure 15.17

- Can you explain how amplification of a signal occurs and why it is important in a cell's response to changes in its environment? See p. 346 and Figure 15.18

We have described how signals from a cell's environment can influence that cell. But the environment of a cell in a multicellular organism is more than the extracellular medium—it includes neighboring cells as well. In the next section we'll look at specialized junctions between cells that allow them to signal one another directly.

15.5 How Do Cells Communicate Directly?

Most cells are in contact with their neighbors. Section 5.2 described various ways in which cells adhere to one another, such as recognition proteins protruding from the cell surface, tight junctions, and desmosomes. However, as we know from our own experience with our neighbors (and roommates), just being in proximity does not necessarily mean that there is functional communication. Neither tight junctions nor desmosomes are specialized for intercellular communication. However, many multicellular organisms have specialized cell junctions that allow their cells to communicate directly. In animals, these structures are *gap junctions*; in plants, they are *plasmodesmata*.

Animal cells communicate by gap junctions

Gap junctions are channels between adjacent cells that occur in many animals, occupying up to 25 percent of the area of the plasma membrane (**Figure 15.19**). Gap junctions traverse the narrow space between the plasma membranes of two cells (the "gap") by means of channel proteins called **connexons**. The walls of these channels are composed of six subunits of an integral membrane protein. In two cells close to each other, two connexons come together, forming a channel that links the two cytoplasms. There may be hundreds of these channels between a cell and its neighbors. The channel pores are about 1.5 nm in diameter—far too narrow for the passage of large molecules such as proteins. But they are wide enough to allow small signal molecules and ions to pass between the cells. Experiments in which a labeled signal molecule or ion is injected into one cell show that it can readily pass into the adjacent cells if the cells are connected by gap junctions.

Gap junctions permit metabolic cooperation among the linked cells. Such cooperation ensures the sharing of important small molecules such as ATP, metabolic intermediates, amino acids, and coenzymes between cells. It may also ensure that concentrations of ions and small molecules are similar in linked cells, thereby maintaining equivalent regulation of metabolism. It is not clear how important this function is in many tissues, but it is known to be vital in some. In the lens of the mammalian eye, for example, only the cells at the periphery are close enough to the blood supply to allow diffusion of nutrients and wastes. But because lens cells are connected by large numbers of gap junctions, material can diffuse between them rapidly and efficiently.

There is evidence that signal molecules such as hormones and second messengers such as cAMP can move through gap junctions. If this is true, then only a few cells would need to have receptors for a signal in order for the signal to spread throughout the tissue. In this way, a tissue could have a coordinated response to the signal.

Plant cells communicate by plasmodesmata

Instead of gap junctions, plants have **plasmodesmata** (singular *plasmodesma*), which are membrane-lined bridges spanning the thick cell walls that separate plant cells from one another. A typical plant cell has several thousand plasmodesmata.

Plasmodesmata differ from gap junctions in one fundamental way: unlike gap junctions, in which the wall of the channel is made of integral proteins from the adjacent plasma membranes, plasmodesmata are lined by the fused plasma membranes themselves. Plant biologists are so familiar with the notion of a tissue as cells interconnected in this way that they refer to these continuous cytoplasms as a *symplast* (see Section 35.1).

The diameter of a plasmodesma is about 6 nm, far larger than a gap junction channel. But the actual space available for diffusion is about the same—1.5 nm. A look at the interior of the plasmodesma gives the reason for this reduction in pore size: a tubule called the **desmotubule**, apparently derived from the endoplasmic reticulum, fills up most of the opening of the plasmodesma (**Figure 15.20**). So, typically, only small metabolites and ions move between plant cells. This fact is important physiologically to plants, which lack the tiny circulatory vessels (capillaries) many animals use to bring gases and nutrients to every cell.

Diffusion from cell to cell through plasma membranes is probably inadequate for hormonal responses in plants. Instead, they rely on more rapid diffusion through plasmodesmata to ensure that all cells of a tissue respond to a signal at the same time. In C$_4$ plants (see Section 8.4), abundant plasmodesmata help to move the carbon fixed in the mesophyll rapidly to the bundle sheath cells. A similar transport system, found at the junctions of nonvascular tissues and phloem, conducts organic solutes throughout the plant.

Plasmodesmata are not merely passive channels, but can be regulated. Plant viruses may infect cells at one location, then spread rapidly through a plant organ by plasmodesmata until they reach the plant's vascular tissue (circulatory system). These viruses, and even their RNA, would appear to be many times too large to pass through the desmotubules, but they get through, apparently by making "movement proteins" that increase the pore size temporarily while attached to the viral genome. Similar movement proteins made by the plants themselves are involved in transporting mRNAs and even proteins such as transcription factors between

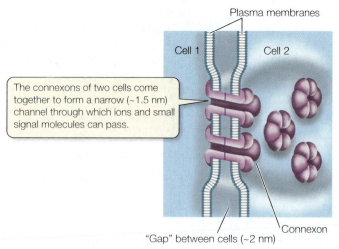

Plasma membranes

Cell 1   Cell 2

The connexons of two cells come together to form a narrow (~1.5 nm) channel through which ions and small signal molecules can pass.

Connexon

"Gap" between cells (~2 nm)

15.19 Gap Junctions Connect Animal Cells An animal cell may contain hundreds of gap junctions connecting it to neighboring cells. Gap junctions are too small for proteins, but small molecules such as ATP, metabolic intermediates, amino acids, and coenzymes can pass through them.

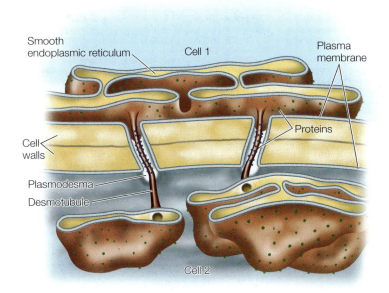

Smooth endoplasmic reticulum
Cell 1
Plasma membrane
Proteins
Cell walls
Plasmodesma
Desmotubule
Cell 2

15.20 Plasmodesmata Connect Plant Cells The desmotubule, derived from the smooth endoplasmic reticulum, fills up most of the space inside a plasmodesma, leaving a tiny gap through which small metabolites and ions can pass.

plant cells. This finding opens up the possibility of long-distance regulation of transcription and translation.

15.5 RECAP

Cells can communicate with their neighbors through specialized cell junctions. In animals, these structures are gap junctions; in plants, they are plasmodesmata.

■ What roles do gap junctions and plasmodesmata play in cell signaling?

CHAPTER SUMMARY

15.1 What are signals, and how do cells respond to them?

Cells receive many signals from the physical environment and from other cells. **Autocrine** signals affect the cells that make them; **paracrine** signals diffuse to and affect nearby cells. Review Figure 15.1

A **signal transduction pathway** involves the interaction of a signal molecule with a **receptor**; the transduction of the signal via a **responder** within the cell; and an effect on the function of the cell. Review Figure 15.2, Web/CD Activity 15.1

15.2 How do signal receptors initiate a cellular response?

Cells respond to signals only if they have specific receptor proteins that can bind those signals. Depending on the nature of its signaling **ligand**, a receptor may be located in the plasma membrane or in the cytoplasm of the target cell. Review Figure 15.4

Receptors located in the plasma membrane include **ion channels**, **protein kinases**, and **G protein-linked receptors**.

Ion channel receptors are "gated": the gate "opens" when the three-dimensional structure of the channel protein is altered by ligand binding. Review Figure 15.5

G proteins have two important binding sites: one for a G protein-linked receptor, and another for GDP/GTP. G proteins can either activate or inhibit an **effector protein**. Review Figure 15.7, Web/CD Tutorial 15.1

When bound by a ligand, **cytoplasmic receptors** change their shape and enter the cell nucleus. Review Figure 15.8

15.3 How is a response to a signal transduced through the cell?

Direct transduction is a function of the receptor itself and occurs at the plasma membrane. **Indirect transduction** involves a **second messenger**. Review Figure 15.9

Protein kinase cascades amplify a response to receptor binding. Review Figure 15.10

Second messengers include **cyclic AMP (cAMP)**; **inositol triphosphate (IP_3)** and **diacylglycerol (DAG)**, which are derived from the phospholipid **phosphatidyl inositol-bisphosphate (PIP2)**; calcium ions; and the gas **nitric oxide (NO)**.

Signal transduction can be regulated in several ways. The balance between making and breaking down the enzymes involved determines the ultimate cellular response to a signal.

15.4 How do cells change in response to signals?

The ultimate cellular response to a signal may be the opening of ion channels, the alteration of enzyme activities, or changes in gene transcription. Review Figure 15.17

Protein kinases covalently add phosphate groups to target proteins; cAMP allosterically binds target proteins. Whether by covalent or noncovalent action, changes in the target protein's conformation expose previously inaccessible active sites.

Activated enzymes may either inhibit or activate other enzymes in a signal transduction pathway, leading to impressive amplification of a signal. Review Figure 15.18

Lipid-soluble signals, such as steroid hormones, can diffuse through the plasma membrane and meet their receptors in the cytoplasm; the ligand–receptor complex may then enter the nucleus to affect gene transcription.

15.5 How do cells communicate directly?

Most animal cells can communicate with one another directly through small pores in their plasma membranes called **gap junctions**. Channel proteins called **connexons** form thin channels between two adjacent cells through which small signal molecules and ions can pass. Review Figure 15.19

Plant cells are connected by somewhat larger pores called **plasmodesmata**, which traverse both plasma membranes and cell walls. The **desmotubule** narrows the opening of the plasmodesma. Review Figure 15.20

See Web/CD Activity 15.2 for a concept review of this chapter.

SELF-QUIZ

1. What is the correct order for the following events in the inter-action of a cell with a signal? (1) alteration of cell function; (2) signal binds to receptor; (3) signal released from source; (4) signal transduction.
   a. 1234
   b. 2314
   c. 3214
   d. 3241

2. Why do some signals ("first messengers") trigger a "second messenger" to activate a target cell?
   a. The first messenger requires activation by ATP.
   b. The first messenger is not water soluble.
   c. The first messenger binds to many types of cells.
   d. The first messenger cannot cross the plasma membrane.
   e. There are no receptors for the first messenger.

3. Steroid hormones such as estrogen act on target cells by
   a. initiating second messenger activity.
   b. binding to membrane proteins.
   c. initiating DNA transcription.
   d. activating enzymes.
   e. binding to membrane lipids.

4. The major difference between a cell that responds to a signal and one that does not is the presence of a
   a. DNA sequence that binds to the signal.
   b. nearby blood vessel.
   c. receptor.
   d. second messenger.
   e. transduction pathway.

5. Which of the following is *not* a consequence of signal binding to a receptor?
   a. Activation of receptor enzyme activity
   b. Diffusion of receptor in the plasma membrane
   c. Change in conformation of the receptor protein
   d. Breakdown of the receptor to amino acids
   e. Release of the signal from the receptor

6. A nonpolar molecule such as a steroid hormone usually binds to a
   a. cytoplasmic receptor.
   b. protein kinase.
   c. ion channel.
   d. phospholipid.
   e. second messenger.

7. Which of the following is *not* a common type of receptor?
   a. Ion channel
   b. Protein kinase
   c. G protein-linked
   d. Transcription factor
   e. Adenylyl cyclase

8. Which of the following is *not* true of the protein kinase cascade?
   a. The signal is amplified.
   b. A second messenger is formed.
   c. Target proteins are phosphorylated.
   d. The cascade ends up in the nucleus.
   e. The cascade begins at the plasma membrane.

9. Which of the following is *not* a second messenger for signal transduction?
   a. Calcium ions
   b. Nitric oxide gas
   c. ATP
   d. Cyclic AMP
   e. Diacylglycerol

10. Plasmodesmata and gap junctions
    a. allow small molecules and ions to pass rapidly between cells.
    b. are both membrane-lined channels.
    c. are channels about 1 mm in diameter.
    d. are present only once per cell.
    e. are involved in cell recognition in signaling.

FOR DISCUSSION

1. Like the Ras protein itself, the various components of the Ras signaling pathway were discovered when cancer cells showed mutations in one or another of their components. What might be the biochemical consequences of mutations in the genes coding for (A) Raf and (B) MAP kinase that resulted in rapid cell division?

2. Cyclic AMP is a second messenger in many different responses. How can the same messenger act in different ways in different cells?

3. Compare direct communication via plasmodesmata or gap junctions with receptor-mediated communication between cells. What are the advantages of one method over the other?

4. The tiny invertebrate *Hydra* has an apical region, which has tentacles, and a long, slender body. *Hydra* can reproduce asexually when cells on the body wall differentiate and form a bud, which then breaks off as a new organism. Buds form only at certain distances from the apex, leading to the idea that the apex releases a signal molecule that diffuses down the body and, at high concentrations (i.e., near the apex), inhibits bud formation. *Hydra* lacks a circulatory system, so the inhibitor must diffuse from cell to cell. If you had an antibody that binds to connexons to plug up the gap junctions, how would you show that *Hydra*'s inhibitory factor passes *through* these junctions?

FOR INVESTIGATION

Endosymbiotic bacteria in the marine invertebrate *Begula neritina* synthesize bryostatins, a name derived from *Begula*'s animal group Ectoprocta, once known as *bryozoans* ("moss animals"); and *stat* (stop). Bryostatins curtail cell division in many cell types, including several cancers. It is proposed that bryostatins inhibit protein kinase C (see Figure 15.13). How would you investigate this hypothesis, and how would you relate this inhibition to cell division?

16 Recombinant DNA and Biotechnology

Baby 81

The tsunami of December 26, 2004, struck the coastal town of Kalmunai, Sri Lanka, with such force that 4-month-old Abilass Jeyarajah was torn from his mother's arms and swept away. Hours later, while his parents desperately searched the devastated town, their tiny son washed up on the beach a kilometer away, alive. A local schoolteacher found him and brought him to the hospital—the eighty-first patient admitted that day. The hospital was overwhelmed with 1,000 bodies, many of them children. Since Abilass was alive and healthy, he was dubbed "Baby 81, the miracle baby" and became an instant celebrity among the staff as they went about their grim duties.

Meanwhile, the parents kept looking. Two days later, they met the schoolteacher, who told them about the baby he had found. Rushing to the hospital, the Jeyarajahs were elated to find their son but were in for a rude shock. Eight other couples who had also lost infants were claiming Baby 81 as theirs. The baby remained in the hospital while the case went to court.

Judge M. P. Mohaideen faced a situation not unlike King Solomon, who 3,000 years ago was asked to decide which of two women was the mother of an infant. Solomon's method of determining parentage is told in a famous biblical passage (I Kings 3:16–28). The Sri Lankan judge had a different method: he called in molecular biologists.

With 6 billion base pairs of DNA packaged in 46 chromosomes, each one of us is unique. Although our protein-coding sequences are similar (after all, our phenotypes are similar), only a few percent of the total base pairs in the human genome actually code for genes. As explained in Chapter 14, the eukaryotic genome contains many repeated sequences, and between individuals the repeat frequency may differ, offering one way to differentiate individuals. Differences in a single base pair due to DNA replication errors or random mutations also distinguish individuals from one another, and these differences are inherited.

After the Tsunami In December of 2004, a tsunami originating in the Indian Ocean struck over a broad region that encompassed many nations in Southeast Asia. The result was an unprecedented humanitarian disaster that left almost a quarter of a million people dead and many more homeless.

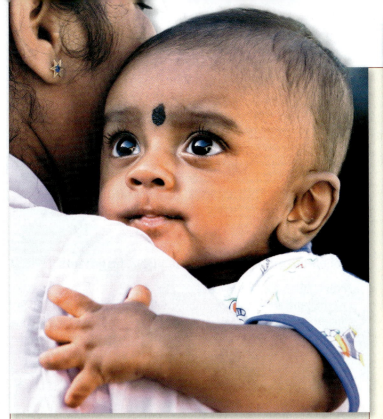

"Baby 81" Abilass Jeyarajah survived the tsunami and was reunited with his parents by court order after DNA testing proved that he is indeed their son.

It is now possible to analyze these differences in DNA (amplified by PCR) to identify people. When DNA from the nine sets of contesting parents was analyzed and compared with that of Baby 81, the only parents whose sequences were consistent with being passed on to the baby were the Jeyarajas. On February 14, 2005, the judge ruled that the Jeyarajas are the true parents, and Baby 81 got his real name and parents back.

IN THIS CHAPTER we will describe some of the techniques that are used to manipulate DNA. We begin with a description of how DNA molecules are cut into smaller fragments and how these fragments are spliced together to create recombinant DNA. We will then describe how recombinant (or any other) DNA is introduced into a suitable host cell. Then we'll look at some other techniques for copying, storing, and altering DNA. Finally, we'll see how some of these DNA manipulation techniques have been applied by scientists.

16.1 How Are Large DNA Molecules Analyzed?

Scientists have long realized that the chemical reactions that living cells employ for one purpose may be applied in the laboratory for other, novel purposes. Similarly, the manipulation of DNA molecules is based on an understanding of the chemical structure of DNA, the properties of certain enzymes, and the rules of complementary base pairing.

In this section we will see how some of the numerous naturally occurring enzymes that cleave and repair DNA are used in the laboratory to manipulate and recombine DNA. We'll learn how enzymes are used to cut DNA into fragments and how these fragments are analyzed.

Restriction enzymes cleave DNA at specific sequences

All organisms, including bacteria, must have ways of dealing with their enemies. As Section 13.1 describes, bacteria are attacked by viruses called bacteriophage that inject their genetic material into the host cell. Some bacteria defend themselves against such invasions by producing **restriction enzymes** (also known as *restriction endonucleases*), which cut double-stranded DNA molecules—such as those injected by phage—into smaller, noninfectious fragments (**Figure 16.1**). These enzymes break the bonds of the DNA backbone between the 3′ hydroxyl group of one nucleotide and the 5′ phosphate group of the next nucleotide. This cutting process is called **restriction digestion**.

There are many such restriction enzymes, each of which cleaves DNA at a specific sequence of bases, called a **recognition sequence** or a **restriction site**. Most recognition sequences are 4–6 base pairs long. The sequence is recognized through the principles of protein–DNA interactions (see Section 11.2): the base pairs inside the double helix of DNA have small differences, so that they fit differently into the three-dimensional structure of an enzyme.

Why doesn't a restriction enzyme cut the DNA of the bacterial cell that makes it? One way that the cell protects itself is

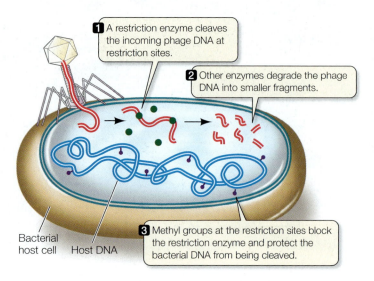

1 A restriction enzyme cleaves the incoming phage DNA at restriction sites.

2 Other enzymes degrade the phage DNA into smaller fragments.

3 Methyl groups at the restriction sites block the restriction enzyme and protect the bacterial DNA from being cleaved.

Bacterial host cell   Host DNA

16.1 Bacteria Fight Invading Viruses with Restriction Enzymes Bacteria produce restriction enzymes that degrade phage DNA by cleaving it into fragments. Methylation protects the bacteria's own DNA from being cleaved.

by modifying the restriction site on its own DNA after the sequence has been made during DNA replication. Specific modifying enzymes called *methylases* add methyl ($-CH_3$) groups to certain bases at the restriction sites of the host's DNA after it has been replicated. The methylation of the host's bases makes the recognition sequence unrecognizable to the restriction enzyme. But unmethylated phage DNA is efficiently recognized and cleaved.

Bacterial restriction enzymes can be isolated from the cells that make them and used as biochemical reagents in the laboratory. If DNA from any organism is incubated in a test tube with a restriction enzyme, that DNA will be cut wherever the restriction site occurs. A specific sequence of bases defines each restriction site. For example, the enzyme *EcoRI* (named after its source, a strain of the bacterium *E. coli*) cuts DNA only where it encounters the following paired sequence in the DNA double helix:

$$5' \ldots \text{GAATTC} \ldots 3'$$
$$3' \ldots \text{CTTAAG} \ldots 5'$$

Notice that this sequence reads the same in the 5'-to-3' direction on both strands. It is *palindromic*, like the word "mom," in the sense that it is the same in both directions from the 5' end. The *EcoRI* enzyme has two identical active sites on its two subunits, which cleave the two strands simultaneously between the G and the A of each strand.

The *EcoRI* recognition sequence occurs, on average, about once in every 4,000 base pairs in a typical prokaryotic genome, or about once per four prokaryotic genes. So *EcoRI* can chop a large piece of DNA into smaller pieces containing, on average, just a few genes. Using *EcoRI* in the laboratory to cut small genomes, such as those of viruses that have tens of thousands of base pairs, may result in a few fragments. For a huge eukaryotic chromosome with tens of millions of base pairs, a very large number of fragments will be created.

Of course, "on average" does not mean that the enzyme cuts all stretches of DNA at regular intervals. The *EcoRI* recognition se-

quence does not occur even once in the 40,000 base pairs of the genome of a phage called T7—a fact that is crucial to the survival of this virus, since its host is *E. coli*. Fortunately for *E. coli*, the *EcoRI* recognition sequence does appear in the DNA of other phage.

Hundreds of restriction enzymes have been purified from various microorganisms. In the test tube, different restriction enzymes that recognize different restriction sites can be used to cut the same sample of DNA. Thus restriction enzymes can be used as "knives" for genetic "surgery" to cut a sample of DNA in many different, specific places.

Gel electrophoresis separates DNA fragments

After a laboratory sample of DNA has been cut with a restriction enzyme, the DNA is in fragments, which must be separated. Because the recognition sequence does not occur at regular intervals, the fragments are not all the same size, and this property provides a way to separate them from one another. Separating the fragments is necessary to determine the number and molecular sizes (in base pairs) of the fragments produced or to identify and purify an individual fragment of particular interest.

A convenient way to separate or purify DNA fragments is by **gel electrophoresis** (**Figure 16.2**). A mixture of DNA fragments is placed in a well in a porous gel, and an electric field (with positive and negative ends) is applied across the gel. Because of its phosphate groups, DNA is negatively charged at neutral pH; therefore, because opposite charges attract, the DNA fragments move toward the positive end of the field. The porous gel acts like a sieve, and the smaller molecules move faster than the larger ones. After a fixed time, and while all the fragments are still in the gel, the electric power is shut off. The separated fragments can be visualized by staining them with a fluorescent dye. Under ultraviolet light, they can then be seen as bars or spots in the gel and can be examined or removed individually. Electrophoresis gives us two types of information:

- *The sizes of the fragments.* DNA fragments of known size are often placed in a well in the gel next to the sample to provide a standard of comparison.

- *The presence of specific DNA sequences.* A specific DNA sequence can be revealed with a single-stranded DNA probe (**Figure 16.3**). The sample DNA is denatured (unwound and separated into single strands) while still in the gel, then the gel is treated so that the DNA is transferred to a nylon filter to make a "blot" (called a *Southern blot*, after the scientist who developed the method). The filter is then exposed to a single-stranded DNA probe with a sequence complementary to the one that is being sought. If the sequence of interest is present in the sample DNA, the probe will hybridize with it. The probe can be labeled with a radioisotope. After hybridization, spots of radioactivity indicate that the probe has hybridized with its target sequence at that location. Unbound probes stay in solution.

The gel region containing the desired fragment (in size or sequence) can be cut out as a lump of gel, and the pure DNA fragment can then be removed from the gel by diffusion into a small volume of water.

RESEARCH METHOD

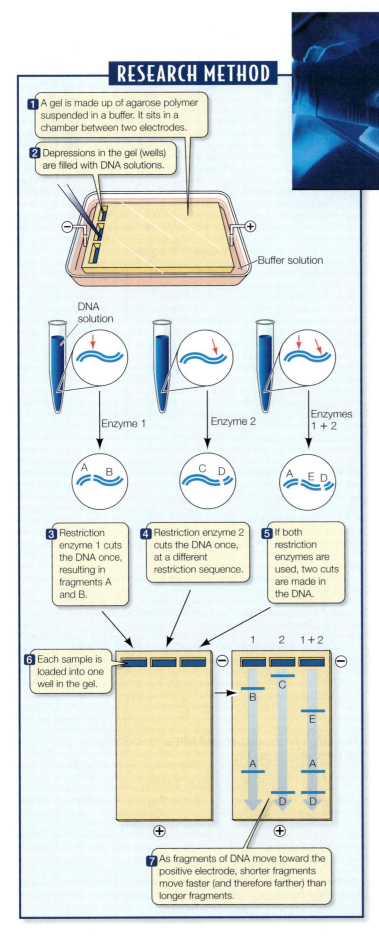

1 A gel is made up of agarose polymer suspended in a buffer. It sits in a chamber between two electrodes.

2 Depressions in the gel (wells) are filled with DNA solutions.

Buffer solution

DNA solution

Enzyme 1

Enzyme 2

Enzymes 1 + 2

A B

C D

A E D

3 Restriction enzyme 1 cuts the DNA once, resulting in fragments A and B.

4 Restriction enzyme 2 cuts the DNA once, at a different restriction sequence.

5 If both restriction enzymes are used, two cuts are made in the DNA.

6 Each sample is loaded into one well in the gel.

1 2 1 + 2

B
C
E
A   A
D   D

7 As fragments of DNA move toward the positive electrode, shorter fragments move faster (and therefore farther) than longer fragments.

16.2 Separating Fragments of DNA by Gel Electrophoresis
A mixture of DNA fragments is placed in a gel and an electric field is applied across the gel. The negatively charged DNA moves toward the positive end of the field, with smaller molecules moving faster than larger ones. When the electric power is shut off, the separated fragments can be analyzed.

DNA fingerprinting uses restriction analysis and electrophoresis

The two methods we have just described—restriction digestion to cut DNA into fragments and gel electrophoresis to separate them by size—are used in **DNA fingerprinting**, a DNA-based technique for identifying individuals. DNA fingerprinting works best with genes that are highly *polymorphic*—that is, genes that have multi-

RESEARCH METHOD

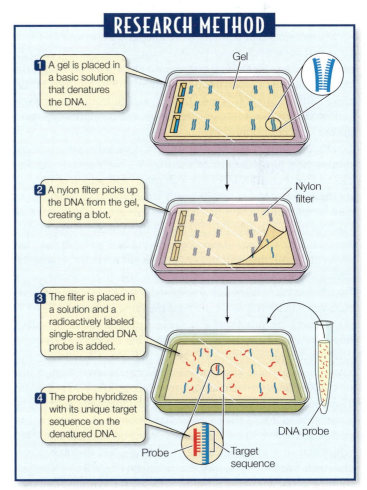

1 A gel is placed in a basic solution that denatures the DNA.

Gel

2 A nylon filter picks up the DNA from the gel, creating a blot.

Nylon filter

3 The filter is placed in a solution and a radioactively labeled single-stranded DNA probe is added.

4 The probe hybridizes with its unique target sequence on the denatured DNA.

DNA probe

Probe   Target sequence

16.3 Analyzing DNA Fragments by Southern Blotting
A probe can be used to locate a specific DNA fragment on an electrophoresis gel.

ple alleles and are therefore likely to be different in different individuals. Two types of polymorphisms are especially informative:

- **Single nucleotide polymorphisms (SNPs)** are inherited variations involving a single nucleotide base. These polymorphisms have been mapped for many organisms. If one parent is homozygous for A at a certain point on the genome, for example, and the other parent has a G at that point, the offspring will be heterozygous: one chromosome will have A at that point and the other, G.

- **Short tandem repeats (STRs)** are short, moderately repetitive DNA sequences that occur side by side on the chromosomes. These repeat patterns are also inherited. For example, an individual might inherit a chromosome 15 with a short sequence repeated six times from her mother and a chromosome 15 with the same sequence repeated two times from her father.

STRs are easily detectable if they lie between two recognition sequences for a restriction enzyme. If the DNA from the heterozygous individual described above is cut with a restriction enzyme, it will form two different-sized fragments: one larger (the one from the mother) and the other smaller (the one from the father). These patterns can be easily seen by the use of gel electrophoresis (**Figure 16.4**). With several different STRs (as many as eight are used), each with numerous alleles, an individual's unique pattern becomes apparent.

DNA fingerprinting methods require at least 1 μg of DNA, or the DNA content of about 100,000 human cells, but this amount is not always available. The polymerase chain reaction (PCR), the powerful technique described in Section 11.5 (see Figure 11.23), can multiply ("amplify") the DNA of interest from even a single cell, producing in a few hours the necessary 1 μg for restriction digestion and electrophoresis.

DNA fingerprinting can be used in forensics (crime investigation) to help prove the innocence or guilt of a suspect. For example, in a rape case, DNA can be extracted from semen or hair left by the attacker and compared with DNA from a suspect. So far, this method more often has been used to prove innocence (the DNA patterns are different) than guilt (the DNA patterns are the same). It is easy to exclude someone on the basis of these tests, but given that DNA fingerprinting examines just a small fragment the genome, two people could theoretically have the same sequence for that fragment, although if several different STRs are used, the probability that two people will have the same alleles becomes very small. Therefore, proof that a suspect is guilty cannot rest on DNA fingerprinting alone, but must rely on other evidence as well.

A fascinating example demonstrates the use of DNA fingerprinting in the analysis of historical events. Three hundred years of rule by the Romanov dynasty in Russia ended on July 16, 1918, when Tsar Nicholas II, his wife, and their five children were executed by a firing squad during the Communist revolution. A report that the bodies had been burned to ashes was never questioned until 1991, when a shallow grave with several skeletons was discovered several miles from the presumed execution site. Recent DNA fingerprinting of bone fragments found in this grave indicated that they came from an older man and woman and three female children, who were clearly related to one another (**Figure 16.5**) and were also related to several living descendants of the Tsar.

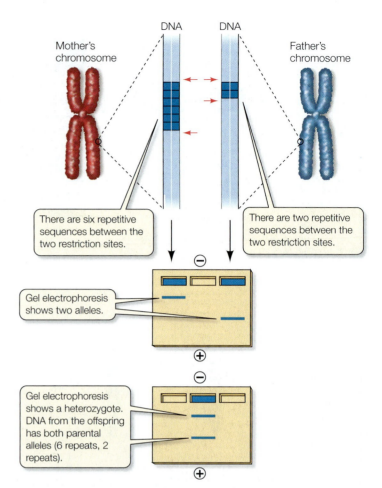

16.4 DNA Fingerprinting with Short Tandem Repeats The number of STRs inherited by an individual from each parent can be used to make a DNA fingerprint. The two alleles can be identified in an electrophoresis gel on the basis of their sizes.

Beluga caviar, the roe (eggs) from Beluga sturgeon, is highly prized and highly priced. Scientists at the American Museum of Natural History in New York developed DNA identification tests for various sturgeon species and went on a gourmet caviar shopping spree. To their surprise, they found that a quarter of the tins labeled Beluga caviar, weren't.

The DNA barcode project aims to identify all organisms on Earth

One of the most exciting aspects of DNA technology for biologists is its potential to identify the species with which they are working. While they can be certain of the species of an organism raised in a pure culture (say, *E. coli* cultivated in the laboratory), different organisms can look very much alike in nature. About 1.7 million species have been named and described, but about ten times that number probably have yet to be identified. A proposal to use DNA technology to identify known species and detect the ones we don't know has been endorsed by a large group of scientific organiza-

16.5 DNA Fingerprinting the Russian Royal Family   The skeletal remains of Tsar Nicholas II, his wife Alexandra, and three of their children were found in 1991 and subjected to DNA fingerprinting. Five STRs were tested. The results can be interpreted as follows: Using STR-2 as an example, the parents had genotypes 8,8 (homozygous) and 7,10 (heterozygous). The three children all inherited type 8 from the Tsarina and either type 7 or type 10 from the Tsar.

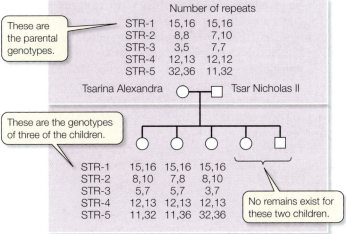

	Number of repeats	
STR-1	15,16	15,16
STR-2	8,8	7,10
STR-3	3,5	7,7
STR-4	12,13	12,12
STR-5	32,36	11,32

These are the parental genotypes.

Tsarina Alexandra ○——□ Tsar Nicholas II

These are the genotypes of three of the children.

STR-1	15,16	15,16	15,16
STR-2	8,10	7,8	8,10
STR-3	5,7	5,7	3,7
STR-4	12,13	12,13	12,13
STR-5	11,32	11,36	32,36

No remains exist for these two children.

the targeted gene fragment has been sequenced for all species, a simple device for conducting field analyses can be developed. The barcode project has the potential to advance biological research on evolution, to track species diversity in ecologically significant areas, to help identify new species, and even to detect undesirable microbes or bioterrorism agents.

16.1 RECAP

Large DNA molecules can be cut into smaller pieces by restriction digestion and then sorted by gel electrophoresis. DNA fingerprinting uses these two techniques to analyze DNA polymorphisms for the purpose of identifying individuals. Scientists hope to identify species using DNA analyses.

- Do you understand how a restriction enzyme recognizes a restriction site on DNA? See p. 355

- On what principle does gel electrophoresis separate DNA fragments? See p. 355 and Figure 16.2

- What are STRs and how are they used to identify individuals? See p. 357

tions known as the Consortium for the Barcode of Life (CBOL).

Paul Hebert at the University of Guelph in Ontario, Canada, has proposed to identify each species with a "DNA barcode" using a short sequence from a single gene. The gene he chose is the cytochrome oxidase gene, which is present in most cells. Because the gene mutates readily, there should be many allelic differences between species. A fragment of 650–750 base pairs in this gene is being sequenced for all organisms, and so far, sufficient variation has been detected to make it diagnostic of each species (**Figure 16.6**). Once the DNA of

16.6 A DNA Barcode   DNA from a 650–750-base pair region of the cytochrome oxidase gene from any organism can be amplified using PCR and then sequenced to identify a species. Each color represents one of the four DNA bases.

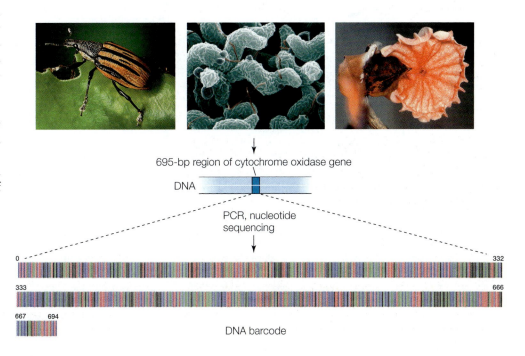

695-bp region of cytochrome oxidase gene

DNA

PCR, nucleotide sequencing

0   332
333   666
667   694

DNA barcode

We've just seen how scientists can cut DNA into fragments, and how they can use those fragments to identify polymorphisms. One of the most exciting ways of using DNA fragments, however, is to put them together in different combinations—to create recombinant DNA.

16.2 What Is Recombinant DNA?

Restriction enzymes cut the DNA backbone into fragments. These fragments can be joined by a second type of enzyme, DNA ligase (the same enzyme that joins Okazaki fragments during DNA replication; see Section 11.3). Once they had isolated restriction enzymes and DNA ligase, scientists realized that they could use these enzymes to generate and splice together any two DNA sequences. Stanley Cohen and Herbert Boyer did just that in 1973. They used restriction enzymes to cut and then DNA ligase to join DNA sequences from two *E. coli* plasmids (see Figure 13.14) containing different antibiotic resistance genes. The resulting plasmid, when inserted into new *E. coli* cells, gave those cells resistance to both antibiotics (**Figure 16.7**). The era of **recombinant DNA** was born.

Some restriction enzymes cleave DNA cleanly, cutting both strands exactly opposite one another. Others, such as *Eco*RI, make staggered cuts in a palindromic recognition sequence, cutting one strand of the double helix several bases away from where they cut the other (**Figure 16.8**). After *Eco*RI makes its two cuts in the complementary strands, the ends of the strands are held together only by the hydrogen bonds between four base pairs. These hydrogen bonds are too weak to persist at warm temperatures (above room temperature), so the fragments of DNA separate when they are warmed. As a result, there are single-stranded "tails" at the location of each cut. These tails are called **sticky ends** because they have a specific base sequence that can bind by base pairing with complementary sticky ends. If *n* restriction sites for a given restriction enzyme are present in a linear DNA molecule, then *n* + 1 fragments will be made, all with the same complementary sequences at their sticky ends.

After a DNA molecule has been cut with a restriction enzyme, complementary sticky ends can form hydrogen bonds with one

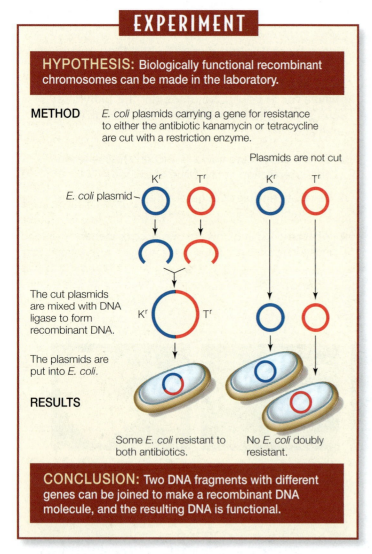

EXPERIMENT

HYPOTHESIS: Biologically functional recombinant chromosomes can be made in the laboratory.

METHOD *E. coli* plasmids carrying a gene for resistance to either the antibiotic kanamycin or tetracycline are cut with a restriction enzyme.

The cut plasmids are mixed with DNA ligase to form recombinant DNA.

The plasmids are put into *E. coli*.

RESULTS

Some *E. coli* resistant to both antibiotics.

No *E. coli* doubly resistant.

CONCLUSION: Two DNA fragments with different genes can be joined to make a recombinant DNA molecule, and the resulting DNA is functional.

16.7 Recombinant DNA   Stanley Cohen and Herbert Boyer performed the first experiment in which two different DNA sequences were combined in the laboratory to make a new, functional DNA molecule. FURTHER RESEARCH: Only one cell in 10,000 took up the plasmid in the experiment. The spontaneous mutation rate to T^r or K^r is one cell in 10^6. How would you distinguish between genetic transformation and spontaneous mutation in this experiment?

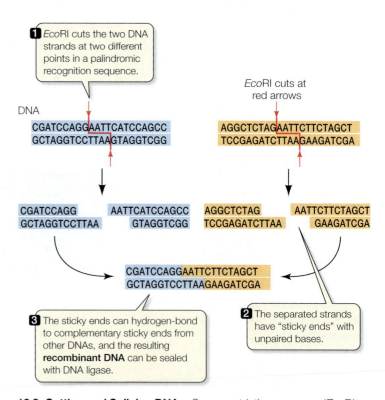

1. *Eco*RI cuts the two DNA strands at two different points in a palindromic recognition sequence.

*Eco*RI cuts at red arrows

DNA

CGATCCAGGAATTCATCCAGCC
GCTAGGTCCTTAAGTAGGTCGG

AGGCTCTAGAATTCTTCTAGCT
TCCGAGATCTTAAGAAGATCGA

CGATCCAGG   AATTCATCCAGCC
GCTAGGTCCTTAA   GTAGGTCGG

AGGCTCTAG   AATTCTTCTAGCT
TCCGAGATCTTAA   GAAGATCGA

CGATCCAGGAATTCTTCTAGCT
GCTAGGTCCTTAAGAAGATCGA

3. The sticky ends can hydrogen-bond to complementary sticky ends from other DNAs, and the resulting **recombinant DNA** can be sealed with DNA ligase.

2. The separated strands have "sticky ends" with unpaired bases.

16.8 Cutting and Splicing DNA   Some restriction enzymes (*Eco*RI is shown here) make staggered cuts in DNA. *Eco*RI can be used to cut DNA from two different sources (blue and yellow). The cuts leave sticky ends, exposed bases that can hybridize with complementary fragments. Sticky ends from different DNAs can bind to each other, forming recombinant DNA.

another. The original ends may rejoin, or an end may pair with a complementary end from another fragment. Furthermore, because the ends of all fragments cut by the same restriction enzyme are the same, fragments from one source, such as a human, can be joined to fragments from another source, such as a bacterium.

When the temperature is lowered, the fragments *anneal* (come together by hydrogen bonding) at their sticky ends, but these associations are unstable because they are held together by only a few hydrogen bonds. The associated sticky ends can be permanently spliced together by DNA ligase.

Many restriction enzymes do not produce sticky ends. Instead, they cut both DNA strands at the same base pair within the recognition sequence, making "blunt" ends. DNA ligase can also connect blunt-ended fragments.

With these two enzyme tools—restriction enzymes and DNA ligase—scientists can cut and rejoin different DNA molecules from *any and all sources* to form recombinant DNA.

16.2 RECAP

DNA fragments from different biological sources can be linked together to make recombinant DNA.

- Describe how Cohen and Boyer made the first recombinant DNA. See Figure 16.7

- Do you understand how a staggered cut in DNA creates a "sticky end"? See pp. 358–359 and Figure 16.8

Recombinant DNA has no biological significance until it is installed inside a living cell, which can replicate and transcribe the transplanted genetic information. How can that be accomplished?

16.3 How Are New Genes Inserted into Cells?

One goal of recombinant DNA technology is to **clone** (produce many copies of) a particular gene, either for analysis or to produce its protein product in quantity. If recombinant DNA is to be used to make a protein, it must be inserted, or **transfected**, into host cells. Such altered hosts are referred to as **transgenic** cells or organisms. The choice of a host cell—prokaryotic or eukaryotic—is important to this endeavor.

Once the host species has been selected, the recombinant DNA is brought together with a population of host cells and, under specific conditions, enters some of them. Because all the host cells proliferate—not just the few that receive the recombinant DNA—the scientist must be able to determine which cells actually contain the sequence of interest. One common method of identifying cells with recombinant DNA is to tag the inserted sequence with **reporter genes**, whose phenotypes are easily observed. These phenotypes serve as *genetic markers* for the sequence of interest. Antibiotic resistance genes were the markers used in Cohen and Boyer's experiment (see Figure 16.7)

Genes can be inserted into prokaryotic or eukaryotic cells

The initial successes of recombinant DNA technology were achieved using bacteria as hosts. As we have seen in preceding chapters, bacterial cells are easily grown and manipulated in the laboratory. Much of their molecular biology is known, especially for certain well-studied bacteria such as *E. coli*, and they have numerous genetic markers that can be used to select for cells harboring the recombinant DNA. Furthermore, bacteria contain plasmids, small circular chromosomes that are easily manipulated to carry recombinant DNA into the cell.

In some important ways, however, bacteria are not ideal organisms for studying and expressing eukaryotic genes. Consider how differently the processes of transcription and translation proceed in prokaryotes and eukaryotes, and recall that DNA often contains the signals for these specific functions. Bacteria, for example, lack the splicing machinery to excise introns from the initial RNA transcript of a eukaryotic gene. Recall also that many eukaryotic proteins are extensively modified after translation by processes such as glycosylation and phosphorylation, and that these modifications are often essential for the protein's activity.

When the expression of a new gene in a eukaryote is the point of the experiment—that is, its aim is to produce a transgenic organism—a eukaryotic host is selected. This host may be a mouse, a wheat plant, a yeast, or even a human. Yeasts such as *Saccharomyces* are common eukaryotic hosts for recombinant DNA studies.

The advantages of using yeasts include rapid cell division (a life cycle completed in 2–8 hours), ease of growth in the laboratory, and a relatively small genome size (about 12 million base pairs and 6,000 genes). The yeast genome is several times larger than that of *E. coli* (see Table 14.2), but has only one-fourth as many genes as the human genome. Nevertheless, yeasts have most of the characteristics of other eukaryotes, except for those characteristics involved in multicellularity (see Table 14.3).

Plant cells can also be used as hosts, especially if the desired result is a transgenic plant. The property that makes plant cells good hosts is their *totipotency*—that is, the ability of a differentiated plant cell to act like a fertilized egg and produce an entire new organism. Isolated plant cells grown in culture can take up recombinant DNA, and by manipulation of the growth medium, these transgenic cells can be induced to form an entire new plant, which can then be reproduced naturally in the field.

Whatever host is chosen, a vehicle for carrying the DNA into the host cell is needed.

Vectors carry new DNA into host cells

The challenge of inserting new DNA into a cell lies not just in getting it into the host cell, but in getting it to replicate as the host cell divides. DNA polymerase, the enzyme that catalyzes DNA replication, does not bind to just any sequence of DNA. If the new DNA is to be replicated, it must become part of a segment of DNA that contains an origin of replication, called a **replicon**, or *replication unit*.

There are two general ways in which the newly introduced DNA can become part of a replicon:

■ It can be inserted near an origin of replication in a host chromosome after entering the host cell. Although such insertion is often a random event, it is nevertheless a common method of integrating a new gene into a host cell.

■ It can enter the host cell as part of a carrier DNA sequence—a **vector**— that already has the appropriate origin of replication.

A vector should have four characteristics:

■ The ability to *replicate independently* in the host cell
■ A *recognition sequence* for a restriction enzyme that will allow the vector to be cut and combined with the new DNA
■ A *reporter gene* that will announce its presence in the host cell
■ A *small size* in comparison to the host chromosomes

Several types of vectors fit this profile, including plasmids, viruses, and artificial chromosomes.

PLASMIDS AS VECTORS Plasmids have all four of the characteristics of a useful vector (**Figure 16.9A**). First, they are small (an *E. coli* plasmid has 2,000–6,000 base pairs, as compared with the main *E. coli* chromosome, which has more than 4.6 million base pairs). Furthermore, because they are so small, many plasmids have only a single recognition sequence for a given restriction enzyme. When such a plasmid is cut with a restriction enzyme, it is transformed into a linear molecule with sticky ends. The sticky ends of another DNA fragment cut with the same restriction enzyme can pair with the sticky ends of the plasmid, resulting in a circular plasmid containing the new DNA at a known location.

Two other characteristics make plasmids good vectors. As we have seen, many plasmids contain genes that confer resistance to antibiotics, which can serve as reporter genes (genetic markers). Finally, plasmids have an origin of replication (*ori*) and can replicate independently of the host chromosome. It is not uncommon for a bacterial cell with a single main chromosome to contain hundreds of copies of a recombinant plasmid.

The plasmids commonly used as vectors in the laboratory have been extensively altered by recombinant DNA technology, and most are combinations of genes and other sequences from several sources. Many of these plasmids have a single marker for antibiotic resistance.

VIRUSES AS VECTORS Constraints on plasmid replication limit the size of the new DNA that can be inserted into a plasmid to about 10,000 base pairs. Although many prokaryotic genes may be smaller than this, 10,000 base pairs is much smaller than most eukaryotic genes, with their introns and extensive flanking sequences. A vector that accommodates larger DNA inserts is needed.

Both prokaryotic and eukaryotic viruses are often used as vectors for eukaryotic DNA. Bacteriophage λ, which infects *E. coli*, has a DNA genome of about 45,000 base pairs. If the genes that cause the host cell to die and lyse—about 20,000 base pairs—are eliminated, the virus can still attach to a host cell and inject its DNA. The deleted 20,000 base pairs can be replaced with DNA from another organism.

Because viruses infect cells naturally, they offer a great advantage over plasmids, which often require artificial means to coax them to enter host cells. As we will see in Section 17.5, viruses are important vectors in human gene therapy.

ARTIFICIAL CHROMOSOMES AS VECTORS Bacterial plasmids are not good vectors for eukaryotic hosts such as yeasts because prokaryotic and eukaryotic DNA sequences use different origins of replication. To remedy this problem, scientists have created a "minimalist chromosome" called the **yeast artificial chromosome**, or **YAC** (**Figure 16.9B**). This artificial DNA molecule contains not only the yeast origin of replication, but the yeast centromere and telomere sequences as well, making it a true eukaryotic chromosome. YACs also contain artificially synthesized restriction sites and useful reporter genes (for yeast nutritional requirements). YACs are only about 10,000 base pairs in size, but can accommodate 50,000 to 1.5 million base pairs of inserted DNA. These artificial chromosomes carry out eukaryotic DNA replication and gene expression normally in yeast cells.

PLASMID VECTORS FOR PLANTS An important vector for carrying new DNA into many types of plants is a plasmid found in *Agrobacterium tumefaciens*. This bacterium lives in the soil and causes a plant disease called crown gall, which is characterized by the pres-

16.9 Vectors for Carrying DNA into Cells (A) A plasmid with reporter genes for antibiotic resistance can be incorporated into an *E. coli* cell. (B) A DNA molecule synthesized in the laboratory constitutes an artificial chromosome that can carry its inserted DNA into yeasts. (C) The Ti plasmid, isolated from the bacterium *Agrobacterium tumefaciens*, is used to insert DNA into many types of plants.

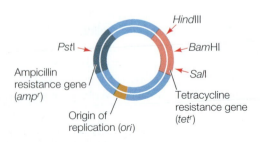

(A) Plasmid pBR322
Host: *E. coli*

- HindIII
- Pstl
- BamHI
- Sall
- Ampicillin resistance gene (*amp*^r)
- Tetracycline resistance gene (*tet*^r)
- Origin of replication (*ori*)

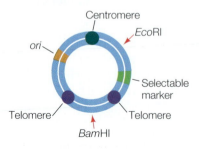

(B) Yeast artificial chromosome (YAC)
Host: yeast

- Centromere
- EcoRI
- *ori*
- Selectable marker
- Telomere
- Telomere
- BamHI

(C) Ti plasmid
Hosts: *Agrobacterium tumefaciens* (plasmid) and infected plants (T DNA)

- T DNA
- Sites for several restriction enzymes
- *ori*

↓ Recognition sites for restriction enzymes

ence of growths, or tumors, in the plant. *A. tumefaciens* contains a plasmid called Ti (for tumor-inducing) (**Figure 16.9C**). The Ti plasmid contains a region called T DNA, which inserts copies of itself into the chromosomes of infected plant cells. The T DNA contains recognition sequences for several restriction enzymes, so that new DNA can be inserted into it. When the T DNA is thus altered, the plasmid no longer produces tumors, but the transposon, with its new DNA, can still be inserted into the host plant cell's chromosomes. A plant cell containing this DNA can then be grown in culture or induced to form a new, transgenic plant.

Whatever vector is effective, the problem of identifying those host cells which have actually taken up the recombinant DNA remains.

Reporter genes identify host cells containing recombinant DNA

Even when a population of host cells interacts with an appropriate vector, only a small proportion of the cells actually take up the vector. In addition, since the process of making recombinant DNA is far from perfect, only a few of the vectors that have moved into the host cells will actually contain the DNA sequence of interest. How can we select only the host cells that contain that sequence?

One common procedure uses *E. coli* bacteria as hosts and the pBR322 plasmid (see Figure 16.9A), which carries the genes for resistance to the antibiotics ampicillin and tetracycline, as a vector. When the plasmid is incubated with the restriction enzyme *Bam*HI, the enzyme encounters its recognition sequence, GGATCC, only once, at a site within the gene for tetracycline resistance. If foreign DNA is inserted at this restriction site, the presence of those "extra" base pairs within the tetracycline resistance gene inactivates it. So plasmids containing the inserted DNA will carry an intact gene for ampicillin resistance, but not for tetracycline resistance (**Figure 16.10**). This difference is the key to the selection of the host bacteria that contain the recombinant plasmid.

The cutting and insertion process results in three types of DNA, all of which can be taken up by host bacteria:

- The recombinant plasmid—the one we want—turns out to be the rarest type of DNA. Its uptake confers resistance to ampicillin, but not to tetracycline, on host *E. coli*.

- More common are bacteria that take up plasmids that have sealed their own ends back together. These plasmids retain intact genes for resistance to both ampicillin and tetracycline.

- Even more common are bacteria that take up the foreign DNA sequence alone, without the plasmid; since it is not part of a replicon, it does not survive as the bacteria divide. These host cells remain susceptible to both antibiotics.

The vast majority (more than 99.9 percent) of host cells take up no DNA at all and remain susceptible to both antibiotics. So the unique drug-resistant phenotype of the cells with recombinant DNA (tetracycline-sensitive and ampicillin-resistant) marks them in a way that can be detected by simply adding ampicillin and/or tetracycline to the medium surrounding the cells.

In addition to genes for antibiotic resistance, several other reporter genes are used to detect recombinant DNA in host cells:

- Artificial vectors include restriction sites within the *lac* operon (see Figure 13.18). When the *lac* operon is inactivated by the insertion of foreign DNA, the vector no longer carries its function into the host cell.

- Green fluorescent protein, which normally occurs in the jellyfish *Aequopora victoriana*, does not require a substrate, but

16.10 Marking Recombinant DNA by Inactivating a Gene　Scientists can inactivate reporter genes within plasmid vectors to mark the host cells that have incorporated recombinant DNA. The host bacteria in this experiment could display any of the three phenotypes indicated in the table.

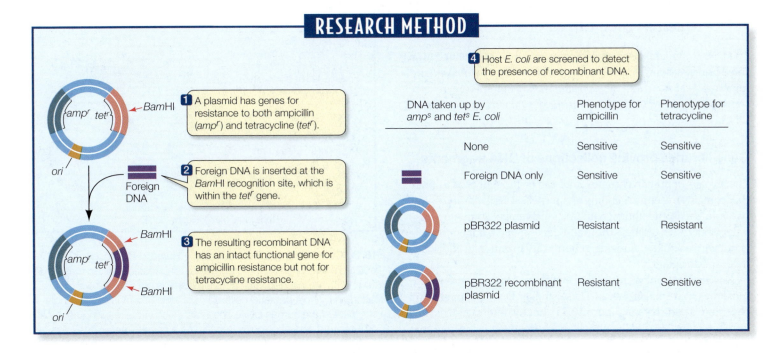

RESEARCH METHOD

1 A plasmid has genes for resistance to both ampicillin (*amp^r*) and tetracycline (*tet^r*).

2 Foreign DNA is inserted at the *Bam*HI recognition site, which is within the *tet^r* gene.

3 The resulting recombinant DNA has an intact functional gene for ampicillin resistance but not for tetracycline resistance.

4 Host *E. coli* are screened to detect the presence of recombinant DNA.

DNA taken up by *amp^s* and *tet^s E. coli*	Phenotype for ampicillin	Phenotype for tetracycline
None	Sensitive	Sensitive
Foreign DNA only	Sensitive	Sensitive
pBR322 plasmid	Resistant	Resistant
pBR322 recombinant plasmid	Resistant	Sensitive

emits visible light when exposed to ultraviolet light. It is now widely used as a reporter gene.

After exposure to the vector, the host cells are grown on a solid medium. If the concentration of cells dispersed on the solid medium is low, each cell will divide and grow into a distinct bacterial colony. The colonies that contain recombinant DNA can be identified by reporter gene expression and removed from the medium, then grown in large amounts in liquid culture. A quick examination of a plasmid can confirm whether the cells of the colony actually have the recombinant DNA. The power of bacterial transformation to amplify a gene is indicated by the fact that a 1-liter culture of bacteria harboring the human β-globin gene in the pBR322 plasmid has as many copies of that gene as the sum total of all the cells in a typical adult human being (10^{14}).

16.3 RECAP

Recombinant DNA can be cloned by using a vector to insert it into a suitable host cell. The vector often has genetic markers that give the host cell a phenotype by which recombinant cells can be identified.

- What are the characteristics of a good vector for introducing new DNA into a host cell? See p. 360

- Do you understand how cells harboring a vector that carries recombinant DNA can be selected? See p. 361 and Figure 16.10

Now that we have described how DNA can be cut, inserted into a vector, and transfected into host cells, and how host cells carrying recombinant DNA can be identified, let's pause briefly to consider where the genes or DNA fragments used in these procedures come from.

16.4 What Are the Sources of DNA Used in Cloning?

The DNA fragments used in cloning procedures are obtained from three principal sources: random fragments of chromosomes maintained as *gene libraries, complementary DNA* obtained by reverse transcription from mRNA, and *artificial synthesis* or *mutation* of DNA.

Gene libraries provide collections of DNA fragments

The 23 pairs of human chromosomes can be thought of as a library that contains the entire genome of our species. Each chromosome, or "volume" in the library, contains, on average, 80 million base pairs of DNA, encoding a thousand genes. Such a huge molecule is not very useful for studying genomic organization or for isolating a specific gene.

Researchers can use restriction enzymes to break human chromosomes into smaller pieces. These smaller DNA fragments still constitute a **gene library** (**Figure 16.11**), but the information is now in many more, smaller "volumes." Each fragment is inserted into

a vector, which is then taken up by a host cell. When bacteria are used as hosts, proliferation of one cell produces a colony of recombinant cells, each of which harbors many copies of the same fragment of human DNA.

When plasmids are used as vectors, about 200,000 separate fragments are required to make a library of the human genome. By using phage λ, which can carry four times as much DNA as a plasmid, the number of volumes can be reduced to about 50,000.

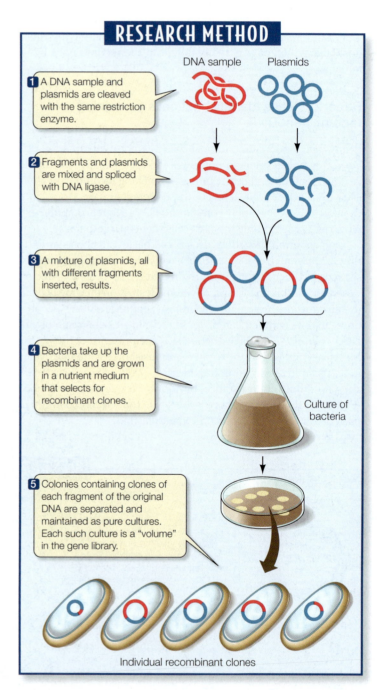

RESEARCH METHOD

DNA sample   Plasmids

1 A DNA sample and plasmids are cleaved with the same restriction enzyme.

2 Fragments and plasmids are mixed and spliced with DNA ligase.

3 A mixture of plasmids, all with different fragments inserted, results.

4 Bacteria take up the plasmids and are grown in a nutrient medium that selects for recombinant clones.

Culture of bacteria

5 Colonies containing clones of each fragment of the original DNA are separated and maintained as pure cultures. Each such culture is a "volume" in the gene library.

Individual recombinant clones

16.11 Constructing a Gene Library Human chromosomal DNA is isolated and broken up into fragments using restriction enzymes. The fragments are inserted into vectors (plasmids are shown here) and taken up by host bacterial cells, each of which then harbors a single fragment of the human DNA. The information in the resulting bacterial cultures and sets of colonies constitutes a gene library.

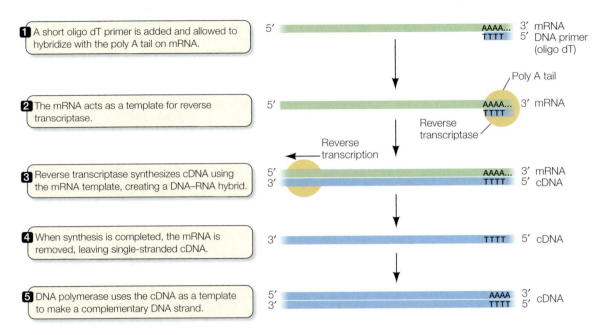

1 A short oligo dT primer is added and allowed to hybridize with the poly A tail on mRNA.

2 The mRNA acts as a template for reverse transcriptase.

3 Reverse transcriptase synthesizes cDNA using the mRNA template, creating a DNA–RNA hybrid.

4 When synthesis is completed, the mRNA is removed, leaving single-stranded cDNA.

5 DNA polymerase uses the cDNA as a template to make a complementary DNA strand.

5' ‖ AAAA... 3' mRNA
TTTT 5' DNA primer (oligo dT)

Poly A tail

5' ‖ AAAA... 3' mRNA
TTTT

Reverse transcriptase

Reverse transcription

5' ‖ AAAA... 3' mRNA
3' TTTT 5' cDNA

3' TTTT 5' cDNA

5' AAAA 3' cDNA
3' TTTT 5'

16.12 Synthesizing Complementary DNA Gene libraries that include only genes transcribed in a particular tissue at a particular time can be made from complementary DNA. cDNA synthesis is especially useful for identifying mRNAs that are present only in a few copies, and is often a starting point for gene cloning.

Although this seems like a large number, a single petri plate can hold up to 80,000 phage colonies, or *plaques*, and is easily screened for the presence of a particular DNA sequence by denaturing the phage DNA and applying a particular probe.

cDNA libraries are constructed from mRNA transcripts

A much smaller DNA library—one that includes only the genes transcribed in a particular tissue—can be made from **complementary DNA**, or **cDNA** (**Figure 16.12**). Recall that most eukaryotic mRNAs have a poly A tail—a string of adenine bases at their 3' end (see Figure 14.10). The first step in cDNA production is to extract mRNA from a tissue and allow its poly A tail to hybridize with a molecule called *oligo dT*, which consists of a string of thymine bases. The oligo dT serves as a primer, and the mRNA as a template, for the enzyme reverse transcriptase, which synthesizes DNA from RNA. In this way, a cDNA strand complementary to the mRNA is formed.

A collection of cDNAs from a particular tissue at a particular time in the life cycle of an organism is called a *cDNA library*. Messenger RNAs do not last long in the cytoplasm and are often present in small amounts, so a cDNA library is a "snapshot" that preserves the transcription pattern of the cell. Complementary DNA libraries have been invaluable in comparisons of gene expression in different tissues at different stages of development. Their use has shown, for example, that up to one-third of all the genes of an animal are expressed only during prenatal development. Complementary DNA is also a good starting point for the cloning of eukaryotic genes. It is especially useful for cloning genes expressed at low levels in only a few cell types.

DNA can be synthesized chemically in the laboratory

If we know the amino acid sequence of a protein, we can use the genetic code to figure out what DNA sequence codes for each amino acid and assemble the corresponding fragments of DNA.

Such artificial DNA synthesis is now fully automated, and a special service laboratory can make short to medium-length sequences overnight for any number of investigators.

Determining the base sequence of the gene that codes for a given protein is just one step in the design of a synthetic gene. Other sequences must be added, such as flanking sequences for transcription initiation, termination, and regulation and start and stop codons for translation initiation and termination. Of course, these noncoding DNA sequences must be the ones actually recognized by the host cell if the synthetic gene is to be transcribed. It does no good to have a prokaryotic promoter sequence near a gene if that gene is to be inserted into a yeast cell for expression. Appropriate selection of the codon for a given amino acid is another important consideration: many amino acids are encoded by more than one codon (see Figure 12.6), and host organisms vary in their use of synonymous codons.

DNA mutations can be created in the laboratory

Mutations that occur in nature have been important in demonstrating cause-and-effect relationships in biology. Recombinant DNA technology allows us to ask "What if?" questions without having to find mutations in nature. Because synthetic DNA can be made in any desired sequence, it can be manipulated to create specific mutations, the consequences of which can be observed when the mutant DNA is expressed by a host cell. These *mutagenesis techniques* have revealed many cause-and-effect relationships. For example, it was hypothesized that the signal sequence at the beginning of a secreted protein is essential to its passage through the membrane of the endoplasmic reticulum (see Section 12.5). A gene coding for such a protein, but with the codons for the signal sequence deleted, was synthesized. Sure enough, when this gene was expressed in yeast cells, the protein did not cross the ER membrane. When the signal sequence codons were added to an unrelated gene encoding a soluble cytoplasmic protein, that protein did cross the ER membrane.

As we have just seen, artificial mutations provide an excellent means of investigating questions about the role of a gene in cell function. Let's look at some ways in which the tools described in this section can be used to study the effects of a gene, beginning with two techniques for blocking a gene's function altogether.

16.5 What Other Tools Are Used to Manipulate DNA?

Section 11.5 describes DNA sequencing and the polymerase chain reaction, two important techniques arising from our understanding of DNA replication. In this section we will examine three additional techniques for manipulating DNA:

- *Knockout experiments*: the use of genetic recombination to create an inactive, or "knocked-out," gene

- *Gene silencing*: the creation of artificial antisense RNA and interference RNA that can block the translation of specific mRNAs

- *DNA chips*: microarrays that detect the presence of many different sequences simultaneously

Genes can be inactivated by homologous recombination

A technique called *homologous recombination* can be used to replace a gene inside a cell with an inactivated form of that gene to see what happens to a living organism lacking that gene. Such a manipulation is called a **knockout** experiment.

Mice are frequently used in knockout experiments (**Figure 16.13**). The normal allele of the mouse gene to be tested is inserted into a plasmid. Restriction enzymes are then used to insert a fragment containing a reporter gene into the middle of the normal gene. This addition of extra DNA plays havoc with the targeted gene's transcription and translation; a functional mRNA is seldom made from a gene whose sequence has been thus interrupted. Next, the plasmid is transfected into a stem cell from an early mouse embryo. (A **stem cell** is an undifferentiated cell that divides and differentiates into specialized cells.)

Because much of the targeted gene is still present in the plasmid (although in two separated regions), homologous sequence recognition takes place between the inactive allele on the plasmid and the active (normal) allele in the mouse genome. The sequence in the plasmid lines up with the homologous sequence in the

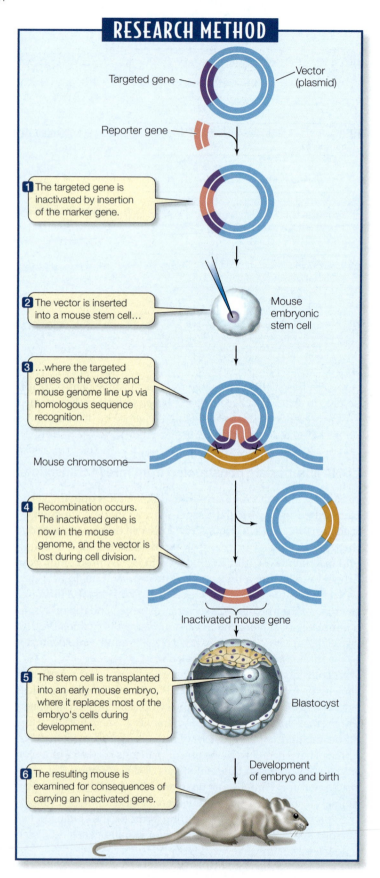

RESEARCH METHOD

Targeted gene — Vector (plasmid)

Reporter gene

1 The targeted gene is inactivated by insertion of the marker gene.

2 The vector is inserted into a mouse stem cell... — Mouse embryonic stem cell

3 ...where the targeted genes on the vector and mouse genome line up via homologous sequence recognition.

Mouse chromosome —

4 Recombination occurs. The inactivated gene is now in the mouse genome, and the vector is lost during cell division.

Inactivated mouse gene

5 The stem cell is transplanted into an early mouse embryo, where it replaces most of the embryo's cells during development. — Blastocyst

Development of embryo and birth

6 The resulting mouse is examined for consequences of carrying an inactivated gene.

16.13 Making a Knockout Mouse   Homologous recombination is used to replace a normal mouse gene with an inactivated copy of that gene, thus "knocking out" the gene. Discovering what happens to a mouse with an inactive gene tells us much about the normal role of that gene.

mouse chromosome, and sometimes recombination occurs, in which case the plasmid's inactive allele can be "swapped" with the functional allele in the host cell. Neither allele can be expressed, however; the allele inserted into the mouse chromosome is an interrupted fragment of the normal gene, and the gene inserted into the plasmid usually lacks its promoter. The reporter gene in the insert is, however, functional, and it is used to identify those stem cells carrying the inactivated gene.

A transfected stem cell is now transplanted into an early mouse embryo, and (through some clever tricks beyond the scope of this discussion) a knockout mouse carrying the inactivated gene in homozygous form is produced. The mutant mouse can then be observed for phenotypic changes providing clues to the function of the non-inactivated gene in the normal, wild-type animal. The knockout technique has been important in assessing the roles of certain genes during development.

Antisense RNA and interference RNA can prevent the expression of specific genes

The expression of a gene can also be blocked by stopping the translation of mRNA. As is often the case, this technique is an example of scientists imitating nature. As described in Section 14.5 (see Figure 14.22), gene expression is occasionally controlled by the production of a small RNA molecule (micro RNA) that is complementary to mRNA. Such a complementary molecule is called **antisense RNA** because it binds by base pairing to the "sense" bases on the mRNA that code for a protein. The resulting double-stranded RNA hybrid inhibits translation of the mRNA, and the hybrid tends to be broken down rapidly in the cytoplasm. Although the gene continues to be transcribed, translation does not take place. After determining the sequence of a gene and its mRNA in the laboratory, scientists can make a specific antisense RNA and add it to a cell to prevent translation of that gene's mRNA (**Figure 16.14**, left).

A related technique takes advantage of **interference RNA** (**RNAi**), a rare natural mechanism for inhibiting mRNA translation. In this case, a short (about 20 nucleotides) double-stranded RNA is unwound to single strands by a protein complex that guides this RNA to a complementary region on mRNA. The protein complex catalyzes the breakdown of the targeted mRNA.

Armed with this knowledge, scientists can synthesize a *small interfering RNA* (siRNA) to inhibit the translation of *any* known gene (Figure 16.14, right). Because these double-stranded siRNAs are more stable than antisense RNAs, RNAi is preferred over antisense RNA as a means of blocking RNA translation.

Antisense RNA and RNAi have been widely used to test cause-and-effect relationships. For example, when antisense RNA was used to block the synthesis of a protein essential for the growth of cancer cells, the cells reverted to a normal phenotype. Such gene silencing techniques offer great potential for the development of drugs to treat diseases that are the result of the inappropriate expression of specific genes.

> In macular degeneration, near-blindness results when blood vessels proliferate in the eye. The signaling molecule for vessel proliferation is a growth factor. An RNAi has been developed that targets this growth factor's mRNA. Without the signal to turn it on, vessel growth stops, and sometimes reverses.

DNA chips can reveal DNA mutations and RNA expression

The emerging science of genomics must deal with two major quantitative realities. First, there are a very large number of genes in eukaryotic genomes. Second, the pattern of gene expression in different tissues at different times is quite distinctive. For example, a cell from a skin cancer at its early stage may have a unique mRNA "fingerprint" that differs from that of both normal skin cells and the cells of a more advanced skin cancer.

To find such patterns, scientists could isolate the mRNA from a cell and test it by hybridization with each gene in the genome, one gene at a time. But it would be far simpler to do these hybridizations all in one step. This is possible with **DNA chip** technology, which provides large arrays of sequences for hybridization experiments.

The development of DNA chips was inspired by methods used for decades by the semiconductor industry. You may be familiar with the silicon microchip, in which an array of microscopic electric circuits is etched onto a tiny chip. In the same way, a series of DNA sequences can be attached to a glass slide in a precise order

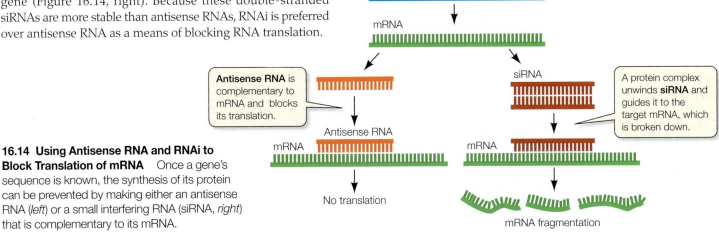

16.14 Using Antisense RNA and RNAi to Block Translation of mRNA Once a gene's sequence is known, the synthesis of its protein can be prevented by making either an antisense RNA (*left*) or a small interfering RNA (siRNA, *right*) that is complementary to its mRNA.

DNA

mRNA

Antisense RNA is complementary to mRNA and blocks its translation.

Antisense RNA

mRNA

No translation

siRNA

A protein complex unwinds **siRNA** and guides it to the target mRNA, which is broken down.

mRNA

mRNA fragmentation

RESEARCH METHOD

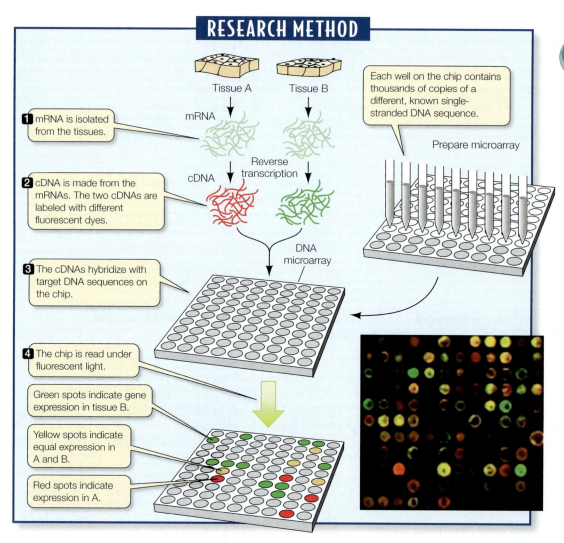

1 mRNA is isolated from the tissues.

2 cDNA is made from the mRNAs. The two cDNAs are labeled with different fluorescent dyes.

3 The cDNAs hybridize with target DNA sequences on the chip.

4 The chip is read under fluorescent light.

Green spots indicate gene expression in tissue B.

Yellow spots indicate equal expression in A and B.

Red spots indicate expression in A.

Tissue A Tissue B

mRNA

Reverse transcription

cDNA

DNA microarray

Each well on the chip contains thousands of copies of a different, known single-stranded DNA sequence.

Prepare microarray

16.15 DNA on a Chip Thousands of known DNA sequences can be attached to a glass slide in an organized grid pattern and hybridized with cDNA derived from mRNA from tissue samples to find out what genes are being expressed in the tissues.

(**Figure 16.15**). The slide is divided into a grid of microscopic spots, or "wells," each of which contains thousands of copies of a particular sequence up to 20 nucleotides long. A computer controls the addition of the sequences in a predetermined pattern. Each 20-base-long sequence can hybridize with only one genomic DNA (or cDNA or RNA) sequence, and thus is a unique identifier of a gene. Up to 60,000 different sequences can be placed on a single chip.

If cellular mRNA is to be analyzed, it is usually incubated with reverse transcriptase (RT) to make cDNA (see Figure 16.12), and the cDNA is amplified by the polymerase chain reaction (PCR) prior to hybridization (see Figure 11.23). This technique, called **RT-PCR**, ensures that mRNA sequences naturally present in only a few copies (or in a small sample, such as a cancer biopsy) will be numerous enough to form a signal. The amplified cDNA is tagged with a fluorescent dye and used to probe the DNA on the chip. Complementary DNA sequences that form hybrids with the DNA on the chip can be located by a sensitive scanner under fluorescent light.

A clinical use of DNA chips was developed by Laura van'T Veer and her colleagues at the Netherlands Cancer Institute. Most women with breast cancer are treated with surgery to remove the tumor, and then treated with radiation soon afterward to kill cancer cells that the surgeon may have missed. In some patients, however, overlooked breast cancer cells eventually form tumors, either

in the breast or elsewhere in the body. The challenge for physicians is to develop criteria to identify those patients and treat them aggressively with tumor-killing chemotherapy. Over the years, doctors have used such characteristics as original tumor size, tumor cell morphology, and tumor spread to lymph nodes (see Section 17.4) to predict whether a patient's breast cancer will recur. These methods have been only moderately successful.

Enter DNA chips. van'T Veer's group looked at the expression of about 1,000 genes in tumors from patients whose prognosis they already knew (the tumors were stored specimens). They found 70 genes whose expression differed dramatically between tumors from patients with poor prognoses (that is, their cancers recurred) and with good prognoses, developing what is called a "gene expression signature." This "signature on a chip" is useful in clinical decision making: patients with a good signature can avoid unnecessary chemotherapy, while those with a poor signature can receive aggressive treatment.

16.5 RECAP

Researchers can study the function of a gene by knocking out that gene in a living organism. Antisense RNA and RNAi silence genes by selectively blocking mRNA translation. DNA chips allow the simultaneous analysis of many different mRNA transcripts.

■ Do you understand how a gene can be "knocked out" in a living organism? See p. 364 and Figure 16.13

■ What is the difference between RNAi and antisense RNA? See p. 365

Now that we've seen how DNA can be fragmented, recombined, manipulated, and put back into living organisms, let's look at some examples of how these techniques can be put together to make useful products.

16.6 What Is Biotechnology?

Biotechnology is the use of living cells to produce materials useful to people, such as foods, medicines, and chemicals. People have been doing this for a very long time. For example, the use of yeasts to brew beer and wine dates back at least 8,000 years, and the use of bacterial cultures to make cheese and yogurt is a technique many centuries old. For a long time, however, people were not aware of the cellular bases of these biochemical transformations.

About 100 years ago, thanks largely to Louis Pasteur's work, it became clear that specific bacteria, yeasts, and other microbes could be used as biological converters to make certain products. Alexander Fleming's discovery that the mold *Penicillium* makes the antibiotic penicillin led to the large-scale commercial culture of microbes to produce antibiotics as well as other useful chemicals. Today, microbes are grown in vast quantities to make much of the industrial-grade alcohol, glycerol, butyric acid, and citric acid that are used by themselves or as starting materials in the manufacture of other products.

But the harvesting of proteins, including hormones and enzymes, was limited to the minuscule amounts that could be extracted from organisms that produce them naturally. Yields were low, and purification was difficult and costly. All this has changed with the advent of gene cloning. The ability to insert almost any gene into bacteria or yeasts, along with methods to induce the gene to make its product in large amounts and export it from the cells, has turned these microbes into versatile factories for important products. Key to this boom in biotechnology has been the development of specialized vectors that not only carry genes into cells, but make those cells express them at high levels.

Expression vectors can turn cells into protein factories

If a eukaryotic gene is inserted into a typical plasmid (see Figure 16.9) and transformed into *E. coli*, little, if any, of the product of the gene will be made by the host cell unless other key sequences are also included. The bacterial promoter for RNA polymerase binding, the terminator for transcription, and a special sequence on mRNA that is necessary for ribosome binding are all needed if the gene is to be expressed and its product synthesized in the bacterial cell.

To solve this kind of problem, scientists make **expression vectors** that have all the characteristics of typical vectors as well as the extra sequences needed for the foreign gene (also called a *transgene*) to be expressed in the host cell. For bacterial hosts, these additional sequences include the elements named above (**Figure 16.16**); for eukaryotes, they include the poly A addition sequence, transcription factor binding sites, and enhancers. Once these sequences are placed at the appropriate location in the vector, a gene

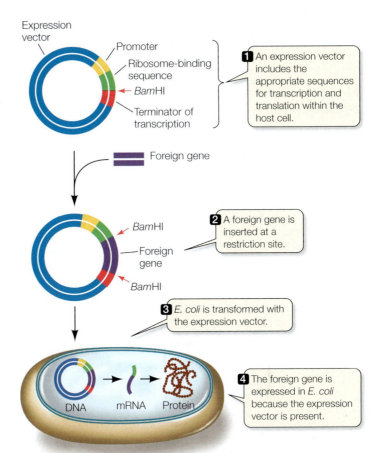

1 An expression vector includes the appropriate sequences for transcription and translation within the host cell.

2 A foreign gene is inserted at a restriction site.

3 *E. coli* is transformed with the expression vector.

4 The foreign gene is expressed in *E. coli* because the expression vector is present.

16.16 An Expression Vector Allows a Transgene to Be Expressed in a Host Cell A transformed eukaryotic gene may not be expressed in *E. coli* if it lacks the necessary bacterial sequences for promotion, termination, and ribosome binding. Expression vectors contain these additional sequences, enabling the eukaryotic protein to be synthesized in the prokaryotic cell.

transfected by that vector can be expressed in almost any kind of host cell.

An expression vector can be modified in various ways.

- An *inducible promoter*, which responds to a specific signal, can be made part of an expression vector. For example, a specific promoter that responds to hormonal stimulation can be used so that the transgene will transcribe its mRNA when the hormone is added. An enhancer that responds to hormonal stimulation can also be added so that transcription and protein synthesis will occur at high rates—a goal of obvious importance in the manufacture of an industrial product.

- A *tissue-specific promoter*, which is expressed only in a certain tissue at a certain time, can be used if localized expression is desired. For example, many seed proteins are expressed only in the plant embryo. Coupling a gene to a seed-specific promoter will allow the gene to be expressed only as a seed protein.

- *Signal sequences* can be added to the expression vector so that the product of the gene is directed to an appropriate destination. For example, when yeast or bacterial cells making a protein are to be maintained in a large vessel, it is economical to include a signal directing the protein to be secreted into the extracellular medium for easier recovery.

Medically useful proteins can be made by biotechnology

Many medically useful products are being made by biotechnology (**Table 16.1**), and hundreds more are in various stages of development. The manufacture of tissue plasminogen activator provides a good illustration of a medical application of biotechnology.

When a wound begins bleeding, a blood clot soon forms to stop the flow. Later, as the wound heals, the clot dissolves. How does the blood perform these conflicting functions at the right times? Mammalian blood contains an enzyme called plasmin that catalyzes the dissolution of the clotting proteins. But plasmin is not always active; if it were, a blood clot would dissolve as soon as it formed. Instead, plasmin is "stored" in the blood in an inactive form called plasminogen. The conversion of plasminogen to plasmin is activated by an enzyme, appropriately called *tissue plasminogen activator* (TPA), that is produced by cells lining the blood vessels:

$$\text{plasminogen} \xrightarrow{\text{TPA}} \text{plasmin}$$
$$\text{(inactive)} \qquad\qquad \text{(active)}$$

Heart attacks and strokes can be caused by blood clots that form in major blood vessels leading to the heart or the brain, respectively. During the 1970s, a bacterial enzyme called streptokinase was found to stimulate the dissolution of clots in some patients. Treatment with this enzyme saved lives, but its use had side effects. Streptokinase was a protein foreign to the body, so patients' immune systems reacted against it. More important, the drug sometimes prevented

clotting throughout the entire circulatory system, leading to an almost hemophilia-like condition in some patients.

The discovery of TPA and its isolation from human tissues led to the hope that this enzyme might be used to treat heart attack and stroke victims—that it would bind specifically to clots, and that it would not provoke an immune reaction. But the amounts

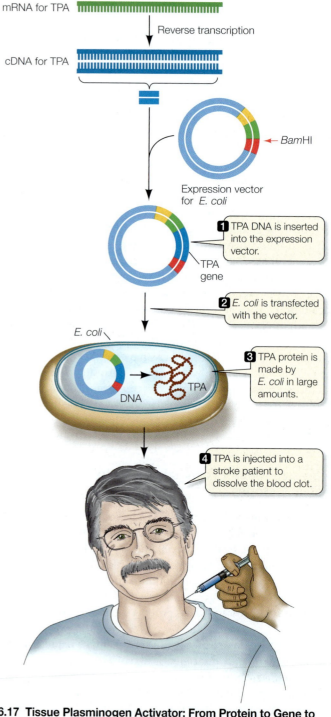

16.17 Tissue Plasminogen Activator: From Protein to Gene to Drug TPA is a naturally occurring human protein involved in dissolving blood clots. Its manufacture and use for treating patients suffering from blood clotting in the heart or brain—in other words, heart attacks or strokes—was made possible by recombinant DNA technology.

TABLE 16.1	
Some Medically Useful Products of Biotechnology	
PRODUCT	**USE**
Colony-stimulating factor	Stimulates production of white blood cells in patients with cancer and AIDS
Erythropoietin	Prevents anemia in patients undergoing kidney dialysis and cancer therapy
Factor VIII	Replaces clotting factor missing in patients with hemophilia A
Growth hormone	Replaces missing hormone in people of short stature
Insulin	Stimulates glucose uptake from blood in people with insulin-dependent (Type I) diabetes
Platelet-derived growth factor	Stimulates wound healing
Tissue plasminogen activator	Dissolves blood clots after heart attacks and strokes
Vaccine proteins: Hepatitis B, herpes, influenza, Lyme disease, meningitis, pertussis, etc.	Prevent and treat infectious diseases

of TPA that could be harvested from human tissues were tiny, certainly not enough to inject at the site of a clot in the emergency room.

Recombinant DNA technology solved this problem. TPA mRNA was isolated and used to make a cDNA copy, which was then inserted into an expression vector and transfected into *E. coli* (**Figure 16.17**). The transgenic bacteria made the protein in quantity, and it soon became available commercially. This drug has had considerable success in dissolving blood clots in people undergoing heart attacks and, especially, strokes.

Another way of making medically useful products in large amounts is what has come to be called **pharming**: the production of proteins in milk. A transgene coding for a useful protein product can be transfected into the eggs of a female domestic animal, such as a sheep, goat, or cow, next to the promoter for lactoglobulin, a protein made in large amounts in milk. The resulting transgenic animal secretes large amounts of the protein in its milk. These natural "bioreactors" can produce a large supply of the protein, which can be easily separated from the other components of the milk (**Figure 16.18**). Products being produced by pharming include blood clotting factors for treating hemophilia and antibodies for treating colon cancer.

DNA manipulation is changing agriculture

The cultivation of plants and husbanding of animals that constitute *agriculture* give us the world's oldest examples of biotechnology, dating back more than 8,000 years. Over the centuries, people have adapted crops and farm animals to their needs. Through cultivation and selective breeding (artificial selection) of these organisms, desirable characteristics, such as ease of cooking seeds or fat content of meat, have been imparted and improved. In addition, people have developed crops with desirable growth characteristics, such as high yield, a reliable ripening season, and resistance to diseases.

Until recently, the most common way to improve crop plants and farm animals was to select and breed varieties with desirable phenotypes that existed in nature through mutational variation. The advent of genetics a century ago was followed by its application to plant and animal breeding. A crop plant or animal with desirable genes could be identified, and through deliberate crosses, those genes could be introduced into a widely used variety of that organism.

Despite some spectacular successes, such as the breeding of "supercrops" of wheat, rice, and corn, such deliberate crossing remains a hit-or-miss affair. Many desirable traits are complex in their genetics, and it is hard to predict the results of a cross or to maintain a prized combination as a pure-breeding variety year after year. In sexual reproduction, combinations of unlinked genes are quickly separated in meiosis during gamete formation. More-over, traditional crop plant breeding takes a long time: many plants can reproduce only once or twice a year—a far cry from the rapid reproduction of bacteria.

Modern recombinant DNA technology has several advantages over traditional methods of breeding:

■ *The ability to target specific genes.* Allowing a breeder to select for specific genes makes the breeding process more precise and less likely to fail as a result of the incorporation of unforeseen genes.

■ *The ability to introduce any gene from any organism into a plant or animal species.* This ability, combined with mutagenesis techniques, vastly expands the range of possible new traits.

■ *The ability to generate new organisms quickly.* Manipulating cells in the laboratory and regenerating a whole plant by cloning is much faster than traditional breeding.

Consequently, recombinant DNA technology has found many applications in agriculture (**Table 16.2**). We will describe a few examples here to demonstrate the approaches that have been used.

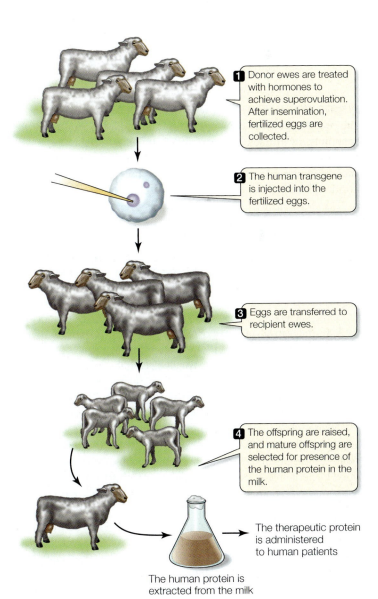

1 Donor ewes are treated with hormones to achieve superovulation. After insemination, fertilized eggs are collected.

2 The human transgene is injected into the fertilized eggs.

3 Eggs are transferred to recipient ewes.

4 The offspring are raised, and mature offspring are selected for presence of the human protein in the milk.

The therapeutic protein is administered to human patients

The human protein is extracted from the milk

16.18 Pharming An expression vector carrying a desired gene can be put into an animal egg and, using reproductive technology, the egg can be made to form an early embryo, which is implanted into a surrogate mother. The transgenic offspring produce the new protein in their milk.

TABLE 16.2

Agricultural Applications of Biotechnology under Development

PROBLEM	TECHNOLOGY/GENES
Improving the environmental adaptations of plants	Genes for drought tolerance, salt tolerance
Improving nutritional traits	High-lysine seeds
Improving crops after harvest	Delay of fruit ripening; sweeter vegetables
Using plants as bioreactors	Plastics, oils, and drugs produced in plants
Controlling crop pests	Herbicide tolerance; resistance to viruses, bacteria, fungi, insects

On the site of a nineteenth-century hat factory in Connecticut, where toxic mercury used in the curing process still contaminates the soil, 160 transgenic cottonwood trees have been planted to clean up the soil. The trees contain a bacterial gene called *MerA*, whose enzyme product converts toxic ionic mercury to volatile and less harmful metallic mercury.

PLANTS THAT MAKE THEIR OWN INSECTICIDES Humans are not the only species that consumes crop plants. Plants are subject to infections by viruses, bacteria, and fungi, but probably the most important crop pests are herbivorous insects. From the locusts of biblical (and modern) times to the cotton boll weevil, insects have continually eaten the crops people grow.

The development of insecticides has improved the situation somewhat, but insecticides have their own problems. Most, such as the organophosphates, are relatively nonspecific, killing not only pests in the field but beneficial insects in the broader ecosystem as well. Some even have toxic effects on other groups of organisms, including people. What's more, many insecticides persist in the environment for a long time.

Some bacteria have solved their own pest problem by producing proteins that can kill insects. For example, there are dozens of strains of *Bacillus thuringiensis*, each of which produces a protein toxic to the insect larvae that prey on it. The toxicity of this protein is 80,000 times that of the usual commercial insecticides. When a hapless larva eats the bacteria, the toxin becomes activated, binding specifically to the insect's gut to produce holes, and the insect starves to death.

Dried preparations of *B. thuringiensis* have been sold for decades as a safe insecticide that breaks down rapidly in the environment. But this *biodegradation* is their limitation, because it means that the dried bacteria must be applied repeatedly during the growing season. A more permanent approach would be to have the crop plants make the toxin themselves.

The toxin genes from different strains of *B. thuringiensis* have been isolated and cloned, and they have been extensively modified by the addition of plant promoters and terminators, plant poly A addition sequences, and plant regulatory elements. These modified genes have been introduced into plant cells in the labora-

tory using the Ti plasmid vector (see Figure 16.9C), and transgenic plants have been grown and tested for insect resistance in the field. Corn, cotton, soybeans, tomatoes, and other crops are being grown successfully with this added gene. Pesticide usage by farmers growing these transgenic crops is greatly reduced.

CROPS THAT ARE RESISTANT TO HERBICIDES Herbivorous insects are not the only threat to agriculture. Weeds may grow in fields and compete with crop plants for water and soil nutrients. Glyphosate (Roundup®) is a widely used and effective *herbicide*, or weed killer. It works only on plants, by inhibiting an enzyme system in the chloroplast that is involved in the synthesis of amino acids. Glyphosate is truly a "miracle herbicide" killing 76 of the world's 78 most prevalent weeds. Unfortunately, it also kills crop plants, so it is best used to rid a field of weeds before the crop plant starts to grow. But, as any gardener knows, when the crop begins to grow, the weeds reappear. If the crop were not affected by the herbicide, the herbicide could be applied to the field at any time and would kill only the weeds.

Scientists have used expression vectors to make plants that synthesize so much of the target enzyme for glyphosate that they are unaffected by it. The gene has been inserted into corn, cotton, and soybean plants, making them resistant to glyphosate. This technology expanded so rapidly in the late 1990s that half of the U.S. crops of these three plants now contain this high-expressing gene.

GRAINS WITH IMPROVED NUTRITIONAL CHARACTERISTICS To remain healthy, humans must eat foods (or supplements) containing an adequate amount of β-carotene, which the body converts into vitamin A (see Figure 3.21). About 400 million people worldwide suffer from vitamin A deficiency, which makes them susceptible to infections and blindness. One reason is that rice grains, which do not contain β-carotene, but only a precursor molecule for it, make up a large part of their diet. Other organisms, such as the bacterium *Erwinia* and daffodil plants, have enzymes that can convert the precursor into β-carotene. The genes for this biochemical pathway are present in the bacterial and daffodil genomes, but not in the rice genome.

Scientists isolated one of the genes for the β-carotene pathway from the bacterium and the other two from daffodil plants. They added promoter signals for expression in the developing rice grain, and then added each gene to rice plants by using the Ti plasmid vector from *Agrobacterium tumefaciens* (see Figure 16.9C). The resulting rice plants produce grains that look yellow because of their high β-carotene content (**Figure 16.19**). About 300 grams of this cooked rice a day can supply all the β-carotene a person needs. This new transgenic strain has been crossed with more locally adapted strains in the hope of improving the diets of millions of people.

CROPS THAT ADAPT TO THE ENVIRONMENT Throughout human history, agriculture has depended on ecological management—tailoring the environment to the needs of crop plants. A farm field is an unnatural, human-designed system, and when conditions in

(A)

(B)

16.19 Transgenic Rice Is Rich in β-Carotene   (A) The grains from a new transgenic strain of rice are yellow because they make the pigment β-carotene, which is converted by humans into vitamin A. (B) Normal rice does not contain β-carotene.

More important, this example illustrates what could become a fundamental shift in the relationship between crop plants and the environment. Instead of manipulating the environment to suit the plant, biotechnology may allow us to adapt the plant to the environment. As a result, some of the negative effects of agriculture, such as water pollution, could be lessened.

There is public concern about biotechnology

Concerns have been raised by the general public about the safety and wisdom of genetically modifying crops. These concerns are centered on three claims:

- Genetic manipulation is an unnatural interference with nature.
- Genetically altered foods are unsafe to eat.
- Genetically altered crop plants are dangerous to the environment.

Advocates of biotechnology tend to agree with the first claim. However, they point out that *all* crops are unnatural in the sense that they come from artificially bred plants growing in a manipulated environment (a farmer's field). Recombinant DNA technology just adds another level of sophistication to these techniques.

Biotechnology advocates counter the concern about whether genetically engineered crops are safe for human consumption by pointing out that only single genes are added and that these genes are specific for plant function. For example, the *B. thuringiensis* toxin produced by transgenic plants has no effect on people. However, as plant biotechnology moves from adding genes to improve plant growth to adding genes that affect human nutrition, such concerns will become more pressing.

that field become intolerable, the crops die. The Fertile Crescent, the region between the Tigris and Euphrates rivers in the Middle East where agriculture probably originated 10,000 years ago, is no longer fertile. It is now a desert, largely because the soil has a high salt concentration. Few plants can grow on salty soils, primarily because the environment is hypertonic to plant roots, and water leaves them, resulting in wilting.

Recently, a gene was discovered in *Arabidopsis thaliana* that allows this tiny weed to thrive in salty soils. The gene codes for a protein that transports sodium ions into the central vacuole. When this gene was added to tomato plants, they grew in soils four times as salty as the normal lethal level (**Figure 16.20**). This finding raises the prospect of growing useful crops on what were previously unproductive soils.

(A)

(B)

16.20 Salt-Tolerant Tomato Plants Transgenic plants containing a gene for salt tolerance thrive in salty soils (A), while plants without the transgene die (B).

The third concern, about environmental effects, centers on the possible "escape" of transgenes from crops to other species. If the gene for herbicide resistance, for example, were inadvertently transferred from a crop plant to a nearby weed, that weed could thrive in herbicide-treated areas. Or beneficial insects could eat plant materials containing *B. thuringiensis* toxin and die. Transgenic plants undergo extensive field testing before they are approved for use, but the complexity of the biological world makes it impossible to predict all potential environmental effects of transgenic organisms. Because of the potential benefits of agricultural biotechnology (see Table 16.2), scientists believe that it is wise to proceed with caution.

16.6 RECAP

Expression vectors maximize the expression of transgenes inserted into host cells. Biotechnology has been used to produce medicines and to develop transgenic plants with improved agricultural and nutritional characteristics.

■ Do you understand how expression vectors work? See p. 367 and Figure 16.16

■ What are some of the concerns that people might have about agricultural biotechnology? See pp. 371–372

CHAPTER SUMMARY

16.1 How are large DNA molecules analyzed?

Restriction enzymes, which are made by bacteria as a defense against viruses, bind to and cut DNA at specific **recognition sequences** (also called **restriction sites**). These enzymes can be used to produce small fragments of DNA for study, a technique known as **restriction digestion**. Review Figure 16.1

DNA fragments can be separated by size using **gel electrophoresis**. Review Figure 16.2, Web/CD Tutorial 16.1

Specific DNA sequences can be identified in a gel by probes with a complementary sequence in a procedure known as Southern blotting. Review Figure 16.3

DNA fingerprinting can distinguish between specific individuals, or reveal which individuals are most closely related, by detecting polymorphisms in their genes, particularly **single nucleotide polymorphisms (SNPs)** and **short tandem repeats (STRs)**. Review Figure 16.4

The DNA barcoding project uses sequencing of a single region of DNA to identify species.

16.2 What is recombinant DNA?

Recombinant DNA is formed by the combination of two DNA sequences from different sources. Review Figure 16.7

Many restriction enzymes make staggered cuts in the two strands of DNA, creating fragments that have **sticky ends** with unpaired bases.

DNA fragments with sticky ends can be used to create recombinant DNA if DNA molecules from different sources are cut with the same restriction enzyme and spliced together with DNA ligase. Review Figure 16.8

16.3 How are new genes inserted into cells?

One goal of recombinant DNA technology is to **clone** a particular gene, either for analysis or to produce its protein product in quantity.

Bacteria, yeasts, and cultured plant cells are commonly used as hosts for recombinant DNA. Host cells into which recombinant DNA is inserted, or **transformed**, are called **transgenic cells**.

To identify host cells that have taken up a foreign gene, the inserted sequence can be tagged with **reporter genes**, genetic markers with easily identifiable phenotypes.

Expression of the foreign gene in the host cell requires that it become part of a segment of DNA that contains a **replicon** (origin and terminus of replication).

Vectors are DNA sequences that can carry new DNA into host cells. Plasmids, viruses, and **yeast artificial chromosomes** are all used as vectors. Review Figure 16.9

16.4 What are the sources of DNA used in cloning?

DNA fragments from a genome can be inserted in host cells to create a **gene library**. Review Figure 16.11

The mRNAs produced in a certain tissue at a certain time can be extracted and used to create **complementary DNA (cDNA)** by reverse transcription. Review Figure 16.12

Synthetic DNA containing any desired sequence can be made and mutated in the laboratory.

16.5 What other tools are used to manipulate DNA?

Homologous recombination can be used to "**knock out**" a gene in a living organism. Review Figure 16.13

Gene silencing techniques can be used to inactivate the mRNA transcript of a gene, which may provide clues to the gene's function. Artificially created **antisense** RNA or **interference RNA (RNAi)** can be added to a cell to prevent translation of a specific mRNA. Review Figure 16.14

DNA chip technology permits the screening of mRNA for thousands of sequences at the same time. Review Figure 16.15, Web/CD Tutorial 16.2

In a technique called **RT-PCR**, mRNA from a specific tissue is made into cDNA by reverse transcription and then amplified. This cDNA can be used as a probe on a DNA chip to explore the pattern of gene expression in the tissue.

16.6 What is biotechnology?

Biotechnology is the use of living cells to produce materials useful to people. Recombinant DNA technology has resulted in a boom in biotechnology.

Expression vectors allow a transgene to be expressed in a host cell. Review Figure 16.16, Web/CD Activity 16.1

Recombinant DNA techniques have been used to make medically useful proteins. Review Figure 16.17

Pharming uses transgenic animals that produce useful products in their milk. Review Figure 16.18

Because recombinant DNA technology has several advantages over traditional agricultural biotechnology, it is being extensively applied to agriculture.

Transgenic crop plants can be adapted to their environment, instead of vice versa.

There is public concern about the application of recombinant DNA technology to food production.

SELF-QUIZ

1. Restriction enzymes
   a. play no role in bacteria.
   b. cleave DNA at highly specific recognition sequences.
   c. are inserted into bacteria by bacteriophage.
   d. are made only by eukaryotic cells.
   e. add methyl groups to specific DNA sequences.

2. When fragments of DNA of different sizes are placed in an electric field,
   a. the smaller pieces move most rapidly toward the positive pole.
   b. the larger pieces move most rapidly toward the positive pole.
   c. the smaller pieces move most rapidly toward the negative pole.
   d. the larger pieces move most rapidly toward the negative pole.
   e. the smaller and larger pieces move at the same rate.

3. From the list below, select the sequence of steps for inserting a piece of foreign DNA into a plasmid vector, introducing the plasmid into bacteria, and verifying that the plasmid and the foreign gene are present:
   (1) Transfect host cells.
   (2) Select for the lack of plasmid reporter gene 1 function.
   (3) Select for the plasmid reporter gene 2 function.
   (4) Digest vector and foreign DNA with a restriction enzyme, which inactivates plasmid reporter gene 1.
   (5) Ligate the digested plasmid together with the foreign DNA.
   a. 45132
   b. 45123
   c. 13425
   d. 32145
   e. 13254

4. Possession of which feature is not desirable in a vector for gene cloning?
   a. An origin of DNA replication
   b. Genetic markers for the presence of the vector
   c. Multiple recognition sequences for the restriction enzyme to be used
   d. One recognition sequence each for one to several different restriction enzymes
   e. Genes other than the target for transfection

5. RNA interference (RNAi) inhibits
   a. DNA replication.
   b. transcription of specific genes.
   c. recognition of the promoter by RNA polymerase.
   d. transcription of all genes.
   e. translation of specific mRNAs.

6. Complementary DNA (cDNA)
   a. is produced from ribonucleoside triphosphates.
   b. is produced by reverse transcription.
   c. is the "other strand" of single-stranded DNAs in a virus.
   d. requires no template for its synthesis.
   e. cannot be placed into a vector because it has the opposite base sequence of the vector DNA.

7. In a gene library of frog DNA in *E. coli* bacteria,
   a. all bacterial cells have the same sequences of frog DNA.
   b. all bacterial cells have different sequences of frog DNA.
   c. each bacterial cell has a random fragment of frog DNA.
   d. each bacterial cell has many fragments of frog DNA.
   e. the frog DNA is transcribed into mRNA in the bacterial cells.

8. An expression vector requires all of the following except
   a. genes for ribosomal RNA.
   b. a reporter gene.
   c. a promoter of transcription.
   d. an origin of DNA replication.
   e. restriction enzyme recognition sequences.

9. "Pharming" is a term that describes
   a. the use of animals in transgenic research.
   b. plants making genetically altered foods.
   c. synthesis of recombinant drugs by bacteria.
   d. large-scale production of cloned animals.
   e. synthesis of a drug by a transgenic animal in its milk.

10. In DNA fingerprinting,
    a. a positive identification can be made.
    b. a gel blot is all that is required.
    c. multiple restriction enzymes generate unique fragments.
    d. the polymerase chain reaction amplifies finger DNA.
    e. the variation in repeated sequences between two restriction sites is evaluated.

FOR DISCUSSION

1. In the recombinant DNA experiment in Figure 16.7, would you expect any *E. coli* to be doubly resistant (to both antibiotics) if the plasmids carrying antibiotic resistance genes were introduced into the cell at the same time but not recombined in the test tube before the experiment?

2. Compare PCR (see Section 11.5) and cloning as methods to amplify a gene. What are the requirements, benefits, and drawbacks of each method?

3. As specifically as you can, outline the steps you would take to (A) insert and express the gene for a new, nutritious seed protein in wheat, and (B) insert and express a gene for a human enzyme in sheep's milk.

4. Compare traditional genetic methods with molecular methods for producing genetically altered plants. For each case, describe (a) sources of new genes; (b) numbers of genes transferred; and (c) how long the process takes.

FOR INVESTIGATION

Green fluorescent protein (GFP) from a jellyfish can be incorporated into a vector as a reporter gene to signal the presence of the vector in a host cell (the cell glows under UV light). How would you alter the technique in Figure 16.10 to substitute GFP for one (or both) of the antibiotic resistance markers?

CHAPTER 17

Genome Sequencing, Molecular Biology, and Medicine

Genomes of the founders of Quebec

In 1535, the French explorer Jacques Cartier sailed up Canada's St. Lawrence River, arriving at a town called Stadacona by the native Iroquois already living there. When Samuel de Champlain arrived with 28 farmers 70 years later, the Algonquin then residing there called it "Quebecq," for "the place where the river narrows."

Over the next 150 years, 15,000 people came to join the first settlers. Half returned home because of the harsh conditions, and thousands more left for greener pastures in the south and west. About 2,600 stayed, had children, and formed the basis of the vibrant French-Canadian society of what is now Quebec Province of Canada. Cultural and religious traditions were very strong, and French Canadians tended to marry within their own society. Even today, about two-thirds of the current population of 6 million Quebecois is descended from that founding group of 2,600. For instance, the marriage of Pierre and Anne Tremblay in 1657 produced 12 children, and there are now 280,000 of their descendants living in Quebec!

French Canadians are a genomic gold mine. Because so many people in this population came from just a few ancestors, alleles that were present in the founding fathers and mothers are more common than in a population derived from a larger set of founders. In addition, because Quebec has a socialized health care system and a centralized database, getting medical records is straightforward.

Scientists are mapping the genomes of thousands of French Canadians, focusing on people whose four grandparents are all French Canadian. This "genome prospecting" involves looking for single nucleotide polymorphisms (SNPs). Scientists are trying to correlate the presence of these SNPs with diseases that may have been inherited from the founders. Such diseases would be indicated by a high frequency in the population along with the presence of a particular SNP. Linkage of genotype and phenotype may suggest genes involved in either the disease itself or susceptibility to it. So far, the researchers have evidence for several genes with some involvement in Crohn's disease, a serious inflammation of the intestines, as well as psoriasis, a skin disorder. Having identified a gene associated with a disease, the next steps include determining its function and seeking possible cures.

Until now, treatment of diseases has been—and largely continues to be—a "top-down" affair, with alle-

Samuel de Champlain In 1608, Champlain led a small group of French settlers to what is now Quebec.

Les Fêtes de la Nouvelle France   This family has dressed in period costume for the annual celebration of the founding of Quebec. Scientists are mapping the genomes of many French Canadians.

viating symptoms the primary goal. For some infectious diseases, knowledge of the microbe causing the disease and the availability of specific antibiotics has led to a cure. But for many complex diseases that still plague humans, such as cancer and heart disease, cures are elusive. For many diseases there can be more than one cause. Molecular medicine aims to identify causes of disease through better understanding of the interactions of genes, proteins, and the environment. The "bottom-up" approach taken in Quebec—identifying genes first, then causes—represents a new era in the understanding, prevention, and cure of disease.

IN THIS CHAPTER we first discuss the kinds of abnormal proteins that can result from an abnormal allele of a gene, whether that allele is inherited or has its origin in a mutation. We then consider how abnormal proteins can cause human genetic diseases, including cancer, and how the alleles that produce them can be detected. Then we'll see how this knowledge has been applied to the development of new treatments. The end of the chapter discusses the promise of molecular biology and medicine embodied in the sequencing of the human genome.

17.1 How Do Defective Proteins Lead to Diseases?

Biochemical genetics, the science that relates genotype (DNA) and phenotype (proteins), has been clearly described in viruses, prokaryotes, and eukaryotic model organisms, but it applies to humans, as well.

Genetic mutations may make proteins dysfunctional

Genetic mutations are often expressed phenotypically as proteins that differ from normal (wild-type) proteins. In principle, a mutation in any gene encoding a protein could result in a genetic disease. Abnormalities in enzymes, receptor proteins, transport proteins, structural proteins, and most of the other functional classes of proteins have all been implicated in genetic diseases.

DYSFUNCTIONAL ENZYMES   In 1934, the urine of two mentally retarded young siblings was found to contain phenylpyruvic acid, an unusual by-product of the metabolism of the amino acid phenylalanine. It was not until two decades later, however, that the complex clinical phenotype of the disease that afflicted these children, called *phenylketonuria* (PKU), was traced back to its molecular phenotype. The disease resulted from an abnormality in a single enzyme, phenylalanine hydroxylase (**Figure 17.1**). This enzyme normally catalyzes the conversion of dietary phenylalanine to tyrosine, but it was not active in PKU patients' livers. Lack of this conversion led to excess phenylalanine in the blood and explained the accumulation of phenylpyruvic acid. Later, the amino acid sequences of phenylalanine hydroxylase in normal people were compared with those in individuals with PKU. Many people with PKU had tryptophan instead of arginine in position 408 of this long polypeptide chain of 451 amino acids.

The exact cause of the mental retardation in PKU remains elusive, although, as we will see later in this chapter, it can be prevented. We can, however, understand why most people with

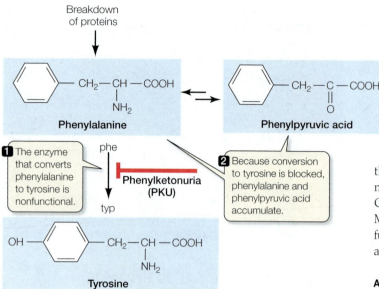

17.1 One Gene, One Enzyme Phenylketonuria is caused by an abnormality in a specific enzyme in the metabolic pathway that metabolizes the amino acid phenylalanine. Knowing the molecular causes of such single-gene, single-enzyme metabolic diseases can aid researchers in developing screening tests as well as treatments.

PKU have light skin and hair color. The pigment melanin, which is responsible for dark skin and hair, is made from tyrosine, which people with PKU cannot synthesize adequately.

Hundreds of human genetic diseases that result from enzyme abnormalities have been discovered, many of which lead to mental retardation and premature death. Most of these diseases are rare; PKU, for example, shows up in one newborn out of every 12,000. But these diseases are just the tip of the iceberg. Some mutations result in amino acid changes that have no obvious clinical effects. In fact, at least 30 percent of all proteins whose sequences are known show detectable amino acid differences among individuals. Thus polymorphism does not necessarily mean disease. There can be numerous normal alleles of a gene, each producing normally functioning forms of its protein.

ABNORMAL HEMOGLOBIN The first human genetic disease known to be caused by an amino acid abnormality was *sickle-cell disease*. This blood disorder most often afflicts people whose ancestors came from the tropics or from the Mediterranean. About 1 in 655 African-Americans are homozygous for the sickle allele and have the disease. The abnormal allele produces abnormal hemoglobin that results in sickle-shaped red blood cells (see Figure 12.18). These cells tend to block narrow blood capillaries, especially when the oxygen concentration of the blood is low. The result is tissue damage and eventually death by organ failure.

Recall that human hemoglobin is a protein with quaternary structure, containing four globin chains—two α chains and two β chains—as well as the pigment heme (see Figure 3.9). In sickle-cell disease, one of the 146 amino acids in the β-globin chain is abnormal: at position 6, the normal glutamic acid has been re-

placed by valine. This replacement changes the charge of the protein (glutamic acid is negatively charged and valine is neutral), causing it to form long, needle-like aggregates in the red blood cells. The result is *anemia*, a deficiency of normal red blood cells and an impaired ability of the blood to carry oxygen.

Because hemoglobin is easy to isolate and study, its variations in the human population have been extensively documented (**Figure 17.2**). Hundreds of single amino acid alterations in β-globin have been reported. For example, at the same position that is mutated in sickle-cell disease, the normal glutamic acid may be replaced by lysine, causing hemoglobin C disease. In this case, the resulting anemia is usually not severe. Many alterations of hemoglobin have no effect on the protein's function, and thus no clinical phenotype. That is fortunate, because about 5 percent of all humans are carriers for one of these variants.

ALTERED MEMBRANE PROTEINS Some of the most common human genetic diseases show their phenotypes as altered membrane receptors or transport proteins. About one person in 500 is born with *familial hypercholesterolemia* (FH), in which levels of cholesterol in the blood are several times higher than normal. The excess cholesterol can accumulate on the inner walls of blood vessels (a condition called *atherosclerosis*), leading to complete blockage if a blood clot forms. If a clot forms in a major vessel serving the heart, the heart becomes starved of oxygen, and a heart attack results. If a clot forms in the brain, the result is a stroke. People with FH often die of heart attacks before the age of 45.

Unlike PKU, which is characterized by the inability to convert phenylalanine to tyrosine, the problem in FH is not an inability to convert cholesterol to other products. People with FH have all the machinery needed to metabolize cholesterol. The problem is that they are unable to transport cholesterol into the liver and other cells that use it.

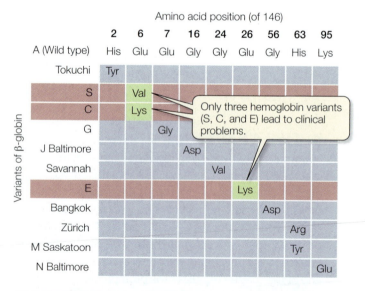

17.2 Hemoglobin Polymorphism Each of these mutant alleles changes a single amino acid in the 146–amino acid chain of β-globin. Only three of the hundreds of known variants of β-globin are known to lead to clinical abnormalities.

Cholesterol travels through the bloodstream in protein-containing particles called *lipoproteins*. One type of lipoprotein, low-density lipoprotein (LDL), carries cholesterol to the liver cells (**Figure 17.3A**). After binding to a specific receptor on the plasma membrane of a liver cell, LDL is taken up by endocytosis and delivers its cholesterol to the interior of the cell. People with FH lack a functional version of the receptor protein. Of the 840 amino acids that make up the receptor, only one may be abnormal, but that is enough to change its structure so that it cannot bind to LDL.

Among Caucasians, about one baby in 2,500 is born with *cystic fibrosis*. The clinical phenotype of this genetic disease is an unusually thick and dry mucus that lines surface tissues such as the airways of the respiratory system and the ducts of glands. In the respiratory passageways, this thick mucus obstructs the passage of air and also prevents the cilia on the surfaces of the epithelial cells from working efficiently to clear out the bacteria and fungal spores that we all take in with every breath. The results are recurrent infections as well as liver, pancreatic, and digestive failures, causing malnutrition and poor growth. People with cystic fibrosis often die in their thirties.

The cause of the thick mucus is a lack of a functional membrane protein, the chloride transporter (**Figure 17.3B**). In normal cells, this ion channel opens to release Cl⁻ to the outside of an epithelial cell. The resulting imbalance, in which there are more Cl⁻ ions outside the cell than inside, causes water to leave the cell by osmosis, resulting in a moist, thin mucus outside the cell. A single amino acid change in the channel protein prevents it from reaching the plasma membrane and functioning properly. As a consequence, water does not lubricate the mucus, and thick mucus and associated clinical problems are the result.

ALTERED STRUCTURAL PROTEINS In some genetic diseases, the defective protein is not an enzyme or receptor, but one involved in biological structure. *Duchenne muscular dystrophy* and *hemophilia* are cases in point.

About one boy in 3,000 is born with Duchenne muscular dystrophy. People with this disease show progressive muscle weakness and are wheelchair-bound by their teenage years. They usually die in their twenties, when the muscles that serve their respiratory system fail. Normal people have a protein in their skeletal muscles called *dystrophin*, which connects the actin filaments of the muscle cells to the extracellular matrix. People with Duchenne muscular dystrophy do not have a working copy of dystrophin, so their muscle cells become structurally disorganized, and the muscles stop working.

Hemophilia is the consequence of the absence of one of the blood clotting proteins. In normal people, inactive blood clotting proteins are always present in the blood and become active only at a wound. Some people with hemophilia risk death from even minor cuts, since they cannot stop bleeding.

We have thus far described disease-causing alterations in wild-type genes. Some abnormal proteins that result in disease, however, arise from other causes.

(A) Hypercholesterolemia

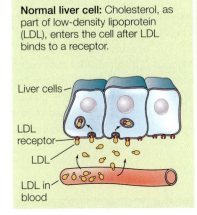

Normal liver cell: Cholesterol, as part of low-density lipoprotein (LDL), enters the cell after LDL binds to a receptor.

Familial hypercholesterolemia: Absence of a functional LDL receptor prevents cholesterol from entering the cells, and it accumulates in the blood.

Liver cells

LDL receptor

LDL

LDL in blood

(B) Cystic fibrosis

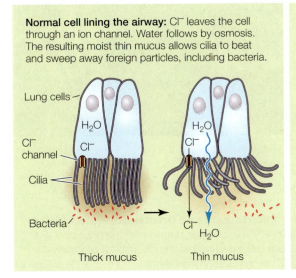

Normal cell lining the airway: Cl⁻ leaves the cell through an ion channel. Water follows by osmosis. The resulting moist thin mucus allows cilia to beat and sweep away foreign particles, including bacteria.

Cystic fibrosis: Lack of a Cl⁻ channel causes a thick, viscous mucus to form. Cilia cannot beat properly and remove bacteria; infections can easily take hold.

Lung cells

H₂O

Cl⁻ channel

Cl⁻

Cilia

Bacteria

Cl⁻
H₂O

Thick mucus   Thin mucus   Thick mucus   Thicker mucus

17.3 Genetic Diseases of Membrane Proteins The two left panels illustrate normal cell function, while the two right panels show the abnormalities caused by (A) familial hypercholesterolemia and (B) cystic fibrosis.

King George III of England suffered from porphyria, a metabolic disorder resulting in nonfunctioning enzymes in the porphyrin pathway that synthesizes heme, the pigment in hemoglobin. Its symptoms—intermittent bouts of high fever, pain, and delirium—may have led to George III's poor administrative decisions, which resulted in the American Revolution and England's loss of its North American colonies.

Prion diseases are disorders of protein conformation

Transmissible spongiform encephalopathies (TSEs) are degenerative brain diseases that occur in many mammals, including humans, in which the brain gradually develops holes, leaving it looking like a sponge. Scrapie, a TSE that causes affected sheep and goats to rub the wool off their bodies, has been known for 250 years. Chronic wasting disease is a TSE that causes elk and deer to become emaciated. In the 1980s, a TSE that appeared in cows in Britain was traced to the cows having eaten products from sheep that had scrapie. These cows would shake and rub their bodies against fences, and their staggering led farmers to dub them "mad cows."

It turns out that TSEs are caused by defective proteins, but the defects arise *not from mutations* in the genes that express them, but from errors in *conformation*—the folding of proteins into the proper three-dimensional shape. This was discovered in the 1990s after some people who had eaten beef from cows with bovine spongiform encephalopathy (BSE) got a human version of the disease (dubbed "mad cow disease" by the media). There was a time delay of several years between when the cows got the disease and its

emergence in humans; thus there is clearly a period of several years between the consumption of meat and the development of symptoms (**Figure 17.4**).

The disease *kuru* offers another example of how humans can acquire a TSE by consuming an infective agent. Kuru is a TSE that was observed in the 1950s among members of the Fore tribe of New Guinea who had consumed the brains of people who had died of it. When this ritual cannibalism stopped, so did the epidemic of kuru.

Researchers found that TSEs can be transmitted from one animal species to another via brain extracts from a diseased animal. At first, a virus was suspected. But when Tikva Alper at Hammersmith Hospital, London, treated infectious extracts with high doses of ultraviolet light to inactivate nucleic acids, they still caused TSEs. She proposed that the causative agent for TSEs was a protein, not a virus. Later, Stanley Prusiner at the University of California purified the protein responsible and showed it to be free of DNA or RNA. He called it a *proteinaceous infective particle*, or **prion**.

Normal brain cells contain a membrane protein called PrP^c. A protein with the same amino acid sequence is present in TSE-affected brain tissues, but that protein, called PrP^{sc}, has a different three-dimensional shape (**Figure 17.5**). Thus TSEs are not caused by a mutated gene (the primary structures of the two proteins are the same), but are somehow caused by an alteration in protein conformation. The altered three-dimensional structure of the protein has profound effects on its function in the cell. PrP^{sc} is insoluble, and it piles up as fibers in brain tissue, causing cell death.

How can the exposure of a normal cell to material containing PrP^{sc} result in a TSE? The abnormal PrP^{sc} protein seems to induce a conformational change in the normal PrP^c protein so that it, too, becomes abnormal, just as one rotten apple results in a whole barrel full of rotten apples. Just how the conversion occurs, and how it causes a TSE, are unclear.

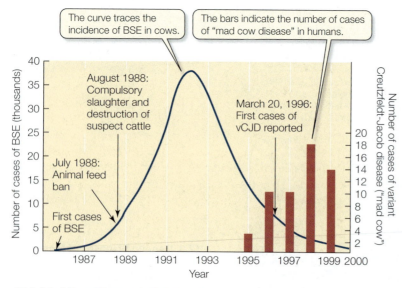

17.4 Mad Cow Disease in Britain There was a lag time of several years between the development of bovine spongiform encephalopathy (BSE) in cows and its equivalent in humans (vCJD).

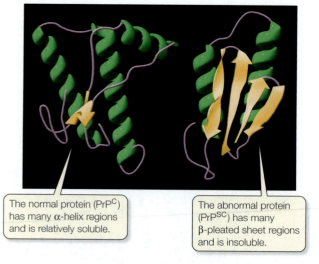

17.5 Prion Diseases are Disorders of Protein Conformation A normal membrane protein in brain cells (PrP^c, left) can be converted into the disease-causing form (PrP^{sc}, right), which has a different three-dimensional structure.

Prions appear to represent a highly unusual phenomenon in human disease. The vast majority of inherited diseases are understood in terms of proteins that are products of functional or dysfunctional genes. But the expression of these genes, like that of all genes, is influenced by the environment.

Most diseases are caused by both genes and environment

The human diseases for which clinical phenotypes can be traced to a single altered protein and its altered gene may number in the thousands, and in most cases they are dramatic evidence of a one-gene, one-polypeptide relationship. Taken together, these diseases have a frequency of about 1 percent in the total human population.

Far more common, however, are diseases that are **multifactorial**; that is, diseases that are caused by the interactions of many genes and proteins and the environment. Although we tend to call individuals either normal (wild-type) or abnormal (mutant), the totality of our genetic makeup is what determines, for example, who among us can eat a high-fat diet and not experience a heart attack and which of us will succumb to disease when exposed to infectious bacteria. Estimates suggest that up to 60 percent of all people are affected by diseases that are genetically influenced.

Human genetic diseases have several patterns of inheritance

As in any human genetic system, the alleles that cause genetic diseases may be inherited in a dominant or recessive pattern, and may be carried on autosomes or on sex chromosomes (see Section 10.4). In addition, some human diseases are caused by more extensive chromosomal abnormalities (see Section 9.5). Different inheritance patterns can be seen when genetic diseases are followed over several human generations.

AUTOSOMAL RECESSIVE PATTERN PKU, sickle-cell disease, and cystic fibrosis are all caused by autosomal recessive mutant alleles. Typically, both parents of an affected person are carriers (with a normal phenotype and a heterozygous genotype). Each time they conceive a child, they have a 25 percent (one in four) chance of having an affected (homozygous) son or daughter.

In the cells of a person who is homozygous for a harmful autosomal recessive mutant allele, a nonfunctional, mutant version of the protein it encodes is made. Thus a biochemical pathway or important cell function is disrupted, and disease results. Heterozygotes, with one normal and one mutant allele, often have 50 percent of the normal level of functional protein. For example, people who are heterozygous for the PKU allele have half as many active molecules of phenylalanine hydroxylase in their liver cells as individuals who carry two normal alleles for this enzyme, but this 50 percent suffices for normal cellular function.

AUTOSOMAL DOMINANT PATTERN Familial hypercholesterolemia is caused by an abnormal autosomal dominant allele. In this case, the presence of only one mutant allele is enough to produce the clinical phenotype. In people who are heterozygous for familial hypercholesterolemia, having half the normal number of functional receptors for low-density lipoprotein on the surfaces of liver cells is simply not enough to clear cholesterol from the blood. In autosomal dominance, direct transmission from an affected parent to offspring is the rule.

X-LINKED RECESSIVE PATTERN Hemophilia is an X-linked recessive condition (see pp. 206–207); that is, the gene responsible is on the X chromosome. Thus a son who inherits a mutant allele on the X chromosome from his mother will have the disease, because his Y chromosome does not contain a normal allele. However, a daughter who inherits one mutant allele will be an unaffected heterozygous carrier, since she has two X chromosomes, and hence two alleles. Because, until recently, few males with these diseases lived to reproduce, the most common pattern of inheritance has been from carrier mother to son, and all rare X-linked diseases are much more common in males than in females.

CHROMOSOMAL ABNORMALITIES Chromosomal abnormalities also cause human diseases. Such abnormalities include a gain or loss of one or more chromosomes (aneuploidy; see Figure 9.20), loss of a piece of a chromosome (deletions), and the transfer of a piece of one chromosome to another chromosome (translocations). About one newborn in 200 is born with a chromosomal abnormality. While some of these abnormalities are inherited, many are the result of meiotic events such as nondisjunction.

One common cause of mental retardation is *fragile-X syndrome* (**Figure 17.6**). About one male in 1,500 and one female in 2,000 is affected. These people have a constriction near the tip of the X chromosome that tends to break during preparation for microscopy, giving the syndrome its name. Although the basic pattern of inheritance is that of an X-linked recessive trait, there are departures from this pattern. Not all people with the fragile-X chromosomal abnormality are mentally retarded, as we will see in Section 17.2.

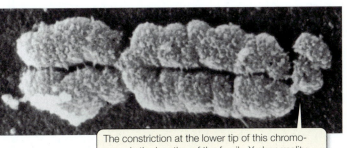

The constriction at the lower tip of this chromosome is the location of the fragile-X abnormality.

17.6 A Fragile-X Chromosome at Metaphase The chromosomal abnormality associated with fragile-X syndrome shows up under the microscope as a constriction in the chromosome.

Many genetic mutations are expressed as non-functional enzymes, structural proteins, or membrane proteins. Human genetic diseases may be inherited in dominant, recessive, or X-linked patterns.

- Can you describe an example of an abnormal protein in humans that results from a genetic mutation ? See pp. 376–377

- How is the brain cell membrane protein PrP^c related to diseases caused by prions? See p. 378

- How are autosomal recessive and autosomal dominant human genetic diseases distinguished? See p. 379

What kinds of changes in DNA result in nonfunctional proteins? An important task of molecular medicine is to find and identify such genetic changes.

17.2 What Kinds of DNA Changes Lead to Diseases?

The isolation and description of human mutations has proceeded rapidly since the modern techniques described in Chapter 16 were developed. When the protein phenotype is known, as in the case of abnormal hemoglobins, cloning the gene responsible has been straightforward, although time-consuming. In other cases, such as Duchenne muscular dystrophy, a chromosome deletion associated with the disease in a patient has pointed the way to the missing gene. In still other cases, such as cystic fibrosis, only a subtle molecular marker was available to lead investigators to the defective gene. In both of the latter examples, the primary phenotype—the defective protein—was unknown; only when the gene was isolated was the protein found.

One way to identify a gene is to start with its protein

The primary phenotype for sickle-cell disease was described in the 1950s as a single amino acid change in β-globin. On the basis of the clinical picture of sickled red blood cells, β-globin was certainly the right protein to examine. By the 1970s, researchers were able to isolate β-globin mRNA from immature red blood cells, which transcribe the globins as their major gene product. A cDNA copy of this mRNA was made and used to probe a human gene library to find the β-globin gene (**Figure 17.7A**). DNA sequencing was

17.7 Two Strategies for Isolating Human Genes (A) Starting with the normal β-globin protein, researchers were able to isolate the mRNA from which it was translated. Once cDNA was made by reverse transcription from the isolated mRNA, it could be used to probe a gene library to isolate the gene. (B) Researchers found a chromosome deletion in some patients affected with Duchenne muscular dystrophy. They then compared the affected persons' chromosomes with normal chromosomes to locate the DNA sequence that was missing.

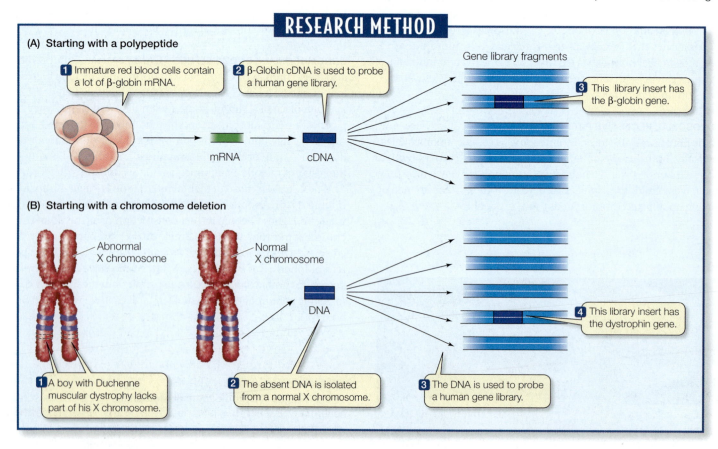

RESEARCH METHOD

(A) Starting with a polypeptide

1 Immature red blood cells contain a lot of β-globin mRNA.

2 β-Globin cDNA is used to probe a human gene library.

mRNA → cDNA

Gene library fragments

3 This library insert has the β-globin gene.

(B) Starting with a chromosome deletion

Abnormal X chromosome

Normal X chromosome

DNA

4 This library insert has the dystrophin gene.

1 A boy with Duchenne muscular dystrophy lacks part of his X chromosome.

2 The absent DNA is isolated from a normal X chromosome.

3 The DNA is used to probe a human gene library.

then used to compare the normal gene with the gene from people with sickle-cell disease. As previously described, it was found that a point mutation had changed only one base pair in the entire β-globin gene.

Chromosome deletions can lead to gene and then protein isolation

The inheritance pattern of Duchenne muscular dystrophy is consistent with an X-linked recessive trait. But until the late 1980s, neither the abnormal protein involved nor the gene encoding it had been described. This failure was not from lack of effort: almost every known muscle protein had been tested without success. Then several boys with the disease were found to have a small deletion in their X chromosome. Comparison of the affected X chromosomes with normal ones made possible the isolation of the gene that was missing in the boys (**Figure 17.7B**).

Genetic markers can point the way to important genes

In cases in which no candidate protein nor chromosome deletion has been identified, a technique called **positional cloning** has been invaluable. To understand this method, imagine an astronaut looking down from space, trying to find her son on a park bench on Chicago's North Shore. The astronaut first picks out reference points—landmarks that will lead her to the park. She recognizes the shape of North America, then moves to Lake Michigan, the Sears Tower, and so on. Once she has zeroed in on the North Shore park, she can use advanced optical instruments to find her son.

The reference points for positional cloning are genetic markers. These markers can be located anywhere in the DNA; the only requirement is that they be polymorphic (have more than one allele).

RFLPs As Section 16.1 describes, restriction enzymes cut DNA molecules at specific recognition sequences. On a particular human chromosome, a given restriction enzyme may make hundreds of cuts, producing many DNA fragments. The enzyme *Eco*RI, for example, cuts DNA at

$$5' \dots \text{GAATTC} \dots 3'$$

Suppose this recognition sequence exists in a certain stretch of human chromosome 7. The restriction enzyme will cut this stretch once and make two fragments of DNA. Now suppose that, in some people, this sequence is mutated as follows:

$$5' \dots \text{GAGTTC} \dots 3'$$

This sequence will not be recognized by the restriction enzyme; thus it will remain intact and yield one larger fragment of DNA.

Differences in the DNA sequence of designated segments of homologous chromosomes are called **restriction fragment length polymorphisms**, or **RFLPs** (**Figure 17.8**). They can be easily seen as bands on an electrophoresis gel. A RFLP band pattern is inherited in a Mendelian fashion and can be followed through a pedigree. Thousands of such markers have been described for the human genome.

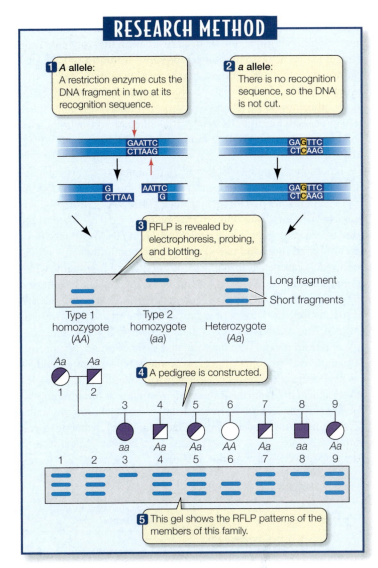

RESEARCH METHOD

1 *A* **allele:** A restriction enzyme cuts the DNA fragment in two at its recognition sequence.

2 *a* **allele:** There is no recognition sequence, so the DNA is not cut.

3 RFLP is revealed by electrophoresis, probing, and blotting.

Long fragment
Short fragments

Type 1 homozygote (*AA*)   Type 2 homozygote (*aa*)   Heterozygote (*Aa*)

4 A pedigree is constructed.

5 This gel shows the RFLP patterns of the members of this family.

17.8 RFLP Mapping Restriction fragment length polymorphisms are differences in DNA sequences that serve as genetic markers. Thousands of such markers have been described for the human genome.

SNPs As noted in Section 16.1, single nucleotide polymorphisms (SNPs) are widespread in the eukaryotic genome. SNP maps indicate that there is roughly one SNP for every 1,330 base pairs in the human genome. SNPs can be detected by direct sequence comparisons or by special chemical methods such as mass spectrometry (see Section 17.6). They differ from RFLPs in that they do not necessarily alter the recognition sequence for a restriction enzyme.

Genetic markers such as RFLPs and SNPs can be used as landmarks to find genes of interest if the genes, too, are polymorphic. The key to this method is the well-known observation that if two genes are located near each other on the same chromosome, they are usually passed on together from parent to offspring. The same holds true for any pair of genetic markers.

To narrow down the location of a gene, a scientist must find a marker and a gene that are *always inherited together*. To do this, family medical histories are taken and pedigrees are constructed. If a genetic marker and a genetic disease are inherited together, then they must be near each other on the same chromosome. This assumption is the basis of the genome prospecting in Quebec that we described at the beginning of this chapter. Unfortunately, "near each other" might be as much as several million base pairs apart. The process of locating the gene is thus similar to the astronaut focusing on Chicago: the first landmarks lead to only an approximate location.

How can the gene be isolated? Many methods are available for narrowing the search. For example, the neighborhood around the RFLP can be screened for further RFLPs using other restriction enzymes. With luck, one of them may be more closely linked to the disease-causing gene. Once a relatively short DNA sequence (several hundred thousand bases) thought to contain the gene is pinpointed, it can be cut into fragments, and those fragments, when denatured, can be used to probe mRNA expressed in cells affected by the disease. If one of the fragments hybridizes with the mRNA, it means that the fragment is part of a gene that is expressed as mRNA. The candidate gene is then sequenced from normal people and from people who have the disease in question. If appropriate mutations are found, the gene of interest has been isolated.

The isolation of genes responsible for genetic diseases has led to spectacular advances in the understanding of human biology. Before the genes, and then the proteins, for Duchenne muscular dystrophy and for cystic fibrosis were isolated, dystrophin and

the chloride transporter had never been described. Thus the identification of mutant genes has opened up new vistas in our understanding of how the human body works.

Disease-causing mutations may involve any number of base pairs

Disease-causing mutations may involve a single base pair (as we saw in the case of hemophilia), a long stretch of DNA, multiple segments of DNA, or even entire chromosomes. Some mutations have no effect at all, as we saw in Chapter 12, while others have many deleterious effects.

DNA sequencing has revealed that mutations occur most often at certain base pairs. These "hot spots" are often located where cytosine has been methylated to 5-methylcytosine (see Section 14.4). Either spontaneously or with chemical prodding, *un*methylated cytosine can lose its amino group and form uracil (**Figure 17.9A**). This type of error, however, is usually detected by the cell and repaired. A DNA repair mechanism recognizes this uracil as inappropriate for DNA (after all, uracil occurs only in RNA) and replaces the uracil with cytosine.

When 5-methylcytosine loses its amino group, however, the product is thymine, a natural base for DNA. The repair mechanism that recognized the inappropriateness of uracil ignores this thymine (**Figure 17.9B**). The mismatch repair mechanism, however, recognizes that GT is a mismatched pair (which should be GC). It cannot tell, though, which base was incorrectly inserted into the sequence. So, half of the time, it matches a new C to the G, but the other half of the time, it matches a new A to the T, resulting in a mutation.

Larger mutations may involve many base pairs of DNA. For example, some deletions in the X chromosome that result in Duchenne muscular dystrophy cover only part of the dystrophin gene, leading to an incomplete protein and a mild form of the disease. In other cases, however, deletions span the entire sequence of the gene, so that the protein is missing entirely from muscle, resulting in the severe form of the disease. In yet other cases, deletions involve millions of base pairs and cover not only the dystrophin gene but adjacent genes as well; the result may be several diseases simultaneously.

Expanding triplet repeats demonstrate the fragility of some human genes

About one-fifth of all males that have the fragile-X chromosomal abnormality are phenotypically normal, as are most of their daughters. But many of those daughters' sons are mentally retarded. In

(A)

When cytosine loses its amino group, uracil is formed.

A DNA repair system removes this abnormal base and replaces it with cytosine.

GGATCACTC / CCTAGTGAG → GGAT␣ACTC / CCTA␣TGAG → GGATCACTC / CCTAGTGAG

Cytosine   Uracil   Cytosine

(B)

When 5-methylcytosine loses its amino group, thymine results. Since thymine is a normal DNA base, it is not removed.

Methyl group

GGATCACTC / CCTAGTGAG → GGAT␣ACTC / CCTA␣TGAG

5-Methylcytosine   Thymine

Replication

50% → GGATTACTC / CCTAATGAG

When DNA replicates, half the daughter DNA is mutant and half is normal.

50% → GGATCACTC / CCTAGTGAG

17.9 5-Methylcytosine in DNA Is a "Hot Spot" for Mutations (A) Cytosine can lose an amino group either spontaneously or because of exposure to certain chemical mutagens. Such mutations are usually repaired. (B) If the cytosine has been methylated to 5-methylcytosine, however, the mutation is unlikely to be repaired and a C-G base pair is replaced with a T-A pair.

a family in which the fragile-X syndrome appears, later generations tend to show earlier onset and more severe symptoms of the disease. It is almost as if the abnormal allele itself is changing—and getting worse. And that's exactly what is happening.

The gene responsible for fragile-X syndrome (*FMR1*) contains a repeated triplet, CGG, at a certain point in the promoter region. In normal people, this triplet is repeated 6 to 54 times (the average is 29). In mentally retarded people with fragile-X syndrome, the CGG sequence is repeated 200 to 2,000 times.

Males carrying a moderate number of repeats (55–200) show no symptoms and are called *premutated*. These repeats become more numerous as the daughters of these men pass the chromosome on to their children (**Figure 17.10**). With more than 200 repeats, increased methylation of the cytosines in the CGG triplets is likely, accompanied by transcriptional inactivation of the *FMR1* gene. The normal role of the protein product of this gene is to bind to mRNAs involved in neuron function and regulate their translation at the ribosome. When the FMR1 protein is not made in adequate amounts, these mRNAs are not properly translated, and nerve cells die. Their loss often results in mental retardation.

This phenomenon of **expanding triplet repeats** has been found in over a dozen other diseases, such as myotonic dystrophy (involving repeated CTG triplets) and Huntington's disease (in which CAG is repeated). Such repeats, which may be found within a protein-coding region or outside it, appear to be present in many other genes without causing harm. How the repeats expand is not known; one theory is that DNA polymerase may slip after copying a repeat and then fall back to copy it again.

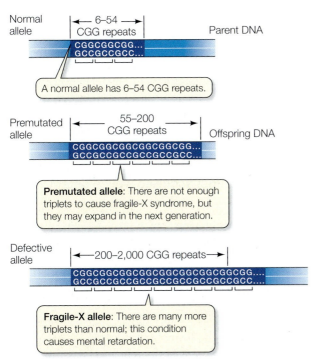

17.10 The CGG Repeats in the *FMR1* Gene Expand with Each Generation The genetic defect in fragile-X syndrome is caused by 200 or more repeats of the CGG triplet.

DNA changes in males and females can have different consequences

Just after fertilization in a mammalian egg, before the nuclei from the egg and from the sperm have fused, there are two haploid *pronuclei*—one from the sperm and the other from the egg. The two pronuclei can be distinguished from each other, and they can be carefully removed with a micropipette and placed in other eggs. It is possible to make mouse zygotes in the laboratory with two male or two female pronuclei, but these diploid cells do not go on to develop into mice. Invariably, if the two sets of chromosomes come from only one sex, development begins, but is quickly terminated. The same thing happens in those rare instances when this occurs in humans—for instance, if two sperm enter an enucleated egg. Again, a fetus never develops.

In addition to showing the obvious need for two sexes, these observations raise the possibility that the male and female genomes are *not functionally equivalent*. In fact, there are groups of genes that differ in their phenotypic effects depending on which parent they came from. This phenomenon is called **genomic imprinting**.

A dramatic example of genomic imprinting is the inheritance and phenotype of a certain small deletion on human chromosome 15:

- A deletion on the mother's chromosome 15 results in a thin child with a wide mouth and prominent jaw (*Angelman syndrome*).

- The same deletion on the father's chromosome 15 results in a short and obese child with small hands and feet (*Prader-Willi syndrome*).

The other alleles for this region on chromosome 15 are not deleted in these people; that is, they are heterozygotes, with one normal and one deleted allele. But the "normal" alleles must differ in males and females in these cases. They must be imprinted in very different ways in the two sexes to result in such different phenotypes. How this happens is not clear.

17.2 RECAP

Genes involved in disease have been identified by first detecting the abnormal DNA sequence and then the protein that the wild-type allele encodes. Linkage between genetic markers (such as RFLPs and SNPs) and a gene of interest is also useful in isolating human genes. Unusual features such as expanding triplet repeats have been detected in the human genome.

- How have gene libraries been instrumental in identifying abnormal gene sequences associated with diseases? See p. 380 and Figure 17.7

- How are genetic markers such as RFLPs and SNPs used to narrow the search for genes of interest, and why must these genes be polymorphic? See p. 381 and Figure 17.8

- Why do many genetic mutations involve GC base pairing? See p. 382

The determination of the precise molecular phenotypes and genotypes of various human genetic diseases has made it possible to diagnose these diseases even before symptoms first appear. Let's take a detailed look at some of these genetic screening techniques.

17.3 How Does Genetic Screening Detect Diseases?

Genetic screening is the use of a test to identify people who have, are predisposed to, or are carriers of a genetic disease. It can be done at many times of life and used for many purposes.

- *Prenatal screening* can identify an embryo or fetus with a disease so that medical intervention can be applied or decisions about continuing the pregnancy can be made.

- *Newborn babies* can be screened so that proper medical intervention can be initiated quickly for those babies who need it.

- *Asymptomatic people who have a relative with a genetic disease* can be screened to determine whether they are carriers of the disease or are likely to develop the disease themselves.

Genetic screening can be done at the level of either the phenotype or the genotype.

Screening for disease phenotypes can make use of protein expression

At the level of the phenotype, genetic screening involves examining a protein for abnormal structure or function. Since many proteins are enzymes, low enzyme activity is strongly suggestive of a mutation, as we saw in Section 17.1. Perhaps the best example of this kind of protein screening is a test for phenylketonuria that has made it possible to identify the disease in newborns so that treatments that prevent the development of mental retardation can be started.

Babies born with phenylketonuria have a normal phenotype because excess phenylalanine in their blood before birth diffuses across the placenta to the mother's circulatory system. Since the mother is almost always heterozygous, and therefore has adequate phenylalanine hydroxylase activity, her body metabolizes the excess phenylalanine from the fetus. After birth, however, the baby begins to consume protein-rich food (milk) and to break down some of its own proteins. Phenylalanine enters the baby's blood and accumulates. After a few days, the phenylalanine level in its blood may be ten times higher than normal. Within days, the developing brain is damaged, and untreated children with PKU become severely mentally retarded. If detected early, PKU can be treated with a special diet low in phenylalanine to avoid the brain damage that would otherwise result. Thus early detection is imperative.

A simple and inexpensive screening test for PKU was devised in 1963 by Robert Guthrie (**Figure 17.11**). This elegant method uses auxotrophic bacteria that require phenylalanine to grow well. If blood samples from newborns are added to a plate containing these bacteria, bacterial growth will be observed around those samples that are high in phenylalanine. The test can be automated so that a screening laboratory can process many samples in a day. This

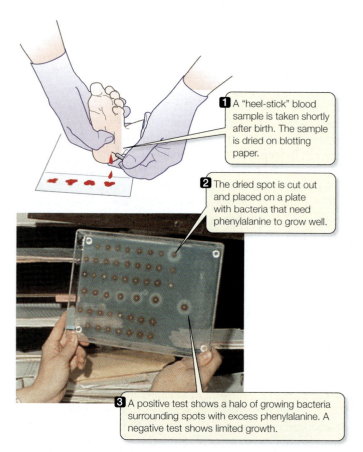

1 A "heel-stick" blood sample is taken shortly after birth. The sample is dried on blotting paper.

2 The dried spot is cut out and placed on a plate with bacteria that need phenylalanine to grow well.

3 A positive test shows a halo of growing bacteria surrounding spots with excess phenylalanine. A negative test shows limited growth.

17.11 Genetic Screening of Newborns for Phenylketonuria A simple test devised by Robert Guthrie in 1963 is used to screen newborns for phenylketonuria. Early detection means that the symptoms of the condition can be prevented by putting the baby on a therapeutic diet.

test is gradually being replaced by more accurate chemical tests that measure phenylalanine directly.

Genetic screening using newborn babies' blood is now done for up to 25 diseases, some rare (occurring in 1 infant in over 100,000) and others more common (occurring in 1 in 3,500). With early intervention, many of these infants can be successfully treated. So it is not surprising that newborn screening is legally mandatory in many countries, including the United States and Canada.

DNA testing is the most accurate way to detect abnormal genes

The blood level of phenylalanine is an indirect measure of phenylalanine hydroxylase activity in the liver. But how can we screen for genetic diseases that are not reflected in the blood? What if blood is difficult to obtain, as it is in a fetus? How are genetic abnormalities in heterozygotes, who express the normal protein at some level, identified?

DNA testing offers the most direct and accurate way of detecting an abnormal allele. With the molecular description of so many of the genetic mutations responsible for human diseases, it has become possible to examine directly any cell in the body at any time during the life span for mutations. These methods work best for diseases caused by only one or a few different mutations. However, with the amplification power of PCR, only one or a few cells are needed for testing.

Consider, for example, two parents who are both heterozygous for the cystic fibrosis allele, have had a child with the disease, and want a normal child. If treated with the appropriate hormones, the mother can be induced to "superovulate," releasing several eggs. One of the eggs can be injected with a single sperm from her husband and the resulting zygote allowed to divide to the 8-cell stage. If one of these embryonic cells is removed, it can be tested for the presence of the cystic fibrosis allele(s). If the test is negative, the remaining 7-cell embryo can be implanted in the mother's womb and go on to develop normally.

Such *preimplantation screening* is performed only rarely. More typical are analyses of fetal cells after implantation in the womb. Fetal cells can be analyzed at about the tenth week of pregnancy by *chorionic villus sampling* or during the thirteenth to seventeenth weeks by *amniocentesis*, two sampling methods described in Section 43.5. In either case, only a few fetal cells are required.

Newborns can also be screened for genetic mutations. The blood samples used for screening for PKU and other disorders contain enough of the baby's blood cells to permit extraction of the DNA, its amplification by PCR, and testing. Pilot studies of screening methods for sickle-cell disease and cystic fibrosis are under way, and other diseases will surely follow.

DNA testing is also widely used to test adults for heterozygosity. For example, a sister or female cousin of a boy with Duchenne muscular dystrophy can determine whether she is a carrier of the X chromosome deletion that results in the disease.

Of the numerous methods of DNA testing, two are the most widespread. We will describe their use to detect the mutation in the β-globin gene that results in sickle-cell disease.

SCREENING FOR ALLELE-SPECIFIC CLEAVAGE DIFFERENCES There is a difference between the normal and sickle alleles of the β-globin gene with respect to a restriction enzyme recognition sequence. Around codon position 6 in the normal gene is the sequence

$$5'\ldots \text{CCTGAGGAG}\ldots 3'$$

This sequence is recognized by the restriction enzyme *Mst*II, which will cleave DNA at

$$5'\ldots \text{CCTNAGGAG}\ldots 3'$$

where *N* is any base.

In the sickle allele, the DNA sequence is

$$5'\ldots \text{CCTGTGGAG}\ldots 3'$$

The point mutation at codon position 6 makes this sequence unrecognizable by *Mst*II. When *Mst*II fails to make the cut in the mutant allele, gel electrophoresis detects a larger DNA fragment (**Figure 17.12**).

This *allele-specific cleavage* method of DNA testing is the same as the use of RFLPs in positional cloning (see Figure 17.8). It works only if a restriction enzyme exists that can recognize the sequence of either the normal or the mutant allele.

SCREENING BY ALLELE-SPECIFIC OLIGONUCLEOTIDE HYBRIDIZATION
The *allele-specific oligonucleotide hybridization* method uses short artificial DNA strands called *oligonucleotides* that will hybridize either (in this example) with the denatured normal β-globin DNA sequence or with the sickle mutant sequence. Usually, an oligonucleotide probe of at least a dozen bases is needed to form a stable double helix with the target DNA. If the probe is radioactively or fluorescently labeled, hybridization can be readily detected (**Figure 17.13**). This method is easier and faster than allele-specific cleavage, and will work no matter what the sequence of the normal or mutant allele is.

17.12 DNA Testing by Allele-Specific Cleavage Allele-specific cleavage, a technique identical to RFLP analysis, can be used to detect mutations such as the one that causes sickle-cell disease.

RESEARCH METHOD

1 DNA from the normal β-globin allele has a recognition sequence for the restriction enzyme *Mst*II.

2 Normal β-globin DNA is cut into two fragments.

Normal β-globin allele
5′ — CCTNAGGAG — 3′
3′ — GGANTCCTA — 5′

Sickle allele
5′ — CCTGTGGAG — 3′
3′ — — 5′

Cut with *Mst*II

3 DNA from the sickle β-globin allele lacks an *Mst*II recognition sequence.

4 Sickle β-globin DNA is not cut, and a larger fragment results.

5 The fragments can be identified by gel electrophoresis on the basis of their sizes.

Normal   Sickle

RESEARCH METHOD

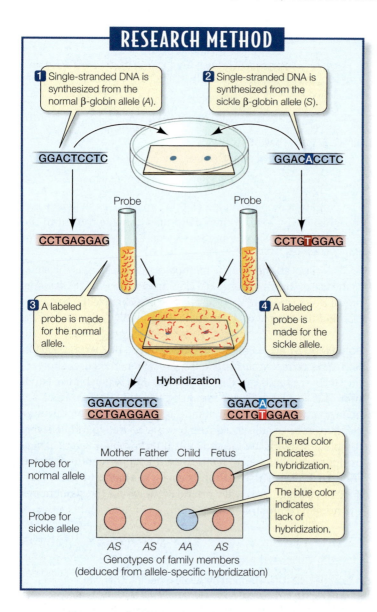

1. Single-stranded DNA is synthesized from the normal β-globin allele (*A*).

2. Single-stranded DNA is synthesized from the sickle β-globin allele (*S*).

GGACTCCTC

GGACACCTC

Probe

Probe

CCTGAGGAG

CCTGTGGAG

3. A labeled probe is made for the normal allele.

4. A labeled probe is made for the sickle allele.

Hybridization

GGACTCCTC
CCTGAGGAG

GGACACCTC
CCTGTGGAG

The red color indicates hybridization.

The blue color indicates lack of hybridization.

Probe for normal allele

Probe for sickle allele

Mother Father Child Fetus

AS   AS   AA   AS
Genotypes of family members
(deduced from allele-specific hybridization)

17.13 DNA Testing by Allele-Specific Oligonucleotide Hybridization Testing of this family reveals that three of them are heterozygous carriers of the sickle allele. The first child, however, has inherited two normal alleles and is neither affected by the disease nor a carrier.

17.3 RECAP

Genetic screening can identify people who have, are predisposed to, or are carriers of a genetic disease. It can be done at the level of the phenotype by identifying an abnormal protein or its function. It can also be done at the level of the genotype by direct testing of DNA.

- Can you describe how newborn babies are screened for phenylketonuria? See p. 384

- What is the advantage of screening for genetic mutations by allele-specific oligonucleotide hybridization relative to screening for allele-specific cleavage differences? See p. 385

We usually think of genetic diseases as inherited, but we'll now turn our attention to a genetic disease that most often affects somatic cells: cancer.

17.4 What Is Cancer?

Perhaps no malady affecting people in the industrialized world instills more fear than cancer. One in three Americans will have some form of cancer in their lifetime, and at present, one in four will die of it. With a million new cases and half a million deaths in the United States annually, cancer ranks second only to heart disease as a killer. Cancer was less common a century ago; then, as now in many regions of the world, people died of infectious diseases and did not live long enough to get cancer. Cancer tends to be a disease of the later years of life; children are much less frequently afflicted.

Since the U.S. government declared "war on cancer" in 1970, a tremendous amount of information on cancer cells—on their growth and spread and on their molecular changes—has been obtained. Perhaps the most remarkable discovery is that cancer is caused primarily by genetic changes. These changes are mostly mutations in the DNA of somatic cells that are propagated by mitosis.

Cancer cells differ from their normal counterparts

Cancer cells differ from the normal cells from which they originate in two major ways.

CANCER CELLS LOSE CONTROL OVER CELL DIVISION Most cells in the body divide only if they are exposed to extracellular influences, such as growth factors or hormones. Cancer cells do not respond to these controls, and instead divide more or less continuously, ultimately forming **tumors** (large masses of cells). By the time a physician can feel a tumor or see one on an X ray or CAT scan, it already contains millions of cells.

Benign tumors resemble the tissue they came from, grow slowly, and remain localized where they develop. A lipoma, for example, is a benign tumor of fat cells that may arise in the armpit and remain there. Benign tumors are not cancers, but they must be removed if they impinge on an important organ, such as the brain.

Malignant tumors, on the other hand, do not look like their parent tissue at all. A flat, specialized epithelial cell in the lung wall may turn into a relatively featureless, round, malignant lung cancer cell (**Figure 17.14**). Malignant cells often have irregular structures, such as variable nucleus sizes and shapes. Many malignant cells express the gene for telomerase and thus do not shorten the ends of their chromosomes after each DNA replication.

CANCER CELLS SPREAD TO OTHER TISSUES The second, and most fearsome, characteristic of cancer cells is their ability to invade surrounding tissues and spread to other parts of the body. This spreading, called **metastasis**, occurs in several stages. First, the cancer cells extend into the tissue that surrounds them by secreting digestive enzymes that disintegrate the surrounding cells and extracellular materials, working their way toward a blood vessel. Then, some

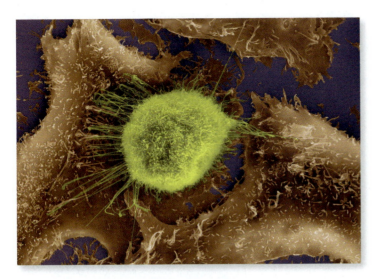

17.14 A Cancer Cell with Its Normal Neighbors This small-cell lung cancer cell (yellow-green) is quite different from the surrounding lung epithelial cells from which it came. This particular form of cancer is very lethal, with a 5-year survival rate of 10 percent. Most cases are caused by tobacco smoking.

TABLE 17.1

Human Cancers Known To Be Caused by Viruses

CANCER	ASSOCIATED VIRUS
Liver cancer	Hepatitis B virus
Lymphoma, nasopharyngeal cancer	Epstein–Barr virus
T cell leukemia	Human T cell leukemia virus (HTLV-I)
Anogenital cancers	Papillomavirus
Kaposi's sarcoma	Kaposi's sarcoma herpesvirus

of the cancer cells enter the bloodstream or the lymphatic system. The journey through these vessels is perilous, and few of the cancer cells survive—perhaps one in 10,000. When by chance a cancer cell arrives at an organ suitable for its further growth, it expresses cell surface proteins that allow it to bind to and invade the new host tissue. Finally, the tumor at the new site secretes chemical signals that cause blood vessels to grow to the tumor and supply it with oxygen and nutrients. This process is called *angiogenesis*.

Different forms of cancer affect different parts of the body. About 85 percent of all human tumors are *carcinomas*—cancers that arise in surface tissues such as the skin and the epithelial cells that line the organs. Lung cancer, breast cancer, colon cancer, and liver cancer are all carcinomas. *Sarcomas* are cancers of tissues such as bone, blood vessels, and muscle. *Leukemias* and *lymphomas* affect the cells that give rise to blood cells.

Some cancers are caused by viruses

In 1909, Peyton Rous, a young physician beginning his research career at the Rockefeller University in New York City, received a phone call from panicked chicken farmers on Long Island. Their chickens were coming down with a strange illness that caused their muscles to deteriorate before they died. The disease was spreading rapidly through the flocks: when one chicken in a coop got it, soon the others did, too. After some medical—or, more correctly, veterinary—detective work, Rous diagnosed the disease as a sarcoma—a muscle tumor—and the cause as a virus. This was a key discovery, because it showed that cancer could be a transmissible disease.

Unfortunately, Rous's discovery was not immediately followed up, and a search for human tumor viruses was initially fruitless. It was only when scientists returned to tumor virology in animals in the 1960s that the importance of Rous's work was fully appre-

ciated. He was awarded the Nobel prize in 1966, 57 years after his discovery! For a time in the 1960s, it was thought that much of human cancer came from viruses. But careful investigations have shown this not to be the case. About 15 percent of human cancers are virally induced. At least five types of human cancer are probably caused by viruses (**Table 17.1**).

Hepatitis B, a liver disease that affects people all over the world, is caused by the hepatitis B virus, which contaminates blood or is carried from mother to child during birth. The viral infection can be long-lasting and may flare up numerous times. The hepatitis B virus is associated with liver cancer, especially in Asia and Africa, where millions of people are infected. But it does not cause cancer by itself. Some gene mutations that are necessary for tumor formation occur in the infected cells of Asians and Africans, although apparently not in those of Europeans and North Americans.

An important group of virally induced cancers among North Americans and Europeans is the various anogenital cancers caused by papillomaviruses. The genital and anal warts that these viruses cause often develop into tumors. These viruses seem to be able to act on their own, not needing mutations in host cells for tumors to arise. Normally, the viral genome is circular, and it undergoes a lytic cycle, producing more viruses. Occasionally, the circle is broken and the virus integrates itself into the chromosome of a host cell in the uterine cervix, disrupting a gene that normally stimulates viral replication and blocks cell division and thus stimulating the cell cycle. Sexual transmission of these papillomaviruses is unfortunately widespread.

> Cancers caused by viruses can be prevented and treated with antiviral vaccines. Large vaccination programs in Asia are already reducing the incidence of liver cancer caused by hepatitis B. Recently, an effective vaccine was developed for the papillomaviruses that cause cervical cancer.

Most cancers are caused by genetic mutations

What causes the 85 percent of cancers that are not caused by viruses? Because most cancers develop in older people, it is reasonable to assume that one must live long enough for a series of events to occur. This assumption turns out to be correct, and the events are genetic mutations.

DNA can be damaged in many ways. As Section 12.6 describes, spontaneous mutations arise because of chemical changes in the nucleotides. In addition, certain mutagens, called **carcinogens**, can cause mutations that lead to cancer. Familiar carcinogens include the chemicals that are present in tobacco smoke and meat preservatives, ultraviolet light from the sun, and ionizing radiation from sources of radioactivity. Less familiar, but just as harmful, are thousands of chemicals that are naturally present in the foods people eat. According to one estimate, these "natural" carcinogens account for well over 80 percent of human exposure.

Both natural and synthetic carcinogens damage DNA, usually by causing changes from one base to another. In somatic cells that divide often, such as epithelial and bone marrow stem cells, there is less time for DNA repair mechanisms to work before replication occurs again. Therefore, such cells are especially susceptible to cancer.

Two kinds of genes are changed in many cancers

The changes in the control of cell division that lie at the heart of cancer can be likened to the controls of an automobile. To make a car move, two things must happen: the gas pedal must be pressed, and the brake must be released. In the human genome, some genes act as *oncogenes*, which "press the gas pedal" to stimulate cell division, and some act as *tumor suppressor genes*, which "put the brake on" to inhibit it.

ONCOGENES The first hint that **oncogenes** (from the Greek *onco*, "mass") were necessary for cells to become cancerous came with the identification of animal cancers induced by viruses. In many cases, these viruses bring a new gene into their host cell that stimulates cell division when it is expressed in the viral genome. It soon became apparent that the viral oncogenes had counterparts that are not usually transcribed in the genomes of host cells. So the search for genes that are damaged by carcinogens quickly zeroed in on these cellular oncogenes. Several dozen such genes were soon found.

Oncogenes are genes that have the capacity to stimulate cell division but are normally "turned off" in differentiated, nondividing cells. Many of them are involved in the pathways by which growth factors stimulate cell division (**Figure 17.15**). Some remarkable oncogenes control *apoptosis* (programmed cell death; see Section 9.6). Activation of these genes by mutation causes them to prevent apoptosis, allowing cells that normally die to continue dividing.

Some oncogenes can be activated by point mutations, others by chromosome changes such as translocations, and still others by gene amplification. Whatever the mechanism, the result is the same: the oncogene becomes activated, and the "gas pedal" for cell division is pressed.

TUMOR SUPPRESSOR GENES About 10 percent of all cancers are inherited. Often the inherited form of a cancer is clinically similar to a noninherited form that occurs later in life, called the *sporadic* form. However, the inherited form strikes much earlier in life, and it usually shows up as multiple tumors.

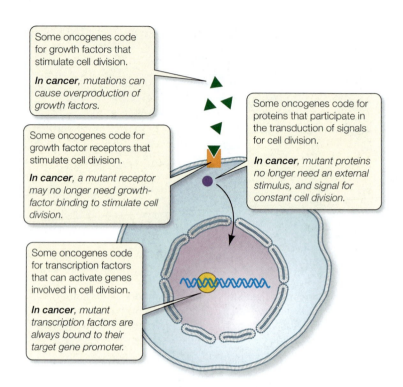

17.15 Oncogene Products Stimulate Cell Division Mutations can affect any of the several ways in which oncogenes can stimulate cell division, thus causing cancer.

In 1971, Alfred Knudson used these observations to predict that for a cancer to occur, a **tumor suppressor gene**, which normally acts as a "brake" on cell division, must be inactivated. But in contrast to oncogenes, in which one mutant allele is all that is needed for activation, the full inactivation of a tumor suppressor gene requires that both alleles be turned off, which requires two mutational events. It takes a long time for both alleles in a single cell to mutate and cause sporadic cancer. But people with inherited cancer are born with one mutant allele for a tumor suppressor gene, and need just one more mutational event for its full inactivation (**Figure 17.16**).

The isolation of various tumor suppressor genes has confirmed Knudson's "two-hit" hypothesis. Some of these genes are involved in inherited forms of rare childhood cancers such as retinoblastoma (a tumor of the eye) and Wilms' tumor of the kidney as well as in inherited breast and prostate cancers.

An inherited form of breast cancer demonstrates the effect of tumor suppressor genes. The 9 percent of women who inherit one mutant allele of the gene *BRCA1* have a 60 percent chance of having breast cancer by age 50 and an 82 percent chance of developing it by age 70. The comparable figures for women who inherit two normal alleles of the gene are 2 percent and 7 percent, respectively.

How do tumor suppressor genes act in the cell? Like the oncogenes, they are normally involved in cell division (**Figure 17.17**). Some regulate progress through the cell cycle (which we described in Section 9.2). The *Rb* gene, which was first described for its contribution to retinoblastoma, is active during the G1 phase. In its active form, it encodes a protein that binds to and inactivates tran-

(A)

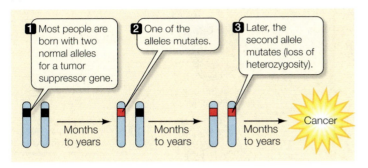

① Most people are born with two normal alleles for a tumor suppressor gene.

② One of the alleles mutates.

③ Later, the second allele mutates (loss of heterozygosity).

Months to years → Months to years → Months to years → Cancer

(B)

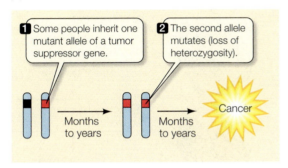

① Some people inherit one mutant allele of a tumor suppressor gene.

② The second allele mutates (loss of heterozygosity).

Months to years → Months to years → Cancer

17.16 The "Two-Hit" Hypothesis for Cancer (A) Although a single mutation can activate an oncogene, two mutations are needed to inactivate a tumor suppressor gene. (B) An inherited predisposition to cancer occurs in people born with one allele already mutated.

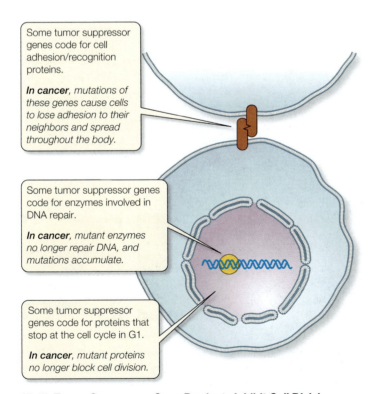

Some tumor suppressor genes code for cell adhesion/recognition proteins.

In cancer, mutations of these genes cause cells to lose adhesion to their neighbors and spread throughout the body.

Some tumor suppressor genes code for enzymes involved in DNA repair.

In cancer, mutant enzymes no longer repair DNA, and mutations accumulate.

Some tumor suppressor genes code for proteins that stop at the cell cycle in G1.

In cancer, mutant proteins no longer block cell division.

17.17 Tumor Suppressor Gene Products Inhibit Cell Division Mutations can affect any of the several ways in which tumor suppressor genes inhibit cell division, allowing cells to divide and form a tumor.

scription factors that are necessary for progress to the S phase and the rest of the cell cycle. In nondividing cells, *Rb* remains active, preventing cell division until the proper growth factor signals are present. When *Rb* is inactivated by mutation, the cell cycle moves forward independently of growth factors.

The protein product of another widespread tumor suppressor gene, *p53*, also stops the cell cycle at G1. It does this by acting as a transcription factor, stimulating the production of (among other things) a protein that blocks the interaction of a cyclin and a protein kinase needed for moving the cell cycle beyond G1. This gene is mutated in many types of cancers, including lung cancer and colon cancer.

Several events must occur to turn a normal cell into a malignant cell

The "gas pedal" and "brake" analogies we have been using for oncogenes and tumor suppressor genes, respectively, are elegant but simplified. There are many oncogenes and tumor suppressor genes, some of which act only in certain cells at certain times. Therefore, a complex sequence of events must occur before a normal cell becomes malignant, as demonstrated by an experiment in which mutations of oncogenes and tumor suppressor genes were introduced in different combinations (**Figure 17.18**). When normal rat cells growing in culture were transfected with an expression vector containing a mutant form of an oncogene (such as *ras*, which encodes a protein involved in signaling cell division), this alone was not enough to turn them into cancer cells. Likewise, if an expression vector containing a mutant form of a tumor suppressor gene (such as *p53*) was introduced in such a way that it overshadowed the normal form already in the cells, that alone was not enough to turn normal cells into cancer cells. *Both* an active gas pedal and a mutant brake are needed for malignant transformation.

More than two gene mutations are usually needed for full-blown cancer. Because colon cancer progresses to full malignancy slowly, it is possible to describe the oncogene and tumor suppressor gene mutations at each stage in great molecular detail. **Figure 17.19** outlines the progress of this form of cancer. At least four tumor suppressor genes and one oncogene must be mutated in sequence for an epithelial cell in the colon to become metastatic. Although the occurrence of all these events in a single cell might appear unlikely, remember that the colon has millions of cells, that the cells giving rise to colon epithelial cells are constantly dividing, and that these changes take place over many years of exposure to natural and synthetic carcinogens as well as spontaneous mutations.

The characterization of the molecular changes in tumor cells has opened up the possibility of genetic diagnosis and screening for cancer. Many cancers are now commonly diagnosed in part by specific oligonucleotide probes for oncogene or tumor suppressor gene alterations. It is also possible to detect early in life whether an individual has inherited a mutant tumor suppressor gene. A person who inherits mutated copies of the tumor suppressor genes involved in colon cancer, for example, normally would have a high probability of developing this cancer by age 40. Surgical removal of the colon would prevent a metastatic tumor from arising.

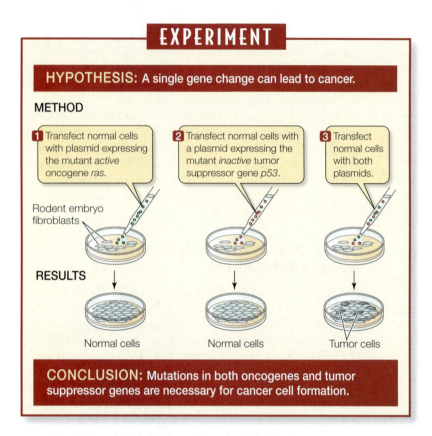

EXPERIMENT

HYPOTHESIS: A single gene change can lead to cancer.

METHOD

1 Transfect normal cells with plasmid expressing the mutant *active* oncogene *ras*.

2 Transfect normal cells with a plasmid expressing the mutant *inactive* tumor suppressor gene *p53*.

3 Transfect normal cells with both plasmids.

Rodent embryo fibroblasts

RESULTS

Normal cells

Normal cells

Tumor cells

CONCLUSION: Mutations in both oncogenes and tumor suppressor genes are necessary for cancer cell formation.

17.18 Cancer Is the Result of Multiple Genetic Alterations A complex series of events leads to malignancy in a normal cell. FURTHER RESEARCH: What would happen if you performed this experiment using the *normal* alleles for *ras* and *p53*?

17.4 RECAP

Uncontrolled cell division and metastasis (spread) are the hallmarks of cancer. Although some cancers are caused by viruses or inherited, most are due to the accumulation of genetic mutations in two classes of genes: oncogenes and tumor suppressor genes.

- What are some viruses that can cause cancer? See p. 387 and Table 17.1
- Can you describe the functions of oncogenes and tumor suppressor genes? See p. 388 and Figures 17.15 and 17.17
- What is the two-hit hypothesis? See p. 388 and Figure 17.16

Ongoing research has resulted in the development of increasingly accurate diagnostic tests and a better understanding of various genetic diseases at the molecular level. This knowledge is now being applied to develop new treatments for genetic diseases. In the next section we will survey various approaches to treatment, ranging from modifications of the mutant phenotype to gene therapy, in which the normal version of a mutant gene is supplied.

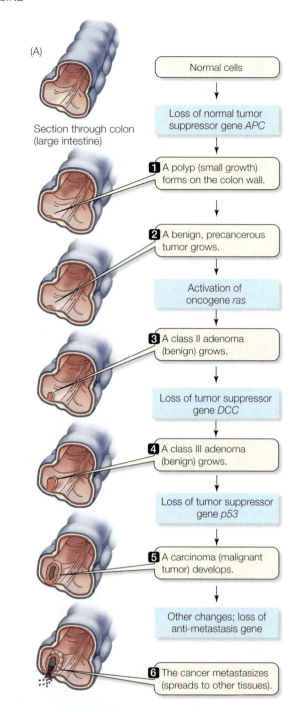

(A)

Section through colon (large intestine)

Normal cells

Loss of normal tumor suppressor gene *APC*

1 A polyp (small growth) forms on the colon wall.

2 A benign, precancerous tumor grows.

Activation of oncogene *ras*

3 A class II adenoma (benign) grows.

Loss of tumor suppressor gene *DCC*

4 A class III adenoma (benign) grows.

Loss of tumor suppressor gene *p53*

5 A carcinoma (malignant tumor) develops.

Other changes; loss of anti-metastasis gene

6 The cancer metastasizes (spreads to other tissues).

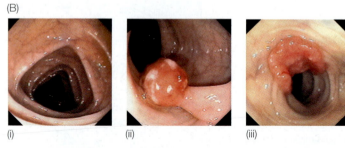

(B)

(i)   (ii)   (iii)

17.19 Multiple Mutations Transform a Normal Colon Epithelial Cell into a Cancer Cell (A) At least five genes must be mutated in a single cell to produce metastatic colon cancer. (B) Colonoscopy is the best screening test for colon cancer. These views reveal (i) normal colon tissue, (ii) a benign adenoma (stalked polyp), and (iii) adenocarcinoma (a malignant tumor).

17.5 How Are Genetic Diseases Treated?

Most treatments for genetic diseases simply try to alleviate the symptoms that affect the patient. But to effectively treat these diseases—whether they affect all cells, as in inherited disorders such as PKU, or only somatic cells, as in cancer—physicians must be able to diagnose the disease accurately, understand how the disease works at the molecular level, and intervene early, before the disease ravages or kills the individual. There are two main approaches to treating genetic diseases: modifying the disease phenotype or replacing the defective gene.

Genetic diseases can be treated by modifying the phenotype

Altering the phenotype of a genetic disease so that it no longer harms an individual is commonly done in one of three ways: by restricting the substrate of a deficient enzyme, by inhibiting a harmful metabolic reaction, or by supplying a missing protein product.

RESTRICTING THE SUBSTRATE Restricting the substrate of a deficient enzyme is the approach taken when a newborn is diagnosed with PKU. In this case, the deficient enzyme is phenylalanine hydroxylase, and the substrate is phenylalanine. The infant's inability to break down the phenylalanine in food leads to a buildup of the substrate, which causes the clinical symptoms. So the infant is immediately put on a special diet that contains only enough phenylalanine for immediate use. Lofenelac, a milk-based product that is low in phenylalanine, is fed to these infants just like formula. Later, certain fruits, vegetables, cereals, and noodles low in phenylalanine can be added to the diet. Meat, fish, eggs, dairy products, and bread, which contain high amounts of phenylalanine, must be avoided, especially during childhood, when brain development is most rapid. The artificial sweetener aspartame must also be avoided because it is made of two amino acids, one of which is phenylalanine.

People with PKU are generally advised to stay on a low-phenylalanine diet for life. Although maintaining these dietary restrictions may be difficult, it is effective. Numerous follow-up studies since newborn screening was initiated have shown that people with PKU who stay on the diet are no different from the rest of the population in terms of mental ability. This is an impressive achievement in public health, given the severity of mental retardation in untreated patients.

METABOLIC INHIBITORS As described in Section 17.1, people with familial hypercholesterolemia accumulate dangerous levels of cholesterol in their blood. These people are not only unable to metabolize dietary cholesterol, but also synthesize a lot of it. One effective treatment for people with this disease is a statin drug, which blocks the patient's own cholesterol synthesis. Patients who receive this drug need only worry about cholesterol in the diet, and not about the cholesterol their cells are making.

Metabolic inhibitors also form the basis of chemotherapy for cancer. The strategy is to kill rapidly dividing cells, since rapid cell division is the hallmark of malignancy. But such a strategy is not

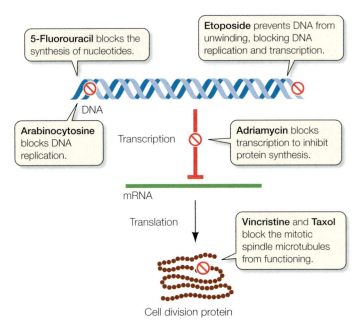

17.20 Strategies for Killing Cancer Cells The medications used in chemotherapy attack rapidly dividing cancer cells in several ways. Unfortunately, most of them also affect rapidly dividing noncancerous cells.

selective for tumor cells. Many drugs can kill cancer cells (**Figure 17.20**), but most of those drugs also damage other, noncancerous, dividing cells in the body. Therefore, it is not surprising that people undergoing chemotherapy suffer side effects such as loss of hair (due to damage to the skin epithelium), digestive upsets (gut epithelial cells), and anemia (bone marrow stem cells). The effective dose of these highly toxic drugs for treating the cancer is often just below the dose that would kill the patient, so they must be used with utmost care. Often they can control the spread of cancer, but not cure it.

SUPPLYING THE MISSING PROTEIN An obvious way to treat a disease phenotype in which a functional protein is missing is to supply that protein. This approach is the basis of treatment of hemophilia, in which the missing blood clotting protein is supplied in pure form. The production of human clotting proteins by recombinant DNA technology has made it possible to provide a pure protein instead of blood products, which can be contaminated with viruses or other pathogens.

Unfortunately, the phenotypes of many diseases caused by genetic mutations are very complex. Simple interventions like those we have described do not work for most such diseases. Indeed, a recent survey showed that current therapies for 351 diseases caused by single-gene mutations improved patients' life spans by only 15 percent.

Gene therapy offers the hope of specific treatments

Clearly, if a cell lacks a functional allele, it would be optimal to provide that allele, which is the aim of gene therapy. Diseases ranging from the rare inherited disorders caused by single-gene mutations to cancer and atherosclerosis are under intensive investigation in an effort to develop gene therapy treatments.

The object of **gene therapy** is to insert a new gene that will be expressed in the host. The new DNA is often attached to a promoter that will be active in human cells. The physicians who are developing such treatments are confronted by all the challenges of recombinant DNA technology: they must find effective vectors and ensure efficient uptake, precise insertion into the host DNA, appropriate expression and processing of mRNA and protein, and selection within the body for the cells that contain the recombinant DNA.

Which human cells should be the targets of gene therapy? The best approach would be to replace the nonfunctional allele with a functional one in every cell of the body. But vectors that could do this are simply not available, and delivery to every cell poses a formidable challenge. Until recently, attempts at gene therapy have used *ex vivo* techniques. That is, physicians have taken cells from the patient's body, added the new gene to those cells in the laboratory, and then returned the cells to the patient in the hope that the correct gene product would be made (**Figure 17.21**). Two examples demonstrate this technique:

- *Adenosine deaminase* is needed for maturation of white blood cells, and people without this enzyme have severe immune system deficiencies. A functional gene for adenosine deaminase was introduced via a viral vector into the white blood cells of a girl with a genetic deficiency of this enzyme. Unfortunately, mature white blood cells were used, and although they survived for a time in the girl and provided some therapeutic benefit, they eventually died, as is the normal fate of such cells. Further clinical trials have used bone marrow stem cells, which constantly divide to produce white blood cells.

- People with *hemophilia* had some skin cells removed and transfected with a plasmid containing a normal allele of the gene coding for the blood clotting protein they were missing. The cells were then reintroduced into the patients' body fat, where they produced adequate protein for normal clotting.

The other approach to gene therapy is to insert the gene directly into cells in the body of the patient. This *in vivo* approach is being attempted for various types of cancer. Lung cancer cells, for example, are accessible to such treatment if the DNA or vector is given as an aerosol through the respiratory system. Vectors carrying functional alleles of the tumor suppressor genes that are mutated in the patient's tumors, as well as vectors expressing antisense RNAs targeting oncogene mRNAs, have been successfully introduced in this way into patients with lung cancer, with some clinical improvement.

Several thousand patients, over half of them with cancer, have undergone gene therapy. Most of these clinical trials have been at a preliminary level, in which people are given the therapy to see whether it has any toxicity and whether the new gene is actually incorporated into the patient's genome. More ambitious trials are under way, in which a larger number of patients will receive the therapy with the hope that their disease will disappear, or at least improve.

17.21 Gene Therapy: The *Ex Vivo* Approach New genes are added to somatic cells taken from a patient's body. These transgenic cells are then returned to the body to make the missing gene product.

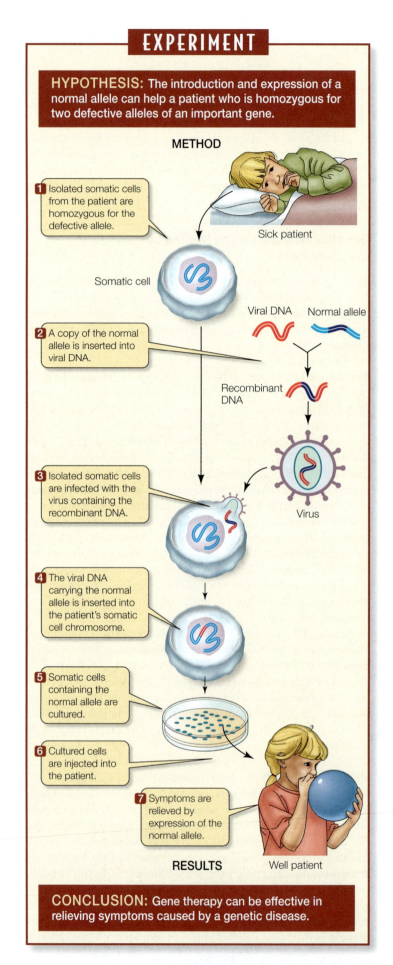

EXPERIMENT

HYPOTHESIS: The introduction and expression of a normal allele can help a patient who is homozygous for two defective alleles of an important gene.

METHOD

1 Isolated somatic cells from the patient are homozygous for the defective allele.

Sick patient

Somatic cell

Viral DNA   Normal allele

2 A copy of the normal allele is inserted into viral DNA.

Recombinant DNA

3 Isolated somatic cells are infected with the virus containing the recombinant DNA.

Virus

4 The viral DNA carrying the normal allele is inserted into the patient's somatic cell chromosome.

5 Somatic cells containing the normal allele are cultured.

6 Cultured cells are injected into the patient.

7 Symptoms are relieved by expression of the normal allele.

RESULTS   Well patient

CONCLUSION: Gene therapy can be effective in relieving symptoms caused by a genetic disease.

Therapies for simple genetic diseases remain a challenge. But most diseases are far more complex, involving many genes interacting with the environment. Knowledge of these genes is coming from the Human Genome Project.

17.6 What Have We Learned from the Human Genome Project?

In 1984, the United States government sponsored a conference on the detection of DNA damage in people exposed to significant levels of radiation, such as those who had survived the atomic bombs in Japan 39 years earlier. Scientists attending this conference quickly realized that the ability to detect such damage would also be useful in evaluating environmental mutagens. But in order to detect changes in the human genome, scientists first needed to know its normal sequence.

In 1986, Renato Dulbecco, who had won the Nobel prize for his pioneering work on cancer-causing viruses, suggested that determining the normal sequence of human DNA could also be a boon to cancer research. He proposed that the scientific community be mobilized for the task. The result was the publicly funded **Human Genome Project**, an international effort. In the 1990s, private industry launched its own sequencing effort.

There are two approaches to genome sequencing

Because of their differing sizes, the 46 human chromosomes can be separated from one another and identified (see Figure 9.15). So it is possible to isolate the DNA of each chromosome for sequencing. The straightforward approach would be to start at one end of a chromosome and simply sequence the entire 120 million base pairs. Unfortunately, this approach is not practical, since only about 700 base pairs can be sequenced at a time (see Figure 11.24 for a description of the DNA sequencing technique).

To sequence an entire genome, chromosomal DNA is first cut into fragments about 500 base pairs long, then each fragment is sequenced. For the haploid human genome, which has about 3.2 billion base pairs, there are more than 6 million such fragments. The problem then becomes putting these millions of fragments

back together. This task is accomplished by assembling smaller portions of DNA sequence. These portions of DNA sequence almost always overlap and must be aligned so that the positions of all of the bases are properly accounted for. Two methods were used to align the DNA fragments: *hierarchical sequencing* and *shotgun sequencing*.

HIERARCHICAL SEQUENCING   The publicly funded sequencing team used a method known as **hierarchical sequencing**. First, they systematically identified short marker sequences along the chromosomes, ensuring that every fragment of DNA to be sequenced would contain a marker (**Figure 17.22A**). This method can be compared to making a road map, showing towns with the mileage separating them. The "towns" are the marker sequences, and the "mileage" is in base pairs. The simplest markers are the recognition sequences for restriction enzymes.

Some restriction enzymes recognize 8–12 base pairs in DNA, not just the usual 4–6 base pairs. A DNA molecule with several million base pairs will have relatively few of these larger sites, and thus the enzyme will generate a small number of relatively large fragments. These large fragments can be added to a vector called a **bacterial artificial chromosome** (**BAC**), which can carry about 250,000 base pairs of inserted DNA, and inserted into bacteria to create a gene library.

The fragments in this library are arranged in the proper order along the chromosome map by using the marker sequences. To arrange the DNA fragments on the map, libraries made with different restriction enzymes are compared. If two large fragments of DNA cut with different enzymes have the same marker, they must overlap. This method works, but is slow.

SHOTGUN SEQUENCING   Instead of finding markers, fragmenting the DNA, and then sequencing it, the **shotgun sequencing** method cuts the DNA at random into small, sequencing-ready fragments and lets powerful computers search for markers that overlap (**Figure 17.22B**). The fragments can then be aligned.

The shotgun approach, which has been used by private industry, is much faster than the hierarchical approach because there is no need to make a map. At first this method was greeted with considerable skepticism. One concern was that without rigorous prior mapping of marker sites on the chromosomes, the computer might pick out repetitive sequences common to many DNA fragments and line the fragments up incorrectly. But the rapid development of sophisticated computers and software for DNA sequencing and analysis (the advent of **bioinformatics**) allowed the shotgun method to be refined so that inaccurate alignment is not a major problem. The entire 180 million-base-pair fruit fly genome was sequenced by the shotgun method in little over a year. This success proved that the shotgun method might work for the much larger human genome, and in fact, it did.

The sequence of the human genome contained many surprises

The two teams of scientists announced a draft human genome sequence in June 2000 to great fanfare, and published their data simultaneously in February 2001. By the start of 2005, the final

RESEARCH METHOD

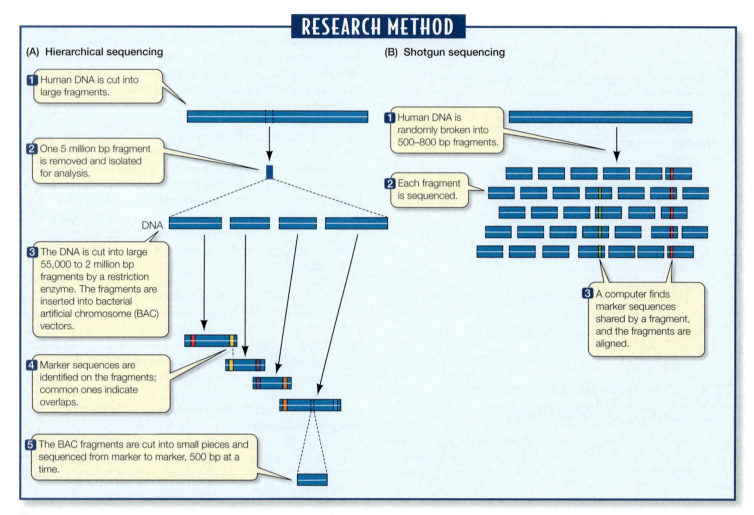

(A) Hierarchical sequencing

1 Human DNA is cut into large fragments.

2 One 5 million bp fragment is removed and isolated for analysis.

DNA

3 The DNA is cut into large 55,000 to 2 million bp fragments by a restriction enzyme. The fragments are inserted into bacterial artificial chromosome (BAC) vectors.

4 Marker sequences are identified on the fragments; common ones indicate overlaps.

5 The BAC fragments are cut into small pieces and sequenced from marker to marker, 500 bp at a time.

(B) Shotgun sequencing

1 Human DNA is randomly broken into 500–800 bp fragments.

2 Each fragment is sequenced.

3 A computer finds marker sequences shared by a fragment, and the fragments are aligned.

17.22 Two Approaches to Sequencing DNA (A) In the hierarchical approach to genome sequencing, genetic markers are mapped, and DNA fragments are then aligned by matching overlapping sites with the same markers. (B) In the shotgun approach, the DNA is fragmented and a computer is used to find overlapping markers.

sequence was completed, two years ahead of the schedule set more than a decade previously and well under budget.

Many interesting and surprising facts about the human genome were revealed upon its sequencing.

- *Of its 3.2 billion base pairs, fewer than 2 percent make up coding regions, containing a total of about 24,000 genes.* Before sequencing began, estimates of the number of human genes ranged from 80,000 to 100,000. This lower number of genes—not many more than the fruit fly's—means that the observed diversity of proteins, which led to the 100,000 estimate, must be produced posttranscriptionally. That is, the average human gene must code for several different proteins.

- *The average gene has 27,000 base pairs.* Gene sizes vary greatly, from 1,000 to 2.4 million base pairs. Variation in gene size is to be expected, given that human proteins (and RNAs) vary in size. Human proteins range from 100 to about 5,000 amino acids per polypeptide chain.

- *Virtually all human genes have many introns.* As we noted previously, only 2 percent of the human genome is coding DNA.

- *Over 50 percent of the genome is made up of highly repetitive sequences.* Repetitive sequences near genes are GC-rich, while those farther away from genes are AT-rich.

- *Almost all (99.9 percent) of the genome is the same in all people.* Despite this apparent homogeneity, there are, of course, many individual differences. Scientists have mapped over 2 million single nucleotide polymorphisms (SNPs)—bases that differ in at least 1 percent of people.

- *Genes are not evenly distributed over the genome.* The small chromosome 19 is packed densely with genes, while chromosome 8 has long stretches of "gene desert," with no coding regions. The Y chromosome has the fewest genes (231), while chromosome 1 has the most (2,968).

- *Many genes exist whose functions are unknown.* There are 740 genes coding for RNAs that are not translated into proteins. Of these RNAs, several dozen are tRNAs, and a few are rRNAs and splicing RNAs. The roles of the rest are not clear. Nor are the roles of the hundreds of genes encoding protein kinases, although it is a good bet that they are involved in cell signaling.

The ENCODE project (*Encyclopedia of DNA Elements*) has been set up to identify all of the functional sequences in the human genome—not just the protein-coding sequences, but others as well, such as those coding for small RNAs involved in gene regulation. This project will make use of comparative data from closely related species, such as the chimpanzee.

The human genome sequence has many applications

Reading the human "book of life" is an achievement that ranks high among other recent great events in scientific exploration. The human genome sequence, and the tools developed to read it, are changing biology in many ways.

- Mapping technology and SNPs have made the isolation of human genes by positional cloning much easier because of the huge number of genetic markers now available. Many disease-related genes have been identified in this way.

- Genetic variation in drug metabolism has been a medical problem for a long time. The emerging field of **pharmacogenomics** is identifying the genes responsible for this variation and developing tests to predict who will react best to which medications.

- DNA chips (see Figure 16.15) are being used to analyze the expression of thousands of genes in different cells in different biochemical states. For example, a Cancer Genome Anatomy Project is seeking to make an mRNA "fingerprint" of a tumor at each stage of its development. Finding out which genes are expressed at which stage will be important not only in diagnosis, but also in identifying targets for therapy.

- "Genome prospecting" refers to the search for important polymorphisms in specific human populations. For example, the Pima Indians in Arizona have a high frequency of extreme obesity and diabetes. A search of their genomes might reveal genes predisposing them to these conditions.

The end result of all this knowledge may be a new approach to medical care, in which each person's genome will be used to prescribe lifestyle changes and treatments that can maximize that person's genetic potential.

The use of genetic information poses ethical questions

When the genetic defect that causes cystic fibrosis was discovered, many people predicted a "tidal wave" of genetic testing for heterozygous carriers. Everyone, it was thought, would want the test—especially the relatives of people with the disease. But this tidal wave has not developed. To find out why, a team of psychologists, ethicists, and geneticists interviewed 20,000 people in the United States. What the researchers found surprised them. Most people are simply not very interested in their genetic makeup, unless they have a close relative with a genetic disease and are involved in a decision about pregnancy.

There are other people, however, who might be very interested in the results of genetic testing. For example, people who test positive for genetic abnormalities, from hypercholesterolemia to cancer, might be denied employment or health insurance. Consequently, there are laws that prohibit discrimination on the basis of genetic information.

The search for valuable genes in diverse human populations has raised many concerns about exploitation and commercialization of a person's DNA sequence. Is a gene that confers resistance to cancer, for example, the property of an individual, an ethnic group in which it may be frequent, the pharmaceutical company that finds it, or humanity at large?

We described at the beginning of this chapter how the French Canadian population of Quebec is being used to discover genes relevant to human diseases. People there have been happy to cooperate with the laboratories doing the investigations. Several other populations that came from a small group of founders are being used by human geneticists for genome analyses:

- 2.5 million Costa Ricans who came from about 4,000 founders 12 generations ago

- 10 million Ashkenazic Jews (from Eastern Europe) who came from about 1,500 founders 30 generations ago

- 1.6 million Sardinians who came from about 500 founders 400 generations ago

Perhaps the best-known population under intensive study is in Iceland, where 280,000 people are descended from about 10,000 settlers who came from Europe 1,100 years ago. A company has been set up, with government support, to sell the knowledge that comes from analyzing the genomes of Iceland's people. The company's approach to mining this genetic lode is illustrated by its search for genes that predispose people to asthma, a respiratory disease:

- The names of Icelanders with asthma were searched for in a genealogy database.

- One group of 104 patients was descended from a single ancestor, born in 1710 (11 generations ago). It was considered likely that the same genes for predisposition to asthma were present in all 104 patients.

- Marker sequences were sought that would identify alleles that all 104 patients had in common, and a small number of such genes were subsequently identified.

The characterization of these genes will lead to a greater understanding of how a group of genes interacts to produce a complex phenotype.

The proteome is more complex than the genome

As mentioned above, many genes encode more than a single protein (**Figure 17.23A**). Alternative splicing leads to different combinations of exons in the mature mRNAs transcribed from a single gene (see Figure 14.21). Posttranslational modifications also increase the number of proteins that derive from one gene (see Figure 12.17). Therefore, the sum total of the proteins produced by an organism—its **proteome**—is more complex than its genome.

Two technologies are commonly used to analyze the proteome:

- Because of their unique amino acid compositions (primary structures), most proteins have unique electric charges and sizes. On the basis of these two properties, they can be separated by *two-dimensional gel electrophoresis*. Thus isolated, individual proteins can be analyzed, sequenced, and studied (**Figure 17.23B**).

(A)

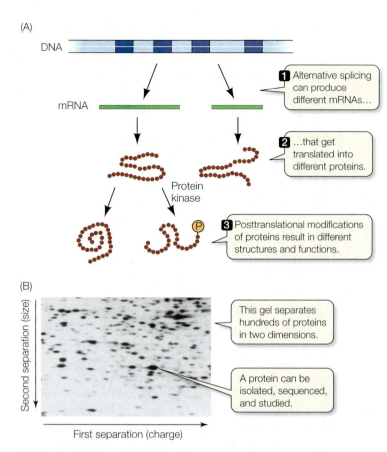

DNA

mRNA

1 Alternative splicing can produce different mRNAs...

2 ...that get translated into different proteins.

Protein kinase

3 Posttranslational modifications of proteins result in different structures and functions.

(B)

Second separation (size)

First separation (charge)

This gel separates hundreds of proteins in two dimensions.

A protein can be isolated, sequenced, and studied.

17.23 Proteomics (A) A small number of genes can code for a large number of proteins. (B) A cell's proteins can be separated on the basis of charge and size by gel electrophoresis. The two separations can distinguish most proteins from one another.

■ *Mass spectrometry* uses electromagnets to identify proteins by the masses of their atoms and displays them as peaks on a graph.

The ultimate aim of proteomics is just as ambitious as that of genomics. While genomics seeks to describe the genome and its expression, proteomics seeks to describe the phenotypes of the expressed proteins.

An amazing example of proteomics, combined with DNA chip technology, is the recent comparison of brain proteins in chimpanzees and humans. Comparative DNA sequencing has shown that humans and chimpanzees differ by no more than 3 percent at the DNA level. Svante Pääbo and his colleagues examined gene expression in the "thinking" part (the cortex) of the brains of three chimpanzees and three humans who had died of natural causes. Of 12,000 DNA sequences tested for expression as mRNA, only 175 (1.4 percent) differed between the two species—a truly humbling result. But proteomics showed that the *specific proteins* expressed by those sequences differed by 7.4 percent, probably due to alternative splicing. The *amounts* of those proteins differed even more (31.4 percent). So what makes our brains different from a chimpanzee's is more quantitative than qualitative. This finding suggests that the control of gene expression may be the key to human evolution.

Systems biology integrates data from genomics and proteomics

With the ease of generating DNA sequences, mRNA expression profiles, and proteomic profiles, a huge amount of data about biological systems is accumulating. How does it all fit together? These kinds of studies are essentially *reductionist*, dissecting biology into ever smaller parts, as has been most of biological science up to now. In contrast, Charles Darwin looked at all of nature, and fit it all together by taking a "top-down" approach in his theory of natural selection. Can we do the same with our molecular data?

Systems biology aims to integrate molecular data into a coherent picture of life. A *system* is a group of parts that interact with one another, forming a whole that is often greater than the sum of the parts. An electric light bulb provides an analogy. You can take three parts—a filament made of tungsten, a metal cup, and a thin glass bowl—and study them individually, but only when the three are put together, with the filament inside the glass bowl and the metal cup at the end, does a "system," with properties that none of the parts possess by themselves, emerge.

Systems biologists strive to discover such *emergent properties* by studying the interactions in a biological system. If the interactions within a system are understood, predictions of how events will proceed in that system under new physiological conditions may be made. For example, when a drug is designed to inhibit one enzyme in a biological pathway, systems biology may predict what the effects will be on the many other pathways that interact with that one. This approach could make drug side effects more predictable and better controlled, and even, if care is taken, eliminate them.

Figure 17.24 is a diagram of a metabolic pathway. It shows interactions between and amounts of some mRNA transcripts, proteins, and metabolites involved in fat metabolism in two different genetic strains of mice. One strain has a mutation that results in a tendency to get atherosclerosis and the other (the wild type) does not. The redder a dot, the more of the molecule it represents in the mutant strain compared with the wild type; the greener a dot, the less of a molecule in the mutant strain. Looking at the left of the diagram, you can see that protein A is clearly up-regulated in the mouse strain that gets atherosclerosis, while protein B is down-regulated in that mutant strain.

Systems biologists can try to use such information to understand how increasing or decreasing the amounts of one molecule might affect the levels of others. For example, what if a drug were designed to increase the level of protein B in the mutant strain of mice, so that it was now at the level seen in normal mice? Looking at Figure 17.24, you can see that protein B has several lines of interaction with other molecules, which in turn interact with still others. A whole new field of computational biology is being developed to help scientists make these kinds of predictions.

The objective of molecular medicine, as in all medicine, is to improve the health of the patient. Scientists have made great progress in understanding and treating diseases caused by single molecular events, such as phenylketonuria. The new approach of systems biology holds promise for understanding and treating the more complex diseases, such as heart disease and cancer, that continue to plague humankind.

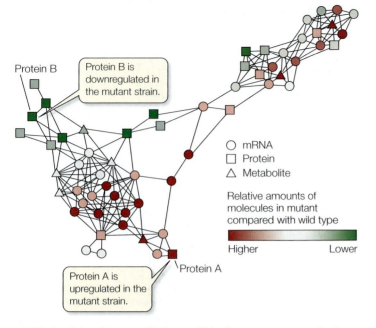

Protein B

Protein B is downregulated in the mutant strain.

○ mRNA
□ Protein
△ Metabolite

Relative amounts of molecules in mutant compared with wild type

Higher   Lower

Protein A is upregulated in the mutant strain.

Protein A

17.24 Applying Systems Biology This diagram compares the levels and interactions of mRNA transcripts, proteins, and metabolites in the pathways involving fat metabolism in two strains of mice. One strain (mutant) tends to get atherosclerosis; the other (wild type) does not. The intensities of the colors indicate amounts of molecules in the mutant strain compared with the wild-type strain (redder means increased and greener means decreased). Lines indicate interactions between molecules.

17.6 RECAP

Advances in DNA sequencing technology made it possible to map the entire human genome. Having a map of the human genome makes it possible to isolate specific genes, to compare genomic sequences between individuals and between species, and to identify mutations that cause disease. Next steps include analysis of the proteome and integrating molecular knowledge into the framework of systems biology.

■ Can you explain the difference between hierarchical and shotgun DNA sequencing? See p. 393 and Figure 17.22

■ What is the proteome and how does it relate to the genome? See pp. 395–396 and Figure 17.23

■ What is systems biology and what is its importance in molecular medicine? See p. 396

CHAPTER SUMMARY

17.1 How do defective proteins lead to diseases?

Abnormalities in nearly all classes of proteins, including enzymes, transport proteins, receptor proteins, and structural proteins, have been implicated in genetic diseases.

While a single amino acid difference can be the cause of disease, amino acid variations have been detected in many functional proteins. Review Figure 17.2

Transmissible spongiform encephalopathies (TSEs) are degenerative brain diseases that can be transmitted from one animal to another by consumption of infected tissues. The infective agent is a **prion**, a protein with an abnormal conformation.

Multifactorial diseases are caused by the interactions of many genes and proteins with the environment. They are much more common than diseases caused by mutations in a single gene.

Predictable patterns of inheritance are associated with some human genetic diseases. Autosomal recessive, autosomal dominant, and X-linked patterns are common.

17.2 What kinds of DNA changes lead to human diseases?

It is possible to isolate both the mutant genes and the abnormal proteins responsible for human diseases.

Positional cloning is a method of locating disease-causing genes by finding polymorphic genetic markers that are inherited with the disease.

Restriction fragment length polymorphisms (**RFLPs**) and single nucleotide polymorphisms (SNPs) are commonly used as genetic markers in positional cloning. Review Figure 17.8, Web/CD Activity 17.1

Mutations occur often where cytosine has been methylated to 5-methylcytosine. Review Figure 17.9

The effects of fragile-X syndrome worsen with each generation. This pattern is the result of an **expanding triplet repeat**. Review Figure 17.10

Genomic imprinting results in a gene being expressed differently depending on the sex of the parent from which it comes.

17.3 How does genetic screening detect human diseases?

Genetic screening can detect human genetic diseases, alleles predisposing people to those diseases, or carriers of those diseases.

Genetic screening can be done by looking for abnormal protein expression.

DNA testing is the direct identification of mutant alleles by sequence comparison. Any cell can be tested at any time in the life cycle.

The two predominant methods of DNA testing are the allele-specific cleavage method and allele-specific oligonucleotide hybridization method. Review Figures 17.12 and 17.13, Web/CD Tutorial 17.1

17.4 What is cancer?

Cancer cells fail to respond to the normal controls on cell division, and divide continuously.

Tumors may be **benign** or **malignant**. Malignant tumors can spread to other parts of the body through a process called **metastasis**.

Some types of human cancers are caused by viruses, but 85 percent of human cancers are caused by genetic mutations of somatic cells. These mutations occur most commonly in dividing cells. **Carcino-**

CHAPTER SUMMARY

gens can cause mutations that lead to cancer, but some mutations arise spontaneously.

Normal cells contain **oncogenes** that stimulate cell division. When mutated, these genes may become active in a cell in which they are normally turned off. Review Figure 17.15

Normal cells also contain **tumor suppressor genes** that inhibit cell division. When mutated, they may become inactive. Review Figure 17.17

The two-hit hypothesis describes the difference between inherited cancer, in which an individual inherits one mutant allele of a tumor suppressor gene and a somatic mutation occurs later in the second allele, and sporadic cancer, in which two mutational events must occur in the same somatic cell to produce cancer. Review Figure 17.16

In most cases, mutations must activate several oncogenes and inactivate several tumor suppressor genes to produce a malignant tumor. Review Figure 17.19

17.5 How are genetic diseases treated?

There are three ways to modify the phenotype of a genetic disease: restrict the substrate of a deficient enzyme, inhibit a harmful metabolic reaction, or supply a missing protein.

In **gene therapy**, a mutant gene is replaced with a normal gene. Both *ex vivo* and *in vivo* therapies are being developed. Review Figure 17.21

17.6 What have we learned from the Human Genome Project?

The entire human genome has been sequenced using both **hierarchical** and **shotgun** methods. Review Figure 17.22, Web/CD Tutorial 17.2

The human genome has only about 24,000 genes.

Humans make many more proteins than predicted by their number of genes. Thus the **proteome** is more complex than the genome. Review Figure 17.23

Systems biology attempts to unify data from genomics and proteomics into a coherent, predictive whole. Review Figure 17.24

See Web/CD Activity 17.2 for a concept review of this chapter.

SELF-QUIZ

1. Phenylketonuria is an example of a genetic disease in which
   a. a single enzyme is not functional.
   b. inheritance is sex-linked.
   c. two parents without the disease cannot have a child with the disease.
   d. mental retardation always occurs, regardless of treatment.
   e. a transport protein does not work properly.

2. Mutations of the gene for β-globin
   a. are usually lethal.
   b. occur only at amino acid position 6.
   c. number in the hundreds.
   d. always result in sickling of red blood cells.
   e. can always be detected by gel electrophoresis.

3. Multifactorial (complex) diseases
   a. are less common than single-gene diseases.
   b. involve the interaction of many genes with the environment.
   c. affect less than 1 percent of humans.
   d. involve the interactions of several mRNAs.
   e. are exemplified by sickle-cell disease.

4. In fragile-X syndrome,
   a. females are affected more severely than males.
   b. a short sequence of DNA is repeated many times to create the fragile site.
   c. both the X and Y chromosomes tend to break when prepared for microscopy.
   d. all people who carry the gene that causes the syndrome are mentally retarded.
   e. the basic pattern of inheritance is autosomal dominant.

5. Most genetic diseases are rare because
   a. each person is unlikely to be a carrier for harmful alleles.
   b. genetic diseases are usually sex-linked and so uncommon in females.
   c. genetic diseases are always dominant.
   d. two parents probably do not carry the same recessive alleles.
   e. mutation rates in humans are low.

6. Mutational "hot spots" in human DNA
   a. always occur in genes that are transcribed.
   b. are common at cytosines that have been modified to 5-methylcytosine.
   c. involve long stretches of nucleotides.
   d. occur where there are long repeats.
   e. are very rare in genes that code for proteins.

7. Newborn genetic screening for PKU
   a. is very expensive.
   b. detects phenylketones in urine.
   c. has not led to the prevention of mental retardation resulting from this disorder.
   d. must be done during the first day of an infant's life.
   e. uses bacterial growth to detect excess phenylalanine in blood.

8. Genetic diagnosis by DNA testing
   a. detects only mutant and not normal alleles.
   b. can be done only on eggs or sperm.
   c. involves hybridization to rRNA.
   d. utilizes restriction enzymes and a polymorphic site.
   e. cannot be done with PCR.

9. Most human cancers
   a. are caused by viruses.
   b. are in blood cells or their precursors.
   c. involve mutations of somatic cells.
   d. spread through solid tissues rather than by the blood or lymphatic system.
   e. are inherited.

10. Current treatments for genetic diseases include all of the following *except*
    a. restricting a dietary substrate.
    b. replacing the mutant gene in all cells.
    c. alleviating the patient's symptoms.
    d. inhibiting a harmful metabolic reaction.
    e. supplying a protein that is missing.

1. How do oncogenes and tumor suppressor genes and their functions change in tumor cells? Propose targets for cancer therapy involving their gene products.

2. In the past, it was common for people with phenylketonuria (PKU) who were placed on a low-phenylalanine diet after birth to be allowed to return to a normal diet during their teenage years. Although the levels of phenylalanine in their blood were high, their brains were thought to be beyond the stage of being harmed. If a woman with PKU becomes pregnant, however, a problem arises. Typically, the fetus is heterozygous, but is unable at early stages of development to metabolize the high levels of phenylalanine that arrive from the mother's blood. Why is the fetus heterozygous? What do you think would happen to the fetus during this "maternal PKU" situation? What would be your advice to a woman with PKU who wants to have a child?

3. Cystic fibrosis is an autosomal recessive disease in which thick mucus is produced in the lungs and airways. The gene responsible for this disease codes for a protein composed of 1,480 amino acids. In most patients with cystic fibrosis, the protein has 1,479 amino acids: a phenylalanine is missing at position 508. A baby is born with cystic fibrosis. He has an older brother. How would you test the DNA of the older brother to determine whether he is a carrier for cystic fibrosis? How would you design a gene therapy protocol to "cure" the cells in the younger brother's lungs and airways?

4. A number of efforts are under way to identify human genetic polymorphisms that correlate with multifactorial diseases such as diabetes, heart disease, and cancer. What would be the uses of such information? What concerns do you think are being raised by the people whose genomes are being analyzed?

We have seen that human cells in culture can be transformed from normal to tumor cells (see Figure 17.18), and that human genes can be isolated by positional cloning (see Figure 17.8). A group of related prostate cancer patients do not have mutations in any of the known tumor suppressor genes. You suspect that a new, as yet undiscovered tumor suppressor gene has been mutated. How would you use these methods to isolate the new gene from these patients?

18 Immunology: Gene Expression and Natural Defense Systems

The most dangerous foe

On January 6, 1777, George Washington, commander of the Revolutionary army of the fledgling United States, wrote to his chief physician: "Finding smallpox to be spreading much, and fearing that no precaution can prevent it from running through the whole of our army, I have determined that the troops shall be inoculated. Should the disease rage with its usual virulence, we should have more to dread from it than the sword of the enemy."

Washington was speaking from experience. He himself had survived smallpox as a teenager, in 1751. And during the year 1776 his army lost 1,000 men in battle—and 10,000 men to smallpox. After Washington's men were inoculated, the death rate due to smallpox in the Revolutionary army plummeted.

Inoculation in Washington's army meant introducing a small amount of fluid obtained from a smallpox pustule on a recent victim to a healthy person. For reasons not understood at the time, most people inoculated in this way became immune to the full ravages of smallpox—as Washington himself, having survived the disease, was immune.

Two decades later, the English country doctor Edward Jenner observed that milkmaids frequently contracted a mild infection called cowpox, and that these women seemed *not* to be affected by the smallpox epidemics that ravaged England. Reasoning that there was some kind of cross-resistance between the two infections, Jenner performed an experiment that modern medical science would forbid as unethical.

Jenner rubbed powdered remnants from cowpox lesions into scratches on the arm of a young boy named James Phipps. Not unexpectedly, James came down with a mild case of cowpox. Six weeks later, Jenner infected James with the residue of smallpox lesions, discovering that, as Jenner expected (and, we presume, hoped), the boy did not contract smallpox.

Because of the connection with cows, for which the Latin word is *vacca*, the term coined for Jenner's procedure was *vaccination*. A much safer approach than inoculation, the practice of vaccination spread quickly. More potent vaccines were developed, and massive programs were instituted to vaccinate much of the world's population. By 1980, smallpox had literally disappeared.

Why was George Washington immune to smallpox? Why did inoculation save his soldiers? How could a vaccination program rid the world of smallpox? The answers lie in the cells and molecules of

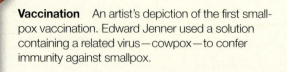

Vaccination An artist's depiction of the first smallpox vaccination. Edward Jenner used a solution containing a related virus—cowpox—to confer immunity against smallpox.

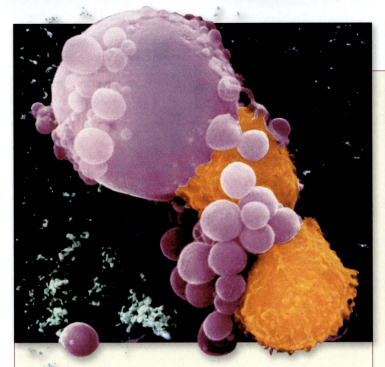

T Cells in Action Two cytotoxic T cells (orange) have come into contact with virus-infected cells (pink). The infected cell at the top left has begun to die, as indicated by membrane blisters. The process is complete in the cell in the center, which has disintegrated into small blebs.

our immune systems. When Washington caught smallpox in 1751, specialized white blood cells called macrophages engulfed and destroyed some of the smallpox viruses and displayed fragments of the viruses on their surfaces. Other specialized white blood cells, T cells, recognized the fragments and become "activated." Descendants of activated T cells attacked Washington's virus-infected cells, preventing the lethal spread of the disease. Other descendants of the T cells persisted in his body as "memory cells," which defended him when he was exposed to the disease later. Inoculation and vaccination against a pathogen stimulate the formation of these memory cells, which then protect the body against infection.

IN THIS CHAPTER we'll see how the nonspecific immune system attempts to prevent pathogens from entering the body. We then describe how the specific immune system targets invaders such as smallpox virus for destruction. We'll see how reshuffling genetic material helps animals fight off a mind-boggling diversity of potential invaders. We conclude by considering what happens when this complex and crucial system malfunctions.

18.1 What Are the Major Defense Systems of Animals?

Animals have a number of ways of defending themselves against **pathogens**—harmful organisms and viruses that can cause disease. These defense systems are based on the distinction between *self*—the animal's own molecules—and *nonself*, or foreign, molecules. In this section we will consider the mechanisms by which animals recognize and mount defenses against nonself molecules. Many of these mechanisms are based on the principles of genetics and molecular biology that have been discussed in earlier chapters.

There are two general types of defense mechanisms:

- **Nonspecific defenses**, or *innate defenses*, are inherited mechanisms that protect the body from many kinds of pathogens. Nonspecific defenses, which typically act very rapidly, include barriers such as the skin, molecules that are toxic to invaders, and phagocytic cells that ingest invaders. Most animals—including both invertebrates and vertebrates—as well as plants have nonspecific defenses.

- **Specific defenses** are *adaptive* mechanisms aimed at a specific pathogen. For example, these defense systems can make an antibody protein that will recognize, bind to, and destroy a certain virus if that virus ever enters the bloodstream. Specific defense mechanisms are found in vertebrate animals. They are typically slow to develop and long-lasting.

In animals that have both kinds of defense mechanisms, nonspecific and specific defenses operate together as a coordinated defense system. Because the specific defenses often require days or even weeks to become effective, the nonspecific defenses are the body's first line of defense. Considering that a bacterium can produce 20 million progeny a day, the nonspecific defenses are tremendously important in keeping disease at bay.

Blood and lymph tissues play important roles in defense systems

The components of the mammalian defense system are dispersed throughout the body and interact with almost all of its other tissues and organs. The *lymphoid tissues*, which include the thymus, bone marrow, spleen, and lymph nodes, are essential parts of the defense system (**Figure 18.1**). Central to their functioning are the blood and lymph.

Blood and lymph are both fluid tissues that consist of water, dissolved solutes, and cells.

- **Blood plasma** is a yellowish solution containing ions, small molecular solutes, and soluble proteins. Suspended in the plasma are red blood cells, white blood cells, and platelets (cell fragments essential to blood clotting). While red blood cells are normally confined to the *closed circulatory system* (the heart, arteries, capillaries, and veins), white blood cells and platelets are also found in the lymph.

- **Lymph** is a fluid derived from the blood and other tissues that accumulates in intercellular spaces throughout the body. From these spaces, the lymph moves slowly into the vessels of the *lymphatic system*. Tiny lymph capillaries conduct this fluid to larger ducts that eventually join together, forming one large vessel, the *thoracic duct*, which joins a major vein (the

left subclavian vein) near the heart. By this system of vessels, the lymph is eventually returned to the blood and the circulatory system.

At many sites along the lymph vessels are small, roundish structures called **lymph nodes**, which contain a variety of white blood cells. As lymph passes through a lymph node, it is filtered and "inspected" for nonself materials by these defensive cells.

White blood cells play many defensive roles

One milliliter of human blood typically contains about 5 billion red blood cells and 7 million of the larger white blood cells. All of these cells originate from *pluripotent stem cells* in the bone marrow, which constantly divide and can differentiate into a wide variety of blood cell types (**Figure 18.2**). **White blood cells** (also called *leukocytes*) have nuclei and are colorless, unlike mammalian red blood cells, which lose their nuclei during development. White blood cells can leave the closed circulatory system and enter intercellular spaces where nonself cells or substances are present. The number of white blood cells in the blood and lymph may rise sharply in response to invading pathogens, providing medical professionals with a useful clue for detecting an infection.

Several types of white blood cells are important in the body's defenses. White blood cells fall into two broad groups:

- **Granular cells** include histamine-producing signaling cells as well as **phagocytes**, which engulf and digest foreign cells and cellular debris. Among the most important phagocytes are dendritic cells and **macrophages**. In addition to engulfing nonself materials, macrophages have the important function of presenting partly digested nonself materials to the T cells.

- **Lymphocytes** participate in specific defenses against nonself or altered cells, such as virus-infected cells and tumor cells. There are two types of lymphocytes, B cells and T cells.

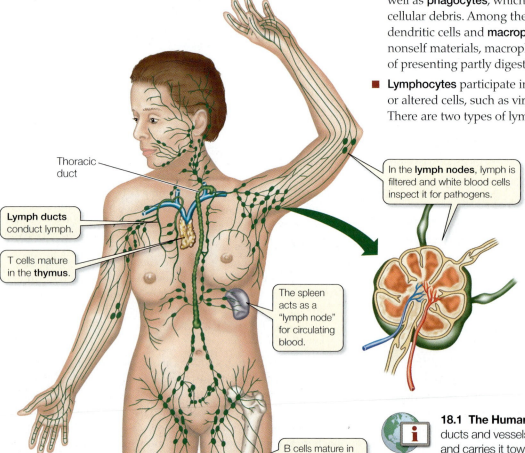

Thoracic duct

Lymph ducts conduct lymph.

T cells mature in the thymus.

The spleen acts as a "lymph node" for circulating blood.

In the **lymph nodes**, lymph is filtered and white blood cells inspect it for pathogens.

B cells mature in the **bone marrow.**

18.1 The Human Lymphatic System A network of ducts and vessels collects lymph from body tissues and carries it toward the heart, where it mixes with blood to be pumped back to the tissues. Other lymphoid tissues, including the thymus, spleen, and bone marrow, are also essential to the body's defense system.

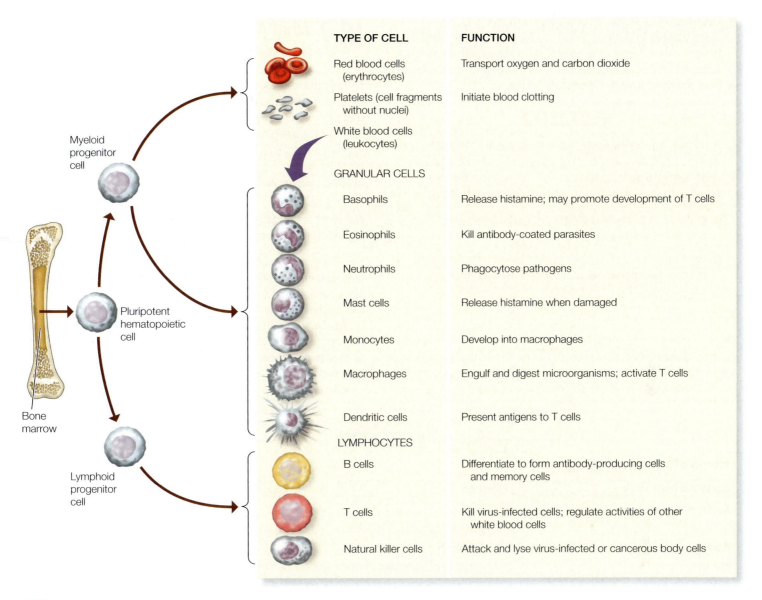

TYPE OF CELL	FUNCTION
Red blood cells (erythrocytes)	Transport oxygen and carbon dioxide
Platelets (cell fragments without nuclei)	Initiate blood clotting
White blood cells (leukocytes)	
GRANULAR CELLS	
Basophils	Release histamine; may promote development of T cells
Eosinophils	Kill antibody-coated parasites
Neutrophils	Phagocytose pathogens
Mast cells	Release histamine when damaged
Monocytes	Develop into macrophages
Macrophages	Engulf and digest microorganisms; activate T cells
Dendritic cells	Present antigens to T cells
LYMPHOCYTES	
B cells	Differentiate to form antibody-producing cells and memory cells
T cells	Kill virus-infected cells; regulate activities of other white blood cells
Natural killer cells	Attack and lyse virus-infected or cancerous body cells

Myeloid progenitor cell

Pluripotent hematopoietic cell

Bone marrow

Lymphoid progenitor cell

18.2 Blood Cells Pluripotent stem cells in the bone marrow can differentiate into red blood cells, platelets, and the various types of white blood cells.

- Immature **T cells** migrate from the bone marrow via the blood to the thymus, where they mature.

- Mature **B cells** leave the bone marrow and circulate through the blood and lymph vessels. B cells make specialized proteins called *antibodies* that enter the blood and bind to nonself substances.

Fundamental to the interactions, control, and defensive functioning of these white blood cells are defensive proteins and other signals.

Immune system proteins bind pathogens or signal other cells

The cells that defend mammalian bodies work together like cast members in a drama, interacting with one another and with the cells of invading pathogens. These cell–cell interactions are accomplished by a variety of key proteins, including receptors, other cell surface proteins, and signaling molecules. While these proteins will be discussed in more detail later in the chapter, four of the major players warrant mention here:

- **Antibodies** are proteins that bind specifically to certain substances identified by the immune system as nonself or altered self, thereby denaturing the invading nonself substance. They are secreted by B cells as defensive weapons.

- **T cell receptors** are integral membrane proteins on the surfaces of T cells. They recognize and bind to nonself substances on the surfaces of other cells.

- **Major histocompatibility complex (MHC)** proteins protrude from the surfaces of most cells in the mammalian body. They are important self-identifying labels and play major parts in coordinating interactions between lymphocytes and macrophages.

■ **Cytokines** are soluble signal proteins released by T cells, macrophages, and other cells. They bind to and alter the behavior of their target cells. Different cytokines activate or inactivate B cells, macrophages, and T cells. Some cytokines limit tumor growth by killing tumor cells.

18.1 RECAP

Animals have nonspecific and specific defenses against pathogens, both based on their ability to differentiate self from nonself. The specific defenses target specific invaders for destruction.

■ Do you understand the differences between specific and nonspecific defenses? See p. 401

■ What are the two classes of white blood cells, and how do they function in vertebrate defense systems? See pp. 402–403 and Figure 18.2

The role that the specific defenses may be called upon to play in fighting disease often depends on the success of the nonspecific responses to invading pathogens. We turn now to these nonspecific defenses that protect vertebrates from disease.

18.2 What Are the Characteristics of the Nonspecific Defenses?

Nonspecific defenses are general protection mechanisms that attempt to stop pathogens from invading the body. As noted above, they are the first lines of the body's defense system, in both time and location. While the specific defense systems of vertebrates evolved about 500 million years ago, the nonspecific defenses are far older. In humans, they include physical barriers as well as cellular and chemical defenses (**Table 18.1**).

Barriers and local agents defend the body against invaders

Skin is a primary nonspecific defense against invasion. Fungi, bacteria, and viruses rarely penetrate healthy, unbroken skin. But damage to the skin or to the internal surface tissues greatly increases the risk of infection by pathogens.

The bacteria and fungi that normally live and reproduce in great numbers on our body surfaces without causing disease are referred to as **normal flora**. These natural occupants of our bodies compete with pathogens for space and nutrients and are thus a form of nonspecific defense.

TABLE 18.1

Human Nonspecific Defenses

DEFENSIVE MECHANISM	FUNCTION
Surface barriers	
Skin	Prevents entry of pathogens and foreign substances
Acid secretions	Inhibit bacterial growth on skin
Mucus	Prevents entry of pathogens; produces defensins that kill pathogens
Mucous secretions	Trap bacteria and other pathogens in digestive and respiratory tracts
Nasal hairs	Filter bacteria in nasal passages
Cilia	Move mucus and trapped materials away from respiratory passages
Gastric juice	Concentrated HCl and proteases destroy pathogens in stomach
Acid in vagina	Limits growth of fungi and bacteria in female reproductive tract
Tears, saliva	Lubricate and cleanse; contain lysozyme, which destroys bacteria
Nonspecific cellular, chemical, and coordinated defenses	
Normal flora	Compete with pathogens; may produce substances toxic to pathogens
Fever	Body-wide response inhibits microbial multiplication and speeds body repair processes
Coughing, sneezing	Expels pathogens from upper respiratory passages
Inflammatory response (involves leakage of blood plasma and phagocytes from capillaries)	Limits spread of pathogens to neighboring tissues; concentrates defenses; digests pathogens and dead tissue cells; released chemical mediators attract phagocytes and lymphocytes to site
Phagocytes (macrophages and neutrophils)	Engulf and destroy pathogens that enter body
Natural killer cells	Attack and lyse virus-infected or cancerous body cells
Antimicrobial proteins	
Interferons	Released by virus-infected cells to protect healthy tissue from viral infection; mobilize specific defenses
Complement proteins	Lyse microorganisms, enhance phagocytosis, and assist in inflammatory and antibody responses

Skin is a major organ of the vertebrate body. In an "average" adult human, skin accounts for about 15 percent of the body's weight. A typical square inch of living human skin holds over 500 sweat glands, 20 blood vessels, and more than a thousand nerve endings.

The mucous membranes found at the surfaces of the visual, respiratory, digestive, excretory, and reproductive systems have other defenses against pathogens. Tears, nasal mucus, and saliva contain an enzyme called **lysozyme** that attacks the cell walls of many bacteria. Mucus in the nose traps airborne microorganisms, and most of those that get past this filter end up trapped in mucus deeper in the respiratory tract. Mucus and trapped pathogens are removed by the beating of cilia in the respiratory passageway, which continuously move a sheet of mucus and the debris it contains up toward the nose and mouth. Sneezing is another way to remove microorganisms from the respiratory tract.

Finally, the mucous membranes produce **defensins**, peptides that consist of 29–42 amino acids and contain hydrophobic domains. They are toxic to a wide range of pathogens, including bacteria, microbial eukaryotes, and enveloped viruses. Defensins insert themselves into the plasma membranes of these organisms and, by some unknown mechanism, kill the invaders. They are also produced in phagocytes, where they kill pathogens trapped by phagocytosis.

Pathogens that reach the digestive tract (stomach, small intestine, and large intestine) are met by other defenses. The gastric juice in the stomach is a deadly environment for many bacteria because of the hydrochloric acid and proteases (protein-digesting enzymes) that are secreted into it. The intact lining of the small intestine is not normally penetrated by bacteria, and some pathogens are killed by bile salts secreted into this part of the digestive tract. The large intestine harbors many bacteria, which multiply freely; however, they are usually removed quickly with the feces. Most of the bacteria in the large intestine are normal flora that provide benefits to their host. We probably add to this beneficial flora when we eat foods such as active-culture yogurt and various cheeses.

All of these barriers and local agents are *nonspecific* defenses because they act on all invading pathogens in the same way. More complex nonspecific defenses await any pathogens that manage to elude this first line of defense.

Other nonspecific defenses include specialized proteins and cellular processes

Pathogens that penetrate the body's outer and inner surfaces encounter more complex nonspecific defenses that involve the secretion of various defensive proteins as well as defensive cells.

COMPLEMENT PROTEINS Vertebrate blood contains about 30 different antimicrobial proteins that make up the **complement system**. These proteins, in different combinations, provide three types of defenses. In each type of defense, the complement proteins act in a characteristic sequence, or *cascade*, with each protein activating the next:

- First they attach to microbes, which helps phagocytes recognize and destroy them.
- Then they activate the inflammation response and attract phagocytes to the site of infection.
- Finally, they lyse (burst) invading cells (such as bacteria).

INTERFERONS When cells are infected by a virus, they produce small amounts of antimicrobial proteins called **interferons** that increase the resistance of neighboring cells to infection by the same *or other* viruses. Interferons have been found in many vertebrates and are one of the body's first lines of nonspecific defense against the internal spread of viral infection.

Interferons differ from species to species, and each vertebrate species produces at least three different interferons. All interferons are glycoproteins (proteins with attached carbohydrate groups) consisting of about 160 amino acids. By binding to receptors in the plasma membranes of uninfected cells, interferons stimulate a signaling pathway that inhibits viral reproduction inside the infected cells. They also stimulate lysosome activity that digests viral proteins into peptides, which, when brought to the cell surface, stimulate the specific immune system (see Section 18.5).

PHAGOCYTES Some phagocytes travel freely in the circulatory and lymphatic systems; others can move out of blood vessels and adhere to certain tissues. Pathogenic cells, viruses, or fragments of these invaders can become attached to the plasma membrane of a phagocyte (**Figure 18.3**), which ingests them by phagocytosis. Defensins inside these phagocytes then kill the pathogens.

NATURAL KILLER CELLS One class of lymphocytes, known as **natural killer cells**, can distinguish virus-infected cells and some tumor cells from their normal counterparts and initiate the lysis of these target cells. In addition to this nonspecific action, natural killer cells interact with the specific defenses.

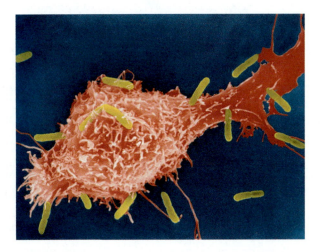

18.3 A Phagocyte and Its Bacterial Prey Several bacteria (yellow in this artificially colored micrograph) have become attached to the surface of a phagocyte in the human bloodstream. The bacteria will be engulfed by the phagocyte and destroyed. A single phagocyte can digest many bacteria.

Inflammation is a coordinated response to infection or injury

The body employs the **inflammation** response in dealing with infection or with any other process that causes tissue injury, either on the surface of the body or internally. The damaged body cells initiate the inflammation response by releasing small molecules and enzymes. Cells adhering to the skin and the linings of organs, called **mast cells**, release a chemical signal, called **histamine**, when they are damaged, as do white blood cells called **basophils**.

You have no doubt experienced the symptoms of inflammation: redness and swelling, accompanied by heat and pain. The redness and heat of inflammation result from histamine-induced dilation of blood vessels in the infected or injured area (**Figure 18.4**). Histamine also causes the capillaries (the smallest blood vessels) to become leaky, allowing blood plasma, along with complement proteins and phagocytes, to escape into the tissue, causing the characteristic swelling. The pain of inflammation results from increased pressure (from the swelling) and from the action of leaked enzymes on sensory nerve endings.

In damaged or infected tissue, complement proteins and other chemical signals attract phagocytes—neutrophils first, and then monocytes, which develop into macrophages. The phagocytes, which engulf the invaders and dead tissue cells, are responsible for most of the healing associated with inflammation. They produce several cytokines, which (among other functions) signal the brain to produce a fever. This rise in body temperature inhibits the growth of the invading pathogens.

Cytokines may also stimulate the cells lining the blood vessels to produce cell adhesion molecules that allow phagocytes in the capillaries to attach to the lining, leave the vessel, and enter the site of the injury. The arrival of phagocytes initiates a specific response to the pathogen.

Calor, dolor, rubor, tumor—the diagnosis of inflammation remains unchanged since 40 A.D. when the Roman medical scribe Aulus Cornelius Celsus described the syndrome in his text *De medicina.*

In some severe bacterial infections, the inflammation response does not remain local. Instead, it is disseminated throughout the bloodstream in a condition called *sepsis*. As in a local infection or injury, blood vessels dilate, but they do so throughout the body. This situation is a medical emergency and can be lethal.

Following inflammation, *pus* may accumulate. Pus is a collection of dead cells (bacteria, neutrophils and the damaged body cells) and leaked fluid. A normal result of inflammation, pus is gradually consumed and digested by macrophages.

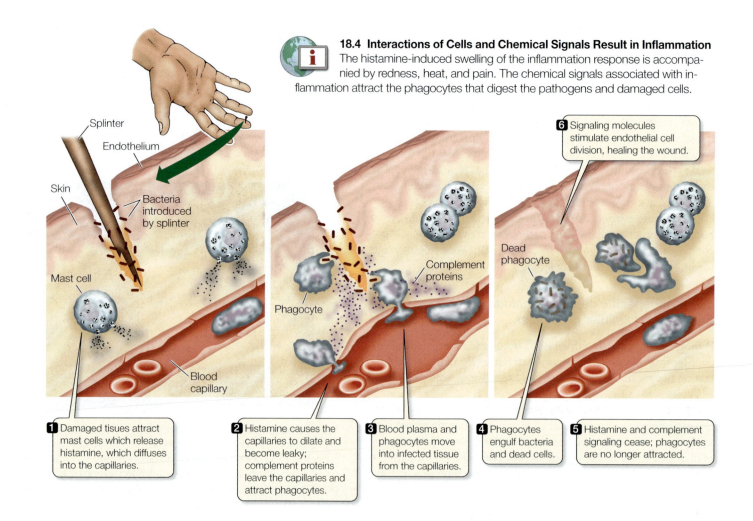

18.4 Interactions of Cells and Chemical Signals Result in Inflammation The histamine-induced swelling of the inflammation response is accompanied by redness, heat, and pain. The chemical signals associated with inflammation attract the phagocytes that digest the pathogens and damaged cells.

6 Signaling molecules stimulate endothelial cell division, healing the wound.

1 Damaged tisues attract mast cells which release histamine, which diffuses into the capillaries.

2 Histamine causes the capillaries to dilate and become leaky; complement proteins leave the capillaries and attract phagocytes.

3 Blood plasma and phagocytes move into infected tissue from the capillaries.

4 Phagocytes engulf bacteria and dead cells.

5 Histamine and complement signaling cease; phagocytes are no longer attracted.

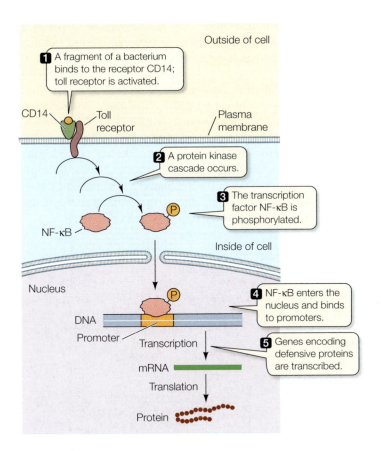

18.5 Cell Signaling and Defense Binding of a pathogenic molecule to the toll receptor initiates a signal transduction pathway which results in the transcription of genes whose products are involved in specific and nonspecific defenses. CD14 is expressed on the surface of white blood cells, including macrophages and monocytes.

A cell signaling pathway stimulates the body's defenses

An invading pathogen such as a bacterium can be regarded as a signal. In response to that signal, the body produces molecules such as complement proteins, interferons, and cytokines that regulate phagocytosis and other defense processes. Not surprisingly, the link between signal and response is a signal transduction pathway, similar to the ones we considered in Section 15.3. The receptor in this pathway is a membrane protein called **toll**. This receptor was originally discovered in fruit flies, in which it plays an essential role in sensing infection by fungi. Comparative genomics has revealed at least ten similar receptors in humans.

Toll is part of a protein kinase cascade that ultimately results in the transcription of at least 40 genes involved in both nonspecific and specific defenses (**Figure 18.5**). The molecules that stimulate this pathway are made only by microbes and include some bacterial and fungal cell wall fragments. Binding of these molecules to toll sets in motion a cascade that results in the phosphorylation of the transcription factor NF-κB. As a result, the transcription factor's conformation changes, allowing it to enter the nucleus, bind to the promoters of genes encoding defensive proteins, and activate their transcription.

Nonspecific defenses are the first line of defense against pathogens. Barriers such as the skin, defensive proteins, and coordinated responses such as inflammation are important nonspecific defenses.

- How do complement proteins and interferons defend the body against microbes? See p. 405
- Can you describe the inflammation response? See p. 406 and Figure 18.4

How does the body deal with pathogens that get by these nonspecific defenses? The next section describes the development of immunity to specific pathogens.

18.3 How Does Specific Immunity Develop?

The body's nonspecific defenses are numerous and effective, but some invaders elude them. Vertebrates animals deal with these pathogens by means of defenses targeted against specific threats. In this section we outline the main features of the specific immune system and consider the two major types of specific responses: the *humoral immune response*, which produces antibodies, and the *cellular immune response*, which destroys infected cells.

The specific immune system has four key traits

The four characteristic features of the specific immune system are specificity; the ability to respond to an enormous diversity of foreign molecules and organisms; the ability to distinguish self from nonself; and immunological memory.

SPECIFICITY Lymphocytes (B cells and T cells) are crucial to specific immunity. T cell receptors and the antibodies produced by B cells recognize and bind to specific nonself or altered-self substances (**antigens**), and this interaction initiates a specific immune response. The specific sites on antigens that the immune system recognizes are called **antigenic determinants** or *epitopes*:

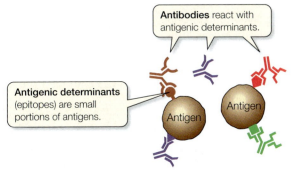

Chemically, an antigenic determinant is a specific portion of a large molecule, such as a certain sequence of amino acids that may be present in several proteins. A large antigen, such as a whole cell, may have many different antigenic determinants on its surface, each capable of being bound by a specific antibody or T cell. Even a single protein has multiple, different antigenic determinants. Some epitopes provoke a more powerful immune response than others and are referred to as *immunodominant*. The host animal responds to the presence of an antigen by producing highly specific defenses: T cells or antibodies that are complementary to, or fit, the antigenic determinants of that antigen. Each T cell and each antibody is *specific for a single antigenic determinant*.

DISTINGUISHING SELF FROM NONSELF The human body contains tens of thousands of different proteins, each with a specific three-dimensional structure capable of generating immune responses. Thus every cell in the body bears a tremendous number of antigenic determinants. A crucial requirement of an individual's immune system is that it recognize the body's own antigenic determinants and not attack them.

DIVERSITY Challenges to the immune system are numerous. Pathogens take many forms: individual foreign molecules, viruses, bacteria, protists, fungi, and multicellular parasites. Furthermore, each pathogenic species usually exists in many subtly differing genetic strains, and each strain possesses multiple surface features. Estimates vary, but a reasonable guess is that humans can respond *specifically* to 10 million different antigenic determinants. Upon recognizing an antigenic determinant, the immune system responds by activating lymphocytes of the appropriate specificity.

IMMUNOLOGICAL MEMORY After responding to a particular type of pathogen once, the immune system "remembers" that pathogen and can usually respond more rapidly and powerfully to the same threat in the future. This **immunological memory** usually saves us from repeats of childhood diseases such as chicken pox. Vaccination against specific diseases works because the immune system "remembers" the antigenic determinants that are introduced into the body.

Two types of specific immune responses interact

The specific immune system mounts two types of responses against invaders: the *humoral immune response* and the *cellular immune response*. These two responses operate in concert—simultaneously and cooperatively, sharing many mechanisms.

HUMORAL IMMUNE RESPONSE In the **humoral immune response** (from the Latin *humor*, "fluid"), antibodies react with antigenic determinants on pathogens in blood, lymph, and tissue fluids. An animal can produce a staggering diversity of antibodies capable of binding to almost any conceivable antigen the animal encounters.

Some antibodies are soluble and travel free in the blood and lymph; others exist as integral membrane proteins on B cells. The first time a specific antigen invades the body, it may be detected and bound by a B cell whose membrane antibody recognizes one of its antigenic determinants. This binding activates the B cell, which makes and secretes multiple copies of an antibody with the same specificity as its membrane antibody.

CELLULAR IMMUNE RESPONSE The **cellular immune response** is directed against antigens that have become established within a cell of the host animal. It detects and destroys virus-infected or mutated cells.

The cellular immune response is carried out by T cells within the lymph nodes, the bloodstream, and the intercellular spaces. These T cells have integral membrane proteins—T cell receptors—that recognize and bind to antigenic determinants. T cell receptors are rather similar to antibodies in structure and function, each including specific molecular configurations that bind to specific antigenic determinants. Once a T cell is bound to an antigenic determinant, it initiates an immune response that typically results in the total destruction of the nonself or altered self cell.

Genetic changes and clonal selection generate the specific immune response

How does the tremendous diversity of the specific immune response arise? How do lymphocytes specific for certain antigenic determinants proliferate? The answers lie in the process of **clonal selection**.

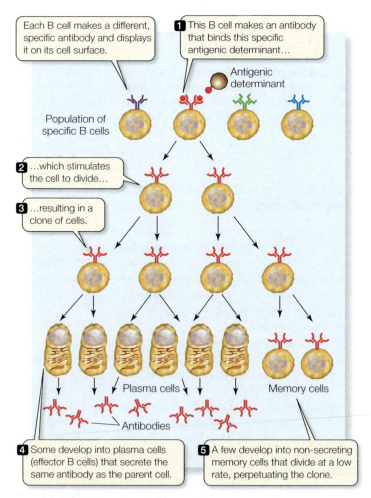

Each B cell makes a different, specific antibody and displays it on its cell surface.

1 This B cell makes an antibody that binds this specific antigenic determinant…

Antigenic determinant

Population of specific B cells

2 …which stimulates the cell to divide…

3 …resulting in a clone of cells.

Plasma cells

Memory cells

Antibodies

4 Some develop into plasma cells (effector B cells) that secrete the same antibody as the parent cell.

5 A few develop into non-secreting memory cells that divide at a low rate, perpetuating the clone.

18.6 Clonal Selection in B Cells The binding of an antigen to a specific antibody on the surface of a B cell stimulates that cell to divide, producing a clone of genetically identical cells to fight that invader.

■ *Diversity is generated primarily by DNA changes*—chromosomal rearrangements and mutations—that occur just after the B and T cells are formed in the bone marrow. Each B cell is able to produce *only one kind of antibody*. Thus there are millions of different B cells, each one producing a particular antibody. Similarly, there are millions of different T cells, each of which has a unique type of T cell receptor that binds to a particular antigenic determinant on a target cell.

■ *Antigen binding "selects" a particular B or T cell for proliferation.* When an antigen that fits the surface antibody on a B cell binds to it, that B cell is activated. It divides to form a *clone* of cells (a genetically identical group derived from a single cell), all of them producing that particular antibody (**Figure 18.6**). In the same way, a foreign or abnormal cell *selects* for the proliferation of a T cell expressing a particular T cell receptor on its surface (hence, *clonal selection*).

Immunity and immunological memory result from clonal selection

An activated lymphocyte produces two types of daughter cells, *effector cells* and *memory cells*.

■ **Effector cells** carry out the attack on the antigen. Effector B cells, called **plasma cells**, secrete antibodies. Effector T cells release cytokines, which initiate reactions that destroy nonself or altered cells. Effector cells live only a few days.

■ **Memory cells** are long-lived cells that retain the ability to start dividing on short notice to produce more effector and more memory cells. Memory B and possibly T cells may survive in the body for decades, dividing at a low rate.

Between them, these two types of lymphocytes can respond to an antigen in two different ways:

■ When the body first encounters a particular antigen, a **primary immune response** is activated, in which the "naive" lymphocytes that recognize that antigen proliferate to produce clones of effector and memory cells. The effector cells destroy the invaders and then die, but one or more clones of memory cells have now been added to the immune system and provide immunological memory.

■ After a primary immune response to a particular antigen, subsequent encounters with the same antigen will trigger a much more rapid and powerful **secondary immune response**. The memory cells that bind with that antigen proliferate, launching a huge army of plasma cells and effector T cells.

The first time a vertebrate animal is exposed to a particular antigen, there is a time lag (usually several days) before the number of antibody molecules and T cells specific to that antigen slowly increases. But for years afterward—sometimes for life—the immune system "remembers" that particular antigen. The secondary immune response is characterized by a shorter lag time, a greater rate of antibody production, and a larger total production of antibodies or T cells than the primary immune response.

TABLE 18.2

Some Human Pathogens for Which Vaccines are Available

INFECTIOUS AGENT	DISEASE	VACCINATED POPULATION
Bacteria		
Bacillus anthracis	Anthrax	Those at risk in biological warfare
Bordetella pertussis	Whooping cough	Children and adults
Clostridium tetani	Tetanus	Children and adults
Corynebacterium diphtheriae	Diphtheria	Children
Haemophilus influenzae	Meningitis	Children
Mycobacterium tuberculosis	Tuberculosis	All people
Salmonella typhi	Typhoid fever	Areas exposed to agent
Streptococcus pneumoniae	Pneumonia	Elderly
Vibrio cholerae	Cholera	People in areas exposed to agent
Viruses		
Adenovirus	Respiratory disease	Military personnel
Hepatitis A	Liver disease	Areas exposed to agent
Hepatitis B	Liver disease, cancer	All people
Influenza virus	Flu	All people
Measles virus	Measles	Children and adolescents
Mumps virus	Mumps	Children and adolescents
Poliovirus	Polio	Children
Rabies virus	Rabies	Persons exposed to agent
Rubella virus	German measles	Children
Vaccinia virus	Smallpox	Laboratory workers, military personnel
Varicella-zoster virus	Chicken pox	Children

Vaccines are an application of immunological memory

Thanks to immunological memory, recovery from many diseases, such as chicken pox, provides a *natural immunity* to those diseases. However, it is possible to provide *artificial immunity* against many life-threatening diseases by *inoculation*: the introduction of antigenic determinants into the body. **Immunization** is inoculation with antigenic proteins, pathogen fragments, or other molecular antigens. **Vaccination** is inoculation with whole pathogens that have been modified so that they cannot cause disease.

Immunization or vaccination initiates a primary immune response, generating memory cells without making the person ill. Later, if the same or very similar pathogens attack, specific memory cells already exist. They recognize the antigen and quickly overwhelm the invaders with a massive production of lymphocytes and antibodies.

Because the antigens used for immunization or vaccination are either parts of or toxic proteins produced by a pathogenic organism, they must be altered so that they cannot cause disease but are still able to provoke an immune response. There are four principal ways to do this:

- *Inactivation* involves treating the antigen—in this case, usually a whole organism such as a bacterium—with heat or chemicals so that it is killed.
- *Attenuation* involves reducing the virulence of a virus by repeatedly infecting cells with it in the laboratory.
- *Recombinant DNA technology* can be used to produce peptide fragments that bind to and activate lymphocytes but do not have the harmful part of a protein toxin.
- *DNA vaccines* are being developed that will introduce a gene encoding an antigen into the body.

Although smallpox vaccination programs have seemingly rid the world of this virus (with everyone vaccinated, the virus simply had nowhere to go), it still exists in several national laboratories under strict security. Because other supplies of the virus may exist and might be used in biological warfare, the known supplies of the virus are being used to generate new vaccines.

For most of the 70 bacteria, viruses, fungi, and parasites known to cause serious human diseases, vaccines are already available or will be in the next few years (**Table 18.2**). Vaccination has completely or almost completely wiped out some deadly diseases, such as smallpox, diphtheria, and polio, in industrialized countries.

Animals distinguish self from nonself and tolerate their own antigens

Normally, the body is tolerant of its own molecules—the same molecules that would generate an immune response in another individual. **Immunological tolerance** is based on two mechanisms: *clonal deletion* and *clonal anergy*.

CLONAL DELETION   Clonal deletion removes certain immature B and T cells from the immune system early in their differentiation. Immature B cells in the bone marrow and T cells in the thymus may encounter self antigens. Any immature B or T cell that shows the potential to mount an immune response against self antigens undergoes programmed cell death (apoptosis) within a short time.

CLONAL ANERGY   Clonal anergy is the suppression of the immune response to self antigens and occurs after lymphocytes mature. A mature T cell, for example, may encounter and recognize a self antigen on the surface of a body cell. But before it sends out the cytokines

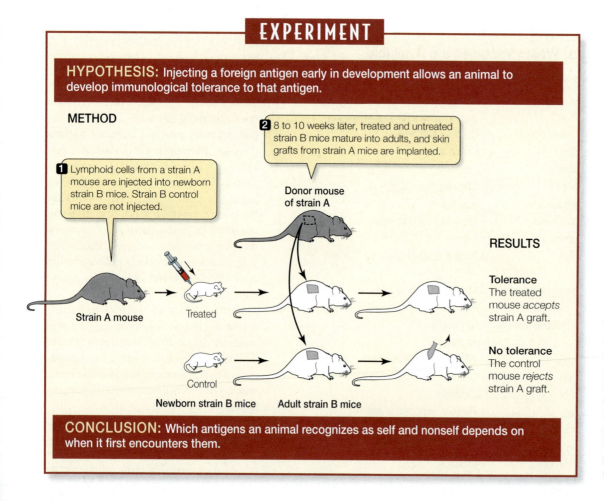

EXPERIMENT

HYPOTHESIS: Injecting a foreign antigen early in development allows an animal to develop immunological tolerance to that antigen.

METHOD

1. Lymphoid cells from a strain A mouse are injected into newborn strain B mice. Strain B control mice are not injected.

2. 8 to 10 weeks later, treated and untreated strain B mice mature into adults, and skin grafts from strain A mice are implanted.

Donor mouse of strain A

Strain A mouse

Treated

Control

Newborn strain B mice   Adult strain B mice

RESULTS

Tolerance
The treated mouse *accepts* strain A graft.

No tolerance
The control mouse *rejects* strain A graft.

CONCLUSION: Which antigens an animal recognizes as self and nonself depends on when it first encounters them.

18.7 Immunological Tolerance   A mouse can tolerate a skin graft from a genetically different mouse if lymphoid cells from the second mouse are injected into the first one as a newborn. FURTHER RESEARCH: What would happen if the lymphoid cells were heated in a boiling water bath prior to injection into strain B mice?

that signal the initiation of an immune response, the T cell must encounter not only an antigen, but also a second molecule, CD28, on the cell surface. Most body cells lack CD28 and thus will not be attacked by the cellular immune response. CD28 is an example of a *co-stimulatory signal* that is expressed only on certain antigen-presenting cells. Such cells "present" nonself antigens on their surfaces, thus stimulating the cellular immune response, as we will see in Section 18.5. They include the macrophages that wander through the body's fluids and the *dendritic cells* that reside among the linings of the respiratory and digestive tracts, as well as B cells.

Immunological tolerance was discovered through the observation that some *nonidentical* twin cattle with different blood types contained some of each other's red blood cells. Why didn't these "foreign" blood cells cause immune responses resulting in their elimination? One hypothesis was that the blood cells had passed between the fetal animals in the womb before the lymphocytes had matured, so that each calf regarded the other's red blood cells as self. This hypothesis was confirmed when it was shown that injecting a foreign antigen into an animal early in its fetal development caused that animal henceforth to recognize that antigen as self (**Figure 18.7**).

Now that we understand the general features of the specific immune system, let's focus in more detail on the B lymphocytes and the humoral response.

 ## 18.4 What Is the Humoral Immune Response?

Every day, billions of B cells survive the test of clonal deletion and are released from the bone marrow into the circulation. B cells are the basis for the humoral immune response.

Some B cells develop into plasma cells

B cells begin by making an antibody that is expressed as a receptor protein on the cell surface. As we have seen, if a B cell is activated by antigen binding to this receptor, it becomes a plasma cell,

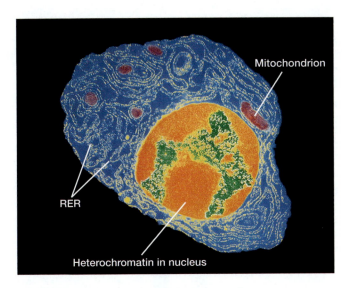

18.8 A Plasma Cell The prominent nucleus with large amounts of heterochromatin (orange) and the cytoplasm (bright blue) crowded with rough endoplasmic reticulum (RER) are features of a cell that is actively synthesizing and exporting proteins—in this case, a specific antibody. Whole blocks of genes not needed for this specialized function are kept turned off in the heterochromatin.

making an antibody protein that is secreted into the bloodstream, and gives rise to a clone of plasma cells as well as memory cells (see Figure 18.6).

Usually, for a B cell to develop into an antibody-secreting plasma cell, a *helper T cell* (T_H) with the same specificity must also bind to the antigen. Thus the B cell also functions as an antigen-presenting cell, as we will see in Section 18.5. The division and differentiation of the B cell is stimulated by the receipt of chemical signals from the T_H cell.

As plasma cells develop, the number of ribosomes and the amount of endoplasmic reticulum in their cytoplasm increase greatly (**Figure 18.8**). These increases allow the cells to synthesize and secrete large amounts of antibody proteins, up to 2,000 per second! All the plasma cells arising from a given B cell produce antibodies that are specific for the antigenic determinant that originally bound to the parent B cell. Thus antibody specificity is maintained as B cells proliferate.

Different antibodies share a common structure

Antibodies belong to a class of proteins called **immunoglobulins**. There are several types of immunoglobulins, but all contain a tetramer consisting of four polypeptide chains. In each immunoglobulin molecule, two of these polypeptides are identical *light chains*, and two are identical *heavy chains*. Disulfide bonds hold the chains together. Each polypeptide chain has a *constant region* and a *variable region* (**Figure 18.9**):

- The amino acid sequences of the **constant regions** are similar among the immunoglobulins. They determine the destination and function—the *class*—of the antibody.

(A)

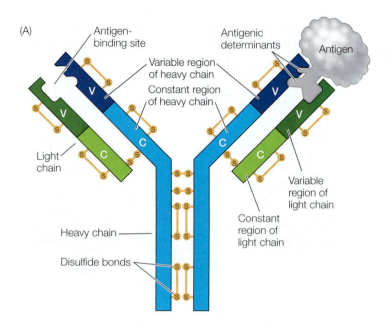

Antigen-binding site
Antigenic determinants
Antigen
Variable region of heavy chain
Constant region of heavy chain
Light chain
Variable region of light chain
Constant region of light chain
Heavy chain
Disulfide bonds

(B)

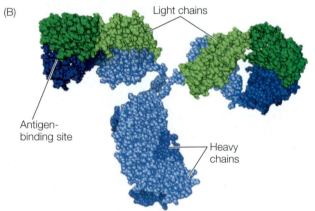

Light chains
Antigen-binding site
Heavy chains

18.9 The Structure of Immunoglobulins (A) The four polypeptide chains (two light, two heavy) of an immunoglobulin molecule. (B) A three-dimensional space-filling model of an antibody molecule in roughly the same orientation as (A).

■ The amino acid sequences of the **variable regions** are different for each specific immunoglobulin. Differences in their secondary structure result in differences in the three-dimensional antigen-binding site of each immunoglobulin and are responsible for antibody specificity.

The two antigen-binding sites on each immunoglobulin molecule are identical, making the antibody *bivalent* (*bi*, "two"; *valent*, "binding"). This ability to bind two antigen molecules at once permits the antibody to form a large complex with antigen and other antibody molecules. Such a complex is an easy target for ingestion and breakdown by phagocytes and complement.

There are five classes of immunoglobulins

While the variable regions are responsible for the *specificity* of an immunoglobulin, the constant regions of the heavy chain determine the *class* of the antibody—for example, whether it will be an integral membrane receptor or a soluble antibody that is secreted into the bloodstream. The five immunoglobulin classes are described in **Table 18.3**. The most abundant immunoglobulin class is IgG; these soluble antibody proteins make up about 80 percent of the total immunoglobulin content of the bloodstream. They are made in greatest quantity during a secondary immune response. IgG defends the body in several ways. For example, after some IgG

TABLE 18.3

Antibody Classes

CLASS	GENERAL STRUCTURE		LOCATION	FUNCTION
IgG	Monomer		Free in blood plasma; about 80 percent of circulating antibodies	Most abundant antibody in primary and secondary immune responses; crosses placenta and provides passive immunization to fetus
IgM	Pentamer		Surface of B cell; free in blood plasma	Antigen receptor on B cell membrane; first class of antibodies released by B cells during primary response
IgD	Monomer		Surface of B cell	Cell surface receptor of mature B cell; important in B cell activation
IgA	Dimer		Saliva, tears, milk, and other body secretions	Protects mucosal surfaces; prevents attachment of pathogens to epithelial cells
IgE	Monomer		Secreted by plasma cells in skin and tissues lining gastrointestinal and respiratory tracts	When bound to antigens, binds to mast cells and basophils to trigger release of histamine that contributes to inflammation and some allergic responses

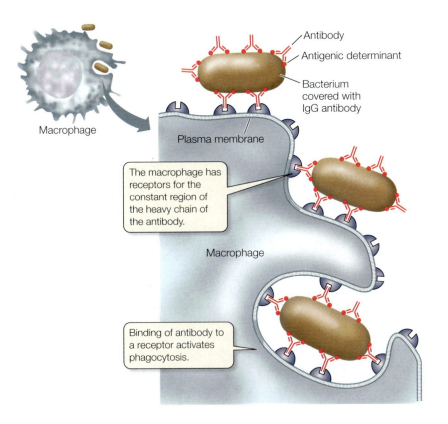

Antibody

Antigenic determinant

Bacterium covered with IgG antibody

Plasma membrane

The macrophage has receptors for the constant region of the heavy chain of the antibody.

Macrophage

Binding of antibody to a receptor activates phagocytosis.

Macrophage

18.10 IgG Antibodies Promote Phagocytosis When IgG antibodies bind to a bacterium, receptors on a macrophage can recognize and bind to those antibodies.

molecules bind to antigens, they become attached by their heavy chains to macrophages. This attachment permits the macrophages to destroy the antigens by phagocytosis (**Figure 18.10**).

Monoclonal antibodies have many uses

The specificity of antibodies suggested to scientists that they might be useful for detecting a specific substance in a fluid. An initial challenge to scientists seeking to accomplish this was that the immune response to a complex antigen is *polyclonal*—that is, because most antigens carry many different antigenic determinants, a single antigen will produce a complex mixture of antibodies, each made by a different clone of B cells.

Suppose that a physician wishes to measure the level of the hormone estrogen in a woman's blood. This could be done by adding an antibody specific for estrogen to a sample of her blood and observing how much antigen–antibody complex formed. But, as emphasized in our study of biochemistry, many biological molecules share regions of similar structure—all human steroid hormones, for example, have a similar multi-ring structure (see Figure 3.22). A polyclonal group of antibodies targeted to estrogen would be uninformative because the antibodies would bind not just to estrogen, but to any steroid hormone present in the blood sample. What is needed is a clone of B cells to produce large amounts of an antibody that binds to *only one* antigenic determinant—a **monoclonal antibody**. How could such a clone be produced?

A clone of cells that produce a single antibody can be made artificially by fusing a B cell (which has a finite lifetime and makes a lot of antibody) with a tumor cell (which has an infinite lifetime and can be grown in culture). The resulting hybrid cell, called a **hybridoma**, makes a specific monoclonal antibody and proliferates in culture (**Figure 18.11**).

A Canadian company called "Toxin Alert" has developed a plastic food wrap impregnated with antibodies against *Listeria, E. coli,* and *Salmonella* bacteria, which together account for 80 percent of food-related deaths in the developed world. The antibodies are distributed in an "X" pattern on the wrap, so if bacteria are present, the antibodies bind and a colored "X" appears on the package.

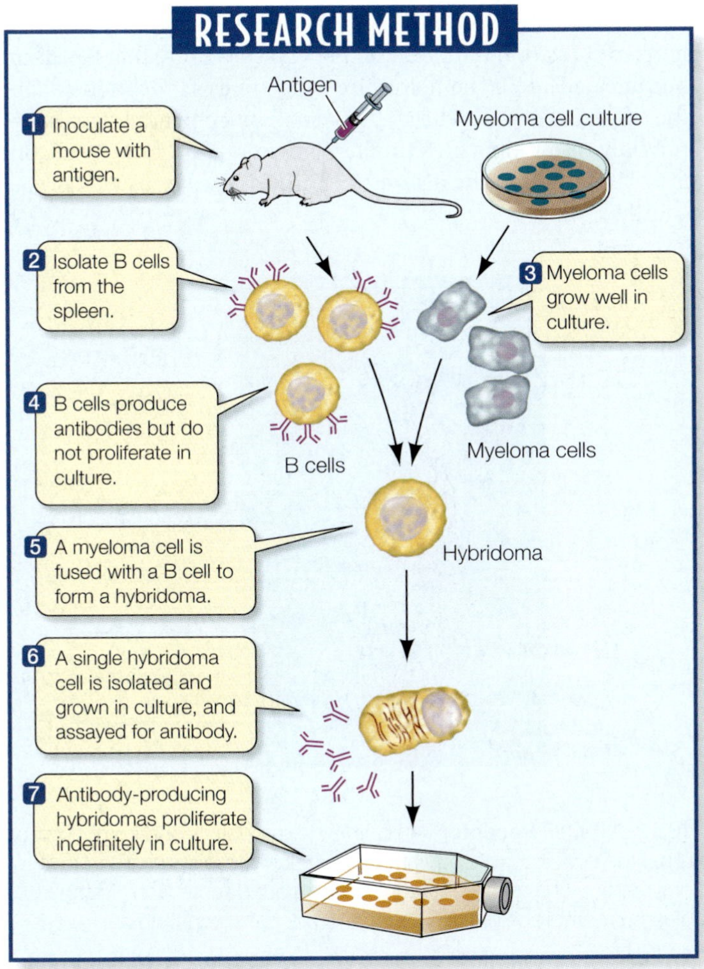

RESEARCH METHOD

Antigen

Myeloma cell culture

1 Inoculate a mouse with antigen.

2 Isolate B cells from the spleen.

3 Myeloma cells grow well in culture.

4 B cells produce antibodies but do not proliferate in culture.

B cells

Myeloma cells

5 A myeloma cell is fused with a B cell to form a hybridoma.

Hybridoma

6 A single hybridoma cell is isolated and grown in culture, and assayed for antibody.

7 Antibody-producing hybridomas proliferate indefinitely in culture.

18.11 Creating Hybridomas for the Production of Monoclonal Antibodies Cancerous myeloma cells and normal B cells can be fused so that the proliferative properties of the myeloma cells are merged with the specificity of the antibody-producing B cells.

Monoclonal antibodies have many practical applications:

- *Immunoassays* use the specificity of monoclonal antibodies to detect tiny amounts of molecules in tissues and fluids. This technique is used, for example, in pregnancy tests to detect the hormone made by the developing embryo.

- *Immunotherapy* uses monoclonal antibodies targeted against antigenic determinants on the surfaces of cancer cells. The coupling of a radioactive ligand or toxin to the antibody makes it into a medical "smart bomb." In some cases, binding of the antibody itself is enough to trigger a cellular immune response that destroys the cancer. (This is the case with Herceptin,® a monoclonal antibody that binds to a growth factor receptor on breast cancer cells.

- *Passive immunization* is inoculation with an immediately acting, but not long-lasting, monoclonal antibody. This approach is necessary when therapy must be effective quickly (within hours). Examples of such life-threatening situations include the early symptoms of rabies infection, rattlesnake bites, and babies born with hepatitis B virus infection—all cases in which the toxic nature of the infection is so serious that there is not enough time to allow the person's immune system to mount its own defense.

18.4 RECAP

The humoral immune response is based on the synthesis by B cells of specific antibodies directed against specific antigens. The specificity of an antibody derives from the amino acid sequence of its variable regions. Monoclonal antibodies are specific to one antigenic determinant and can be produced artificially for use in diagnostics and therapy.

- Can you describe the B cell's response to an antigen? See p. 411

- How are the structure and function of an antibody molecule related? See p. 412 and Figure 18.9

- Can you explain what a monoclonal antibody is and how are they used? See pp. 413–414 and Figure 18.11

Both T cells and B cells are involved in the humoral immune response. T cells are also the effectors of the cellular immune response, which we explore in the next section.

 ## 18.5 What Is the Cellular Immune Response?

The effector molecules of the humoral immune response are the antibodies secreted by plasma cells that develop from activated B cells. The *cellular immune response*, which is directed against any factor (e.g., a virus or mutation) that changes a normal cell into an abnormal cell, is directed by T cells.

Two types of effector T cells (*helper T cells* and *cytotoxic T cells*) are involved in the cellular immune response, along with the *MHC*

(*major histocompatibility complex*) proteins, which underlie the immune system's tolerance for the body's own cells.

T cell receptors are found on two types of T cells

Like B cells, T cells possess specific membrane receptors. T cell receptors are not immunoglobulins, however, but glycoproteins with molecular weights about half that of an IgG. They are made up of two polypeptide chains, each encoded by a separate gene (**Figure 18.12**). Thus the two chains are nearly always different in their amino acid sequence, especially in their variable regions.

The genes that code for T cell receptors are similar to those that code for immunoglobulins, suggesting that both are derived from a single, evolutionarily more ancient group of genes. Like the immunoglobulins, T cell receptors include both variable and constant regions, and the variable regions are responsible for their specificity. There is one major difference: whereas antibodies bind to an intact antigen, T cell receptors bind to a piece of an antigen displayed on the surface of an antigen-presenting cell.

When a T cell is activated by contact with a specific antigenic determinant, it proliferates and forms a clone. Its descendants differentiate into two types of effector T cells:

- **Cytotoxic T cells**, or T_C cells, recognize virus-infected or mutated cells and kill them by inducing lysis (see page 401).

- **Helper T cells**, or T_H cells, assist both the cellular and humoral immune responses.

As mentioned in Section 18.4, a specific T_H cell must bind to an antigenic determinant presented on a B cell before that B cell can become activated. The helper T cell becomes the "conductor" of the "immunological orchestra" as it sends out chemical signals that not only result in its own proliferation and that of the B cell, but also set in motion the actions of cytotoxic T cells, as we will see shortly.

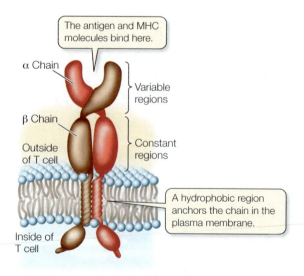

18.12 A T Cell Receptor In both T cell receptors and immunoglobulins, the specificity of each antigen-binding site is determined by two polypeptide chains. T cell receptors are bound more firmly to the plasma membrane of the T cell that produces them than antibodies are to B cells.

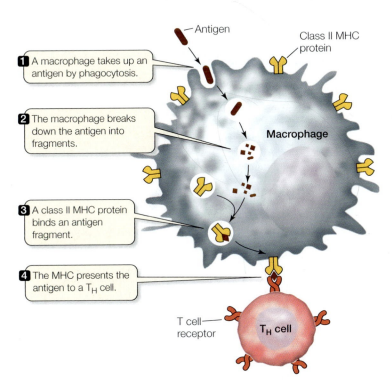

1 A macrophage takes up an antigen by phagocytosis.

2 The macrophage breaks down the antigen into fragments.

3 A class II MHC protein binds an antigen fragment.

4 The MHC presents the antigen to a T_H cell.

Antigen

Class II MHC protein

Macrophage

T cell receptor

T_H cell

18.13 Macrophages Are Antigen-Presenting Cells A fragment of an antigen is displayed by MHC II on the surface of a macrophage. T cell receptors on a specific helper T cell can then bind to and interact further with the antigen–MHC II complex.

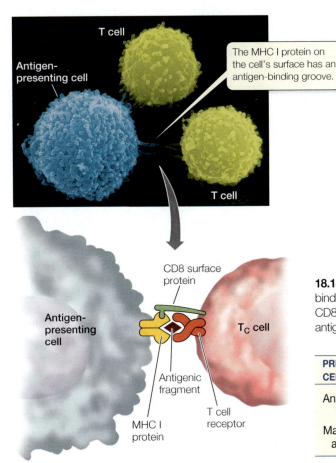

T cell

Antigen-presenting cell

The MHC I protein on the cell's surface has an antigen-binding groove.

T cell

CD8 surface protein

Antigen-presenting cell

Antigenic fragment

T_C cell

T cell receptor

MHC I protein

Now that we are familiar with the two types of T cells, we can address the question of how T cells recognize their antigenic determinants and the role the MHC proteins play in the process.

The MHC encodes proteins that present antigens to the immune system

We have seen that an animal's immune system recognizes the body's own cells by their surface proteins. Several types of mammalian cell surface proteins are involved in this process, but we will focus here on the products of a cluster of genes called the **major histocompatibility complex**, or **MHC**. These proteins have important roles in the cellular and humoral immune responses as well as in self-tolerance.

The MHC gene products are plasma membrane glycoproteins. In humans, the MHC proteins are called *human leukocyte antigens* (HLA), while in mice they are called *H-2 proteins*. Their major role is to present antigens to a T cell receptor in such a way that it can distinguish between self and nonself antigens. There are two classes of MHC proteins:

■ *Class I MHC proteins* are present on the surface of every nucleated cell in the animal body. When cellular proteins are degraded into small peptide fragments by a proteasome (see Section 14.6), an MHC I protein may bind to a fragment and travel to the plasma membrane. There, the MHC I protein "presents" the cellular peptide to T_C cells. The T_C cells have a surface protein called CD8 that recognizes and binds to MHC I.

■ *Class II MHC proteins* are found mostly on the surfaces of B cells, macrophages, and other antigen-presenting cells. When an antigen-presenting cell ingests a nonself antigen, such as a virus, the antigen is broken down in a phagosome. An MHC II molecule may bind to one of the fragments and carry it to the cell surface, where it is presented to a T_H cell (**Figure 18.13**). T_H cells have a surface protein called CD4 that recognizes and binds to MHC II.

To accomplish their roles in antigen presentation, both MHC I and MHC II proteins have an antigen-binding site, which can hold a peptide of about 10–20 amino acids (**Figure 18.14**). The T cell receptor recognizes not just the antigenic fragment, but the fragment *bound to an MHC I or MHC II molecule*. The table in Figure 18.14 summarizes the relationships of T cells and antigen-presenting cells.

In humans, there are three genetic loci for MHC I and three for MHC II; each of these six loci has as many as 100 different alleles.

18.14 The Interaction between T Cells and Antigen-Presenting Cells An antigen-binding site in the MHC I protein holds an antigen, which it presents to cytotoxic T cells. CD8 surface proteins on T_C cells bind to MHC I, and specific T cell receptors bind to the antigen. The binding of MHC II protein by T_H cells works in a similar manner.

PRESENTING CELL TYPE	ANTIGEN PRESENTED	MHC CLASS	T CELL TYPE	T CELL SURFACE PROTEIN
Any cell	Intracellular protein fragment	Class I	Cytotoxic T cell (T_C)	CD8
Macrophages and B cells	Fragments from extracellular proteins	Class II	Helper T cell (T_H)	CD4

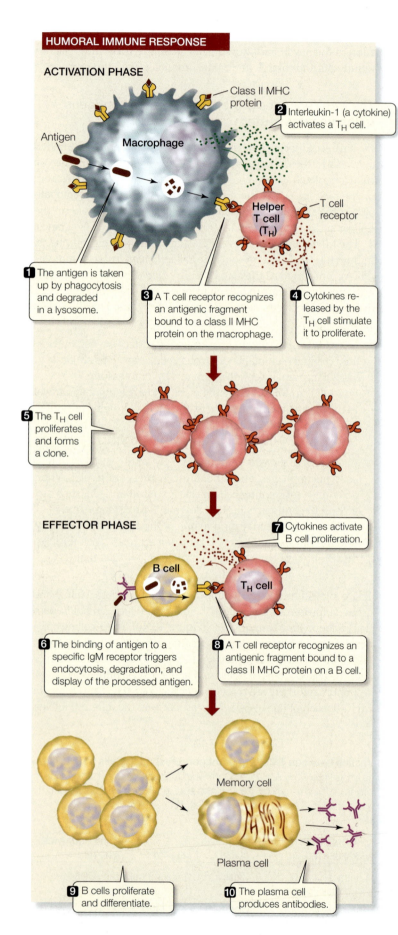

HUMORAL IMMUNE RESPONSE

ACTIVATION PHASE

Class II MHC protein

2 Interleukin-1 (a cytokine) activates a T$_H$ cell.

Antigen

Macrophage

Helper T cell (T$_H$)

T cell receptor

1 The antigen is taken up by phagocytosis and degraded in a lysosome.

3 A T cell receptor recognizes an antigenic fragment bound to a class II MHC protein on the macrophage.

4 Cytokines released by the T$_H$ cell stimulate it to proliferate.

5 The T$_H$ cell proliferates and forms a clone.

EFFECTOR PHASE

7 Cytokines activate B cell proliferation.

B cell

T$_H$ cell

6 The binding of antigen to a specific IgM receptor triggers endocytosis, degradation, and display of the processed antigen.

8 A T cell receptor recognizes an antigenic fragment bound to a class II MHC protein on a B cell.

Memory cell

Plasma cell

9 B cells proliferate and differentiate.

10 The plasma cell produces antibodies.

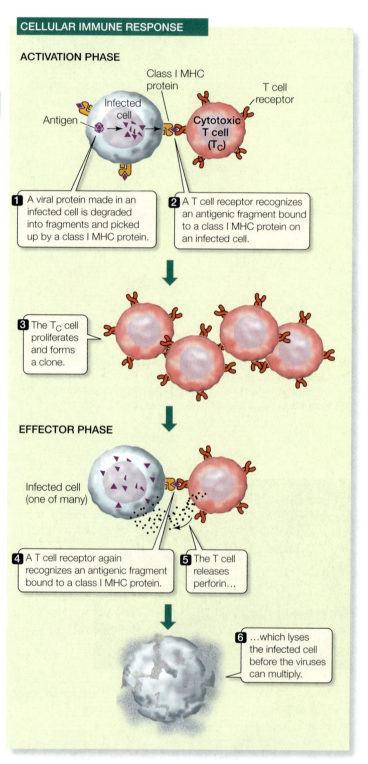

CELLULAR IMMUNE RESPONSE

ACTIVATION PHASE

Class I MHC protein

Infected cell

T cell receptor

Antigen

Cytotoxic T cell (T$_C$)

1 A viral protein made in an infected cell is degraded into fragments and picked up by a class I MHC protein.

2 A T cell receptor recognizes an antigenic fragment bound to a class I MHC protein on an infected cell.

3 The T$_C$ cell proliferates and forms a clone.

EFFECTOR PHASE

Infected cell (one of many)

4 A T cell receptor again recognizes an antigenic fragment bound to a class I MHC protein.

5 The T cell releases perforin…

6 …which lyses the infected cell before the viruses can multiply.

18.15 Phases of the Humoral and Cellular Immune Responses
Both the humoral and the cellular immune responses have activation and effector phases, all of which involve T cells.

With so many possible allele combinations, it is not surprising that different people are very likely to have different MHC genotypes. Similarities in base sequences between the MHC genes and the genes coding for antibodies and T cell receptors suggest that all three may have descended from the same ancestral genes and are part of a gene "superfamily." Thus major aspects of the immune system in vertebrates seem to be woven together by a common evolutionary thread.

Helper T cells and MHC II proteins contribute to the humoral immune response

When a T_H cell binds to an antigen-presenting macrophage, the T_H cell releases cytokines, which activate the T_H cell to produce a clone of T_H cells with the same specificity. The steps to this point constitute the *activation phase* of the humoral immune response, and they occur in the lymphoid tissues. Next comes the *effector phase*, in which the T_H cells activate B cells with the same specificity to produce antibodies (**Figure 18.15, left**).

B cells are also antigen-presenting cells. B cells take up antigens bound to their surface immunoglobulin receptors by endocytosis, break them down, and display antigenic fragments on class II MHC proteins. When a T_H cell binds to the displayed antigen–MHC II complex, it releases cytokines, which cause the B cell to produce a clone of plasma cells. Finally, the plasma cells secrete antibodies, completing the effector phase of the humoral immune response.

Cytotoxic T cells and MHC I proteins contribute to the cellular immune response

Class I MHC proteins play a role in the cellular immune response that is similar to the role played by class II MHC proteins in the humoral immune response. In a virus-infected or mutated cell, foreign or abnormal proteins or peptide fragments combine with MHC I molecules. The resulting complex is displayed on the cell surface and presented to T_C cells. When a T_C cell recognizes and binds to this –antigen–MHC I complex, it is activated to proliferate (**Figure 18.15, right**).

In the effector phase of the cellular immune response, T_C cells recognize and bind to cells bearing the same antigen–MHC I complex. These T_C cells produce a substance called perforin, which lyses the target cell. In addition, the T_C cells can bind to a specific receptor (called Fas) on the target cell that initiates apoptosis in that cell. These two mechanisms, cell lysis and programmed cell death, work in concert to eliminate the altered host cell.

Because T_C cells recognize self MHC proteins complexed with *nonself* antigens, they help rid the body of its own virus-infected cells. Because they also recognize MHC proteins complexed with *altered self* antigens (altered as a result of mutations), they help eliminate tumor cells, since most tumor cells have been altered by mutations.

In addition to the binding of an antigen–MHC complex to their receptors, T cells must receive a second signal for activation. This co-stimulatory signal occurs after the initial specific binding and involves the interaction of additional proteins on the T cell with the CD28 protein on certain antigen-presenting cells, as we saw above. This second binding event leads to T cell activation, including cytokine production and proliferation. It also sets in motion the production of an *inhibitor* of these events, so that the response is appropriately terminated. This inhibitor, a cell surface protein called CTLA4 competes with CD28, blocking the activation process, especially for self antigens.

MHC proteins underlie the tolerance of self

MHC proteins play a key role in establishing self-tolerance, without which an animal would be destroyed by its own immune system. Throughout the animal's life, developing T cells are tested in the thymus. This "test" consists of two "questions":

1. Can this cell recognize the body's MHC proteins? A T cell unable to recognize self MHC proteins would be useless to the animal because it could not participate in any immune reactions. Such a T cell fails the test and dies within about 3 days.

2. Does this cell bind to self MHC proteins *and* to one of the body's own antigens? A T cell that satisfied both of these criteria would be harmful or lethal to the animal; it also fails the test and undergoes apoptosis.

T cells that survive this test mature into either T_C cells or T_H cells.

In humans, one consequence of the major histocompatibility complex became important with the development of organ transplant surgery. Because the proteins produced by the MHC are specific to each individual, they act as nonself antigens if transplanted into another individual. An organ or a piece of tissue transplanted from one person to another is recognized as nonself and soon provokes an immune response; the tissue is then killed, or "rejected," by the host's cellular immune system. But if the transplant is performed immediately after birth, or if it comes from a genetically identical person (an identical twin), the material is recognized as self and is not rejected.

The rejection problem can be overcome by treating a patient with drugs, such as *cyclosporin*, that suppress the immune system. Cyclosporin works by blocking the activation of a transcription factor essential for T cell development. This approach, however, compromises the ability of transplant recipients to defend themselves against pathogens. These risks must be managed by the use of antibiotics and other drugs.

18.5 RECAP

The cellular immune response acts against antigens expressed on the surfaces of virus-infected or mutated body cells. Specific receptors on T cells bind to antigens displayed on the cell surface by MHC proteins. MHC proteins are also involved in self-tolerance.

- What are the roles of a T cell receptor in cellular immunity? See p. 414

- Can you describe the events of the cellular immune response to a virus-infected cell? See p. 415 and Figure 18.15

- What is the role of MHC proteins in the cellular immune response? See p. 417

We have alluded to genes that encode various components of the immune system and to the tremendous diversity in the immune response. We will now consider the genetic mechanisms that make this diversity possible.

18.6 How Do Animals Make So Many Different Antibodies?

Each cell of a newborn mammal possesses a full set of genetic information for immunoglobulin synthesis. At each of the loci coding for the heavy and light antibody chains, it has one allele from its mother and one from its father. Throughout the animal's life, each of its cells begins with the same full set of immunoglobulin genes. However, as B cells develop, their genomes become modified in such a way that each mature B cell can produce one—and only one—specific type of immunoglobulin. In other words, different B cells develop *slightly different genomes* encoding different antibody specificities. How can a single organism produce millions of different genomes?

One hypothesis was that mammals simply have millions of antibody genes. However, a simple calculation (the number of base pairs needed per antibody gene multiplied by millions of antibodies) shows that if this were true, the entire human genome would be taken up by antibody genes! More than 30 years ago, an alternative hypothesis was proposed: a relatively small number of genes recombine to produce many unique combinations, and it is this shuffling of the genetic deck, plus the random pairing of light and heavy chains, that produces antibody diversity. This second hypothesis is now the accepted molecular genetic theory.

In this section we will describe the unusual genetic events that generate the enormous antibody diversity that normally characterizes each individual mammal. Then we will see how similar events produce five classes of antibodies with slightly different functions.

Antibody diversity results from DNA rearrangement and other mutations

Each gene encoding an immunoglobulin chain is in reality a "supergene" assembled from several clusters of smaller genes scattered along part of a chromosome (**Figure 18.16**). Every cell in the body has hundreds of genes, located in separate clusters, that are potentially capable of participating in the synthesis of the variable and constant regions of immunoglobulin chains. In most body cells and tissues, these genes remain intact and separated from one another. During B cell development, however, these genes are cut out, rearranged, and joined together. One gene from each cluster is chosen randomly for joining, and the others are deleted (**Figure 18.17**).

In this manner, a unique antibody supergene is assembled from randomly selected "parts." Each B cell precursor in the animal assembles its own two specific antibody supergenes, one for a specific heavy chain and the other, assembled independently, for a specific light chain. This remarkable example of essentially irreversible cell differentiation generates an enormous diversity of antibody specificities from the same starting genome, one for each individual B cell.

In both humans and mice, the gene clusters coding for immunoglobulin heavy chains are on one pair of chromosomes and those for light chains are on others. The variable region of the light chain is encoded by two families of genes; the variable region of the heavy chain is encoded by three families.

Figure 18.16 illustrates the gene families coding for the heavy-chain constant and variable regions in mice. There are multiple genes coding for each of the four kinds of regions in the polypeptide chain: 100 *V*, 30 *D*, 6 *J*, and 8 *C*. Each B cell that becomes committed to making an antibody randomly selects *one* gene for each of these clusters to make the final heavy-chain coding sequence, *VDJC*. So the number of *different* heavy chains that can be made through this random recombination process is quite large (144,000 possible combinations in mice).

Now consider that the light chains are similarly constructed, with a similar amount of diversity made possible by random recombination. If we assume that light-chain diversity is the same as heavy-chain diversity, the number of possible combinations of light and heavy chains is 144,000 different light chains × 144,000 different heavy chains = 21 *billion* possibilities! But there are other mechanisms that generate even more diversity:

- When the DNA sequences coding for the *V*, *J*, and *C* regions are rearranged so that they are next to one another, the re-

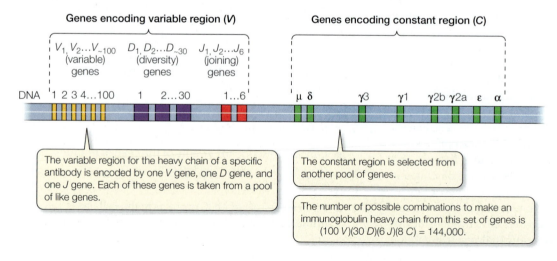

Genes encoding variable region (*V*)

Genes encoding constant region (*C*)

$V_1, V_2...V_{\sim100}$ (variable) genes $D_1, D_2...D_{\sim30}$ (diversity) genes $J_1, J_2...J_6$ (joining) genes

μ δ   γ3   γ1   γ2b γ2a ε α

DNA 1 2 3 4...100   1   2...30   1...6

The variable region for the heavy chain of a specific antibody is encoded by one *V* gene, one *D* gene, and one *J* gene. Each of these genes is taken from a pool of like genes.

The constant region is selected from another pool of genes.

The number of possible combinations to make an immunoglobulin heavy chain from this set of genes is (100 *V*)(30 *D*)(6 *J*)(8 *C*) = 144,000.

18.16 Heavy-Chain Genes
Mouse immunoglobulin heavy chains have four regions, each of which is coded for by one of a number of possible genes selected from a cluster of similar genes.

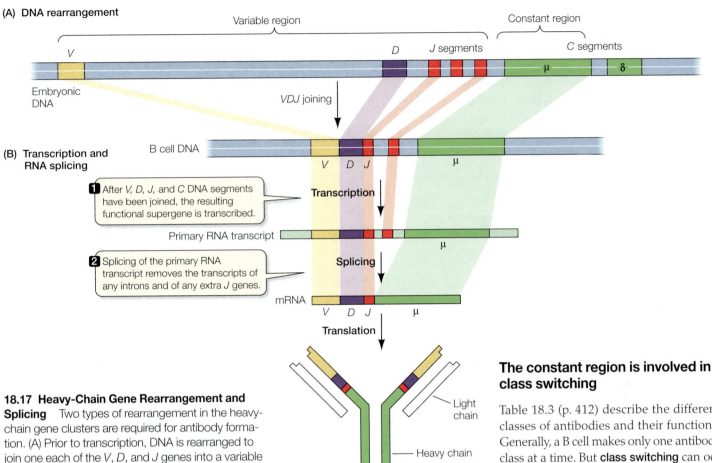

18.17 Heavy-Chain Gene Rearrangement and Splicing   Two types of rearrangement in the heavy-chain gene clusters are required for antibody formation. (A) Prior to transcription, DNA is rearranged to join one each of the *V*, *D*, and *J* genes into a variable region supergene. (B) After transcription, RNA splicing joins the *VDJ* region to the constant region.

1 After *V*, *D*, *J*, and *C* DNA segments have been joined, the resulting functional supergene is transcribed.

2 Splicing of the primary RNA transcript removes the transcripts of any introns and of any extra *J* genes.

The constant region is involved in class switching

Table 18.3 (p. 412) describe the different classes of antibodies and their functions. Generally, a B cell makes only one antibody class at a time. But **class switching** can occur, in which a B cell changes the antibody class it synthesizes. For example, a B cell making IgM can switch to making IgG.

Early in its life, a B cell produces IgM molecules, which are the receptors responsible for its recognition of a specific antigenic determinant. At this time, the constant region of the antibody's heavy chain is encoded by the first constant region gene, the μ gene (see Figure 18.17). If the B cell later becomes a plasma cell during a humoral immune response, another deletion commonly occurs in the cell's DNA, positioning the heavy-chain variable region genes (consisting of the same *V*, *D*, and *J* genes) next to a constant region gene farther down the original DNA (**Figure 18.18**). Such a DNA deletion results in the production of an antibody with a different constant region of the heavy chain, and therefore a different function. However, the antibody produced has *the same variable regions*, and therefore the same antigen specificity, as before. The new antibody falls into one of the four other immunoglobulin classes (IgA, IgD, IgE, or IgG), depending on which of the constant region genes is placed adjacent to the variable region genes.

After switching classes, the plasma cell cannot go back to making the previous immunoglobulin class, because that part of the DNA has been lost. On the other hand, if additional constant region genes are still present, the cell may switch classes again.

What triggers class switching, and what determines the class to which a given B cell will switch? T_H cells direct the course of an immune response and determine the nature of the attack on the antigen. These T cells induce class switching by sending cytokine signals. The cytokines bind to receptors on the target B cells,

combination event is not precise, and errors occur at the junctions. This *imprecise recombination* can create new codons at the junctions, with resulting amino acid changes.

- After the DNA sequences are cut out and before they are joined, an enzyme, *terminal transferase*, often adds some nucleotides to the free ends of the DNAs. These additional bases create *insertion mutations*.

- There is a relatively high spontaneous *mutation rate* in immunoglobulin genes. Once again, this process creates many new alleles and adds to antibody diversity.

When we add these possibilities to the billions of combinations that can be made by random DNA rearrangements, it is not surprising that the immune system can mount a response to almost any natural or artificial substance.

Once this pretranscriptional processing in completed, pre-mRNA can be transcribed from each supergene. Posttranscriptional processing removes the remaining introns, so that the mature mRNA contains a continuous coding sequence for an immunoglobulin light chain or heavy chain. Translation then produces the polypeptide chains, which combine to form an active antibody protein.

This genetic system is capable of still other kinds of changes, as seen when a B cell or plasma cell switches the immunoglobulin class it produces, but retains its antibody specificity.

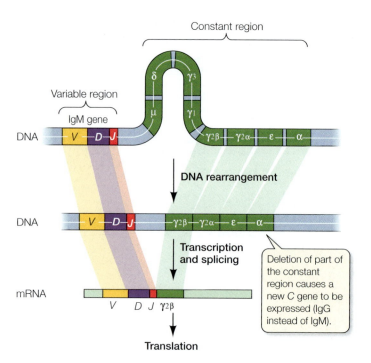

18.18 Class Switching The supergene produced by joining *V*, *D*, *J*, and *C* genes (see Figure 18.19) may later be modified, causing a different *C* region to be transcribed. This modification, known as class switching, is accomplished by deletion of part of the constant region gene cluster. Shown here is class switching from IgM to IgG.

generating a signal transduction cascade that results in altered transcription of the immunoglobulin genes.

18.6 RECAP

The immune system can make millions of antibodies with different specificities by rearranging the *V*, *D*, and *J* genes that code for variable regions in the immature B cell. Additional posttranscriptional processing and mRNA splicing result in a unique immunoglobulin chain. The class of the immunoglobulin molecule can be changed by deletion of a gene coding for the constant region of the heavy chain.

■ How can millions of antibodies with different specificities be generated from a relatively small number of genes? See pp. 418–419 and Figures 18.16 and 18.17

■ What is the role of the constant region of the immunoglobulin in class switching? See p. 419 and Figure 18.18

Given the numerous and complex cellular interactions that activate the immune system and generate antibody diversity, you may perceive many points at which the immune system can fail. We now turn to several situations in which one or more components of this complex system malfunction.

18.7 What Happens When the Immune System Malfunctions?

Sometimes the immune system fails us in one way or another. It may overreact, as in an *allergic reaction*; it may attack self antigens, as in an *autoimmune disease*; or it may function weakly or not at all, as in an *immune deficiency* disease.

Allergic reactions result from hypersensitivity

An **allergic reaction** arises when the human immune system overreacts to (is *hypersensitive* to) a dose of antigen. Although the antigen itself may present no danger to the host, the inappropriate immune response may produce inflammation and other symptoms, which can cause serious illness or death. Allergic reactions are the most familiar examples of this phenomenon. There are two types of allergic reactions: *immediate hypersensitivity* and *delayed hypersensitivity*.

IMMEDIATE HYPERSENSITIVITY **Immediate hypersensitivity** arises when an individual exposed to an antigen in a food, pollen, or the venom of an insect, referred to as an *allergen*, makes large amounts of IgE. When this happens, mast cells in tissues and basophils in blood bind the constant end of the IgE. If that individual is exposed to that allergen again, binding of the allergen to the IgE causes the mast cells and basophils to rapidly release a large amount of histamine (**Figure 18.19**). This results in symptoms such as dilation of blood vessels, inflammation, and difficulty breathing. If not treated with antihistamines, a severe allergic reaction can lead to death. Why an initial exposure to an allergen stimulates IgE production in some people is not known. There is some evidence for genetic factors predisposing people to an allergic response.

Allergy to pollen can be treated by a process called *desensitization*. The process involves injecting small amounts of the allergen (typically just an extract of the offending plant tissue) into the skin—enough to stimulate IgG production but not enough to stimulate IgE production. So the next time the person is exposed to the allergen, IgG binds to it, tying it up before IgE can bind it and exert its harmful effects. Desensitization does not work for food allergens because the IgE response to those substances is so strong that even a small amount of antigen provokes it.

One of the most common food allergens is the peanut. Some people are so sensitive that kissing someone who has recently eaten a peanut can cause a life-threatening allergic reaction.

DELAYED HYPERSENSITIVITY **Delayed hypersensitivity** is an allergic reaction that does not begin until hours after exposure to an antigen. In this case, the antigen is taken up by antigen-presenting cells and a T cell response is initiated. An example is the rash that develops after exposure to poison ivy.

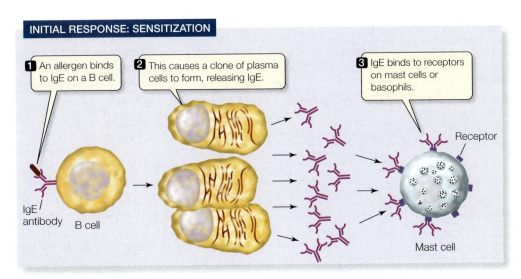

INITIAL RESPONSE: SENSITIZATION

1 An allergen binds to IgE on a B cell.

2 This causes a clone of plasma cells to form, releasing IgE.

3 IgE binds to receptors on mast cells or basophils.

IgE antibody

B cell

Receptor

Mast cell

18.19 An Allergic Reaction An allergen is an antigen that stimulates B cells to make large amounts of IgE antibodies that bind to mast cells and basophils. When the body encounters the allergen again, these cells produce large amounts of histamine that have harmful physiological effects.

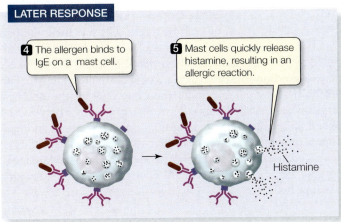

LATER RESPONSE

4 The allergen binds to IgE on a mast cell.

5 Mast cells quickly release histamine, resulting in an allergic reaction.

Histamine

Autoimmune diseases are caused by reactions against self antigens

Sometimes, people produce one or more "forbidden clones" of B and T cells directed against self antigens. Although the precise origin of **autoimmunity** is not known, there are several hypotheses:

- *Failure of clonal deletion:* A clone of lymphocytes making antibodies against self antigens that should have been destroyed by apoptosis is not.
- *Viral infection:* A virus that has an antigenic determinant that resembles a self antigen infects a person, and because of the polyclonal response to complex antigens, the body generates some antibodies against the self antigen.
- *Molecular mimicry:* T cells that recognize a nonself antigen also recognize something on the self that has a similar structure.

Analyses of human pedigrees show that autoimmune diseases tend to "run in families," indicating a genetic component. Genome scans for SNPs in patients with these diseases have found a transcription factor involved in B cell development, called RUNX1, that may be involved. Some alleles of *MHC II* are strongly linked to certain autoimmune diseases.

Autoimmunity does not always result in disease, but a number of autoimmune diseases are common:

- People with *systemic lupus erythematosis* (SLE) have antibodies to many cellular components, including DNA and nuclear proteins. These antinuclear antibodies can cause serious damage when they bind to normal tissue antigens to form large circulating antigen–antibody complexes, which become stuck in tissues and provoke inflammation.
- People with *rheumatoid arthritis* have difficulty in shutting down a T cell response. We mentioned earlier that the inhibitor CTLA4 blocks T cells from reacting to self antigens. People with rheumatoid arthritis may have low CTLA4 activity, which results in inflammation of the joints due to the infiltration of excess white blood cells.
- *Hashimoto's thyroiditis* is the most common autoimmune disease in women over 50. Immune cells attack thyroid secretions resulting in fatigue, depression, weight gain, and other symptoms.
- *Insulin-dependent diabetes mellitus*, or type I diabetes, occurs most often in children. It is caused by an immune reaction against several proteins in the cells of the pancreas that manufacture the protein hormone insulin. This reaction kills the insulin-producing cells, so people with type I diabetes must take insulin daily in order to survive.

AIDS is an immune deficiency disorder

There are a number of inherited and acquired *immune deficiency disorders*. In some individuals, T or B cells never form; in others, B cells lose the ability to give rise to plasma cells. In either case, the affected individual is unable to mount an immune response and thus lacks a major line of defense against pathogens.

Because of its essential roles in both the humoral and cellular immune responses, the T_H cell is perhaps the most central of all the components of the immune system—a significant cell to lose to an immune deficiency disorder. This cell is the target of **HIV** (*human immunodeficiency virus*), the retrovirus that eventually results in **AIDS** (*acquired immune deficiency syndrome*).

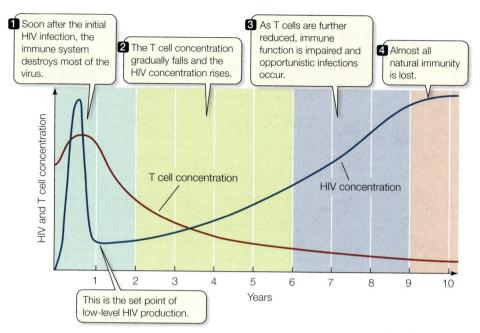

1 Soon after the initial HIV infection, the immune system destroys most of the virus.

2 The T cell concentration gradually falls and the HIV concentration rises.

3 As T cells are further reduced, immune function is impaired and opportunistic infections occur.

4 Almost all natural immunity is lost.

T cell concentration

HIV concentration

This is the set point of low-level HIV production.

18.20 The Course of an HIV Infection An HIV infection may be carried, unsuspected, for many years before the onset of symptoms. This long "dormant" period means that the infection is often spread by people who are unaware that they are carrying the virus.

HIV can be transmitted from person to person in several ways:

■ Through blood, such as by a needle contaminated with the virus after being used to inject an infected individual

■ Through the exposure of broken skin, an open wound, or mucous membranes to body fluids, such as blood or semen, from an infected individual

■ Through the blood of an infected mother to her baby during birth

HIV initially infects macrophages, T_H cells, and dendritic cells in blood and tissues. These infected cells carry the virus to the lymph nodes and spleen, where T and B cells are present. Normally, the dendritic cells present their captured antigens to T_H cells in the lymph nodes, and this causes the T_H cells to proliferate and form a clone (see Figure 18.15). But HIV preferentially infects activated—not resting—T_H cells. So the HIV arriving in the lymph nodes proceeds to infect the many activated T_H cells that are already responding to other antigens there. These two processes—the transport of the virus to the nodes and the presence in the nodes of cells already receptive to viral infection—combine to ensure that HIV reproduces vigorously. Up to 10 billion viruses are made every day during this initial phase of infection. The numbers of T_H cells quickly drop, and infected people show symptoms similar to mononucleosis, such as enlarged lymph nodes and fever.

These symptoms abate within 3 weeks, however, as T cells recognize infected cells, an immune response is mounted, and antibodies specific to HIV appear in the blood (**Figure 18.20**). By this time, the patient has a high level of circulating HIV complexed with antibodies, which is gradually removed by the action of dendritic cells over the next several months. But before they are removed, these antibody-complexed viruses can still infect T_H cells that come into contact with them. This secondary infection process reaches a low, steady-state level called the "set point." This point varies among individuals and is a strong predictor of the rate of progres-

sion of the disease. For most people, it takes 8–10 years, even without treatment, for the more severe manifestations of AIDS to develop. In some, it can take as little as a year; in others, 20 years.

During this dormant period, people carrying HIV generally feel fine, and their T_H cell levels are adequate for them to mount immune responses. Eventually, however, the virus destroys the T_H cells, and their numbers fall to dangerous levels. At this point, the infected person is considered to have *full-blown AIDS* and is susceptible to infections that the T_H cells would normally eliminate (**Figure 18.21**). Most notable among these infections are the otherwise rare skin tumor called Kaposi's sarcoma, caused by a herpesvirus; pneumonia, caused by the fungus *Pneumocystis carinii*; and lymphoma tumors, caused by the Epstein–Barr virus. These conditions are called *opportunistic infections* because they take advantage of the crippled immune system of the host. They lead to death within a year or two.

HIV INFECTION AND REPLICATION As the AIDS epidemic has grown, so has our knowledge of the molecular biology of HIV infection. We described the life cycle of HIV in Section 13.1 (see Figure 13.6). Briefly, as an enveloped retrovirus, the viral core of HIV is surrounded by a membrane, and its genome is RNA.

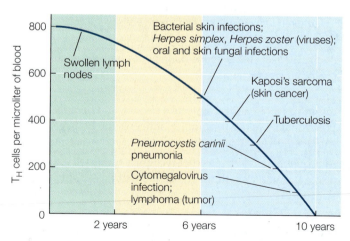

Swollen lymph nodes

Bacterial skin infections; *Herpes simplex*, *Herpes zoster* (viruses); oral and skin fungal infections

Kaposi's sarcoma (skin cancer)

Tuberculosis

Pneumocystis carinii pneumonia

Cytomegalovirus infection; lymphoma (tumor)

18.21 Relationship between T_H Cell Count and Opportunistic Infections As HIV kills more and more T_H cells, the immune system is less and less able to defend the body against various pathogens, including many that are not usually infectious to healthy people.

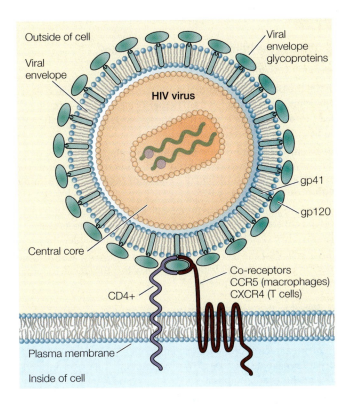

Outside of cell

Viral envelope glycoproteins

Viral envelope

HIV virus

gp41

gp120

Central core

Co-receptors
CCR5 (macrophages)
CXCR4 (T cells)

CD4+

Plasma membrane

Inside of cell

18.22 Two Receptors for HIV Two proteins on the cell surface are needed for HIV to enter a cell. Some people lack the co-receptor CCR5 and so are resistant to HIV infection.

Several virally encoded proteins are especially vital for HIV infection and replication:

- The virus uses the membrane proteins *gp120* and *gp41* to attach to human cells (*gp* stands for *glyco*protein). These proteins target two proteins on the host cell surfaces, CD4 and a co-receptor (CCR5 in macrophages and CXCR4 in T cells) (**Figure 18.22**). Binding of the virus is followed by membrane fusion and the viral genome with attached proteins enters the cell.

- *Reverse transcriptase* catalyzes the synthesis of cDNA from viral RNA. Unfortunately, it lacks the proofreading function of many DNA polymerases, and makes about 10 errors in every round of cDNA synthesis. This creates a pool of mutant viruses.

- *Integrase* catalyzes the insertion of cDNA into the host chromosome.

- *Protease* is necessary to complete the formation of individual viral proteins from larger initial products of translation.

Additional proteins encoded by HIV are needed to complete the transcription of its cDNA (Tat protein) and the splicing of RNA (Rev protein).

TREATING HIV INFECTION Armed with a staggering amount of detailed knowledge of HIV, scientists have been attempting to develop treatments for AIDS. The general therapeutic strategy is to try to block stages in the viral life cycle without damaging the host cell. Potential therapeutic agents that interfere with the major steps

of the life cycle are being tested. It is crucial to block only steps that are unique to the virus so that drug therapies do not harm the patient by blocking a step in the patient's own metabolism.

*H*ighly *a*ctive *a*ntiretroviral *t*herapy (**HAART**), developed in the late 1990s, has had considerable success in delaying the onset of AIDS symptoms in people infected with HIV, and in prolonging the lives of people with AIDS. The logic of HAART comes from cancer treatment: employ a combination of drugs that act at different parts of the viral life cycle. Generally, the HAART regimen uses a protease inhibitor and two reverse transcriptase inhibitors. The two reverse transcriptase inhibitors can be given as a single pill. These drug regimens may eliminate HIV entirely in some people, especially in those treated within the first few days after infection, before the virus has arrived in the lymph nodes. Most patients, however, face a lifetime of anti-HIV therapy.

Unfortunately, many patients who take HAART develop mutant strains of HIV that are resistant to this regimen; this is a consequence of the error rate of reverse transcriptase. Thus there is a never-ending race to modify HAART by adding new and/or different drug combinations, and today there are around 150 different HAART treatments. In short, we seem trapped in an evolutionary struggle with HIV. How can we gain a lasting advantage? The greatest hope is for the development of a vaccine against HIV. The first clinical trials of vaccines directed against the HIV membrane protein gp120 were not successful, but other vaccines are under development.

The worldwide AIDS epidemic began in the early 1980s and continues to expand, especially in Africa and Southeast Asia. About 40 million people are currently infected with HIV. Some 4 million new infections occur each year, along with 3 million deaths. At least half of the people who have been infected since the epidemic began are already dead. What can be done until biomedical science can bring the epidemic to an end?

Above all, people must recognize that they are in danger whenever they have sex with a partner whose *total* sexual history they do not know. The danger rises as the number of sexual partners rises, and the danger is much greater if partners participating in sexual intercourse are not protected by a latex condom. The danger that heterosexual intercourse will transmit HIV rises significantly if either partner has another sexually transmitted disease.

18.7 RECAP

Failures of the immune system include allergic reactions (caused by hypersensitivity to antigens), autoimmune diseases (caused by reactions against self antigens), and immune deficiency disorders.

- How does immediate hypersensitivity develop? See p. 420 and Figure 18.19

- What is an autoimmune disease? Give an example. See p. 421

- Can you describe the course of events in the human immune system during HIV infection? See pp. 422–423 and Figure 18.20

CHAPTER SUMMARY

18.1 What are the major defense systems of animals?

Animal defenses against **pathogens** are based on the body's ability to distinguish between self and nonself.

Nonspecific (innate) **defenses** are inherited mechanisms that protect the body from many kinds of pathogens and typically act rapidly.

Specific defenses are adaptive mechanisms that respond to a specific pathogen. They develop slowly but are long-lasting.

Many defenses are implemented by cells and proteins carried in the **blood plasma** and **lymph**. Review Figure 18.1, Web/CD Activity 18.1

White blood cells fall into two broad groups. **Phagocytes** include **macrophages** that engulf pathogens by phagocytosis. **Lymphocytes**, which include **B cells** and **T cells**, participate in specific responses. Review Figure 18.2, Web/CD Tutorial 18.1

18.2 What are the characteristics of the nonspecific defenses?

An animal's nonspecific defenses include physical barriers such as the skin and competing resident microorganisms known as **normal flora**. Review Table 18.1

The **complement system** consists of about 30 different antimicrobial proteins.

Circulating defensive cells, such as **phagocytes** and **natural killer cells**, eliminate invaders.

The **inflammation** response calls on several cells and proteins that act against invading pathogens. **Mast cells** release **histamines**, which cause blood vessels to dilate and become "leaky." Review Figure 18.4, Web/CD Activity 18.2

A cell signaling pathway involving the **toll** receptor stimulates the body's defenses.

18.3 How does specific immunity develop?

See Web/CD Tutorial 18.2

The specific immune response is characterized by recognition of specific **antigens**, mechanisms for developing a response to an enormous diversity of **antigenic determinants**, the ability to distinguish self from nonself, and **immunological memory** of the antigens it has encountered.

Each antibody and each T cell is specific for a single antigenic determinant. **T cell receptors** bind to antigens on the surface of virus-infected cells.

The **humoral immune response** is directed against pathogens in the blood, lymph, and tissue fluids. The **cellular immune response** is directed against an antigen established within a host cell. Both responses are mediated by antigenic fragments being presented on a cell surface along the proteins of the **major histocompatibility complex**.

Clonal selection accounts for the specificity and diversity of the immune response as well as immunological memory and tolerance to self. Review Figure 18.6

An activated lymphocyte produces **effector cells** that carry out an attack on the antigen and **memory cells** that retain the ability to divide to produce more effector and memory cells. Effector B cells are called **plasma cells**.

Immunization is inoculation with antigenic molecules, and **vaccination** is inoculation with modified pathogens. Both techniques prevent disease by stimulating immunological memory.

18.4 What is the humoral immune response?

See Web/CD Tutorial 18.3

B cells are the basis of the humoral immune response. Activated B cells, stimulated by signals from helper T cells with the same specificity, form plasma cells, which synthesize and secrete specific antibodies.

The basic unit of an antibody, or **immunoglobulin**, is a tetramer of four polypeptides: two identical light chains and two identical heavy chains, each consisting of a **constant region** and a **variable region**. Review Figure 18.9, Web/CD Activity 18.3

The variable regions determine the specificity of an immunoglobulin, and the constant regions of the heavy chain determine its class. There are five classes of immunoglobulins. Review Table 18.3

A clone of cells that can be grown in culture to produce a **monoclonal antibody** can be made by fusing a B cell with a tumor cell to form a **hybridoma**. Review Figure 18.11

18.5 What is the cellular immune response?

See Web/CD Tutorial 18.4

T cells are the effectors of the cellular immune response. T cell receptors are similar in structure to the immunologlobulins, having variable and constant regions. Review Figure 18.12

There are two types of T cells. **Cytotoxic T cells** recognize and kill virus-infected cells or mutated cells. **Helper T cells** direct both the cellular and humoral immune responses.

The genes of the major histocompatibility complex (MHC) encode membrane proteins that bind antigenic fragments and present them to T cells. Review Figure 18.13

Organ transplants are rejected when the host's immune system recognizes MHC proteins on transplanted tissue as nonself.

18.6 How do animals make so many different antibodies?

See Web/CD Tutorial 18.5

B cell genomes undergo changes as the cell develops so that each cell can produce a specific antibody protein. The immunoglobulin chains derive from "supergenes" that are constructed from one each of numerous V, D, J, and C genes. This DNA rearrangement and rejoining yields a unique immunoglobulin chain. Review Figures 18.16 and 18.17

Once a B cell becomes a plasma cell, it may undergo **class switching**, in which a deletion of one or more constant region genes results in the production of an antibody with a different constant region and a different function. Review Figure 18.18

18.7 What happens when the immune system malfunctions?

An **allergic reaction** is an inappropriate immune response caused by **immediate** or **delayed hypersensitivity** to certain antigens. Review Figure 18.19

Autoimmune diseases result when the immune system produces B and T cells that attack self antigens.

Immune deficiency disorders result from failure of one or another part of the immune system. **AIDS** is an acquired immune deficiency disorder arising from depletion of the T_H cells as a result of infection with **HIV**. Review Figure 18.20

SELF-QUIZ

1. Phagocytes kill harmful bacteria by
   a. endocytosis.
   b. producing antibodies.
   c. complement proteins.
   d. T cell stimulation.
   e. inflammation.

2. Which statement about immunoglobulins is true?
   a. They help antibodies do their job.
   b. They recognize and bind antigenic determinants.
   c. They encode some of the most important genes in an animal.
   d. They are the chief participants in nonspecific defense mechanisms.
   e. They are a specialized class of white blood cells.

3. Which statement about an antigenic determinant is *not* true?
   a. It is a specific chemical grouping.
   b. It may be part of many different molecules.
   c. It is the part of an antigen to which an antibody binds.
   d. It may be part of a cell.
   e. A single protein has only one on its surface.

4. T cell receptors
   a. are the primary receptors for the humoral immune system.
   b. are carbohydrates.
   c. cannot function unless the animal has previously encountered the antigen.
   d. are produced by plasma cells.
   e. are important in combating viral infections.

5. According to the clonal selection theory,
   a. an antibody changes its shape according to the antigen it meets.
   b. an individual animal contains only one type of B cell.
   c. an individual animal contains many types of B cells, each producing one kind of antibody.
   d. each B cell produces many types of antibodies.
   e. many clones of antiself lymphocytes appear in the bloodstream.

6. Immunological tolerance
   a. depends on exposure to an antigen.
   b. develops late in life and is usually life-threatening.
   c. disappears at birth.
   d. results from the activities of the complement system.
   e. results from DNA splicing.

7. The extraordinary diversity of antibodies results in part from
   a. the action of monoclonal antibodies.
   b. the splicing of protein molecules.
   c. the action of cytotoxic T cells.
   d. the rearrangement of genes.
   e. their remarkable nonspecificity.

8. Which of the following play(s) no role in the antibody response?
   a. Helper T cells
   b. Growth factors
   c. Macrophages
   d. Reverse transcriptase
   e. Products of class II MHC genes

9. The major histocompatibility complex
   a. codes for specific proteins found on the surfaces of cells.
   b. plays no role in T cell immunity.
   c. plays no role in antibody responses.
   d. plays no role in skin graft rejection.
   e. is encoded by a single locus with multiple alleles.

10. Which of the following plays no role in HIV reproduction?
    a. Integrase
    b. Reverse transcriptase
    c. gp120
    d. Interleukin-1
    e. Protease

FOR DISCUSSION

1. Describe the part of an antibody molecule that interacts with an antigenic determinant. How is it similar to the active site of an enzyme? How does it differ from the active site of an enzyme?

2. Contrast immunoglobulins and T cell receptors with respect to their structure and function.

3. Discuss the diversity of antibody specificities in an individual in relation to the diversity of enzymes. Does every cell in an animal contain genetic information for all the organism's enzymes? Does every cell contain genetic information for all the organism's immunoglobulins?

4. The gene family determining MHC on the cell surface in humans is on a single chromosome. A father's MHC type is A1, A3, B5, B7, D9, D11. A mother's phenotype is A2, A4, B6, B7, D11, D12 Their child is A1, A4, B6, B7, D11, D12. What are the parents' haplotypes—that is, which alleles are linked on the diploid chromosomes of each parent? Assuming there is no recombination among the genes determining the MHC type, can these same two parents have a child who is A1, A2, B7, B8, D10, D11?

FOR INVESTIGATION

Development of an effective HIV vaccine requires that the person being vaccinated develop both cellular and humoral immunity against HIV. Explain what this means in terms of studies you would do on people given a potential new vaccine.

19 Differential Gene Expression in Development

Stem cells from fat

Dr. Marc Hedrick was concerned about fat—other people's fat. As a plastic surgeon in Los Angeles, part of his medical practice involved performing liposuction, a technique to remove unwanted fat. He wondered if this excess fat could be put to some use. Looking at globs of fat under the microscope, Hedrick was surprised to see cells that resembled bone cells. How did bone cells come to be found in excised fat tissue?

Hedrick hypothesized that fat stores harbor a population of *stem cells*—dividing, unspecialized cells that have the potential to produce many different cell types on demand. In this case, the potential of the stem cells would be limited to types of cells whose origins in the embryo are related to fat; this would include bone, cartilage, blood vessel, and muscle cells.

The surgeon set out to isolate these stem cells. There was no shortage of discarded fat from Los Angeles cosmetic surgery practices, and soon he had a cell population that, when provided with growth factors and hormones, could differentiate into several types of cells in a laboratory dish.

But can these fat-derived adult stem cells differentiate inside an animal? Experiments on rats and mice have now shown that these stem cells can be implanted into organs and that they differentiate into cells appropriate for the tissue they are in. Thus they could be used to treat damaged hearts, injured bones, and damaged blood vessels. An advantage of using fat stem cells in human medicine is that the patient's own cells can be used. The immune system, as we saw in Chapter 18, recognizes and rejects nonself tissues. With fat stem cells, there would be no rejection implanted tissue.

Hedrick has developed the machinery to separate stem cells from fat quickly enough to be done in the operating room while patients await treatment for a damaged heart or other tissue. Two hundred million stem cells are enough for therapy, and they can be obtained from 450 grams (about a pound) of fat. Recently, three women in Japan had stem cells from their own fat implanted to help reconstruct and heal wounded breast tissue following mastectomy. The era of personalized medicine has begun.

Behind all of this impressive medical technology lies a great deal of basic developmental genetics. The processes that underlie the differentiation of stem cells in an adult are the same ones that occur in the embryo. Much of our knowledge of developmental

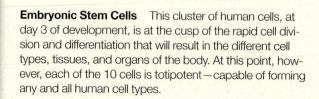

Embryonic Stem Cells This cluster of human cells, at day 3 of development, is at the cusp of the rapid cell division and differentiation that will result in the different cell types, tissues, and organs of the body. At this point, however, each of the 10 cells is totipotent—capable of forming any and all human cell types.

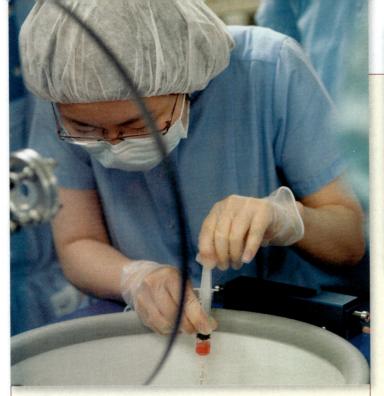

Fat As a Source of Stem Cells This centrifuge separates dense fatty tissues from the lighter stem cells. Stem cells from fat have been found to be capable of differentiating into several specialized cell types.

genetics has come from studies on certain model organisms, such as the fruit fly *Drosophila melanogaster*, the nematode *Caenorhabditis elegans*, frogs, sea urchins, and a flowering plant, the thale cress, *Arabidopsis thaliana*. As we saw in Section 14.1, the genomes of all eukaryotes are surprisingly similar, and the cellular and molecular principles underlying their development also turn out to be similar. Thus discoveries from one organism aid us in understanding other organisms, including ourselves.

IN THIS CHAPTER we will see that every body cell in a multicellular organism contains all of the genes present in the zygote that gave rise to that organism—even though only some of those genes are expressed in any particular cell type. We will see how cellular changes during development result from the differential expression of genes. We will describe the various mechanisms of transcriptional control and chemical signaling that work together to produce a complex organism.

19.1 What Are the Processes of Development?

Development is the process by which a multicellular organism undergoes a series of progressive changes, taking on the successive forms that characterize its life cycle (**Figure 19.1**). In its earliest stages of development, a plant or animal is called an **embryo**. Sometimes the embryo is contained within a protective structure, such as a seed coat, an eggshell, or a uterus. An embryo does not photosynthesize or feed actively; instead, it obtains its food from its mother directly or indirectly (by way of nutrients stored in a seed or egg). A series of embryonic stages may precede the birth of the new, independent organism. Most organisms continue to develop throughout their life cycle; development ceases only with death.

Development proceeds via determination, differentiation, morphogenesis, and growth

Four processes are responsible for the developmental changes an organism undergoes as it progresses from an embryo to mature adulthood:

- **Determination** sets the developmental *fate* of a cell—what type of cell it will become—even before any characteristics of that cell type are observable.

- **Differentiation** is the process by which different types of cells arise; that is, differentiation leads to cells with the specific structures and functions of that cell's determined fate.

- **Morphogenesis** (Greek for "origin of form") is the shaping of differentiated cells into the multicellular body and its organs.

- **Growth** is the increase in size of the body and its organs by cell division and cell expansion.

Mitosis produces daughter cells that are chromosomally and genetically identical to the cell that divides to produce them (see Section 9.1). Why, then, are the cells of a multicellular organism not all identical in structure or function? Cells differ from one another because different genes are expressed in different cells, a phenomenon called **differential gene expression**.

19.1 From Fertilized Egg to Adult Stages of development from zygote to maturity are shown for an animal and for a plant.

In an early embryo that consists of only a few cells, each cell has the potential to develop in many different ways. As development proceeds, regulation of the expression of genes results in the production of different proteins and thus in cells with progressively different features and functions.

Morphogenesis, the process by which recognizable tissues and organs develop in an organism, can occur in several ways. In plant development, cells are constrained by cell walls and do not move around the body, so organized division and expansion of cells are the major processes that build the plant body. In animals, cell movements are very important in morphogenesis, as we will see in Section 43.2. And in both plants and animals, apoptosis (programmed cell death) is essential to orderly development. Like differentiation, morphogenesis results ultimately from the closely regulated activities of genes and their products as well as from the interplay of signals secreted by one group of cells and their effects on other (target) cells.

In all multicellular organisms, repeated mitotic divisions generate the multicellular body. Increases in the sizes of cells also contribute to growth. In plants, cell expansion begins shortly after the first divisions of the fertilized egg, or *zygote*. In animals, on the other hand, cell expansion is often slow to begin: the animal embryo may

consist of thousands of cells before it becomes larger than the original zygote. Growth continues throughout the individual's life in some species, but reaches a more or less stable end point in others.

Cell fates become more and more restricted

A cell's **fate**—that is, the type of cell into which it will ultimately differentiate—is a function of both differential gene expression and morphogenesis. The role of morphogenesis in determining cell fate can be revealed in experiments in which specific cells of an early embryo are grafted into new positions on another embryo. The cells are marked with stains so that their development into adult structures can be traced.

For instance, we know that the blue-shaded area of the early frog embryo shown in **Figure 19.2** normally becomes part of the skin of the tadpole (the larva of a frog). However, if we cut out a piece from this region and transplant it to another location on another early frog embryo, it does not become skin. The type of tissue it does become is determined by its new location in the embryo. The developmental potential of these early embryonic cells—that is, their range of possible fates—is thus greater than their actual fate.

The developmental potential of cells becomes restricted fairly early in normal development. Tissue from a later-stage frog embryo, for example, if taken from a region that normally develops

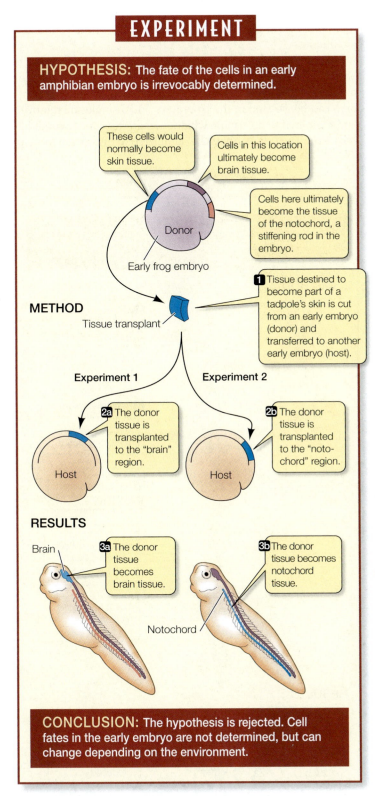

EXPERIMENT

HYPOTHESIS: The fate of the cells in an early amphibian embryo is irrevocably determined.

These cells would normally become skin tissue.

Cells in this location ultimately become brain tissue.

Cells here ultimately become the tissue of the notochord, a stiffening rod in the embryo.

Donor

Early frog embryo

METHOD

Tissue transplant

1 Tissue destined to become part of a tadpole's skin is cut from an early embryo (donor) and transferred to another early embryo (host).

Experiment 1

Experiment 2

2a The donor tissue is transplanted to the "brain" region.

Host

2b The donor tissue is transplanted to the "noto-chord" region.

Host

RESULTS

Brain

3a The donor tissue becomes brain tissue.

3b The donor tissue becomes notochord tissue.

Notochord

CONCLUSION: The hypothesis is rejected. Cell fates in the early embryo are not determined, but can change depending on the environment.

19.2 Developmental Potential in Early Frog Embryos Cells that would normally form one type of tissue can form completely different tissue types when they are experimentally moved to another location. In this experiment, epithelial (skin) tissue from an early-stage frog embryo is transplanted from a donor to a host embryo. The tissue that develops in the host tadpole is not skin, but is consistent with the location to which the tissue is transplanted. FURTHER RESEARCH: What would happen if tissue from an adult were transplanted into an early embryo?

into the brain, becomes brain tissue even if transplanted to a part of an early-stage embryo destined to become another structure.

Determination is influenced by the action of both the extracellular environment and the contents of the cell on the cell's genome. Determination is not something that is visible under the microscope—cells do not change their appearance when they become determined. Determination is followed by differentiation—the actual changes in biochemistry, structure, and function that result in cells of different types. Differentiation often involves a change in appearance as well as function. *Determination is a commitment; the final realization of that commitment is differentiation.*

19.1 RECAP

Development takes place via the processes of determination, differentiation, morphogenesis, and growth. Cells in the early embryo have not yet had their differentiated fates determined; as development proceeds, their potential fates become more and more restricted.

- Can you name and describe the four processes of development? See p. 427

- Do you understand how the experiment in Figure 19.2 illuminates how cell fates become determined? See pp. 428–429

Is a photosynthetic cell in the mesophyll of a leaf or a liver cell in a human being irrevocably committed to that specialization? Under the right experimental circumstances, differentiation is reversible in many cells. The next section describes how the genome of a cell can be induced to express a different pattern of differentiation.

19.2 Is Cell Differentiation Irreversible?

A zygote has the ability to give rise to every type of cell in the adult body; in other words, it is **totipotent**. Its genome contains instructions for all of the structures and functions that will arise throughout the life cycle of the organism. Later in development, the cellular descendants of the zygote lose their totipotency and become determined. These determined cells then differentiate into specific types of specialized cells. A liver cell in a human or a mesophyll cell in the leaf of a green plant generally retains its differentiated form and function throughout its life. But this does not necessarily mean that these cells have irrevocably lost their totipotency.

Most of the differentiated cells of an animal or plant have nuclei containing the entire genome and certainly have the genetic capacity for totipotency. We explore here several examples of how this capacity has been demonstrated experimentally.

Plant cells are usually totipotent

A food storage cell in a carrot root normally faces a dark future. It cannot photosynthesize or give rise to new carrot plants. However, if we isolate that cell from the root, maintain it in a suitable nutrient medium, and provide it with appropriate chemical cues, we can "fool" the cell into acting as if it were a zygote. It can divide and give rise to a mass of undifferentiated cells, called a *callus*, and eventually to a complete plant (**Figure 19.3**). Since the new plant is genetically identical to the cell from which it came, we call the plant a **clone**.

The ability of scientists to clone an entire carrot plant from a differentiated root cell indicates that the cell contains the entire carrot genome and that it can express the appropriate genes in the right sequence. Many types of cells from other plant species show similar behavior in the laboratory. This ability to generate a whole plant from a single cell has been invaluable in agricultural biotechnology. For example, forest products companies harvest trees for use in making paper, lumber, and other products. To replace the trees reliably, they remove fragments of leaves and, in a procedure just like the one used on carrot root cells, form embryos and then small plants, which can be planted in the forest. The yield of trees from these clones is higher and more predictable than that from seeds produced by fertilization.

Among animals, the cells of early embryos are totipotent

Animal somatic cells cannot be fooled into acting like zygotes as easily as plant cells can, but nuclear transplantation experiments have demonstrated that animal somatic cells retain their totipotency. Such experiments were first done on frogs by Robert Briggs and Thomas King, who asked whether the nuclei of early frog embryos had retained the ability to do what the totipotent zygote nucleus could do. They first removed the nucleus from an unfertilized egg, forming an *enucleated* egg. Then, with a very fine glass tube, they punctured a cell from an early embryo and drew up part of its contents, including the nucleus, which they injected into the enucleated egg. They stimulated the eggs to divide, and many went on to form embryos, tadpoles, and eventually, frogs. These experiments led to two important conclusions:

- No information is lost from the nuclei of cells as they pass through the early stages of embryonic development. This fundamental principle of developmental biology is known as **genomic equivalence**.

- The cytoplasmic environment around a nucleus can modify its fate.

Similar experiments have been performed on rhesus monkeys, in which a single cell can be removed from an 8-cell embryo and fused with an enucleated egg. This *cell fusion* technique causes the nucleus of the embryonic cell to enter the egg cytoplasm. The resulting cell acts like a zygote, forming an embryo, which can be implanted into a foster mother, who ultimately gives birth to a normal monkey. Each of the remaining 7 cells from the orig-

inal embryo can give rise to offspring by the same cell fusion technique.

In humans, the totipotency of early embryonic cells permits both genetic screening (see Section 17.3) and certain assisted reproductive technologies (see Section 42.4). An 8-cell human embryo can be isolated in the laboratory and a single cell removed and examined to determine whether a harmful genetic condition is present. The cells can then be stimulated to divide and form an

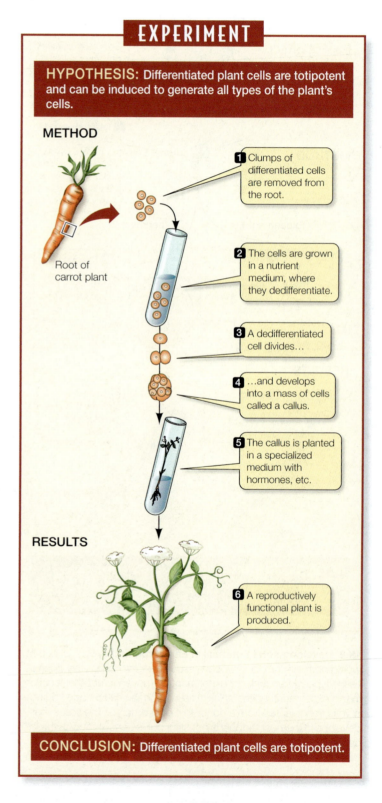

EXPERIMENT

HYPOTHESIS: Differentiated plant cells are totipotent and can be induced to generate all types of the plant's cells.

METHOD

Root of carrot plant

1 Clumps of differentiated cells are removed from the root.

2 The cells are grown in a nutrient medium, where they dedifferentiate.

3 A dedifferentiated cell divides…

4 …and develops into a mass of cells called a callus.

5 The callus is planted in a specialized medium with hormones, etc.

RESULTS

6 A reproductively functional plant is produced.

CONCLUSION: Differentiated plant cells are totipotent.

19.3 Cloning a Plant Differentiated, specialized food storage cells from the root of a carrot can be induced to dedifferentiate by placing them in a new chemical environment. These cells can then act like early embryonic cells and form a new plant.

embryo, which can be implanted into the mother's uterus, where it develops into an infant.

The somatic cells of adult animals retain the complete genome

The experimental reproductive cloning of animals from adult somatic cells was revolutionized in the late 1990s when Ian Wilmut and his colleagues at a biotechnology company in Scotland used the cell fusion technique to clone sheep. Previous attempts to produce mammals by this method had worked, as in the rhesus monkey case, only if the donor nucleus was from an early embryo. Problems had arisen when mammalian donor cells that were in the G2 phase of the cell cycle (see Figure 9.3) were fused with the cytoplasm of eggs that were also in G2; extra DNA replication took place that created havoc with the cell cycle in the egg when it attempted to divide. Wilmut found a way to ensure that the cells used in animal cloning experiments were in the G1 phase of the cell cycle. His procedure is outlined in **Figure 19.4**.

Wilmut took differentiated cells from a ewe's udder and starved them of nutrients for a week, thus halting the cells in G1 phase of the cell cycle. One of these cells was fused with an enucleated egg from a different breed of ewe. When mitotic inducers in the egg cytoplasm were stimulated, the donor nucleus entered S phase, and the rest of the cell cycle proceeded normally. After several cell divisions, the resulting early embryo was transplanted into the womb of a surrogate mother.

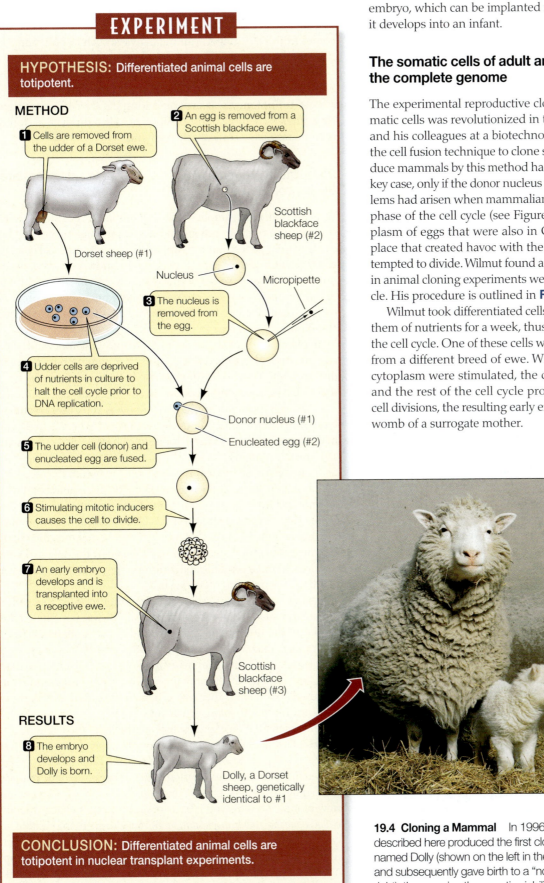

EXPERIMENT

HYPOTHESIS: Differentiated animal cells are totipotent.

METHOD

1 Cells are removed from the udder of a Dorset ewe.

Dorset sheep (#1)

2 An egg is removed from a Scottish blackface ewe.

Scottish blackface sheep (#2)

Nucleus

Micropipette

3 The nucleus is removed from the egg.

4 Udder cells are deprived of nutrients in culture to halt the cell cycle prior to DNA replication.

Donor nucleus (#1)

Enucleated egg (#2)

5 The udder cell (donor) and enucleated egg are fused.

6 Stimulating mitotic inducers causes the cell to divide.

7 An early embryo develops and is transplanted into a receptive ewe.

Scottish blackface sheep (#3)

RESULTS

8 The embryo develops and Dolly is born.

Dolly, a Dorset sheep, genetically identical to #1

CONCLUSION: Differentiated animal cells are totipotent in nuclear transplant experiments.

19.4 Cloning a Mammal In 1996, the experimental procedure described here produced the first cloned mammal, a Dorset sheep named Dolly (shown on the left in the photo). As an adult, Dolly mated and subsequently gave birth to a "normal" offspring (the lamb on the right), thus proving the genetic viability of cloned mammals.

Out of 277 successful attempts to fuse adult cells with enucleated eggs, one lamb survived to be born; she was named Dolly, and she became world-famous overnight. DNA analyses confirmed that Dolly's nuclear genes were identical to those of the ewe from whose udder the donor nucleus had been obtained. Dolly grew to adulthood, mated, and produced offspring in the classic manner (see Figure 19.4), thus proving her status as a fully functioning adult animal. Dolly died in 2003 at the age of six—middle age for a sheep. Whether her cloned genome contributed to her relatively early death is still being debated.

One goal of Wilmut's experiments was to develop a method of cloning transgenic sheep—sheep into which genes with therapeutic properties have been introduced. For example, transgenic sheep have been developed that are capable of expressing pharmaceutical products in their milk (see Section 16.6). The hope was that the cloning procedure could make multiple, identical transgenic sheep that are all reliable producers of drugs.

The trick of starving donor cells for cloning has been applied to other mammals. Mice have been cloned using the somatic cells surrounding the egg as a source of donor nuclei (**Figure 19.5**). Cattle have been cloned to preserve a rare breed in New Zealand. Genetically engineered goats that produce several useful proteins in their milk have now been cloned to expand the numbers of these valuable animals. The recent successful cloning of horses may be a prelude to the cloning of valuable racing and show animals.

Cloning might be useful in preserving endangered species. More than 25 years ago, geneticists at the San Diego Zoo began freezing cells from endangered species, creating a modern-day Noah's Ark in anticipation of the emergence of new knowledge about animal development and its application to reproductive cloning. With that knowledge now at hand, they have begun thawing out cells and using them as nuclear donors to expand the populations of endangered species by cloning. The banteng, an endangered relative of the cow, was the first animal cloned in this way, using a cow enucleated egg and a cow surrogate mother. Meanwhile, in China, scientists are using rabbits as surrogate mothers for cloned pandas (which are only about 6 centimeters long when they are born). Even pets such as cats and dogs have been cloned. The evidence that clones are genetically identical to their nuclear donor "parent" comes from DNA fingerprinting analyses, such as SNP and STR typing (see Section 16.1).

The advent of cloning has been surrounded by controversy and ethical concerns, but cloning is not a new scientific concept. The idea of totipotency was accepted long before Dolly was born. Nevertheless, demonstrating totipotency via reproductive cloning is an impressive technical achievement.

Pluripotent stem cells can be induced to differentiate by environmental signals

Genomic equivalence implies that differentiated cells stay specialized because of environment and developmental history, not because of their genes, and thus appropriate environmental changes could result in a new pattern of differentiation. In normal development, a complex series of timed signals results in the patterns of differentiation we see in a newborn organism. If these signals could be described in enough detail, we should be able to understand how any cell type might become any other.

In plants, the growing regions at the tips of the roots and stems contain *meristems*, which are clusters of undifferentiated, rapidly dividing cells. These cells can give rise to the specialized cell types that make up roots and stems, respectively. Plants have far fewer (15–20) cell types than animals (as many as 200). Most plant cell types differ in the structure of their cell walls, whereas most animal cell types have specific cytoplasmic characteristics and many cell-specific proteins.

In mammals, **stem cells** are found in adult tissues that need frequent cell replacement, such as the skin, the inner lining of the intestine, and the blood system. Canadians Ernest McCulloch and James Till discovered stem cells in 1960 while doing bone marrow transplants in mice. They noticed that sometimes the recipient mice developed small clumps of tissue in the spleen. When they looked more carefully at the clumps, they found that each was composed of undifferentiated cells.

As they divide, stem cells produce daughter cells that differentiate to replace dead cells and maintain the tissues. These adult stem cells are not totipotent, but they do have limited abilities to differentiate into certain cell types; in other words, they are **pluripotent**. For example, there are two types of stem cells in bone marrow. One type produces only the various types of red and white blood cells, while the other type produces the cells that make bone and surrounding tissues, such as muscle.

The differentiation of pluripotent stem cells is "on demand." The blood cells that differentiate in the bone marrow do so in response to specific signals known as growth factors. If blood cells are removed from the circulatory system and put back into bone marrow, the signals will still be present and the bone marrow will generate new blood cells. This is the basis of an important cancer therapy called *bone marrow transplantation*. Because some therapies that kill cancer cells also kill other dividing cells (see Section 17.4), bone marrow stem cells in patients will die if exposed to the therapy agents. To prevent this, stem cells are removed from the patient's bone marrow and stored during therapy, then added back to the

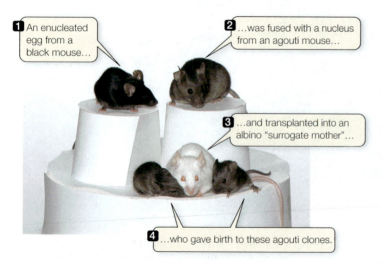

1 An enucleated egg from a black mouse…

2 …was fused with a nucleus from an agouti mouse…

3 …and transplanted into an albino "surrogate mother"…

4 …who gave birth to these agouti clones.

19.5 Cloned Mice Because so much is known about mouse genetics and molecular biology, cloned mice may be useful in studies of basic biology.

bone marrow when therapy is over. The stem cells retain their ability to differentiate in the bone marrow environment.

Adjacent cells can also influence stem cell differentiation. For example, the bone marrow stem cells that can form muscle will do so if implanted into the heart (**Figure 19.6**). This has been demonstrated in animal experiments, in which the stem cells were used to repair a damaged heart, and clinical uses in human patients are beginning.

Embryonic stem cells are potentially powerful therapeutic agents

As stated earlier, totipotent stem cells are found only in early embryos. In laboratory mice, these embryonic stem cells can be removed from an early embryo (the *blastocyst*; see Figure 43.4) and grown almost indefinitely. When injected back into a mouse blastocyst, the stem cells mix with the resident cells and differentiate to form all the cell types of the mouse. This kind of experiment shows that blastocyst cells do not lose any of their developmental potential while growing in the laboratory.

Embryonic stem cells growing in the laboratory also can be induced to differentiate in a particular way if the right signal is provided (**Figure 19.7**). For example, treatment of mouse embryonic stem cells with a derivative of vitamin A causes them to form neu-

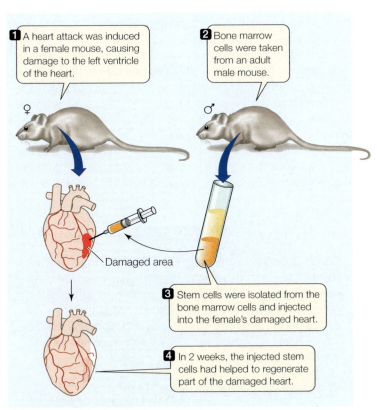

19.6 Repairing a Damaged Heart Pluripotent stem cells from the bone marrow of a male mouse were used successfully to repair the damaged heart of a female mouse. Mice of different sexes were used so that the Y chromosome could act as a genetic marker for the donor cells.

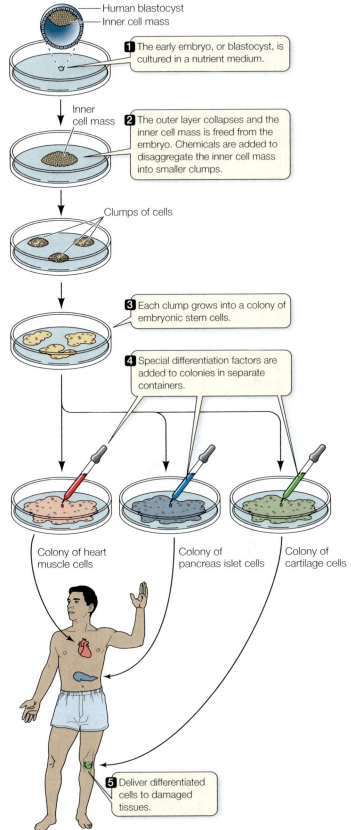

19.7 The Potential Use of Embryonic Stem Cells in Medicine Totipotent human embryonic stem cells can be cultured in the laboratory and induced to differentiate into a particular cell type. The use of these cells to replace tissues damaged by injury or disease is under intensive investigation.

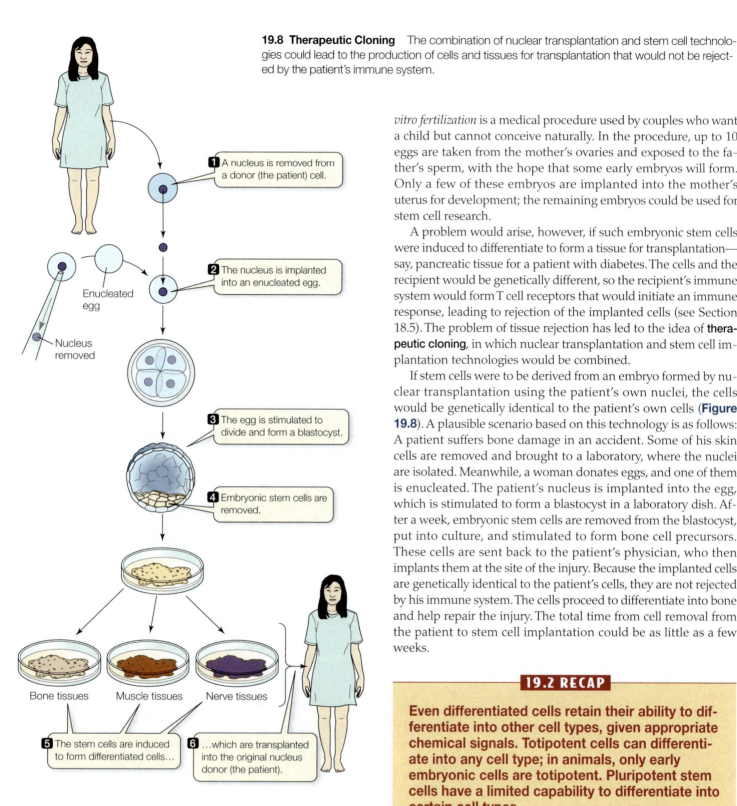

19.8 Therapeutic Cloning The combination of nuclear transplantation and stem cell technologies could lead to the production of cells and tissues for transplantation that would not be rejected by the patient's immune system.

1 A nucleus is removed from a donor (the patient) cell.

Enucleated egg

2 The nucleus is implanted into an enucleated egg.

Nucleus removed

3 The egg is stimulated to divide and form a blastocyst.

4 Embryonic stem cells are removed.

Bone tissues   Muscle tissues   Nerve tissues

5 The stem cells are induced to form differentiated cells...

6 ...which are transplanted into the original nucleus donor (the patient).

vitro fertilization is a medical procedure used by couples who want a child but cannot conceive naturally. In the procedure, up to 10 eggs are taken from the mother's ovaries and exposed to the father's sperm, with the hope that some early embryos will form. Only a few of these embryos are implanted into the mother's uterus for development; the remaining embryos could be used for stem cell research.

A problem would arise, however, if such embryonic stem cells were induced to differentiate to form a tissue for transplantation—say, pancreatic tissue for a patient with diabetes. The cells and the recipient would be genetically different, so the recipient's immune system would form T cell receptors that would initiate an immune response, leading to rejection of the implanted cells (see Section 18.5). The problem of tissue rejection has led to the idea of **therapeutic cloning**, in which nuclear transplantation and stem cell implantation technologies would be combined.

If stem cells were to be derived from an embryo formed by nuclear transplantation using the patient's own nuclei, the cells would be genetically identical to the patient's own cells (**Figure 19.8**). A plausible scenario based on this technology is as follows: A patient suffers bone damage in an accident. Some of his skin cells are removed and brought to a laboratory, where the nuclei are isolated. Meanwhile, a woman donates eggs, and one of them is enucleated. The patient's nucleus is implanted into the egg, which is stimulated to form a blastocyst in a laboratory dish. After a week, embryonic stem cells are removed from the blastocyst, put into culture, and stimulated to form bone cell precursors. These cells are sent back to the patient's physician, who then implants them at the site of the injury. Because the implanted cells are genetically identical to the patient's cells, they are not rejected by his immune system. The cells proceed to differentiate into bone and help repair the injury. The total time from cell removal from the patient to stem cell implantation could be as little as a few weeks.

rons, while other growth factors induce them to form blood cells, again demonstrating their developmental potential and the roles of environmental signals. This finding raises the possibility of using embryonic stem cell cultures as sources of differentiated cells for clinical medicine. A key advance toward this use has been the ability to grow human embryonic stem cells in the laboratory.

Embryonic stem cells could be harvested from human embryos made for in vitro fertilization with the consent of the donors. *In*

Cloning experiments and observations of stem cells have shown that a differentiated cell still has all of the genes of every other cell type. But not all genes are expressed in every cell. What turns gene expression on and off as cells differentiate? In the next section we explore several of the controls of gene expression that lead to cell differentiation.

19.3 What Is the Role of Gene Expression in Cell Differentiation?

Although every cell contains all the genes needed to produce every protein encoded by its genome, each cell synthesizes only selected proteins. For example, certain cells in our hair follicles continuously produce keratin, the protein that makes up hair, while other cell types in the body do not produce keratin. What determines whether a cell will produce keratin or not?

Chapter 14 describes a number of ways in which cells regulate gene expression—and hence the production of proteins. These pathways included transcriptional, translational, and posttranslational controls. The major controls of the gene expression that results in cell differentiation are transcriptional.

Differential gene transcription is a hallmark of cell differentiation

The gene for β-globin, one of the protein components of hemoglobin, is expressed in red blood cells as they form in the bone marrow of mammals. That this same gene is also present—but unexpressed—in neurons in the brain (which do not make hemoglobin) can be demonstrated by nucleic acid hybridization. Recall that in nucleic acid hybridization, a probe made of single-stranded DNA or RNA of known sequence is added to denatured DNA to reveal complementary coding regions on the DNA template strand (see Figure 14.6). A probe for the β-globin gene can be applied to DNA from both brain cells and immature red blood cells (recall that mature mammalian red blood cells lose their nuclei during development). In both cases, the probe finds its complement, showing that the β-globin gene is present in both types of cells. On the other hand, if the probe is applied to mRNA, rather than DNA, from the two cell types, it finds β-globin mRNA only in the red blood cells, not in the brain cells. This result shows that the gene is expressed in only one of the two cell types.

What leads to this differential gene expression? One well-studied example of cell differentiation is the conversion of undifferentiated muscle precursor cells, called *myoblasts*, into the large, multinucleated *muscle fibers* that make up mammalian skeletal muscles. The key event that starts this conversion is the expression of a gene called *MyoD* (*myo*blast-*d*etermination gene). The protein product of this gene is a *transcription factor* (MyoD) with a helix-loop-helix domain (see Figure 14.15), which not only binds to the promoters of muscle-determining genes to stimulate their transcription, but also acts on its own promoter to keep its levels high in the myoblasts and in their descendants.

Strong evidence for the controlling role of *MyoD* in muscle fiber differentiation comes from experiments in which an artificial DNA sequence containing an active promoter adjacent to *MyoD* was transfected into the precursors of other cell types. For example, when this sequence was added to fat cell precursors, the fat cells were reprogrammed to become muscle cells. Genes such as *MyoD* that direct the most fundamental decisions in development (often by regulating other genes on other chromosomes) usually encode transcription factors. Such genes, which act as a kind of molecular on-off switch, are called *developmental genes*.

Tools of molecular biology are used to investigate development

As the *MyoD* system demonstrates, transcription factors are important regulators of cell differentiation in the embryo. Determination and differentiation are not carried out by the actions of single genes, however, but rather by a complex series of interactions of many genes and their products. Eric Davidson at the California Institute of Technology leads a team of scientists who have explored these interactions in the early stages of development in the sea urchin, which has long been a favorite model organism of developmental biologists.

Davidson and his colleagues have described dozens of genes that are turned on and off during sea urchin development. Indeed, they estimate that at least one-third of the eukaryotic genome is used *only* during development. Besides studying transcription using DNA chips, they inactivated the expression of single genes using RNAi (see Section 16.5). The researchers looked at the effects of "silenced" genes not only on overall phenotype, but also on other proteins. For example, for a single gene called *endo16* that codes for a cell membrane protein, they were able to describe where and when in the embryo it is transcribed; how much of the protein was made; how long the transcript and protein lasted in the cell; and what molecules interacted with the *endo16* promoter to regulate its transcription during development. The result is a complex network of hundreds of RNAs and proteins interacting. The computational tools of systems biology are important in setting up and analyzing this network.

> ### 19.3 RECAP
>
> **Differentiation involves selective gene expression. In some cases, a single transcription factor can cause a cell to differentiate in a certain way. In others, complex interactions between genes and proteins determine a sequence of transcriptional events that leads to differential gene expression.**
>
> - Do you understand how differential gene expression underlies cell differentiation? See pp. 435
> - What is a transcription factor? Do you understand the role of transcription factors in differentiation? See pp. 435

We have seen how cell differentiation involves the extensive transcriptional regulation of genes. But what causes a cell to express one set of genes, and not some other set? In other words, how is a cell's fate determined?

19.4 How Is Cell Fate Determined?

The intricate networks of transcriptional controls that lead to cell differentiation are stimulated by chemical signals. In general, there are two mechanisms for producing such signals:

■ **Cytoplasmic segregation.** A factor within an egg, zygote, or precursor cell may be unequally distributed in the cytoplasm. After cell division, the factor ends up in some daughter cells or regions of cells, but not others.

■ **Induction.** A factor is actively produced and secreted by certain cells to induce other cells to differentiate.

Cytoplasmic segregation can determine polarity and cell fate

Some differences in patterns of gene expression are the result of *cytoplasmic* differences between cells. The emergence of **polarity**—the difference between the "top" and "bottom" ends of an organism or structure—is one such phenomenon. Polarity is obvious throughout development. Our heads are distinct from our rear ends, and the distal ends of our arms and legs (wrists, ankles, fingers, toes) differ from the proximal ends (shoulders and hips). Polarity may develop early; even within the fertilized egg, yolk and other factors are often distributed asymmetrically. During early development, polarity is specified by an *animal pole* at the top ("north pole") of the zygote and a *vegetal pole* at the bottom (the "south pole").

A famous series of experiments by Hans Driesch demonstrated the effects of cytoplasmic segregation on development (**Figure 19.9**). Very early development in sea urchins occurs by equal mitotic divisions of the fertilized egg; there is no increase in size at this stage. If an 8-cell embryo is cut vertically, both halves develop into normal (albeit small) embryos. But if an 8-cell embryo is cut horizontally, the top half does not develop at all, while the bottom half develops into a small, abnormal embryo.

Clearly, then, there must be at least one factor essential for development that is segregated in the vegetal half of the sea urchin egg, such that the bottom cells of the 8-cell embryo have it and the top cells do not. This and many other experiments have established that certain materials, called **cytoplasmic determinants**, are distributed unequally in the egg cytoplasm. These materials play a role in directing the embryonic development of many organisms (**Figure 19.10**).

In the fertilized eggs of many animals, the nucleus is positioned near the animal pole, and yolk molecules that will nourish the embryo accumulate in the vegetal half. The presence of yolk can slow down cell division, so the top halves of such eggs undergo many more cell divisions than the bottom halves do.

The cytoskeleton contributes to the asymmetrical distribution of cytoplasmic determinants in the egg. Recall from Section 4.3 that an important function of the microtubules and microfilaments

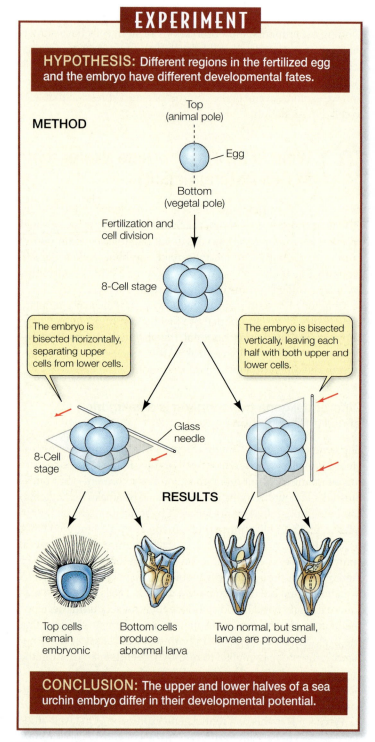

EXPERIMENT

HYPOTHESIS: Different regions in the fertilized egg and the embryo have different developmental fates.

METHOD

Top (animal pole)

Egg

Bottom (vegetal pole)

Fertilization and cell division

8-Cell stage

The embryo is bisected horizontally, separating upper cells from lower cells.

The embryo is bisected vertically, leaving each half with both upper and lower cells.

8-Cell stage

Glass needle

RESULTS

Top cells remain embryonic

Bottom cells produce abnormal larva

Two normal, but small, larvae are produced

CONCLUSION: The upper and lower halves of a sea urchin embryo differ in their developmental potential.

19.9 Asymmetry in the Early Sea Urchin Embryo The top and bottom halves of an 8-cell sea urchin differ in the cytoplasmic determinants they contain. Cells from both halves are necessary to produce a normal larva.

in the cytoskeleton is to help move materials in the cell. Two properties allow these structures to accomplish this:

■ Microtubules and microfilaments have polarity— they grow by adding subunits to the plus (+) end.

■ Cytoskeletal elements can bind specific proteins.

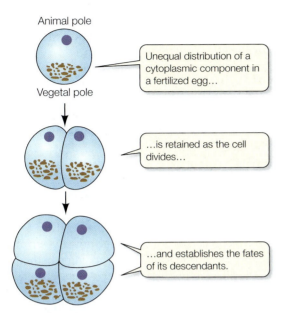

19.10 The Principle of Cytoplasmic Segregation The unequal distribution of some component in the cytoplasm of a cell may determine the fates of its descendants.

For example, in the sea urchin egg, a protein binds to both the growing (+) end of a microfilament and to an mRNA encoding a cytoplasmic determinant. As the microfilament grows toward one end of the cell, it carries the mRNA along with it. This asymmetrical distribution of the mRNA leads to a similar distribution of the protein it encodes.

Inducers passing from one cell to another can determine cell fates

Experimental work on developing embryos has established that in many cases, the fates of particular cells and tissues are determined by interactions with other specific tissues in the embryo. Many such instances of induction, in which one tissue causes an adjacent tissue to develop in a particular manner, have been observed in developing animal embryos. These effects are mediated by inter-

cellular communication—that is, by chemical signals and signal transduction mechanisms. We will describe two examples of differentiation by embryonic induction: one in the developing vertebrate eye, and the other in a developing reproductive structure in the nematode *C. elegans*.

LENS DIFFERENTIATION IN THE VERTEBRATE EYE The development of the lens of the vertebrate eye is a classic example of induction. In a frog embryo, the developing forebrain bulges out at both sides to form the *optic vesicles*, which expand until they come into contact with the cells at the surface of the head (**Figure 19.11**). The surface tissue in the region of contact with the optic vesicles thickens, forming a *lens placode*. The lens placode bends inward, folds over on itself, and ultimately detaches from the surface tissue to produce a structure that will develop into the lens. If the growing optic vesicle is cut away before it contacts the surface cells, no lens forms. Placing an impermeable barrier between the optic vesicle and the surface cells also prevents the lens from forming. These observations suggest that the surface tissue begins to develop into a lens when it receives a signal—an **inducer**—from the optic vesicle.

A chain of inductive interactions results in the development of the eye. There is a "dialogue" between the developing optic vesicle and the surface tissue. The optic vesicle induces lens development, and the developing lens determines the size of the *optic cup* that forms from the optic vesicle. If head surface tissue from a frog species with small eyes is grafted over the optic vesicle of one with large eyes, both lens and optic cup will have an intermediate size. The developing lens also induces the surface tissue over it to develop into a *cornea*, a specialized layer that allows light to pass through and enter the eye.

As this example shows, tissues do not induce themselves; rather, different tissues interact and induce one another. Embryonic inducers trigger a sequence of gene expression in the responding cells. How cells switch on different sets of genes that govern development and direct the formation of body plans is a subject of great interest to both developmental and evolutionary biologists. We will look at embryonic induction in more detail in Chapter 43.

VULVAL DIFFERENTIATION IN THE NEMATODE The tiny nematode *Caenorhabditis elegans* is a favorite model organism for studying

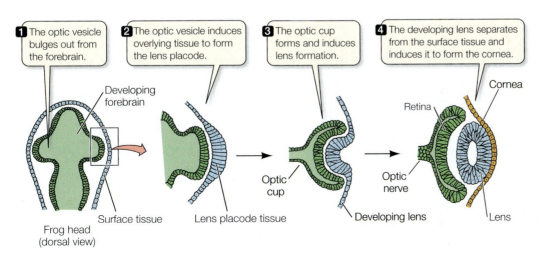

19.11 Embryonic Inducers in the Vertebrate Eye The eye of a frog develops as different tissues take their turns inducing one another.

development. It normally lives in the soil, where it feeds on bacteria, but can also grow in the laboratory if supplied with its food source. The process of development from fertilized egg to larva takes only about 8 hours, and the worm reaches the adult stage in just 3.5 days. The process is easily observed using a low-magnification dissecting microscope because the body covering is transparent (**Figure 19.12A**). The development of *C. elegans* does not vary from one individual to the next, so it has been possible to identify the source of each of the 959 somatic cells of the adult worm.

The adult nematode is *hermaphroditic*, containing both male and female reproductive organs. It lays eggs through a pore called the *vulva* on the ventral (belly) surface. During development, a single cell, called the *anchor cell*, induces the vulva to form. If the anchor cell is destroyed by laser surgery, no vulva forms; the eggs are fertilized inside the parent, and ultimately the baby worms consume the parent. This striking phenotype has allowed geneticists to identify the genes that have a role in the development of the vulva.

The anchor cell controls the fates of six cells on the developing worm's ventral surface through two molecular signals, the *primary inducer* and the *secondary inducer*. Each of these cells has three pos-

sible fates: it may become a primary vulval precursor cell, a secondary vulval precursor cell, or simply become part of the worm's surface—an epidermal cell (**Figure 19.12B**).

The anchor cell produces the primary inducer, which diffuses out of the cell and interacts with adjacent cells. Cells that receive enough primary inducer become vulval precursor cells; cells slightly farther from the anchor cell become epidermal cells. Thus the anchor cell, by releasing the primary inducer, determines whether a cell takes the "track" toward becoming part of the vulva or the track toward becoming part of the epidermis.

The cell closest to the anchor cell, having received the most primary inducer, differentiates into the primary vulval precursor cell. It produces its own inducer (the secondary inducer), which acts on the two neighboring cells and directs them to become secondary vulval precursor cells. Thus the primary vulval precursor cell produces a second signal, determining whether a vulval precursor cell will take the primary track or the secondary track. The two inducers control the activation or inactivation of specific genes through a signal transduction cascade in the responding cells (**Figure 19.13**).

Nematode development illustrates the important observation that *much of development is controlled by molecular switches that allow a cell to proceed down one of two alternative tracks.* One challenge for developmental biologists is to find these switches and determine how they work. The primary inducer released by the *C. elegans* anchor cell appears to be a growth factor homologous to a mammalian

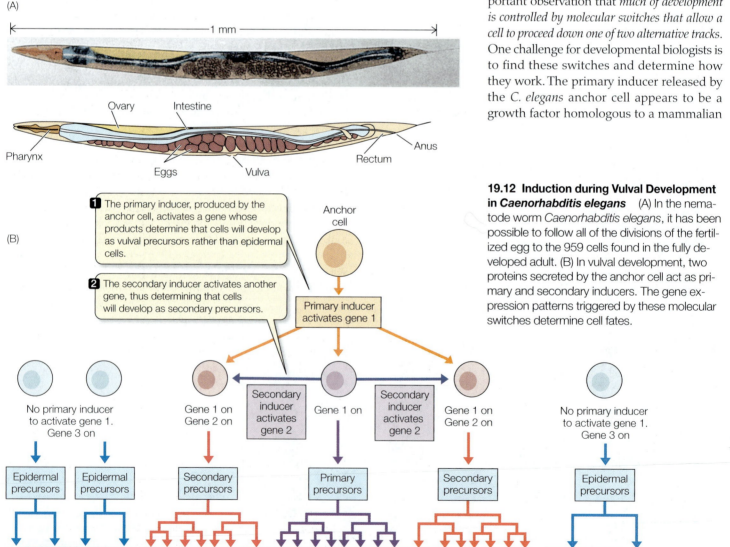

19.12 Induction during Vulval Development in *Caenorhabditis elegans* (A) In the nematode worm *Caenorhabditis elegans*, it has been possible to follow all of the divisions of the fertilized egg to the 959 cells found in the fully developed adult. (B) In vulval development, two proteins secreted by the anchor cell act as primary and secondary inducers. The gene expression patterns triggered by these molecular switches determine cell fates.

(A)

1 mm

Ovary   Intestine

Pharynx

Anus

Rectum

Eggs   Vulva

(B)

1 The primary inducer, produced by the anchor cell, activates a gene whose products determine that cells will develop as vulval precursors rather than epidermal cells.

2 The secondary inducer activates another gene, thus determining that cells will develop as secondary precursors.

Anchor cell

Primary inducer activates gene 1

No primary inducer to activate gene 1. Gene 3 on

Gene 1 on Gene 2 on

Secondary inducer activates gene 2

Gene 1 on

Secondary inducer activates gene 2

Gene 1 on Gene 2 on

No primary inducer to activate gene 1. Gene 3 on

Epidermal precursors

Epidermal precursors

Secondary precursors

Primary precursors

Secondary precursors

Epidermal precursors

Epidermis

Vulva

Epidermis

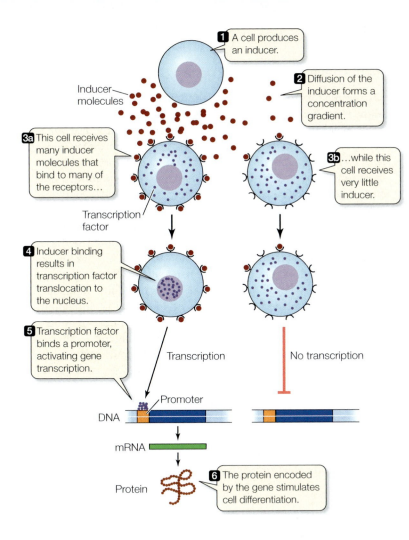

1 A cell produces an inducer.

Inducer molecules

2 Diffusion of the inducer forms a concentration gradient.

3a This cell receives many inducer molecules that bind to many of the receptors...

3b ...while this cell receives very little inducer.

Transcription factor

4 Inducer binding results in transcription factor translocation to the nucleus.

5 Transcription factor binds a promoter, activating gene transcription.

Transcription

No transcription

Promoter

DNA

mRNA

Protein

6 The protein encoded by the gene stimulates cell differentiation.

19.13 Embryonic Induction The concentration of an inducer directly affects the degree to which a transcription factor is activated. The inducer acts by binding to a receptor on the target cell. This binding is followed by signal transduction involving transcription factor translocation from the cytoplasm to the nucleus.

The differentiation of cells is beginning to be understood in terms of molecular and cellular events, but how do these events result in the organization of multitudes of cells into specific body parts, such as a leaf, a flower, a shoulder blade, or a tear duct? We now turn to the final phase of development, the formation of organs.

19.5 How Does Gene Expression Determine Pattern Formation?

Pattern formation is the process that results in the spatial organization of a tissue or organism. It is inextricably linked to morphogenesis, the creation of body form. You might expect morphogenesis to involve a lot of cell division, followed by differentiation—and it does. But what you might not expect is the degree of programmed cell death—apoptosis—that occurs during morphogenesis.

Some genes determine programmed cell death during development

We noted in Section 9.6 that apoptosis is used extensively during development to "sculpt" organs. For example, in the early human embryo, the hands and feet look like tiny paddles: the fingers and toes are linked by connective tissue. Between days 41 and 56 of development, the cells between the digits die, freeing the individual fingers and toes (**Figure 19.14**). As we will see when we discuss mammalian development in Chapter 43, many structures (such as the primitive structure in the back of the human embryo called the notochord) form and then disappear; this is also an example of apoptosis.

Model organisms have been very useful in studying the genes involved in apoptosis. For example, the nematode *C. elegans* produces precisely 1,090 somatic cells as it develops from a fertilized egg into an adult, but 131 of those cells die (leaving 959 cells to develop into an adult worm, as described in the previous section). The sequential expression of two genes, called *ced-4* and *ced-3* (for *ce*ll *d*eath), appears to control this cell death.

In the nematode nervous system, 302 neurons come from 405 precursors; thus 103 neural precursor cells undergo apoptosis. If the protein encoded by either *ced-3* or *ced-4* is nonfunctional, all 405 cells form neurons, resulting in abnormal brain development. Experiments have identified a third gene, *ced-9*, that codes for an inhibitor of apoptosis—that is, it codes for a protein that blocks the function of the *ced-3* and *ced-4* genes. Where apoptosis is required, *ced-3* and *ced-4* are active and *ced-9* is inactive; if apoptosis is not appropriate, *ced-9* is active and blocks *ced-3* and *ced-4*.

A similar system of cell death genes acts in humans. The proteins—a class of enzymes called *caspases*—that stimulate apopto-

growth factor called EGF (*e*pidermal *g*rowth *f*actor). The nematode growth factor, called LIN-3, binds to a receptor on the surface of a potential vulval precursor cell, setting in motion a signal transduction cascade involving the Ras protein and MAP kinases (see Figure 15.10). The end result is increased transcription of the genes involved in the differentiation of vulval cells.

19.4 RECAP

Cytoplasmic segregation is the unequal distribution of molecular signaling factors in the egg, zygote, or early embryo. Embryonic induction occurs when one cell or tissue sends a chemical signal to another. Both types of signaling trigger transcription factor activity, and thus cellular differentiation.

- Can you describe how cytoplasmic segregation results in polarity in a fertilized egg, and how polarity affects cell differentiation? See p. 436 and Figure 19.10

- Do you understand how embryonic induction forms the tissues of a vertebrate's eyes? See p. 437 and Figure 19.11

- How do inducer molecules interact with transcription factors to produce differentiated cells? See p. 438 and Figure 19.13

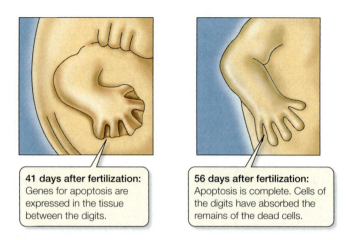

41 days after fertilization: Genes for apoptosis are expressed in the tissue between the digits.

56 days after fertilization: Apoptosis is complete. Cells of the digits have absorbed the remains of the dead cells.

19.14 Apoptosis Removes the Tissue between Human Fingers Early in the second month of human development, the tissue connecting the fingers is removed by apoptosis, freeing the individual fingers.

sis are similar in amino acid sequence to the protein encoded by *ced-3*, and a human protein (BCL-2) that inhibits apoptosis is similar to the product of *ced-9*. So humans and nematodes, two creatures separated by more than 600 million years of evolutionary history, have similar genes controlling programmed cell death.

Plants have organ identity genes

Like animals, plants have organs—for example, leaves and roots. Many plants form flowers, and many flowers are composed of four types of organs: sepals, petals, stamens, and carpels. These floral organs occur in *whorls*, which are groups of each organ type stacked around a central axis. The whorls develop from meristems in the shape of domes, which develop at growing points on the stem (**Figure 19.15A**). How is the identity of a particular whorl determined? A group of genes, called **organ identity genes**, code for proteins that act in combination to produce specific whorl features.

Organ identity genes have been well described in the thale cress, *Arabidopsis thaliana*. This model organism is very useful for studies of development because of its small size (about 25 cm), abundant seed production (over 1,000 seeds per plant), rapid development (from seed to plant to seed in 6 weeks), and small genome (see Section 14.1). Finally, it is easy to produce mutations in this plant by treating the seeds with mutagens.

The development of the flower begins with the meristem, which contains about 700 undifferentiated cells. Within this seemingly homogeneous cell population, individual cells "sense" their position and differentiate into the whorls. This happens through the expression of three classes of organ identity genes, which code for proteins that act in combination with one another:

■ Genes in class A are expressed in whorls 1 and 2 (which form sepals and petals, respectively).

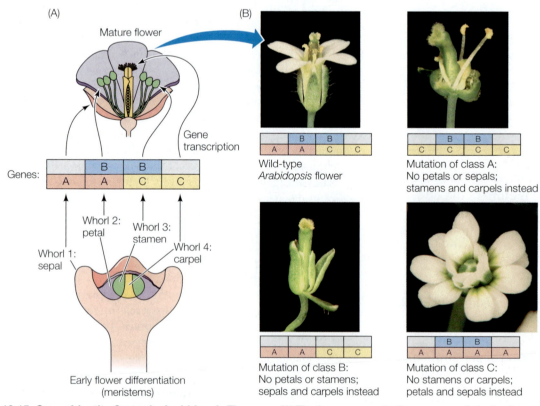

19.15 Organ Identity Genes in *Arabidopsis* Flowers (A) The four organs of a flower—carpels (yellow), stamens (green), petals (purple), and sepals (pink)—grow in whorls that develop from meristems. (B) When a mutation in one of the three organ identity genes occurs, one type of organ replaces another. Such mutations helped scientists decipher the pattern of gene expression that gives rise to normal flowers.

- Genes in class B are expressed in whorls 2 and 3 (which form petals and stamens, respectively).
- Genes in class C are expressed in whorls 3 and 4 (which form stamens and carpels, respectively).

There are two lines of experimental evidence for this model of organ identity gene function (**Figure 19.15B**):

- *Loss-of-function mutations:* for example, a mutation in gene class A results in no sepals or petals.
- *Gain-of-function mutations:* for example, the promoter for gene class C can be artificially coupled to gene class A. In this case, A is expressed in all four whorls, resulting in only sepals and petals.

Gene classes A, B, and C code for subunits of transcription factors, which are active as dimers. Gene regulation in these cases is *combinatorial*—that is, the composition of the dimer determines which genes will be activated by the transcription factor. For example, a dimer made up of two monomers of transcription factor A would activate transcription of the genes that make sepals; a dimer made up of A and B monomers would result in petals, and so forth. A common feature of the A, B, and C proteins, as well as many other plant transcription factors, is a DNA-binding domain called the **MADS box**. These 200-amino-acid proteins also have domains that can bind to other proteins in a *transcription initiation complex.*

The MADS box gets its name from homologous amino acid sequences found in four proteins in widely divergent species: *MCMI* (from yeast), *AGAMOUS* (*Arabidopsis*), *DEFICIENS* (snapdragons), and *SRF* (humans).

In addition to being fascinating to biologists, plant organ identity genes have caught the eye of horticultural and agricultural scientists. Flowers filled with petals instead of stamens and carpels often have mutations of the C genes. Many of the foods that make up the human diet, such as the grains of wheat, rice, and corn, come from fruits and seeds. These fruits and seeds form from the carpels (the female reproductive organs) of the flower. Genetically modifying the number of carpels on a particular plant could increase the amount of grain a crop could produce.

A gene called *leafy* codes for a protein that controls the transcription of the organ identity genes. Plants with a mutation that causes the underexpression of *leafy* are just that—they make leaves, but no flowers. The protein product of this gene acts as a transcription factor, stimulating gene classes A, B, and C so that they produce flowers (**Figure 19.16**). This finding, too, has practical applications. It usually takes 6–20 years for a citrus tree to produce flowers and fruits. Scientists have made orange trees transgenic for *leafy* coupled to a strongly expressed promoter; the transgenic trees flower and fruit years earlier than normal trees.

Morphogen gradients provide positional information

During development, the key cellular question, "What am I (or what will I be)?" is often answered in part by "Where am I?" Think of a cell in the apical meristem of a developing *Arabidopsis*: it needs to "know" in what whorl it is located. This spatial "sense" is called **positional information**. Positional information usually comes in the form of a signal, called a **morphogen**, that diffuses from one group of cells down a body axis, setting up a concentration gradient. There are two requirements for a signal to be considered a morphogen:

- It must directly affect target cells, rather than triggering a secondary signal that affects target cells.
- Different concentrations of the signal must cause different effects.

The development of the vertebrate limb provides us with an example of a morphogen in action. The limb develops from a paddle-shaped *limb bud*. The cells that become the bones and muscles of the limb must receive positional information. If they do not, the limbs will be totally disorganized (imagine a hand with only thumbs or only little fingers). A group of cells at the posterior base of the limb bud, just where it joins the body wall, is called the *zone of polarizing activity* and makes a morphogen called BMP2, which forms a gradient. This gradient determines the posterior–anterior ("little finger to thumb") axis of the developing limb. The cells getting the highest dose of BMP2 make the little finger, and those getting the lowest dose make the thumb.

Wild type

Leafy mutant

19.16 A Nonflowering Mutant Mutations in the *leafy* gene of *Arabidopsis* prevent the transcription of the organ identity genes, and the resulting plant does not produce any flowers.

Different concentrations of morphogens act through differential regulation of gene expression in their target cells. The model organism most often used for studying this process has been the fruit fly.

In the fruit fly, a cascade of transcription factors establishes body segmentation

Insects such as the fruit fly *Drosophila melanogaster* develop a highly modular body composed of different types of segments. Complex interactions of different sets of genes underlie pattern formation in segmented bodies.

Unlike the body segments of segmented worms such as earthworms, which are all essentially alike, the segments of the *Drosophila* body are clearly different from one another. The adult fly has an anterior *head* (composed of several fused segments), three different *thoracic* segments, and eight *abdominal* segments at the posterior end. In the *Drosophila* larva, the thoracic and abdominal segments all appear to be similar, but at this point it has *already been determined* that they will form these specialized adult segments. Several types of genes are expressed sequentially in the embryo to define these segments:

- First, a set of genes from the mother's cells adjacent to the egg sets up anterior–posterior and dorsal–ventral axes in the egg.

- Next, a series of genes in the embryo successively define the position of each cell in a segment relative to these axes. The end result is that a cell "knows" precisely where it is in the embryo; for example, that it is part of the head at the very front.

- Finally, a set of genes called *Hox genes* control the ultimate identity of each segment; for example, determining that the cells at the very front of the head will make antennae.

The genes involved in each of these steps code for transcription factors, which in turn control the synthesis of transcription factors acting on the next set of genes. This cascade of events may remind you of a signal transduction cascade (see Section 15.3), only in this case it is a cascade of events over time and location, rather than in a single cell. The final expressed genes are the ones familiar to you: they code for protein kinases, receptors, and other proteins that carry out the functions of the cell.

The description of these events in fruit fly development is one of the great achievements in modern biology. We will only skim the surface of the process here, but keep in mind the basic principle of a transcriptional cascade. As we will see in Chapter 20, the fruit fly has been a true model organism in this case, as the basic pattern of events is similar in many other organisms, including humans.

MATERNAL EFFECT GENES Like those of the sea urchin, *Drosophila* eggs and larvae are characterized by unevenly distributed cytoplasmic determinants (see Figure 19.10). These molecules, which include both mRNAs and proteins, are the products of specific **maternal effect genes**. These genes are transcribed in the cells of the mother's ovary that surround what will be the anterior portion of the egg. Two maternal effect genes, called *bicoid* and *nanos*, determine the anterior–posterior axis of the egg. Other maternal effect genes that will not be described here determine the dorsal–ventral axis.

The mRNAs for *bicoid* and *nanos* diffuse from the mother's cells into what will be the anterior end of the egg through cytoplasmic bridges. The *bicoid* mRNA stays where it enters the egg and is translated to produce Bicoid protein, which diffuses away from the anterior end, establishing a gradient in the egg (**Figure 19.17**). Where it is present in sufficient concentration, Bicoid acts as a transcription factor to stimulate the transcription of the *hunchback* gene in the egg. A gradient of the latter protein establishes the head, or anterior, region.

Meanwhile, *nanos* mRNA is transported by the cytoskeleton to the posterior end of the egg, where it enters the egg and is translated and its protein forms a gradient. The Nanos protein is not a morphogen; instead, it is an inhibitor of the translation of *hunchback* mRNA. It is the low level of *hunchback* expression in the posterior because of this "double whammy"—inhibition of its translation by Nanos and a lack of stimulation of its transcription by Bicoid—that results in the establishment of the posterior region.

How do we know all of this? Throughout this book, we have described two approaches to demonstrating cause and effect: we study mutations that *prevent* an event from occurring, and we do experiments to *cause* that event to occur. Scientists demonstrated that maternal effect genes specify the axes of the *Drosophila* embryo by causing mutations in those genes, and by experiments in which cytoplasm was transferred from one egg to another:

- Females that are homozygous for a particular mutation of *bicoid* produce larvae with no head and no thorax; thus the Bicoid protein must be needed for the anterior structures to develop.

- If the eggs of these mutant females are inoculated at the anterior end with cytoplasm from the anterior region of a wild-type egg, the "rescued" eggs develop into normal larvae; this experiment also shows that Bicoid protein is involved in the development of anterior structures.

- If Bicoid protein from a wild-type egg is injected into the posterior region of an egg, anterior structures develop there. The degree of induction depends on how much cytoplasm is injected.

- Eggs from homozygous *nanos* mutant females develop into larvae with missing abdominal segments.

- Injecting cytoplasm from the posterior region of a wild-type egg into a *nanos* mutant egg allows normal development.

The events involving *bicoid*, *nanos*, and *hunchback* begin before fertilization and continue after it. Before the *Drosophila* egg is fertilized, *bicoid* and *nanos* mRNAs are made by the maternal cells and enter the egg. After the egg is fertilized and laid, nuclear divisions begin. In *Drosophila*, cytokinesis does not begin right away; until the thirteenth nuclear division, the embryo is a single, multinucleated cell called a *syncytium*. At this early stage, *bicoid* and *nanos* mRNAs are translated and establish gradients.

After the axes of the embryo are determined, the next step in pattern formation is the determination of the locations of the segments.

SEGMENTATION GENES The number, boundaries, and polarity of the larval segments are determined by proteins encoded by the **segmentation genes**. These genes are expressed when there are about 6,000 nuclei in the embryo. These nuclei all look the same, but in terms of gene expression, they are not equivalent. Their fates are sealed early in development.

19.17 Bicoid Protein Provides Positional Information The anterior–posterior axis of *Drosophila* arises from the gradient of a morphogen encoded by *bicoid*, a maternal effect gene. Bicoid protein acts as a transcription factor to activate a gene that specifies that this region will produce the structures of the head. Other maternal effect genes in the posterior portion of the embryo inhibit Bicoid, thus limiting its region.

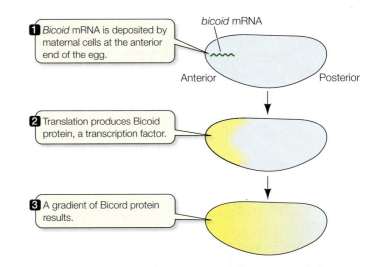

1 *Bicoid* mRNA is deposited by maternal cells at the anterior end of the egg.

bicoid mRNA

Anterior   Posterior

2 Translation produces Bicoid protein, a transcription factor.

3 A gradient of Bicord protein results.

Three classes of segmentation genes act one after the other to regulate finer and finer details of the segmentation pattern (**Figure 19.18**):

- **Gap genes** organize broad areas along the anterior–posterior axis. Mutations in gap genes result in gaps in the body plan—the omission of several larval segments.

- **Pair rule genes** divide the embryo into units of two segments each. Mutations in pair rule genes result in embryos missing every other segment.

- **Segment polarity genes** determine the boundaries and anterior–posterior organization of the individual segments. Mutations in segment polarity genes can result in segments in which posterior structures are replaced by reversed (mirror-image) anterior structures.

4 High concentrations of Bicoid stimulate the head-specifying genes.

The expression of these genes is sequential. The maternal effect protein Bicoid, which begins the cascade, acts as a morphogen and transcription factor to stimulate the expression in the egg of genes such as *hunchback* that set up the anterior–posterior axis. As a result, a nucleus in the egg "knows" where it is. The Hunchback protein stimulates gap gene transcription; the products of the gap genes stimulate pair rule genes; and the products of the pair rule genes stimulate segment polarity genes. By the end of this cascade,

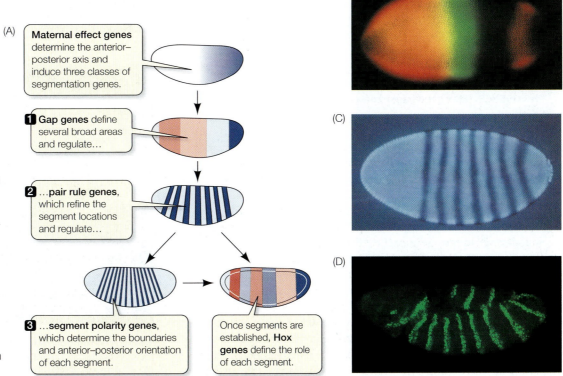

(A)

Maternal effect genes determine the anterior–posterior axis and induce three classes of segmentation genes.

1 **Gap genes** define several broad areas and regulate…

2 …**pair rule genes**, which refine the segment locations and regulate…

3 …**segment polarity genes**, which determine the boundaries and anterior–posterior orientation of each segment.

Once segments are established, **Hox genes** define the role of each segment.

19.18 A Gene Cascade Controls Pattern Formation in the *Drosophila* Embryo (A) Maternal effect genes (see Figure 19.17) induce gap, pair rule, and segment polarity genes—collectively referred to as segmentation genes. (B) Two gap genes, *hunchback* (orange) and *Krüppel* (green) overlap; both genes are transcribed in the yellow area. (C) The pair rule gene *fushi tarazu* is transcribed in the dark blue areas. (D) The segment polarity gene *engrailed* (bright green) is seen here at a slightly more advanced stage than is depicted in (A).

(B)

(C)

(D)

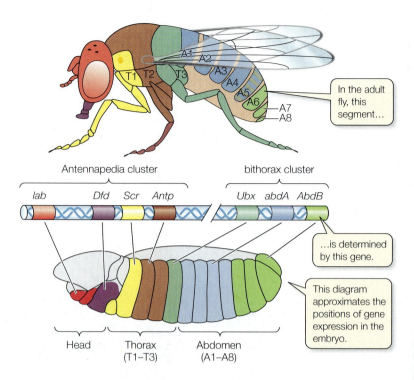

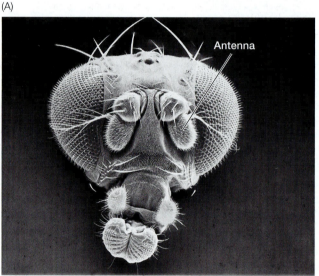

19.19 Hox Genes in *Drosophila* Two clusters of genes on chromosome 3 (center) determine segment function in the adult fly (top). These genes are expressed in the embryo (bottom) long before the structures of the segments actually appear.

In the adult fly, this segment…

…is determined by this gene.

This diagram approximates the positions of gene expression in the embryo.

Antennapedia cluster   bithorax cluster

lab *Dfd* *Scr* *Antp*   *Ubx* *abdA* *AbdB*

Head   Thorax (T1–T3)   Abdomen (A1–A8)

ward Lewis at Caltech found that *Antennapedia* and *bithorax* mutations resulted from changes in Hox genes.

The first cluster of Hox genes, the bithorax cluster, specifies anterior segments, starting with genes for the different head segments and ending with thoracic segments. The second cluster, the Antennapedia cluster, begins with a gene specifying the last thoracic segment, followed by a gene for the anterior abdominal segments, and ends with a gene for the posterior abdominal segments. Lewis hypothesized that all of the *Drosophila*

a group of nuclei at the very anterior of the egg, for example, "knows" that they are going to become part of the first anterior segment in the adult fly.

The next set of genes in the cascade determines the form and function of each segment.

Drosophila genes are often named for their mutant phenotype. Thus the *hunchback* gene received its name because flies who lack this gene have deformed or missing heads. Likewise, *fushi tarazu* is Japanese for "not enough parts"—and, indeed, mutants for this pair rule gene have too few segments.

HOX GENES Hox genes are expressed in different combinations along the length of the embryo and tell each segment what to become. Hox gene expression tells the cells of a segment in the head to make eyes, those of a segment in the thorax to make wings, and so on. Remarkably, the *Drosophila* Hox genes map on chromosome 3, in the same order as the segments whose function they determine (**Figure 19.19**). By the time the fruit fly larva hatches, its segments are completely determined. Hox genes are analogous to the organ identity genes of plants. The maternal effect, segmentation, and Hox genes interact to "build" a *Drosophila* larva step by step, beginning with the unfertilized egg.

Once again, how do we know that the Hox genes determine segment identity? An important clue came from bizarre mutations observed in *Drosophila* that were called *homeotic* (Greek *homeos*, "a being resembling something else"). In the *Antennapedia* mutation, legs grow on the head in place of antennae (**Figure 19.20**), and in the *bithorax* mutation, an extra pair of wings grows in a thoracic segment where wings do not normally grow (see Figure 20.2). Ed-

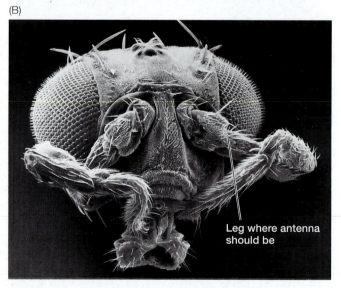

(A)

Antenna

(B)

Leg where antenna should be

19.20 A Homeotic Mutation in *Drosophila* Mutations of the Hox genes cause body parts to form on inappropriate segments. (A) A wild-type fruit fly. (B) An *Antennapedia* mutant fruit fly.

Hox genes might have come from the duplication of a single gene in an ancestral, unsegmented organism.

Homeobox-containing genes encode transcription factors

Molecular biologists confirmed Lewis's hypothesis using nucleic acid hybridization. Several scientists found that a probe for a sequence found in one of the genes in the bithorax cluster bound not only to its own gene, but also to the adjacent genes in its cluster and to the genes in the Antennapedia cluster. In other words, this DNA sequence is common to all the Hox genes in both clusters. It is also found in several of the segmentation genes, as well as other genes that encode transcription factors.

This 180-base-pair DNA sequence is called the **homeobox**. It encodes a 60-amino acid sequence, called the *homeodomain*, that binds to DNA. The homeodomain recognizes a specific DNA sequence in the promoter of its target genes, but that recognition is usually not sufficient to allow the transcription factor to bind fully to a promoter and turn the target gene on or off. Other transcription factors are also involved.

Genes containing the homeobox are found in many animals, including humans. They play a role in development similar to the role the MADS box genes play in plants. The evolutionary significance of these common pathways for development will be discussed in the next chapter.

19.5 RECAP

A cascade of transcription factors governs pattern formation and the subsequent development of animal and plant organs. In plants, organ identity genes code for transcription factors that determine which organ a group of cells will form. In animals, morphogen gradients provide positional information.

- Do you see how apoptosis is crucial in shaping the developing embryo? See p. 440
- Can you describe how organ identity genes act in *Arabidopsis*? See p. 441 and Figure 19.15
- What are the key attributes of a morphogen? How does Bicoid protein fit this definition? See pp. 441–442 and Figure 19.17

CHAPTER SUMMARY

19.1 What are the processes of development?

A multicellular organism begins its **development** as an **embryo**. A series of embryonic stages may precede the birth of an independent organism. Review Figure 19.1

The processes of development are **determination**, **differentiation**, **morphogenesis**, and **growth**.

Differential gene expression is responsible for the differences between cell types. A cell's **fate** is determined by environmental factors, such as its position in the embryo, as well as by intracellular influences.

Determination is followed by differentiation, the actual changes in biochemistry, structure, and function that result in cells of different types. Determination is a commitment; differentiation is the realization of that commitment.

19.2 Is cell differentiation irreversible?

See Web/CD Tutorial 19.1

The zygote is **totipotent**; it is capable of forming every type of cell in the adult body.

The ability to create **clones** from differentiated cells demonstrates the principle of **genomic equivalence**. Review Figures 19.3 and 19.4

Stem cells produce daughter cells that differentiate when provided with appropriate intercellular signals. Some **pluripotent** stem cells in the adult body can differentiate into a limited number of cell types to replace dead cells and maintain tissues.

Embryonic stem cells are totipotent and can be cultured in the laboratory. With suitable environmental stimulation, these cells can be induced to differentiate. A technique called **therapeutic cloning** could combine nuclear transplantation and stem cell technologies to replace cells or tissues damaged by injury or disease. Review Figures 19.7 and 19.8

19.3 What is the role of gene expression in cell differentiation?

Differential gene expression results in cell differentiation. Transcription factors are especially important in regulating gene expression during differentiation.

Complex interactions of many genes and their products are responsible for differentiation during development.

19.4 How is cell fate determined?

Cytoplasmic segregation—the unequal distribution of **cytoplasmic determinants** in the egg, zygote, or early embryo—can establish **polarity** and lead to determination of a cell's descendants. Review Figures 19.9 and 19.10, Web/CD Tutorial 19.2

Induction is a process by which embryonic animal tissues direct the development of neighboring cells and tissues by secreting chemical signals, called **inducers**. Review Figure 19.11

The induction of the vulva in the nematode *Caenorhabditis elegans* offers an example of how inducers act as molecular switches to direct a cell down one of two differentiation paths. Review Figures 19.12 and 19.13

19.5 How does gene expression determine pattern formation?

Pattern formation is the process that results in the spatial organization of a tissue or organism.

During development, selective elimination of cells by apoptosis results from the expression of specific genes.

Combinatorial interactions of transcription factors coded for by **organ identity genes** in plants causes the formation of sepals, petals, stamens, and carpels. Review Figure 19.15

CHAPTER SUMMARY

The transcription factors encoded by floral organ identity genes contain an amino acid sequence called the **MADS box** that can bind to DNA.

Both plants and animals use **positional information** as a basis for pattern formation. Positional information usually comes in the form of a signal called a **morphogen**. Different concentrations of the morphogen cause different effects.

In the fruit fly *Drosophila melanogaster*, a cascade of transcription factors sets up the axes of the embryo and then causes the cells of each body segment to differentiate into particular organs. The cascade involves the sequential expression of maternal effect genes, gap genes, pair rule genes, segment polarity genes, and Hox genes. Review Figures 19.17 and 19.18, Web/CD Tutorial 19.3

The **homeobox** is a DNA sequence found in Hox genes and other genes that code for transcription factors. The sequence of amino acids encoded by the homeobox is called the homeodomain. Review Figure 19.19

SELF-QUIZ

1. Which statement about determination is true?
   a. Differentiation precedes determination.
   b. All cells are determined after two cell divisions in most organisms.
   c. A determined cell will keep its determination no matter where it is placed in an embryo.
   d. A cell changes its appearance when it becomes determined.
   e. A differentiated cell has the same pattern of transcription as a determined cell.

2. Cloning experiments on sheep, frogs, and mice have shown that
   a. nuclei of adult cells are totipotent.
   b. nuclei of embryonic cells can be totipotent.
   c. nuclei of differentiated cells have different genes than zygote nuclei have.
   d. differentiation is fully reversible in all cells of a frog.
   e. differentiation involves permanent changes in the genome.

3. The term "induction" describes a process in which a cell or group of cells
   a. influences the development of another group of cells.
   b. triggers the cell movements in an embryo.
   c. stimulates the transcription of their own genes.
   d. organizes the egg cytoplasm before fertilization.
   e. inhibits the movement of the embryo.

4. The term "therapeutic cloning" describes a method for
   a. modification of a clone by a transgene.
   b. combining nuclear transplantation and stem cell technologies.
   c. making clones that produce useful drugs.
   d. producing embryonic stem cells for transplantation.
   e. making many identical copies of an organism.

5. Which statement about cytoplasmic determinants in *Drosophila* is *not* true?
   a. They specify the dorsal–ventral and anterior–posterior axes of the embryo.
   b. Their positions in the embryo are determined by cytoskeletal action.
   c. Some are products of specific genes in the mother fruit fly.
   d. They do not produce gradients.
   e. They have been studied by the transfer of cytoplasm from egg to egg.

6. In fruit flies, the following genes are used to determine segment polarity: (k) gap genes; (l) Hox genes; (m) maternal effect genes; (n) pair rule genes. In what order are these genes expressed during development?
   a. klmn
   b. lknm
   c. mknl
   d. nkml
   e. nmkl

7. Which statement about embryonic induction is *not* true?
   a. One group of cells induces adjacent cells to develop in a certain way.
   b. It triggers a sequence of gene expression in target cells.
   c. Single cells cannot form an inducer.
   d. A tissue may be induced as well as an inducer.
   e. The chemical identification of specific inducers has not been achieved.

8. In the process of pattern formation in the *Drosophila* embryo,
   a. the first steps are specified by Hox genes.
   b. mutations in pair rule genes result in embryos missing every other segment.
   c. mutations in gap genes result in the insertion of extra segments.
   d. segment polarity genes determine the dorsal–ventral axes of segments.
   e. segmentation is the same as in earthworms.

9. Homeotic mutations
   a. are often severe and result in structures at inappropriate places.
   b. cause subtle changes in the forms of larvae or adults.
   c. occur only in prokaryotes.
   d. do not affect the animal's DNA.
   e. are confined to the zone of polarizing activity.

10. Which statement about the homeobox is *not* true?
    a. It is transcribed and translated.
    b. It is found only in animals.
    c. Proteins containing the homeodomain bind to DNA.
    d. It is a sequence of DNA shared by many genes.
    e. It occurs only in Hox genes.

FOR DISCUSSION

1. Molecular biologists can attach genes to active promoters and insert them into cells (see Section 16.3). What would happen if the following were inserted and overexpressed? Explain your answers.
   a. *ced-9* in embryonic neuron precursors in *C. elegans*
   b. *MyoD1* in undifferentiated myoblasts
   c. the gene for BMP2 in a chick limb bud
   d. *nanos* at the anterior end of the *Drosophila* embryo

2. A powerful method to test for the function of a gene in development is to generate a "knockout" organism, in which the gene in question is inactivated (see Section 16.5). What do you think would happen in each of the following cases?
   a. a knocked-out *ced-9* in *C. elegans*
   b. a knocked-out *nanos* in *Drosophila*

3. Look at the chart of organ identity mutations in Figure 19.15. What pattern do you perceive in the results of these mutations, and what might this pattern mean?

4. During development, the potential of a cell becomes ever more limited, until, in the normal course of events, its potential is the same as its original prospective fate. On the basis of what you have learned in this chapter, discuss possible mechanisms for the progressive limitation of the cell's potential.

5. How were biologists able to obtain such a complete accounting of all the cells in *Caenorhabditis elegans*? What major conclusions came from these studies?

FOR INVESTIGATION

Cloning involves considerable reprogramming of gene expression in a differentiated cell so that it acts like an egg cell.

How would you investigate this reprogramming?

CHAPTER 20

Development and Evolutionary Change

The eyes have it

Eyes are not essential for survival; many animals and all plants get by just fine without them. However, over 90 percent of all animals *do* have eyes or some type of light-sensing organs, and having eyes can confer a selective advantage.

About a dozen different kinds of eyes are found among the different animals, including the camera-like eyes of humans and the compound eyes of insects, with their thousands of individual units. In trying to understand the origin of this variety, scientists—starting with Charles Darwin—proposed that eyes evolved independently many times in different animal groups, and that each improvement in the ability of eyes to gather light and form images conferred a selective advantage on their possessor.

Our understanding of the evolution of eyes remained at this level until 1915, when a mutant fruit fly without eyes was found and the gene involved, appropriately called *eyeless*, was mapped on one of its chromosomes. This mutant fly remained a laboratory curiosity until the 1990s, when the Swiss developmental biologists Rebecca Quiring and Walter Gehring began looking for transcription factors that might be involved in fly development. They found one, and when they mapped its gene, it was at the *eyeless* locus. The product of the *eyeless* gene is a transcription factor that controls the formation of the eye. Quiring and Gehring demonstrated this by transplanting early embryonic tissue that was overexpressing the *eyeless* gene to other places in the embryo. It did not matter where they placed the tissue; even on legs, as long as the gene was active, an eye developed.

When they did comparative genomic studies, Quiring and Gehring were in for a big surprise: the *eyeless* gene sequence was quite similar to that of *Pax6*, a gene in mice that, when mutated, leads to the development of very small eyes. Could the very different eyes of flies and mice just be variations on a common developmental theme?

To test the similarity in function between the insect and mammalian genes, Quiring and Gehring inserted the mouse gene *Pax6* into the fly genome and repeated their transplantation experiments. Once again, eyes developed. A gene whose expression normally leads to the development of a mammalian "camera" eye now led to the development of

Eye of the Fly   Unlike the single-lensed eyes of vertebrates, the compound eyes of flies and some other insects are composed of thousands of individual lenses, or ommatidia.

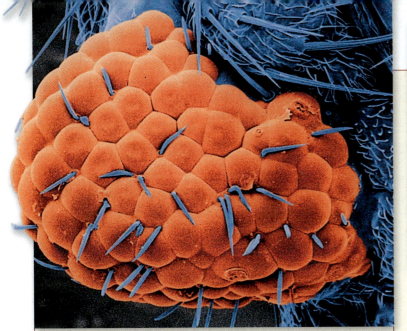

A Mouse Gene Can Produce a Fly's Eye When the mouse *Pax6* eye-specifying gene was implanted in the limb disk of a fruit fly, ommatidia emerged in place of a leg.

an insect's "compound" eye—a very different eye type. Thus a single transcription factor appears to function as a molecular switch that turns on eye development. Although eyes evolved many times during animal evolution, all of them may depend on the same gene. The special features of the many different eyes in diverse animals all evolved from a common developmental process.

The discovery that the same genes govern development in a wide variety of animals led to the rapid proliferation of a new discipline: evolutionary developmental biology, often known as "evo-devo." Evolutionary developmental biologists compare the genes that regulate development in many different multicellular organisms to understand how a single gene can do so many different things.

IN THIS CHAPTER we will see that the genes that control pattern formation are shared by a diverse array of organisms. We will discover how changes can occur in some parts of an organism without causing undesirable changes in other parts, and how a common set of genes can produce such a great variety of body forms. We will also see how some organisms can modulate their development by responding to signals from their environment. Finally, we will see how developmental processes constrain evolution.

20.1 How Does a Molecular Tool Kit Govern Development?

Biologists have long known that a fruit fly looks like a fruit fly and a mouse looks like a mouse largely because of their genes; clearly, certain genes are unique to each type of organism. But we have also discussed how the genomes of complex organisms share numerous homologous coding sequences and proteins (see Section 14.1). When developmental biologists began to describe the events that specify differentiation, morphogenesis, and pattern formation at the molecular level, they found common themes that encompassed both phenomena.

Developmental genes in diverse organisms are similar, but have different results

As we saw in Chapter 19, many of the genes that control development encode transcription factors that act on proteins in a cascade of events. These *developmental genes* were not discovered until the mid 1980s because geneticists had previously concentrated their attention on the transmission of inherited characteristics from adult organisms to their offspring. Until that time, genetics was almost exclusively the study of that part of DNA encoding structural proteins and enzymes. The processes of development remained a "molecular black box."

The discovery of developmental mutants in *Drosophila* eventually led to the identification of genes and gene products that are responsible for some of normal insect development (see Section 19.5). Using the tools of comparative genomics, scientists discovered that a similar set of genes exists in vertebrates. This discovery of a common set of developmental genes in organisms as evolutionarily distant as fruit flies and mice led to a major conclusion of evolutionary developmental biology: *the amazing diversity of organisms is produced by a modest number of regulatory genes.* Differences in body form result from differences in where and when these genes are turned on and off.

The transcription factors and extracellular signals that govern pattern formation in multicellular organisms, and the genes that encode them, can be thought of as a **molecular tool kit**, in the same sense that a few tools in a carpenter's tool kit can be used to build many different structures.

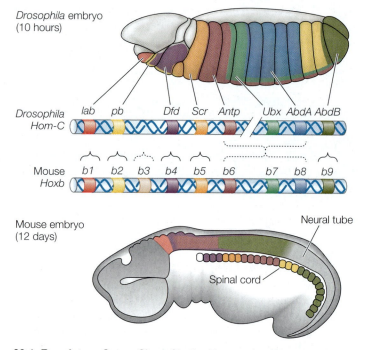

20.1 Regulatory Genes Show Similar Expression Patterns
Homologous genes encoding similar transcription factors are expressed in similar patterns along the anterior–posterior axes of both insects and vertebrates.

Plants and animals share many regulatory genes. We saw in Section 19.5 that members of two families of genes that encode transcription factors, the MADS box genes of plants and the Hox genes of fruit flies, regulate important developmental processes. In *Drosophila*, these genes govern the development of structures such as legs, antennae, and wings, whereas in plants they govern the development of flowers, fruits, and seeds. So even though the structures produced are unique and dissimilar, the "rules" governing cell differentiation during development are quite similar in animals and plants. This similarity is remarkable, given that the evolutionary paths of these two lineages diverged in the far distant past. The regulatory genes have been *conserved*—that is, they have changed little—over many millions of years.

Most animals that move through their environment under their own power have bilaterally symmetrical bodies with a head (anterior) and a tail (posterior) end; the bodies of many of them are divided into segments, although these segments may not be visible externally (see Chapter 31). The same kinds of homeobox-containing genes encoding transcription factors provide positional information to cells along the anterior–posterior axis of the body in both insect and mammalian embryos. For example, certain *Drosophila* genes and the homologous genes of vertebrates both are expressed in the anterior regions of the brain (**Figure 20.1**). When certain Hox genes are mutated in *Drosophila*, the segments differentiate in the wrong way (see Figure 19.20). The *bithorax* mutation, which is caused by a deletion of the *Ubx* gene, causes the developing insect to form two sets of forewings rather than the normal one pair (**Figure 20.2A**). In vertebrates, altering the expression patterns of some Hox genes can change lumbar (abdominal)

vertebrae into thoracic (ribbed) vertebrae (**Figure 20.2B**). Altering the expression of other genes can replace neck bones with duplications of the ear bones and jaw.

20.1 RECAP

A molecular tool kit consisting of highly conserved regulatory genes encoding transcription factors and chemical signals governs pattern formation in multicellular organisms.

- How does the story of the mouse *Pax6* gene demonstrate the existence of common developmental genes in different organisms? See pp. 448–449

- Do you understand the effects of mutations in developmental genes, and why the existence of mutant phenotypes led to the discovery of these genes? See pp. 449–450

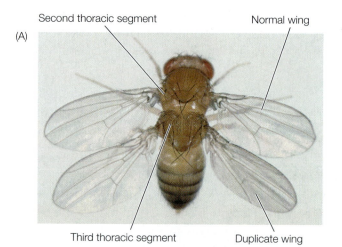

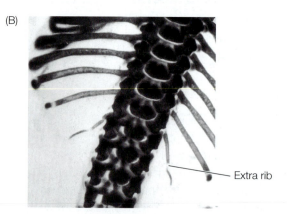

20.2 Altering Homeotic Genes Changes Morphology (A) Deletion of the *Ubx* gene in *Drosophila* converts the third thoracic segment, which does not normally bear wings, into a duplication of the second thoracic (forewing-bearing) segment. (B) Deletion of the *Hoxc-8* gene in mice transforms a lumbar (abdominal) vertebra into a copy of a thoracic (ribbed) vertebra.

The striking abnormalities that first alerted geneticists to the existence of regulatory genes in *Drosophila* transformed one body part into another part normally found elsewhere. Surprisingly, these mutations affected only single structures. A fly with a leg coming out of its head was, in all other respects, normal. How was it possible that a mutation with such a massive effect on one structure did not cause deformities in any other part of the fly's body?

20.2 How Can Mutations with Large Effects Change Only One Part of the Body?

It was by studying homeotic mutations that geneticists discovered that embryos, like adults, are made up of **modules**—functional entities encompassing genes and the various signaling pathways the genes stimulate or inhibit, as well as the physical structures that result from these signaling cascades.

The form of each module in an organism may be changed independently of the other modules because many developmental genes exert their effects on only a single module. The form of a developing animal's heart, for example, can change independently of changes in its limbs because the genes that govern heart formation do not affect limb formation, and vice versa. If this were not true, a single mutation in a developmental gene would likely result in an adult with multiple, widely different deformities. Such an adult would be unlikely to function well in any environment.

Genetic switches govern how the molecular tool kit is used

Different structures can evolve within a single organism using a common set of genetic instructions because there are components of DNA, called *genetic switches*, that control how the molecular tool kit is used. These DNA sequences and the signal cascades that converge and act on these sequences determine when and where genes will be turned on and off. Multiple switches control each gene by influencing its expression at different times and in different places. In turn, most elements of the molecular tool kit influence more than one developmental process. They are able to do so because multiple switches allow them to be used in many contexts.

Genetic switches integrate positional information in the developing embryo and play a key role in making different modules develop differently. We have seen that different Hox genes are expressed in different segments and appendages of developing fruit flies. The pattern and functioning of each segment depends on the unique Hox gene or combination of Hox genes that are expressed in it. Genetic switches control this activity by activating each Hox gene in different zones of the body.

As an example, consider the formation of *Drosophila* wings. A fruit fly has three thoracic segments, the first of which bears no wings. The second segment bears the large forewings, and the third segment bears small hind wings that function as balancing organs. Hox proteins are not expressed in forewing cells, but all hind wing cells express the Hox gene *Ultrabithorax* (*Ubx*) because a set of genetic switches activates the *Ubx* gene in the third thoracic segment.

Ubx turns off genes that promote the formation of the veins and other structures of the forewing, and it turns on genes that promote the formation of hind wing features (**Figure 20.3**). In butterflies, on the other hand, *Ubx* does not bind in the third-segment cells, so full wings develop.

Modularity allows differences in the timing and spatial pattern of gene expression

Modularity also allows the relative *timing* of different developmental processes to shift independently of one another, a process called **heterochrony**. That is, the genes regulating the development of one module (say, the eyes of vertebrates) may be expressed at different times in different species, relative to the genes regulating the development of other modules.

Two salamander species of the genus *Bolitoglossa* illustrate how heterochrony can result in new morphology. The webbing between the toes of the larvae of most salamander species, including *Bolitoglossa rostratus* (**Figure 20.4A**), disappears as the animals mature, resulting in feet with separated toes, well suited to moving on the ground. But what if a mutation slowed or stopped expression of the genes that trigger apoptosis in the webbing? The resulting "juvenile" webbed feet would act like suction cups, allowing the animal to adhere to tree branches (**Figure 20.4B**). This is exactly what happened to *Bolitoglossa occidentalis*, opening up a new, arboreal way of life for this salamander species.

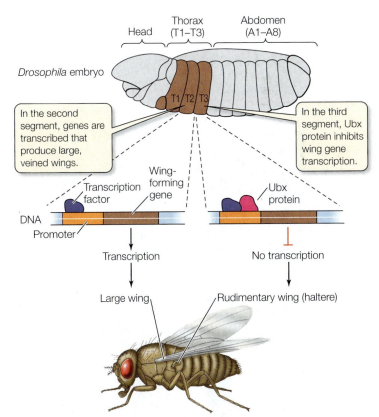

20.3 Segments Differentiate under Control of Genetic Switches
The binding of a single protein, Ultrabithorax, determines whether a thoracic segment produces full wings or the reduced wings known as halteres (balancers).

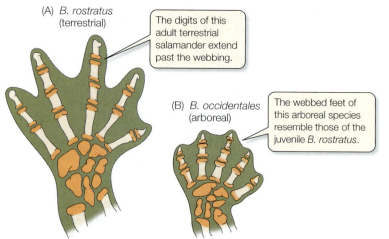

(A) *B. rostratus*
(terrestrial)

> The digits of this adult terrestrial salamander extend past the webbing.

(B) *B. occidentales*
(arboreal)

> The webbed feet of this arboreal species resemble those of the juvenile *B. rostratus*.

20.4 Heterochrony Created an Arboreal Salamander (A) The foot of an adult *Bolitoglossa rostratus*, a terrestrial salamander. (B) The foot of *B. occidentalis*, a closely related salamander, does not lose its webbing; this species uses the suction of its webbed feet in an arboreal lifestyle.

The webbed feet of ducks are an example of an evolutionary change resulting from an altered *spatial* expression pattern of a developmental gene. The feet of ducks have webs that connect their toes, but those of chickens and most other birds do not. The developing feet of early embryos of both ducks and chickens have webs (as do those of humans; see Figure 19.14). A particular gene is expressed in the spaces between the developing bones of the toes. This gene encodes a protein called bone morphogenetic protein 4 (BMP4), which instructs the cells between the developing toes to undergo apoptosis. The death of these cells destroys the webbing between the toes.

Embryonic duck and chicken hind limbs both express BMP4 in the webbing between the toes, but they differ in the expression of a BMP *inhibitor* protein, called Gremlin (**Figure 20.5**). Gremlin expression occurs around the digits in both chicken and duck hind limbs. In ducks, but not chickens, the *gremlin* gene is also expressed in the webbing cells. The Gremlin protein prevents the BMP4 protein from signaling for apoptosis in the webbing. The result is a webbed foot. Experimental application of the Gremlin protein to chicken hind limbs converts them into ducklike feet (**Figure 20.6**).

20.5 Changes in *gremlin* Expression Correlate with Changes in Hindlimb Structure The upper row of photos shows the development of a chicken's foot; the lower row shows foot development in a duck. Gremlin protein in the webbing of the duck foot inhibits BMP4 signaling, thus preventing the embryonic webbing from undergoing apoptosis.

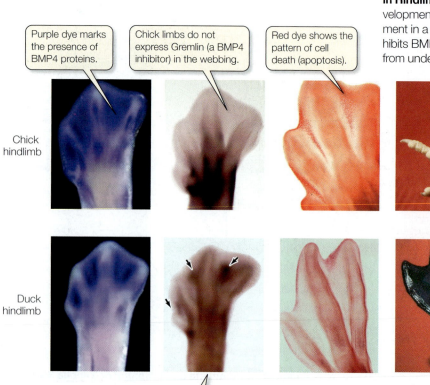

> Purple dye marks the presence of BMP4 proteins.

> Chick limbs do not express Gremlin (a BMP4 inhibitor) in the webbing.

> Red dye shows the pattern of cell death (apoptosis).

> In the chicken, webbing undergoes apoptosis, resulting in the separated toes of the adult.

Chick hindlimb

Duck hindlimb

> Webbing in the adult duck's foot remains intact.

> Duck limbs express Gremlin in the webbing (arrows).

EXPERIMENT

HYPOTHESIS: Adding Gremlin protein (a BMP4 inhibitor) to a developing chicken foot will transform it into a ducklike foot.

METHOD

Open chicken eggs and carefully add Gremlin-secreting beads to the webbing of embryonic chicken hindlimbs. Add beads that do not contain Gremlin to other hindlimbs (controls). Close the eggs and observe limb development.

RESULTS

In the hindlimbs in which Gremlin was secreted, the webbing does not undergo apoptosis, and the hindlimb resembles that of a duck. The control hindlimbs develop the normal chicken form.

Control

Gremlin added

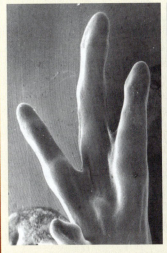
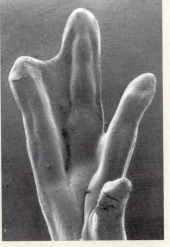

CONCLUSION: Differences in *gremlin* gene expression cause differences in morphology, allowing duck hindlimbs to retain their webbing.

Studies of pattern formation within embryos and experimental insertions of genes have shown that the different structures that develop on different parts of an individual organism can result from similar genetic instructions. Such information suggests that the processes that generate multiple structures *within* an organism might also explain how different structures develop in different species.

20.3 How Can Differences among Species Evolve?

Can the processes that allow different structures to develop in different regions of an embryo also explain the evolution of the striking differences in form among species? Apparently they can. The

20.6 Changing the Form of an Appendage In this experiment, chicken hindlimbs exposed to Gremlin-secreting beads developed ducklike webbed feet. FURTHER RESEARCH: In what other body parts might the *gremlin* gene be expressed?

action of genes controlled by genetic switches that determine where and when they will be expressed or suppressed appears to underlie both the transformation of an individual from egg to adult and the evolution of differences among species. Arthropods provide good examples of how morphological changes in species can evolve through mutations in the genes that regulate the differentiation of segments.

The arthropods (which include crustaceans, centipedes, spiders, and insects) all possess the Hox gene *Ubx*. The insect *Ubx* gene, however, has a mutation not found in the other arthropods. The Ubx protein transcribed from this mutated gene represses expression of the *distal-less* gene (*dll*), which is essential for leg formation. Insect *Ubx* is expressed in the abdomen (see Figure 19.19), and the resulting Ubx protein represses *dll*. As a result, insects have only six legs, none of which grow from the abdominal segments (**Figure 20.7**). In contrast, the Ubx protein of other arthropods—such as millipedes, centipedes, spiders, mites, and crustaceans—does not repress the expression of *dll*. Consequently, those animals have legs on their abdominal segments.

> Their distinctive limb anatomy gives the insects and their relatives their formal phylogenetic name. This arthropod group is known as Hexapoda—Greek for "six legs."

Similar processes govern the development of differences in the segments of the vertebral column. The vertebral column consists of a set of anterior-to-posterior regions: cervical (neck), thoracic (chest), lumbar (back), sacral, and caudal (tail). The transitions from one region to another correspond to the transitions between zones of expression of particular Hox genes. The anterior boundary of expression of *Hoxc-6*, for example, always falls at the boundary between the cervical and thoracic vertebrae of mice, chickens, and geese, even though all of these animals have different numbers of thoracic vertebrae. This observation suggests that the characteristic numbers of different vertebrae seen among different species result from genetic changes that expand or contract the expression domains of different Hox genes.

20.3 RECAP

The action of genes controlled by genetic switches that determine where and when they will be expressed or suppressed underlies the evolution of the striking differences in form among species.

- Can you explain why insects, unlike other arthropods, lack abdominal limbs? See p. 453 and Figure 20.7

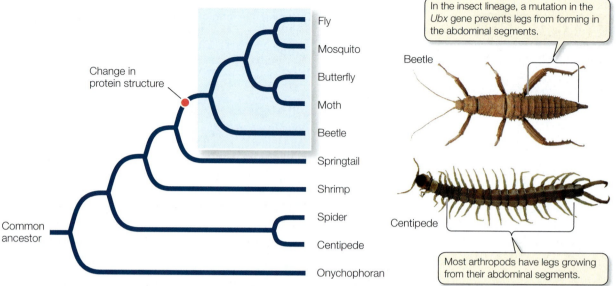

20.7 Mutation in a Hox Gene Changed the Number of Legs in Insects In the insect lineage (blue) of the arthropods, a mutation in the *Ubx* gene resulted in a protein that inhibits the *dll* gene, which is required for legs to form. Because insects express *Ubx* in their abdominal segments, no legs grow from these segments. Other arthropods, such as centipedes, do grow legs from their abdominal segments.

Our discussion so far might suggest that the form of an adult organism is determined entirely by its genes. Yet we know that the form of an adult organism is also influenced by environmental conditions during its development. We also know that no single body form is best for all environments. How are developmental processes modified so that the adult organism is adapted for the environment in which it will live?

 20.4 How Does the Environment Modulate Development?

The environment in which an individual will live may differ from the one in which its parents lived. It would be advantageous for a developing individual to "sense" the kind of environment in which it will function as an adult and modify its development accordingly. As it turns out, the development of many organisms is modified by environmental conditions. The ability of an organism to modify its development in response to environmental conditions is called **developmental plasticity**. In other words, a single genotype may produce a range of phenotypes, and signals from the environment may determine which phenotype is expressed. But how can organisms respond to signals from the environment?

No single way of responding to signals from the environment would always result in adaptation because environmental signals tell organisms many different things. We can divide signals from the environment into two major types, based on their significance and how organisms should respond to them:

- *Some environmental signals are accurate predictors of future conditions.* Some of these signals always occur, but organisms may develop without ever encountering others. In either case, the developmental processes of organisms should respond adaptively to these signals when they occur.

- *Some environmental signals are poorly correlated with future conditions.* Organisms should fail to respond to such signals.

Do developing organisms respond to these different types of signals as we expect them to?

Organisms respond to signals that accurately predict the future

Organisms *do* respond to signals that are accurate predictors of future environmental conditions. For example, seasonal changes in day length occur reliably at the same time every year. Increasing day length accurately predicts the approach of summer; decreasing day length signals winter is coming. In most tropical regions, wet and dry seasons alternate in a regular pattern during the year. Developing organisms respond to these signals in such a way that the adults they become are adapted to the future conditions.

The West African butterfly *Bicyclus anynana* has two color forms. The dry-season form matches the dead brown leaves on the dry-season forest floor, where the butterfly typically rests. The more active wet-season form has a white line along the wing and conspicuous ventral eyespots on its hind wings. These eyespots deceive predatory lizards and birds into attacking the wing, rather than the butterfly's actual eye, increasing the butterfly's chances of escape.

Temperature during pupation determines the color form of the adult butterfly. Pupae that develop in the soil under cooler temperatures (less than 20°C) produce the dry-season color form; temperatures above 24°C produce the wet-season color form. In the late larval stages, transcription of the *distal-less* gene is restricted to several small areas of the wing that have the potential to become the centers of eyespots. During pupal development, the area over which distal-less protein is expressed increases with temper-

20.8 Eyespot Development Responds to Temperature Warm temperatures during pupation increase the expression of the *distal-less* gene, resulting in eyespot formation on the wet-season morph of the adult butterfly. At cooler temperatures, *distal-less* expression is decreased, and conspicuous eyespots fail to form in the dry-season adult.

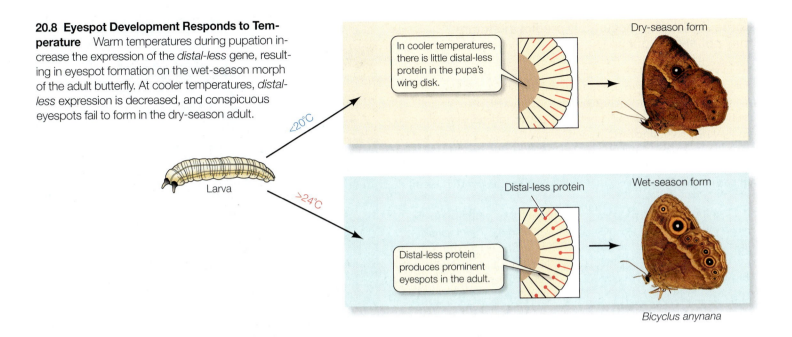

Larva

<20°C

In cooler temperatures, there is little distal-less protein in the pupa's wing disk.

Dry-season form

>24°C

Distal-less protein

Wet-season form

Distal-less protein produces prominent eyespots in the adult.

Bicyclus anynana

ature, resulting in conspicuous eyespots on adults that pupate while exposed to warm temperatures (**Figure 20.8**). Thus, by responding to an environmental signal—temperature—the pupae develop into adults with a form that is adapted to the conditions under which they will live.

Another example of developmental plasticity in response to seasonal changes occurs in the larvae of the moth *Nemoria arizonaria*. This moth produces two generations each year. Larvae (caterpillars) that hatch from eggs in spring feed on oak flowers (catkins). These larvae complete their development, form pupae, and transform themselves into adult moths in summer. These moths lay their eggs on oak trees. The larvae that hatch from these eggs in summer eat oak leaves, complete their development, and lay their eggs on oak branches. These eggs overwinter and hatch the following spring. The spring caterpillars resemble the catkins on which they feed (**Figure 20.9A**); the summer caterpillars that feed on oak leaves resemble small, year-old oak branches (**Figure 20.9B**). Both types of caterpillars are well camouflaged in the environment in which they feed. An experimenter was able to convert spring caterpillars into summer caterpillars by feeding them oak leaves. A chemical in the oak leaves induces them to develop the twiglike summer form.

Another example of developmental plasticity resulting in phenotypic change in response to the environment is found among the venomous tiger snakes (genus *Notechis*) of Australia. On the Australian mainland, tiger snakes eat small frogs and mice. On nearby Carnac Island, they eat much larger prey, the chicks of silver gulls. Island snakes grow larger, and have larger heads relative to their body length, than mainland snakes. In the laboratory, juvenile island snakes fed large mice developed larger heads and larger jaws than their siblings that ate only small prey. The heads and jaws of juvenile mainland snakes changed much less with diet. Thus head development in island snakes shows marked developmental plasticity, but that of mainland snakes does not.

Catkins

Nemoria arizonaria caterpillars

(A)

(B)

20.9 Spring and Summer Forms of a Caterpillar Differ (A) Spring caterpillars of the moth *Nemoria arizonaria* resemble the oak catkins on which they feed. (B) Summer caterpillars of the same species resemble oak twigs.

Some signals that accurately predict the future may not always occur

We have discussed developmental responses to environmental signals, such as changes in day length, that always occur. However, many other changes in an organism's environment occur sporadically: predators may or may not be present; an individual may live under crowded or uncrowded conditions; or the sexes and ages of its associates may change. Nevertheless, if such changes have occurred frequently during the evolution of a species, developmental plasticity may evolve so that individuals can respond to them when they do occur.

Developmental responses to the presence of predators have evolved in numerous species. When water fleas (*Daphnia cucullata*) encounter predatory larvae of the fly *Chaoborus*, the "helmets" on the top of their heads grow to twice their normal size (**Figure 20.10**). The fly larvae have difficulty ingesting *Daphnia* with large helmets. Helmet induction also occurs if *Daphnia* are exposed to water in which fly larvae have been swimming. Offspring that develop in the abdomens of mothers with induced large helmets are born with large helmets. Why don't all individuals develop large helmets just in case? Because there is a cost in terms of allocation of limited energy: *Daphnia* with large helmets produce fewer eggs than do *Daphnia* with small helmets.

20.11 Light Seekers The bean plants on the left were grown experimentally under low light levels. The plant's cells have elongated in response to low light, and the plants have become spindly. The control plants on the right were grown under normal light conditions.

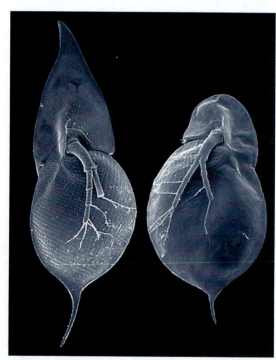

Daphnia cucullata

20.10 Predator-Induced Developmental Plasticity in *Daphnia*
This scanning electron micrograph shows the predator-induced form of *Daphnia* (left), with an enlarged helmet, and the normal form of this tiny crustacean (right). These two *Daphnia* are genetically identical individuals from a single asexually produced clone.

Light exerts a powerful influence on plant development. Low light levels stimulate the elongation of cells, so that plants growing in the shade become spindly (**Figure 20.11**). This developmental plasticity is adaptive because a spindly plant is more likely to reach a patch of brighter light than a plant that remains compact. And because they have undifferentiated tissues called meristems, plants can continue to respond to light as long as they grow.

Among some animals, learning is a developmental response to environmental change. As you know from your struggles to absorb the contents of this book, learning is costly. Learning takes much effort and time, during which an individual cannot do other useful things. But learning can continue throughout adult life, allowing an individual to adjust its behavior to the physical, biological, and social environment in which it matures. As we will see in Chapter 53, learning is particularly important in species with complex social systems. Individuals of these species must learn the identities and individual characteristics of many associates and adjust their behavior accordingly. Meanwhile, bear in mind that, as difficult as learning may be, ignorance is even more costly.

Organisms do not respond to signals that are poorly correlated with future conditions

Just as we expect organisms to respond to environmental signals that accurately predict the future, we expect them to ignore signals that are poorly correlated with future conditions, because responding to such signals would not help them adapt to the environment in which they will live. Consider, for example, seed production by plants. Plants respond to changing environmental conditions by varying their size, shape, and number of their flowers, as well as the *number* of seeds they produce; but the *size* of the seeds they produce remains nearly constant.

The amount of energy a growing plant has available to allocate to seed production depends, among other things, on temperature, rainfall, and the sizes and numbers of its neighbors. But these environmental conditions in any one year are poor predictors of what those conditions will be in the next year. A seedling that germinates from a large seed will survive better under conditions of intense competition than a seedling that germinates from a small seed because it can grow larger using the greater energy reserves stored within the seed. But, for a given amount of energy, a mature plant can produce far fewer large seeds than small seeds. So if the next generation of plants grows under more favorable conditions, plants that produced a larger number of smaller seeds in the previous year will have more surviving offspring. Thus plants do not change the sizes of the seeds they produce in response to the conditions under which they grow.

Instead, many plants have genes for seed dormancy, so that the seeds made in one year germinate at staggered intervals over future years that will have different and unknowable rainfall patterns and densities of neighbors. Seed size has evolved in response to the average conditions encountered by plants over many generations, and thus remains relatively consistent.

Organisms may lack appropriate responses to new environmental signals

Although organisms can respond adaptively to environmental signals that have occurred frequently during their recent evolutionary histories, organisms typically lack adaptive responses to environmental signals that they have not encountered before. Chapter 1 of this book gave an example in describing the effects of a particular parasite on frog limb development. The lack of useful developmental responses to newly encountered environmental events is an important problem today because human societies have changed the environment in so many ways.

For example, humans today release thousands of chemical compounds into the environment, some of which disrupt normal development. In 1962, Rachel Carson's classic book *Silent Spring* focused attention on the devastating effects the widely used chemical pesticide DDT had on bird populations, in part by interfering with the development of the eggshell. Research over the years has confirmed the malignant effects of DDT on the reproductive systems and development of many birds and mammals.

Another event that made headlines in 1962 underscored the unforeseen effects of environmental agents, this time affecting humans. More than 7,000 infants with missing or underdeveloped limbs were born to women who had taken a drug called thalidomide. Nobody in the medical profession had expected thalidomide, which was believed to be a safe mild sedative, to affect the expression of limb development genes in the fetus.

20.4 RECAP

Developmental plasticity enables developing organisms to adjust their forms to fit the environment in which they live. Organisms respond to environmental signals that are accurate predictors of future conditions, but fail to respond to signals that are poorly correlated with future conditions.

■ Describe several examples of how an organism's phenotype can be a response to environmental signals. See pp. 454–455 and Figure 20.8

■ Can you explain how learning allows an individual to adjust its behavior to the physical and social environment in which it matures? See p. 456

■ Why do organisms typically lack adaptive responses to environmental signals that their ancestors never encountered? See p. 457

Appropriate responses to new environmental conditions are likely to evolve over time, but what are the limits of such evolution? Do developmental genes dictate what structures and forms are possible?

20.5 How Do Developmental Genes Constrain Evolution?

Three decades ago, the French geneticist François Jacob made the analogy that evolution works like a tinker, assembling new structures by combining and modifying the available materials, and not like an engineer, who is free to develop dramatically different designs (say, a jet engine to replace a propeller-driven engine). The evolution of form has not been governed by the appearance of radically new genes, but by modifications of what the existing genes do. Developmental genes and their expression constrain evolution in two major ways:

■ Nearly all evolutionary innovations are modifications of previously existing structures.

■ The genes that control development are highly conserved; that is, the regulatory genes themselves do not greatly change over the course of evolution.

Evolution proceeds by changing what's already there

The features of organisms almost always evolve from pre-existing features in their ancestors. To cite a classic example, insects, birds, and bats did not "invent" their wing genes. Wings arose as modifications of an existing structure. Wings evolved independently in insects and vertebrates—only once in the insects, but in three equally independent instances among the vertebrates (**Figure 20.12**). But in all four cases, the wings are modified limbs.

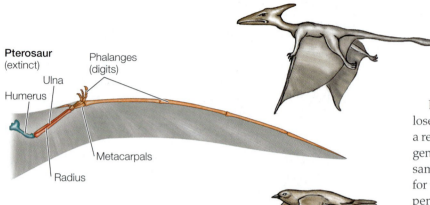

Pterosaur
(extinct)

Phalanges
(digits)

Ulna

Humerus

Metacarpals

Radius

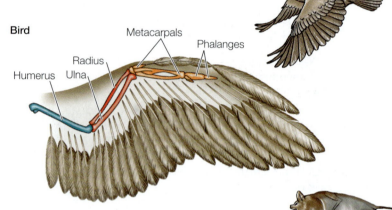

Bird

Metacarpals

Phalanges

Radius

Humerus Ulna

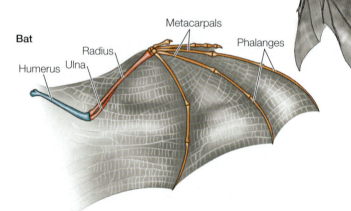

Metacarpals

Phalanges

Bat

Radius

Humerus Ulna

20.12 Wings Evolved Three Times in Vertebrates The wings of pterosaurs, birds, and bats are all modified forelimbs constructed of the same skeletal components. However, the components have different forms in the different groups.

Dr. Seuss was not alone in conceiving animals that evolution never produced. Storytellers and writers have envisioned many imaginary beings with wings sprouting from their backs—think of fairies, flying horses, and angels. Such beings are likely to remain confined to the world of human imagination, because wings probably cannot evolve from any structure other than an already existing limb.

Developmental controls also influence how organisms lose structures. The ancestors of snakes lost their limbs as a result of changes in the body segments in which the Hox genes that suppress limb formation express themselves. The same process, involving the same genes, was responsible for the dramatic evolutionary changes in the number of appendages arthropods have and on which body segments they are borne (see Figure 20.7). But in the case of snakes, suppression is not necessarily complete, and some snakes occasionally develop rudimentary limbs (**Figure 20.13**).

Conserved developmental genes can lead to parallel evolution

The nucleotide sequences of many of the genes that govern development have been highly conserved throughout the evolution of multicellular organisms—in other words, these genes exist in similar form across a broad spectrum of species. We made mention of this fact in relation to the *Pax6* eye-specifying gene described at the start of this chapter.

The existence of highly conserved developmental genes makes it likely that similar traits will evolve repeatedly, es-

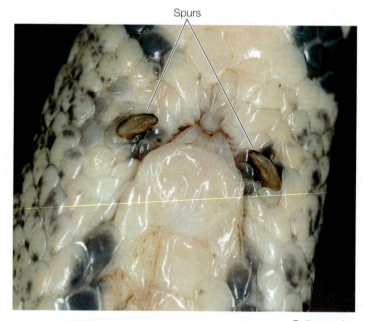

Spurs

Python regius

20.13 Developmental Genes and Lost Structures In snakes, Hox gene expression in the developing vertebral column suppresses limb formation. However, in a number of snake species, mutations occasionally result in the development of tiny legs. These spurs, or claws, are the outward projections of a royal python's vestigial hind legs.

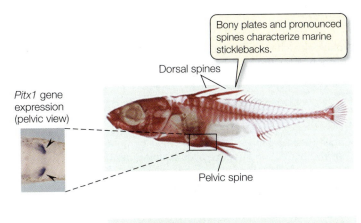

Bony plates and pronounced spines characterize marine sticklebacks.

Dorsal spines

Pitx1 gene expression (pelvic view)

Pelvic spine

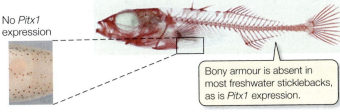

No *Pitx1* expression

Bony armour is absent in most freshwater sticklebacks, as is *Pitx1* expression.

20.14 Parallel Phenotypic Evolution in Sticklebacks Three-spined sticklebacks are small fish (often no more than 3 cm) that are widespread in the oceans. Marine sticklebacks (top) have sharp spines and a bony skeleton. In the many independent cases in which freshwater sticklebacks have arisen from marine ancestors, the spines and bony armor are greatly reduced or disappear; the phenotypic change from marine to freshwater forms seems to be consistently parallel, the result of the absence of *Ptx1* gene expression.

pecially among closely related species—a process called **parallel phenotypic evolution**. A good example of parallel phenotypic evolution is provided by a small fish, the three-spined stickleback (*Gasterosteus aculeatus*)

Sticklebacks are widely distributed across the Atlantic and Pacific Oceans and are found in many freshwater lakes. Marine populations of this species spend most of their lives at sea but return to fresh water to breed. Members of freshwater populations live in lakes and never journey to salt water. Genetic evidence shows that freshwater populations have arisen independently many times from adjacent marine ones. Marine sticklebacks have several structures that protect them from predators: well-developed pelvic bones, long dorsal and pelvic spines, and bony plates. In the freshwater populations descended from them, this body armor is greatly reduced, and dorsal and pelvic spines are much shorter or even lacking (**Figure 20.14**).

When researchers began to investigate the genetics of stickleback evolution, they expected that different genes would govern the changes in different populations. Instead, wherever they looked—Japan, British Columbia, California, Iceland—investigators have found that the same single gene, or small set of nearby genes, is responsible for the loss of armor in freshwater populations, as well as for changes in a suite of other bony parts, including jaws and bones that protect the gills. That gene, *Pitx1*, is inactive in the bone-forming regions of freshwater stickleback embryos, but active in those regions in marine sticklebacks. (Note that *mod-*

ularity is important to the results of gene expression in this example; the *Pitx1* gene is expressed in other developing organs in both marine and freshwater sticklebacks; it is only in the developing bone that structural differences arise.) The consistency of these changes in gene expression that have arisen multiple times in freshwater populations strongly suggests that the changes are adaptive. Most of the lakes in which freshwater sticklebacks live formed less than 10,000 years ago, after the glaciers that covered them retreated. These lakes lack predatory fish, but they contain predatory dragonfly larvae that can more easily capture sticklebacks with long spines. In addition, calcium is typically in short supply in freshwater lakes, making armor more costly for the fish to produce, and freshwater lakes have abundant vegetation in which sticklebacks can hide, so they have less need of an armored defense.

20.5 RECAP

Developmental controls constrain evolution because nearly all evolutionary innovations are modifications of previously existing structures, and because the conservation of many genes makes it likely that similar traits will evolve repeatedly.

- Do you understand how diverse body forms have evolved by means of modifications in the functioning of existing genes? See p. 457 and Figure 20.12

- Do you understand how the differences between marine and freshwater sticklebacks exemplify phenotypic evolution by gene regulation? See p. 459 and Figure 20.14

Many novel traits have arisen that failed to persist beyond a single generation. Part Five of this book examines the processes of evolution—the powerful forces that influence the different survival and reproductive success of various life forms—and how different adaptations become prevalent in different environments, resulting over evolutionary time in the luminous diversity of life.

20.1 **How does a molecular tool kit govern development?**

Evolutionary diversity is produced using a modest number of regulatory genes. Review Figure 20.1

The transcription factors and chemical signals that govern pattern formation in the bodies of multicellular organisms, and the genes that encode them, can be thought of as a "molecular tool kit."

Regulatory genes have been highly conserved during evolution. Review Figure 20.2

20.2 **How can mutations with large effects change only one part of the body?**

See Web/CD Tutorial 20.1

The bodies of both adults and embryos are organized into self-contained units called **modules** that can be modified independently. Modularity allows the timing of different developmental processes to shift independently, a process called **heterochrony**. Review Figure 20.4

Alterations in the spatial expression patterns of regulatory genes can also result in evolutionary changes. Review Figures 20.5 and 20.6

20.3 **How can differences among species evolve?**

The action of genes controlled by genetic switches that determine where and when they will be expressed or suppressed underlies both the transformation of an individual from an egg to an adult and the evolution of differences among species.

Morphological changes in species can evolve through mutations in the genes that regulate the differentiation of segments. Review Figure 20.7

20.4 **How does the environment modulate development?**

See Web/CD Activity 20.1

The ability of an organism to modify its development in response to environmental conditions is called **developmental plasticity**.

During development, organisms respond to environmental signals that are accurate predictors of future conditions. Some of these signals, such as changes in day length, always occur. Review Figure 20.8

Other kinds of changes in an organism's environment may or may not occur, but if such changes have occurred frequently during the evolution of a species, the ability to respond to them may evolve.

Organisms tend not to respond to environmental signals that are poorly correlated with future conditions, or to environmental signals that they have not encountered before.

20.5 **How do developmental genes constrain evolution?**

Virtually all evolutionary innovations are modifications of preexisting structures. Review Figure 20.12

Because many genes that govern development have been highly conserved, similar traits are likely to evolve repeatedly, especially among closely related species, a process called **parallel phenotypic evolution**.

SELF-QUIZ

1. Plants and animals share many regulatory genes that
   a. govern only the early development of the embryo.
   b. are inherited from a shared multicellular ancestor.
   c. govern the development of similar structures in both groups.
   d. govern the development of very different structures in the two groups.
   e. operate by very different rules.

2. Geneticists discovered regulatory genes only recently because
   a. geneticists did not believe that development was genetically controlled.
   b. geneticists concentrated their attention on genes that produced small effects.
   c. geneticists competed with developmental biologists for research funds and therefore de-emphasized the importance of development for understanding evolution.
   d. regulatory genes were particularly difficult to discover.
   e. geneticists concentrated their attention on the transmission of inherited characteristics from adults to their offspring.

3. Which of the following is *not* true of genetic switches?
   a. They control how a molecular tool kit is used.
   b. They integrate positional information in an embryo.
   c. A single switch controls each gene.
   d. They allow different structures to evolve within an individual organism.
   e. They determine when and where a gene is turned on or off.

4. Ducks have webbed feet and chickens don't because
   a. ducks need webbed feet to swim, whereas terrestrial chickens do not.

b. both duck and chicken embryos express BMP4 in the webbing between the toes, but the *gremlin* gene is expressed in the webbing cells only in ducks.
c. both duck and chicken embryos express BMP4 in the webbing between the toes, but the *gremlin* gene is expressed in the webbing cells only in chickens.
d. only duck embryos express BMP4 in the webbing between the toes.
e. only chicken embryos express BMP4 in the webbing between the toes.

5. Modularity is important for development because it
   a. guarantees that all units of a developing embryo will change in a coordinated way.
   b. coordinates the establishment of the anterior–posterior axis of the developing embryo.
   c. allows changes in the genes to change one part of the body without affecting other parts.
   d. guarantees that the timing of gene expression is the same in all parts of a developing embryo.
   e. allows organisms to be built up one module at a time.

6. Organisms often respond developmentally to regularly occurring environmental signals that accurately predict future conditions by
   a. stopping development until the signal changes.
   b. altering their development such that the resulting adult is adapted to the future environment.
   c. altering their development such that the resulting adult can produce offspring adapted to the future environment.
   d. producing new mutants.
   e. developing normally because the predicted conditions may not last long.

7. The process whereby changes in the relative timing of different developmental events can change the form of an organism is called
   a. heterochrony.
   b. developmental plasticity.
   c. adaptation.
   d. modularity.
   e. mutation.

8. *Daphnia* with large helmets are more difficult for some predators to capture and eat, but not all *Daphnia* produce large helmets because
   a. individuals with large helmets cannot feed efficiently.
   b. individuals with large helmets have trouble mating.
   c. individuals with large helmets produce fewer eggs than individuals with small helmets.
   d. individuals with large helmets become ensnared in vegetation.
   e. some individuals lack the genes that govern helmet formation.

9. Which of the following plant structures does *not* change in response to the conditions under which a plant grows?
   a. Roots
   b. Seeds
   c. Leaves
   d. Stems
   e. Branches

10. Parallel phenotypic evolution is common because
   a. closely related organisms typically face similar problems.
   b. the conservation of regulatory genes during evolution means that similar traits are likely to evolve repeatedly.
   c. many different phenotypes can be produced by a given genotype.
   d. phenotypic plasticity, which generates parallel phenotypic evolution, is widespread.
   e. evolutionary biologists have looked especially hard to find evidence of it.

FOR DISCUSSION

1. What components of environmental influences on development would probably be missed if investigations were confined to simple organisms such as bacteria and single-celled eukaryotes?

2. A spadefoot toad tadpole that develops in a rapidly drying pond is likely to eat many of its brothers and sisters. How can eating its siblings, which share half of an individual's genes, be favored by natural selection?

3. If evolutionary innovations can result from rather simple changes in the timing of expression of a few genes, why have such innovations arisen relatively infrequently during evolution?

4. François Jacob stated that evolution was more like tinkering than engineering. Does the observation that developmental genes have changed little over evolutionary time support his assertion? Why?

5. Despite their major differences, plants and animals share many of the genes that regulate development. What are the implications of this observation for the ways in which humans can respond to the adverse effects of the many substances we release into the environment that cause developmental abnormalities in plants and animals? What kinds of substances are most likely to have such effects? Why?

FOR INVESTIGATION

Figure 20.6 describes an experiment in which the protein Gremlin, which inhibits expression of the *BMP4* gene, was introduced into the foot of a developing chicken. What result would you expect from introducing Gremlin protein into other parts of the body of a developing chicken? Why? Into what other body parts would it be most informative to introduce Gremlin? If you were particularly interested in parallel phenotypic evolution, into what other organisms would you want to introduce Gremlin?

Appendix A: The Tree of Life

Phylogeny is the organizing principle of modern biological taxonomy, and a guiding principle of modern phylogeny is *monophyly*: a monophyletic group is considered to be one that contains an ancestral lineage and *all* of its descendants. Any such a group can be extracted from a phylogenetic tree with a single cut. The tree shown here provides a guide to the relationships among the major groups of the extant (living) organisms in the Tree of Life as we have presented them throughout this book. We do include three groups that are not believed to be monophyletic; these are designated with quotation marks.

The position of the branching "splits" indicates the relative branching order of the lineages of life, but the timing of splits in different groups is not drawn on a comparable time scale. In addition, the groups appearing at the branch tips do not necessarily carry equal phylogenetic "weight." For example, the ginkgo [55] is indeed at the apex of its lineage; this gymnosperm group consists of a single living species.

In contrast, a phylogeny of the angiosperms [52] would continue on from this point to fill many more trees the size of this one.

The glossary entries that follow are informal descriptions of some major features of the organisms described in Part Six of this book. Each entry gives the group's common name, followed by the formal scientific name of the group (in parentheses). Numbers in square brackets reference the location of the respective groups on the tree.

It is sometimes convenient to use an informal name to refer to a collection of organisms that are not monophyletic but nonetheless all share (or all lack) some common attribute. We call these "convenience terms"; such groups are indicated in these entries by quotation marks, and we do not give them formal scientific names. Examples include "prokaryotes," "protists," and "algae." Note that these groups cannot be removed with a single cut; they represent a collection of distantly related groups that appear in different parts of the tree.

– A –

acorn worms (*Enteropneusta*) Benthic marine hemichordates [109] with an acorn-shaped proboscis, a short collar (neck), and a long trunk.

"algae" A convenience term encompassing various distantly related groups of aquatic, photosynthetic chromalveolates [5] and certain members of the Plantae [8].

alveolates (*Alveolata*) [7] Unicellular eukaryotes with a layer of flattened vesicles (alveoli) supporting the plasma membrane. Major alveolate groups include the dinoflagellates [49], apicomplexans [50], and ciliates [51].

ambulacrarians (*Ambulacraria*) [27] The echinoderms [108] and hemichordates [109].

amniotes (*Amniota*) [34] Mammals, reptiles, and their extinct close relatives. Characterized by many adaptations to terrestrial life, including an amniotic egg (with a unique set of membranes—the amnion, chorion, and allantois), a water-repellant epidermis (with epidermal scales, hair, or feathers), and, in males, a penis that allows internal fertilization.

amoebozoans (*Amoebozoa*) [76] A group of eukaryotes [4] that use lobe-shaped pseudopods for locomotion and to engulf food. Major amoebozoan groups include the loboseans, plasmodial slime molds, and cellular slime molds.

amphibians (*Amphibia*) [118] Tetrapods [33] with glandular skin that lacks epidermal scales, feathers, or hair. Many amphibian species undergo a complete metamorphosis from an aquatic larval form to a terrestrial adult form, although direct development is also common. Major amphibian groups include frogs and toads (anurans), salamanders, and caecilians.

amphipods (*Amphipoda*) Small crustaceans [106] that are abundant in many marine and freshwater habitats. They are important herbivores, scavengers, and micropredators, and are an important food source for many aquatic organisms.

angiosperms (*Anthophyta* or *Magnoliophyta*) [52] The flowering plants. Major angiosperm groups include the monocots, eudicots, and magnoliids.

animals (*Animalia* or *Metazoa*) [19] Multicellular heterotrophic eukaryotes. The majority of animals are bilaterians [21]. Other major groups are the cnidarians [87], ctenophores [86], calcareous sponges [85], demosponges [84], and glass sponges [83]. The closest living relatives of the animals are the choanoflagellates [82].

annelids (*Annelida*) [94] Segmented worms, including earthworms, leeches, and polychaetes. One of the major groups of lophotrochozoans [23].

anthozoans (*Anthozoa*) One of the major groups of cnidarians [87]. Includes the sea anemones, sea pens, and corals.

anurans (*Anura*) Comprising the frogs and toads, this is the largest group of living amphibians [118]. They are tailless, with a shortened vertebral column and elongate hind legs modified for jumping. Many species have an aquatic larval form known as a tadpole.

apicomplexans (*Apicomplexa*) [50] Parasitic alveolates [7] characterized by the possession of an apical complex at some stage in the life cycle.

arachnids (*Arachnida*) Chelicerates [104] with a body divided into two parts: a cephalothorax that bears six pairs of appendages (four pairs of which are usually used as legs) and an abdomen that bears the genital opening. Familiar arachnids include spiders, scorpions, mites and ticks, and harvestmen.

archaeans (*Archaea*) [3] Unicellular organisms lacking a nucleus and lacking peptidoglycan in the cell wall. Once grouped with the bacteria, archaeans possess distinctive membrane lipids.

archosaurs (*Archosauria*) [36] A group of reptiles [35] that includes dinosaurs and crocodilians [123]. Most dinosaur groups became extinct at the end of the Cretaceous; birds [122] are the only surviving dinosaurs.

arrow worms (*Chaetognatha*) [96] Small planktonic or benthic predatory marine worms with fins and a pair of hooked, prey-grasping spines on each side of the head.

arthropods (*Arthropoda*) The largest group of ecdysozoans [24]. Arthropods are characterized by a stiff exoskeleton, segmented bodies, and jointed appendages. Includes the chelicerates [104], myriapods [105], crustaceans [106], and hexapods (insects and their relatives) [107].

ascidians (*Ascidiacea*) "Sea squirts"; the largest group of urochordates [110]. Also known as tunicates, they are sessile (as adults), marine, sac-like filter feeders.

ascomycetes (*Ascomycota*) [78] Fungi that bear the products of meiosis within sacs (asci) if the organism is multicellular. Some are unicellular.

– B –

bacteria (*Eubacteria*) [2] Unicellular organisms lacking a nucleus, possessing distinctive ribosomes and initiator tRNA, and generally containing peptidoglycan in the cell wall. Different bacterial groups are distinguished primarily on nucleotide sequence data.

barnacles (*Cirripedia*) Crustaceans [106] that undergo two metamorphoses—first from a feeding planktonic larva to a nonfeeding swimming larva, and then to a sessile adult that forms a "shell" composed of four to eight plates cemented to a hard substrate.

basidiomycetes (*Basidiomycota*) [77] Fungi [17] that, if multicellular, bear the products of meiosis on club-shaped basidia and possess a long-lasting dikaryotic stage. Some are unicellular.

bilaterians (*Bilateria*) [21] Those animal groups characterized by bilateral symmetry and three distinct tissue types (endoderm, ectoderm, and mesoderm). Includes the protostomes [22] and deuterostomes [26].

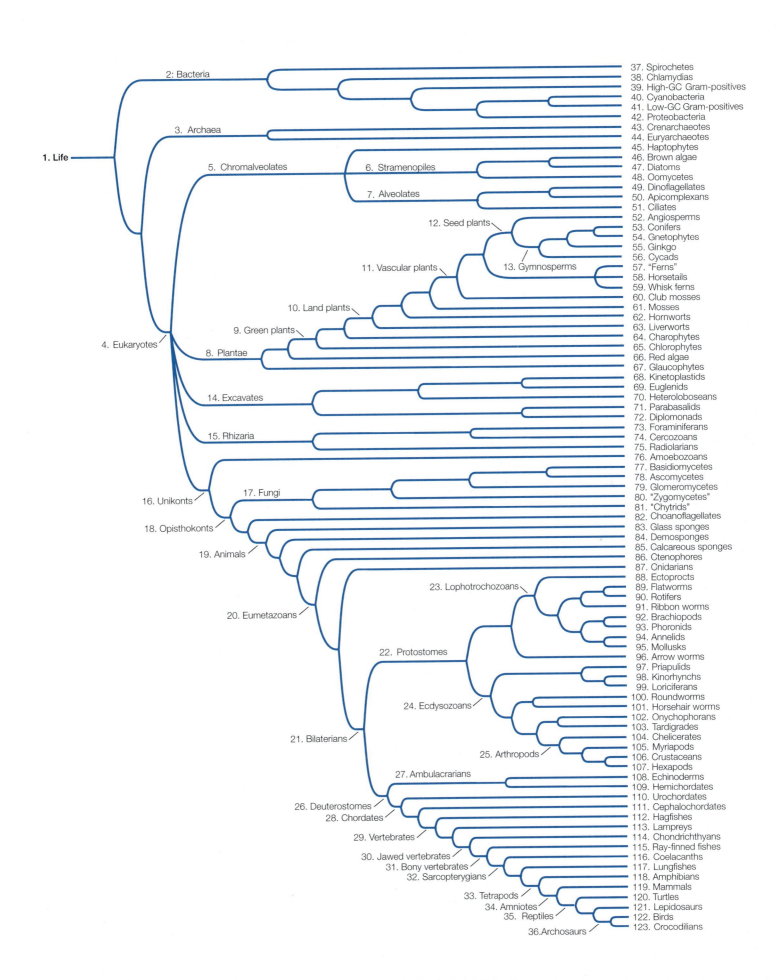

1. Life
2: Bacteria
 37. Spirochetes
 38. Chlamydias
 39. High-GC Gram-positives
 40. Cyanobacteria
 41. Low-GC Gram-positives
 42. Proteobacteria
3. Archaea
 43. Crenarchaeotes
 44. Euryarchaeotes
4. Eukaryotes
5. Chromalveolates
6. Stramenopiles
 45. Haptophytes
 46. Brown algae
 47. Diatoms
 48. Oomycetes
7. Alveolates
 49. Dinoflagellates
 50. Apicomplexans
 51. Ciliates
8. Plantae
9. Green plants
10. Land plants
11. Vascular plants
12. Seed plants
13. Gymnosperms
 52. Angiosperms
 53. Conifers
 54. Gnetophytes
 55. Ginkgo
 56. Cycads
 57. "Ferns"
 58. Horsetails
 59. Whisk ferns
 60. Club mosses
 61. Mosses
 62. Hornworts
 63. Liverworts
 64. Charophytes
 65. Chlorophytes
 66. Red algae
 67. Glaucophytes
14. Excavates
 68. Kinetoplastids
 69. Euglenids
 70. Heteroloboseans
 71. Parabasalids
 72. Diplomonads
15. Rhizaria
 73. Foraminiferans
 74. Cercozoans
 75. Radiolarians
16. Unikonts
17. Fungi
 76. Amoebozoans
 77. Basidiomycetes
 78. Ascomycetes
 79. Glomeromycetes
 80. "Zygomycetes"
 81. "Chytrids"
18. Opisthokonts
 82. Choanoflagellates
19. Animals
 83. Glass sponges
 84. Demosponges
 85. Calcareous sponges
 86. Ctenophores
20. Eumetazoans
 87. Cnidarians
21. Bilaterians
22. Protostomes
23. Lophotrochozoans
 88. Ectoprocts
 89. Flatworms
 90. Rotifers
 91. Ribbon worms
 92. Brachiopods
 93. Phoronids
 94. Annelids
 95. Mollusks
 96. Arrow worms
24. Ecdysozoans
 97. Priapulids
 98. Kinorhynchs
 99. Loriciferans
 100. Roundworms
 101. Horsehair worms
 102. Onychophorans
 103. Tardigrades
25. Arthropods
 104. Chelicerates
 105. Myriapods
 106. Crustaceans
 107. Hexapods
26. Deuterostomes
27. Ambulacrarians
 108. Echinoderms
 109. Hemichordates
28. Chordates
 110. Urochordates
 111. Cephalochordates
29. Vertebrates
 112. Hagfishes
 113. Lampreys
30. Jawed vertebrates
31. Bony vertebrates
32. Sarcopterygians
 114. Chondrichthyans
 115. Ray-finned fishes
 116. Coelacanths
 117. Lungfishes
33. Tetrapods
34. Amniotes
35. Reptiles
36. Archosaurs
 118. Amphibians
 119. Mammals
 120. Turtles
 121. Lepidosaurs
 122. Birds
 123. Crocodilians

birds (*Aves*) [122] Feathered, flying (or secondarily flightless) tetrapods [33].

bivalves (*Bivalvia*) Major mollusk [95]group; clams and mussels. Bivalves typically have two similar hinged shells that are each asymmetrical across the midline.

bony vertebrates (*Osteichthyes*) [31] Vertebrates [29] in which the skeleton is usually ossified to form bone. Includes the ray-finned fishes [115], coelacanths [116], lungfishes [117], and tetrapods [33].

brachiopods (*Brachiopoda*) [92] Lophotrochozoans [23] with two similar hinged shells that are each symmetrical across the midline. Superficially resemble bivalve mollusks, except for the shell symmetry.

brittle stars (*Ophiuroidea*) Echinoderms [108] with five long, whip-like arms radiating from a distinct central disk that contains the reproductive and digestive organs.

brown algae (*Phaeophyta*) [46] Multicellular, almost exclusively marine stramenopiles [6] generally containing the pigment fucoxanthin as well as chlorophylls *a* and *c* in their chloroplasts.

– C –

caecilians (*Gymnophiona*) A group of burrowing or aquatic amphibians [118]. They are elongate, legless, with a short tail (or none at all), reduced eyes covered with skin or bone, and a pair of sensory tentacles on the head.

calcareous sponges (*Calcarea*) [85] Filter-feeding marine sponges with spicules composed of calcium carbonate.

cellular slime molds (*Dictyostelida*) Amoebozoans [76] in which individual amoebas aggregate under stress to form a multicellular pseudoplasmodium.

cephalochordates (*Cephalochordata*) [111] A group of weakly swimming, eel-like benthic marine chordates [28]; also called lancelets. Thought to be the closest living relatives of the vertebrates [29].

cephalopods (*Cephalopoda*) Active, predatory mollusks [95] in which the molluscan foot has been modified into muscular hydrostatic arms or tentacles. Includes octopuses, squids, and nautiluses.

cercozoans (*Cercozoa*) [74] Unicellular eukaryotes [4] that feed by means of threadlike pseudopods. Together with foraminiferans [73] and radiolarians [75], the cercozoans comprise the group *Rhizaria* [15].

charophytes (*Charales*) [64] Multicellular green algae with branching, apical growth and plasmodesmata between adjacent cells. The closest living relatives of the land plants [10], they retain the egg in the parent organism.

chelicerates (*Chelicerata*) [104] A major group of arthropods [25] with pointed appendages (chelicerae) used to grasp food (as opposed to the chewing mandibles of most other arthropods). Includes the arachnids, horseshoe crabs, pycnogonids, and extinct sea scorpions.

chimaeras (*Holocephali*) A group of bottom-dwelling, marine, scaleless chondrichthyan fishes [114] with large, permanent, grinding tooth plates (rather than the replaceable teeth found in other chondrichthyans).

chitons (*Polyplacophora*) Flattened, slow-moving mollusks [95] with a dorsal protective calcareous covering made up of eight articulating plates.

chlamydias (*Chlamydiae*) [38] A group of very small Gram-negative bacteria; they live as intracellular parasites of other organisms.

chlorophytes (*Chlorophyta*) [65] The most abundant and diverse group of green algae, including freshwater, marine, and terrestrial forms; some are unicellular, others colonial, and still others multicellular. Chlorophytes use chlorophylls *a* and *c* in their photosynthesis.

choanoflagellates (*Choanozoa*) [82] Unicellular eukaryotes [4] with a single flagellum surrounded by a collar. Most are sessile, some are colonial. The closest living relatives of the animals [19]

chondrichthyans (*Chondrichthyes*) [114] One of the two main groups of jawed vertebrates [30]; includes sharks, rays, and chimaeras. They have cartilaginous skeletons and paired fins.

chordates (*Chordata*) [28] One of the two major groups of deuterostomes [26], characterized by the presence (at some point in development) of a notochord, a hollow dorsal nerve cord, and a post-anal tail. Includes the urochordates [110], cephalochordates [111], and vertebrates [29].

chromalveolates (*Chromalveolata*) [5] A contested group, said to have arisen from a common ancestor with chloroplasts derived from a red alga and supported by molecular evidence. Major chromalveolate groups include the alveolates [7] and stramenopiles [6].

"chytrids" [81] A convenience term used for a paraphyletic group of mostly aquatic, microscopic fungi [17] with flagellated gametes. Some exhibit alternation of generations.

ciliates (*Ciliophora*) [51] Alveolates [7] with numerous cilia and two types of nuclei (micronuclei and macronuclei).

clitellates (*Clitellata*) Annelids [94] with gonads contained in a swelling (called a clitellum) toward the head of the animal. Includes earthworms (oligochaetes) and leeches.

club mosses (*Lycophyta*) [60] Vascular plants [11] characterized by microphylls.

cnidarians (*Cnidaria*) [87] Aquatic, mostly marine eumetazoans [20] with specialized stinging organelles (nematocysts) used for prey capture and defense, and a blind gastrovascular cavity. The closest living relatives of the bilaterians [21].

coelacanths (*Actinista*) [116] A group of marine sarcopterygians [32] that was diverse from the Middle Devonian to the Cretaceous, but is now known from just two living species. The pectoral and anal fins are on fleshy stalks supported by skeletal elements, so they are also called lobe-finned fishes.

conifers (*Pinophyta* or *Coniferophyta*) [53] Cone-bearing, woody seed plants.

copepods (*Copepoda*) Small, abundant crustaceans [106] found in marine, freshwater, or wet terrestrial habitats. They have a single eye, long antennae, and a body shaped like a teardrop.

craniates (*Craniata*) Some biologist exclude the hagfishes [112] from the vertebrates [29], and use the term craniates to refer to the two groups combined.

crenarchaeotes (*Crenarchaeota*) [43] A major and diverse group of archaeans [3], defined on the basis of rRNA base sequences. Many are extremophiles (inhabit extreme environments), but the group may also be the most abundant archaeans in the marine environment.

crinoids (*Crinoidea*) Echinoderms [108] with a mouth surrounded by feeding arms, and a U-shaped gut with the mouth next to the anus. They attach to the substratum by a stalk or are free-swimming. Crinoids were abundant in the middle and late Paleozoic, but only a few hundred species have survived to the present. Includes the sea lilies and feather stars.

crocodilians (*Crocodylia*) [123] A group of large, predatory, aquatic archosaurs [36]. The closest living relatives of birds [122]. Includes alligators, caimans, crocodiles, and gharials.

crustaceans (*Crustacea*) [106] Major group of marine, freshwater, and terrestrial arthropods [25] with a head, thorax, and abdomen (although the head and thorax may be fused), covered with a thick exoskeleton, and with two-part appendages. Crustaceans undergo metamorphosis from a nauplius larva. Includes decapods, isopods, krill, barnacles, amphipods, copepods, and ostracods.

ctenophores (*Ctenophora*) [86] Radially symmetrical, diploblastic marine animals [19], with a complete gut and eight rows of fused plates of cilia (called ctenes).

cyanobacteria (*Cyanobacteria*) [40] A group of unicellular, colonial, or filamentous bacteria that conduct photosynthesis using chlorophyll *a*.

cycads (*Cycadophyta*) [56] Palmlike gymnosperms with large, compound leaves.

cyclostomes (*Cyclostomata*) This term refers to the possibly monophyletic group of lampreys [113] and hagfishes [112]. Molecular data support this group, but morphological data suggest that lampreys are more closely related to jawed vertebrates [30] than to hagfishes.

– D –

decapods (*Decapoda*) A group of marine, freshwater, and semiterrestrial crustaceans [106] in which five of the eight pairs of thoracic appendages function as legs (the other three pairs, called maxillipeds, function as mouthparts). Includes crabs, lobsters, crayfishes, and shrimps.

demosponges (*Demospongiae*) [84] The largest of the three groups of sponges, accounting for 90 percent of all sponge species. Demosponges have spicules made of silica, spongin fiber (a protein), or both.

deuterostomes (*Deuterostomia*) [26] One of the two major groups of bilaterians [21], in which the mouth forms at the opposite end of the embryo from the blastopore in early development (contrast with protostomes). Includes the ambulacrarians [27] and chordates [28].

diatoms (*Bacillariophyta*) [47] Unicellular, photosynthetic stramenopiles [6] with glassy cell walls in two parts.

dinoflagellates (*Dinoflagellata*) [49] A group of alveolates [7] usually possessing two flagella, one in an equatorial groove and the other in a longitudinal groove; many are photosynthetic.

diplomonads (*Diplomonadida*) [72] A group of eukaryotes [4] lacking mitochondria; most have two nuclei, each with four associated flagella.

– E –

ecdysozoans (*Ecdysozoa*) [24] One of the two major groups of protostomes [22], characterized by periodic molting of their exoskeletons. Roundworms [100] and arthropods [25] are the largest ecdysozoan groups.

echinoderms (*Echinodermata*) [108] A major group of marine deuterostomes [26] with fivefold radial symmetry (at some stage of life) and an endoskeleton made of calcified plates and spines. Includes sea stars, crinoids, sea urchins, sea cucumbers, and brittle stars.

ectoprocts (*Ectoprocta*) [88] A group of marine and freshwater lophotrochozoans [23] that live in colonies attached to substrata. Also known as bryozoans or moss animals.

elasmobranchs (*Elasmobranchii*) The largest group of chondrichthyan fishes [114]. Includes sharks, skates, and rays. In contrast to the other group of living chondrichthyans (the chimaeras), they have replaceable teeth.

eudicots (*Eudicotyledones*) A group of angiosperms [52] with pollen grains possessing three openings. Typically with two cotyledons, net-veined leaves, taproots, and floral organs typically in multiples of four or five.

euglenids (*Euglenida*) [69] Flagellate excavates characterized by a pellicle composed of spiraling strips of protein under the plasma membrane; the mitochondria have disk-shaped cristae. Some are photosynthetic.

eukaryotes (*Eukarya*) [4] Organisms made up of one or more complex cells in which the genetic material is contained in nuclei. Contrast with archaeans [3] and bacteria [2].

eumetazoans (*Eumetazoa*) [20] Those animals [19] characterized by body symmetry, a gut, a nervous system, specialized types of cell junctions, and well-organized tissues in distinct cell layers (although there have been secondary losses of some of these characteristics in some eumetazoans).

euphyllophytes (*Euphyllophyta*) This clade is sister to the club mosses [60] and includes all plants with megaphylls.

euryarchaeotes (*Euryachaeota*) [44] A major group of archaeans [3], diagnosed on the basis of rRNA sequences. Includes many methanogens, extreme halophiles, and thermophiles.

eutherians (*Eutheria*) A group of viviparous mammals [119], eutherians are well developed at birth (contrast to prototherians and marsupials, the other two groups of mammals). Most familiar mammals outside the Australian and South American regions are eutherians (see Table 33.1).

excavates (*Excavata*) [14] Diverse group of unicellular, flagellate eukaryotes, many of which possess a feeding groove; some lack mitochondria.

– F –

"ferns" [57] Vascular plants [11] usually possessing large, frond-like leaves that unfold from a "fiddlehead." Not a monophyletic group, although most fern species are encompassed in a monophyletic clade, the leptosporangiate ferns.

flatworms (*Platyhelminthes*) [89] A group of dorsoventrally flattened and generally elongate soft-bodied lophotrochozoans [23]. May be free-living or parasitic, found in marine, freshwater, or damp terrestrial environments. Major flatworm groups include the tapeworms, flukes, monogeneans, and turbellarians.

flowering plants *See* angiosperms.

flukes (*Trematoda*) A group of wormlike parasitic flatworms [89] with complex life cycles that involve several different host species. May be paraphyletic with respect to tapeworms.

foraminiferans (*Foraminifera*) [73] Amoeboid organisms with fine, branched pseudopods that form a food-trapping net. Most produce external shells of calcium carbonate.

fungi (*Fungi*) [17] Eukaryotic heterotrophs with absorptive nutrition based on extracellular digestion; cell walls contain chitin. Major fungal groups include the "chytrids" [81], "zygomycetes" [80], glomeromycetes [79], ascomycetes [78], and basidiomycetes [77].

– G –

gastropods (*Gastropoda*) The largest group of mollusks [95]. Gastropods possess a well-defined head with two or four sensory tentacles (often terminating in eyes) and a ventral foot. Most species have a single coiled or spiraled shell. Common in marine, freshwater, and terrestrial environments.

ginkgo (*Ginkgophyta*) [55] A gymnosperm [13] group with only one living species. The ginkgo seed is surrounded by a fleshy tissue not derived from an ovary wall and hence not a fruit.

glass sponges (*Hexactinellida*) [83] Sponges with a skeleton composed of four- and/or six-pointed spicules made of silica.

glaucophytes (*Glaucophyta*) [67] Unicellular freshwater algae with chloroplasts containing traces of peptidoglycan, the characteristic cell wall material of bacteria.

glomeromycetes (*Glomeromycota*) [79] A group of fungi [17] that form arbuscular mycorrhizae.

gnathostomes (*Gnathostomata*) *See* jawed vertebrates.

gnetophytes (*Gnetophyta*) [54] A gymnosperm [13] group with three very different lineages; all have wood with vessels, unlike other gymnosperms.

green plants (*Viridiplantae*) [9] Organisms with chlorophylls *a* and *b*, cellulose-containing cell walls, starch as a carbohydrate storage product, and chloroplasts surrounded by two membranes.

gymnosperms (*Gymnospermae*) [13] Seed plants [12] with seeds "naked" (i.e., not enclosed in carpels). Probably monophyletic, but status still in doubt. Includes the conifers [53], gnetophytes [54], ginkgo [55], and cycads [56].

– H –

hagfishes (*Myxini*) [112] Elongate, slimy-skinned vertebrates [29] with three small accessory hearts, a partial cranium, and no stomach or paired fins. *See also* craniata; cyclostomes.

haptophytes (*Haptophyta*) [45] Unicellular, photosynthetic stramenopiles [6] with two slightly unequal, smooth flagella. Abundant as phytoplankton, some form marine algal blooms.

hemichordates (*Hemichordata*) [109] One of the two primary groups of ambulacrarians [27]; marine wormlike organisms with a three-part body plan.

heteroloboseans (*Heterolobosea*) [70] Colorless excavates [14] that can transform among amoeboid, flagellate, and encysted stages.

hexapods (*Hexapoda*) [107] Major group of arthropods [25] characterized by a reduction (from the ancestral arthropod condition) to six walking appendages, and the consolidation of three body segments to form a thorax. Includes insects and their relatives (see Table 32.2).

high-GC Gram-positives (*Actinobacteria*) [39] Gram-positive bacteria with a relatively high G+C/A+T ratio of their DNA, with a filamentous growth habit.

hornworts (*Anthocerophyta*) [62] Nonvascular plants with sporophytes that grow from the base. Cells contain a single large, platelike chloroplast.

horsehair worms (*Nematomorpha*) [101] A group of very thin, elongate, wormlike freshwater ecdysozoans [24]. Largely nonfeeding as adults, they are parasites of insects and crayfish as larvae.

horseshoe crabs (*Xiphosura*) Marine chelicerates [104] with a large outer shell in three parts: a carapace, an abdomen, and a tail-like telson. Only five living species remain, but many additional species are known from fossils.

horsetails (*Sphenophyta* or *Equisetophyta*) [58] Vascular plants [11] with reduced megaphylls in whorls.

hydrozoans (*Hydrozoa*) A group of cnidarians [87]. Most species go through both polyp and mesuda stages, although one stage or the other is eliminated in some species.

– I –

insects (*Insecta*) The largest group within the hexapods [107]. Insects are characterized by exposed mouthparts and one pair of antennae containing a sensory receptor called a Johnston's organ. Most have two pairs of wings as adults. There are more described species of insects than all other groups of life [1] combined, and many species remain to be discovered. The major insect groups are described in Table 32.2.

isopods (*Isopoda*) Crustaceans [106] characterized by a compact head, unstalked compound eyes, and mouthparts consisting of four pairs of appendages. Isopods are abundant and widespread in salt, fresh, and brackish water, although some species (the sow bugs) are terrestrial.

– J –

jawed vertebrates (*Gnathostomata*) [30] A major group of vertebrates [29] with jawed mouths. Includes chondrichthyans [114], ray-finned fishes [115], and sarcopterygians [32].

– K –

kinetoplastids (*Kinetoplastida*) [68] Unicellular, flagellate organisms characterized by the presence in their single mitochondrion of a kinetoplast (a structure containing multiple, circular DNA molecules).

kinorhynchs (*Kinorhyncha*) [98] Small (< 1 mm) marine ecdysozoans [24] with bodies in 13 segments and a retractable proboscis.

korarchaeotes (*Korarchaeota*) A group of archaeans [3] known only by evidence from nucleic acids derived from hot springs. Its phylogenetic relationships within the Archaea are unknown.

krill (*Euphausiacea*) A group of shrimplike marine crustaceans [106] that are important components of the zooplankton.

– L –

lampreys (*Petromyzontiformes*) [113] Elongate, eel-like vertebrates [29] that often have rasping and sucking disks for mouths.

lancelets (*Cephalochordata*) *See* cephalochordates.

land plants (*Embryophyta*) [10] Plants with embryos that develop within protective structures; sporophytes and gametophytes are multicellular. Land plants possess a cuticle. Major groups are the liverworts [63], hornworts [62], mosses [61], and vascular plants [11].

larvaceans (*Larvacea*) Solitary, planktonic urochordates [110] that retain both notochords and nerve cords throughout their lives.

lepidosaurs (*Lepidosauria*) [121] Reptiles [35] with overlapping scales. Includes tuataras and squamates (lizards, snakes, and amphisbaenians).

life (*Life*) [1] The monophyletic group that includes all known living organisms. Characterized by a nucleic-acid based genetic system (DNA or RNA), metabolism, and cellular structure. Some parasitic forms, such as viruses, have secondarily lost some of these features and rely on the cellular environment of their host.

liverworts (*Hepatophyta*) [63] Nonvascular plants lacking stomata; stalk of sporophyte elongates along its entire length.

loboseans (*Lobosea*) A group of unicellular amoebozoans [76]; includes the most familiar amoebas (e.g., *Amoeba proteus*).

"lophophorates" Not a monophyletic group. A convenience term used to describe several groups of lophotrochozoans [23] that have a feeding structure called a lophophore (a circular or U-shaped ridge around the mouth that bears one or two rows of ciliated, hollow tentacles).

lophotrochozoans (*Lophotrochozoa*) [23] One of the two main groups of protostomes [22]. This group is morphologically diverse, and is supported primarily on information from gene sequences. Includes ectoprocts [88], flatworms [89], rotifers [90], ribbon worms [91], brachiopods [92], phoronids [93], annelids [94], and mollusks [95].

loriciferans (*Loricifera*) [99] Small (< 1 mm) ecdysozoans [24] with bodies in four parts, covered with six plates.

low-GC Gram-positives (*Firmicutes*) [41] A diverse group of bacteria [2] with a relatively low G+C/A+T ratio of their DNA, often but not always Gram-positive, some producing endospores.

lungfishes (*Dipnoi*) [117] A group of aquatic sarcopterygians [32] that are the closest living relatives of the tetrapods [33]. They have a modified swim bladder used to absorb oxygen from air, so some species can survive the temporary drying of their habitat.

– M –

magnoliids Major group of angiosperms [52] possessing two cotyledons and pollen grains with a single opening. The group is defined primarily by nucleotide sequence data; it is more closely related to the eudicots and monocots than to three other small angiosperm groups.

mammals (*Mammalia*) [119] A group of tetrapods [33] with hair covering all or part of their skin; females produce milk to feed their developing young. Includes the prototherians, marsupials, and eutherians.

marsupials (*Marsupialia*) Mammals [119] in which the female typically has a marsupium (a pouch for rearing young, which are born at an extremely early stage in development). Includes such familiar mammals as opossums, koalas, and kangaroos.

metazoans (*Metazoa*) *See* animals.

mollusks (*Mollusca*) [95] One of the major groups of lophotrochozoans [23], mollusks have bodies composed of a foot, a mantle (which often secretes a hard, calcareous shell), and a visceral mass. Includes monoplacophorans, chitons, bivalves, gastropods, and cephalopods.

monocots (*Monocotyledones*) Angiosperms [52] characterized by possession of a single cotyledon, usually parallel leaf veins, a fibrous root system, pollen grains with a single opening, and floral organs usually in multiples of three.

monogeneans (*Monogenea*) A group of ectoparasitic flatworms [89].

monoplacophorans (*Monoplacophora*) Mollusks [95] with segmented body parts and a single, thin, flat, rounded, bilateral shell.

mosses (*Bryophyta*) [61] Nonvascular plants with true stomata and erect, "leafy" gametophytes; sporophytes elongate by apical cell division.

multicellular eukaryotes *See* "protists."

myriapods (*Myriapoda*) [105] Arthropods [25] characterized by an elongate, segmented trunk with many legs. Includes centipedes and millipedes.

– N –

nanoarchaeotes (*Nanoarchaeota*) A hypothetical group of extremely small, thermophilic archaeans [3] with a much-reduced genome. The only described example can survive only when attached to a host organism.

nematodes (*Nematoda*) [100] A very large group of elongate, unsegmented ecdysozoans [24] with thick, multilayer cuticles. They are among the most abundant and diverse animals, although most species have not yet been described. Include free-living predators and scavengers, as well as parasites of most species of land plants [10] and animals [19].

neognaths (*Neognathae*) The main group of birds [122], including all living species except the ratites (ostrich, emu, rheas, kiwis, cassowaries) and tinamous (*see* palaeognaths).

– O –

oligochaetes (*Oligochaeta*) An annelid [94] group whose members lack parapodia, eyes, and anterior tentacles, and have few setae. Earthworms are the most familiar oligochaetes.

onychophorans (*Onychophora*) [102] Elongate, segmented ecdysozoans [24] with many pairs of soft, unjointed, claw-bearing legs. Also known as velvet worms.

oomycetes (*Oomycota*) [48] Water molds and relatives; absorptive heterotrophs with nutrient-absorbing, filamentous hyphae.

opisthokonts (*Opisthokonta*) [18] A group of unikonts [16] in which the flagellum on motile cells, if present, is posterior. The opisthokonts include the fungi [17], animals [19], and choanoflagellates [82].

ostracods (*Ostracoda*) Marine and freshwater crustaceans [106] that are laterally compressed and protected by two clam-like calcareous or chitinous shells.

– P –

palaeognaths (*Palaeognathae*) A group of secondarily flightless or weakly flying birds [122]. Includes the flightless ratites (ostrich, emu, rheas, kiwis, cassowaries) and the weakly flying tinamous.

parabasalids (*Parabasalia*) [71] A group of unicellular eukaryotes [4] that lack mitochondria; they possess flagella in clusters near the anterior of the cell.

phoronids (*Phoronida*) [93] A small group of sessile, wormlike marine lophotrochozoans [23] that secrete chitinous tubes and feed using a lophophore.

placoderms (*Placodermi*) An extinct group of jawed vertebrates [30] that lacked teeth. Placoderms were the dominant predators in Devonian oceans.

Plantae [8] The most broadly defined plant group. In most parts of this book, we use the word "plant" as synonymous with "land plant" [10], a more restrictive definition.

plasmodial slime molds (*Myxogastrida*) Amoebozoans [76] that in their feeding stage consist of a coenocyte called a plasmodium.

pogonophorans (*Pogonophora*) Deep-sea annelids [94] that lack a mouth or digestive tract; they feed by taking up dissolved organic matter, facilitated by endosymbiotic bacteria in a specialized organ (the trophosome).

polychaetes (*Polychaeta*) A group of mostly marine annelids [94] with one or more pairs of eyes and one or more pairs of feeding tentacles; parapodia and setae extend from most body segments. May be paraphyletic with respect to the clitellates.

priapulids (*Priapulida*) [97] A small group of cylindrical, unsegmented, wormlike marine ecdysozoans [24] that takes its name from its phallic appearance.

"progymnosperms" Paraphyletic group of extinct vascular plants [11] that flourished from the Devonian through the Mississippian periods. The first truly woody plants, and the first with vascular cambium that produced both secondary xylem and secondary phloem, they reproduced by spores rather than by seeds.

"prokaryotes" Not a monophyletic group; as commonly used, includes the bacteria [2] and archaeans [3]. A term of convenience encompassing all cellular organisms that are not eukaryotes.

proteobacteria (*Proteobacteria*) [42] A large and extremely diverse group of Gram-negative bacteria that includes many pathogens, nitrogen fixers, and photosynthesizers. Includes the alpha, beta, gamma, delta, and epsilon proteobacteria.

"protists" This term of convenience does not describe a monophyletic group but is used to encompass a large number of distinct and distantly related groups of eukaryotes, many but far from all of which are microbial and unicellular. Essentially a "catch-all" term for any eukaryote group not contained within the land plants [10], fungi [17], or animals [19].

protostomes (*Protostomia*) [22] One of the two major groups of bilaterians [21]. In protostomes, the mouth typically forms from the blastopore (if present) in early development (contrast with deuterostomes). The major protostome groups are the lophotrochozoans [23] and ecdysozoans [24].

prototherians (*Prototheria*) A mostly extinct group of mammals [119], common during the Cretaceous and early Cenozoic. The three living species—the echidnas and the duck-billed platypus—are the only extant egg-laying mammals.

pterobranchs (*Pterobranchia*) A small group of sedentary marine hemichordates [109] that live in tubes secreted by the proboscis. They have one to nine pairs of arms, each bearing long tentacles that capture prey and function in gas exchange.

pteridophytes (*Pteridophyta*) A group of vascular plants [11], sister to the seed plants [12], characterized by overtopping and possession of megaphylls. The pteridophytes include the horsetails [58], whisk ferns [59], and "ferns" [57].

pycnogonids (*Pycnogonida*) Treated in this book as a group of chelicerates [104], but sometimes considered an independent group of arthropods [25]. Pycnogonids have reduced bodies and very long, slender legs. Also called sea spiders.

– R –

radiolarians (*Radiolaria*) [75] Amoeboid organisms with needle-like pseudopods supported by microtubules. Most have glassy internal skeletons.

ray-finned fishes (*Actinopterygii*) [115] A highly diverse group of freshwater and marine bony vertebrates [31]. They have reduced swim bladders that often function as hydrostatic organs and fins supported by soft rays (lepidotrichia). Includes most familiar fishes.

red algae (*Rhodophyta*) [66] Mostly multicellular, marine algae characterized by the presence of phycoerythrin in their chloroplasts.

reptiles (*Reptilia*) [35] One of the two major groups of extant amniotes [34], supported on the basis of similar skull structure and gene sequences. The term "reptiles" traditionally excluded the birds [122], but the resulting group is then clearly paraphyletic. As used in this book, the reptiles include turtles [120], lepidosaurs [121], birds [122], and crocodilians [123].

rhizaria (*Rhizaria*) [15] Mostly amoeboid unicellular eukaryotes with pseudopods, many with external or internal shells. Includes the foraminiferans [73], cercozoans [74], and radiolarians [75].

rhyniophytes (*Rhyniophyta*) A group of early vascular plants [11] that appeared in the Silurian and became extinct in the Middle Devonian. Possessed dichotomously branching stems with terminal sporangia but no true leaves or roots.

ribbon worms (*Nemertea*) [91] A group of unsegmented lophotrochozoans [23] with an eversible proboscis used to capture prey. Mostly marine, but some species live in fresh water or on land.

rotifers (*Rotifera*) [90] Tiny (< 0.5 mm) lophotrochozoans [23] with a pseudocoelomic body cavity that functions as a hydrostatic organ and a ciliated feeding organ called the corona that surrounds the head. They live in freshwater and wet terrestrial habitats.

roundworms (*Nematoda*) [100] *See* nematodes.

– S –

salamanders (*Caudata*) A group of amphibians [118] with distinct tails in both larvae and adults and limbs set at right angles to the body.

salps *See* thaliaceans

sarcopterygians (*Sarcopterygii*) [32] One of the two major groups of bony vertebrates [31], characterized by jointed appendages (paired fins or limbs).

scyphozoans (*Scyphozoa*) Marine cnidarians [87] in which the medusa stage dominates the life cycle. Commonly known as jellyfish.

sea cucumbers (*Holothuroidea*) Echinoderms [108] with an elongate, cucumber-shaped body and leathery skin. They are scavengers on the ocean floor.

sea spiders *See* pycnogonids.

sea squirts *See* ascidians.

sea stars (*Asteroidea*) Echinoderms [108] with five (or more) fleshy "arms" radiating from an indistinct central disk. Also called starfishes.

sea urchins (*Echinoidea*) Echinoderms [108] with a test (shell) that is covered in spines. Most are globular in shape, although some groups (such as the sand dollars) are flattened.

"seed ferns" A paraphyletic group of loosely related, extinct seed plants that flourished in the Devonian and Carboniferous. Characterized by large, frond-like leaves that bore seeds.

seed plants (*Spermatophyta*) [12] Heterosporous vascular plants [11] that produce seeds; most produce wood; branching is axillary (not dichotomous). The major seed plant groups are gymnosperms [13] and angiosperms [52].

sow bugs *See* isopods.

spirochetes (*Spirochaetes*) [37] Motile, Gram-negative bacteria with a helically coiled structure and characterized by axial filaments.

"sponges" A term of convenience used for a paraphyletic group of relatively asymmetric, filter-feeding animals that lack a gut or nervous system and generally lack differentiated tissues. (*See* glass sponges [83], demosponges [84], and calcareous sponges [85].)

springtails (*Collembola*) Wingless hexapods [107] with springing structures on the third and fourth segments of their bodies. Springtails are extremely abundant in some environments (especially in soil, leaf litter, and vegetation).

squamates (*Squamata*) The major group of lepidosaurs [121], characterized by the possession of movable quadrate bones (which allow the upper jaw to move independently of the rest of the skull) and hemipenes (a paired set of eversible penises, or penes) in males. Includes the lizards (a paraphyletic group), snakes, and amphisbaenians.

starfish (*Asteroidea*) *See* sea stars

stramenopiles (*Heterokonta* or *Stramenopila*) [6] Organisms having, at some stage in their life cycle, two unequal flagella, the longer possessing rows of tubular hairs. Chloroplasts, when present, surrounded by four membranes. Major stramenopile groups include the brown algae [46], diatoms [47], and oomycetes [48].

– T –

tapeworms (*Cestoda*) Parasitic flatworms [89] that live in the digestive tracts of vertebrates as adults, and usually in various other species of animals as juveniles.

tardigrades (*Tardigrada*) [103] Small (< 0.5 mm) ecdysozoans [24] with fleshy, unjointed legs and no circulatory or gas exchange organs. They live in marine sands, in temporary freshwater pools, and on the water films of plants. Also called water bears.

tetrapods (*Tetrapoda*) [33] The major group of sarcopterygians [32]; includes the amphibians [118] and the amniotes [34]. Named for the presence of four jointed limbs (although limbs have been secondarily reduced or lost completely in several tetrapod groups).

thaliaceans (*Thaliacea*) A group of solitary or colonial planktonic marine urochordates [110]. Also called salps.

therians (*Theria*) Mammals [119] characterized by viviparity (live birth). Includes eutherians and marsupials.

theropods (*Theropoda*) Archosaurs [36] with bipedal stance, hollow bones, a furcula ("wishbone"), elongated metatarsals with three-fingered feet, and a pelvis that points backwards. Includes many well-known extinct dinosaurs (such as *Tyrannosaurus rex*), as well as the living birds [122].

trilobites (*Trilobita*) An extinct group of arthropods [25] related to the chelicerates [104]. Trilobites flourished from the Cambrian through the Permian.

tuataras (*Rhyncocephalia*) A group of lepidosaurs [121] known mostly from fossils; there are just two living tuatara species. The quadrate bone of the upper jaw is fixed firmly to the skull. Sister group of the squamates.

tunicates *See* ascidians.

turbellarians (*Turbellaria*) A group of free-living, generally carnivorous flatworms [89]. Their monophyly is questionable.

turtles (*Testudines*) [120] A group of reptiles [35] with a bony carapace (upper shell) and plastron (lower shell) that encase the body.

– U –

unikonts (*Unikonta*) [16] A group of eukaryotes [4] whose motile cells possess a single flagellum. Major unikont groups include the amoebozoans [76], fungi [17], and animals [19].

urochordates (*Urochordata*) [110] A group of chordates [28] that are mostly saclike filter feeders as adults, with motile larvae stages that resemble a tadpole.

– V –

vascular plants (*Tracheophyta*) [11] Plants with xylem and phloem. Major groups include the club mosses [60] and euphyllophytes.

vertebrates (*Vertebrata*) [29] The largest group of chordates [28], characterized by a rigid endoskeleton supported by the vertebral column and an anterior skull encasing a brain. Includes hagfishes [112], lampreys [113], and the jawed vertebrates [30], although some biologists exclude the hagfishes from this group (see craniates).

– W –

water bears *See* tardigrades.

whisk ferns (*Psilotophyta*) [59] Vascular plants [11] lacking leaves and roots.

– Y –

"yeasts" A convenience term for several distantly related groups of unicellular fungi [17].

– Z –

"zygomycetes" [80] A convenience term for a paraphyletic group of fungi [17] in which hyphae of differing mating types conjugate to form a zygosporangium.

Appendix B: Some Measurements Used in Biology

MEASURES OF	UNIT	EQUIVALENTS	METRIC → ENGLISH CONVERSION
Length	meter (m)	base unit	1 m = 39.37 inches = 3.28 feet
	kilometer (km)	1 km = 1000 (10^3) m	1 km = 0.62 miles
	centimeter (cm)	1 cm = 0.01 (10^{-2}) m	1 cm = 0.39 inches
	millimeter (mm)	1 mm = 0.1 cm = 10^{-3} m	1 mm = 0.039 inches
	micrometer (μm)	1 μm = 0.001 mm = 10^{-6} m	
	nanometer (nm)	1 nm = 0.001 μm = 10^{-9} m	
Area	square meter (m^2)	base unit	1 m^2 = 1.196 square yards
	hectare (ha)	1 ha = 10,000 m^2	1 ha = 2.47 acres
Volume	liter (L)	base unit	1 L = 1.06 quarts
	milliliter (mL)	1 mL = 0.001 L = 10^{-3} L	1 mL = 0.034 fluid ounces
	microliter (μL)	1 μL = 0.001 mL = 10^{-6} L	
Mass	gram (g)	base unit	1 g = 0.035 ounces
	kilogram (kg)	1 kg = 1000 g	1 kg = 2.20 pounds
	metric ton (mt)	1 mt = 1000 kg	1 mt = 2,200 pounds = 1.10 ton
	milligram (mg)	1 mg = 0.001 g = 10^{-3} g	
	microgram (μg)	1 μg = 0.001 mg = 10^{-6} g	
Temperature	degree Celsius (°C)	base unit	°C = (°F − 32)/1.8
			0°C = 32°F (water freezes)
			100°C = 212°F (water boils)
			20°C = 68°F ("room temperature")
			37°C = 98.6°F (human internal body temperature)
	Kelvin (K)*	°C + 273	0 K = −460°F
Energy	joule (J)		1 J ≈ 0.24 calorie = 0.00024 kilocalorie[†]

*0 K (–273°C) is "absolute zero," a temperature at which molecular oscillations approach 0—that is, the point at which motion all but stops.

[†]A *calorie* is the amount of heat necessary to raise the temperature of 1 gram of water 1°C. The *kilocalorie*, or nutritionist's calorie, is what we commonly think of as a calorie in terms of food.

Glossary

– A –

abdomen (ab' duh mun) [L. *abdomin*: belly]   In arthropods, the posterior segments of the body; in mammals, the part of the body containing the intestines and most other internal organs, posterior to the thorax.

abiotic (a' bye ah tick) [Gk. *a*: not + *bios*: life] Nonliving. (Contrast with biotic.)

abscisic acid (ab sighs' ik)   A plant growth substance having growth-inhibiting action. Causes stomata to close.

abscission (ab sizh' un) [L. *abscissio*: break off] The process by which leaves, petals, and fruits separate from a plant.

absorption   (1) Of light: complete retention, without reflection or transmission. (2) Of liquids: soaking up (taking in through pores or cracks).

absorption spectrum   A graph of light absorption versus wavelength of light; shows how much light is absorbed at each wavelength.

absorptive period   When there is food in the gut and nutrients are being absorbed.

abyssal zone (uh biss' ul) [Gk. *abyssos*: bottomless]   The deep ocean, below the point that light can penetrate.

accessory pigments   Pigments that absorb light and transfer energy to chlorophylls for photosynthesis.

acetylcholine   A neurotransmitter substance that carries information across vertebrate neuromuscular junctions and some other synapses.

acetylcholinesterase   An enzyme that breaks down acetylcholine.

acetyl coenzyme A (acetyl CoA)   Compound that reacts with oxaloacetate to produce citrate at the beginning of the citric acid cycle; a key metabolic intermediate in the formation of many compounds.

acid [L. *acidus*: sharp, sour]   A substance that can release a proton in solution. (Contrast with base.)

acid precipitation   Precipitation that has a lower pH than normal as a result of acid-forming precursors introduced into the atmosphere by human activities.

acidic   Having a pH of less than 7.0 (a hydrogen ion concentration greater than 10^{-7} molar).

acoelomate   Lacking a coelom.

Acquired Immune Deficiency Syndrome   *See* AIDS.

acrosome (a' krow soam) [Gk. *akros*: highest + *soma*: body]   The structure at the forward tip of an animal sperm which is the first to fuse with the egg membrane and enter the egg cell.

ACTH (adrenocorticotropin)   A pituitary hormone that stimulates the adrenal cortex.

actin [Gk. *aktis*: ray]   One of the two major proteins of muscle; it makes up the thin filaments. Forms the microfilaments found in most eukaryotic cells.

action potential   An impulse in a neuron taking the form of a wave of depolarization or hyperpolarization imposed on a polarized cell surface.

action spectrum   A graph of a biological process versus light wavelength; shows which wavelengths are involved in the process.

activating enzymes   Enzymes that catalyze the addition of amino acids to their appropriate tRNAs. Also called aminoacyl-tRNA synthetases.

activation energy (E_a)   The energy barrier that blocks the tendency for a set of chemical substances to react.

active site   The region on the surface of an enzyme where the substrate binds, and where catalysis occurs.

active transport   The energy-dependent transport of a substance across a biological membrane against a concentration gradient—that is, from a region of low concentration (of that substance) to a region of high concentration. (See also primary active transport, secondary active transport; contrast with facilitated diffusion.)

adaptation (a dap tay' shun)   (1) In evolutionary biology, a particular structure, physiological process, or behavior that makes an organism better able to survive and reproduce. Also, the evolutionary process that leads to the development or persistence of such a trait. (2) In sensory neurophysiology, a sensory cell's loss of sensitivity as a result of repeated stimulation.

adaptive radiation   The proliferation of members of a single clade into a variety of different adaptive forms.

adenine (A) (a' den een)   A nitrogen-containing base found in nucleic acids, ATP, NAD, and other compounds.

adenosine triphosphate   *See* ATP.

adenylate cyclase   Enzyme catalyzing the formation of cyclic AMP (cAMP) from ATP.

adrenal (a dree' nal) [L. *ad*: toward + *renes*: kidneys]   An endocrine gland located near the kidneys of vertebrates, consisting of two glandular parts, the cortex and medulla.

adrenaline   *See* epinephrine.

adrenocorticotropin   *See* ACTH.

adsorption   Binding of a gas or a solute to the surface of a solid.

aerobic (air oh' bic) [Gk. *aer*: air + *bios*: life]   In the presence of oxygen; requiring oxygen.

afferent (af' ur unt) [L. *ad*: toward + *ferre*: to carry]   Carrying to, as in a neuron that carries impulses to the central nervous system, or a blood vessel that carries blood to a structure. (Contrast with efferent.)

AIDS (acquired immune deficiency syndrome) Condition caused by a virus (HIV) in which the body's helper T lymphocytes are reduced, leaving the victim subject to opportunistic diseases.

alcoholic fermentation   Breakdown of glucose in cells under anaerobic conditions to produce alcohol.

aldehyde (al' duh hide)   A compound with a —CHO functional group. Many sugars are aldehydes. (Contrast with ketone.)

aldosterone (al dohs' ter own)   A steroid hormone produced in the adrenal cortex of mammals. Promotes secretion of potassium and reabsorption of sodium in the kidney.

allantois (al lan' to is)   A sac-like extraembryonic membrane that contains nitrogen waste from embryo.

allele (a leel') [Gk. *allos*: other]   The alternate forms of a genetic character found at a given locus on a chromosome.

allele frequency   The relative proportion of a particular allele in a specific population.

allergy [Ger. *allergie*: altered reaction]   An overreaction to amounts of an antigen that do not affect most people; often involves IgE antibodies.

allometric growth   A pattern of growth in which some parts of the body of an organism grow faster than others, resulting in a change in body proportions as the organism grows.

allopatric speciation (al' lo pat' rick) [Gk. *allos*: other + *patria*: homeland]   The formation of two species from one when reproductive isolation occurs because of the interposition of (or crossing of) a physical geographic barrier such as a river. Also called geographic speciation. (Contrast with parapatric speciation, sympatric speciation.)

allopolyploidy   The possession of more than two entire chromosomes sets that are derived from a single species.

allostery (al' lo steer y) [Gk. *allos*: other + *stereos*: structure]   Regulation of the activity of a protein by the binding of an effector molecule at a site other than the active site.

alpha (α) helix   A prevalent type of secondary protein structure; a right-handed spiral.

alternation of generations   The succession of multicellular haploid and diploid phases in some sexually reproducing organisms, notably plants.

alternative splicing   A process for generating different mature mRNAs from a single gene by splicing together different sets of exons during RNA processing.

altruism   Behavior that harms the individual who performs it but benefits other individuals.

alveolus (al ve' o lus) (plural: alveoli) [L. *alveus*: cavity]   A small, baglike cavity, especially the blind sacs of the lung.

amensalism (a men' sul ism) Interaction in which one animal is harmed and the other is unaffected. (Contrast with commensalism, mutualism.)

amine An organic compound with an amino group. (Compare with amino acid.)

amino acid Organic compounds containing both NH_2 and COOH groups. Proteins are polymers of amino acids.

amino acid replacement A change in a protein sequence in which one amino acid is replaced by another.

ammonotelic (am moan' o teel' ic) [Gk. *telos*: end] Describes an organism in which the final product of breakdown of nitrogen-containing compounds (primarily proteins) is ammonia. (Contrast with ureotelic, uricotelic.)

amnion (am' nee on) The fluid-filled sac in which the embryos of reptiles, birds, and mammals develop.

amniote egg A shelled egg surrounding four extraembryonic membranes and embryo-nourishing yolk. This adaptation allowed animals to colonize the terrestrial environment.

amphipathic (am' fi path' ic) [Gk. *amphi*: both + *pathos*: emotion] Of a molecule, having both hydrophilic and hydrophobic regions.

amylase (am' ill ase) Any of a group of enzymes that digest starch.

anabolic reaction A single reaction that participates in anabolism.

anabolism (an ab' uh liz' em) [Gk. *ana*: upward + *ballein*: to throw] Synthetic reactions of metabolism, in which complex molecules are formed from simpler ones. (Contrast with catabolism.)

anaerobic (an ur row' bic) [Gk. *an*: not + *aer*: air + *bios*: life] Occurring without the use of molecular oxygen, O_2.

anagenesis Evolutionary change in a single lineage over time.

analogy (a nal' o jee) [Gk. *analogia*: resembling] A resemblance in function, and often appearance as well, between two structures that is due to convergent evolution rather than to common ancestry. (Contrast with homology.)

anaphase (an' a phase) [Gk. *ana*: upward progress] The stage in nuclear division at which the first separation of sister chromatids (or, in the first meiotic division, of paired homologs) occurs.

anaphylactic shock A precipitous drop in blood pressure caused by loss of fluid from capillaries because of an increase in their permeability stimulated by an allergic reaction.

ancestral trait The trait originally present in the ancestor of a given group; may be retained or changed in the descendants of that ancestor.

androgens (an' dro jens) The male sex steroids.

aneuploidy (an' you ploy dee) A condition in which one or more chromosomes or pieces of chromosomes are either lacking or present in excess.

angiotensin (an' jee oh ten' sin) A peptide hormone that raises blood pressure by causing peripheral vessels to constrict. Also maintains glomerular filtration by constricting efferent vessels and stimulates thirst and the release of aldosterone.

animal hemisphere The metabolically active upper portion of some animal eggs, zygotes, and embryos; does not contain the dense nutrient yolk. (Contrast with vegetal hemisphere.)

anion (an' eye on) [Gk. *ana*: upward progress] A negatively charged ion. (Contrast with cation.)

anisogamy (an eye sog' a mee) [Gk. *aniso*: unequal + *gamos*: marriage] The existence of two dissimilar gametes (egg and sperm).

annual Referring to a plant whose life cycle is completed in one growing season. (Contrast with biennial, perennial.)

antenna system In photosynthesis, a group of different molecules that cooperate to absorb light energy and transfer it to a reaction center.

anterior pituitary The portion of the vertebrate pituitary gland that derives from gut epithelium and produces tropic hormones.

anther (an' thur) [Gk. *anthos*: flower] A pollen-bearing portion of the stamen of a flower.

antheridium (an' thur id' ee um) [Gk. *antheros*: blooming] The multicellular structure that produces the sperm in nonvascular plants and ferns.

antibody One of the millions of proteins produced by the immune system that specifically binds to a foreign substance and initiates its removal from the body.

anticodon The three nucleotides in transfer RNA that pair with a complementary triplet (a codon) in messenger RNA.

antidiuretic hormone (ADH) A hormone that promotes water reabsorption by the kidney. ADH is produced by neurons in the hypothalamus and released from nerve terminals in the posterior pituitary. Also called vasopressin.

antigen (an' ti jun) Any substance that stimulates the production of an antibody or antibodies in the body of a vertebrate.

antigenic determinant A specific region of an antigen, which is recognized by and binds to a specific antibody.

antiparallel Pertaining to molecular orientation in which a molecule or parts of a molecule have opposing directions.

antipodal cell At one end of the megagametophyte, one of the three cells which eventually degenerate.

antiport A membrane transport process that carries one substance in one direction and another in the opposite direction. (Contrast with symport.)

antisense nucleic acid A single-stranded RNA or DNA complementary to and thus targeted against the mRNA transcribed from a harmful gene such as an oncogene.

anus (a' nus) Opening through which digestive wastes are expelled, located at the posterior end of the gut.

aorta (a or' tah) [Gk. *aorte*: aorta] The main trunk of the arteries leading to the systemic (as opposed to the pulmonary) circulation.

aortic body A nodule of modified muscle tissue on the aorta that is sensitive to the oxygen supply provided by arterial blood.

apex (a' pecks) The tip or highest point of a structure, as the apex of a growing stem or root.

apical (a' pi kul) Pertaining to the apex, or tip, usually in reference to plants.

apical dominance Inhibition by the apical bud of the growth of axillary buds.

apical meristem The meristem at the tip of a shoot or root; responsible for the plant's primary growth.

apomixis (ap oh mix' is) [Gk. *apo*: away from + *mixis*: sexual intercourse] The asexual production of seeds.

apoplast (ap' oh plast) in plants, the continuous meshwork of cell walls and extracellular spaces through which material can pass without crossing a plasma membrane. (Contrast with symplast.)

apoptosis (ap uh toh' sis) A series of genetically programmed events leading to cell death.

aquaporin A transport protein in plant and animal cells through which water passes in osmosis.

aquatic (a kwa' tic) [L. *aqua*: water] Living in water. (Compare with marine, terrestrial.)

aqueous (a' kwee us) Pertaining to water or a watery solution.

archegonium (ar' ke go' nee um) [Gk. *archegonos*: first, foremost] The multicellular structure that produces eggs in nonvascular plants, ferns, and gymnosperms.

archenteron (ark en' ter on) [Gk. *archos*: first + *enteron*: bowel] The earliest primordial animal digestive tract.

area phylogenies Phylogenies in which the names of the taxa are replaced with the names of the places where those taxa live or lived.

arteriosclerosis *See* atherosclerosis.

artery A muscular blood vessel carrying oxygenated blood away from the heart to other parts of the body. (Contrast with vein.)

artficial selection The selection by plant and animal breeders of individuals with certain desirable traits.

ascus (ass' cuss) [Gk. *askos*: bladder] In ascomycete fungi (sac fungi), the club-shaped sporangium within which spores (ascospores) are produced by meiosis.

asexual Without sex.

assortative mating A breeding system in which mates are selected on the basis of a particular trait or group of traits.

atherosclerosis (ath' er oh sklair oh' sis) [Gk. *athero*: gruel, porridge + *skleros*: hard] A disease of the lining of the arteries characterized by fatty, cholesterol-rich deposits in the walls of the arteries. When fibroblasts infiltrate these deposits and calcium precipitates in them, the disease become arteriosclerosis, or "hardening of the arteries."

atmosphere The gaseous mass surrounding our planet. Also a unit of pressure, equal to the normal pressure of air at sea level.

atom [Gk. *atomos*: indivisible] The smallest unit of a chemical element. Consists of a nucleus and one or more electrons.

atomic mass The average mass of an atom of an element; the average depends on the relative amounts of ifferent isotopes of the element on Earth. Also called atomic weight.

atomic number The number of protons in the nucleus of an atom; also equals the number of electrons around the neutral atom. Determines the chemical properties of the atom.

ATP (adenosine triphosphate) An energy-storage compound containing adenine, ribose, and three phosphate groups. When it is formed from ADP, useful energy is stored; when it is broken down (to ADP or AMP), energy is released to drive endergonic reactions.

ATP synthase An integral membrane protein that couples the transport of proteins with the formation of ATP.

atrioventricular node A modified node of cardiac muscle that organizes the action potentials that control contraction of the ventricles.

atrium (a' tree um) [L. *atrium*: central hall] An internal chamber. In the hearts of vertebrates, the thin-walled chamber(s) entered by blood on its way to the ventricle(s). Also, the outer ear.

autocrine Referring to a cell signaling mechanism in which the signal binds to and affects the cell that makes it. An autocrine hormone, for example, is one that influences the cell that releases it. (Compare with endocrine gland;, paracrine)

autoimmune disease A disorder in which the immune system attacks the animal's own antigens.

autonomic nervous system The system that controls such involuntary functions as those of guts and glands. (Compare with central nervous system.)

autopolyploidy The possession of more than two chromosome sets that are derived from more than one species.

autosome Any chromosome (in a eukaryote) other than a sex chromosome.

autotroph (au' tow trow' fik) [Gk. *autos*: self + *trophe*: food] An organism that is capable of living exclusively on inorganic materials, water, and some energy source such as sunlight or chemically reduced matter. (Contrast with heterotroph.)

auxin (awk' sin) [Gk. *auxein*: to grow] In plants, a substance (the most common being indoleacetic acid) that regulates growth and various aspects of development.

auxotroph (awks' o trofe) [Gk. *auxein*: to grow + *trophe*: food] A mutant form of an organism that requires a nutrient or nutrients not usually required by the wild type. (Contrast with prototroph.)

Avogadro's number The number of atoms or molecules in a mole (weighed out in grams) of a substance, calculated to be 6.022×10^{23}.

Avr genes (avirulence genes) Genes in a pathogen that may trigger defenses in plants. *See* gene-for-gene resistance.

axillary bud A bud occurring in the upper angle (axil) between a leaf and stem.

axon [Gk. *axon*: axle] The part of a neuron that conducts action potentials away from the cell body.

axon hillock The junction between an axon and its cell body, where action potentials are generated.

axon terminals The endings of an axon; they form synapses and release neurotransmitter.

axoneme (ax' oh neem) The complex of microtubules and their crossbridges that forms the motile apparatus of a cilium.

– B –

bacillus (bah sil' us) [L: little rod] Any of various rod-shaped bacteria.

bacterial artificial chromosome (BAC) A DNA cloning vector used in bacteria that can carry up to 150,000 base pairs of foreign DNA.

bacteriophage (bak teer' ee o fayj) [Gk. *bakterion*: little rod + *phagein*: to eat] One of a group of viruses that infect bacteria and ultimately cause their disintegration.

bacteroids Nitrogen-fixing organelles that develop from endosymbiotic bacteria.

balanced polymorphism [Gk. *polymorphos*: many forms] The maintenance of more than one form, or the maintenance at a given locus of more than one allele, at frequencies of greater than 1 percent in a population. Often results when heterozygotes are more fit than either homozygote.

bark All tissues outside the vascular cambium of a plant.

baroreceptor [Gk. *baros*: weight] A pressure-sensing cell or organ.

Barr body In female mammals, an inactivated X chromosome.

basal Pertaining to one end—the base—of an axis.

basal body Centriole found at the base of a eukaryotic flagellum or cilium.

basal metabolic rate (BMR) The minimum rate of energy turnover in an awake (but resting) bird or mammal that is not expending energy for thermoregulation.

base (1) A substance tha can accept a hydrogen ion in solution. (Contrast with acid.) (2) In nucleic acids, the purine or pyrimidine that is attached to each sugar in the backbone.

base pairing *See* complementary base pairing.

basic Having a pH greater than 7.0 (i.e., having a hydrogen ion concentration lower than 10^{-7} molar).

basidium (bass id' ee yum) In basidiomycete fungi, the characteristic sporangium in which four spores are formed by meiosis and then borne externally before being shed.

basophils One type of phagocytic white blood cell that releases histamine and may promote T cell development.

Batesian mimicry The convergence in appearance of an edible species (mimic) with an unpalatable species (model).

B cell A type of lymphocyte involved in the humoral immune response of vertebrates. Upon recognizing an antigenic determinant, a B cell develops into a plasma cell, which secretes an antibody. (Contrast with T cell.)

benefit An improvement in survival and reproductive success resulting from performing a behavior or having a trait. (Contrast with cost.)

benign (be nine') Referring to a tumor that grows to a certain size and then stops, uaually with a fibrous capsule surrounding the mass of

cells. Benign tumors do not spread (metastasize) to other organs. (Contrast with malignant.)

benthic zone [Gk. *benthos*: bottom] The bottom of the ocean.

beta (β) pleated sheet Type of protein secondary structure; results from hydrogen bonding between polypeptide regions running antiparallel to each other.

biennial Referring to a plant whose life cycle includes vegetative growth in the first year and flowering and senescence in the second year. (Contrast with annual, perennial.)

bilateral symmetry The condition in which only the right and left sides of an organism, divided exactly down the back, are mirror images of each other. (Contrast with biradial symmetry.)

bilayer In membranes, a structure that is two lipid layers in thickness.

bile A secretion of the liver delivered to the small intestine via the common bile duct. In the intestine, bile emulsifies fats.

binary fission Reproduction by cell division of a single-celled organism.

binding domain The region of a receptor molecule where its ligand attaches.

binocular cells Neurons in the visual cortex that respond to input from both retinas; involved in depth perception.

binomial (bye nome' ee al) Consisting of two; for example, the binomial nomenclature of biology in which each species has two names (the genus name followed by the species name).

biodiversity crisis The current high rate of loss of species, caused primarily by human activities.

biofilm A community of microorganisms embedded in a polysaccharide matrix, forming a highly resistant coating on almost any moist surface.

biogeochemical cycle Movement of elements through living organisms and the physical environment.

biogeographic region One of several defined, continental-scale regions of Earth, each of which has a biota distinct from that of the others. (Contrast with biome.)

bioinformatics The use of computers and/or mathematics to analyze complex biological information, such as DNA sequences.

bioluminescence The production of light by biochemical processes in an organism.

biomass The total weight of all the living organisms, or some designated group of living organisms, in a given area.

biome (bye' ome) A major division of the ecological communities of Earth, characterized primarily by distinctive vegetation. A given biogeographic region contains many different biomes.

biosphere (bye' oh sphere) All regions of Earth (terrestrial and aquatic) and Earth's atmosphere in which organisms can live.

biota (bye oh' tah) All of the organisms—animals, plants, fungi, and microorganisms—found in a given area. (Contrast with flora, fauna.)

biotic (bye ah' tick) [Gk. *bios*: life] Alive. (Contrast with abiotic.)

biradial symmetry Radial symmetry modified so that only two planes can divide the animal into similar halves.

blastocoel (blass' toe seal) [Gk. *blastos*: sprout + *koilos*: hollow] The central, hollow cavity of a blastula.

blastocyst (blass' toe cist) An early embryo formed by the first divisions of the fertilized egg (zygote). In mammals, a hollow ball of cells.

blastodisc (blass' toe disk) A disk of cells forming on the surface of a large yolk mass, comparable to a blastula, but occurring in animals such as birds and reptiles, in which the massive yolk restricts cleavage to one side of the egg only.

blastomere A cell produced by the division of a fertilized egg.

blastopore The opening from the archenteron to the exterior of a gastrula.

blastula (blass' chu luh) An early stage in animal embryology; in many species, a hollow sphere of cells surrounding a central cavity, the blastocoel. (Contrast with blastodisc.)

blood–brain barrier A property of the blood vessels of the brain that prevents most chemicals from diffusing from the blood into the brain.

blood plasma (plaz' muh) [Gk. *plassein*: to mold] The liquid portion of blood, in which blood cells and other particulates are suspended.

blue light receptors Pigments in plants that absorbs blue light (400–500 nm). These pigments mediate many plant responses including phototropism, stomatal movements, and expression of some genes.

body cavity Membrane-lined, fluid-filled compartment that lies between the cell layers of many animals.

Bohr effect The fact that low pH decreases the affinity of hemoglobin for oxygen.

bond *See* chemical bond.

bottleneck A stressful period during which only a few individuals of a once large population survive.

Bowman's capsule An elaboration of kidney tubule cells that surrounds a know of capillaries (the glomerulus). Blood is filtered across the walls of these capillaries and the filtrate is collected into Bowman's capsule.

brain stem The portion of the vertebrate brain between the spinal cord and the forebrain.

brassinosteroids Plant steroid hormones that mediate light effects promoting the elongation of stems and pollen tubes.

bronchioles The smallest airways in a vertebrate lung, branching off the bronchi.

bronchus (plural: bronchi) The major airway(s) branching off the trachea into the vertebrate lung.

brown fat Fat tissue in mammals that is specialized to produce heat. It has many mitochondria and capillaries, and a protein that uncouples oxidative phosphorylation.

browser An animal that feeds on the tissues of woody plants.

budding Asexual reproduction in which a more or less complete new organism simply grows from the body of the parent organism and eventually detaches itself.

buffer A substance that can transiently accept or release hydrogen ions and thereby resist changes in pH.

bundle of His Fibers of modified cardiac muscle that conduct action potentials from the atria to the ventricular muscle mass.

bundle sheath cell Part of a tissue that surrounds the veins of plants; contains chloroplasts in C_4 plants.

- C -

C_3 photosynthesis Form of photosynthesis in which 3-phosphoglycerate is the first stable product, and ribulose bisphosphate is the CO_2 receptor.

C_4 photosynthesis Form of photosynthesis in which oxaloacetate is the first stable product, and phosphoenolpyruvate is the CO_2 acceptor. C_4 plants also perform the reactions of C_3 photosynthesis.

calcitonin A hormone produced by the thyroid gland; it lowers blood calcium and promotes bone formation. (Compare with parathyroid hormone.)

calmodulin (cal mod' joo lin) A calcium-binding protein found in all animal and plant cells; mediates many calcium-regulated processes.

calorie [L. *calor*: heat] The amount of heat required to raise the temperature of one gram of water by one degree Celsius (1°C) from 14.5°C to 15.5°C. Calorie spelled with a capital C refers to the kilocalorie (1 kcal = 1,000 cal).

Calvin cycle The stage of photosynthesis in which CO_2 reacts with RuBP to form 3PG, 3PG is reduced to a sugar, and RuBP is regenerated, while other products are released to the rest of the plant. Also known as the Calvin–Benson cycle.

calyx (kay' licks) [Gk. *kalyx*: cup] All of the sepals of a flower, collectively.

CAM *See* crassulacean acid metabolism.

cambium (kam' bee um) [L. *cambiare*: to exchange] A meristem that gives rise to radial rows of cells in stem and root, increasing them in girth; commonly applied to the vascular cambium which produces wood and phloem, and the cork cambium, which produces bark.

Cambrian explosion The rapid diversification of life that took place during the Cambrian period.

cAMP (cyclic AMP) A compound formed from ATP that mediates the effects of numerous animal hormones.

canopy The leaf-bearing part of a tree. Collectively, the aggregate of the leaves and branches of the larger woody plants of an ecological community.

capillaries [L. *capillaris*: hair] Very small tubes, especially the smallest blood-carrying vessels of animals between the termination of the arteries and the beginnings of the veins. Capillaries are the site of exchange of materials between the blood and the interstitial fluid.

capsid The outer shell of a virus that encloses its nucleic acid.

carbohydrates Organic compounds containing carbon, hydrogen, and oxygen in the ratio 1:2:1 (i.e., with the general formula $C_nH_{2n}O_n$). Common examples are sugars, starch, and cellulose.

carboxylase An enzyme that catalyzes the addition of carboxyl groups to a substrate.

carboxylic acid (kar box sill' ik) An organic acid containing the carboxyl group, —COOH, which dissociates to the carboxylate ion, —COO⁻.

carcinogen (car sin' oh jen) A substance that causes cancer.

cardiac (kar' dee ak) [Gk. *kardia*: heart] Pertaining to the heart and its functions.

cardiac muscle One of the three types of muscle tissue, it makes up the vertebrate heart. Characterized by branching cells with single nuclei and striated (striped) appearance. (Contrast with smooth muscle, striated muscle.)

carnivore [L. *carn*: flesh + *vovare*: to devour] An organism that eats animal tissues. (Contrast with detritivore, herbivore, omnivore.)

carotenoid (ka rah' tuh noid) A yellow, orange, or red lipid pigment commonly found as an accessory pigment in photosynthesis; also found in fungi.

carpel (kar' pel) [Gk. *karpos*: fruit] The organ of the flower that contains one or more ovules.

carrier (1) In facilitated diffusion, a membrane protein that binds a specific molecule and transports it through the membrane. (2) In respiratory and photosynthetic electron transport, a participating substance such as NAD that exists in both oxidized and reduced forms. (3) In genetics, a person heterozygous for a recessive trait.

carrier protein *See* Carrier (1).

cartilage In vertebrates, a tough connective tissue found in joints, the outer ear, and elsewhere. Forms the entire skeleton in some animal groups.

Casparian strip A band of cell wall containing suberin and lignin, found in the endodermis. Restricts the movement of water across the endodermis.

caspase A member of a group of proteases that catalyze cleavage of target proteins and are active in apoptosis.

catabolic reaction A single reaction that participates in catabolism.

catabolism [Gk. *kata*: to break down + *ballein*: to throw] Degradational reactions of metabolism, in which complex molecules are broken down. (Contrast with anabolism.)

catabolite repression In the presence of abundant glucose, the diminished synthesis of catabolic enzymes for other energy sources.

catalyst (cat' a list) [Gk. *kata*: to break down] A chemical substance that accelerates a reaction without itself being consumed in the overall course of the reaction. Catalysts lower the activation energy of a reaction. Enzymes are biological catalysts.

catalytic subunit The polypeptide in an enzyme protein with quaternary structure that contains the active site for the enzyme. (Contrast regulatory subunit.)

cation (cat' eye on) An ion with one or more positive charges. (Contrast with anion.)

caudal [L. *cauda*: tail] Pertaining to the tail, or to the posterior part of the body.

cDNA *See* complementary DNA.

cecum (see' cum) [L. *caecus*: blind]   A blind branch off the large intestine. In many nonruminant mammals, the cecum contains a colony of microorganisms that contribute to the digestion of food.

cell adhesion molecules   Molecules on animal cell surfaces that affect the selective association of cells during development of the embryo.

cell cycle   The stages through which a cell passes between one division and the next. Includes all stages of interphase and mitosis.

cell division   The reproduction of a cell to produce two new cells. In eukaryotes, this process involves nuclear division (mitosis) and cytoplasmic division (cytokinesis).

cell junctions   Specialized structures associated with the plasma membranes of epithelial cells. Some contribute to cell adhesion, others to intercellular communication.

cell plate   A structure that forms at the equator of the spindle in dividing plant cells.

cell recognition   Binding of cells to one another mediated by membrane proteins or carbohydrates.

cell wall   A relatively rigid structure that encloses cells of plants, fungi, many protists, and most prokaryotes. Gives these cells their shape and limits their expansion in hypotonic media.

cellular immune response   Action of the immune system based on the activities of T cells. Directed against parasites, fungi, intracellular viruses, and foreign tissues (grafts). (Contrast with humoral immune system.)

cellular respiration   *See* respiration.

cellulose (sell' you lowss)   A straight-chain polymer of glucose molecules, used by plants as a structural supporting material.

centimorgan   In genetic mapping, a recombinant frequency of 0.01; a map unit.

central dogma   The statement that information flows from DNA to RNA to polypeptide (in retroviruses, there is also information flow from RNA to cDNA).

central nervous system   That part of the nervous system which is condensed and centrally located, e.g., the brain and spinal cord of vertebrates; the chain of cerebral, thoracic and abdominal ganglia of arthropods. (Compare with autonomic nervous system.)

centrifuge [L. *centrum*: center + *fugere*: to flee]   A laboratory device in which a sample is spun around a central axis at high speed. Used to separate suspended materials of different densities.

centriole (sen' tree ole)   A paired organelle that helps organize the microtubules in animal and protist cells during nuclear division.

centromere (sen' tro meer) [Gk. *centron*: center + *meros*: part]   The region where sister chromatids join.

centrosome (sen' tro soam)   The major microtubule organizing center of an animal cell.

cephalization (sef ah luh zay' shun) [Gk. *kephale*: head]   The evolutionary trend toward increasing concentration of brain and sensory organs at the anterior end of the animal.

cerebellum (sair uh bell' um) [L.: diminutive of *cerebrum*, brain]   The brain region that controls muscular coordination; located at the anterior end of the hindbrain.

cerebral cortex   The thin layer of gray matter (neuronal cell bodies) that overlays the cerebrum.

cerebrum (su ree' brum) [L. *cerebrum*: brain]   The dorsal anterior portion of the forebrain, making up the largest part of the brain of mammals. In mammals, the chief coordination center of the nervous system; consists of two cerebral hemispheres.

cervix (sir' vix) [L. *cervix*: neck]   The opening of the uterus into the vagina.

cGMP (cyclic guanosine monophosphate) An intracellular messenger that is part of signal transmission pathways involving G proteins. (*See* G protein.)

channel protein   A membrane protein that forms an aqueous passageway though which specific solutes may pass.

chemical bond   An attractive force stably linking two atoms.

chemical equilibrium   A state reached in a reversible chemical reaction when the forward and reverse reactions balance each other and there is no net change.

chemical evolution   The theory that life originated through the chemical transformation of inanimate substances.

chemical reaction   The change in the composition or distribution of atoms of a substance with consequent alterations in properties.

chemiosmosis   The formation of ATP in mitochondria and chloroplasts, resulting from a pumping of protons across a membrane (against a gradient of electrical charge and of pH), followed by the return of the protons through a protein channel with ATPase activity.

chemoautotroph   *See* chemolithotroph.

chemoheterotroph   An organism that must obtain both carbon and energy from organic substances. (Contrast with chemolithotroph, photoautotroph, photoheterotroph.)

chemolithotroph [Gk. *lithos*: stone, rock]   An organism that uses carbon dioxide as a carbon source and obtains energy by oxidizing inorganic substances from its environment. (Contrast with chemoheterotroph, photoautotroph, photoheterotroph.)

chemoreceptor   A sensory receptor cell or tissue that senses specific substances in its environment.

chemosynthesis   Synthesis of food substances, using the oxidation of reduced materials from the environment as a source of energy.

chiasma (kie az' muh) (plural: chiasmata) [Gk. *chiasmata*: cross]   An X-shaped connection between paired homologous chromosomes in prophase I of meiosis. A chiasma is the visible manifestation of crossing over between homologous chromosomes.

chitin (kye' tin) [Gk. *kiton*: tunic]   The characteristic tough but flexible organic component of the exoskeleton of arthropods, consisting of a complex, nitrogen-containing polysaccharide. Also found in cell walls of fungi.

chlorophyll (klor' o fill) [Gk. *kloros*: green + *phyllon*: leaf]   Any of a few green pigments associated with chloroplasts or with certain bacterial membranes; responsible for trapping light energy for photosynthesis.

chloroplast [Gk. *kloros*: green + *plast*: a particle]   An organelle bounded by a double membrane containing the enzymes and pigments that perform photosynthesis. Chloroplasts occur only in eukaryotes.

choanocyte (ko' an uh site)   The collared, flagellated feeding cells of sponges.

cholecystokinin (ko' luh sis tuh kai' nin)   A hormone produced and released by the lining of the duodenum when it is stimulated by undigested fats and proteins. It stimulates the gallbladder to release bile and slows stomach activity.

chorion (kor' ee on) [Gk. *khorion*: afterbirth]   The outermost of the membranes protecting mammal, bird, and reptile embryos; in mammals it forms part of the placenta.

chromatid (kro' ma tid)   Each of a pair of new sister chromosomes from the time at which the molecular duplication occurs until the time at which the centromeres separate at the anaphase of nuclear division.

chromatin   The nucleic acid–protein complex found in eukaryotic chromosomes.

chromatin remodeling   Changes in chromatin structure to allow transcription, translation, and chromosome condensation.

chromatophore (krow mat' o for) [Gk. *kroma*: color + *phoreus*: carrier]   A pigment-bearing cell that expands or contracts to change the color of the organism.

chromosomal mutation   Loss of or changes in position/direction of a DNA segment on a chromosome.

chromosome (krome' o sowm) [Gk. *kroma*: color + *soma*: body]   In bacteria and viruses, the DNA molecule that contains most or all of the genetic information of the cell or virus. In eukaryotes, a structure composed of DNA and proteins that bears part of the genetic information of the cell.

chylomicron (ky low my' cron)   Particles of lipid coated with protein, produced in the gut from dietary fats and secreted into the extracellular fluids.

chyme (kime) [Gk. *kymus*, juice]   Created in the stomach; a mixture of ingested food with the digestive juices secreted by the salivary glands and the stomach lining.

cilium (sil' ee um) (plural: cilia) [L.: eyelash]   Hairlike organelle used for locomotion by many unicellular organisms and for moving water and mucus by many multicellular organisms. Generally shorter than a flagellum.

circadian rhythm (sir kade' ee an) [L. *circa*: approximately + *dies*: day]   A rhythm in behavior, growth, or some other activity that recurs about every 24 hours under constant conditions.

circannual rhythm [L. *circa*: approximately + *annus*: year]   A rhythm of behavior, growth, or some other activity that recurs on a yearly basis.

citric acid cycle   A set of chemical reactions in cellular respiration, in which acetyl CoA is oxidized to carbon dioxide, and hydrogen atoms are stored as NADH and $FADH_2$. Also called the Krebs cycle.

clade [Gk. *klados*: branch] A monophyletic group made up of an ancestor and all of its descendants.

class I MHC molecules These cell surface proteins participate in the cellular immune response directed against virus-infected cells.

class II MHC molecules These cell surface proteins participate in the cell-cell interactions (of helper T cells, macrophages, and B cells) of the humoral immune response.

class switching The process whereby a plasma cell changes the class of immunoglobulin that it synthesizes by changing the DNA region coding for the C segment.

clathrin A fibrous protein on the inner surfaces of animal cell membranes that strengthens coated vesicles and thus participates in receptor-mediated endocytosis.

cleavage First divisions of the fertilized egg of an animal.

climate The average of the atmospheric conditions (temperature, precipitation, wind direction and velocity) found in a region over time.

cline A gradual change in the traits of a species over a geographical gradient.

cloaca (klo ay' kuh) [L. *cloaca*: sewer] In some invertebrates, the posterior part of the gut; in many vertebrates, a cavity receiving material from the digestive, reproductive, and excretory systems.

clonal anergy Prevention of the synthesis of antibodies against the body's own antigens. When a T cell binds to a self-antigen, it does not receive signals from an antigen-presenting cell; thus the T cell dies (becomes anergic) rather than yielding a clone of active cells.

clonal deletion The inactivation or destruction of lymphocyte clones that would produce immune reactions against the animal's own body.

clonal selection The mechanism by which exposure to antigen results in the activation of selected T- or B-cell clones, resulting in an immune response.

clone [Gk. *klon*: twig, shoot] Genetically identical cells or organisms produced from a common ancestor by asexual means.

cnidocytes (nye' duh sites) The feeding cells of cnidarians, within which nematocysts are housed.

coacervate (ko as' er vate) [L. *coacervare*: to heap up] An aggregate of colloidal particles in suspension.

coated vesicle Cytoplasmic vesicle containing distinctive proteins, including clathrin.

coccus (kock' us) [Gk. *kokkos*: berry, pit] Any of various spherical or spheroidal bacteria.

cochlea (kock' lee uh) [Gk. *kokhlos* snail] A spiral tube in the inner ear of vertebrates; it contains the sensory cells involved in hearing.

codominance A condition in which two alleles at a locus produce different phenotypic effects and both effects appear in heterozygotes.

codon Three nucleotides in messenger RNA that direct the placement of a particular amino acid into a polypeptide chain. (Contrast with anti-codon.)

coelom (see' lum) [Gk. *koiloma*: cavity] The body cavity of certain animals; the coelom is lined with cells of mesodermal origin.

coelomate Having a coelom.

coenocyte (seen' a sight) [Gk. *koinos*: common + *kytos*: container] A "cell" enclosed by a single plasma membrane but containing many nuclei.

coenzyme A nonprotein organic molecule that plays a role in catalysis by an enzyme.

cofactor An inorganic ion that is weakly bound to an enzyme and required for its activity.

cohesin Proteins involved in binding chromatids together.

cohesion The tendency of molecules (or any substances) to stick together.

cohort (co' hort) [L. *cohors*: company of soldiers] A group of similar-aged organisms, considered as it passes through time.

coleoptile A sheath that surrounds and protects the shoot apical meristem and young primary leaves of a grass seedling as they move through the soil.

collagen [Gk. *kolla*: glue] A fibrous protein found extensively in bone and connective tissue.

collecting duct In vertebrates, a tubule that receives urine produced in the nephrons of the kidney and delivers that fluid to the ureter for excretion.

collenchyma (cull eng' kyma) [Gk. *kolla*: glue + *enchyma*: infusion] A type of plant cell, living at functional maturity, which lends flexible support by virtue of primary cell walls thickened at the corners. (Contrast with parenchyma, sclerenchyma.)

colon [Gk. *kolon*] The large intestine.

common bile duct A single duct that delivers bile from the gallbladder and secretions from the pancreas into the small intestine.

communication A signal from one organism (or cell) that alters the functioning or behavior of another organism (or cell).

community Any ecologically integrated group of species of microorganisms, plants, and animals inhabiting a given area.

companion cell A specialized cell found adjacent to a sieve tube element in flowering plants.

comparative genomics Computer-aided comparison of DNA sequences between different organisms to reveal genes with related functions.

comparative method An experimental design in which two samples or populations exposed to different conditions or treatments are compared to each other.

compensation point The light intensity at which the rates of photosynthesis and of cellular respiration are equal.

competition In ecology, use of the same resource by two or more species, when the resource is present in insufficient supply for the combined needs of the species.

competitive exclusion A result of competition between species for a limiting resource in which one species completely eliminates the other.

competitive inhibitor A nonsubstrate that binds to the active site of an enzyme and thereby inhibits binding of substrate and reaction from part of the environment.

complement system A group of eleven proteins that play a role in some reactions of the immune system. The complement proteins are not immunoglobulins.

complementary base pairing The AT (or AU), TA (or UA), CG, and GC pairing of bases in double-stranded DNA, in transcription, and between tRNA and mRNA.

complementary DNA (cDNA) DNA formed by reverse transcriptase acting with an RNA template; essential intermediate in the reproduction of retroviruses; used as a tool in recombinant DNA technology; lacks introns.

complete metamorphosis A change of state during the life cycle of an organism in which the body is almost completely rebuilt to produce an individual with a very different body form. Characteristic of insects such as butterflies, moths, beetles, ants, wasps, and flies.

compound (1) A substance made up of atoms of more than one element. (2) A structure made up of many units, as the compound eyes of arthropods.

concerted evolution The common evolution of a family of repeated genes, such that changes in one copy of the gene family are replicated in other copies of the gene family.

condensation reaction A reaction in which two molecules become connected by a covalent bond and a molecule of water is released. ($AH + BOH \rightarrow AB + H_2O$.)

condensing A protein complex involved in chromosome condensation during mitosis and meiosis.

conditional mutations Mutations that show characteristic phenotype only under certain environmental conditions such as temperature.

conduction The transfer of heat from one object to another through direct contact. In neurophysiology, the progression of an action potential along an axon.

cone cells (1) In the vertebrate retina, photoreceptor cells that are responsible for color vision. (2) In gymnosperms, reproductive structures consisting of spore-bearing scales inserted on a short axis; the scales are modified branches. (Contrast with strobilus.)

conformation The three-dimensional shape of a protein or other macromolecule.

conidium (ko nid' ee um) [Gk. *konis*: dust] An asexual fungus spore borne singly or in chains either apically or laterally on a hypha.

conifer (kahn' e fer) [Gr. *konos*: cone + *phero*: carry] One of the cone-bearing gymnosperms, mostly trees, such as pines and firs.

conjugation (kon ju gay' shun) [L. *conjugare*: yoke together] The close approximation of two cells during which they exchange genetic material, as in *Paramecium* and other ciliates, or during which DNA passes from one to the other, as in bacteria.

conjugation tube Cytoplasmic connection between two bacterial cells through which DNA passes during conjugation.

connective tissue An animal tissue that connects or surrounds other tissues; its cells are embedded in a collagen-containing matrix.

connexon In a gap junction, a protein channel linking adjacent animal cells.

consensus sequences Short stretches of DNA that appear, with little variation, in many different genes.

conservation biology An applied, normative science that carries out investigations designed to help preserve the diversity of life on Earth.

constant region For a particular class of immunoglobulin molecules, the region with identical amino acid composition.

constitutive enzyme An enzyme that is present in approximately constant amounts in a system, whether its substrates are present or absent. (Contrast with inducible enzyme.)

consumer An organism that eats the tissues of some other organism.

continental drift The gradual movements of the world's continents that have occurred over billions of years.

contraception The act of preventing union of sperm and egg.

controlled experiment An experimental design in which a sample or population is divided into two groups; one group is exposed to a manipulated variable while the other group serves as a nontreated control. The two groups are compared to see if there are changes in a "dependent" variable as a result of the experimental manipulation.

convection The transfer of heat to or from a surface via a moving stream of air or fluid.

convergent evolution The independent evolution of similar features from different ancestral traits.

copulation Reproductive behavior that results in a male depositing sperm in the reproductive tract of a female.

coprophagy Ingesting ones own feces.

corepressor A low-molecular-weight compound that unites with a protein (the repressor) to prevent transcription in a repressible operon.

cork A waterproofing tissue in plants, with suberin-containing cell walls. Produced by a cork cambium.

cornea The clear, transparent tissue that covers the eye and allows light to pass through to the retina.

corolla (ko role' lah) [L. *corolla*: a small crown] All of the petals of a flower, collectively.

coronary (kor' oh nair ee) [L. *corona*: crown] Referring to the blood vessels of the heart.

coronary thrombosis A fibrous clot that blocks a coronary vessel.

corpus luteum (kor' pus loo' tee um) [L.: yellow body] A structure formed from a follicle after ovulation; it produces hormones important to the maintenance of pregnancy.

cortex [L. *cortex*: covering, rind] (1) In plants, the tissue between the epidermis and the vascular tissue of a stem or root. (2) In animals, the outer tissue of certain organs, such as the adrenal cortex and cerebral cortex.

corticosteroids Steroid hormones produced and released by the cortex of the adrenal gland.

cortisol A corticosteroid that mediates stress responses.

cost–benefit analysis An approach for explaining why the traits of organisms evolve as they do; assumes that an organism has a limited amount of time and energy to devote to its activities, and that each activity has costs (e.g., risks, expenditure of energy) as well as benefits (e.g., obtaining food, mating and successful reproduction). (*See also* trade-off.)

cotyledon (kot' ul lee' dun) [Gk. *kotyledon*: hollow space] A "seed leaf." An embryonic organ that stores and digests reserve materials; may expand when seed germinates.

countercurrent exchange An adaptation that promotes maximum exchange of heat or any diffusible substance between two fluids by the fluids flow in opposite directions through parallel tubes close together

countercurrent multiplier The mechanism of increasing the concentration of the interstitial fluid in the mammalian kidney as a result of countercurrent exchange in the loops of Henle and selective permeability and active transport of ions by segments of the loops of Henle.

covalent bond A chemical bond that arises from the sharing of electrons between two atoms. Usually a strong bond.

crassulacean acid metabolism (CAM) A metabolic pathway enabling the plants that possess it to store carbon dioxide at night and then perform photosynthesis during the day with stomata closed.

crista (plural: cristae) A small, shelflike projection of the inner membrane of a mitochondrion; the site of oxidative phosphorylation.

critical night length In the photoperiodic flowering response of short-day plants, the length of night above which flowering occurs and below which the plant remains vegetative. (The reverse applies in the case of long-day plants.)

critical period The age during which some particular type of learning must take place or during which it occurs much more easily than at other times. Typical of song learning among birds.

cross section A section taken perpendicular to the longest axis of a structure. Also called a transverse section.

crossing over The mechanism by which linked markers undergo recombination. In general, the term refers to the reciprocal exchange of corresponding segments between two homologous chromatids.

cryptic [Gk. *kryptos*: hidden] The resemblance of an animal to some part of its environment, which helps it to escape detection by predators.

cryptochromes [Gk. *kryptos*: hidden + *kroma*: color] Photoreceptors mediating some blue-light effects in plants and animals.

culture (1) A laboratory association of organisms under controlled conditions. (2) The collection of knowledge, tools, values, and rules that characterize a human society.

cuticle A waxy layer on the outer surface of a plant or an insect, tending to retard water loss.

cyanobacteria (sigh an' o bacteria) [Gr. *kuanos*: blue] A group of photosynthetic bacteria, formerly referred to as blue-green algae; they use chlorophyll *a* in photosynthesis.

cyclic AMP *See* cAMP.

cyclic electron transport In photosynthetic light reactions, the flow of electrons that produces ATP but no NADPH or O_2.

cyclins Proteins that activate cyclin-dependent kinases, bringing about transitions in the cell cycle.

cyclin-dependent kinase (Cdk) A kinase whose target proteins are involved in transitions in the cell cycle and which is active only when complexed with additional protein subunits, called cyclins.

cytochromes (sy' toe chromes) [Gk. *kytos*: container + *chroma*: color] Iron-containing red proteins, components of the electron-transfer chains in photophosphorylation and respiration.

cytokine A regulatory protein made by immune system cells that affects other target cells in the immune system.

cytokinesis (sy' toe kine ee' sis) [Gk. *kytos*: container + *kinein*: to move] The division of the cytoplasm of a dividing cell. (Compare with mitosis.)

cytokinin (sy' toe kine' in) A member of a class of plant growth substances playing roles in senescence, cell division, and other phenomena.

cytoplasm The contents of the cell, excluding the nucleus.

cytoplasmic determinants In animal development, gene products whose spatial distribution may determine such things as embryonic axes.

cytoplasmic domain The portion of a membrane bound receptor molecule that projects into the cytoplasm.

cytoplasmic segregation The asymmetrical distribution of cytoplasmic determinants in a developing animal embryo.

cytosine (C) (site' oh seen) A nitrogen-containing base found in DNA and RNA.

cytoskeleton The network of microtubules and microfilaments that gives a eukaryotic cell its shape and its capacity to arrange its organelles and to move.

cytosol The fluid portion of the cytoplasm, excluding organelles and other solids.

cytotoxic T cells (T_C) Cells of the cellular immune system that recognize and directly eliminate virus-infected cells. (Compare with helper T cells.)

– D –

DAG *See* diacylglycerol.

daughter chromosomes During mitosis, the separated chromatids from the beginning of anaphase onward.

deciduous [L. *deciduus*: falling off] Refers to a woody plant that sheds it leaves but does not die.

decomposers Organisms that metabolize organic compounds in debris and dead organisms, releasing inorganic material; found among the bacteria, protists, and fungi. *See also* detritivore.

defensin A type of protein made by phagocytes that kills bacteria and enveloped viruses by insertion into their plasma membranes.

degeneracy The situation in which a single amino acid may be represented by any of two or more different codons in messenger RNA. Most of the amino acids can be represented by more than one codon.

delayed hypersensitivity An increased immune reaction against an antigen that does not appear for 1–2 days after exposure. (Contrast with immediate hypersensitivity.)

deletion A mutation resulting from the loss of a continuous segment of a gene or chromosome. Such mutations never revert to wild type. (Contrast with duplication, point mutation.)

deme (deem) [Gk. *demos*: the populace] Any local population of individuals belonging to the same species that interbreed with one another.

demographic processes Events (births, deaths, immigration, and emigration) that change the number of individuals and the age structure of a population.

demographic stochasticity Random variations in the factors influencing the size, density, and distribution of a population.

denaturation Loss of activity of an enzyme or nucleic acid molecule as a result of structural changes induced by heat or other means.

dendrite [Gk. *dendron*: tree] A fiber of a neuron which often cannot carry action potentials. Usually much branched and relatively short compared with the axon, and commonly carries information to the cell body of the neuron.

denitrification Metabolic activity by which nitrate and nitrite ions are reduced to form nitrogen gas; carried on by certain soil bacteria.

density dependence The state in which changes in the severity of action of agents affecting birth and death rates within populations are directly or inversely related to population density.

density independence The state in which the severity of action of agents affecting birth and death rates within a population does not change with the density of the population.

deoxyribonucleic acid *See* DNA.

deoxyribose A five-carbon sugar found in nucleotides and DNA.

depolarization A change in the electric potential across a membrane from a condition in which the inside of the cell is more negative than the outside to a condition in which the inside is less negative, or even positive, with reference to the outside of the cell. (Contrast with hyperpolarization.)

derived trait A trait that differs from the ancestral trait. (Compare with shared derived trait.)

dermal tissue system The outer covering of a plant, consisting of epidermis in the young plant and periderm in a plant with extensive secondary growth. (Contrast with ground tissue system and vascular tissue system.)

desmosome (dez' mo sowm) [Gk. *desmos*: bond + *soma*: body] An adhering junction between animal cells.

desmotubule A membrane extension connecting the endoplasmic retitulum of two plant cells that traverses the plasmodesma.

determination Process whereby an embryonic cell or group of cells becomes fixed into a predictable developmental pathway.

determined The state of a cell prior to its differentiation but after its developmental fate has been channeled.

detritivore (di try' ti vore) [L. *detritus*: worn away + *vorare*: to devour] An organism that obtains its energy from the dead bodies and/or waste products of other organisms.

development Progressive change, as in structure or metabolism; in most kinds of organisms, development continues throughout the life of the organism.

developmental plasticity The capacity of an organism to alter its pattern of development in response to environmental conditions.

diacylglycerol (DAG) In hormone action, the second messenger produced by hydrolytic removal of the head group of certain phospholipids.

diaphragm (dye' uh fram) [Gk. *diaphrassein*: barricade] (1) A sheet of muscle that separates the thoracic and abdominal cavities in mammals; responsible for breathing. (2) A method of birth control in which a sheet of rubber is fitted over the woman's cervix, blocking the entry of sperm.

diastole (dye ass' toll ee) [Gk. dilation] The portion of the cardiac cycle when the heart muscle relaxes. (Contrast with systole.)

diencephalons The portion of the vertebrate forebrain that becomes the thalamus and hypothalamus.

differential gene expression The hypothesis that, given that all cells contain all genes, what makes one cell type different from another is the difference in transcription and translation of those genes.

differentiation Process whereby originally similar cells follow different developmental pathways. The actual expression of determination.

diffusion Random movement of molecules or other particles, resulting in even distribution of the particles when no barriers are present.

digestion Enzyme-catalyzed process by which large, usually insoluble, molecules (foods) are hydrolyzed to form smaller molecules of soluble substances.

dihybrid cross A mating in which the parents differ with respect to the alleles of two loci of interest.

dikaryon (di care' ee ahn) [Gk. *di*: two + *karyon*: kernel] A cell or organism carrying two genetically distinguishable nuclei. Common in fungi.

dioecious (die eesh' us) [Gk.: *di*: two + *oikos*: house] Refers to organisms in which the two sexes are "housed" in two different individuals, so that eggs and sperm are not produced in the same individuals. Examples: humans, fruit flies, date palms. (Contrast with monoecious.)

diploblastic Having two cell layers. (Contrast with triploblastic.)

diploid (dip' loid) [Gk. *diplos*: double] Having a chromosome complement consisting of two copies (homologs) of each chromosome. Designated 2*n*.

diplontic A type of life cycle in which gametes are the only haploid cells and mitosis occurs only in diploid cells. (Contrast with haplontic.)

direct transduction A cell signaling mechanism in which the receptor acts as the effector in

the cellular response. (Contrast with indirect transduction.)

directional selection Selection in which phenotypes at one extreme of the population distribution are favored. (Contrast with disruptive selection, stabilizing selection.)

disaccharide A carbohydrate made up of two monosaccharides (simple sugars).

displacement activity Apparently irrelevant behavior performed by an animal under conflict situations, especially when tendencies to attack and escape are closely balanced.

display A behavior that has evolved to influence the actions of other individuals.

disruptive selection Selection in which phenotypes at both extremes of the population distribution are favored. (Contrast with directional selection; stabilizing selection.)

distal Away from the point of attachment or other reference point. (Contrast with proximal.)

disturbance A short-term event that disrupts populations, communities, or ecosystems by changing the environment.

disulfide bridge The covalent bond between twosulfur atoms (–S—S–) linking to molecules or remote parts of the same molecule.

diverticulum (di ver tik' u lum) [L. *divertere*: turn away] A small cavity or tube that connects to a major cavity or tube.

division A term used by some microbiologists and formerly by botanists, corresponding to the term phylum.

DNA (deoxyribonucleic acid) The fundamental hereditary material of all living organisms. In eukaryotes, stored primarily in the cell nucleus. A nucleic acid using deoxyribose rather than ribose.

DNA chip A small glass or plastic square onto which thousands of single-stranded DNA sequences are fixed. Hybridization of cell-derived RNA or DNA to the target sequences can be performed.

DNA fingerprint An individual's unique pattern of DNA fragments produced by action of restriction endonucleases and separated by electrophoresis.

DNA helicase An enzyme that functions during DNA replication to unwind the double helix.

DNA ligase Enzyme that unites Okazaki fragments of the lagging strand during DNA replication; also mends breaks in DNA strands. It connects pieces of a DNA strand and is used in recombinant DNA technology.

DNA methylation The addition of methyl groups to DNA. Methylation plays a role in regulation of gene expression. Among bacteria, it protects DNA against restriction endonucleases.

DNA polymerase Any of a group of enzymes that catalyze the formation of DNA strands from a DNA template.

DNA sequencing Determining the precise sequence of nucleotides, and thus the sequence of amino acids encoded in a segment DNA.

DNA topoisomerase Enzymes that introduce positive or negative supercoils into the double-stranded DNA of continuous (circular) chromosomes.

docking protein A receptor protein that binds (docks) a ribosome to the membrane of the endoplasmic reticulum by binding the signal sequence attached to a new protein being made at the ribosome.

domain (1) Independent structural elements within proteins that affect the protein's function. Encoded by recognizable nucleotide sequences, a domain often folds separately from the rest of the protein. Similar domains can appear in a variety of different proteins across phylogenetic groups (e.g., "homeobox domain"; "calcium-binding domain.") (2) In phylogenetics, the three monophyletic branches of Life. Members of the three domains (Bacteria, Archaea, and Eukarya) are believed to have been evolving independently of each other for at least a billion years.

dominance In genetics, the ability of one allelic form of a gene to determine the phenotype of a heterozygous individual in which the homologous chromosomes carry both it and a different (recessive) allele. (Contrast with recessive.)

dormancy A condition in which normal activity is suspended, as in some seeds and buds.

dorsal [L. *dorsum*: back] Pertaining to the back or upper surface. (Contrast with ventral.)

dorsal lip In amphibian embryos, the dorsal segment of the blastopore. Also called the "organizer," this region directs the development of nearby embryonic regions.

double fertilization Virtually unique to angiosperms, a process in which the nuclei of two sperm fertilize one egg. One sperm's nucleus combines with the egg nucleus to produce a zygote, while the other combines with the same egg's two polar nuclei to produce the first cell of the triploid endosperm (the tissue that will nourish the growing plant embryo).

double helix In DNA, the natural, right-handed coil configuration of two complementary, antiparallel strands.

duodenum (do' uh dee' num) The beginning portion of the vertebrate small intestine. (Compare with ileum, jejunum.)

duplication A mutation in which a segment of a chromosome is duplicated, often by the attachment of a segment lost from its homolog. (Contrast with deletion.)

dynein [Gk. *dynamis*: power] A protein that plays a part in the movement of eukaryotic flagella and cilia by means of conformational changes.

– E –

ecdysone (eck die' sone) [Gk. *ek*: out of + *dyo*: to clothe] In insects, a hormone that induces molting.

ecological community The species living together at a particular site.

ecological niche (nitch) [L. *nidus*: nest] The functioning of a species in relation to other species and its physical environment.

ecological succession The sequential replacement of one assemblage of populations by another in a habitat following some disturbance.

ecology [Gk. *oikos*: house + *logos*: study] The scientific study of the interaction of organisms with their living (biotic) and nonliving (abiotic) environment.

ecosystem (eek' oh sis tum) The organisms of a particular habitat, such as a pond or forest, together with the physical environment in which they live.

ectoderm [Gk. *ektos*: outside + *derma*: skin] The outermost of the three embryonic tissue layers first delineated during gastrulation. Gives rise to the skin, sense organs, nervous system, etc.

ectotherm [Gk. *ektos*: outside + *thermos*: heat] An animal unable to control its body temperature. (Contrast with endotherm.)

edema (i dee' mah) [Gk. *oidema*: swelling] Tissue swelling caused by the accumulation of fluid.

edge effect The changes in ecological processes in a community caused by physical and biological factors originating in an adjacent community.

effector Any organ, cell, or organelle that moves the organism through the environment or else alters the environment; for example, muscle, exocrine glands, chromatophores.

effector cell A cell responsible for the effector phase of the immune response.

effector phase Stage of the immune response, when cytotoxic T cells attack virus-infected cells, and helper T cells assist B cells to differentiate into plasma cells.

effector protein In cell signaling, a protein responsible for the cellular reponse to a signal transduction pathway.

efferent [L. *ex*: out + *ferre*: to bear] In physiology, conducting outward or away from an organ or structure. (Contrast with afferent.)

egg In all sexually reproducing organisms, the female gamete; in birds, reptiles, and some other vertebrates, a structure within which early embryonic development occurs. (Compare with amniote egg.)

elasticity The property of returning quickly to a former state after a disturbance.

electrocardiogram (ECG or EKG) A graphic recording of electrical potentials from the heart.

electroencephalogram (EEG) A graphic recording of electrical potentials from the brain.

electromagnetic radiation A self-propagating wave that travels though space and has both electrical and magnetic properties.

electromyogram (EMG) A graphic recording of electrical potentials from muscle.

electron A subatomic particle outside the nucleus carrying a negative charge and very little mass.

electron shell The region surrounding the atomic nucleus at a fixed energy level in which electrons orbit.

electron transport The passage of electrons through a series of proteins with a release of energy which may be captured in a concentration gradient or chemical form such as NADH or ATP.

electronegativity The tendency of an atom to attract electrons when it occurs as part of a compound.

electrostatic Pertaining to the attraction and repulsion of negative and positive charges on atoms due to the number and distribution of electrons.

electrophoresis (e lek' tro fo ree' sis) [L. *electrum*: amber + Gk. *phorein*: to bear] A separation technique in which substances are separated from one another on the basis of their electric charges and molecular weights.

element A substance that cannot be converted to simpler substances by ordinary chemical means.

elongation (1) In molecular biology, the addition of monomers to make a longer RNA or protein during synthesis. (2) Growth of a plant axis or cell primarily in the longitudinal direction.

embolus (em' buh lus) [Gk. *embolos*: inserted object; stopper] A circulating blood clot. Blockage of a blood vessel by an embolus or by a bubble of gas is referred to as an embolism. (Contrast with thrombus.)

embryo [Gk. *en*: within + *bryein*: to grow] A young animal, or young plant sporophyte, while it is still contained within a protective structure such as a seed, egg, or uterus.

embryophyte A photosynthetic organism that develops from a protected embryo; synonymous with "land plant." In this book, "plant," when unmodified, is synonymous with "embryophyte."

embryo sac In angiosperms, the female gametophyte. Found within the ovule, it consists of eight or fewer cells, membrane bounded, but without cellulose walls between them.

emergent property A property of a complex system that is not exhibited by its individual component parts.

emigration The deliberate and usually oriented departure of an organism from the habitat in which it has been living.

3' end (3 prime) The end of a DNA or RNA strand that has a free hydroxyl group at the 3' carbon of the sugar (deoxyribose or ribose).

5' end (5 prime) The end of a DNA or RNA strand that has a free phosphate group at the 5' carbon of the sugar (deoxyribose or ribose).

embolism A blockage of a coronary vessel resulting from a circulating blood clot.

endemic (en dem' ik) [Gk. *endemos*: native, dwelling in] Confined to a particular region, thus often having a comparatively restricted distribution.

endergonic reaction A chemical reaction that requires the input of energy in order to proceed. (Contrast with exergonic reaction.)

endocrine gland (en' doh krin) [Gk. *endo*: within + *krinein*: to separate] Any gland, such as the adrenal or pituitary gland of vertebrates, that secretes certain substances, especially hormones, into the body through the blood. The *endocrine system* consists of all cells and glands in the body that produce and release hormones.

endocytosis A process by which liquids or solid particles are taken up by a cell through invagination of the plasma membrane. (Contrast with exocytosis.)

endoderm [Gk. *endo*: within + *derma*: skin] The innermost of the three embryonic tissue layers delineated during gastrulation. Gives rise to the digestive and respiratory tracts and structures associated with them.

endodermis In plants, a specialized cell layer marking the inside of the cortex in roots and some stems. Frequently a barrier to free diffusion of solutes.

endomembrane system Endoplasmic reticulum plus Golgi apparatus; also lysosomes, when present. A system of membranes that exchange material with one another.

endoplasmic reticulum (ER) [Gk. *endo*: within + L. *plasma*: form + L. *reticulum*: net] A system of membranous tubes and flattened sacs found in the cytoplasm of eukaryotes. Exists in two forms: rough ER, studded with ribosomes; and smooth ER, lacking ribosomes.

endorphins, enkephalins Naturally occurring, opiate-like substances in the mammalian brain.

endoskeleton [Gk. *endo*: within + *skleros*: hard] An internal skeleton covered by other, soft body tissues. (Contrast with exoskeleton.)

endosperm [Gk. *endo*: within + *sperma*: seed] A specialized triploid seed tissue found only in angiosperms; contains stored nutrients for the developing embryo.

endosymbiosis [Gk. *endo*: within + *sym*: together + *bios*: life] Two species living together, with one living inside the body (or even the cells) of the other.

endosymbiotic theory The theory that the eukaryotic cell evolved via the engulfing of one prokaryotic cell by another.

endotherm [Gk. *endo*: within + *thermos*: heat] An animal that can control its body temperature by the expenditure of its own metabolic energy. (Contrast with ectotherm.)

end product inhibition A control capacity of some metabolic pathways in which the final product produced inhibits an early enzyme in the pathway.

energetic cost The difference between the energy an animal expends in performing a behavior and the energy it would have expended had it rested.

energy The capacity to do work or move matter against an opposing force. The capacity to accomplish change.

energy budget A quantitative description of all paths of energy exchange between an animal and its environment.

enhancer In eukaryotes, a DNA sequence, lying on either side of the gene it regulates, that stimulates a specific promoter.

enterocoelous development A pattern of development in which the coelum is formed by an outpocketing of the embryonic gut (enteron).

enterokinase (ent uh row kine' ase) An enzyme secreted by the mucosa of the duodenum. It activates the zymogen trypsinogen to create the active digestive enzyme trypsin.

enthalpy The sum of the internal energy of a system; the product of the volume multiplied by the pressure.

entrainment With respect to circadian rhythms, the process whereby the period is adjusted to match the 24-hour environmental cycle.

entropy (en' tro pee) [Gk. *tropein*: to change] A measure of the degree of disorder in any system. Spontaneous reactions in a closed system are always accompanied by an increase in disorder and entropy.

environment Whatever surrounds and interacts with a population, organism, or cell. May be external or internal.

environmental carrying capacity (K) The number of individuals in a population that the resources of a habitat can support.

enzyme (en' zime) [Gk. *zyme*: to leaven (as in yeast bread)] A protein, on the surface of which are chemical groups so arranged as to make the enzyme a catalyst for a chemical reaction.

enzyme–substrate complex The complex that forms when an enzyme binds to its substrate(s).

eosinophils Phagocytic white blood cells that attack multicellular parasites once they have been coated with antibodies.

epi- [Gk.: upon, over] A prefix used to designate a structure located on top of another; for example: epidermis, epiphyte.

epiblast The upper or overlying portion of the avian blastula which is joined to the hypoblast at the margins of the blastodisc.

epiboly The movement of cells over the surface of the blastula toward the forming blastopore.

epicotyl (epp' i kot' il) [Gk. *epi*: over + *kotyle*: something hollow] That part of a plant embryo or seedling that is above the cotyledons.

epidermis [Gk. *epi*: over + *derma*: skin] In plants and animals, the outermost cell layers. (Only one cell layer thick in plants.)

epididymis (epuh did' uh mus) [Gk. *epi*: over + *didymos*: testicle] Coiled tubules in the testes that store sperm and conduct sperm from the seminiferous tubules to the vas deferens.

epinephrine (ep i nef' rin) [Gk. *epi*: over + *nephros*: kidney] The "fight or flight" hormone produced by the medulla of the adrenal gland; it also functions as a neurotransmitter. (Also known as adrenaline.)

epiphyte (ep' e fyte) [Gk. *epi*: over + *phyton*: plant] A specialized plant that grows on the surface of other plants but does not parasitize them.

epistasis Interaction between genes in which the presence of a particular allele of one gene determines whether another gene will be expressed.

epithelium Sheets of cells that line or cover organs, make up tubules, and cover the surface of the body.

equatorial plate In a cell undergoing mitosis, the region in the middle of a cell where the centromeres will align during metaphase.

equilibrium Any state of balanced opposing forces and no net change.

ER *See* endoplasmic reticulum.

erythrocyte (ur rith' row site) [Gk. *erythros*: red + *kytos*: container] A red blood cell.

erythropoietin A hormone produced by the kidney in response to lack of oxygen. It stimulates the production of red blood cells.

esophagus (i soff' i gus) [Gk. *oisophagos*: gullet] That part of the gut between the pharynx and the stomach.

essential element A mineral nutrient element required in order for a seed to develop and complete the plant's life cycle, producing viable new seeds.

essential acids Amino acids or fatty acids that an animal cannot synthesize for itself.

ester linkage A condensation (water-releasing) reaction in which the carboxyl group of a fatty acid reacts with the hydroxyl group of an alcohol. Lipids are formed in this way.

estivation (ess tuh vay' shun) [L. *aestivalis*: summer] A state of dormancy and hypometabolism that occurs during the summer; usually a means of surviving drought and/or intense heat. (Contrast with hibernation.)

estrogen Any of several steroid sex hormones; produced chiefly by the ovaries in mammals.

estrus (es' truss) [L. *oestrus*: frenzy] The period of heat, or maximum sexual receptivity, in some female mammals. Ordinarily, the estrus is also the time of release of eggs in the female.

ethology An approach to the study of animal behavior developed in Europe. Emphasizes the causes of the evolution of behavior.

ethylene One of the plant growth hormones, the gas $H_2C=CH_2$. Involved in fruit ripening and other growth and developmental responses.

euchromatin Chromatin that is diffuse and non-staining during interphase; may be transcribed. (Contrast with heterochromatin.)

eukaryotes (yew car' ree oats) [Gk. *eu*: true + *karyon*: kernel or nucleus] Organisms whose cells contain their genetic material inside a nucleus. Includes all life other than the viruses, archaea, and bacteria.

eutrophication (yoo trofe' ik ay' shun) [Gk. *eu*: truly + *trephein*: to flourish] The addition of nutrient materials to a body of water, resulting in changes in ecological processes and species composition therein.

evaporation The transition of water from the liquid to the gaseous phase.

evolution Any gradual change. Organic or Darwinian evolution, often referred to as evolution, is any genetic and resulting phenotypic change in organisms from generation to generation. (*See* macroevolution, microevolution; compare with speciation.)

evolutionary agent Any factor that influences the direction and rate of evolutionary change.

evolutionarily conserved Refers to traits that have evolved very slowly and are similar or even identical in individuals of highly divergent groups.

evolutionary radiation The proliferation of species within a single evolutionary lineage.

evolutionary reversal The reappearance of an ancestral trait in a group that had previously acquired a derived trait.

excision repair The removal and damaged DNA and its replacement by the appropriate nucleotides.

excited state The state of an atom or molecule when, after absorbing energy, it has more energy than in its normal, ground state. (Compare with ground state.)

excretion Release of metabolic wastes by an organism.

exergonic reaction A reaction in which free energy is released. (Contrast with endergonic reaction.)

exocrine gland (eks' oh krin) [Gk. *exo*: outside + *krinein*: to separate]   Any gland, such as a salivary gland, that secretes to the outside of the body or into the gut. (Contrast with endocrine gland.)

exocytosis   A process by which a vesicle within a cell fuses with the plasma membrane and releases its contents to the outside. (Contrast with endocytosis.)

exon   A portion of a DNA molecule, in eukaryotes, that codes for part of a polypeptide. (Contrast with intron.)

exoskeleton (eks' oh skel' e ton) [Gk. *exos*: outside + *skleros*: hard]   A hard covering on the outside of the body to which muscles are attached. (Contrast with endoskeleton.)

exotoxins   Highly toxic proteins released by living, multiplying bacteria.

expanding triplet repeat   A three-base-pair sequence in a human gene that is unstable and can be repeated a few to hundreds of times. Often, the more the repeats, the less the activity of the gene involved. Expanding triplet repeats occur in some human diseases such as Huntington's disease and fragile-X syndrome.

experiment   A testing process to support or disprove hypotheses and to answer questions. The basis of the scientific method. *See* comparative experiment, controlled experiment.

expiratory reserve volume   The amount of air that can be forcefully exhaled beyond the normal tidal expiration. (Compares with inspiratory reserve volume, tidal volume, vital capacity.)

exploitation competition   Competition in which individuals reduce the quantities of their shared resources. (Compare with interference competition.)

exponential growth   Growth, especially in the number of organisms in a population, which is a geometric function of the size of the growing entity: the larger the entity, the faster it grows. (Contrast with logistic growth.)

expression vector   A DNA vector, such as a plasmid, that carries a DNA sequence that includes the adjacent sequences for its expression into mRNA and protein in a host cell.

expressivity   The degree to which a genotype is expressed in the phenotype; may be affected by the environment.

extensor   A muscle the extends an appendage.

extinction   The termination of a lineage of organisms.

extrinsic protein   A membrane protein found only on the surface of the membrane. (Contrast with intrinsic protein.)

extracellular matrix   In animal tissues, a material of heterogeneous composition surrounding cells and performing many functions including adhesion of cells.

extraembryonic membranes   Four membranes that support but are not part of the developing embryos of reptiles, birds, and mammals, defining these groups phylogenetically as amniotes. (*See* amnion, allantois, chorion, and yolk sac.)

- F -

F₁ (first filial generation)   The immediate progeny of a parental (P) mating.

F₂ (second filial generation)   The immediate progeny of a mating between members of the F₁ generation.

facilitated diffusion   Passive movement through a membrane involving a specific carrier protein; does not proceed against a concentration gradient. (Contrast with active transport, diffusion.)

facultative anaerobes (alternatively, facultative aerobes)   Prokaryotes that can shift their metabolism between anaerobic and aerobic operations depending on the presence or absence of O_2.

FAD   *See* flavin adenine dinucleotide.

fast-twitch fibers   Skeletal muscle fibers that can generate high tension rapidly, but fatigue rapidly ("sprinter" fibers). Characterized by an abundance of enzymes of glycolysis.

fat   A triglyceride that is solid at room temperature. (Contrast with oil.)

fate   In an embryo, the type of cell that a particular cell will become in the adult.

fate map   A diagram of the blastula showing which cells (blastomeres) are "fated" to contribute to specific tissues and organs in the mature body.

fatty acid   A molecule with a long hydrocarbon tail and a carboxyl group at the other end. Found in many lipids.

fauna (faw' nah)   All of the animals found in a given area. (Contrast with flora.)

feces [L. *faeces*: dregs]   Waste excreted from the digestive system.

feedback information   Information relevant to the rate of a process that can be used by a control system to regulate that process at a particular level.

feedforward information   Information that can be used to alter the setpoint of a regulatory process.

fermentation (fur men tay' shun) [L. *fermentum*: yeast]   The anaerobic degradation of a substance such as glucose to smaller molecules with the extraction of energy.

ferredoxin   A protein containing iron that mediates the transfer of electrons in a number of pathways, including the light reactions of photosynthesis.

fertility factor (F factor)   A plasmid that confers on a bacterium the ability to act as a DNA donor during conjugation.

fertilization   Union of gametes. Also known as syngamy.

fertilization membrane   A membrane surrounding an animal egg which becomes rapidly raised above the egg surface within seconds after fertilization, serving to prevent entry of a second sperm.

fetus   Medical and legal term for the latter stages of a developing human embryo, from about the eighth week of pregnancy (the point at which all major organ systems have formed) to the moment of birth.

fiber   An elongated, tapering cell of flowering plants, usually with a thick cell wall. Serves a support function.

fibrin   A protein that polymerizes to form long threads that provide structure to a blood clot.

fibrinogen   A circulating protein that can be stimulated to fall out of solution and provide the structure for a blood clot.

Fick's law of diffusion   An equation that describes the factors that determine the rate of diffusion of a molecule from an area of higher concentration to an area of lower concentration.

filter feeder   An organism that feeds upon much smaller organisms, that are suspended in water or air, by means of a straining device.

filtration   In the excretory physiology of some animals, the process by which the initial urine is formed; water and most solutes are transferred into the excretory tract, while proteins are retained in the blood or hemolymph.

first law of thermodynamics   Energy can be neither created nor destroyed.

fission   Reproduction of a prokaryote by division of a cell into two comparable progeny cells.

fitness   The contribution of a genotype or phenotype to the genetic composition of subsequent generations, relative to the contribution of other genotypes or phenotypes. (*See* also inclusive fitness.)

flagellum (fla jell' um) (plural: flagella) [L. *flagellum*: whip]   Long, whiplike appendage that propels cells. Prokaryotic flagella differ sharply from those found in eukaryotes.

flavin adenine dinucleotide (FAD)   A coenzyme involved in redox reactions and containing the vitamin riboflavin (B_2).

flexor   A muscle that flexes an appendage.

flora (flore' ah)   All of the plants found in a given area. (Contrast with fauna.)

floral meristem   Meristem that forms the sexual parts of flowering plants (sepals, petals, stamens, and carpels).

florigen   A plant hormone involved in the conversion of a vegetative shoot apex to a flower.

flower   The total reproductive structure of an angiosperm; its basic parts include the calyx, corolla, stamens, and carpels.

fluid mosaic model   A molecular model for the structure of biological membranes consisting of a fluid phospholipid bilayer in which suspended proteins are free to move in the plane of the bilayer.

fluorescence   The emission of a photon of visible light by an excited atom or molecule.

follicle [L. *folliculus*: little bag]   In female mammals, an immature egg surrounded by nutritive cells.

follicle-stimulating hormone   A gonadotropic hormone produced by the anterior pituitary.

food chain   A portion of a food web, most commonly a simple sequence of prey species and the predators that consume them.

food vacuole   Membrane enclosed structure formed by phagocytosis in which engulfed food particles are digested by the action of lysosomal enzymes.

food web   The complete set of food links between species in a community; a diagram indicating which ones are the eaters and which are eaten.

forb   Any broad-leaved herbaceous plant. Especially applied to such plants growing in grasslands.

fossil Any recognizable structure originating from an organism, or any impression from such a structure, that has been preserved over geological time.

founder effect Random changes in allele frequencies resulting from establishment of a population by a very small number of individuals.

fovea [L. *fovea*; a small pit] In the vertebrate retina, the area of most distinct vision.

frame-shift mutation A mutation resulting from the addition or deletion of one or two consecutive base pairs in the DNA sequence of a gene, resulting in misreading mRNA during translation and production of a nonfunctional protein. (Compare with missense substitution, nonsense substitution, synonymous substitution.)

Frank–Starling law The stroke volume of the heart increases with increased return of blood to the heart.

free energy That energy which is available for doing useful work, after allowance has been made for the increase or decrease of disorder.

freeze-fracturing Method of tissue preparation for transmission and scanning electron microscopy in which a tissue is frozen and a knife is then used to crack open the tissue; the fracture often occurs in the path of least resistance, within a membrane.

frequency In population genetics, the proportion of all alleles or genotypes in a population composed of a particular allele or genotype.

frequency-dependent selection Selection that changes in intensity with the proportion of individuals in a population having the trait.

fruit In angiosperms, a ripened and mature ovary (or group of ovaries) containing the seeds. Sometimes applied to reproductive structures of other groups of plants.

fruiting body Any structure that bears spores.

functional genomics The assignment of functional roles to genes first identified by sequencing entire genomes.

functional group A characteristic combination of atoms that contribute specific properties when attached to larger molecules.

functional mRNA Eukaryotic mRNA that has been modified after transcription by the removal of introns and the addition of a 5' cap and a 3' poly(A) tail.

– G –

G cap A chemically modified GTP added to the 5' end of mRNA; facilitates binding of mRNA to ribosome and prevents mRNA breakdown.

G$_1$ phase In the cell cycle, the gap between the end of mitosis and the onset of the S phase.

G$_2$ phase In the cell cycle, the gap between the S (synthesis) phase and the onset of mitosis.

G protein A membrane protein involved in signal transduction; characterized by binding GDP or GTP.

gametangium (gam uh tan' gee um) [Gk. *gamos*: marriage + *angeion*: vessel] Any plant or fungal structure within which a gamete is formed.

gamete (gam' eet) [Gk. *gamete/gametes*: wife, husband] The mature sexual reproductive cell: the egg or the sperm.

gametogenesis (ga meet' oh jen' e sis) The specialized series of cellular divisions that leads to the production of sex cells (gametes). (Contrast with oogenesis and spermatogenesis.)

gametophyte (ga meet' oh fyte) In plants and photosynthetic protists with alternation of generations, the multicellular haploid phase that produces the gametes. (Contrast with sporophyte.)

ganglion (gang' glee un) [Gk.: tumor] A cluster of neurons that have similar characteristics or function.

gap genes During insect development, the first step of segmentation genes to act organizing the anterior-posterior axis.

gap junction A 2.7-nanometer gap between plasma membranes of two animal cells, spanned by protein channels. Gap junctions allow chemical substances or electrical signals to pass from cell to cell.

gas exchange In animals, the process of taking up oxygen from the environment and releasing carbon dioxide to the environment.

gastrovascular cavity Serving for both digestion (gastro) and circulation (vascular); in particular, the central cavity of the body of jellyfish and other cnidarians.

gastrula (gas' true luh) [Gk. *gaster*: stomach] An embryo forming the characteristic three cell layers (ectoderm, endoderm, and mesoderm) that give rise to all of the major tissue systems of the adult animal.

gastrulation Development of a blastula into a gastrula.

gated channel A membrane protein that opens and closes in response to binding of specific molecules or to changes in membrane potential. When open, it allows specific ions to move across the membrane.

gene [Gk. *genes*: to produce] A unit of heredity. Used here as the unit of genetic function which carries the information for a single polypeptide or RNA.

gene amplification Creation of multiple copies of a particular gene, allowing the production of large amounts of the RNA transcript (as in rRNA synthesis in oocytes).

gene cloning Formation of a clone of bacteria or yeast containing a particular foreign gene.

gene family A set of identical (or once-identical) genes derived from a single parent gene; need not be on the same chromosomes. The vertebrate globin genes constitute a classic example of a gene family.

gene flow Exchange of genes between different species (an extreme case referred to as hybridization) or between different populations of the same species caused by migration followed by breeding.

gene-for-gene resistance A mechanism for resistance to pathogens, in which resistance is triggered by the specific interaction of the products of pathogens' *Avr* genes and plants' *R* genes.

gene frequency See allele frequency.

gene library All of the cloned DNA fragments generated by action of a restriction endonuclease on a genome or chromosome.

gene pool All of the different alleles of all of the genes existing in all individual of a population.

gene therapy Treatment of a genetic disease by providing patients with cells containing functioning alleles of the genes that are nonfunctional in their bodies.

generative cell In a pollen tube, a haploid nucleus that undergoes mitosis to produce the two sperm nuclei that participate in double fertilization. (Contrast with tube cell.)

genetic code The set of instructions, in the form of nucleotide triplets, that translate a linear sequence of nucleotides in mRNA into a linear sequence of amino acids in a protein.

genetic drift Changes in gene frequencies from generation to generation as a result of random (chance) processes.

genetic map The positions of genes along a chromosome as revealed by recombination frequencies.

genetic screening The application of medical tests to determine whether an individual carries a specific allele.

genetic stochasticity Random variation in the frequencies of alleles and genotypes in a population over time. (Compare with demographic stochasticity.)

genetics The study of the structure, functioning, and inheritance of genes, the units of hereditary information.

genome (jee' nome) All the genes in a complete haploid set of chromosomes. (Compare with proteome.)

genomics The study of entire sets of genes and their interactions.

genomic imprinting When a given gene's phenotype is determined by whether that gene is inherited from the male or the female parent.

genotype (jean' oh type) [Gk. *gen*: to produce + *typos*: impression] An exact description of the genetic constitution of an individual, either with respect to a single trait or with respect to a larger set of traits. (Contrast with phenotype.)

genus (jean' us) (plural: genera) [Gk. *genos*: stock, kind] A group of related, similar species recognized by taxonomists with a distinct name used in binomial nomenclature.

geotropism See gravitropism.

germ cell [L. *germen*: to beget] A reproductive cell or gamete of a multicellular organism. (Contrast with somatic cell.)

germ line mutation A change in the genetic material in a germ cell; such mutations are heritable. (Contrast with somatic mutation.)

germ layers The three embryonic tissue layers formed during gastrulation (ectoderm, mesoderm, endoderm).

germination Sprouting of a seed or spore.

gestation (jes tay' shun) [L. *gestare*: to bear] The period during which the embryo of a mammal develops within the uterus. Also known as pregnancy.

gibberellin (jib er el' lin) A class of plant growth substances playing roles in stem elongation, seed germination, flowering of certain plants, etc. Named for the fungus *Gibberella*.

gill An organ for gas exchange in aquatic organisms.

gill arch A skeletal structure that supports gill filaments and the blood vessels that supply them.

gizzard (giz′ erd) [L. *gigeria*: cooked chicken parts] A muscular port of the stomach of birds that grinds up food, sometimes with the aid of fragments of stone.

gland An organ or group of cells that produces and secretes one or more substances.

glans penis Sexually sensitive tissue at the tip of the penis.

glia (glee′ uh) [Gk. *glia*: glue] Cells, found only in the nervous system, that do not conduct action potentials.

glomerular filtration rate (GFR) The rate at which the blood is filtered in the glomeruli of the kidney.

glomerulus (glo mare′ yew lus) [L. *glomus*: ball] Sites in the kidney where blood filtration takes place. Each glomerulus consists of a knot of capillaries served by afferent and efferent arterioles.

glucagon Hormone produced by alpha cells of the pancreatic islets of Langerhans. Glucagon stimulates the liver to break down glycogen and release glucose into the circulation.

glucocorticoids Steroid hormones produced by the adrenal cortex. Secreted in response to ACTH, they inhibit glucose uptake by many tissues in addition to mediating other stress responses.

gluconeogenesis The biochemical synthesis of glucose from other substances, such as amino acids, lactate, and glycerol.

glucose [Gk. *gleukos*: sugar, sweet] The most common monosaccharide; the monomer of the polysaccharides starch, glycogen, and cellulose.

glycerol (gliss′ er ole) A three-carbon alcohol with three hydroxyl groups; a component of phospholipids and triglycerides.

glycogen (gly′ ko jen) An energy storage polysaccharide found in animals and fungi; a branched-chain polymer of glucose, similar to starch.

glycolipid A lipid to which sugars are attached.

glycolysis (gly kol′ li sis) [Gk. *gleukos*: sugar + *lysis*: break apart] The enzymatic breakdown of glucose to pyruvic acid. One of the evolutionarily oldest of the cellular energy-yielding mechanisms.

glycoprotein A protein to which sugars are attached.

glycosidic linkage Bond between carbohydrate (sugar) molecules through an intervening oxygen atom (–O–).

glycosylation The addition of carbohydrates to another type of molecule, such as a protein.

glyoxysome (gly ox′ ee soam) An organelle found in plants, in which stored lipids are converted to carbohydrates.

Golgi apparatus (goal′ jee) A system of concentrically folded membranes found in the cytoplasm of eukaryotic cells; functions in secretion from cell by exocytosis.

gonad (go′ nad) [Gk. *gone*: seed] An organ that produces sex cells in animals: either an ovary (female gonad) or testis (male gonad).

gonadotropin A hormone that stimulates the gonads.

gonadotropin-releasing hormone (GnRH) Hypothalamic hormone that stimulates the anterior pituitary to secrete growth hormone.

Gondwana The large southern land mass that existed from the Cambrian (540 mya) to the Jurassic (138 mya). Present-day remnants are South America, Africa, India, Australia, and Antarctica.

graft Tissue artificially and viably transplanted from one organism to another. In agriculture, refers to the transfer of bud or stem segment from one plant onto another plant as a form of asexual reproduction.

Gram stain A differential purple stain useful in characterizing bacteria. The peptidoglycan-rich cell walls of Gram-positive bacteria stain purple; cell walls of Gram-negative bacteria generally stain orange.

granum (plural: grana) Within a chloroplast, a stack of thylakoids.

gravitropism A directed plant growth response to gravity.

grazer An animal that eats the vegetative tissues of herbaceous plants.

green gland An excretory organ of crustaceans.

greenhouse effect The heating of Earth's atmosphere by gases such as water vapor, carbon dioxide, and methane; such greenhouse gases are transparent to sunlight but opaque to heat; thus sunlight-engendered heat builds up at Earth's surface and cannot be dissipated into the atmosphere.

gross primary production The total energy captured by plants growing in a particular area.

gross primary productivity (GPP) The rate of assimilation of energy by plants growing in a particular area.

ground meristem That part of an apical meristem that gives rise to the ground tissue system of the primary plant body.

ground state The lowest energy state of an atom of molecule. (Compare with excited state.)

ground tissue system Those parts of the plant body not included in the dermal or vascular tissue systems. Ground tissues function in storage, photosynthesis, and support.

group transfer The exchange of atoms between molecules.

growth Irreversible increase in volume (an accurate definition, but at best a dangerous oversimplification).

growth hormone A peptide hormone of the anterior pituitary that stimulates many anabolic processes.

guanine (G) (gwan′ een) A nitrogen-containing base found in DNA, RNA, and GTP.

guard cells In plants, specialized, paired epidermal cells that surround and control the opening of a stoma (pore). *See* stoma.

gut An animal's digestive tract.

guttation The extrusion of liquid water through openings in leaves, caused by root pressure.

gymnosperm (jim′ no sperm) [Gk. *gymnos*: naked + *sperma*: seed] A plant, such as a pine or other conifer, whose seeds do not develop within an ovary (hence, the seeds are "naked").

gyrus The raised or ridged portion of the convoluted surface of the brain. (Contrast with sulcus.)

– H –

habitat The environment in which an organism lives.

habituation (ha bich′ oo ay shun) The simplest form of learning, in which an animal presented with a stimulus without reward or punishment eventually ceases to respond.

hair cell A type of mechanoreceptor in animals. Detects sound waves and other forms of motion in air or water.

half-life The time required for half of a sample of a radioactive isotope to decay to its stable, non-radioactive form, or for a drug or other substance to reach half its initial dosage.

halophyte (hal′ oh fyte) [Gk. *halos*: salt + *phyton*: plant] A plant that grows in a saline (salty) environment.

haploid (hap′ loid) [Gk. *haploeides*: single] Having a chromosome complement consisting of just one copy of each chromosome; designated $1n$ or n. (Contrast with diploid.)

haplontic A type of life cycle in which the zygote is the only diploid cell and mitosis occurs only in haploid cells. (Contrast with diplontic.)

Hardy–Weinberg equilibrium The allele frequency at a given locus in a sexually reproducing population that is not being acted on by agents of evolution; the conditions that would result in no evolution in a population.

haustorium (haw stor′ ee um) [L. *haustus*: draw up] A specialized hypha or other structure by which fungi and some parasitic plants draw nutrients from a host plant.

Haversian systems Units of organization in compact bone that reflect the action of intercommunicating osteoblasts.

heat of vaporization The energy that must be supplied to convert a molecule from a liquid to a gas at its boiling point.

heat-shock proteins Chaperone proteins expressed in cells exposed to high or low temperatures or other forms of environmental stress.

helical Shaped like a screw or spring; this shape occurs in DNA and proteins.

helper T cells (T_H) T cells that participate in the activation of B cells and of other T cells; targets of the HIV-I virus, the agent of AIDS. (Contrast with cytotoxic T cells.)

hematocrit (heme at′ o krit) [Gk. *heaema*: blood + *krites*: judge] The proportion of 100 cc of blood that consists of red blood cells.

hemizygous (hem′ ee zie′ gus) [Gk. *hemi*: half + *zygotos*: joined] In a diploid organism, having only one allele for a given trait, typically the case for X-linked genes in male mammals and Z-linked genes in female birds. (Contrast with homozygous, heterozygous.)

hemoglobin (hee′ mo glow bin) [Gk. *heaema*: blood + L. *globus*: globe] Oxygen-transporting protein found in the red blood cells of vertebrates (and found in some invertebrates).

Hensen's node In avian embryos, a structure at the anterior end of the primitive groove; determines the fates of cells passing over it during gastrulation.

hepatic (heh pat′ ik) [Gk. *hepar*: liver] Pertaining to the liver.

hepatic duct   Conveys bile from the liver to the gallbladder.

herbivore (ur' bi vore) [L. *herba*: plant + *vorare*: to devour]   An animal that eats plant tissues. (Contrast with carnivore, detritivore, omnivore.)

heritable   Able to be inherited; in biology, refers to genetically influenced traits.

hermaphroditism (her maf' row dite ism) [Gk. Hermes (messenger god) + Aphrodite (goddess of love)]   The coexistence of both female and male sex organs in the same organism.

hertz (abbreviated Hz)   Cycles per second.

hetero- [Gk.: *heteros*: other, different]   A prefix specifying that two or more different conditions are involved; for example, heterotroph, heterozygous.

heterochromatin   Chromatin that retains its coiling during interphase; generally not transcribed. (Contrast with euchromatin.)

heterochrony   Comparing different species, an alternation in the timing of developmental events, leading to different results in the adult.

heterocyst   A large, thick-walled cell in the filaments of certain cyanobacteria; performs nitrogen fixation.

heterogeneous nuclear RNA (hnRNA)   The product of transcription of a eukaryotic gene, including transcripts of introns.

heterokaryon   In fungi, hypha containing two genetically different nuclei.

heteromorphic (het' er oh more' fik) [Gk. *heteros*: different + *morphe*: form]   Having a different form or appearance, as two heteromorphic life stages of a plant. (Contrast with isomorphic.)

heterosporous (het' er os' por us)   Producing two types of spores, one of which gives rise to a female megaspore and the other to a male microspore. (Contrast with homosporous.)

heterosis   Situation in which heterozygous genotypes are superior to homozygous genotypes with respect to growth, survival, or fertility. Also called hybrid vigor.

heterotherm   An animal that regulates its body temperature at a constant level at some times but not others, such as a hibernator.

heterotroph (het' er oh trof) [Gk. *heteros*: different + *trophe*: food]   An organism that requires preformed organic molecules as food. (Contrast with autotroph.)

heterotypic   Referring to adhesion of cells of different types. (Contrast with homotypic.)

heterozygous (het' er oh zie' gus) [Gk. *heteros*: different + *zygotos*: joined]   Of a diploid organism having different alleles of a given gene on the pair of homologs carrying that gene. (Contrast with homozygous.)

hexose [Gk. *hex*: six]   A sugar containing six carbon atoms.

hibernation [L. *hibernum*: winter]   The state of inactivity of some animals during winter; marked by a drop in body temperature and metabolic rate.

hierarchical sequencing   An approach to DNA sequencing in which markers are mapped and DNA sequences are aligned by matching overlapping sites of known sequence.

highly repetitive DNA   Short DNA sequences present in millions of copies in the genome, next to each other (in tandem). In reassociation experiments, denatured highly repetitive DNA reanneals very quickly.

hindbrain   The region of the developing vertebrate brain that gives rise to the medulla, pons, and cerebellum.

hippocampus [Gr.: sea horse]   A part of the forebrain that takes part in long-term memory formation.

histamine (hiss' tah meen)   A substance released by damaged tissue, or by mast cells in response to allergens. Histamine increases vascular permeability, leading to edema (swelling).

histology [Gk. *histos*: weaving]   The study of tissues.

histone   Any one of a group of basic proteins forming the core of a nucleosome, the structural unit of a eukaryotic chromosome. (Compare with nucleosome.)

hierarchical sequencing   An approach to DNA sequencing in which markers are mapped and DNA sequences are aligned by matching overlapping sites of known sequence.

hnRNA   See heterogeneous nuclear RNA.

homeobox   A 180-base-pair segment of DNA found in certain homeotic genes; regulates the expression of other genes and thus controls large-scale developmental processes.

homeostasis (home' ee o sta' sis) [Gk. *homos*: same + *stasis*: position]   The maintenance of a steady state, such as a constant temperature or a stable social structure, by means of physiological or behavioral feedback responses.

homeotherm (home' ee o therm) [Gk. *homos*: same + *thermos*: heat]   An animal that maintains a constant body temperature by its own internal heating and cooling mechanisms. (Contrast with heterotherm, poikilotherm.)

homeotic genes (home ee ot' ic)   Genes that determine the developmental fate of entire segments of an animal.

homeotic mutations   Mutations in homeotic genes that drastically alter the characteristics of a particular body segment, giving it the characteristics of other segments (as when wings grow from a *Drosophila* thoracic segment that should have produced legs).

homing   The ability to return over long distances to a specific site.

homo- [Gk. *homos*: same]   Prefix indicating two or more similar conditions, structures, or processes. (Contrast with hetero-.)

homolog (home' o log') [Gk. *homos*: same + *logos*: word]   In cytogenetics, one of a pair (or larger set) of chromosomes having the same overall genetic composition and sequence. In diploid organisms, each chromosome inherited from one parent is matched by an identical (except for mutational changes) chromosome—its homolog—from the other parent.

homology (ho mol' o jee) [Gk. *homologia*: of one mind; agreement]   A similarity between two or more structures that is due to inheritance from a common ancestor. The structures are said to be homologous, and each is a homolog of the others. (Contrast with analogy.)

homoplasy (home' uh play zee) [Gk. *homos*: same + *plastikos*: shape, mold]   The presence in multiple groups of a trait that is not inherited from the common ancestor of those groups. Can result from convergent evolution, evolutionary reversal, or parallel evolution.

homosporous   Producing a single type of spore that gives rise to a single type of gametophyte, bearing both female and male reproductive organs. (Contrast with heterosporous.)

homotypic   Referring to adhesion of cells of the same type. (Contrast with heterotypic.)

homozygous (home' oh zie' gus) [Gk. *homos*: same + *zygotos*: joined]   In a diploid organism, having identical alleles of a given gene on both homologous chromosomes. An individual may be a homozygote with respect to one gene and a heterozygote with respect to another. (Contrast with heterozygous.)

hormone (hore' mone) [Gk. *hormon*: to excite, stimulate]   A substance produced in minute amount at one site in a multicellular organism and transported to another site where it acts on target cells.

host   An organism that harbors a parasite or symbiont and provides it with nourishment.

Hox genes   Conserved homeotic genes found in vertebrates, *Drosophila*, and other animal groups. Hox genes contain the homeobox domain and specify pattern and axis formation in these animals.

human chorionic gonadotropin (hCG)   A hormone secreted by the placenta which sustains the corpus luteum and helps maintain pregnancy.

Human Genome Project   An effort to determine the DNA sequence of the entire human genome, understand the structures and functions of the genes, make comparisons with other organisms, and understand the social implications of this information.

humoral immune response   The part of the immune system mediated by B cells that produce circulating antibodies active against extracellular bacterial and viral infections.

humus (hew' muss)   The partly decomposed remains of plants and animals on the surface of a soil.

hyaluronidase (high' uh loo ron' uh dase)   An enzyme that digests proteoglycans. In sperm cells, it digests the coatings surrounding an egg so the sperm can enter.

hybrid (high' brid) [L. *hybrida*: mongrel]   (1) The offspring of genetically dissimilar parents. (2) In molecular biology, a double helix formed of nucleic acids from different sources.

hybridize   To combine the genetic material of two distinct species or of two distinguishable populations within a species.

hybrid vigor   See heterosis.

hybridoma   A cell produced by the fusion of an antibody-producing cell with a myeloma cell; it produces monoclonal antibodies.

hybrid zone   A narrow zone where two populations interbreed, producing hybrid individuals.

hydrocarbon   A compound containing only carbon and hydrogen atoms.

hydrogen bond   A weak electrostatic bond which arises from the attraction between the slight positive charge on a hydrogen atom and a slight negative charge on a nearby oxygen or nitrogen atom.

hydrological cycle   The movement of water from the oceans to the atmosphere, to the soil, and back to the oceans.

hydrolysis (high drol' uh sis) [Gk. *hydro*: water + *lysis*: break apart]   A chemical reaction that breaks a bond by inserting the components of water: $AB + H_2O \rightarrow AH + BOH$.

hydrophilic (high dro fill' ik) [Gk. *hydro*: water + *philia*: love]   Having an affinity for water. (Contrast with hydrophobic.)

hydrophobic (high dro foe' bik) [Gk. *hydro*: water + *phobia*: fear]   Having no affinity for water. Uncharged and nonpolar groups of atoms are hydrophobic; for example, fats and the side chain of the amino acid phenylalanine. (Contrast with hydrophilic.)

hydrostatic pressure   Pressure generated by compression of liquid in a confined space. Generated in plants, fungi, and some protists with cell walls by the osmotic uptake of water. Generated in animals with closed circulatory systems by the beating of a heart.

hydrostatic skeleton   The incompressible internal liquids of some animals that transfer forces from one part of the body to another when acted upon by the surrounding muscles.

hydroxyl group   The —OH group found on alcohols and sugars.

hyper- [Gk. *hyper*: above, over]   Prefix indicating above, higher, more.

hyperpolarization   A change in the resting potential of a membrane so the inside of a cell becomes more electronegative. (Contrast with depolarization.)

hypersensitive response   A defensive response of plants to microbial infection; it results in a "dead spot."

hypertension   High blood pressure.

hypertonic   Having a greater solute concentration. Said of one solution compared to another. (Contrast with hypotonic, isotonic.)

hypha (high' fuh) (plural: hyphae) [Gk. *hyphe*: web]   In the fungi and oomycetes, any single filament.

hypo- [Gk. *hypo*: beneath, under]   Prefix indicating underneath, below, less.

hypoblast   The lower tissue portion of the avian blastula which is joined to the epiblast at the margins of the blastodisc.

hypocotyl [Gk. *hypo*: beneath + *kotyledon*: hollow space]   That part of the embryonic or seedling plant shoot that is below the cotyledons.

hypothalamus   The part of the brain lying below the thalamus; it coordinates water balance, reproduction, temperature regulation, and metabolism.

hypothesis   A tentative answer to a question, from which testable predictions can be generated. (Contrast with theory.)

hypotonic   Having a lesser solute concentration. Said of one solution in comparing it to another. (Contrast with hypertonic, isotonic.)

– I –

ileum   The final segment of the small intestine.

imaginal disc [L. *imagos*: image, form]   In insect larvae, groups of cells that develop into specific adult organs.

imbibition   Water uptake by a seed; first step in germination.

immediate hypersensitivity   A rapid, extensive immune reaction against an allergen involving IgE and histamine release. (Contrast with delayed hypersensitivity.)

immune system [L. *immunis*: exempt from]   A system in vertebrates that recognizes and attempts to eliminate or neutralize foreign substances (e.g., bacteria, viruses, pollutants).

immunization   The deliberate introduction of antigen to bring about an immune response.

immunoassay   The use of labeled antibodies to measure the concentration of an antigen in a sample.

immunoglobulins   A class of proteins, with a characteristic structure, active as receptors and effectors in the immune system.

immunological memory   The capacity to more rapidly and massively respond to a second exposure to an antigen than occurred on first exposure.

immunological tolerance   A mechanism by which an animal does not mount an immune response to the antigenic determinants of its own macromolecules.

implantation   The process by which the early mammalian embryo becomes attached to and embedded in the lining of the uterus.

imprinting   (1) In genetics, the differential modification of a gene depending on whether it is present in a male or a female. (2) In animal behavior, a rapid form of learning in which an animal comes to make a particular response, which is maintained for life, to some object or other organism.

inbreeding   Breeding among close relatives.

inclusive fitness   The sum of an individual's genetic contribution to subsequent generations both via production of its own offspring and via its influence on the survival of relatives who are not direct descendants.

incomplete dominance   Condition in which the heterozygous phenotype is intermediate between the two homozygous phenotypes.

incomplete metamorphosis   Insect development in which changes between instars are gradual.

incus (in' kus) [L. *incus*: anvil]   The middle of the three bones that conduct movements of the eardrum to the oval window of the inner ear. (*See* malleus, stapes.)

independent assortment   During meiosis, the random separation of genes carried on nonhomologous chromosomes. Articulated by Mendel as his second law.

indirect transduction   A cell signaling mechanism in which a second messenger mediates the interaction between receptor binding and cellular response. (Contrast with direct transduction.)

individual fitness   That component of inclusive fitness resulting from an organism producing its own offspring. (Contrast with kin selection.)

indoleacetic acid   *See* auxin.

induced fit   A change in enzyme conformation upon binding to substrate with an increase in the rate of catalysis.

induced mutation   A mutation resulting from treatment with a chemical or other agent.

inducer   (1) In enzyme systems, a small molecule which, when added to a growth medium, causes a large increase in the level of some enzyme. (2) In embryology, a substance that causes a group of target cells to differentiate in a particular way.

inducible enzyme   An enzyme that is present in much larger amounts when a particular compound (the inducer) has been added to the system. (Contrast with constitutive enzyme.)

inflammation   A nonspecific defense against pathogens; characterized by redness, swelling, pain, and increased temperature.

inflorescence   A structure composed of several to many flowers.

inflorescence meristem   A meristem that produces floral meristems as well as other small leafy structures (bracts).

inhibitor   A substance that binds to the surface of an enzyme and interferes with its action on its substrates.

initial cells   In plant meristems, undifferentiated cells that retain the capacity to divide producing both undifferentiated cells (initials) and cells committed to differentiation. (Compare with stem cells.)

initiation   In molecular biology, the beginning of transcription or translation.

initiation complex   Combination of a ribosomal light subunit, an mRNA molecule, and the tRNA charged with the first amino acid coded for by the mRNA; formed at the onset of translation.

initiation factors   Proteins that assist in forming the translation initiation complex at the ribosome.

inner cell mass   Derived from the mammalian blastula (bastocyst), the inner cell mass will give rise to the yolk sac (via hypoblast) and embryo (via epiblast).

inositol triphosphate (IP_3)   An intracellular second messenger derived from membrane phospholipids.

inspiratory reserve volume   The amount of air that can be inhaled above the normal tidal inspiration. (Compares with expiratory reserve volume, tidal volume, vital capacity.)

instar (in' star)   An immature stage of an insect between molts.

insulin (in' su lin) [L. *insula*: island]   A hormone synthesized in islet cells of the pancreas that promotes the conversion of glucose into the storage material, glycogen.

integral membrane protein   A membrane protein embedded in the bilayer of the membrane. (Contrast with peripheral membrane protein.)

integrase   An enzyme that integrates retroviral cDNA into the genome of the host cell.

integrated pest management Control of pests by the use of natural predators and parasites in conjunction with sparing use of chemicals; an attempt to limit environmental damage.

integument [L. *integumentum*: covering] A protective surface structure. In gymnosperms and angiosperms, a layer of tissue around the ovule which will become the seed coat.

intercalary meristem A meristematic region in plants which occurs not apically, but between two regions of mature tissue. Intercalary meristems occur in the nodes of grass stems, for example.

intercostal muscles Muscles between the ribs that can augment breathing movements by elevating and suppressing the rib cage.

interference competition Competition in which individuals actively interfere with one another's access to resources. (Compare with exploitation competition.)

interference RNA (RNAi) A mechanism for reducing mRNA translation whereby a double-stranded RNA, made by the cell or synthetically, is processed to a small, single-stranded RNA, and binding of this RNA to a target mRNA results in the latter's breakdown.

interferon A glycoprotein produced by virus-infected animal cells; increases the resistance of neighboring cells to the virus.

interkinesis The period between meiosis I and meiosis II.

interleukins Regulatory proteins, produced by macrophages and lymphocytes, that act upon other lymphocytes and direct their development.

intermediate disturbance hypothesis The hypothesis that explains why species richness is lower in areas with both high and low rates of disturbance than in areas with intermediate rates of disturbance.

intermediate filaments Cytoskeletal component with diameters between the larger microtubules and smaller microfilaments.

internal environment The physical and chemical characteristics of the extracellular fluids of the body.

interneuron A neuron that communicates information between two other neurons.

ionotropic receptors A receptor that that directly alters membrane permeability to a type of ion when it combines with its ligand.

internode The region between two nodes of a plant stem.

interphase The period between successive nuclear divisions during which the chromosomes are diffuse and the nuclear envelope is intact. It is during this period that the cell is most active in transcribing and translating genetic information.

interspecific competition Competition between members of two or more species. (Contrast with intraspecific competition.)

intertropical convergence zone The tropical region where the air rises most strongly; moves north and south with the passage of the sun overhead.

intraspecific competition Competition among members of the same species. (Contrast with interspecific competition.)

intrinsic protein A membrane protein that is embedded in the phospholipid bilayer of the membrane. (Contrast with extrinsic protein.)

intrinsic rate of increase The rate at which a population can grow when its density is low and environmental conditions are highly favorable.

intron A portion of a DNA molecule that, because of RNA splicing, is not involved in coding for part of a polypeptide molecule. (Contrast with exon.)

invagination An infolding of cells during animal embryonic development.

inversion A rare 180° reversal of the order of genes within a segment of a chromosome.

invertebrate A "convenience term" that encompasses any animal that is not a vertebrate—that is, whose nerve cord is not enclosed in a backbone of bony segments.

in vitro [L.: in glass] In a test tube, rather than in a living organism. (Contrast with in vivo.)

in vitro evolution Evolution in the laboratory, as used to produce compounds for industrial and pharmaceutical purposes.

in vivo [L.: in the living state] In a living organism. Many processes that occur in vivo can be reproduced in vitro with the right selection of cellular components. (Contrast with in vitro.)

ion (eye' on) [Gk.: *ion*: wanderer] An atom or group of atoms with electrons added or removed, giving it a negative or positive electrical charge.

ion channel A membrane protein that can let ions diffuse across the membrane. The channel can be ion-selective, and it can be voltage-gated or ligand-gated.

ionic bond An electrostatic attraction between positively and negatively charged ions. Usually a strong bond.

iris (eye' ris) [Gk. *iris*: rainbow] The round, pigmented membrane that surrounds the pupil of the eye and adjusts its aperture to regulate the amount of light entering the eye.

irruption A rapid increase in the density of a population. Often followed by massive emigration.

islets of Langerhans Clusters of hormone-producing cells in the pancreas.

iso- [Gk. *iso*: equal] Prefix used two separate entities that share some element of identity.

isogamous Describes male and female gametes that are morphologically identical.

isolating mechanism Geographical, physiological, ecological, or behavioral mechanisms that lead to a reduction in the frequency of successful matings between individuals in separate populations of a species. Can lead to the eventual evolution of separate species.

isomers Molecules consisting of the same numbers and kinds of atoms, but differing in the bonding patterns by which the atoms are held together.

isomorphic (eye so more' fik) [Gk. *isos*: equal + *morphe*: form] Having the same form or appearance, as when the haploid and diploid life stages of an organism appear identical. (Contrast with heteromorphic.)

isotonic Having the same solute concentration; said of two solutions. (Contrast with hypertonic, hypotonic.)

isotope (eye' so tope) [Gk. *isos*: equal + *topos*: place] Isotopes of a given chemical element have the same number of protons in their nuclei (and thus are in the same position on the periodic table), but differ in the number of neutrons

isozymes Enzymes that have somewhat different amino acid sequences but catalyze the same reaction.

– J –

jasmonates Plant hormones that trigger defenses against pathogens and herbivores.

jejunum (jih jew' num) The middle division of the small intestine, where most absorption of nutrients occurs. (*See* duodenum, ileum.)

joule (jool, or jowl) A unit of energy, equal to 0.24 calories.

juvenile hormone In insects, a hormone maintaining larval growth and preventing maturation or pupation.

– K –

karyogamy The fusion of nuclei of two cells. (Contrast with plasmogamy.)

karyotype The number, forms, and types of chromosomes in a cell.

keratin (ker' a tin) [Gk. keras: horn] A protein which contains sulfur and is part of such hard tissues as horn, nail, and the outermost cells of the skin.

ketone (key' tone) A compound with a C=O group attached to two other groups, neither of which is an H atom. Many sugars are ketones. (Contrast with aldehyde.)

keystone species Species that have a dominant influence on the composition of a community.

kidneys A pair of excretory organs in vertebrates.

kin selection That component of inclusive fitness resulting from helping the survival of relatives containing the same alleles by descent from a common ancestor. (Contrast with individual fitness.)

kinase (kye' nase) An enzyme that transfers a phosphate group from ATP to another molecule. Protein kinases transfer phosphate from ATP to specific proteins, playing important roles in cell regulation.

kinesin Motor protein having the capacity to attach to organelles or vesicles and move them along microtubules of the cytoskeleton.

kinetic energy The energy associated with movement.

kinetochore (kin net' oh core) [Gk. *kinetos*: moving] Specialized structure on a centromere to which microtubules attach.

knockout A molecular genetic method in which a single gene of an organism is permanently inactivated.

Koch's posulates A set of rules for establishing that a particular microorganism causes a particular disease.

Krebs cycle *See* citric acid cycle.

- L -

lactic acid fermentation   Fermentation whose end product is lactic acid (lactate).

lagging strand   In DNA replication, the daughter strand that is synthesized in discontinuous stretches. (*See* Okazaki fragments.)

lamella (la mell' ah) [L. *lamina*: thin sheet]   Layer.

larva (plural: larvae) [L. *lares*: guiding spirits]   An immature stage of any invertebrate animal that differs dramatically in appearance from the adult.

larynx (lar' inks) [Gk. *larynx*: voice box]   A structure between the pharynx and the trachea that includes the vocal cords.

lateral   Pertaining to the side.

lateral gene transfer   The transfer of genes from one species to another, common among bacteria and archaea.

lateral meristems   The vascular cambium and cork cambium, which give rise to secondary tissue in plants.

laticifers (luh tiss' uh furs)   In some plants, elongated cells containing secondary plant products such as latex.

leader sequence   A sequence of amino acids at the amino-terminal end of a newly synthesized protein; determines where the protein will be placed in the cell.

leading strand   In DNA replication, the daughter strand that is synthesized continuously. (Contrast with lagging strand.)

leaf primordium   An outgrowth on the side of the shoot apical meristem that will eventually develop into a leaf.

lenticel (len' ti sill)   Spongy region in a plant's periderm, allowing gas exchange.

leukocyte (loo' ko sight) [Gk. *leukos*: clear + *kytos*: container]   A white blood cell.

lichen (lie' kun)   An organism resulting from the symbiotic association of a true fungus and either a cyanobacterium or a unicellular alga.

life cycle   The entire span of the life of an organism from the moment of fertilization (or asexual generation) to the time it reproduces in turn.

life history   The stages an individual goes through during its life.

life table   A table showing, for a group of equal-aged individuals, the proportion still alive at different times in the future and the number of offspring they produce during each time interval.

ligament   A band of connective tissue linking two bones in a joint.

ligand (lig' and)   Any molecule that binds to a receptor site of another (usually larger) molecule.

light reactions   The initial phase of photosynthesis, in which light energy is converted into chemical energy.

light-independent reactions   The phase of photosynthesis in which chemical energy captured in the light reactions is used to drive the reduction of CO_2 to form carbohydrates.

lignin   The principal noncarbohydrate component of wood, a polymer that binds together cellulose fibrils in some plant cell walls.

limbic system   A group of primitive vertebrate forebrain nuclei that form a network and are involved in emotions, drives, instinctive behaviors, learning, and memory.

limiting resource   The required resource whose supply most strongly influences the size of a population.

linkage   Association between genetic markers on the same chromosome such that they do not show random assortment and seldom recombine; the closer the markers, the lower the frequency of recombination.

lipase (lip' ase; lye' pase)   An enzyme that digests fats.

lipids (lip' ids) [Gk. *lipos*: fat]   Substances in a cell which are easily extracted by organic solvents; fats, oils, waxes, steroids, and other large organic molecules, including those which, with proteins, make up the cell membranes. (Compare with phospholipids.)

littoral zone   The coastal zone from the upper limits of tidal action down to the depths where the water is thoroughly stirred by wave action.

liver   A large digestive gland. In vertebrates, it secretes bile and is involved in the formation of blood.

lobes   Regions of the human cerebral hemispheres; includes the temporal, frontal, parietal, and occipital lobes.

locus   In genetics, a specific location on a chromosome. May be considered to be synonymous with *gene*.

logistic growth   Growth, especially in the size of an organism or in the number of organisms in a population, that slows steadily as the entity approaches its maximum size. (Contrast with exponential growth.)

long-day plant (LDP)   A plant that requires long days (actually, short nights) in order to flower.

long-term potentiation (LTP)   A long-lasting increase in the sensitivity of a neuron resulting from a period of intense stimulation.

loop of Henle (hen' lee)   Long, hairpin loop of the mammalian renal tubule that runs from the cortex down into the medulla, and back to the cortex. Creates a concentration gradient in the interstitial fluids in the medulla.

lophophore   A U-shaped fold of the body wall with hollow, ciliated tentacles that encircles the mouth of animals in several different groups. Used for filtering prey from the surrounding water.

lumen (loo' men) [L. *lumen*: light]   The open cavity inside any tubular organ or structure, such as the gut or a kidney tubule.

luteinizing hormone   A gonadotropin produced by the anterior pituitary. It stimulates the gonads to produce sex hormones.

lymph [L. *lympha*: liquid]   A clear, watery fluid that is formed as a filtrate of blood; it contains white blood cells; it collects in a series of special vessels and is returned to the bloodstream.

lymph nodes   Specialized tissue regions that act as filters for cells, bacteria and foreign matter.

lymphocyte   A major class of white blood cells. Includes T cells, B cells, and other cell types important in the immune response.

lymphoid tissue   Tissues of the immune defense system dispersed throughout the body and consisting of: thymus, spleen, bone marrow, lymph nodes, blood, and lymph.

lysis (lie' sis) [Gk. *lysis*: break apart]   Bursting of a cell.

lysogenic bacteria   Bacteria that harbor a viral chromosome capable of the lysogenic cycle.

lysogenic cycle   A form of viral replication in which the virus becomes incorporated into the bacterial chromosome and the host cell is not killed. (Contrast with lytic cycle.)

lysosome (lie' so soam) [Gk. *lysis*: break away + *soma*: body]   A membrane-enclosed organelle found in eukaryotic cells (other than plants). Lysosomes contain a mixture of enzymes that can digest most of the macromolecules found in the rest of the cell.

lysozyme (lie' so zyme)   An enzyme in saliva, tears, and nasal secretions that attacks bacterial cell walls, as one of the body's nonspecific defense mechanisms.

lytic cycle   A form of viral reproduction that lyses the host bacterium releasing the new viruses. (Contrast with lysogenic cycle.)

- M -

M phase   The portion of the cell cycle in which mitosis takes place.

macroevolution [Gk. *makros*: large, long]   Evolutionary changes occurring over long time spans and usually involving changes in many traits. (Contrast with microevolution.)

macromolecule   A giant polymeric molecule. The macromolecules are proteins, polysaccharides, and nucleic acids.

macronutrient   A mineral element required by plant tissues in concentrations of at least 1 milligram per gram of their dry matter.

macrophage (mac' roh faj)   A type of white blood cell that endocytoses bacteria and other cells.

MADS box   A DNA-binding domain in many plant transcription factors that is active in development.

major histocompatibility complex (MHC)   A complex of linked genes, with multiple alleles, that control a number of cell surface antigens that identify self and can lead to graft rejection.

malignant   Referring to a tumor that can grow indefinitely and/or spread from the original site of growth to other locations in the body. (Contrast with benign.)

malleus (mal' ee us) [L. *malleus*: hammer]   The first of the three bones that conduct movements of the eardrum to the oval window of the inner ear. (*See* incus, stapes.)

Malpighian tubule (mal pee' gy un)   A type of protonephridium found in insects.

mantle   A sheet of specialized tissues that covers most of the viscera of mollusks; provides protection to internal organs and secretes the shell.

mapping   In genetics, determining the order of genes on a chromosome and the distances between them.

map unit   The distance between two genes, a recombinant frequency of 0.01.

marine [L. *mare*: sea, ocean] Pertaining to or living in the ocean. (Contrast with aquatic, terrestrial.)

marker A gene of identifiable phenotype that indicates the presence on another gene, DNA segment, or chromosome fragment.

mass extinctions Periods of evolutionary history during which rates of extinction were much higher than during intervening times.

mass number The sum of the number of protons and neutrons in an atom's nucleus.

mast cells Typically found in connective tissue, mast cells can be provoked by antigens or inflammation to release histamine.

maternal effect genes These genes code for morphogens that determine the polarity of the egg and larva in the fruit fly *Drosophila melanogaster*.

maternal inheritance Inheritance in which the mother's phenotype is exclusively expressed. Mitochondria and chloroplasts are maternally inherited via egg cytoplasm. Also known as cytoplasmic inheritance.

mating type A particular strain of a species that is incapable of sexual reproduction with another member of the same strain but capable of sexual reproduction with members of other strains of the same species.

maximum likelihood A statistical method of determining which of two or more hypotheses (such as phylogenetic trees) best fit the observed data, given an explicit model of how the data were generated.

mechanoreceptor A cell that is sensitive to physical movement and generates action potentials in response.

medulla (meh dull' luh) (1) The inner, core region of an organ, as in the adrenal medulla (adrenal gland) or the renal medulla (kidneys). (2) The portion of the brain stem that connects to the spinal cord.

megaphyll The generally large leaf of a fern, horsetail, or seed plant, with several to many veins. (Contrast with microphyll.)

megaspore [Gk. *megas*: large + *spora*: to sow] In plants, a haploid spore that produces a female gametophyte.

meiosis (my oh' sis) [Gk. *meiosis*: diminution] Division of a diploid nucleus to produce four haploid daughter cells. The process consists of two successive nuclear divisions with only one cycle of chromosome replication. In *meiosis I,* homologous chromosomes separate but retain their chromatids. The second division *meiosis II,* is similar to mitosis, in which chromatids separate.

melatonin A hormone released by the pineal gland that is involved in photoperiodism and circadian rhythms.

membrane potential The difference in electrical charge between the inside and the outside of a cell, caused by a difference in the distribution of ions.

memory cells Long-lived lymphocytes produced by exposure to antigen. They persist in the body and are able to mount a rapid response to subsequent exposures to the antigen.

Mendelian population A local population of individuals belonging to the same species and exchanging genes with one another.

Mendel's first law *See* segregation.

Mendel's second law *See* independent assortment.

menstrual cycle The monthly sloughing off of the uterine lining if fertilization does not occur in the female. Occurs between puberty and menopause.

meristem [Gk. *meristos*: divided] Plant tissue made up of undifferentiated actively dividing cells.

mesenchyme (mez' en kyme) [Gk. *mesos*: middle + *enchyma*: infusion] Embryonic or unspecialized cells derived from the mesoderm.

mesoderm [Gk. *mesos*: middle + *derma*: skin] The middle of the three embryonic tissue layers first delineated during gastrulation. Gives rise to skeleton, circulatory system, muscles, excretory system, and most of the reproductive system.

mesophyll (mez' uh fill) [Gk. *mesos*: middle + *phyllon*: leaf] Chloroplast-containing, photosynthetic cells in the interior of leaves.

mesosome (mez' uh soam') [Gk. *mesos*: middle + *soma*: body] A localized infolding of the plasma membrane of a bacterium.

messenger RNA (mRNA) A transcript of one of the strands of DNA; carries information (as a sequence of codons) for the synthesis of one or more proteins.

meta- [Gk.: between, along with, beyond] A prefix used in biology to denote a change or a shift to a new form or level; for example, as used in metamorphosis.

metabolic compensation Changes in metabolic properties of an organism that render it less sensitive to temperature changes.

metabolic factor In bacteria, a plasmid that carries genes determining unusual metabolic functions, such as the breakdown of hydrocarbons in oil.

metabolic pathway A series of enzyme-catalyzed reactions so arranged that the product of one reaction is the substrate of the next.

metabolism (meh tab' a lizm) [Gk. *metabole*: to change] The sum total of the chemical reactions that occur in an organism, or some subset of that total (as in respiratory metabolism).

metabotropic receptor A receptor that that indirectly alters membrane permeability to a type of ion when it combines with its ligand.

metamorphosis (met' a mor' fo sis) [Gk. *meta*: between + *morphe*: form, shape] A change occurring between one developmental stage and another, as for example from a tadpole to a frog. (*See* complete metamorphosis, incomplete metamorphosis.)

metaphase (met' a phase) The stage in nuclear division at which the centromeres of the highly supercoiled chromosomes are all lying on a plane (the metaphase plane or plate) perpendicular to a line connecting the division poles.

metapopulation A population divided into subpopulations, among which there are occasional exchanges of individuals.

metastasis (meh tass' tuh sis) The spread of cancer cells from their original site to other parts of the body.

methylation The addition of a methyl group (—CH$_3$) to a molecule. Extensive methylation of cytosine in DNA is correlated with reduced transcription.

MHC *See* major histocompatibility complex.

micelle a particle of lipid covered with bile salts that is produced in the duodenum and facilitates digestion and absorption of lipids.

microbiology [Gk. *mikros*: small + *bios*: life + *logos*: discourse] The scientific study of microscopic organisms, particularly bacteria, protists, and viruses.

microevolution The small evolutionary changes typically occurring over short time spans; generally involving a small number of traits and minor genetic changes. (Contrast with macroevolution.)

microfilament Minute fibrous structure generally composed of actin found in the cytoplasm of eukaryotic cells. They play a role in the motion of cells.

micronutrient A mineral element required by plant tissues in concentrations of less than 100 micrograms per gram of their dry matter.

microphyll A small leaf with a single vein, found in club mosses and their relatives. (Contrast with megaphyll.)

micropyle (mike' roh pile) [Gk. *mikros*: small + *pylon*: gate] Opening in the integument(s) of a seed plant ovule through which pollen grows to reach the female gametophyte within.

micro RNA A small RNA, typically about 21 bases long, that binds to mRNA to reduce its translation.

microspore [Gk. *mikros*: small + *spora*: to sow] In plants, a haploid spore that produces a male gametophyte.

microtubules Minute tubular structures found in centrioles, spindle apparatus, cilia, flagella, and cytoskeleton of eukaryotic cells. These tubules play roles in the motion and maintenance of shape of eukaryotic cells.

microvilli (singular: microvillus) The projections of epithelial cells, such as the cells lining the small intestine, that increase their surface area.

middle lamella A layer of polysaccharides that separates plant cells; a shared middle lamella lies outside the primary walls of the two cells.

migration The regular, seasonal movements of animals.

mineral An inorganic substance other than water.

mineral nutrients Inorganic ions required by organisms for normal growth and reproduction.

mismatch repair When a single base in DNA is changed into a different base, or the wrong base inserted during DNA replication, there is a mismatch in base pairing with the base on the opposite strand. A repair system removes the incorrect base and inserts the proper one for pairing with the opposite strand.

missense substitution A change in a gene from one nucleotide to another that also results in a change in the amino acid specified by the corresponding codon. (Compare with frame-shift mutation, nonsense substitution, synonymous substitution.)

mitochondrial matrix　The fluid interior of the mitochondrion, enclosed by the inner mitochondrial membrane.

mitochondrion (my′ toe kon′ dree un) [Gk. *mitos*: thread + *chondros*: grain]　An organelle in eukaryotic cells that contains the enzymes of the citric acid cycle, the respiratory chain, and oxidative phosphorylation.

mitosis (my toe′ sis) [Gk. *mitos*: thread]　Nuclear division in eukaryotes leading to the formation of two daughter nuclei, each with a chromosome complement identical to that of the original nucleus.

mitotic center　Cellular region that organizes the microtubules for mitosis. In animals a centrosome serves as the mitotic center.

moderately repetitive DNA　DNA sequences that appear hundreds to thousands of times in the genome. They include the DNA sequences coding for rRNAs and tRNAs, as well as the DNA at telomeres.

modular organism　An organism which grows by producing additional units of body construction (modules) that are very similar to the units of which it is already composed.

mole　A quantity of a compound whose weight in grams is numerically equal to its molecular weight expressed in atomic mass units. Avogadro′s number of molecules: 6.023×10^{23} molecules.

molecular clock　The theory that macromolecules diverge from one another over evolutionary time at a constant rate; this rate may provide insight into the phylogenetic relationships among organisms.

molecular tool kit　A set of developmental genes and proteins that is common to most animals and is hypothesized to be responsible for the evoltuion of their differing developmental pathways.

molecular weight　The sum of the atomic weights of the atoms in a molecule.

molecule　A particle made up of two or more atoms joined by covalent bonds or ionic attractions.

molting　The process of shedding part or all of an outer covering, as the shedding of feathers by birds or of the entire exoskeleton by arthropods.

monoclonal antibody　Antibody produced in the laboratory from a clone of hybridoma cells, each of which produces the same specific antibody.

monocytes　White blood cells that produce macrophages.

monoecious (mo nee′ shus) [Gk. *mono*: one + *oikos*: house]　Describes organisms in which both sexes are "housed" in a single individual that produces both eggs and sperm. (In some plants, these are found in different flowers within the same plant.) Examples include corn, peas, earthworms, hydras. (Contrast with dioecious.)

monohybrid cross　A mating in which the parents differ with respect to the alleles of only one locus of interest.

monomer [Gk. *mono*: one + *meros*: unit]　A small molecule, two or more of which can be combined to form oligomers (consisting of a few monomers) or polymers (consisting of many monomers).

monophyletic (mon′ oh fih leht′ ik) [Gk. *mono*: one + *phylon*: tribe]　Referring to a group that consists of an ancestor and all of its descendants. (Compare with paraphyletic, polyphyletic.)

monosaccharide　A simple sugar. Oligosaccharides and polysaccharides are made up of monosaccharides.

monosomic　Referring to an organism with one less than the normal diploid number of chromosomes.

monosynaptic reflex　A neural reflex that begins in a sensory neuron and makes a single synapse before activating a motor neuron.

morphogen　A diffusible substances whose concentration gradients determine patterns of development in animals and plants.

morphogenesis (more′ fo jen′ e sis) [Gk. *morphe*: form + *genesis*: origin]　The development of form; the overall consequence of determination, differentiation, and growth.

morphology (more fol′ o jee) [Gk. *morphe*: form + *logos*: study, discourse]　The scientific study of organic form, including both its development and function.

mosaic development　Pattern of animal embryonic development in which each blastomere contributes a specific part of the adult body. (Contrast with regulative development.)

motor end plate　The modified area on a muscle cell membrane where a synapse is formed with a motor neuron.

motor neuron　A neuron carrying information from the central nervous system to an effector such as a muscle fiber.

motor proteins　Specialized proteins that use energy to change shape and move cells or structures within cells. *See* dynein, kinesin.

motor unit　A motor neuron and the set of muscle fibers it controls.

mRNA　*See* messenger RNA.

mucosa (mew koh′ sah)　An epithelial membrane containing cells that secrete mucus. The inner cell layers of the digestive and respiratory tracts.

Müllerian mimicry　The convergence in appearance of two or more unpalatable species.

multifactorial　In medicine, referring to a disease with many interacting causes, both genetic and environmental.

muscle fiber　A single muscle cell. In the case of skeletal (striated) muscle, a syncitial, multinucleate cell.

muscle tissue　Excitable tissue that can contract due to interactions of actin and myosin. Three types are striated, smooth, and cardiac.

muscle tone　The degree of contraction of a muscle.

muscle spindle　Modified muscle fibers encased in a connective sheat and functioning as stretch receptors.

mutagen (mute′ ah jen) [L. *mutare*: change + Gk. *genesis*: source]　Any agent (e.g., chemicals, radiation) that increases the mutation rate.

mutation　A detectable, heritable change in the genetic material not caused by recombination.

mutation pressure　Evolution (change in gene proportions) by different mutation rates alone (i.e., without the influence of natural selection).

mutualism　The type of symbiosis, such as that exhibited by fungi and algae or cyanobacteria in forming lichens, in which both species profit from the association.

mycelium (my seel′ ee yum) [Gk. *mykes*: fungus]　In the fungi, a mass of hyphae.

mycorrhiza (my′ ko rye′ za) [Gk. *mykes*: fungus + *rhiza*: root]　An association of the root of a plant with the mycelium of a fungus.

myelin (my′ a lin)　A material forming a sheath around some axons. Formed by Schwann cells that wrap themselves about the axon, myelin insulates the axon electrically and increases the rate of transmission of a nervous impulse.

myocardial infarction　A blockage of an artery that carries blood to the heart muscle.

myofibril (my′ oh fy′ bril) [Gk. *mys*: muscle + L. *fibrilla*: small fiber]　A polymeric unit of actin or myosin in a muscle.

myogenic (my oh jen′ ik) [Gk. *mys*: muscle + *genesis*: source]　Originating in muscle.

myoglobin (my′ oh globe′ in) [Gk. *mys*: muscle + L. *globus*: sphere]　An oxygen-binding molecule found in muscle. Consists of a heme unit and a single globiin chain, and carrys less oxygen than hemoglobin.

myosin　One of the two major proteins of muscle, it makes up the thick filaments. (*See* actin.)

– N –

NAD (nicotinamide adenine dinucleotide)　A compound found in all living cells, existing in two interconvertible forms: the oxidizing agent NAD^+ and the reducing agent $NADH + H^+$.

NADP (nicotinamide adenine dinucleotide phosphate)　A compound similar to NAD, but possessing another phosphate group; plays similar roles but is used by different enzymes.

natural killer cells　A nonspecific defensive cell (lymphocyte) that attacks tumor cells and virus infected cells.

natural selection　The differential contribution of offspring to the next generation by various genetic types belonging to the same population. The mechanism of evolution proposed by Charles Darwin.

necrosis (nec roh′ sis) [Gk. *nekros*: death]　Tissue damage resulting from cell death.

negative control　The situation in which a regulatory macromolecule (generally a repressor) functions to turn off transcription. In the absence of a regulatory macromolecule, the structural genes are turned on.

negative feedback　Information relevant to the rate of a process that can be used by a control system to return the outcome of that process to an optimal level.

nematocyst (ne mat′ o sist) [Gk. *nema*: thread + *kystis*: cell]　An elaborate, threadlike structure produced by cells of jellyfish and other cnidarians, used chiefly to paralyze and capture prey.

nephridium (nef rid' ee um) [Gk. *nephros*: kidney] An organ which is involved in excretion, and often in water balance, involving a tube that opens to the exterior at one end.

nephron (nef' ron) [Gk. *nephros*: kidney]   The functional unit of the kidney, consisting of a structure for receiving a filtrate of blood, and a tubule that absorbs selected parts of the filtrate back into the bloodstream.

nephrostome (nef' ro stome) [Gk. *nephros*: kidney + *stoma*: opening]   An opening in a nephridium through which body fluids can enter.

Nernst equation   A mathematical statement; calculates the potential across a membrane permeable to a single type of ion that differs in concentration on the two sides of the membrane.

nerve   A structure consisting of many neuronal axons and connective tissue.

net primary production   Total photosynthesis minus respiration by plants.

neural plate   A thickened strip of ectoderm along the dorsal side of the early vertebrate embryo; gives rise to the central nervous system.

neural tube   An early stage in the development of the vertebrate nervous system consisting of a hollow tube created by two opposing folds of the dorsal ectoderm along the anterior–posterior body axis.

neurohormone   A chemical signal produced and released by neurons; the signal then acts as a hormone.

neuromuscular junction   The region where a motor neuron contacts a muscle fiber, creating a synapse.

neuron (noor' on) [Gk. *neuron*: nerve]   A nervous system cell that can generate and conduct action potentials along an axon to a synapse with another cell.

neurotransmitter   A substance produced in and released by one a neuron (the presynaptic cell) that diffuses across a synapse and excites or inhibits another cell (the postsynaptic cell).

neurula (nure' you la)   Embryonic stage during the dorsal nerve cord forms from two ectodermal ridges.

neurulation   A stage in vertebrate development during which the nervous system begins to form.

neutral allele   An allele that does not alter the functioning of the proteins for which it codes.

neutral theory   A view of molecular evolution that postulates that most mutations do not affect the amino acid being coded for, and that such mutations accumulate in a population at rates driven by genetic drift and mutation rates.

neutron (new' tron)   One of the three most fundamental particles of matter, with mass approximately 1 amu and no electrical charge.

neutrophils   Abundant, short-lived phagocytic leukocytes that attack antibody-coated antigens.

nitrate reduction   The process by which nitrate (NO_3^-) is reduced to ammonia (NH_3).

nitric oxide (NO)   An unstable molecule (a gas) that serves as a second messenger causing smooth muscle to relax. In the nervous system it operates as a neurotransmitter.

nitrification   The oxidation of ammonia to nitrite and nitrate ions, performed by certain soil bacteria.

nitrogenase   In nitrogen-fixing organisms, an enzyme complex that mediates the stepwise reduction of atmospheric N_2 to ammonia.

nitrogen fixation   Conversion of nitrogen gas to ammonia, which makes nitrogen available to living things. Carried out by certain prokaryotes, some of them free-living and others living within plant roots.

node [L. *nodus*: knob, knot]   In plants, a (sometimes enlarged) point on a stem where a leaf is or was attached.

node of Ranvier   A gap in the myelin sheath covering an axon; the point where the axonal membrane can fire action potentials.

noncompetitive inhibitor   An inhibitor that binds the enzyme at a site other than the active site. (Contrast with competitive inhibitor.)

noncyclic electron transport   In photosynthesis, the flow of electrons that forms ATP, NADPH, and O_2.

nondisjunction   Failure of sister chromatids to separate in meiosis II or mitosis, or failure of homologous chromosomes to separate in meiosis I. Results in aneuploidy.

nonpolar molecule   A molecule whose electric charge is evenly balanced from one end of the molecule to the other.

nonrandom mating   The selection by individuals of other individuals of particular genotypes as mates.

non-REM sleep   A state of sleep characterized by low muscle tone, but not atonia, quiescence,

nonsense substitution   A change in a gene from one nucleotide to another that prematurely terminates a polypeptide. Termination occurs when a codon that specifies an amino acid is changed to one of the codons (UAG, UAA, or UGA) that signal termination of translation. (Compare with frame-shift mutation, missense substitution, synonymous substitution.)

nonspecific defenses   Immunologic responses directed against any invading agent without reacting to apecific antigens.

nonsynonymous substitution   A change in a gene from one nucleotide to another that changes the amino acid specified by the corresponding codon (i.e., AGC → AGA, or serine → arginine). (Contrast with synonymous substitution.)

nonvascular plants   Those plants lacking well-developed vascular tissue; the liverworts, hornworts, and mosses. (Contrast with vascular plants.)

norepinephrine   A neurotransmitter found in the central nervous system and also at the post-ganglionic nerve endings of the sympathetic nervous system. Also called noradrenaline.

notochord (no' tow kord) [Gk. *notos*: back + *chorde*: string]   A flexible rod of gelatinous material serving as a support in the embryos of all chordates and in the adults of tunicates and lancelets.

nuclear envelope   The surface, consisting of two layers of membrane, that encloses the nucleus of eukaryotic cells.

nuclear lamina   A meshwork of fibers on the inner surface of the nuclear envelope.

nuclear pore complex   A protein structure situated in nuclear pores through which RNA and proteins enter and leave the nucleus.

nucleic acid (new klay' ik)   A long-chain alternating polymer of deoxyribose or ribose and phosphate groups, with nitrogenous bases—adenine, thymine, uracil, guanine, or cytosine (A, T, U, G, or C)—as side chains. DNA and RNA are nucleic acids.

nucleic acid hybridization   A technique in which a single-stranded nucleic acid probe in made that is complementary to, and binds to, a target sequence, either DNA or RNA. The resulting double-stranded molecule is a hybrid.

nucleoid (new' klee oid)   The region that harbors the chromosomes of a prokaryotic cell. Unlike the eukaryotic nucleus, it is not bounded by a membrane.

nucleolar organizer (new klee' o lar)   A region on a chromosome that is associated with the formation of a new nucleolus following nuclear division. The site of the genes that code for ribosomal RNA.

nucleolus (new klee' oh lus)   A small, generally spherical body found within the nucleus of eukaryotic cells. The site of synthesis of ribosomal RNA.

nucleoplasm (new' klee o plazm)   The fluid material within the nuclear envelope of a cell, as opposed to the chromosomes, nucleoli, and other particulate constituents.

nucleoside   A nucleotide without the phosphate group.

nucleosome   A portion of a eukaryotic chromosome, consisting of part of the DNA molecule wrapped around a group of histone molecules, and held together by another type of histone molecule. The chromosome is made up of many nucleosomes.

nucleotide   The basic chemical unit in a nucleic acid. A nucleotide in RNA consists of one of four nitrogenous bases linked to ribose, which in turn is linked to phosphate. In DNA, deoxyribose is present instead of ribose.

nucleotide substitution   A change of one base pair to another in a DNA sequence.

nucleus (new' klee us) [L. *nux*: kernel or nut] (1) In cells, the centrally located compartment of eukaryotic cells that is bounded by a double membrane and contains the chromosomes. (2) In the brain, an identifiable group of neurons that share common characteristics or functions.

null hypothesis   The assertion that an effect proposed by its companion hypothesis does not in fact exist.

nutrient   A food substance; or, in the case of mineral nutrients, an inorganic element required for completion of the life cycle of an organism.

– O –

obligate anaerobe   An anaerobic prokaryote that cannot survive exposure to O_2.

odorant   A molecule that can bind to an olfactory receptor.

oil   A triglyceride that is liquid at room temperature. (Contrast with fat.)

Okazaki fragments   Newly formed DNA making up the lagging strand in DNA replication. DNA ligase links Okazaki fragments together to give a continuous strand.

olfactory [L. *olfacere*: to smell]   Having to do with the sense of smell.

oligomer [Gk.: *oligo*: a few + *meros*: units] A compound molecule of intermediate size, made up of two to a few monomers. (Contrast with monomer, polymer.)

oligosaccharide A polymer containing a small number of monosaccharides.

oligosaccharins Plant hormones, derived from the plant cell wall, that trigger defenses against pathogens.

ommatidium [Gk. *omma*: eye] One of the units which, collected into groups of up to 20,000, make up the compound eye of arthropods.

omnivore [L. *omnis*: everything + *vorare*: to devour] An organism that eats both animal and plant material. (Contrast with carnivore, detritivore, herbivore.)

oncogene [Gk. *onkos*: mass, tumor + *genes*: born] Genes that greatly stimulate cell division, giving rise to tumors.

one-gene, one-polypeptide The principle that each gene codes for a single polypeptide.

oocyte (oh' eh site) [Gk. *oon*: egg + *kytos*: container] The cell that gives rise to eggs in animals.

oogenesis (oh' eh jen e sis) [Gk. *oon*: egg + *genesis*: source] Female gametogenesis, leading to production of the egg.

oogonium (oh' eh go' nee um) In some algae and fungi, a cell in which an egg is produced.

operator The region of an operon that acts as the binding site for the repressor.

operon A genetic unit of transcription, typically consisting of several structural genes that are transcribed together; the operon contains at least two control regions: the promoter and the operator.

opportunity cost The sum of the benefits an animal forfeits by not being able to perform some other behavior during the time when it is performing a given behavior.

opsin (op' sin) [Gk. *opsis*: sight] The protein portion of the visual pigment rhodopsin. (*See* rhodopsin.)

optic chiasm [Gk. *chiasma*: cross] Structure on the lower surface of the vertebrate brain where the two optic nerves come together.

optical isomers Two isomers that are mirror images of each other.

orbital A region in space surrounding the atomic nucleus in which an electron is most likely to be found.

organ [Gk. *organon*: tool] A body part, such as the heart, liver, brain, root, or leaf. Organs are composed of different tissues integrated to perform a distinct function. Organs are in turn often integrated into systems, such as the digestive or reproductive systems.

organ identity genes Plant genes that specify the various parts of the flower. *See* homeotic genes.

organ of Corti Structure in the inner ear that transforms mechanical forces produced from pressure waves ("sound waves") into action potentials that are sensed as sound.

organ system An interrelated and integrated group of tissues and organs that work together in a physiological function.

organelles (or gan els') Organized structures found in or on eukaryotic cells. Examples include ribosomes, nuclei, michrondria, chloroplasts, cilia, and contractile vacuoles.

organic Pertaining to any aspect of living matter, e.g., to its evolution, structure, or chemistry. The term is also applied to any chemical compound that contains carbon.

organism Any living entity.

organizer Region of an early embryo that directs the development of nearby regions. In amphibian early gastrulas, the dorsal lip of the blastopore is the organizer.

organogenesis The formation of organs and organ systems during development.

origin of replication DNA sequence at which helicase unwinds the DNA double helix and DNA polymerase binds to initiate DNA replication.

orthology (or thol' o jee) A type of homology applied to genes in which the divergence of homologous genes can be traced to speciation events. The genes are said to be *orthologous*, and each is an *ortholog* of the others. (Compare with paralogy)

osmoconformer An aquatic animal that maintains an osmotic concentration of its extracellular fluid that is thet same as that of the external environment.

osmolarity The concentration of osmotically active particles in a solution.

osmoregulation Regulation of the chemical composition of the body fluids of an organism.

osmoreceptor Neuron that converts changes in the solute potential of interstial fluids into action potentials.

osmosis (oz mo' sis) [Gk. *osmos*: to push] The movement of water across a differentially permeable membrane, from one region to another region where the water potential is more negative.

ossicle (oss' ick ul) [L. *os*: bone] The calcified construction unit of echinoderm skeletons.

osteoblasts (oss' tee oh blast) [Gk. *osteon*: bone + *blastos*: sprout] Cells that lay down the protein matrix of bone.

osteoclasts (oss' tee oh clast) [Gk. *osteon*: bone + *klastos*: broken] Cells that dissolve bone.

otolith (oh' tuh lith) [Gk. *otikos*: ear + *lithos*: stone[Structures in the vertebrate vestibular apparatus that mechanically stimulate hair cells when the head moves or changes position.

oval window The flexible membrane that, when moved by the bones of the middle ear, produces pressure waves in the inner ear

ovary (oh' var ee) [L. *ovum*: egg] Any female organ, in plants or animals, that produces an egg.

oviduct [L. *ovum*: egg + *ducere*: to lead] In mammals, the tube serving to transport eggs to the uterus or to outside of the body.

oviparity Reproduction in which eggs are released by the female and development is external to the mother's body. (Contrast with viviparous.)

ovoviviparity Reproduction in which fertilized eggs develop and hatch within the body of the mother but are not attached to the mother by means of a placenta.

ovulation The release of an egg from an ovary.

ovule (oh' vule) In plants, a structure that contains a gametophyte and, within the gametophyte, an egg; when it matures, an ovule becomes a seed.

ovum (oh' vum) [L. *ovum*: egg] The egg; the female sex cell.

oxidation (ox i day' shun) Relative loss of electrons in a chemical reaction; either outright removal to form an ion, or the sharing of electrons with substances having a greater affinity for them, such as oxygen. Most oxidation, including biological ones, are associated with the liberation of energy. (Contrast with reduction.)

oxidative phosphorylation ATP formation in the mitochondrion, associated with flow of electrons through the respiratory chain.

oxidizing agent A substance that can accept electrons from another. The oxidizing agent becomes reduced; its partner becomes oxidized.

oxygenase An enzyme that catalyzes the addition of oxygen to a substrate from O_2.

– P –

P generation Parental generation. The individuals that mate in a genetic cross. Their immediate offspring are the F_1 generation.

pacemaker That part of the heart which undergoes most rapid spontaneous contraction, thus setting the pace for the beat of the entire heart. In mammals, the sinoatrial (SA) node. Also, an artificial device, implanted in the heart, that initiates rhythmic contraction of the organ.

Pacinian corpuscle A modified nerve ending that senses touch and vibration.

pair rule genes Segmentation genes that divide the *Drosophila* larva into two segments each.

pancreas (pan' cree us) A gland located near the stomach of vertebrates that secretes digestive enzymes into the small intestine and releases insulin into the bloodstream.

Pangaea (pan jee' uh) [Gk. *pan*: all, every] The single land mass formed when all the continents came together in the Permian period.

para- [Gk. *para*: akin to, beside] Prefix indicating association in being along side or accessory to.

parabronchi Passages in the lungs of birds through which air flows.

paracrine A substance, such as a hormone, that acts locally, near the site of its secretion. (Compare with autocrine, endocrine gland.)

parallel evolution Repeated evolutionary patterns of change that occur independently in multiple lineages.

paralogy (par al' o jee) A type of homology applied to genes in which the divergence of homologous genes can be traced to gene duplication events. The genes are said to be paralogous, and each is an paralog of the others. (Compare with orthology.)

parapatric speciation [Gk. *para*: along side + *patria*: homeland] Reproductive isolation between subpopulations arising from some non-geographic but physical condition, such as soil nutrient content. (Contrast with allopatric speciation, sympatric speciation.)

paraphyletic (par' a fih leht' ik) [Gk. *para*: beside + *phylon*: tribe] Referring to a group that consists of an ancestor and some (but not all) of its descendants. (Compare with monophyletic, polyphyletic.)

parasite   An organism that attacks and consumes parts of an organism much larger than itself. Parasites sometimes, but not always, kill their host.

parasympathetic nervous system   A portion of the autonomic (involuntary) nervous system. (Contrast with sympathetic nervous system.)

parathyroids   Four glands on the posterior surface of the thyroid that produce and release parathormone.

parathyroid hormone   Hormone secreted by the parathyroid glands. Stimulates osteoclast activity and raises blood calcium levels.

parenchyma (pair eng' kyma)   A plant tissue composed of relatively unspecialized cells without secondary walls.

parsimony   The principle of preferring the simplest among a set of plausible explanations of any phenomenon.

parthenocarpy   Formation of fruit from a flower without fertilization.

parthenogenesis (par' then oh jen' e sis) [Gk. parthenos: virgin + genesis: source]   The production of an organism from an unfertilized egg.

partial pressure   The portion of the barometric pressure of a mixture of gases that is due to one component of that mixture. For example, the partial pressure of oxygen at sea level is 20.9% of barometric pressure.

particulate theory   In genetics, the theory that genes are physical entities that retain their identities after fertilization.

passive transport   Diffusion across a membrane; may or may not require a channel or carrier protein. (Contrast with active transport.)

patch clamping   A technique for isolating a tiny patch of membrane to allow the study of ion movement through a particular channel.

pathogen (path' o jen) [Gk. pathos: suffering + genesis: source]   An organism that causes disease.

pattern formation   In animal embryonic development, the organization of differentiated tissues into specific structures such as wings.

pedigree   The pattern of transmission of a genetic trait within a family.

penetrance   Of a genotype, the proportion of individuals with that genotype who show the expected phenotype.

pentose [Gk. penta: five]   A sugar containing five carbon atoms.

PEP carboxylase   The enzyme that combines carbon dioxide with PEP to form a 4-carbon dicarboxylic acid at the start of C_4 photosynthesis or of crassulacean acid metabolism (CAM).

pepsin [Gk. pepsis: digestion]   An enzyme in gastric juice that digests protein.

pepsinogen   Inactive secretory product that is converted into pepsin by low pH or by enzymatic action.

peptide linkage   The bond between amino acids in a protein. Formed between a carboxyl group and amino group (CO—NH⁻) with the loss of water molecules.

peptidoglycan   The cell wall material of many bacteria, consisting of a single enormous molecule that surrounds the entire cell.

perennial (per ren' ee al) [L. per: throughout + annus: year]   Refers to a plant that survives from year to year. (Contrast with annual, biennial.)

perfect flower   A flower with both stamens and carpels, therefore hermaphroditic.

pericycle [Gk. peri: around + kyklos: ring or circle]   In plant roots, tissue just within the endodermis, but outside of the root vascular tissue. Meristematic activity of pericycle cells produces lateral root primordia.

periderm   The outer tissue of the secondary plant body, consisting primarily of cork.

period   (1) A category in the geological time scale. (2) The duration of a single cycle in a cyclical event, such as a circadian rhythm.

peripheral membrane protein   Membrane protein not embedded in the bilayer. (Contrast with integral membrane protein.)

peripheral nervous system   Neurons that transmit information to and from the central nervous system and whose cell bodies reside outside the brain or spinal cord.

peristalsis (pair' i stall' sis) [Gk. peri: around + stellein: place]   Wavelike muscular contractions proceeding along a tubular organ, propelling the contents along the tube.

peritoneum   The mesodermal lining of the body cavity among coelomate animals.

permease   A membrane protein that specifically transports a compound or family of compounds across the membrane.

peroxisome   An organelle that houses reactions in which toxic peroxides are formed. The peroxisome isolates these peroxides from the rest of the cell.

petal [Gk. petalon: spread out]   In an angiosperm flower, a sterile modified leaf, nonphotosynthetic, frequently brightly colored, and often serving to attract pollinating insects.

petiole (pet' ee ole) [L. petiolus: small foot]   The stalk of a leaf.

pH   The negative logarithm of the hydrogen ion concentration; a measure of the acidity of a solution. A solution with pH = 7 is said to be neutral; pH values higher than 7 characterize basic solutions, while acidic solutions have pH values less than 7.

phage (fayj)   Short for bacteriophage. A virus that infects bacteria.

phagocyte [Gk. phagein: to eat + kystos: sac]   A white blood cell that ingests microorganisms by endocytosis.

phagocytosis   Endocytosis by a cell of another cell or large particle.

pharmacogenomics   The relaionship between an individual's genetic makeup and response to drugs.

pharming   The use of genetically modified animals to produce medically useful products in their milk.

pharynx [Gk. pharynx: throat]   The part of the gut between the mouth and the esophagus.

phenotype (fee' no type) [Gk. phanein: to show]   The observable properties of an individual resulting from both genetic and environmental factors. (Contrast with genotype.)

phenotypic plasticity   Refers to the fact that the phenotype of a developing organism is determined by a complex series of processes that are affected by both its genotype and its environment.

pheromone (feer' o mone) [Gk. pheros: carry + hormon: excite, arouse]   A chemical substance used in communication between organisms of the same species.

phloem (flo' um) [Gk. phloos: bark]   In vascular plants, the tissue that transports sugars and other solutes from sources to sinks. It consists of sieve cells or sieve tubes, fibers, and other specialized cells.

phosphate group   The functional group — OPO_3H_2. The transfer of energy from one compound to another is often accomplished by the transfer of a phosphate group.

phosphodiester linkage   The connection in a nucleic acid strand, formed by linking two nucleotides.

phospholipids   Lipids containing a phosphate group; important constituents of cellular membranes. (See lipids.)

phosphorylation   The addition of a phosphate group.

photoautotroph   An organism that obtains energy from light and carbon from carbon dioxide. (Contrast with chemolithotroph, chemoheterotroph, photoheterotroph.)

photoheterotroph   An organism that obtains energy from light but must obtain its carbon from organic compounds. (Contrast with chemolithotroph, chemoheterotroph, photoautotroph.)

photon (foe' ton) [Gk. photos: light]   A quantum of visible radiation; a "packet" of light energy.

photoperiod (foe' tow peer' ee ud)   The duration of a period of light, such as the length of time in a 24-hour cycle in which daylight is present.

photoperiodicity   A condition in which physiological and behavioral changes are induced by changes in day length.

photoreceptor   (1) A pigment that triggers a physiological response when it absorbs a photon. (2) A sensory receptor cell that senses and responds to light energy.

photorespiration   Light-driven uptake of oxygen and release of carbon dioxide, the carbon being derived from the early reactions of photosynthesis.

photosynthesis (foe tow sin' the sis) [literally, "synthesis from light"]   Metabolic processes, carried out by green plants, by which visible light is trapped and the energy used to synthesize compounds such as ATP and glucose.

photosystem [Gk. phos: light + systema: assembly]   A light-harvesting complex in the chloroplast thylakoid composed of pigments and proteins.

photosystem I   In photosynthesis, the reactions that absorb light at 700 nm, passing electrons to ferrodoxin and thence to NADPH. Rich in chlorophyll a.

photosystem II   In photosynthesis, the reactions that absorb light at 660 nm, passing electrons to the electron transport chain in the chloroplast. Rich I chlorphyll b.

phototropins A class of blue light receptors that mediate phototropism and other plant responses.

phototropism [Gk. *photos*: light + *trope*: turning] A directed plant growth response to light.

phycobilin Photosynthetic pigment that absorbs red, yellow, orange, and green light and is found in cyanobacteria and some red algae.

phylogenetic tree A graphic representation of lines of descent among organisms or their genes.

phylogeny (fy loj' e nee) [Gk. *phylon*: tribe, race + *genesis*: source] The evolutionary history of a particular group of organisms or their genes.

physiology (fiz' ee ol' o jee) [Gk. *physis*: natural form + *logos*: discourse, study] The scientific study of the functions of living organisms and the individual organs, tissues, and cells of which they are composed.

phytoalexins Substances toxic to pathogens, produced by plants in response to fungal or bacterial infection.

phytochrome (fy' tow krome) [Gk. *phyton*: plant + *chroma*: color] A plant pigment regulating a large number of developmental and other phenomena in plants.

pigment A substance that absorbs visible light.

pineal gland A gland located between the cerebral hemispheres that secretes melatonin.

pinocytosis Endocytosis by a cell of liquid containing dissolved substances.

pistil [L. *pistillum*: pestle] The structure of an angiosperm flower within which the ovules are borne. May consist of a single carpel, or of several carpels fused into a single structure. Usually differentiated into ovary, style, and stigma.

pith In plants, relatively unspecialized tissue found within a cylinder of vascular tissue.

pituitary A small gland attached to the base of the brain in vertebrates. Its hormones control the activities of other glands. Also known as the hypophysis.

pits Recessed cavities in the cell walls of a plant vascular element where only the primary wall is present. facilitating the movement of sap between cells.

placenta (pla sen' ta) [Gk. *plax*: flat surface] The organ found in most mammals that provides for the nourishment of the fetus and elimination of the fetal waste products.

placental (pla sen' tal) Pertaining to mammals of the subclass Eutheria, a group characterized by the presence of a placenta; contains the majority of living species of mammals.

plankton Free-floating small organisms inhabiting the surface waters of lakes and oceans. Photosynthetic members of the plankton are referred to as phytoplankton.

plant *See* embryophyte.

planula (plan' yew la) [L. *planum*: flat] The free-swimming, ciliated larva of the cnidarians.

plaque (plack) [Fr.: a metal plate or coin] (1) A circular clearing in a layer (lawn) of bacteria growing on the surface of a nutrient agar gel. (2) An accumulation of prokaryotic organisms on tooth enamel. Acids produced by these microorganisms can cause tooth decay. (3) A region of arterial wall invaded by fibroblasts and fatty deposits (*see* atherosclerosis).

plasma *See* blood plasma.

plasma cell An antibody-secreting cell that developed from a B cell. The effector cell of the humoral immune system.

plasma membrane The membrane that surrounds the cell, regulating the entry and exit of molecules and ions. Every cell has a plasma membrane.

plasmid A DNA molecule distinct from the chromosome(s); that is, an extrachromosomal element. May replicate independently of the chromosome.

plasmodesma (plural: plasmodesmata) [Gk. *plassein*: to mold + *desmos*: band] A cytoplasmic strand connecting two adjacent plant cells.

plasmogamy The fusion of the cytoplasm of two cells. (Contrast with karyogamy.)

plasmolysis (plaz mol' i sis) Shrinking of the cytoplasm and plasma membrane away from the cell wall, resulting from the osmotic outflow of water. Occurs only in cells with rigid cell walls.

plastid Organelle in plants that serves for food manufacture (by photosynthesis) or food storage; bounded by a double membrane.

plastoquinone A mobile electron carrier within the thylakoid membrane of the chloroplast linking photosystems I and II of photosynthesis.

platelet A membrane-bounded body without a nucleus, arising as a fragment of a cell in the bone marrow of mammals. Important to blood-clotting action.

pleiotropy (plee' a tro pee) [Gk. *pleion*: more] The determination of more than one character by a single gene.

pleural membrane [Gk. *pleuras*: rib, side] The membrane lining the outside of the lungs and the walls of the thoracic cavity. Inflammation of these membranes is a condition known as pleurisy.

pluripotent Of a stem cell, having the ability to differentiate into any of a limted number of cell types. (Compare with totipotent.)

podocytes Cells of Bowman's capsule of the nephron that cover the capillaries of the glomerulus, forming filtration slits.

poikilotherm (poy' kill o therm) [Gk. *poikilos*: varied + *thermos*: heat] An animal whose body temperature tends to vary with the surrounding environment. (Contrast with homeotherm, heterotherm.)

point mutation A mutation that results from a small, localized alteration in the chemical structure of a gene; can revert to wild type. (Contrast with deletion.)

polar body A nonfunctional nucleus produced by meiosis, accompanied by very little cytoplasm. The meiosis which produces the mammalian egg produces in addition three polar bodies.

polar molecule A molecule in which the electric charge is not distributed evenly in the covalent bonds.

polar nuclei In flowering plants, the two nuclei in the central cell of the megagametophyte; following fertilization they give rise to the endosperm.

polarity In development, the difference between one end and the other. In chemistry, the property that makes a polar molecule.

pollen [L. *pollin*: fine, powdery flour] In seed plants, the microscopic grains containing the male gametophyte (microgametophyte) and gamete (microspore).

pollination The process of transferring pollen from an anther to the stigma of a pistil in an angiosperm or from a strobilus to an ovule in a gymnosperm.

poly- [Gk. *poly*: many] A prefix denoting multiple entities.

poly(A) tail A long sequence of adenine nucleotides (50–250) added after transcription to the 3' end of most eukaryotic mRNAs.

polygenes Multiple loci whose alleles increase or decrease a continuously variable phenotypic trait.

polymer [Gk. *poly*: many + *meros*: unit] A large molecule made up of similar or identical subunits called monomers. (Contrast with monomer, oligomer.)

polymerase chain reaction (PCR) An enzymatic technique for the rapid production of millions of copies of a particular stretch of DNA.

polymerization reactions Chemical reactions that generate polymers by linking monomers.

polymorphic Referring to a gene whose most frequent allele in a population is present less than 99% of the time.

polymorphism (pol' lee mor' fiz um) [Gk. *poly*: many + *morphe*: form, shape] In genetics, the coexistence in the same population of two distinct hereditary types based on different alleles.

polyp The sessile asexual stage in the life cycle of most cnidarians.

polypeptide A large molecule made up of many amino acids joined by peptide linkages. Large polypeptides are called proteins.

polyphyletic (pol' lee fih leht' ik) [Gk. *poly*: many + *phylon*: tribe] Referring to a group that consists of multiple distantly related organisms, and does not include the common ancestor of the group. (Compare with monophyletic, paraphyletic.)

polyploidy (pol' lee ploid ee) The possession of more than two entire sets of chromosomes.

polysaccharide A macromolecule composed of many monosaccharides (simple sugars). Common examples are cellulose and starch.

polysome (polyribosome) A complex consisting of a threadlike molecule of messenger RNA and several (or many) ribosomes. The ribosomes move along the mRNA, synthesizing polypeptide chains as they proceed.

polytene (pol' lee teen) [Gk. *poly*: many + *taenia*: ribbon] An adjective describing giant interphase chromosomes, such as those found in the salivary glands of fly larvae. The characteristic pattern of bands and bulges seen on these chromosomes provided a method for preparing detailed chromosome maps of several organisms.

pons [L. *pons*: bridge] Region of the brain stem anterior to the medulla.

population Any group of organisms coexisting at the same time and in the same place and capable of interbreeding with one another.

population bottleneck   *See* bottleneck.

population density   The number of individuals (or modules) of a population in a unit of area or volume.

population genetics   The study of genetic variation and its causes within populations.

population structure   The proportions of individuals in a population belonging to different age classes (age structure). Also, the distribution of the population in space.

portal blood vessels   Blood vessels that begin and end in capillary beds.

positional cloning   A technique for isolating a gene associated with a disease on the basis of its approximate chromosomal location.

positional information   Signals by which genes regulate cell functions to locate cells in a tissue during development.

positive control   The situation in which a regulatory macromolecule is needed to turn transcription of structural genes on. In its absence, transcription will not occur.

positive cooperativity   Occurs when a molecule can bind several ligands and each one that binds alters the conformation of the molecule so that it can bind the next ligand more easily. The binding of four molecules of O_2 by hemoglobin is an example of positive cooperativity.

post [L. *postere*: behind, following after]   Prefix denoting something that comes after.

postabsorptive period   When there is no food in the gut and no nutrients are being absorbed.

posterior pituitary   The portion of the pituitary gland that is derived from neural tissue.

postsynaptic cell   The cell whose membranes receive neurotransmitter after its release by another cell (the presynaptic cell) at a synapse.

postzygotic reproductive barriers   Barriers to the reproductive process that occur after the union of the nuclei of two gametes. (Contrast with prezygotic reproductive barriers.)

potential energy   "Stored" energy not doing work, such as the energy in chemical bonds.

precapillary sphincter   A cuff of smooth muscle that can shut off the blood flow to a capillary bed.

pre-mRNA (precursor mRNA)   Initial gene transcript before it is modified to produce functional mRNA. Also known as the primary transcript.

predator   An organism that kills and eats other organisms.

pressure flow model   An effective model for phloem transport in angiosperms. It holds that sieve element transport is driven by an osmotically driven pressure gradient between source and sink.

pressure potential   The hydrostatic pressure of an enclosed solution in excess of the surrounding atmospheric pressure. (Contrast with solute potential, water potential.)

presynaptic excitation/inhibition   Occurs when a neuron modifies activity at a synapse by releasing a neurotransmitter onto the presynaptic nerve terminal.

prey [L. *praeda*: booty]   An organism consumed as an energy source.

prezygotic reproductive barriers   Barriers to the reproductive process that occur before the union of two gametes (Contrast with postzygotic reproductive barriers.)

primary active transport   Form of active transport in which ATP is hydrolyzed, yielding the energy required to transport ions against their concentration gradients. (Contrast with secondary active transport.)

primary consumer   An herbivore; an organism that eats plant tissues.

primary embryonic organizer   *See* organizer.

primary growth   In plants, growth produced by the apical meristems. (Contrast with secondary growth.)

primary immune response   The first response of the immune system to an antigen, involving recognition by lymphocytes and the production of effector cells and memory cells. (Contrast with secondary immune response.)

primary motor cortex   The region of the cerebral cortex that contains motor neurons that directly stimulate specific muscle fibers to contract.

primary producer   A photosynthetic or chemosynthetic organism that synthesizes complex organic molecules from simple inorganic ones.

primary sex determination   Genetic determination of gametic sex, male or female. (Contrast with secondary sex determination.)

primary somatosensory cortex   The region of the cerebral cortex that receives input from mechanosensors distributed throughout the body.

primary succession   Succession that begins in an area initially devoid of life, such as on recently exposed glacial till or lava flows. (Contrast with secondary succession.)

primary structure   The specific sequence of amino acids in a protein.

primary wall   Cellulose-rich cell wall layers laid down by a growing plant cell.

primase   An enzyme that catalyzes the synthesis of a primer for DNA replication.

primer   A short, single-stranded segment of DNA that is the necessary starting material for the synthesis of a new DNA strand, which is synthesized from the 3' end of the primer.

primitive streak   A line running axially along the blastodisc, the site of inward cell migration during formation of the three-layered embryo. Formed in the embryos of birds and fish.

primordium [L. *primordium*: origin]   The most rudimentary stage of an organ or other part.

prion   An infectious protein that can proliferate by converting other proteins.

pro- [L.: first, before, favoring]   A prefix often used in biology to denote a developmental stage that comes first or an evolutionary form that appeared earlier than another. For example, prokaryote, prophase.

probe   A segment of single stranded nucleic acid used to identify DNA molecules containing the complementary sequence.

procambium   Primary meristem that produces the vascular tissue.

processive   Referring to an enzyme that catalyzes many reactions each time it binds to a substrate, as DNA polymerase does during DNA replication.

progesterone [L. *pro*: favoring + *gestare*: to bear]   A vertebrate female sex hormone that maintains pregnancy.

prokaryotes (pro kar' ry otes) [L. *pro*: before + Gk. *karyon*: kernel, nucleus]   Organisms whose genetic material is not contained within a nucleus: the bacteria and archaea. Considered an earlier stage in the evolution of life than the eukaryotes.

prometaphase   The phase of nuclear division that begins with the disintegration of the nuclear envelope.

promoter   The region of an operon that acts as the initial binding site for RNA polymerase.

proofreading   The correction of an error in DNA replication just after an incorrectly paired base is added to the growing polynucleotide chain.

prophage (pro' fayj)   The noninfectious units that are linked with the chromosomes of the host bacteria and multiply with them but do not cause dissolution of the cell. Prophage can later enter into the lytic phase to complete the virus life cycle.

prophase (pro' phase)   The first stage of nuclear division, during which chromosomes condense from diffuse, threadlike material to discrete, compact bodies.

prostaglandin   Any one of a group of specialized lipids with hormone-like functions. It is not clear that they act at any considerable distance from the site of their production.

prostate gland   Glandular tissue that surrounds the male urethra at its junction with the vas deferens; contributes an alkaline fluid to the semen.

prosthetic group   Any nonprotein portion of an enzyme.

proteasome   In the eukaryotic cytoplasm, a huge protein structure that binds to and digests cellular proteins that have been tagged by ubiquitin.

protein (pro' teen) [Gk. *protos*: first]   One of the most fundamental building substances of living organisms. A long-chain polymer of amino acids with twenty different common side chains. Occurs with its polymer chain extended in fibrous proteins, or coiled into a compact macromolecule in enzymes and other globular proteins.

protein domain   *See* domain (1)

protein kinase   An enzyme that catalyzes the addition of a phosphate group from ATP to a target protein.

protein kinase cascade   aA series of reactions in response to a molecular signal, in which a series of protein kinases activates one another in sequence, amplifying the signal at each step.

proteoglycan   A glycoprotein containing a protein core with attached long, linear carbohydrate chains.

proteolysis [protein + Gk. *lysis*: break apart]   An enzymatic digestion of a protein or polypeptide.

proteome   The total of the different proteins that can be made by an organism. Because of alternate splicing of pre-mRNA, the number of proteins that can be made is usually much larger than the number of protein-coding genes present in the organism's genome.

protobiont [Gk. *protos*: first, before + *bios*: life] Aggregates of abiotically produced molecules that cannot reproduce but do maintain internal chemical environments that differ from their surroundings.

protoderm Primary meristem that gives rise to the plant epidermis.

proton (pro' ton) [Gk. *protos*: first, before] (1) A subatomic particle with a single positive charge. The number of protons in the nucleus of an atom determine its element. (2) A hydrogen ion, H^+.

proton pump An active transport system that uses ATP energy to move hydrogen ions across a membrane generating an electric potential (voltage).

proton motive force A force generated across a membrane expressed in millivolts having two components: a chemical potential (difference in proton concentration) plus an electrical potential due to the electrostatic charge on the proton.

proto-oncogenes The normal alleles of genes possessing oncogenes (cancer-causing genes) as mutant alleles. Proto-oncogenes encode growth factors and receptor proteins.

prototroph (pro' tow trofe') [Gk. *protos*: first + *trophein*: to nourish] The nutritional wild type, or reference form, of an organism. Any deviant form that requires growth nutrients not required by the prototrophic form is said to be a nutritional mutant, or auxotroph.

proximal Near the point of attachment or other reference point. (Contrast with distal.)

pseudocoelom [Gk. *pseudes*: false] A body cavity not surrounded by a peritoneum. Characteristic of nematodes and rotifers.

pseudogene [Gk. *pseudes*: false] A DNA segment that is homologous to a functional gene but is not expressed because of changes to its sequence or changes to its location in the genome.

pseudopod (soo' do pod) [Gk. *pseudes*: false + *podos*: foot] A temporary, soft extension of the cell body that is used in location, attachment to surfaces, or engulfing particles.

pulmonary [L. *pulmo*: lung] Pertaining to the lungs.

punctuated equilibrium An evolutionary pattern in which periods of rapid change are separated by longer periods of little or no change.

Punnett square A method of predicting the results of a genetic cross by arranging the gametes of each parent at the edges of a square.

pupa (pew' pa) [L. *pupa*: doll, puppet] In certain insects (the Holometabola), the encased developmental stage between the larva and the adult.

pupil The opening in the vertebrate eye through which light passes.

purine (pure' een) One of the types of nitrogenous bases. The purines adenine and guanine are found in nucleic acids. (Contrast with pyrimidine.)

Purkinje fibers Specialized heart muscle cells that conduct excitation throughout the ventricular muscle.

pyrimidine (per im' a deen) A type of nitrogenous base. The pyrimidines cytosine, thymine, and uracil are found in nucleic acids.

pyruvate A three-carbon acid; the end product of glycolysis and the raw material for the citric acid cycle.

pyruvate oxidation Conversion of pyruvate to acetyl CoA and CO_2 that occurs in the mitochondrial matrix in the presence of O_2.

- Q -

Q_{10} A value that compares the rate of a biochemical process or reaction over a 10°C range of temperature. A process that is not temperature-sensitive has a Q_{10} of 1; values of 2 or 3 mean the reaction speeds up as temperature increases.

quantum (kwon' tum) [L. *quantus*: how great] An indivisible unit of energy.

quaternary structure The specific three dimensional arrangement of protein subunits.

quiescent center In root meristem, central region where cells do not divide or divide very slowly.

- R -

R factor (resistance factor) A plasmid that contains one or more genes that encode resistance to antibiotics.

R genes Resistance genes that function in plant defenses against bacteria, fungi, and nematodes. See gene-for-gene resistance.

R group The distinguishing group of atoms of a particular amino acid.

radial symmetry The condition in which two halves of a body are mirror images of each other regardless of the angle of the cut, providing the cut is made along the center line. Thus, a cylinder cut lengthwise down its center displays this form of symmetry. (Contrast with biradial symmetry.)

radioisotope A radioactive isotope of an element. Examples are carbon-14 (^{14}C) and hydrogen-3, or tritium (^3H).

radiometry The use of the regular, known rates of decay of radioisotopes of elements to determine dates of events in the distant past.

rain shadow The relatively dry area on the leeward side of a mountain range.

rapid-eye-movement See REM sleep

reactant A chemical substance that enters into a chemical reaction with another substance.

reaction A chemical change in which changes take place in the kind, number, or position of atoms making up a substance.

reaction center A group of electron transfer proteins that receive energy from light-absorbing pigments and convert it to chemical energy by redox reactions.

receptacle The end of a plant stem to which the parts of the flower are attached.

receptive field The area of visual space that activates a particular cell in the visual system.

receptor A site or protein on the outer surface of the plasma membrane or in the cytoplasm to which a specific ligand from another cell binds.

receptor-mediated endocytosis Endocytosis initiated by macromolecular binding to a specific membrane receptor.

receptor potential The change in the resting potential of a sensory cell when it is stimulated.

recessive In genetics, an allele that does not determine phenotype in the presence of a dominant allele. (Contrast with dominance.)

reciprocal crosses A pair of matings in one of which a female of genotype A mates with a male of genotype B and in the other of which a female of genotype B mates with a male of genotype A.

recognition site See restriction site.

recombinant An individual, meiotic product, or single chromosome in which genetic materials originally present in two individuals end up in the same haploid complement of genes. The reshuffling of genes can be either by independent segregation, or by crossing over between homologous chromosomes.

recombinant DNA DNA generated in vitro, from more than one source.

recombinant DNA technology The application of restriction endonucleases, plasmids, and transformation to alter and assemble recombinant DNA, with the goal of producing specific proteins.

recombinant frequency The proportion of offspring of a genetic cross that have phenotypes different from the parental phenotypes due to crossing over between linked genes during gamete formation.

reconciliation ecology The practice of making exploited lands more biodiversity-friendly.

rectum The terminal portion of the gut, ending at the anus.

redox reaction A chemical reaction in which one reactant becomes oxidized and the other becomes reduced.

reducing agent A substance that can donate electrons to another substance. The reducing agent becomes oxidized, and its partner becomes reduced.

reduction Gain of electrons by a chemical reactant; any reduction is accompanied by an oxidation. (Contrast with oxidation.)

reflex An automatic action, involving only a few neurons (in vertebrates, often in the spinal cord), in which a motor response swiftly follows a sensory stimulus.

refractory period Of a neuron, the time interval after an action potential, during which another action potential cannot be elicited.

regulative development A pattern of animal embryonic development in which the fates of the first blastomeres are not absolutely fixed. (Contrast with mosaic development.)

regulator sequence A DNA sequence to which the protein product of a regulatory gene binds.

regulatory gene A gene that codes for a protein that controls the transcription of another gene(s).

regulatory subunit The polypeptide in an enzyme protein with quaternary structure that does not contain the active site, but instead binds nonsubstrate molecules and changes its structure, in turn changing the structure and function of the active site. (Contrast catalytic subunit.)

regulatory system A system that uses feedback information to maintain a physiological function or parameter at an optimal level.

reinforcement The evolution of enhanced reproductive isolation between populations due to natural selection for greater isolation.

releaser A sensory stimulus that triggers the performance of a stereotyped behavior pattern.

releasing hormone One of several hypothalamic hormones that stimulates the secretion of anterior pituitary hormone.

REM sleep A sleep state characterized by vivid dreams, skeletal muscle relaxation, and rapid eye movements.

renal [L. *renes*: kidneys] Relating to the kidneys.

replication Pertaining to the duplication of genetic material.

replication complex The close association of several proteins operating in the replication of DNA.

replication fork A point at which a DNA molecule is replicating. The fork forms by the unwinding of the parent molecule.

replicon A region of DNA controlled by a single origin of replication.

reporter gene A marker gene included in recombinant DNA to indicate the presence of the recombinant DNA in a host cell.

repressible enzyme An enzyme whose synthesis can be decreased or prevented by the presence of a particular compound. A repressible operon often controls the synthesis of such an enzyme.

repressor A protein coded by the regulatory gene. The repressor can bind to a specific operator and prevent transcription of the operon.

reproductive isolating mechanism Any trait that prevents individuals from two different populations from producing fertile hybrids.

reproductive isolation The condition in which a population is not exchanging genes with other populations of the same species.

rescue effect The process by which a few individuals moving among declining subpopulations of a species and reproducing may prevent their extinction.

resolution Of an optical device such as a microscope, the smallest distance between two lines that allows the lines to be seen as separate from one another.

resource Something in the environment required by an organism for its maintenance and growth that is consumed in the process of being used.

respiration (res pi ra′ shun) [L. *spirare*: to breathe] (1) Cellular respiration; the catabolic pathways by which electrons are removed from various molecules and passed through intermediate electron carriers to O_2, generating H_2O and releasing energy. (2) Breathing.

respiratory chain The terminal reactions of cellular respiration, in which electrons are passed from NAD or FAD, through a series of intermediate carriers, to molecular oxygen, with the concomitant production of ATP.

resting potential The membrane potential of a living cell at rest. In cells at rest, the interior is negative to the exterior. (Contrast with action potential, electrotonic potential.)

restoration ecology The science and practice of restoring damaged or degraded ecosystems.

restriction digestion Use of restriction enzymes to cleave DNA into fragments in a test tube.

restriction endonuclease *See* restriction enzyme.

restriction enzyme Any one of several enzymes, produced by bacteria, that break foreign DNA molecules at specific sites. Some produce "sticky ends." Extensively used in recombinant DNA technology.

restriction fragment length polymorphism *See* RFLP.

restriction point The specific time during G1 of the cell cycle at which the cell becomes committed to undergo the rest of the cell cycle.

restriction site A specific DNA base sequence recognized and acted on by a restriction endonuclease cutting the DNA.

reticular system A central region of the vertebrate brain stem that includes complex fiber tracts conveying neural signals between the forebrain and the spinal cord, with collateral fibers to a variety of nuclei that are involved in autonomic functions, including arousal from sleep.

retina (rett′ in uh) [L. *rete*: net] The light-sensitive layer of cells in the vertebrate or cephalopod eye.

retinal The light-absorbing portion of visual pigment molecules. Derived from β-carotene.

retinoblastoma protein A protein that inhibits an animal cell from passing through the restriction point; inactivation of this protein is necessary for the cell cycle to proceed.

retrovirus An RNA virus that contains reverse transcriptase. Its RNA serves as a template for cDNA production, and the cDNA is integrated into a chromosome of the mammalian host cell.

reversal *See* evolutionary reversal.

reverse transcriptase An enzyme that catalyzes the production of DNA (cDNA), using RNA as a template; essential to the reproduction of retroviruses.

reversible reaction A chemical change that can occur in both the forward and reverse directions.

RFLP (restriction fragment length polymorphism) Coexistence of two or more patterns of restriction fragments (patterns produced by restriction enzymes), as revealed by a probe. The polymorphism reflects a difference in DNA sequence on homologous chromosomes.

rhizoids (rye′ zoids) [Gk. *rhiza*: root] Hairlike extensions of cells in mosses, liverworts, and a few vascular plants that serve the same function as roots and root hairs in vascular plants. The term is also applied to branched, rootlike extensions of some fungi and algae.

rhizome (rye′ zome) A special underground stem (as opposed to root) that runs horizontally beneath the ground.

rhodopsin A photopigment used in the visual process of transducing photons of light into changes in the membrane potential of photoreceptor cells.

ribonucleic acid *See* RNA.

ribose A five-carbon sugar in nucleotides and RNA.

ribosomal RNA (rRNA) Several species of RNA that are incorporated into the ribosome. Involved in peptide bond formation.

ribosome A small organelle that is the site of protein synthesis.

ribozyme An RNA molecule with catalytic activity.

risk cost The increased chance of being injured or killed as a result of performing a behavior, compared to resting.

RNA (ribonucleic acid) An often single stranded nucleic acid whose nucleotides use ribose rather than deoxyribose and in which the base uracil replaces thymine found in DNA. Serves as genome from some viruses. (*See* rRNA, tRNA, mRMA, and ribozyme.)

RNA editing The alteration of bases on mRNA prior to its translation.

RNA polymerase An enzyme that catalyzes the formation of RNA from a DNA template.

RNA primase A replication complex enzyme that makes the primer strand of DNA needed to initiate DNA replication.

RNA splicing The last stage of RNA processing in eukaryotes, in which the transcripts of introns are excised through the action of small nuclear ribonucleoprotein particles (snRNP).

rod cells Light-sensitive cell in the vertebrate retina; these sensory receptor cells are sensitive in extremely dim light and are responsible for dim light, black and white vision.

root The organ responsible for anchoring the plant in the soil, absorbing water and minerals, and producing certain hormones. Some roots are storage organs.

root cap A thimble-shaped mass of cells, produced by the root apical meristem, that protects the meristem; the organ that perceives the gravitational stimulus in root gravitropism.

root hair A long, thin process from a root epidermal cell that absorbs water and minerals from the soil solution.

rough ER That portion of the endoplasmic reticulum whose outer surface has attached ribosomes. (Contrast with smooth ER.)

rRNA *See* ribosomal RNA.

RT-PCR A technique in which RNA is first converted to cDNA by the use of the enzyme reverse transcriptase, then the cDNA is amplified by the polymerase chain reaction.

rubisco (ribulose bisphosphate carboxylase/ oxygenase) Acronym for the enzyme that combines carbon dioxide or oxygen with ribulose bisphosphate to catalyze the first step of the Calvin-Benson cycle.

rumen (rew′ mun) The first division of the ruminant stomach. It stores and initiates bacterial fermentation of food. Food is regurgitated from the rumen for further chewing.

ruminant Herbivorous, cud-chewing mammals such as cows or sheep. The ruminant stomach consists of four compartments.

– S –

S phase In the cell cycle, the stage of interphase during which DNA is replicated. (Contrast with G_1 phase, G_2 phase, M phase.)

saprobe [Gk. *sapros*: rotten] An organism (usually a bacterium or fungus) that obtains its carbon and energy directly from dead organic matter.

sarcomere (sark' o meer) [Gk. *sark*: flesh + *meros*: unit]   The contractile unit of a skeletal muscle.

sarcoplasm   The cytoplasm of a muscle cell.

sarcoplasmic reticulum   The endoplasmic reticulum of a muscle cell.

saturated fatty acid   A fatty acid usually containing from 12 to 18 carbon atoms and no double bonds.

scientific method   A means of gaining knowledge about the natural world by making observations, posing hypotheses, and conducting experiments to test those hypotheses.

schizocoelous development [Gk. *schizo*: split + *koiloma*: cavity]   Formation of a coelom during embryological development by a splitting of mesodermal masses.

Schwann cell   A glial cell that wraps around part of the axon of a peripheral neuron, creating a myelin sheath.

sclereid [Gk. *skleros*: hard]   A type of sclerenchyma cell, commonly found in nutshells, that is not elongated.

sclerenchyma (skler eng' kyma) [Gk. *skleros*: hard + *kymus*: juice]   A plant tissue composed of cells with heavily thickened cell walls, dead at functional maturity. The principal types of sclerenchyma cells are fibers and sclereids.

scrotum   A sac of skin that encloses the testicles in male mammals.

second law of thermodynamics   States that in any real (irreversible) process, there is a decrease in free energy and an increase in entropy.

second messenger   A compound, such as cAMP, that is released within a target cell after a hormone (first messenger) has bound to a surface receptor on a cell; the second messenger triggers further reactions within the cell.

secondary active transport   Form of active transport which does not use ATP as an energy source; rather, transport is coupled to ion diffusion down a concentration gradient established by primary active transport.**secondary consumer** An organism that eat primary consumers.

secondary growth   In plants, growth produced by vascular and cork cambia, contributing to an increase in girth. (Contrast with primary growth.)

secondary immune response   A rapid and intense response to a second or subsequent exposure to an antigen, initiated by memory cells. (Contrast with primary immune response.)

secondary metabolite   A compound synthesized by a plant that is not needed for basic cellular metabolism. Typically has an antiherbivore or antiparasite function.

secondary sex determination   Formation of nongametic features of sex, such as external organs and body hair. (Contrast with primary sex determination.)

secondary structure   Of a protein, localized regularities of structure, such as the α helix and the β pleated sheet.

secondary succession   Ecological succession after a disturbance that did not eliminate all the organisms originally living on the site. (Contrast with primary succession.)

secondary wall   Wall layers laid down by a plant cell that has ceased growing; often impregnated with lignin or suberin.

secretin (si kreet' in)   A peptide hormone secreted by the upper region of the small intestine when acidic chyme is present. Stimulates the pancreatic duct to secrete bicarbonate ions.

section   A thin slice, usually for microscopy, as a tangential section or a transverse section.

seed   A fertilized, ripened ovule of a gymnosperm or angiosperm. Consists of the embryo, nutritive tissue, and a seed coat.

seed plant   Plants in which the embryo is protected and nourished within a seed; the gymnosperms and angiosperms.

seedling   A young plant that has grown from a seed (rather than by grafting or by other means.)

segmentation genes   In insect larvae, genes that determine the number and polarity of larval segments.

segment polarity genes   Genes that determine the boundaries and front-to-back organization of the segments in the *Drosophila* larva.

segregation   In genetics, the separation of alleles, or of homologous chromosomes, from each other during meiosis so that each of the haploid daughter nuclei produced by meiosis contains one or the other member of the pair found in the diploid parent cell, but never both. This principle was articulated by Mendel as his first law.

selective permeability   Allowing certain substances to pass through while other substances are excluded; a characteristic of membranes.

self incompatability   In plants, the rejection of their own pollen; promotes genetic variation and limits inbreeding.

selfish act   A behavioral act that benefits its performer but harms the recipients.

semen (see' men) [L. *semin*: seed]   The thick, whitish liquid produced by the male reproductive organ in mammals, containing the sperm.

semiconservative replication   The way in which DNA is synthesized. Each of the two partner strands in a double helix acts as a template for a new partner strand. Hence, after replication, each double helix consists of one old and one new strand.

seminiferous tubules   The tubules within the testes within which sperm production occurs.

senescence [L. *senescere*: to grow old]   Aging; deteriorative changes with aging; the increased probability of dying with increasing age.

sensory receptor cells   Cells that are responsive to a particular type of physical or chemical stimulation.

sensory transduction   The transformation of environmental stimuli or information into neural signals.

sepal (see' pul) [L. *sepalum*: covering]   One of the outermost structures of the flower, usually protective in function and enclosing the rest of the flower in the bud stage.

septum [L. *saeptum*: wall, fence]   (1) A partition or cross-wall appearing in the hyphae of some fungi. (2) The bony structure dividing the nasal passages.

Sertoli cells   Cells in the seminiferous tubules that nuture the developing sperm.

sessile (sess' ul) [L. *sedere*: to sit]   Permanently attached; not moving.

set point   In a regulatory system, the threshold sensitivity to the feedback stimulus.

sex chromosome   In organisms with a chromosomal mechanism of sex determination, one of the chromosomes involved in sex determination.

sex linkage   The pattern of inheritance characteristic of genes located on the sex chromosomes of organisms having a chromosomal mechanism for sex determination.

sex pilus (pill' us) [L. *pilus*: hair]   A structure on the cell wall that allows one bacterium to adhere to another prior to conjugation.

sexual reproduction   Reproduction involving union of gametes.

sexual selection   Selection by one sex of characteristics in individuals of the opposite sex. Also, the favoring of characteristics in one sex as a result of competition among individuals of that sex for mates.

shared derived trait   A trait that arose in the ancestor of a phylogenetic group and is present (sometimes in modified form) in all of its members, thus helping define that group. Also called a synapomorphy.

shoot system   The aerial parts of a vascular plant, consisting of the leaves, stem(s), and flowers.

short-day plant (SDP)   A plant that requires short days (or long nights) in order to flower.

short tandem repeat (STR)   An inherited, short (2–5 base pairs), moderately repetitive sequence of DNA.

shotgun sequencing   A relatively rapid method of analyzing DNA sequences in which a large DNA molecule is broken up into overlapping fragments, each fragment is sequenced, and computers are used to analyze and realign the fragments.

sieve tube   A column of specialized cells found in the phloem, specialized to conduct organic matter from sources (such as photosynthesizing leaves) to sinks (such as roots). Found principally in flowering plants.

sieve tube element   A single cell of a sieve tube, containing cytoplasm but relatively few organelles, with highly specialized perforated end walls leading to elements above and below.

signal   A chemical (neurotransmitter or hormone) or light message emitted from a cell or cells or organism(s) and received by others to cause some change in function or behavior.

signal recognition particle (SRP)   A complex of RNA and protein that recognizes both the signal sequence on a growing polypeptide and receptor protein on the surface of the ER.

signal sequence   The sequence of a protein that directs the protein through a particular cellular membrane.

signal transduction pathway   The series of biochemical steps whereby a stimulus to a cell (such as a hormone or neurotransmitter binding to a receptor) is translated into a response of the cell.

silencer sequence   A sequence of eukaryotic DNA that binds proteins that inhibit the transcription of an associated gene.

silent substitution   A change in gene sequence that, due to the redundancy of the genetic code, has no effect on the amino acid produced, and thus no effect on the protein phenotype. Also called a synonymous substitution.

similarity matrix   A matrix used to compare the degree of divergence among pairs of objects. For molecular sequences, constructed by summing the number or percentage of nucleotidies or amino acids that are identical in each pair of sequences.

single nucleotide polymorphisms (SNPs)   Inherited variations in a single nucleotide base in DNA.

single-strand binding protein   In DNA replication, a protein that binds to single strands of DNA after they have been separated from each other, keeping the two strands separate for replication.

sinoatrial node (sigh' no ay' tree al) [L. *sinus*: curve + *atrium*: hall, chamber]   The pacemaker of the mammalian heart.

sink   In plants, any organ that imports the products of photosynthesis, such as roots, developing fruits, immature leaves. (Contrast with source.)

sinus (sigh' nus) [L. *sinus*: curve, hollow]   A cavity in a bone, a tissue space, or an enlargement in a blood vessel.

sister chromatid   In the eukaryotic cell, a chromatid resulting from chromosome replication during interphase.

sister groups   Two phylogenetic groups that are each other's closest relatives.

skeletal muscle   *See* striated muscle.

sliding DNA clamp   A protein complex that keeps DNA polymerase bound to DNA during replication.

sliding filament theory   A proposed mechanism of muscle contraction based on formation and breaking of crossbridges between actin and myosin filaments, causing them to slide together.

slow-twitch fibers   Skeletal muscle fibers that generate tension slowly, but are resistant to fatigue ("marathon" fibers). They have abundant mitochondria, enzymes of aerobic metabolism, myoglobin, and blood supply.

slow-wave sleep   A state of mammalian and avian sleep characterized by high amplitude slow waves in the EEG.

small intestine   The portion of the gut between the stomach and the colon; consists of the duodenum, the jejunum, and the ileum.

small nuclear ribonucleoprotein particle (snRNP)   A complex of an enzyme and a small nuclear RNA molecule, functioning in RNA splicing.

smooth muscle   One of three types of muscle tissue. Usually consists of sheets of mononucleated cells innervated by the autonomic nervous system. (Compare with cardiac muscle, striated muscle.)

sodium–potassium pump (Na–K pump)   The complex protein in plasma membranes that is responsible for primary active transport; it pumps sodium ions out of the cell and potassium ions into the cell, both against their concentration gradients.

solute   A substance that is dissolved in a liquid (solvent) to form a solution.

solute potential   A property of any solution, resulting from its solute contents; it may be zero or have a negative value. The more negative the solute potential, the greater the tendency of the solution to take up water through a differentially permeable membrane. (Contrast with pressure potential, water potential.)

solution   A liquid (the solvent) and its dissolved solutes.

somatic [Gk. *soma*: body]   Pertaining to the body. Somatic cells are cells of the body (as opposed to germ cells).

somatic mutation   Permanent genetic change in a somatic cell. These mutations affect the individual only; they are not passed on to offspring. (Contrast with germ line mutation)

somite (so' might)   One of the segments into which an embryo becomes divided longitudinally, leading to the eventual segmentation of the animal as illustrated by the spinal column, ribs, and associated muscles.

source   In plants, an organ exporting photosynthetic products in excess of its own needs. For example, a mature leaf or storage organ. (Contrast with sink.)

spatial summation   In the production or inhibition of action potentials in a postsynaptic neuron, the interaction of depolarizations and hyperpolarizations produced by several terminal boutons.

spawning   The direct release of sex cells into the water.

speciation (spee' shee ay' shun)   The process of splitting one population into two populations that are reproductively isolated from one another.

species (spee' shees) [L. *species*: kind]   The basic lower unit of classification, consisting of an ancestor–descendant lineage of populations of closely related and similar organisms. The more narrowly defined "biological species" consists of individuals capable of interbreeding freely with each other but not with members of other species.

species–area relationship   The relationship between the sizes of areas and the numbers of species they support.

species diversity   A weighted representation of the species of organisms living in a region; large and common species are given greater weight than are small and rare ones. (Contrast with species richness.)

species richness   The total number of species living in a region. (Contrast with species diversity.)

specific defenses   Defensive reactions of the immune system that are based on antibody reaction with a specific antigen.

specific heat   The amount of energy that must be absorbed by a gram of a substance to raise its temperature by one degree centigrade. By convention, water is assigned a specific heat of one.

sperm [Gk. *sperma*: seed]   A male gamete (reproductive cell).

spermatogenesis (spur mat' oh jen' e sis) [Gk. *sperma*: seed + *genesis*: source]   Male gametogenesis, leading to the production of sperm.

sphincter (sfink' ter) [Gk. *sphinkter*: something that binds tightly]   A ring of muscle that can close an orifice, for example at the anus.

spindle apparatus   An array of microtubules stretching from pole to pole of a dividing nucleus and playing a role in the movement of chromosomes at nuclear division. Named for its shape.

spiracle (spy' rih kel) [L. *spirare*: to breathe]   An opening of the treacheal respiratory system of terrestrial arthorpods.

spleen   An organ that serves as a reservoir for venous blood and eliminates old, damaged red blood cells from the circulation.

spliceosome   An RNA–protein complex that splices out introns from eukaryotic pre-mRNAs.

splicing   Removal of introns and connecting of exons in eukaryotic pre-mRNAs.

spontaneous mutation   A genetic change caused by internal cellular mechanisms, such as an error in DNA replication. (Contrast with induced mutation.)

spontaneous reaction   A chemical reaction that will proceed on its own, without any outside influence. A spontaneous reaction need not be rapid.

sporangium (spor an' gee um) [Gk. *spora*: seed + *angeion*: vessel or reservoir]   In plants and fungi, any specialized stucture within which one or more spores are formed.

spore [Gk. *spora*: seed]   Any asexual reproductive cell capable of developing into an adult organism without gametic fusion. In plants, haploid spores develop into gametophytes, diploid spores into sporophytes. In prokaryotes, a resistant cell capable of surviving unfavorable periods.

sporocyte   Specialized cells of the diploid sporophyte that will divide by meiosis to produce four haploid spores. Germination of these spores produces the haploid gametophyte.

sporophyte (spor' o fyte) [Gk. *spora*: seed + *phyton*: plant]   In plants and protists with alternation of generations, the diploid phase that produces the spores. (Contrast with gametophyte.)

stabilizing selection   Selection against the extreme phenotypes in a population, so that the intermediate types are favored. (Contrast with disruptive selection.)

stamen (stay' men) [L. *stamen*: thread]   A male (pollen-producing) unit of a flower, usually composed of an anther, which bears the pollen, and a filament, which is a stalk supporting the anther.

starch [O.E. *stearc*: stiff]   A polymer of glucose; used by plants to store energy.

start codon   The mRNA triplet (AUG) that acts as a signal for the beginning of translation at the ribosome. (Contrast with stop codon.)

stasis [Gk. *stasis*: to stop, stand still]   Period during which little or no evolutionary change takes place within a lineage or groups of lineages.

statocyst (stat' oh sist) [Gk. *statos*: stationary + *kystos*: cell]   An organ of equilibrium in some invertebrates.

statolith (stat' oh lith) [Gk. *statos*: stationary + *lithos*: stone]   A solid object that responds to gravity or movement and stimulates the mechanoreceptors of a statocyst.

stele (steel) [Gk. *stylos*: pillar]   The central cylinder of vascular tissue in a plant stem.

stem   Plant structure that holds leaves and/or flowers; it is the site for transporting and distributing material throughout the plant.

stem cells   In animals, undifferentiated cells that are capable of extensive proliferation. A stem cell generates more stem cells and a large clone of differentiated progeny cells.

steroid   Any of numerous lipids based on a 17-carbon atom ring system.

sticky ends   On a piece of two-stranded DNA, short, complementary, one-stranded regions produced by the action of a restriction endonuclease. Sticky ends allow the joining of segments of DNA from different sources.

stigma [L. *stigma*: mark, brand]   The part of the pistil at the apex of the style that is receptive to pollen, and on which pollen germinates.

stimulus [L. *stimulare*: to goad]   Something causing a response; something in the environment detected by a receptor.

stolon [L. *stolon*: branch, sucker]   A horizontal stem that forms roots at intervals.

stoma (plural: stomata) [Gk. *stoma*: mouth, opening]   Small opening in the plant epidermis that permits gas exchange; bounded by a pair of guard cells whose osmotic status regulates the size of the opening.

stop codon   Any of the thre mRNA codons that signal the end of protein translation at the ribosome: UAG, UGA, UAA.

stratosphere   The upper part of Earth's atmosphere, above the troposphere; extends from approximately 18 kilometers upward to approximately 50 kilometers above the surface.

stratum (plural strata) [L. *stratos*: layer]   A layer of sedimentary rock laid down at a particular time in the past.

striated muscle   Contractile tissue characterized by multinucleated cells containing highly ordered arrangements of actin and myosin microfilaments. Also known as skeletal muscle. (Compare with cardiac muscle, smooth muscle.)

strobilus   A conelike structure consisting of spore-bearing scales (modified leaves) inserted on an axis. (Contrast with cone.)

stroma   The fluid contents of an organelle, such as a chloroplast.

stromatolites   Composite, flat-to-domed structures composed of successive mineral layers produced by the action of cyanobacteria in water; ancient ones provide evidence for early life on the earth.

structural gene   A gene that encodes the primary structure of a protein.

structural isomers   Molecules made up of the same kinds and numbers of atoms, in which the atoms are bonded differently.

style [Gk. *stylos*: pillar or column]   In flowering plants, a column of tissue extending from the tip of the ovary, and bearing the stigma or receptive surface for pollen at its apex.

sub- [L. *sub*: under]   A prefix often used to designate a structure that lies beneath another or is less than another. For example, subcutaneous (beneath the skin); subspecies.

suberin   A waxlike lipid that acts as a barrier to water and solute movement across the Casparian strip of the endodermis. Suberin is the waterproofing element in the cell walls of cork.

submucosa (sub mew koe' sah)   The tissue layer just under the epithelial lining of the lumen of the digestive tract.

substrate (sub' strayte)   The molecule or molecules on which an enzyme exerts catalytic action.

substrate-level phosphorylation   Reaction in which ATP is formed from ADP by the addition of a phosphate group directly from a reactant.

substratum   The base material on which a sessile organism lives.

succession   In ecology, the gradual, sequential series of changes in species composition of a community following a disturbance.

sulcus [L. *sulcare*: to plow]   The valleys or creases between the raised portions of the convoluted surface of the brain. (Contrast with gyrus.)

summation   The ability of a neuron to fire action potentials in response to numerous subthreshold postsynaptic potentials arriving simultaneously at differentiated places on the cell, or arriving at the same site in rapid succession.

surface area-to-volume ratio   For any cell, organism, or geometrical solid, the ratio of surface area to volume; this is an important factor in setting an upper limit on the size a cell or organism can attain.

surface tension   The attractive intermolecular forces at the surface of liquid; especially important in water.

surfactant   A substance that decreases the surface tension of a liquid. Lung surfactant, secreted by cells of the alveoli, is mostly phospholipid and decreases the amount of work necessary to inflate the lungs.

survivorship   The proportion of individuals in a cohort that is alive at some time in the future.

suspensor   In the embryos of seed plants, the stalk of cells that pushes the embryo into the endosperm and is a source of nutrient transport to the embryo.

symbiosis (sim' bee oh' sis) [Gk. *sym*: together + *bios*: living]   The living together of two or more species in a prolonged and intimate ecological relationship. (Compare with parasitism, mutualism.)

symmetry   Describes an attribute of an animal body in which at least one plane can divide the body into similar, mirror-image halves. (*See* bilateral symmetry, biradial symmetry, radial symmetry.)

sympathetic nervous system   A division of the autonomic (involuntary) nervous system. (Contrast with parasympathetic nervous system.)

sympatric speciation (sim pat' rik) [Gk. *sym*: same + *patria*: homeland]   Speciation due to reproductive isolation without any physical separation of the subpopulation. (Contrast with allopatric speciation, parapatric speciation.)

symplast   The continuous meshwork of the interiors of living cells in the plant body, resulting from the presence of plasmodesmata. (Contrast with apoplast.)

symport   A membrane transport process that carries two substances in the same direction across the membrane. (Contrast with antiport.)

synapomorphy   *See* shared derived trait.

synapse (sin' aps) [Gk. *syn*: together + *haptein*: to fasten]   The narrow gap between the terminal bouton of one neutron and the dendrite or cell body of another.

synapsis (sin ap' sis)   The highly specific parallel alignment (pairing) of homologous chromosomes during the first division of meiosis.

synaptic vesicle   A membrane-bounded vesicle containing neurotransmitter; the neurotransmitter is produced in and discharged by a presynaptic neuron.

synergids [Gk. *syn*: together + *ergos*: performing work]   In flowering plants, the two cells accompanying the egg cell at one end of the megmagametophyte.

syngamy (sing' guh mee) [Gk. *syn*: together + *gamos*: marriage]   Union of gametes. Also known as fertilization.

synonymous substitution   A change of one nucleotide in a sequence to another when that change does not affect the amino acid specified (i.e., UUA → UUG, both specifying leucine). (Compare with nonsynonymous substitution, missense substitution, nonsense substitution.)

systematics   The scientific study of the diversity of organisms, and of their relationships.

systemic circulation   The part of the circulatory system serving those parts of the body other than the lungs or gills (which are served by the pulmonary circulation.)

systemic acquired resistance   A general resistance to many plant pathogens following infection by a single agent.

systemin   The only polypeptide plant hormone; participates in response to tissue damage.

systems biology   The study of an organism as an integrated and interacting system of genes, proteins, and biochemical reactions.

systole (sis' tuh lee) [Gk. *systole*: contraction]   Contraction of a chamber of the heart, driving blood forward in the circulatory system.

- T -

T cell   A type of lymphocyte, involved in the cellular immune response. The final stages of its development occur in the thymus gland. (Contrast with B cell; see also cytotoxic T cell, helper T cell, suppressor T cell.)

T cell receptor   A protein on the surface of a T cell that recognizes the antigenic determinant for which the cell is specific.

target cell   A cell with the appropriate receptors to bind and respond to a particular hormone or other chemical mediator.

taste bud   A structure in the epithelium of the tongue that includes a cluster of chemoreceptors innervated by sensory neurons.

TATA box   An eight-base-pair sequence, found about 25 base pairs before the starting point for transcription in many eukaryotic promoters, that binds a transcription factor and thus helps initiate transcription.

taxis (tak' sis) [Gk. *taxis*: arrange, put in order]   The movement of an organism or its part directly toward or away from the stimulus. For example, positive phototaxis is movement toward a light source, negative geotaxis is movement away from gravity).

taxon   A biological group (typically a species or a clade) that is given a name.

telencephalon   The frontmost division of the vertebrate brain; becomes the cerebrum.

telomerase   An enzyme that catalyzes the addition of telomeric sequences lost from chromosomes during DNA replication.

telomeres (tee' lo merz) [Gk. *telos*: end + *meros*: units, segments]   Repeated DNA sequences at the ends of eukaryotic chromosomes.

telophase (tee' lo phase) [Gk. *telos*: end]   The final phase of mitosis or meiosis during which chromosomes became diffuse, nuclear envelopes reform, and nucleoli begin to reappear in the daughter nuclei.

template   (1) In biochemistry, a molecule or surface upon which another molecule is synthesized in complementary fashion, as in the replication of DNA. (2) In the brain, a pattern that responds to a normal input but not to incorrect inputs.

template strand   In double-stranded DNA, the strand that is transcribed to create an RNA transcript that will be processed into a protein. Also refers to a strand of RNA that is used to create a complementary RNA.

temporal summation [L. *tempus*: time; *summus*: highest amount]   In the production or inhibition of action potentials in a postsynaptic neuron, the interaction of depolarizations or hyperpolarizations produced by rapidly repeated stimulation of a single point.

tendon   A collagen-containing band of tissue that connects a muscle with a bone.

termination   The end of protein synthesis triggered by a stop codon which binds release factor that causes the polypeptide to release from the ribosome.

terminator   A sequence at the 3′ end of mRNA that causes the RNA strand to be released from the transcription complex.

terrestrial (ter res' tree al) [L. *terra*: earth]   Pertaining to the land. (Contrast with aquatic, marine.)

territory   A fixed area from which an animal or group of animals excludes other members of the same (and sometimes other) species by aggressive behavior or displays.

tertiary structure   In reference to a protein, the relative locations in three-dimensional space of all the atoms in the molecule. The overall shape of a protein. (Contrast with primary, secondary, and quaternary structures.)

test cross   Mating of a dominant-phenotype individual (who may be either heterozygous or homozygous) with a homozygous-recessive individual.

testis (tes' tis) (plural: testes) [L. *testis*: witness]   The male gonad; the organ that produces the male sex cells.

testosterone (tes toss' tuhr own)   A male sex steroid hormone.

tetanus [Gk. *tetanos*: stretched]   (1) A state of sustained maximal muscular contraction caused by rapidly repeated stimulation. (2) In medicine, an often fatal disease ("lockjaw") caused by the bacterium *Clostridium tetani*.

tetrad [Gk. *tettares*: four]   During prophase I of meiosis, the association of a pair of homologous chromosomes or four chromatids.

thalamus [Gk. *thalamos*: chamber]   A region of the vertebrate forebrain; involved in integration of sensory input.

thallus (thal' us) [Gk. *thallos*: sprout]   Any algal body which is not differentiated into root, stem, and leaf.

therapeutic cloning   The use of cloning by nuclear transfer to produce an embryo that will provide embryonic stem cells to be used in therapy.

theory [Gk. *theoria*: analysis of facts]   A far-reaching explanation of observed facts that is supported by such a wide body of evidence, with no significant contradictory evidence, that it is scientifically accepted as a factual framework. Examples are Newton's theory of gravity and Darwin's theory of evolution. (Contrast with hypothesis.)

thermoneutral zone [Gk. *thermos*: temperature]   The range of temperatures over which an endotherm does not have to expend extra energy to thermoregulate.

thermoreceptor   A cell or structure that responds to changes in temperature.

thoracic cavity [Gk. *thorax*: breastplate]   The portion of the mammalian body cavity bounded by the ribs, shoulders, and diaphragm. Contains the heart and the lungs.

thoracic duct   The connection between the lymphatic system and the circulatory system.

thorax [Gk. *thorax*: breastplate]   In an insect, the middle region of the body, between the head and abdomen. In mammals, the part of the body between the neck and the diaphragm.

thrombin   An enzyme that converts fibrinogen to fibrin, thus triggering the formation of blood clots.

thrombus (throm' bus) [Gk. *thrombos*: clot]   A blood clot that forms within a blood vessel and remains attached to the wall of the vessel. (Contrast with embolus.)

thylakoid (thigh la koid) [Gk. *thylakos*: sack or pouch]   A flattened sac within a chloroplast. Thylakoid membranes contain all of the chlorophyll in a plant, in addition to the electron carriers of photophosphorylation. Thylakoids stack to form grana.

thymine (T)   A nitrogen-containing base found in DNA.

thymus [Gk. *thymos*: warty]   A ductless, glandular portion of the lymphoid system, involved in development of the immune system of vertebrates. In humans, the thymus degenerates during puberty.

thyroid [Gk. *thyreos*: door-shaped]   A two-lobed gland in vertebrates. Produces the hormone thyroxin.

thyrotropin (thyroid-stimulating hormone, TSH)   A hormone of the anterior pituitary that stimulates the thyroid gland to produce and release thyroxin.

thyrotropin-releasing hormone (TRH)   A hypothalamic hormone that stimulates anterior pituitary cells to release TSH.

thyroxine   The hormone produced by the thyroid gland that controls many metabolic processes.

tidal volume   The amount of air that is exchanged during each breath when a person is at rest. (Compare with vital capacity.)

tight junction   A junction between epithelial cells, in which there is no gap whatever between the adjacent cells. Materials may pass through a tight junction only by entering the epithelial cells themselves.

tissue   A group of similar cells organized into a functional unit; usually integrated with other tissues to form part of an organ.

toll   A member of a receptor family that responds to the binding of a molecule from a pathogen by initiating a protein kinase cascade, resulting in the synthesis of defensive proteins.

tonus (toe' nuss) [L. *tonus*: tension]   A low level of muscular tension that is maintained even when the body is at rest.

topsoil   The uppermost soil layer; contains most of the organic matter of soil, but may be depleted of most mineral nutrients.

totipotent [L. *toto*: whole, entire + *potens*: powerful]   Of a cell, possessing all the genetic information and other capacities necessary to form an entire individual. (Compare with pluripotent.)

toxic [L. *toxicum*: poison]   Injurious to the tissues of the host organism.

trachea (tray' kee ah) [Gk. *trakhoia*: tube]   A tube that carries air to the bronchi of the lungs of vertebrates. When plural (*tracheae*), refers to the major airways of insects.

tracheary element   Refers to either or both types of conductive xylem cells: tracheids and vessel elements.

tracheid (tray' kee id)   A distinctive conducting and supporting cell found in the xylem of nearly all vascular plants, characterized by tapering ends and walls that are pitted but not perforated. (Contrast with vessel element.)

tracheophytes [Gk. *trakhoia*: tube + *phyton*: plant]   *See* vascular plants.

trade-off   The relationship between the costs of performing a behavior or other trait and the benefits the individual gains from the trait or behavior. (*See also* cost–benefit analysis.)

trait   In genetics, one form of a character: eye color is a character; brown eyes and blue eyes are traits. (Compare with character.)

transcription   The synthesis of RNA using one strand of DNA as the template.

transcription factors   Proteins that assemble on a eukaryotic chromosome, allowing RNA polymerase II to perform transcription.

transduction   (1) Transfer of genes from one bacterium to another, with a bacterial virus acting as the carrier of the genes. (2) In sensory cells, the transformation of a stimulus (e.g., light energy, sound pressure waves, chemical or electrical stimulants) into action potentials.

transfection   Uptake, incorporation, and expression of recombinant DNA.

transfer cell   A modified parenchyma cell that transports solutes from its cytoplasm into its cell wall, thus moving the solutes from the symplast into the apoplast.

transfer RNA (tRNA)   A family of double-stranded RNA molecules. Each tRNA carries a specific amino acid and anticodon that will pair with the complementary codon in mRNA during translation.

transformation Mechanism for transfer of genetic information in bacteria in which pure DNA extracted from bacteria of one genotype is taken in through the cell surface of bacteria of a different genotype and incorporated into the chromosome of the recipient cell.

transforming principle An early term for the as yet unidentified chemical substance responsible for bacterial tranformation.

transgenic organism An organism containing recombinant DNA incorporated into its genetic material.

transition-state species A short-lived, unstable intermediate with high potential energy in a chemical reaction.

translation The synthesis of a protein (polypeptide). Takes place on ribosomes, using the information encoded in messenger RNA.

transmembrane domain The portion of a protein that lies inside the membrane bilayer.

transmembrane protein An integral membrane protein that spans the lipid bilayer.

translocation (1) In genetics, a rare mutational event that moves a portion of a chromosome to a new location, generally on a nonhomologous chromosome. (2) In vascular plants, movement of solutes in the phloem.

transpiration [L. *spirare*: to breathe] The evaporation of water from plant leaves and stem, driven by heat from the sun, and providing the motive force to raise water (plus mineral nutrients) from the roots.

transposable element A segment of DNA that can move to, or give rise to copies at, another locus on the same or a different chromosome.

transposon Mobile DNA segment that can insert into a chromosome and cause genetic change.

triglyceride A simple lipid in which three fatty acids are combined with one molecule of glycerol.

triplet *See* codon.

triplet repeat The occurrence of repeated triplet of bases in a gene, often leading to genetic disease, as does excessive repetition of CGG in the gene responsible for fragile-X syndrome.

triploblastic Having three cell layers. (Contrast with diploblastic.)

trisomic Containing three rather than two members of a chromosome pair.

tRNA *See* transfer RNA.

trophic cascade The progression over successively lower trophic levels of the indirect effects of a predator.

trophic level [Gk *trophes*: nourishment] A group of organisms united by obtaining their energy from the same part of the food web of a biological community.

trophoblast [Gk *trophes*: nourishment + *blastos*: sprout] At the 32-cell stage of mammalian development, the outer group of cells that will become part of the placenta and thus nourish the growing embryo. (Comtrast with inner cell mass.)

trochophore (troke' o fore) [Gk. *trochos*: wheel + *phoreus*: bearer] The free-swimming larva of some annelids and mollusks, distinguished by a wheel-like band of cilia around the middle.

tropic hormones Hormones of the anterior pituitary that control the secretion of hormones by other endocrine glands.

tropism [Gk. *tropos*: to turn] In plants, growth toward or away from a stimulus such as light (phototropism) or gravity (gravitropism).

tropomyosin [troe poe my' oh sin] A protein that, along with actin, constitutes the thin filaments of myofibrils. It controls the interactions of actin and myosin necessary for muscle contraction.

troposphere The lowest atmospheric zone, reaching upward from the Earth's surface approximately 17 km in the tropics and subtropics but only to about 10 km at higher latitudes. The zone in which virtually all the water vapor in the atmosphere is located.

true-breeding A genetic cross in which the same result occurs every time with respect to the trait(s) under consideration, due to homozygous parents.

trypsin A protein-digesting enzyme. Secreted by the pancreas in its inactive form (trypsinogen), it becomes active in the duodenum of the small intestine.

T tubules A system of tubules that runs throughout the cytoplasm of muscle fibers, through which action potentials spread.

tube cell The larger of the two cells in a pollen grain; responsible for growth of the pollen tube. *See* generative cell.

tubulin A protein that polymerizes to form microtubules.

tumor [L. *tumor*: a swollen mass] A disorganized mass of cells, often growing out of control. Malignant tumors spread to other parts of the body.

tumor suppressor genes Genes which, when homozygous mutant, result in cancer. Such genes code for protein products that inhibit cell proliferation.

turgor pressure [L. *turgidus*: swollen] *See* pressure potential.

tympanic membrane [Gk. *tympanum*: drum] The eardrum.

- U -

ubiquinone (yoo bic' kwi known) [L. *ubique*: everywhere] A mobile electron carrier of the mitochondrial respiratory chain. Similar to plastoquinone found in chloroplasts.

ubiquitin A small protein that is covalently linked to other cellular proteins identified for breakdown by the proteosome.

umbilical cord Tissue made up of embryonic membranes and blood vessels that connects the embryo to the placenta in eutherian mammals.

understory The aggregate of smaller plants growing beneath the canopy of dominant plants in a forest.

unicellular (yoon' e sell' yer ler) [L. *unus*: one + *cella*: chamber] Consisting of a single cell, as in a unicellular organism. (Contrast with multicellular.)

uniport [L. *unus*: one + portal: doorway] A membrane transport process that carries a single substance. (Contrast with antiport, symport.)

unsaturated hydrocarbon A compound containing only carbon and hydrogen atoms, with one or more pairs of carbon atoms that are connected by double bonds.

upwelling The surfaceward movement of nutrient-rich, cooler water from deeper layers of the ocean.

uracil (U) A pyrimidine base found in nucleotides of RNA.

urea A compound serving as the main excreted form of nitrogen by many animals, including mammals.

ureotelic Describes an organism in which the final product of the breakdown of nitrogen-containing compounds (primarily proteins) is urea. (Contrast with ammonotelic, uricotelic.)

ureter (your' uh tur) A long duct leading from the vertebrate kidney to the urinary bladder or the cloaca.

urethra (you ree' thra) In most mammals, the canal through which urine is discharged from the bladder and which serves as the genital duct in males.

uric acid A compound that serves as the main excreted form of nitrogen in some animals, particularly those which must conserve water, such as birds, insects, and reptiles.

uricotelic Describes an organism in which the final product of the breakdown of nitrogen-containing compounds (primarily proteins) is uric acid. (Contrast with ammonotelic, ureotelic.)

urine (you' rin) In vertebrates, the fluid waste product containing the toxic nitrogenous by-products of protein and amino acid metabolism.

uterus (yoo' ter us) [L. *utero*: womb] The uterus or womb is a specialized portion of the female reproductive tract in certain mammals. It receives the fertilized egg and nurtures the embryo in its early development.

V -

vaccination Injection of virus or bacteria or their proteins into the body, to induce immunization. The injected material is usually attenuated (weakened) before injection.

vacuole (vac' yew ole) [Fr.: small vacuum] A liquid-filled, membrane-enclosed compartment in cytoplasm; may function as digestive chambers, storage chambers, waste bins.

vagina (vuh jine' uh) [L.: sheath] In female mammals, the passage leading from the external genital orifice to the uterus; receives the copulatory organ of the male in mating.

van der Waals forces Weak attractions between atoms resulting from the interaction of the electrons of one atom with the nucleus of another. This type of attraction is about one-fourth as strong as a hydrogen bond.

variable regions The part of an immunoglobulin molecule or T-cell receptor that includes the antigen-binding site.

vasa recta Blood vessels that parallel the loops of Henle and the collecting ducts in the renal medulla of the kidney.

vascular (vas' kew lar) [L. *vasculum*: a small vessel]   Pertaining to organs and tissues that conduct fluid, such as blood vessels in animals and phloem and xylem in plants.

vascular bundle   In vascular plants, a strand of vascular tissue, including conducting cells of xylem and phloem as well as thick-walled fibers.

vascular cambium   A lateral meristem giving rise to secondary xylem and phloem.

vascular rays   In vascular plants, radially oriented sheets of cells produced by the vascular cambium, carrying materials laterally between the wood and the phloem.

vascular system   The conductive system of the plant, consisting primarily of xylem and phloem.

vas deferens   The duct that transfers sperm from the epididymis to the urethra.

vasopressin *See* antidiuretic hormone.

vector   (1) An agent, such as an insect, that carries a pathogen affecting another species. (2) A plasmid or virus that carries an inserted piece of DNA into a bacterium for cloning purposes in recombinant DNA technology.

vegetal hemisphere   The lower portion of some animal eggs, zygotes, and embryos, in which the dense nutrient yolk settles. The vegetal pole refers to the very bottom of the egg or embryo. (Contrast with animal hemisphere.)

vegetative   Nonreproductive, nonflowering, or asexual.

vegetative reproduction   Asexual reproduction.

vein [L. *vena*: channel]   A blood vessel that returns blood to the heart. (Contrast with artery.)

ventral [L. *venter*: belly, womb]   Toward or pertaining to the belly or lower side. (Contrast with dorsal.)

ventricle   A muscular heart chamber that pumps blood through the lungs or through the body.

vernalization [L. *vernalis*: spring]   Events occurring during a required chilling period, leading eventually to flowering.

vertebral column [L. *vertere*: to turn]   The jointed, dorsal column that is the primary support structure of vertebrates.

vesicle   A membrane enclosed compartment within the cytoplasm.

vessel element   In plants, a nonliving water-conducting cell with perforated end walls. (Contrast with tracheid.)

vestibular apparatus (ves tib' yew lar) [L. *vestibulum*: an enclosed passage]   Structures associated with the vertebrate ear; these structures sense changes in position or momentum of the head, affecting balance and motor skills.

vestigial (ves tij' ee al) [L. *vestigium*: footprint, track]   The remains of body structures that are no longer of adaptive value to the organism and therefore are not maintained by selection.

vicariant distribution   A population distribution resulting from the disruption of a formerly continuous range by a vicariant event.

vicariant event (vye care' ee unce) [L. *vicus*: change]   The splitting of the range of a taxon by the imposition of some barrier to interchange among its members.

villus (vil' lus) (plural: villi) [L. *villus*: shaggy hair or beard]   A hairlike projection from a membrane; for example, from many gut walls.

virion (veer' e on)   The virus particle, the minimum unit capable of infecting a cell.

viroid (vye' roid)   An infectious agent consisting of a single-stranded RNA molecule with no protein coat; produces diseases in plants.

virulent [L. *virus*: poison, slimy liquid]   Causing or capable of causing disease and death.

virus   Any of a group of ultramicroscopic particles constructed of nucleic acid and protein (and, sometimes, lipid) that require living cells in order to reproduce. Viruses probably evolved from eukaryotic cells, secondarily losing some cellular attributes.

vital capacity   The sum total of the tidal volume and the inspiratory and expiratory reserve volumes.

vitamins [L. *vita*: life]   Organic compounds that an organism cannot synthesize, but nevertheless requires in small quantity for normal growth and metabolism.

viviparous (vye vip' uh rus) [L. *vivus*: alive]   Reproduction in which fertilization of the egg and development of the embryo occur inside the mother's body. (Contrast with oviparous.)

VNTRs (variable number of tandem repeats)   In the human genome, short DNA sequences that are repeated a characteristic number of times in related individuals. Can be used to make a DNA fingerprint.

voltage-gated channels   Ion channels in membranes that change conformation and therefore ion conductance when a certain potential difference exists across the membrane in which they are inserted.

- W -

waggle dance   The running movement of a working honey bee on the hive, during which the worker traces out a repeated figure eight. The dance contains elements that transmit to other bees the location of the food.

water potential   In osmosis, the tendency for a system (a cell or solution) to take up water from pure water through a differentially permeable membrane. Water flows toward the system with a more negative water potential. (Contrast with solute potential, pressure potential.)

wavelength   The distance between successive peaks of a wave train, such as electromagnetic radiation.

wild-type   Geneticists' term for standard or reference type. Deviants from this standard, even if the deviants are found in the wild, are usually referred to as mutant. (Note that this terminology is not usually applied to human genes.)

wood   Secondary xylem tissue.

- X -

xanthophyll (zan' tho fill) [Gk. *xanthos*: yellowish-brown + *phyllon*: leaf]   A yellow or orange pigment commonly found as an accessory pigment in photosynthesis, but found elsewhere as well. An oxygen-containing carotenoid.

X-linked   A character that is coded for by a gene on the X chromosome; a sex-linked trait.

xerophyte (zee' row fyte) [Gk. *xerox*: dry + *phyton*: plant]   A plant adapted to an environment with a limited water supply.

xylem (zy' lum) [Gk. *xylon*: wood]   In vascular plants, the tissue that conducts water and minerals; xylem consists, in various plants, of tracheids, vessel elements, fibers, and other highly specialized cells.

- Y -

yolk [M.E. *yolke*: yellow]   The stored food material in animal eggs, usually rich in protein and lipid.

yolk sac   In reptiles, birds, and mammals, the extraembryonic membrane that forms from the endoderm of the hypoblast; it encloses and digests the yolk.

- Z -

Z-DNA   A form of DNA in which the molecule spirals to the left rather than to the right.

zeaxanthin   A blue-light receptor involved in the opening of plant stomata.

zona pellucida   A jellylike substance that surrounds the mammalian ovum when it is released from the ovary.

zoospore (zoe' o spore) [Gk. *zoon*: animal + *spora*: seed]   In algae and fungi, any swimming spore. May be diploid or haploid.

zygote (zye' gote) [Gk. *zygotos*: yoked]   The cell created by the union of two gametes, in which the gamete nuclei are also fused. The earliest stage of the diploid generation.

zymogen   An inactive precursor of a digestive enzyme secreted into the lumen of the gut, where a protease cleaves it to form the active enzyme.

Answers to Self-Quizzes

Chapter 2
1.	b	6.	a
2.	d	7.	d
3.	c	8.	a
4.	c	9.	c
5.	d	10.	b

Chapter 3
1.	e	6.	a
2.	e	7.	c
3.	c	8.	e
4.	d	9.	a
5.	b	10.	d

Chapter 4
1.	b	6.	e
2.	d	7.	a
3.	c	8.	d
4.	e	9.	b
5.	a	10.	d

Chapter 5
1.	e	6.	c
2.	c	7.	c
3.	a	8.	b
4.	d	9.	e
5.	c	10.	c

Chapter 6
1.	c	6.	e
2.	e	7.	d
3.	b	8.	b
4.	c	9.	d
5.	c	10.	e

Chapter 7
1.	d	6.	d
2.	d	7.	a
3.	e	8.	b
4.	e	9.	a
5.	c	10.	e

Chapter 8
1.	c	6.	c
2.	b	7.	c
3.	d	8.	d
4.	b	9.	d
5.	e	10.	b

Chapter 9
1.	d	6.	d
2.	c	7.	e
3.	b	8.	d
4.	d	9.	c
5.	c	10.	c

Chapter 10*
1.	e	6.	d
2.	a	7.	b
3.	d	8.	b
4.	d	9.	b
5.	d	10.	b

Chapter 11
1.	c	6.	b
2.	a	7.	d
3.	c	8.	d
4.	b	9.	c
5.	e	10.	c

Chapter 12
1.	c	6.	d
2.	d	7.	b
3.	e	8.	d
4.	b	9.	d
5.	a	10.	a

Chapter 13
1.	b	6.	d
2.	e	7.	d
3.	a	8.	c
4.	c	9.	b
5.	c	10.	d

Chapter 14
1.	c	6.	c
2.	c	7.	c
3.	a	8.	b
4.	a	9.	e
5.	c	10.	d

Chapter 15
1.	d	6.	a
2.	d	7.	e
3.	c	8.	b
4.	c	9.	c
5.	d	10.	a

Chapter 16
1.	b	6.	b
2.	a	7.	c
3.	a	8.	a
4.	c	9.	e
5.	e	10.	e

Chapter 17
1.	a	6.	b
2.	c	7.	e
3.	b	8.	d
4.	b	9.	c
5.	d	10.	b

Chapter 18
1.	a	6.	a
2.	b	7.	d
3.	e	8.	d
4.	e	9.	a
5.	c	10.	d

Chapter 19
1.	c	6.	c
2.	a	7.	d
3.	a	8.	b
4.	b	9.	a
5.	b	10.	b

Chapter 20
1.	d	6.	b
2.	e	7.	a
3.	c	8.	c
4.	b	9.	b
5.	c	10.	b

Chapter 21
1.	d	6.	a
2.	b	7.	c
3.	e	8.	b
4.	c	9.	c
5.	a	10.	e

*Answers to Chapter 10 Genetics Problems

1. Each of the eight boxes in the Punnett squares should contain the genotype *Tt*, regardless of which parent was tall and which dwarf.

2. See Figure 10.4, page 212.

3. The trait is autosomal. Mother *dp dp*, father *Dp dp*. If the trait were sex-linked, all daughters would be wild-type and sons would be *dumpy*.

4. Yellow parent = $s^Y s^b$; offspring 3 yellow (s^Y–): 1 black ($s^b s^b$).
Black parent = $s^b s^b$; offspring all black ($s^b s^b$). Orange parent = $s^O s^b$; offspring 3 orange (s^O–): 1 black ($s^b s^b$). Both s^O and s^Y are dominant to s^b.

5. All females wild-type; all males spotted.

6. F_1 all wild-type, *PpSwsw*; F_2 9:3:3:1 in phenotypes. See Figure 10.7, page 214, for analogous genotypes.

7a. Ratio of phenotypes in F_2 is 3:1 (double dominant to double recessive).

7b. The F_1 are *Pby pB*^{Y}; they produce just two kinds of gametes (*Pby* and *pBy*). Combine them carefully and see the 1:2:1 phenotypic ratio fall out in the F_2.

7c. Pink-blistery.

7d. See Figures 9.16 and 9.18 (pages 196–198). Crossing over took place in the F_1 generation.

8. The genotypes are:

PpSwsw
Ppswsw
ppSwsw
ppswsw
Ratio: 1:1:1:1

The phenotypes are:
wild eye, long wing	pink eye, long wing
wild eye, short wing	pink eye, short wing

Ratio: 1:1:1:1

9a. 1 black:2 blue:1 splashed white

9b. Always cross black with splashed white.

10a. $w^+ > w^e > w$

10b. Parents $w^e w$ and $w^+ Y$. Progeny $w+w^e$, $w+w$, $w^e Y$, and wY.

11. All will have normal vision because they inherit dad's wild-type X chromosome, but half of them will be carriers.

12. Agouti parent *AaBb*. Albino offspring *aaBb* and *aabb*; black offspring *Aabb*; agouti offspring *AaBb*.

13. Because the gene is carried on mitochondrial DNA, it is passed through the mother only. Thus if the woman does not have the disease but her husband does, their child will not be affected. On the other hand, if the woman has the disease but her husband does not, their child *will* have the disease.

Chapter 22
1.	d	6.	d
2.	e	7.	b
3.	d	8.	e
4.	c	9.	b
5.	d	10.	c

Chapter 23
1.	c	6.	d
2.	e	7.	a
3.	d	8.	a
4.	c	9.	c
5.	a	10.	e

Chapter 24
1.	a	6.	e
2.	a	7.	e
3.	d	8.	e
4.	a	9.	b
5.	b	10.	e

Chapter 25
1.	b	6.	b
2.	e	7.	e
3.	a	8.	a
4.	b	9.	e
5.	e	10.	d

Chapter 26
1.	e	6.	b
2.	e	7.	d
3.	b	8.	d
4.	c	9.	c
5.	e	10.	d

Chapter 27
1.	a	6.	b
2.	e	7.	d
3.	c	8.	b
4.	d	9.	a
5.	c	10.	d

Chapter 28
1.	d	6.	e
2.	c	7.	c
3.	e	8.	b
4.	b	9.	b
5.	b	10.	d

Chapter 29
1.	d	6.	c
2.	c	7.	a
3.	d	8.	e
4.	a	9.	c
5.	d	10.	a

Chapter 30
1.	b	6.	a
2.	d	7.	e
3.	e	8.	a
4.	c	9.	c
5.	d	10.	c

Chapter 31
1.	c	6.	b
2.	d	7.	c
3.	c	8.	d
4.	d	9.	e
5.	c	10.	d

Chapter 32
1.	b	6.	e
2.	e	7.	b
3.	c	8.	d
4.	d	9.	d
5.	a	10.	e

Chapter 33
1.	b	6.	e
2.	a	7.	a
3.	c	8.	e
4.	c	9.	c
5.	d	10.	b

Chapter 34
1.	d	6.	b
2.	b	7.	b
3.	e	8.	c
4.	e	9.	a
5.	a	10.	d

Chapter 35
1.	c	6.	d
2.	d	7.	d
3.	b	8.	e
4.	b	9.	e
5.	b	10.	a

Chapter 36
1.	d	6.	c
2.	d	7.	e
3.	c	8.	a
4.	a	9.	d
5.	a	10.	e

Chapter 37
1.	a	6.	c
2.	e	7.	e
3.	c	8.	c
4.	d	9.	a
5.	b	10.	b

Chapter 38
1.	d	6.	e
2.	b	7.	a
3.	e	8.	b
4.	b	9.	c
5.	d	10.	d

Chapter 39
1.	e	6.	a
2.	b	7.	b
3.	c	8.	c
4.	c	9.	d
5.	d	10.	a

Chapter 40
1.	c	6.	b
2.	c	7.	e
3.	a	8.	a
4.	d	9.	e
5.	b	10.	b

Chapter 41
1.	b	6.	b
2.	a	7.	d
3.	b	8.	e
4.	e	9.	c
5.	e	10.	c

Chapter 42
1.	c	6.	d
2.	e	7.	d
3.	a	8.	c
4.	d	9.	d
5.	d	10.	a

Chapter 43
1.	a	6.	c
2.	c	7.	b
3.	e	8.	b
4.	a	9.	b
5.	d	10.	a

Chapter 44
1.	d	6.	e
2.	d	7.	c
3.	c	8.	c
4.	c	9.	d
5.	e	10.	d

Chapter 45
1.	d	6.	e
2.	d	7.	b
3.	a	8.	c
4.	b	9.	c
5.	e	10.	d

Chapter 46
1.	c	6.	c
2.	a	7.	a
3.	e	8.	c
4.	d	9.	a
5.	d	10.	c

Chapter 47
1.	e	6.	d
2.	a	7.	e
3.	b	8.	a
4.	c	9.	a
5.	b	10.	e

Chapter 48
1.	e	6.	b
2.	d	7.	c
3.	a	8.	c
4.	b	9.	a
5.	c	10.	d

Chapter 49
1.	d	6.	d
2.	a	7.	b
3.	c	8.	d
4.	d	9.	c
5.	c	10.	e

Chapter 50
1.	b	6.	d
2.	e	7.	a
3.	c	8.	b
4.	a	9.	d
5.	b	10.	d

Chapter 51
1.	d	6.	b
2.	a	7.	e
3.	d	8.	a
4.	a	9.	c
5.	d	10.	e

Chapter 52
1.	c	6.	c
2.	a	7.	d
3.	a	8.	c
4.	c	9.	d
5.	a	10.	a

Chapter 53
1.	c	6.	d
2.	e	7.	c
3.	c	8.	c
4.	a	9.	d
5.	d	10.	e

Chapter 54
1.	c	6.	b
2.	c	7.	d
3.	a	8.	c
4.	d	9.	a
5.	c	10.	d

Chapter 55
1.	a	6.	e
2.	a	7.	c
3.	b	8.	a
4.	c	9.	d
5.	d	10.	e

Chapter 56
1.	d	6.	b
2.	d	7.	a
3.	e	8.	c
4.	e	9.	d
5.	c	10.	b

Chapter 57
1.	b	6.	a
2.	d	7.	d
3.	e	8.	c
4.	e	9.	a
5.	b	10.	c

Illustration Credits

© Garry T. Cole/Biological Photo Service. 9.14 *left*: © Andrew Syred/SPL/Photo Researchers, Inc. 9.14 *center*: David McIntyre. 9.14 *right*: © Gerry Ellis, DigitalVision/PictureQuest. 9.15: Courtesy of Dr. Thomas Ried and Dr. Evelin Schröck, NIH. 9.16: © C. A. Hasenkampf/Biological Photo Service. 9.17: Courtesy of J. Kezer. 9.21A: © Gopal Murti/Photo Researchers, Inc.

Chapter 10 *Bris*: © David H. Wells/CORBIS. *Eye test*: © Brand X Pictures/Alamy. 10.1: © the Mendelianum. 10.11: Courtesy the American Netherland Dwarf Rabbit Club. 10.14: Courtesy of Madison, Hannah, and Walnut. 10.15: Courtesy of Pioneer Hi-Bred International, Inc. 10.16: © Carolyn A. McKeone/Photo Researchers, Inc. 10.17: © Peter Morenus/U. of Connecticut. *Bay scallops*: © Barbara J. Miller/Biological Photo Service.

Chapter 11 *T. rex*: © The Natural History Museum, London. *Earrings*: David McIntyre/Model: Ashley Ying. 11.3: © Biozentrum, U. Basel/SPL/Photo Researchers, Inc. 11.6 *X-ray crystallograph*: Courtesy of Prof. M. H. F. Wilkins, Dept. of Biophysics, King's College, U. London. 11.6 *Franklin*: © CSHL Archives/Peter Arnold, Inc. 11.8A: © A. Barrington Brown/Photo Researchers, Inc. 11.8B: Data from S. Arnott & D. W. Hukins, 1972. *Biochem. Biophys. Res. Commun.* 47(6): 1504. 11.21C: © Dr. Peter Lansdorp/Visuals Unlimited.

Chapter 12 *Ricinus*: © Alan L. Detrick/Photo Researchers, Inc. *Ribosome*: Data from PDB 1GIX and 1G1Y. M. M. Yusupov et al., 2001. *Science* 292: 883. 12.4: Data from PDB 1I3Q. P. Cramer et al., 2001 *Science* 292: 1863. 12.8: Data from PDB 1EHZ. H. Shi & P. B. Moore, 2000. *RNA* 6: 1091. 12.9B: Data from PDB 1EUQ. L. D. Sherlin et al., 2000. *J. Mol. Biol.* 299: 431. 12.10: Data from PDB 1GIX and 1G1Y. M. M. Yusupov et al., 2001. *Science* 292: 883. 12.14: Courtesy of J. E. Edström and *EMBO J.* 12.18: © Stanley Flegler/Visuals Unlimited.

Chapter 13 *Market*: © Derek Brown/Alamy. *Scientist*: © Tek Image/SPL/Photo Researchers, Inc. 13.1A: © Dept. of Microbiology, Biozentrum/SPL/Photo Researchers, Inc. 13.1B: © Dennis Kunkel Microscopy, Inc. 13.2A: © Dr. Harold Fisher/Visuals Unlimited. 13.2B: © E.O.S./Gelderblom/Photo Researchers, Inc. 13.2C: © BSIP Agency/Index Stock Imagery/Jupiter Images. 13.7: Courtesy of Roy French. 13.11: Courtesy of L. Caro and R. Curtiss. 13.22: Based on an illustration by Anthony R. Kerlavage, Institute for Genomic Research. *Science* 269: 449 (1995). *For Investigation*: Adapted from J. J. Perry, J. T. Staley, & S. Lory, 2002. *Microbial Life*. Sinauer Associates.

Chapter 14 *Cheetah*: © John Giustina/ImageState/Jupiter Images. *Panther*: © Lynn M. Stone/Naturepl.com. 14.7: From D. C. Tiemeier et al., 1978. *Cell* 14: 237. 14.18: Courtesy of Murray L. Barr, U. Western Ontario. 14.20: Courtesy of O. L. Miller, Jr.

Chapter 15 *Student*: © Ryan McVay, Photodisc Green/Getty Images. *Ethiopian*: © JTB Photo/photolibrary.com. 15.2: © Biophoto Associates/Photo Researchers, Inc. 15.3: From A. M. de Vos, M. Ultsch, & A. A. Kossiakoff, 1992. *Science* 255: 306. 15.14: © Stephen A. Stricker, courtesy of Molecular Probes, Inc.

Chapter 16 *Destruction*: © Suzanne Plunkett/AP Images. *Baby 81*: © Gemunu Amarasinghe/AP Images. 16.2: © Philippe Plailly/Photo

Researchers, Inc. 16.5: Bettmann/CORBIS. 16.6: Courtesy of Keith Weller, De Wood, Chris Pooley, & Scott Bauer/USDA ARS. 16.19A: Courtesy of Ingo Potrykus, Swiss Federal Institute of Technology. 16.19B: Joan Gemme. 16.20: Courtesy of Eduardo Blumwald.

Chapter 17 *Champlain*: © North Wind Picture Archives/Alamy. *Family*: © David Sanger Photography/Alamy. 17.6: From C. Harrison et al., 1983. *J. Med. Genet.* 20: 280. 17.11: Courtesy of Harvey Levy and Cecelia Walraven, New England Newborn Screening Program. 17.14: © Dennis Kunkel Microscopy, Inc. 17.19B: © David M. Martin, M.D./SPL/Photo Researchers, Inc. 17.23: From P. H. O'Farrell, 1975. High resolution two-dimensional electrophoresis of proteins. *J. Biol. Chem.* 250: 4007-21. Courtesy of Patrick H. O'Farrell. 17.24: After N. M. Morel et al., 2004. *Mayo Clin. Proc.* 79: 651.

Chapter 18 *Vaccination*: Bettmann/CORBIS. *T cell*: © Dr. Andrejs Liepins/SPL/Photo Researchers, Inc. 18.3: © Dennis Kunkel Microscopy, Inc. 18.8: © Dr. Gopal Murti/SPL/Photo Researchers, Inc. 18.14: © David Phillips/Science Source/Photo Researchers, Inc.

Chapter 19 *Embryo*: © Dr. Yorgas Nikas/Photo Researchers, Inc. *Centrifuge*: Courtesy of Cytori Therapeutics. 19.4: © Roddy Field, the Roslin Institute. 19.5: Courtesy of T. Wakayama and R. Yanagimachi. 19.12: From J. E. Sulston & H. R. Horvitz, 1977. *Dev. Bio.* 56: 100. 19.15B, 19.16 *left*: Courtesy of J. Bowman. 19.16 *right*: Courtesy of Detlef Weigel. 19.17: Courtesy of W. Driever and C. Nüsslein-Vollhard. 19.18B: Courtesy of C. Rushlow and M. Levine. 19.18C: Courtesy of T. Karr. 19.18D: Courtesy of S. Carroll and S. Paddock. 19.20: Courtesy of F. R. Turner, Indiana U.

Chapter 20 *Fly head*: © David Scharf/Photo Researchers, Inc. *Mutant leg*: From G. Halder et al., 1995. *Science* 267: 1788. Courtesy of W. J. Gehring and G. Halder. 20.2A: Courtesy of E. B. Lewis. 20.2B: From H. Le Mouellic et al., 1992. *Cell* 69: 251. Courtesy of H. Le Mouellic, Y. Lallemand, and P. Brûlet. 20.5: Courtesy of J. Hurle and E. Laufer. 20.6: Courtesy of J. Hurle. 20.7 *Cladogram*: After R. Galant & S. Carroll, 2002. *Nature* 415: 910. 20.7 *Beetle*: © Stockbyte/PictureQuest. 20.7 *Centipede*: © Burke/Triolo/Brand X Pictures/PictureQuest. 20.8: Courtesy of S. Carroll and P. Brakefield. 20.9: © Erick Greene. 20.10: Photograph by C. Laforsch & R. Tollrian, courtesy of A. A. Agrawal. 20.11: © Nigel Cattlin, Holt Studios International/Photo Researchers, Inc. 20.13: © Simon D. Pollard/Photo Researchers, Inc. 20.14: Courtesy of Mike Shapiro and David Kingsley.

Chapter 21 *Capybara*: © Frans Lanting/Minden Pictures. *Grand Canyon*: © Robert Fried/Tom Stack & Assoc. 21.4A: David McIntyre. 21.4B: © Robin Smith/photolibrary.com. 21.7: © François Gohier/The National Audubon Society Collection/Photo Researchers, Inc. 21.8: © Jeff J. Daly/Visuals Unlimited. 21.9 *left*: © Ken Lucas/Visuals Unlimited. 21.9 *right*: © The Natural History Museum, London. 21.10: © Chip Clark. 21.11: © Hans Steur/Visuals Unlimited. 21.12: © Tom McHugh/Field Museum, Chicago/Photo Researchers, Inc. 21.13: Courtesy of Conrad C. Labandeira, Department of Paleobiology, National Museum of Natural History, Smithsonian Institution. 21.14: © Chase Studios, Cedarcreek, MO. 21.16: © The Natural History Museum, London. 21.18: © K. Simons and David Dilcher.

21.20: David McIntyre. 21.22: © John Worrall. 21.23: © Calvin Larsen/Photo Researchers, Inc.

Chapter 22 *Snake*: © Joseph T. Collins/Photo Researchers, Inc. *Newt*: © Robert Clay/Visuals Unlimited. *Pufferfish*: © Georgette Douwma/Naturepl.com. 22.1A, B: © SPL/Photo Researchers, Inc. 22.2: From W. Levi, 1965. *Encyclopedia of Pigeon Breeds*. T. F. H. Publications, Jersey City, NJ. (A, B: photos by R. L. Kienlen, courtesy of Ralston Purina Company; C, D: photos by Stauber). 22.9A: © S. Maslowski/Visuals Unlimited. 22.9B: © Anthony Cooper/SPL/Photo Researchers, Inc. 22.11, 22.17A: David McIntyre. 22.21A: © Marilyn Kazmers/Dembinsky Photo Assoc. 22.21B: © Paul Osmond/Painet Inc. 22.22 *upper*: © Jeff Foott/Naturepl.com. 22.22 *lower*: © franzfoto.com/Alamy.

Chapter 23 *Hummingbird*: © Tui De Roy/Minden Pictures. *Swift*: © Kim Taylor/Naturepl.com. 23.1A *left*: © Gary Meszaros/Dembinsky Photo Assoc. 23.1A *right*: © Lior Rubin/Peter Arnold. 23.1B: © Fi Rust/Painet Inc. 23.8: © Jan Vermeer/Foto Natura/Minden Pictures. 23.9A: ©Virginia P. Weinland/Photo Researchers, Inc. 23.9B: © Pablo Galán Cela/AGE Fotostock. 23.10A: © J. S. Sira/photolibrary.com. 23.10B: © Daniel L. Geiger/SNAP/Alamy. 23.12: © Boris I. Timofeev/Pensoft. 23.14A: © Tony Tilford/photolibrary.com. 23.14B: © W. Peckover/VIREO. 23.15 *Madia*: © Peter K. Ziminsky/Visuals Unlimited. 23.15 *Argyroxiphium*: © Elizabeth N. Orians. 23.15 *Wilkesia*: © Gerald D. Carr. 23.15 *Dubautia*: © Noble Proctor/The National Audubon Society Collection/Photo Researchers, Inc.

Chapter 24 *Vaccination*: Courtesy of Chris Zahniser, B.S.N., R.N., M.P.H./CDC. *Virus*: © Dr. Tim Baker/Visuals Unlimited. 24.3 *Rice*: data from pdb 1CCR. *Tuna*: data from pdb 5CYT. 24.4: From P. B. Rainey & M. Travisano, 1998. *Nature* 394: 69. © Macmillan Publishers Ltd. 24.7A: © Barrie Britton/Naturepl.com. 24.7B: © M. Graybill/J. Hodder/Biological Photo Service.

Chapter 25 *HIV*: © James Cavallini/Photo Researchers, Inc. *Chimpanzees*: © John Cancalosi/Naturepl.com. 25.6A: © Mark Smith/Photo Researchers, Inc. 25.6B: After E. Verheyen et al., 2003. *Science* 300: 325. 25.7: © Larry Jon Friesen. 25.9: © Alexandra Basolo. 25.10: David M. Hillis and Matthew Brauer. 25.11A: © Helen Carr/Biological Photo Service. 25.11B: © Michael Giannechini/Photo Researchers, Inc. 25.11C: © Skip Moody/Dembinsky Photo Assoc.

Chapter 26 *River*: © Felipe Rodriguez/Alamy. *Salmonella* and *Methanospirillum*: © Kari Lounatmaa/Photo Researchers, Inc. 26.2A: © David Phillips/Photo Researchers, Inc. 26.2B: © R. Kessel & G. Shih/Visuals Unlimited. 26.2C: Courtesy of Janice Carr/NCID/CDC. 26.4: From F. Balagaddé et al., 2005. *Science* 309: 137. Courtesy of Frederick Balagaddé. 26.5A *left*: © David M. Phillips/Visuals Unlimited. 26.5A *right*: Courtesy of Peter Hirsch and Stuart Pankratz. 26.5B *left*: Courtesy of the CDC. 26.5B *right*: Courtesy of Peter Hirsch and Stuart Pankratz. 26.6A: © J. A. Breznak & H. S. Pankratz/Biological Photo Service. 26.6B: © J. Robert Waaland/Biological Photo Service. 26.7: © USDA/Visuals Unlimited. 26.8: © Steven Haddock and Steven Miller. 26.12: Courtesy of David Cox/CDC. 26.13: Courtesy of Randall C. Cutlip. 26.14: © David Phillips/Visuals Unlimited. 26.15A: © Paul W. Johnson/Biological Photo Service. 26.15B: © H. S. Pankratz/Biological Photo Service. 26.15C: © Bill Kamin/Visuals

Unlimited. 26.16: © Dr Kari Lounatmaa/Photo Researchers, Inc. 26.17: © Dr. Gary Gaugler/Visuals Unlimited. 26.18: © Michael Gabridge/Visuals Unlimited. 26.20: © Phil Gates/Biological Photo Service. 26.21: From K. Kashefi & D. R. Lovley, 2003. *Science* 301: 934. Courtesy of Kazem Kashefi. 26.23: © Krafft/Hoa-qui/Photo Researchers, Inc. 26.24: © David Sanger Photography/Alamy. 26.25: From H. Huber et al., 2002. *Nature* 417: 63. © Macmillan Publishers Ltd. Courtesy of Karl O. Stetter.

Chapter 27 *Leishmania*: © Dennis Kunkel Microscopy, Inc. *Macrocystis*: © Karen Gowlett-Holmes/photolibrary.com. 27.1: © Steve Gschmeissner/Photo Researchers, Inc. 27.2: © David Patterson, Linda Amaral Zettler, Mike Peglar, & Tom Nerad/micro*scope. 27.3: © London School of Hygiene/SPL/Photo Researchers, Inc. 27.4A: © Bill Bachman/Photo Researchers, Inc. 27.4B: © Markus Geisen/NHMPL. 27.5: © Astrid & Hanns-Frieder Michler/SPL/Photo Researchers, Inc. 27.9: © Mike Abbey/Visuals Unlimited. 27.12A: © Wim van Egmond/Visuals Unlimited. 27.12B: © Biophoto Associates/Photo Researchers, Inc. 27.18: © Dennis Kunkel Microscopy, Inc. 27.19A: © Mike Abbey/Visuals Unlimited. 27.19B: © Dennis Kunkel Microscopy, Inc. 27.19C: © Paul W. Johnson/Biological Photo Service. 27.19D: © M. Abbey/Photo Researchers, Inc. 27.21: © Manfred Kage/Peter Arnold, Inc. 27.22: After A. Ianora et al., 2005. *Nature* 429: 403. 27.23A: © Duncan McEwan/Naturepl.com. 27.23B, C: © Larry Jon Friesen. 27.23D: © J. N. A. Lott/Biological Photo Service. 27.24: © James W. Richardson/Visuals Unlimited. 27.25A: © Larry Jon Friesen. 27.25B: © Milton Rand/Tom Stack & Assoc. 27.26A: © Carolina Biological/Visuals Unlimited. 27.26B: © Larry Jon Friesen. 27.27A: © J. Paulin/Visuals Unlimited. 27.27B: © Dr. David M. Phillips/Visuals Unlimited. 27.29: © Andrew Syred/SPL/Photo Researchers, Inc. 27.30A: © William Bourland/micro*scope. 27.30B: © David Patterson & Aimlee Laderman/micro*scope. 27.31: © Larry Jon Friesen. 27.32: Courtesy of R. Blanton and M. Grimson.

Chapter 28 *Fossils*: © Sinclair Stammers/SPL/Photo Researchers, Inc. *Rainforest*: © Photo Resource Hawaii/Alamy. 28.2A: © Ronald Dengler/Visuals Unlimited. 28.2B: © Larry Mellichamp/Visuals Unlimited. 28.3: © Brian Enting/Photo Researchers, Inc. 28.5: © J. Robert Waaland/Biological Photo Service. 28.6: After L. E. Graham et al., 2004. *PNAS* 101: 11025. Micrograph courtesy of Patricia Gensel and Linda Graham. 28.8: © U. Michigan Exhibit Museum. 28.10: Courtesy of the Biology Department Greenhouses, U. Massachusetts, Amherst. 28.11: After C. P. Osborne et al., 2004. *PNAS* 101: 10360. 28.13A, B: David McIntyre. 28.13C: © Harold Taylor/photolibrary.com. 28.14: © Daniel Vega/AGE Fotostock. 28.15: © Danilo Donadoni/AGE Fotostock. 28.16A: © Ed Reschke/Peter Arnold, Inc. 28.16B: © Carolina Biological/Visuals Unlimited. 28.17A: © J. N. A. Lott/Biological Photo Service. 28.17B: © David Sieren/Visuals Unlimited. 28.18: Courtesy of the Biology Department Greenhouses, U. Massachusetts, Amherst. 28.19 A: © Rod Planck/Dembinsky Photo Assoc. 28.19B: © Nuridsany et Perennou/Photo Researchers, Inc. 28.19C: Courtesy of the Talcott Greenhouse, Mount Holyoke College. 28.20 inset: David McIntyre.

Chapter 29 *Masada*: © Eddie Gerald/Alamy. *Coconut*: © Ben Osborne/Naturepl.com. 29.1 *Seed fern*: David McIntyre. 29.1 *Cycad*: © Patricio Robles

Gil/Naturepl.com. 29.1 *Magnolia*: © Plantography/Alamy. 29.4: © Natural Visions/Alamy. 29.5A: © Patricio Robles Gil/Naturepl.com. 29.5B: © Dave Watts/Naturepl.com. 29.5C: © M. Graybill/J. Hodder/Biological Photo Service. 29.5D: © Frans Lanting/Minden Pictures. 29.7A *left*: David McIntyre. 29.7A *right*: © Stan W. Elems/Visuals Unlimited. 29.7B *left*: David McIntyre. 29.7B *right*: © Dr. John D. Cunningham/Visuals Unlimited. 29.10A: David McIntyre. 29.10B: © Aflo Foto Agency/Naturepl.com. 29.10C: © Richard Shiell/Dembinsky Photo Assoc. 29.11A: © Plantography/Alamy. 29.11B: © Thomas Photography LLC/Alamy. 29.15A: © Inga Spence/Tom Stack & Assoc. 29.15B: © Holt Studios/Photo Researchers, Inc. 29.15C: © Catherine M. Pringle/Biological Photo Service. 29.15D: © blickwinkel/Alamy. 29.17A: Courtesy of Stephen McCabe, U. California, Santa Cruz, and UCSC Arboretum. 29.17B: David McIntyre. 29.17C: © Rob & Ann Simpson/Visuals Unlimited. 29.17D: © R. C. Carpenter/Photo Researchers, Inc. 29.17E: © Geoff Bryant/Photo Researchers, Inc. 29.17F: © José Antonio Jiménez/AGE Fotostock. 29.18A: © Andrew E. Kalnik/Photo Researchers, Inc. 29.18B: © Ed Reschke/Peter Arnold, Inc. 29.18C: © Adam Jones/Dembinsky Photo Assoc. 29.19A: © Willard Clay/photolibrary.com. 29.19B: © Adam Jones/Dembinsky Photo Assoc. 29.19C: © Alan & Linda Detrick/The National Audubon Society Collection/Photo Researchers, Inc. 29.20: © Diaphor La Phototheque/photolibrary.com.

Chapter 30 *Fusarium*: © Dr. Gary Gaugler/Visuals Unlimited. *Ant*: © L. E. Gilbert/Biological Photo Service. 30.3: © David M. Phillips/Visuals Unlimited. 30.5A: © G. T. Cole/Biological Photo Service. 30.6: © N. Allin & G. L. Barron/Biological Photo Service. 30.7: © Richard Packwood/photolibrary.com. 30.8A: © Amy Wynn/Naturepl.com. 30.8B: © Geoff Simpson/Naturepl.com. 30.8C: © Gary Meszaros/Dembinsky Photo Assoc. 30.10A: © R. L. Peterson/Biological Photo Service. 30.10B: © Ken Wagner/Visuals Unlimited. 30.13A: © J. Robert Waaland/Biological Photo Service. 30.13B: © M. F. Brown/Visuals Unlimited. 30.13C: © Dr. John D. Cunningham/Visuals Unlimited. 30.13D: © Biophoto Associates/Photo Researchers, Inc. 30.14 *left*: © William E. Schadel/Biological Photo Service. 30.14 *right*: © Dr. John D. Cunningham/Visuals Unlimited. 30.15: © John Taylor/Visuals Unlimited. 30.16: © G. L. Barron/Biological Photo Service. 30.17A: © Richard Shiell/Dembinsky Photo Assoc. 30.17B: © Matt Meadows/Peter Arnold, Inc. 30.18: © Andrew Syred/SPL/Photo Researchers, Inc. 30.19A: © Manfred Danegger/Photo Researchers, Inc. 30.19B: © Botanica/photolibrary.com.

Chapter 31 *Fossil*: Photograph by Diane Scott. *Dinosaur*: © Joe Tucciarone/Photo Researchers, Inc. 31.2A: Courtesy of J. B. Morrill. 31.2B: From G. N. Cherr et al., 1992. *Microsc. Res. Tech.* 22: 11. Courtesy of J. B. Morrill. 31.5A: © Jurgen Freund/Naturepl.com. 31.5B: © Henry W. Robison/Visuals Unlimited. 31.6A: © Brian Parker/Tom Stack & Assoc. 31.6B: © Tui De Roy/Minden Pictures. 31.8: © Colin M. Orians. 31.9A: © T. Kitchin & V. Hurst/Photo Researchers, Inc. 31.9B: © Stockbyte/PictureQuest. 31.10: Adapted from F. M. Bayer and H. B. Owre, 1968. *The Free-Living Lower Invertebrates*, Macmillan Publishing Co. 31.11A: © Scott Camazine/Alamy. 31.11B, C: David McIntyre. 31.13A: © Stephen Dalton/Minden Pictures. 31.13B: © Gerald & Buff Corsi/Visuals Unlimited. 31.14A: © Dave Watts/Naturepl.com.

31.14B: © Larry Jon Friesen. 31.16A: © Jurgen Freund/Naturepl.com. 31.16B: © Larry Jon Friesen. 31.16C: © David Wrobel/Visuals Unlimited. 31.17B: © Larry Jon Friesen. 31.18: Adapted from F. M. Bayer and & H. B. Owre, 1968. *The Free-Living Lower Invertebrates*, Macmillan Publishing Co. 31.19A: © Larry Jon Friesen. 31.19B: © Michael Patrick O'Neill/Alamy. 31.19C, D: © Larry Jon Friesen. 31.20A: From J. E. N. Veron, 2000. *Corals of the World*. © J. E. N. Veron. 31.20B: © Jurgen Freund/Naturepl.com. 31.21: Adapted from F. M. Bayer and & H. B. Owre, 1968. *The Free-Living Lower Invertebrates*, Macmillan Publishing Co.

Chapter 32 *Strepsipterans*: Courtesy of Dr. Hans Pohl. *Wasp*: © Paulo de Oliveira/OSF/Jupiter Images. 32.2: © Robert Brons/Biological Photo Service. 32.3A: From D. C. García-Bellido & D. H. Collins, 2004. *Nature* 429: 40. Courtesy of Diego García-Bellido Capdevila. 32.3B: © Piotr Naskrecki/Minden Pictures. 32.6: © Alexis Rosenfeld/Photo Researchers, Inc. 32.7A: © Ed Robinson/photolibrary.com. 32.8B: © Robert Brons/Biological Photo Service. 32.9B: © Larry Jon Friesen. 32.10A: © Stan Elems/Visuals Unlimited. 32.11: © David J. Wrobel/Visuals Unlimited. 32.13A: © Jurgen Freund/Naturepl.com. 32.13B: Courtesy of R. R. Hessler, Scripps Institute of Oceanography. 32.13C: © Roger K. Burnard/Biological Photo Service. 32.13D: © Larry Jon Friesen. 32.15A: © Larry Jon Friesen. 32.15B: © Dave Fleetham/photolibrary.com. 32.15C, D, E: © Larry Jon Friesen. 32.15F: © Orion Press/Jupiter Images. 32.16A: Courtesy of Jen Grenier and Sean Carroll, U. Wisconsin. 32.16B: Courtesy of Graham Budd. 32.16C: Courtesy of Reinhardt Møbjerg Kristensen. 32.17: © R. Calentine/Visuals Unlimited. 32.18B, C: © James Solliday/Biological Photo Service. 32.20A: © Michael Fogden/photolibrary.com. 32.20B: © Diane R. Nelson/Visuals Unlimited. 32.21: © Ken Lucas/Visuals Unlimited. 32.22A: © Frans Lanting/Minden Pictures. 32.22B: © Larry Jon Friesen. 32.22C: © Tom Branch/Photo Researchers, Inc. 32.22D: © Norbert Wu/Minden Pictures. 32.25: © Mark Moffett/Minden Pictures. 32.26: © Oxford Scientific Films/Jupiter Images. 32.27A: © Larry Jon Friesen. 32.27B: © Larry Jon Friesen. 32.27C: © Meul/ARCO/Naturepl.com. 32.27D: © Piotr Naskrecki/Minden Pictures. 32.27E: © Colin Milkins/Oxford Scientific Films/photolibrary.com. 32.27F: © Peter J. Bryant/Biological Photo Service. 32.27G: © Larry Jon Friesen. 32.27H: David McIntyre. 32.29A: © John Mitchell/Jupiter Images. 32.29B: © John R. MacGregor/Peter Arnold, Inc. 32.30A: © Norbert Wu/Minden Pictures. 32.30B: © Frans Lanting/Minden Pictures. 32.31A: © Kelly Swift, www.swiftinverts.com. 32.31B: © Larry Jon Friesen. 32.31C: © W. M. Beatty/Visuals Unlimited. 32.31D: Photo by Eric Erbe; colorization by Chris Pooley/USDA ARS.

Chapter 33 *Homo floresiensis*: © Christian Darkin/Alamy. *Tunicates*: © Larry Jon Friesen. 33.2: From S. Bengtson, 2000. Teasing fossils out of shales with cameras and computers. *Palaeontologia Electronica* 3(1). 33.4A: © Larry Jon Friesen. 33.4B: © Hal Beral/Visuals Unlimited. 33.4C: © Randy Morse/Tom Stack & Assoc. 33.4D: © Mark J. Thomas/Dembinsky Photo Assoc. 33.4E: © Peter Scoones/Photo Researchers, Inc. 33.4F: © Larry Jon Friesen. 33.5A: © C. R. Wyttenbach/Biological Photo Service. 33.6: © Robert Brons/Biological Photo Service. 33.7A: © Jurgen Freund/Naturepl.com. 33.7B: © David Wrobel/Visuals Unlimited. 33.9A: © Brian Parker/Tom Stack &

Courtesy of Thomas Eisner, Cornell U. 48.10A: © SPL/Photo Researchers, Inc. 48.10C: © P. Motta/Photo Researchers, Inc. 48.13: © Berndt Fischer/Jupiter Images. 48.17: After C. R. Bainton, 1972. *J. Appl. Physiol.* 33: 775.

Chapter 49 *Lewis*: © NBAE/Getty Images. *Emergency room*: © BananaStock/Alamy. 49.8A: © Brand X Pictures/Alamy. 49.9: After N. Campbell, 1990. *Biology*, 2nd Ed., Benjamin Cummings. 49.10B: © CNRI/Photo Researchers, Inc. 49.12: © Ed Reschke/Peter Arnold, Inc. 49.15A: Chuck Brown/Science Source/Photo Researchers, Inc. 49.15B: © Biophoto Associates/Science Source/Photo Researchers, Inc. 49.19: © Doc White/Nature Picture Library.

Chapter 50 *Pima*: © Marilyn "Angel" Wynn/Nativestock.com. *Fast food*: © Matt Bowman/FoodPix/Jupiter Images. 50.1A: © Gerry Ellis, DigitalVision/PictureQuest. 50.1B: © Rinie Van Meurs/Foto Natura/Minden Pictures. 50.3: © AP/Wide World Photos. 50.6: © Ace Stock Limited/Alamy. 50.9: © Dennis Kunkel Microscopy, Inc. 50.14: © Eye of Science/SPL/Photo Researchers, Inc. 50.20: © Science VU/Jackson/Visuals Unlimited.

Chapter 51 *Bat*: © Michael & Patricia Fogden/Minden Pictures. *Kangaroo rat*: © Mary McDonald/Naturepl.com. 51.1 *inset*: © Kim Taylor/Naturepl.com. 51.2B: © Rod Planck/Photo Researchers, Inc. 51.8: From R. G. Kessel & R. H. Kardon, 1979. *Tissues and Organs.* W. H. Freeman, San Francisco. 51.11: From L. Bankir & C. de Rouffignac, 1985. *Am. J. Physiol.* 249: R643-R666. Courtesy of Lise Bankir, INSERM Unit, Hôpital Necker, Paris. 51.13: © Hank Morgan/Photo Researchers, Inc.

Chapter 52 *Katrina*: Courtesy of the NOAA Satellite and Information Service. *Swamp*: © Tim Fitzharris/Minden Pictures. 52.1: © Tim Fitzharris/Minden Pictures. *Tundra, upper*: © Tim Acker/Auscape/Minden Pictures. *Tundra, lower*: © Elizabeth N. Orians. *Boreal, upper*: © Carr Clifton/Minden Pictures. *Boreal, lower*: © Robert Harding Picture Library Ltd/Alamy. *Temperate deciduous*: © Paul W. Johnson/Biological Photo Service. *Temperate grasslands, upper*: © Robert & Jean Pollock/Biological Photo Service. *Temperate grasslands, lower*: © Elizabeth N. Orians. *Cold desert, upper*: © Edward Ely/Biological Photo

Service. *Cold desert, lower*: © Robert Harding Picture Library Ltd./photolibrary.com. *Hot desert, left*: © Terry Donnelly/Tom Stack & Assoc. *Hot desert, right*: © Dave Watts/Tom Stack & Assoc. *Chaparral, left*: © Elizabeth N. Orians. *Chaparral, right*: © Larry Jon Friesen. *Thorn forest*: © Frans Lanting/Minden Pictures. *Savanna*: © Nigel Dennis/AGE Fotostock. *Tropical deciduous*: Courtesy of Donald L. Stone. *Tropical evergreen*: © Elizabeth N. Orians. 52.6: © Elizabeth N. Orians. 52.9 *Chameleon*: © Pete Oxford/Naturepl.com. 52.9 *Tenrec*: © Nigel J. Dennis/Photo Researchers, Inc. 52.9 *Fossa*: © Frans Lanting/Minden Pictures. 52.9 *Lemur*: © Wendy Dennis/Dembinsky Photo Associates. 52.9 *Baobab*: © Pete Oxford/Naturepl.com. 52.9 *Pachypodium*: © Michael Leach/Oxford Scientific Films/photolibrary.com. 52.12: Courtesy of E. O. Wilson. 52.13A *left*: © Bill Lea/Dembinsky Photo Associates. 52.13A *right*: © Stephen J. Krasemann/Photo Researchers, Inc. 52.13B *left*: © Frans Lanting/Minden Pictures. 52.13B *center*: © Kenneth W. Fink/Photo Researchers, Inc. 52.13B *right*: © Chris Gomersall/Naturepl.com.

Chapter 53 *Macaques*: © Frans de Waal, Emory U. *Spider*: © Mark Gibson/Jupiter Images. 53.2A: From J. R. Brown et al., 1996. *Cell* 86: 297. Courtesy of Michael Greenberg. 53.4A: © Nina Leen/Time Life Pictures/Getty Images. 53.4B: © Frans Lanting/Minden Pictures. 53.9A: © Anup Shah/Naturepl.com. 53.9B: © Mitsuaki Iwago/Minden Pictures. 53.9C: © Konrad Wothe/Minden Pictures. 53.11: © Frans Lanting/Minden Pictures. 53.13: Courtesy of John Alcock, Arizona State U. 53.16: © Tui De Roy/Minden Pictures. 53.18: © Cyril Ruoso/JH Editorial/Minden Pictures. 53.21: © Nigel Dennis/AGE Fotostock. 53.22: © Piotr Naskrecki/Minden Pictures. 53.23: © Steve & Dave Maslowski/Photo Researchers, Inc. 53.24: Courtesy of John Alcock, Arizona State U. 53.25: © José Fuste Raga/AGE Fotostock.

Chapter 54 *Caterpillars*: Courtesy of John R. Hosking, NSW Agriculture, Australia. *Farmer*: © Thomas Shjarback/Alamy. 54.1A: © Flip Nicklin/Minden Pictures. 54.1B: © PhotoStockFile/Alamy. 54.1C: © Robert McGouey/Alamy. 54.5A: © Frans Lanting/Minden Pictures. 54.5B: © David Nunuk/SPL/Photo Researchers, Inc. 54.6A: © Michael Durham/Minden Pictures. 54.6B: Courtesy of Colin Chapman. 54.7: © Larry Jon Friesen. 54.11: After P. A. Marquet, 2000. *Science* 289: 1487. 54.12:

© Ed Reschke/Peter Arnold, Inc. 54.13: © Adam Jones/Photo Researchers, Inc. 54.14: © T. W. Davies/California Academy of Sciences. 54.18: © Kathie Atkinson/OSF/Jupiter Images.

Chapter 55 *Ant on spine*: © Piotr Naskrecki/Minden Pictures. *Ant with Beltian bodies*: © Mark Moffett/Minden Pictures. 55.1: After R. H. Whittaker, 1960. *Ecological Monographs* 30: 279. 55.4: © Alan & Sandy Carey/Photo Researchers, Inc. 55.6: © Dave Watts/Naturepl.com. 55.7: © Lawrence E. Gilbert/Biological Photo Service. 55.8: © Shehzad Noorani/Peter Arnold, Inc. 55.11: © Mitsuaki Iwago/Minden Pictures. 55.12A: © Kim Taylor/Naturepl.com. 55.12B: © Perennou Nuridsany/Photo Researchers, Inc. 55.13D: Courtesy of William W. Dunmire/National Park Service. 55.14A: David McIntyre. 55.15: © Jim Zipp/Photo Researchers, Inc. 55.16: Courtesy of Jim Peaco/National Park Service. 55.18: After M. Begon, J. Harper, & C. Townsend, 1986. *Ecology.* Blackwell Scientific Publications.

Chapter 56 *Crane*: © Mark Moffett/Minden Pictures. *Researchers*: Courtesy of Christian Koerner, U. Basel. 56.1: After M. C. Jacobson et al., 2000. *Introduction to Earth System Science.* Academic Press. 56.9A: © Corbis Images/PictureQuest.

Chapter 57 *Flight*: © Tom Hugh-Jones/Naturepl.com. *Crane suit*: © Mark Payne-Gill/Naturepl.com. 57.1: © Michael Long/NHMPL. 57.2B: © Mark Godfrey/The Nature Conservancy. 57.6: Richard Bierregaard, Courtesy of the Smithsonian Institution, Office of Environmental Awareness. 57.7: © Mr_Jamsey/iStockphoto.com. 57.8A: © Terry Whittaker/Alamy. 57.8B: © Paul Johnson/Naturepl.com. 57.9: © Michael & Patricia Fogden/Minden Pictures. 57.10A: After R. B. Aronson et al., 2000. *Nature* 405: 36. 57.10B: © Fred Bavendam/Minden Pictures. 57.12: Courtesy of WWF-US, GIS Map provided by J. Morrison. 57.14B: © Jim Brandenburg/Minden Pictures. 57.14C: Courtesy of Jesse Achtenberg/U.S. Fish and Wildlife Service. 57.15A: Courtesy of Christopher Baisan and the Laboratory of Tree-Ring Research, U. Arizona, Tucson. 57.15B: © Karen Wattenmaker/Painet, Inc. 57.17: After S. H. Reichard & C. W. Hamilton, 1997. *Conservation Biology* 11: 193. 57.18A: © Elizabeth N. Orians. 57.20: © Tom Vezo/Nature Picture Library.

Index